Advances in Total Hip Arthroplasty

Yasuharu Nakashima
Editor

Advances in Total Hip Arthroplasty

JSRA
JAPANESE SOCIETY FOR REPLACEMENT ARTHROPLASTY

Springer

Editor
Yasuharu Nakashima
Department of Orthopedic Surgery
Kyushu University
Fukuoka, Japan

ISBN 978-981-97-2460-4 ISBN 978-981-97-2461-1 (eBook)
https://doi.org/10.1007/978-981-97-2461-1

Translation from the Japanese language edition: "Jinkou Kokansetsu Chikan Shujutsu" by Yasuharu Nakashima, © The Japanese Society for Replacement Arthroplasty 2022. Published by Nankodo Co., Ltd.. All Rights Reserved.

The translation was completed using an artificial intelligence translation tool. Following this, a human revision was conducted to further refine the work, ensuring that the translation is accurate in terms of content and scientific correctness.

This Springer imprint is published by the registered company Springer Nature Singapore Pte Ltd.
The registered company address is: 152 Beach Road, #21-01/04 Gateway East, Singapore 189721, Singapore

Preface

More than 60 years have passed since Total Hip Arthroplasty (THA) first became popular in the 1960s. Various innovations have dramatically improved the long-term durability of THA and have allowed for consistent clinical results. This progress has been based on basic and clinical research, with many notable achievements centered on biomaterials, hip morphology research, and computer-assisted technology.

The Japanese Society for Replacement Surgery (JSRA) planned to produce scientific books covering the entire field of artificial hip and knee joint replacement, reviewing those achievements conducted in Japan and around the world. As the person responsible for THA volume, I would like to express my sincere appreciation to all the authors for their understanding of the purpose of this book. This book is unique in that it covers from preoperative planning to surgical techniques, postoperative care, and countermeasures for complications.

The history of THA is introduced first, and then the concepts and clinical results of hip implants developed in Japan are introduced by the physicians who were involved in their development. In the section on biomaterials, a number of achievements that Japan has made in the world are introduced. Preoperative planning, surgical techniques, and postoperative therapy are covered in practical terms, including 3D planning, risk assessment, each surgical approach, and aggressive early rehabilitation, and patient-oriented evaluation, which has become indispensable for postoperative assessment. In the latter half of the book, many pages are devoted to difficult cases such as Crowe IV and post-osteotomy, as well as measures to prevent complications, to ensure that the book covers THA as a whole.

We hope that through this publication the readers will learn about the history of progress in THA and that it will lead them to new research for further progress. Finally, we would like to thank all the authors, and the editorial board members for their hard work in editing the book.

Fukuoka, Japan — Yasuharu Nakashima

Contents

Part I

History and Development of THA

History of THA Development

1

Nobuhiko Sugano

1 From Interposition Hip Arthroplasty Using Artificial Materials to Total Hip Replacement Arthroplasty

Historically, interposition arthroplasty was initially introduced, which involved resecting an ankylosed and immobile joint surface and inserting a membrane material to prevent rejection of the bone ends. In 1902, Robert Jones inserted gold foil into the articular surface; Murphy inserted the fascia and adipose tissue [1] to avoid bony ankylosis of the joint. However, these intermediate insertions did not provide stable load-bearing results for the lower limb joints. Smith Petersen in the United States attempted to improve interposition arthroplasty and devised a cup-shaped mold (mold arthroplasty) in 1923; it was inserted into the hip joint such that fibrocartilage was formed spherically in the acetabulum and femoral head [2]. Initially, cup arthroplasty was performed using artificial materials such as glass, Pyrex, and Bakelite; however, short-term failures occurred due to material failure. The use of vitallium (65% cobalt, 30% chromium, and 5% molybdenum), a corrosion-resistant metal developed in 1932, made it possible to avoid implant material failure from 1938 onwards. In this method, the inserted cup-shaped implant was not fixed to the pelvis or femur. Since it was cast to shape the fibrocartilage formation from the subchondral bone into a spherical shape, the cast was removed once the fibrocartilage regenerated. In practice, however, simply inserting the cast provided pain relief and improved mobility; furthermore, patients often refused the procedure to remove it; therefore, it was left in place. In vitallium cup arthroplasty, bone resorption of the femoral head and upper migration of the cup to the acetabulum lead to leg shortening and greater trochanter elevation (Fig. 1.1), resulting in the distal transfer of the greater trochanter and cup replacement required in many cases. In 1957, Aufranc reported the results of 1000 cup arthroplasty cases performed at the Massachusetts General Hospital over 15 years and found a revision rate of 22.5%, which was considered a very good result for arthroplasty at that time [3]. Until the late 1960s, cup arthroplasty was the standard surgical treatment for degenerative hip joint diseases. Although this method differs from total hip arthroplasty (THA), developed in later periods, as the inserted implant was not fixed to either the pelvis or femur, Norman Sharp underwent cup arthroplasty in 1948. On June 28, 2019, 70 years and 209 days later, he was awarded a place in the Guinness Book of World Records for the longest-lasting hip replacement.

The beginning of joint replacement with artificial materials required resection and replacement of the proximal femoral bone tumor (giant cell tumor) with an artificial femoral head and medullary stem implant made of vitallium by Bohlman in 1937, in which the vitallium stem was fixed in the femoral marrow cavity and passed without foreign body reaction; this proved that the human acetabular cartilage could accept the vitallium metallic head and provide adequate joint functioning [4]. In 1950, the Judet brothers examined 300 cases of acrylic femoral neck stem-designed prostheses. In most cases, especially in patients with femoral neck fractures, the results were very good, although they were used only for a short period [5]. Moore used a window in the stem, similar to the design used in Bohlman's case, to allow for fixation by osteogenesis. He called it a self-locking device (Fig. 1.2) based on three-point compression fixation [6]. Thompson used an anteroposterior curved stem design without a window to avoid fracture risk due to the reduction in strength and the difficulty of removal [7]. However, replacement hemiarthroplasty did not provide good results in pain relief, mobility, and support due to friction and bone erosion between the artificial head and acetabulum in patients with loss of articular cartilage and non-spherical sliding surfaces due to osteoarthritis of the hip joint. For THA, placing an artificial acetabulum (cup) in the pelvis and an artificial femoral head is necessary to reconstruct a sliding surface

N. Sugano (✉)
Department of Orthopaedic Medical Engineering, Osaka University Graduate School of Medicine, Suita, Japan
e-mail: n-sugano@umin.net

Y. Nakashima (ed.), *Advances in Total Hip Arthroplasty*, https://doi.org/10.1007/978-981-97-2461-1_1

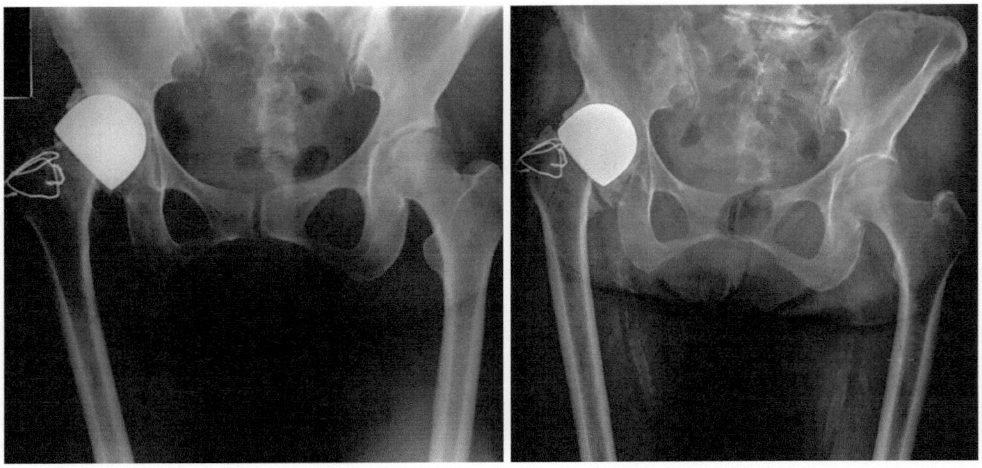

Fig. 1.1 Radiographs of cup arthroplasty with a vitalium cup. The left side is 5 years postoperatively, and the right is 37 years postoperatively. The cup migrated outward and upward from the original acetabulum but was fixed to the pelvis. The femoral head moved slightly within the cup. Since there is almost no pain and difficulty in the sitting position, conversion to THA is unnecessary

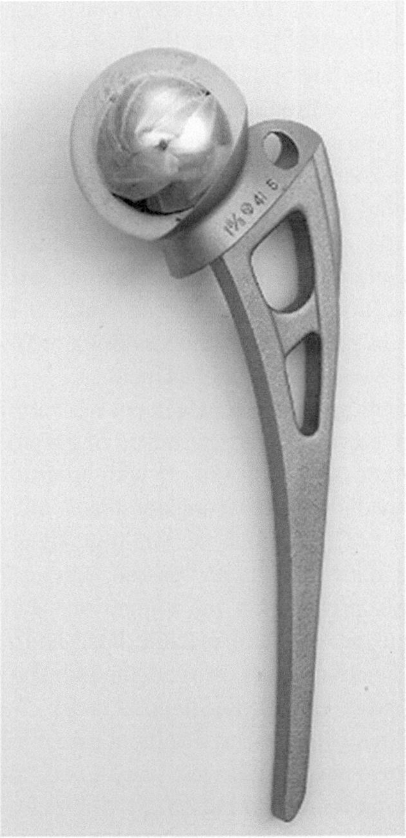

Fig. 1.2 Austin Moore self-locking artificial femoral head prosthesis

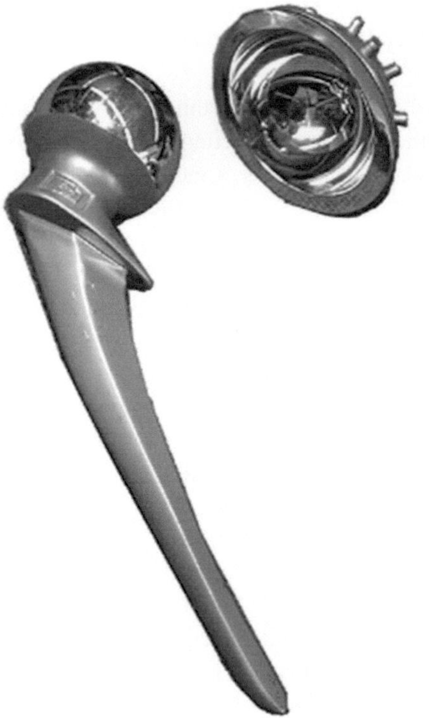

Fig. 1.3 McKee Farrar-type metal on metal hip prosthesis

with less friction, such as a real joint. The stainless-steel hip prosthesis completed by Wiles in 1938 was the world's first THA with an acetabular and femoral head component that had a metal-on-metal (MOM) sliding part. Six patients with Still's disease underwent surgery, and all six could walk despite a nearly bedridden state. However, materials and information were lost during World War II, and only one case was followed up for 13 years. Although the long-term follow-up case showed mobility, the screws and bolts that held the cup and head components in place were radiologically fractured [8]. McKee and Farrar focused on high-strength vitallium and developed a Thompson stem combined with a vitallium acetabular cup (Fig. 1.3) and cemented it in place [9]. The sliding portion of the MOM was lubricated by the lymphatic fluid supplied by the inter-tissue fluid.

Charnley thought that the low friction of the joint was the key to the success of THA because of the squeaking noise problem of the Judet acrylic artificial head. Therefore, he focused on artificial materials with a low friction coefficient; he developed a cup and surface-replacement type femoral head with Teflon and implanted them through a trochanteric approach. However, he was concerned about blood flow disturbance in the femoral head and replaced the surface-type head implant with a Thomson-type cobalt-chromium stemmed femoral head [10]. Subsequently, the torque generated by friction during joint sliding would become a shearing force at the interface between the acetabular cup and bone, and the fixation of the cup would fail. The femoral head diameter was reduced from 13/8 inches (approximately 41 mm) to 7/8 inches (22.225 mm) in stages (Fig. 1.4) and reported as a new procedure to reduce the torque [10]. However, Teflon wore rapidly, and the foreign body reaction caused by the wear particles resulted in osteolysis around the implant, resulting in short-term failure due to fracture and loosening. Therefore, he introduced ultrahigh molecular weight polyethylene, a new material, and implanted pieces of Teflon and polyethylene under Charnley's skin and confirmed that the biological reaction did not occur with polyethylene [11]; this resulted in the change of the cup to polyethylene in 1962, and the Charnley hip prosthesis was considered a success. The concept of low-frictional-torque arthroplasty was successfully demonstrated using a pendulum tester with a longer amplitude than McKee Farrar's MOM hip prosthesis; it has spread rapidly throughout the world since the 1970s.

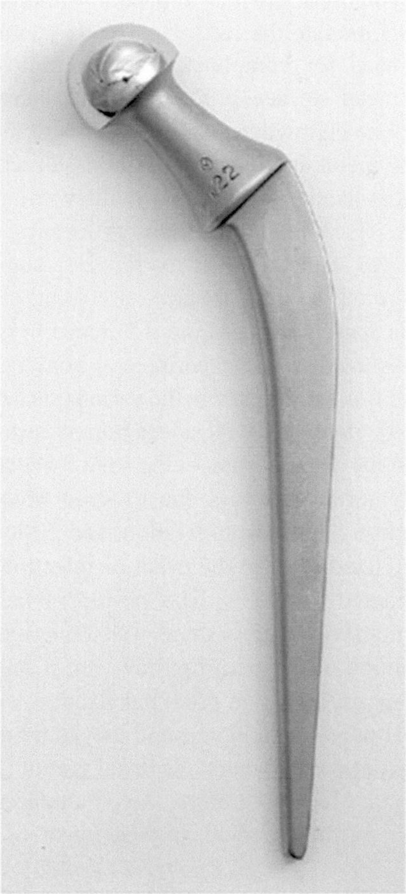

Fig. 1.4 A femoral stem of Charnley low frictional torque arthroplasty

Charnley considered low frictional torque to be the key to the success of THA; therefore, he set the head diameter to approximately 22 mm, which not only favored cup fixation but also resulted in a shorter sliding distance per cycle than a larger head diameter, which leads to lower volumetric wear rate of the polyethylene and good long-term results [12]. Although the linear wear rate was higher with a smaller head diameter owing to increased surface pressure, it was reported that the volumetric wear rate did not significantly differ among head diameters of 22 mm, 26 mm, and 28 mm in relation to the cross-sectional area of the head [13]. However, the results of hip prostheses with femoral head diameters of ≥32 mm (Fig. 1.5), such as a Müller hip implant, were poor [14] until cross-linked polyethylene was available. The concept of a small-head design, such as that by Charnley, was derived from the lack of artificial material with sufficient strength and wear resistance to replace the cartilaginous part of the body. However, the small diameter of the femoral head, which is far from the anatomical shape of the joint, is disadvantageous for maintaining a wide range of motion and stability (dislocation resistance). Stemmed femoral implants cause stress shielding of the proximal femur, bypassing the hip loading stress to the femoral diaphysis via the stem, and bone atrophy is inevitable due to stress shielding [15].

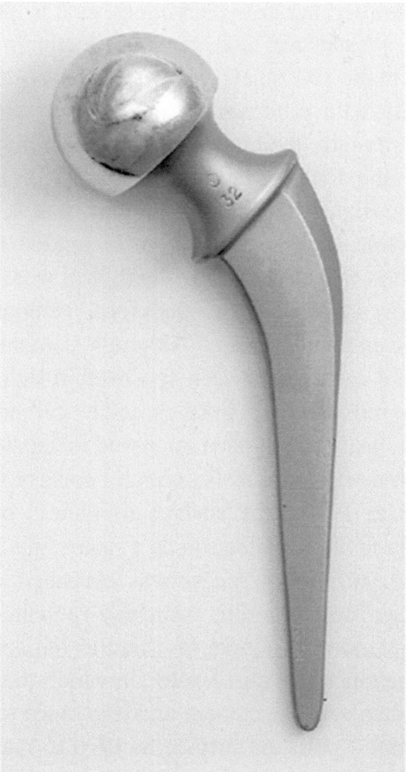

Fig. 1.5 Müller's cemented femoral stem (so-called Charnley Müller type)

2 Progress in Artificial Hip Fixation

Early models of hip prostheses relied on mechanical fixation such as screws, fins, and press fit, which did not yield good long-term results [8, 9, 16]. Charnley suggested that the three-point fixation of a Moore-type prosthetic head stem to provide stability against vertical loading does not consider rotational stability around the stem axis, particularly in the fixation of femoral stems, where stress and shear forces are concentrated distal to the stem. Therefore, if clay-like bone cement is first inserted into the medullary cavity, followed by the insertion of a metal stem and hardening of the cement, the cement and stem would become a metal-cement composite intraoperative custom-made stem suited to the complex medullary cavity geometry and provide very stable rotational fixation [17]. When early cemented THA became popular in the United States, radiographic loosening was 6.5% for the cups and 24% for stems at 5 years and 11.3% and 29.9% at 10 years, respectively, which was not as good as that reported by Charley et al. [18]. Younger and more active patients had even poorer results [19]. Since the cement is not an adhesive but a filler, its fixation force is exerted only when it is introduced within a strong cancellous bone near the cortical bone, and the quality of the cementing technique [20] influences its performance. Femoral cementation was initially finger-packed (first-generation technique); second-generation techniques using cement guns and plugs were introduced in 1975, and third-generation techniques such as pulse lavage, centralizer, pressurizer, and cement vacuum mixing were introduced in the 1990s [21]. There are also different types of bone cement with different physical properties; the higher the tensile strength, the higher the fatigue strength [22], and the higher the long-term survival rate [23]. In terms of cement viscosity, low-viscosity cement had lower tensile strength, while high-viscosity cement tended to leave air bubbles when the monomer and polymer were mixed and had inferior survival rates; therefore, standard-viscosity cement became mainstream. Although cement-containing antimicrobial agents have been approved in Europe because of their risk-reducing infections caused by cement [24], their prophylactic use has not been approved in Japan because of concerns about bacterial resistance and nephrotoxicity.

In the late 1990s, the surface roughness of cemented stems began to attract attention as a factor affecting performance. The surface roughness was classified as textured, rough, matte, satin, smooth, and shiny (polished) [25]; the processes intended to adhere the stem to the cement as the precoat. The taper slip type Exeter, in which the stem slides at the interface with the cement and is allowed to settle, was changed from a polished surface in 1970 to matte between 1976 and 1985 to increase the stem strength; however, polishing was reintroduced in 1986. The matte surface caused less stem subsidence but more frequent loosening [26]. Although the precoat type showed good results when the stem did not subside in the cement, once the cement was detached from the stem, the rough surface caused cement wear, accelerating the breakdown of stem fixation [27, 28]. The results for satin and smooth-surface stems were good, even though they were of the composite beam type [29]. A risk-unadjusted analysis in the British Registry showed that Exeter has recently outperformed other cemented stems in terms of survival [30].

In contrast, in the Nordic registry, the relative risk is significantly higher for Exeter when the risk-adjusted Charnley is set to one. The relative risk was significantly smaller for the composite beam type Lubinus, which is also frequently used [31]. The taper slip type is expected to reduce stress shielding by subsidence in the cement mantle hoop stress [32, 33]; however, reports have suggested that excessive sinking can lead to an increased risk of periprosthetic fracture compared to that from the composite beam type [34].

Cementless fixation, which relies on mechanical fixation during hip joint development, has problems during long-term use. However, research on cementless fixation, in which metal is directly bonded to the bone, has been conducted since the 1970s to improve long-term results in young patients and highly active men with poor results using cement fixation, to eliminate the risk of pulmonary embolism, and avoid the need for complicated cementation procedures. When the metal surface is roughened to a certain degree (Fig. 1.6), bone ingrowth enters the hollows of the irregularities, and the metal and bone matrix bond directly, enabling the biological fixation of artificial joints without the use of cement [35–38]. Although methods of forming porous surface structures using a lost-wax process, sintering metal beads or wire mesh on the surface, or coating metal powder with plasma spray were used, no difference in implant fixation was observed in the experimental bone ingrowth rate even if there was a difference in the various surface treatment methods [39]. On the femoral side, a porous surface was processed up to the distal portion of the stem for secure fixation. Proximal femoral bone loss due to stress shielding, stem fatigue fracture due to distal fixation, and difficulty removing a firmly fixed stem in the event of infection or fracture were considered problems. The postoperative thigh pain characteristic of cementless stems was thought to be caused by micromotion of the stem tip; however, it may be due to poor fixation, especially in proximal fixation without a circumferential porous coating around the stem because polyethylene particles could reach the distal part of the stem and cause osteolysis [40]. Therefore, the circumferential porous coating is now the standard surface finish of cementless stems.

Bone ingrowth requires 6–12 weeks to stabilize because the gap between the bone matrix and the implant is filled with bone from the matrix side. If the mechanical fixation of

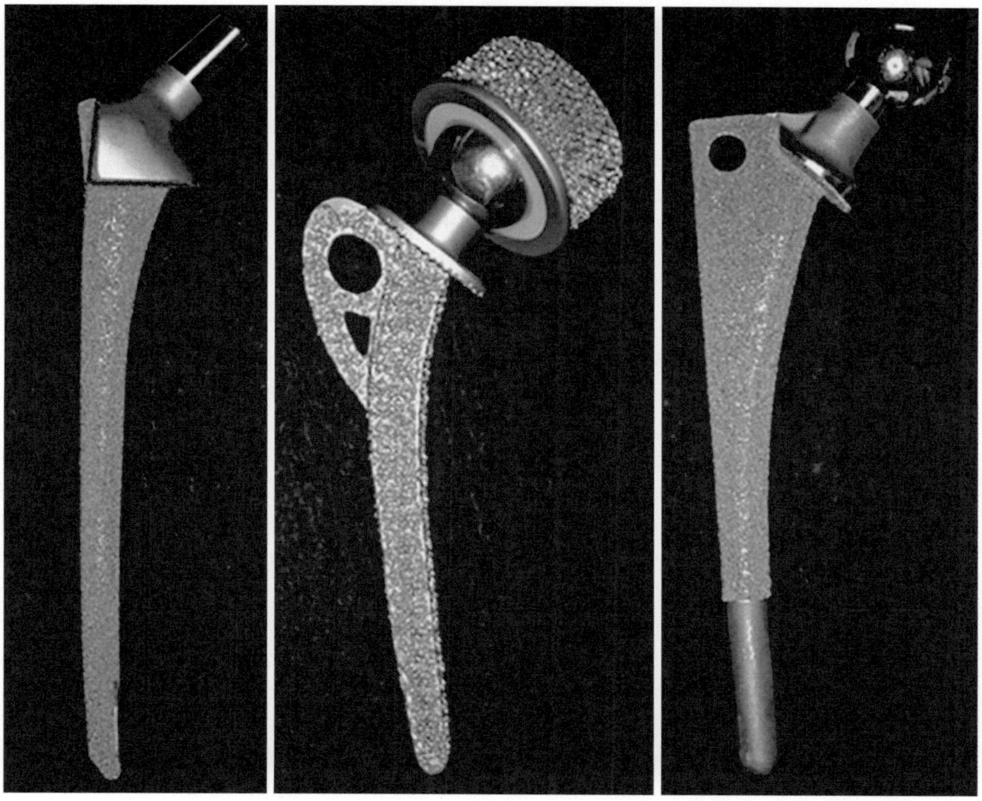

Fig. 1.6 Early design porous cementless femoral prostheses for bone ingrowth. From left to right: Lord madreporique, Judet cementless system, AML stem

the implant is poor during this period, biological fixation fails when there is a micromotion of 50 μm or more between the implant and the matrix. Therefore, bioactive coating techniques, such as hydroxyapatite (HA), have also been developed to increase the osteoconductive activity of the implant surface to double the gap repair rate [41]. It is difficult to prove that the HA coating further increases the survival rate because the long-term survival rate of other porous fabricated stems is also high [42]. In contrast, blasting the titanium alloy surface with a surface roughness of 3–5 μm causes bone growth on the titanium alloy surface in stems with excellent rotational fixation, resulting in strong biological fixation. Since these surface treatment techniques are also compatible with stem shape, they are classified while considering the combination of these techniques [43]. In addition, femurs with hip dysplasia have a different medullary cavity shape than normal and often have abnormal anteversion. To cope with this, offset and anteversion adjustments using modular parts and short custom-made designs with proximal fixation have been attempted [44]. In recent years, extremely short stems have been developed to facilitate placement through small incisions and minimize proximal femoral atrophy due to stress shielding with bone preservation. Still, the risks of size selection, poor alignment, and thigh pain have also been noted [45]. Except for resurfacing [46], it is difficult to avoid osteopenia due to stress shielding in the stem type because it is difficult to physiologically load the femoral neck even if the stem length is shortened. The problem with the extensive porous process up to the distal portion of the stem is that it provides secure stem fixation and causes significant stress shielding. However, full HA coating of the stem has been reported to cause relatively mild stress shielding without distal cortical bone thickening [47], and its use has been increasing in recent years.

In cementless fixation of cups, the method in which the acetabulum is drilled conically and a threaded cup is screwed into the acetabulum did not give positive results, and it became clear in the 1990s that a press-fit fixation method using a hemispherical metal cup with a porous finish was superior to screw fixation [48]. Although the cement technique improved the fixation of cemented fixation stems, there was no breakthrough in the cement fixation of cups, and metal-back cups failed to prove their theoretical advantage [48, 49]. The problems with the early cementless cups were the inadequate locking mechanism of the liner and the thin polyethylene liner that could be attached to the thick metal cup [50]. The former has problems of backside wear of the liner and osteolysis spreading from around the screw hole, in addition to liner deformation [51]. The latter had a risk of breakage but was unrelated to accelerated wear [52]. The locking mechanism of the liner has been improved, and the standard method for metal and ceramic liners is taper locking, and for polyethylene liners, a circumferential groove and a notch are used for rotational fixation of the liner to the cup rim.

Regarding the HA coating of cups, it has been reported that HA coating on the porous process is more advantageous for fixation and gap filling [53]. Porous cups have also changed from beads to plasma spray titanium coatings, which have higher friction with the bone surface, to 3D tantalum porous cups [54] or 3D porous cups made by 3D printing, which have emerged due to superior initial fixation and bone ingrowth [55]; however, whether these new 3D porous cups are superior to plasma-sprayed titanium-coated cups in primary THA needs to be evaluated [56].

3 Progress of Sliding Parts

The first sliding part for THA was the MOM, but Charnley's success in the 1960s led to metal-on-polyethylene (MOP) articulation due to the frictional torque problems faced using MOM. Although Charnley experienced Teflon wear and foreign body reaction failures, he did not believe that osteolysis from polyethylene wear debris could occur. In 1976, Harris et al. reported four cases of localized extensive bone resorption in the femur after cemented THA [57]. They suspected infection or tumor, but the examination results were negative, and they considered it a type of reaction to stem loosening and cement debris. At the same time, Charnley et al. reported that radiographic evaluation of 141 cases of Charnley THAs at an average age of 10 years showed that 66% had demarcation around the cup cement, and 9.6% of them had progressive cup migration [58]. Charnley considered a latent infection even if the bacterial culture test was negative and concluded that the infection was caused by a problem in the surgical technique after examining the surgical records.

On the other hand, in 1977, Willert et al. histologically analyzed the biological reactions to metal, polyethylene, and cement particles generated from hip prostheses. They found that the lymphatic system processed small amounts of foreign particles, but large amounts accumulated in the periarticular synovial membrane, causing phagocytosis and bone resorption by macrophages and foreign giant cells, eventually leading to the breakdown of cement fixation [59]. In the 1980s, osteolysis was referred to as "cement disease" [60], and in young, highly active patients, cement fixation was not strong enough. The osteolytic reaction to cement particles after fixation failure made functional reconstruction difficult, so cementless fixation became common to avoid the "cement disease." However, osteolysis has been reported to occur in the distal femur, even in well-fixed cementless stems, due to the accumulation of macrophages that have phagocytosed polyethylene particles [61]. The problem, recognized in the 1990s, is polyethylene wear debris, which causes osteolysis and prosthesis fixation failure [62, 63]. Analysis of the wear debris showed that the average size was 0.5 µm, more than 90% were less than 1 µm, 70–90% was polyethylene, and the rest were cement and metal particles, and, in many cases, chromium corrosion products from the trunnion [64]. Histologically, macrophage infiltration, fine particles, and rarely foreign body giant cells were found in fibrous tissue, including macrophages, fibroblasts, endothelial cells, rare lymphocytes, osteoblasts, and osteoclasts. Although the factors related to the biological reactivity of the wear debris are still unclear, cytokines such as interleukin-1, interleukin-6, tumor necrosis factor-alpha, matrix metalloproteinases, collagenase, stromelysin, prostaglandin E2, receptor activator of nuclear factor kappa-B ligand, and their single nucleotide polymorphisms and enzymes are involved in increased intra-articular pressure and access between the joint space and osteolytic lesions.

To eradicate osteolysis, the realistic options were to use ceramic heads, non-polyethylene bearing surfaces, or to improve polyethylene quality. Consequently, in the 1990s, there was progress in the improvement and development of new designs for metal-on-metal (MOM) and ceramic-on-ceramic (COC) bearing surfaces. On the other hand, attempts to improve polyethylene by incorporating carbon fibers, using heat press technology, or introducing Hylamer had failed early in clinical use [65–67], and cautiousness was required while introducing new polyethylene. Subsequently, based on sufficient basic experimental data, highly cross-linked polyethylene became available from various manufacturers at the end of the 1990s. First-generation cross-linked polyethylene became commercially available after 2000. Even with 10-year results, the linear wear calculated from the femoral head movement was approximately 0.3 mm, mostly due to initial creep deformation, indicating that the wear issue of polyethylene was being overcome by molecular cross-linking treatment [68]. Furthermore, it should be noted that even if the small head diameter concept of Charnley is not adopted. The larger head is free from wear concerns and simultaneously reduces the risk of dislocation. Various second-generation highly cross-linked polyethylene [69, 70] that is not remelted, reduces tensile strength, and has higher antioxidant performance, are currently available. However, the effect of cross-linking on polyethylene wear resistance is much greater than the negative effect of oxidative degradation. Therefore, it is unclear whether second-generation highly cross-linked polyethylene with enhanced antioxidant performance can be clinically proven to have better wear resistance than first-generation polyethylene.

MOM was considered inferior to Charnley's THA with MOP; however, it was found that THA with a cobalt-chromium MOM maintained good clinical results without osteolysis in a long-term follow-up study [16]. MOM was also suitable for resurfacing design because the metal allows for sufficient strength even with a thinner cup, which allows for a larger head diameter than THA with a polyethylene cup. To minimize the friction and wear rates in MOMs, the metal manufacturing method, clearance at sliding parts, and perfect sphericity are essential. McMinn analyzed the sliding parts of Ring and McKee's clinically successful MOMs, and Birmingham hip resurfacing, which employs a historically successful large clearance (250 microns), was developed using a cast-and-burn process (as-cast) of cobalt-chromium alloy. Different placement techniques have reported excellent long-term success [71–73]. Nevertheless, accelerated wear because of edge loading caused by poor cup placement angles may require early revision due to pseudotumors and local tissue damage called adverse reactions to metal debris (ARMD). ARMD occurs as a result of large amounts of metal wear debris [74]. High reoperation rates were observed in specific models such as ASR (DePuy) and Durom (Zimmer), which became RECALL products. Although ARMD is thought to be mainly caused by metal wear debris from the sliding parts of the MOM, it is also induced by metal debris at the taper junction of the cobalt-chromium femoral head, stem trunnion, and at cobalt-chromium modular necks and titanium stems. ARMD occurs more frequently in MOM stem-type THA with large-diameter heads than in MOM surface replacement [74]; therefore, in recent years, MOM stem-type THAs are no longer used, and since trunnionitis has been reported even in MOP sliding parts, ceramic heads have also been selected for polyethylene-compatible femoral heads.

Alumina COC sliding parts have the advantage of low wear resulting from low friction owing to their hardness, scratch resistance, high lubricity, and lower biological activity of wear debris compared to metals and polyethylene [75]. However, many early alumina COCs were unsuccessful because of their poor design and low strength, resulting in failure and wear. Third-generation alumina ceramics have dramatically improved strength by HIP treatment, high-purity alumina, and finer particles, and reduced frequency of osteolysis and 0% mechanical loosening compared with conventional polyethylene have been reported after 8 years [76]. In contrast, polyethylene sandwich-type COCs have a high frequency of cup-liner dislocations and fractures [77, 78]. As the use of COCs increases, squeak-like noises have been reported with a high frequency of 0.2 (20%) [79]. However, this squeak is not associated with early signs of fracture or functional impairment and does not cause osteolysis or laxity [80]. Furthermore, to improve the strength of ceramic heads, composite ceramics of alumina and zirconia are now employed in the sliding portion, and the risk of head fracture has decreased to one-tenth of that of third-generation alumina. The wear performance of the COC tends to improve slightly with increasing head diameter rather than decreasing, and a one-piece metal cup with a ceramic liner that reduces the risk of cup liner failure allows the use of larger heads. The long-term clinical results of these systems are awaited.

4 Advances in Surgical Techniques

Advances in surgical techniques, complication prevention, and postoperative care have significantly improved THA outcomes. For the surgical approach, Charnley recommended a technique with greater trochanter osteotomy, which allows accurate placement of the implant and lateral trochanteric shift to improve the biomechanics of the hip joint [10]. However, many complications related to greater trochanter fusion have occurred, so Müller used an anterolateral approach to make the surgery easier and faster and eliminate complications related to the greater trochanter [81]. In addition, because insertion of a long, straight stem into the femur from the anterior aspect is not easy, a curved stem design was used to ease insertion. Keggi et al. reported that a smaller incision using the Smith–Petersen anterior approach was possible for THA [82]. Although Hardinge's lateral approach has become popular in the United Kingdom [83], many surgeons use Dall's variant approach to avoid the potential complication of limping due to abductor muscle weakness [84]. However, the posterolateral approach is the most commonly used worldwide. The acetabulum is well exposed, the posterior pelvic column can be exposed, stem insertion is the easiest, stem alignment is well controlled, the surgical wound to the distal femur can be easily expanded, and it is widely applicable from a hemiarthroplasty to revision surgery. If the external rotator muscles are not repaired, the risk of dislocation is said to be higher when the diameter of the femoral head is small than with other approaches [85]; however, when the external rotators are repaired, the risk of dislocation is not significantly different from that of other techniques [86]. Minimally invasive surgery (MIS) has been gaining attention because endoscopic procedures such as cholecystectomy are less painful than conventional laparotomy, the postoperative recovery is faster, and hospital stay is shorter. In THA, MIS, which requires smaller skin incisions and fewer incisions of soft tissues, such as muscle tendons [87], has been proposed and rapidly spread as a marketing method for joint prosthesis companies. However, MIS is prone to complications (bleeding, fracture, poor implant placement, etc.) due to the small surgical field [88], and even if functional recovery is rapid in the short term, no MIS has been proven to be significantly functionally better than conventional techniques after 6 months. Computer-assisted surgery, such as surgical robots and navigation, is superior to conventional methods in terms of the accuracy of implant placement [89–91], but the cost has hindered its widespread use. However, if the long-term effectiveness of computer-assisted surgery is demonstrated and reimbursement can be obtained, the general public may adopt it to improve surgical accuracy.

References

1. Murphy JB. Ankylosis: arthroplasty-clinical and experimental. JAMA. 1905;27:1–77.
2. Smith-Petersen MN. Evolution of mold arthroplasty of the hip joint. J Bone Joint Surg Br. 1948;30B:59–75.
3. Aufranc OE. Constructive hip surgery with a vitallium mold; a report on 1,000 cases of arthroplasty of the hip over a fifteen-year period. Surg Am. 1957;39-A:237–48.
4. Moore AT, et al. Metal hip joint. a case report. J Bone Joint Surg. 1948;25:688.
5. Judet J, et al. The use of an artificial femoral head for arthroplasty of the hip joint. J Bone Joint Surg Br. 1950;32-B:166–73.
6. Moore AT. The self-locking metal hip prosthesis. J Bone Joint Surg Am. 1957;39-A:811–27.
7. Thompson FR. Two and a half years' experience with a vitallium intramedullary hip prosthesis. J Bone Joint Surg Am. 1954;36-A:489–502.
8. Wiles P. The surgery of the osteoarthritic hip. Br J Surg. 1958;45:488–97.
9. McKee GK, et al. Replacement of arthritic hips by the McKee-Farrar prosthesis. J Bone Joint Surg Br. 1966;48:245–59.
10. Charnley J. Arthroplasty of the hip. A new operation. Lancet. 1961;1:1129–32.
11. Charnley J. Tissue reactions to polytetrafluorethylene. Lancet. 1963;282(7322):1379.
12. Schmalzried TP, et al. Wear in total hip and knee replacements. J Bone Joint Surg Am. 1999;81:115–36.
13. Livermore J, et al. Effect of femoral head size on wear of the polyethylene acetabular component. J Bone Joint Surg Am. 1990;72:518–28.
14. Morrey BF, et al. Size of the femoral head and acetabular revision in total hip-replacement arthroplasty. J Bone Joint Surg Am. 1989;71:50–5.
15. Wroblewski BM, et al. Stem fixation in the Charnley low-friction arthroplasty in young patients using an intramedullary bone block. J Bone Joint Surg. 1998;80:273–8.
16. Ring PA. Complete replacement arthroplasty of the hip by a ring prosthesis. J Bone Joint Surg Br. 1968;50:720–31.
17. Charnley J. The bonding of prostheses to bone by cement. J Bone Joint Surg Br. 1964;46:518–29.
18. Stauffer RN. Ten-year follow-up study of total hip replacement. J Bone Joint Surg Am. 1982;64:983–90.
19. Chandler HP, et al. Total hip replacement in patients younger than thirty years old. A five-year follow-up study. J Bone Joint Surg Am. 1981;63:1426–34.
20. Barrack RL, et al. Improved cementing techniques and femoral component loosening in young patients with hip arthroplasty: a 12-year radiographic review. J Bone Joint Surg Br. 1992;74:385–9.
21. Herberts P, et al. How outcome studies have changed total hip arthroplasty practices in Sweden. Clin Orthop Relat Res. 1997;344:44–60.
22. Harper EJ, Bonfield W. Tensile characteristics of ten commercial acrylic bone cements. J Biomed Mater Res. 2000;53:605–16.
23. Espehaug B, et al. The type of cement and failure of total hip replacements. J Bone Joint Surg Br. 2002;84:832–8.
24. Engesaeter LB, et al. Does cement increase the risk of infection in primary total hip arthroplasty? Revision rates in 56,275 cemented and uncemented primary THAs followed for 0-16 years in the Norwegian arthroplasty register. Acta Orthop. 2006;77:351–8.
25. Crowninshield RD, et al. Cemented femoral component surface finish mechanics. Clin Orthop Relat Res. 1998;355:90–102.
26. Howie DW, et al. Loosening of matt and polished cemented femoral stems. J Bone Joint Surg Br. 1998;80:573–6.
27. Sylvain GM, et al. Early failure of a roughened surface, precoated femoral component in total hip arthroplasty. J Arthroplast. 2001;16:141–8.
28. One A, et al. Early failure of precoated femoral components in primary total hip arthroplasty. J Bone Joint Surg Am. 2002;84:786–92.
29. Jaffe WL, et al. Normalized and proportionalized cemented femoral stem survivorship at 15 years. J Arthroplast. 1999;14:708–13.
30. Kazi HA, et al. Not all cemented hips are the same: a register-based (NJR) comparison of taper-slip and composite beam femoral stems. Acta Orthop. 2019;90:214–9.
31. Junnila M, et al. Implant survival of the most common cemented total hip devices from the Nordic arthroplasty register association database. Acta Orthop. 2016;87:546–53.
32. Fowler JL, et al. Experience with the Exeter total hip replacement since 1970. Orthop Clin North Am. 1988;19:477–89.
33. Wroblewski BM, et al. Triple taper polished cemented stem in total hip arthroplasty: rationale for the design, surgical technique, and 7 years of clinical experience. J Arthroplast. 2001;16(8 Suppl 1):37–41.
34. Palan J, et al. The influence of cemented femoral stem choice on the incidence of revision for periprosthetic fracture after primary total hip arthroplasty: an analysis of national joint registry data. Bone Joint J. 2016;98-B:1347–54.
35. Galante J, et al. Sintered fiber metal composites as a basis for attachment of implants to bone. J Bone Joint Surg Am. 1971;53:101–14.
36. Lord GA, et al. An uncemented total hip replacement: experimental study and review of 300 madreporique arthroplasties. Clin Orthop Relat Res. 1979;141:2–16.
37. Judet R, et al. A noncemented total hip prosthesis. Clin Orthop Relat Res. 1978;137:76–84.
38. Engh CA. Hip arthroplasty with a Moore prosthesis with porous coating. A five-year study. Clin Orthop Relat Res. 1983;176:52–66.
39. Turner TM, et al. A comparative study of porous coatings in a weight-bearing total hip-arthroplasty model. J Bone Joint Surg Am. 1986;68:1396–409.
40. Tanzer M, et al. The progression of femoral cortical osteolysis in association with total hip arthroplasty without cement. J Bone Joint Surg Am. 1992;74:404–10.
41. Geesink RG. Hydroxyapatite-coated total hip prostheses. Two-year clinical and roentgenographic results of 100 cases. Clin Orthop Relat Res. 1990;261:39–58.
42. Hailer NP, et al. Hydroxyapatite coating does not improve uncemented stem survival after total hip arthroplasty! Acta Orthop. 2015;86:18–25.
43. Zweymüller KA, et al. Biologic fixation of a press-fit titanium hip joint endoprosthesis. Clin Orthop Relat Res. 1988;235:195–206.
44. Sakai T, et al. Stem length and canal filling in uncemented custom-made total hip arthroplasty. Int Orthop. 1999;23:219–23.
45. Loppini M, et al. Uncemented short stems in primary total hip arthroplasty: the state of the art. EFORT Open Rev. 2018;3:149–59.
46. Kishida Y, et al. Preservation of the bone mineral density of the femur after surface replacement of the hip. J Bone Joint Surg Br. 2004;86(2):185–9.
47. Vidalain JP. Twenty-year results of the cementless corail stem. Int Orthop. 2011;35:189–94.
48. Morscher EW. Current status of acetabular fixation in primary total hip arthroplasty. Clin Orthop Relat Res. 1992;274:172–93.
49. Yahiro MA, et al. Comparison of the results of cemented, porous-ingrowth, and threaded acetabular cup fixation. A meta-analysis of the orthopedic literature. J Arthroplast. 1995;10:339–50.
50. González della Valle A, et al. Dislodgment of polyethylene liners in first- and second-generation Harris-Galante acetabular components. J Bone Joint Surg Am. 2001;83:553–9.

51. Schmalzried TP, et al. The role of acetabular component screw holes and/or screws in the development of pelvic osteolysis. Proc Inst Mech Eng H. 1999;213:147–53.
52. Shih CH, et al. Measurement of polyethylene wear in cementless total hip arthroplasty. J Bone Joint Surg Br. 1997;79:361–5.
53. Thanner J, et al. Porous cups with and without hydroxylapatite-tricalcium phosphate coating: 23 matched pairs evaluated with radiostereometry. Arthroplasty. 1999;14:266–71.
54. Unger AS, et al. Evaluation of a porous tantalum uncemented acetabular cup in revision total hip arthroplasty: clinical and radiological results of 60 hips. J Arthroplast. 2005;20:1002–9.
55. Goldman AH, et al. Does increased coefficient of friction of highly porous metal increase initial stability at the acetabular interface? J Arthroplast. 2016;31:721–6.
56. Carli AV, et al. Short to midterm follow-up of the tritanium primary acetabular component: a cause for concern. J Arthroplast. 2017;32:463–9.
57. Harris WH, et al. Extensive localized bone resorption in the femur following total hip replacement. J Bone Joint Surg Am. 1976;58:612–8.
58. DeLee JG, et al. Radiological demarcation of cemented sockets in total hip replacement. Clin Orthop Relat Res. 1976;121:20–32.
59. Willert HG, et al. Reactions of the articular capsule to wear products of artificial joint prostheses. J Biomed Mater Res. 1977;11:157–64.
60. Jones LC, et al. Cement disease. Clin Orthop Relat Res. 1987;225:192–206.
61. Maloney WJ, et al. Endosteal erosion in association with stable uncemented femoral components. J Bone Joint Surg Am. 1990;72:1025–34.
62. Schmalzried TP, et al. The mechanism of loosening of cemented acetabular components in total hip arthroplasty. Analysis of specimens retrieved at autopsy. Clin Orthop Relat Res. 1992;274:60–78.
63. Harris WH. The problem is osteolysis. Clin Orthop Relat Res. 1995;311:46–53.
64. Jacobs JJ, et al. Osteolysis: basic science. Clin Orthop Relat Res. 2001;393:71–7.
65. Wright TM, et al. Analysis of surface damage in retrieved carbon-fiber-reinforced and plain polyethylene tibial components from posterior stabilized total knee replacements. J Bone Joint Surg Am. 1988;70:1312–9.
66. Bloebaum RD, et al. Investigation of early surface delamination observed in retrieved heat-pressed tibial inserts. Clin Orthop Relat Res. 1991;269:120–7.
67. Livingston BJ, et al. Complications of total hip arthroplasty associated with the use of an acetabular component with a Hylamer liner. J Bone Joint Surg Am. 1997;79:1529–38.
68. Capello WN, et al. Continued improved wear with annealed highly cross-linked polyethylene. Clin Orthop Relat Res. 2011;469:825–30.
69. Dumbleton JH, et al. The basis for a second-generation highly cross-linked UHMWPE. Clin Orthop Relat Res. 2006;453:265–71.
70. Oral E, et al. Wear resistance and mechanical properties of highly cross-linked, ultrahigh-molecular weight polyethylene doped with vitamin E. Arthroplasty. 2006;21:580–91.
71. Treacy RBC, et al. Birmingham hip resurfacing: a minimum follow-up of ten years. J Bone Joint Surg Br. 2011;93:27–33.
72. Van Der Straeten C. Hip resurfacing arthroplasty in young patients: international high-volume centres' report on the outcome of 11,382 metal-on-metal hip resurfacing arthroplasties in patients ⩽50 years at surgery. Hip Int. 2020;32:353–62.
73. Liu F, et al. A safe zone for acetabular component position in metal-on-metal hip resurfacing arthroplasty: winner of the 2012 HAP PAUL award. Arthroplasty. 2013;28:1224–30.
74. Sugano N, et al. Nationwide investigation into adverse tissue reactions to metal debris after metal-on-metal total hip arthroplasty in Japan. Orthop Sci. 2014;19:85–9.
75. Bierbaum BE, et al. Ceramic-on-ceramic bearings in total hip arthroplasty. Clin Orthop. 2002;405:158–63.
76. Capello WN, et al. Ceramic-on-ceramic total hip arthroplasty: update. J Arthroplast. 2008;23(7 Suppl):39–43.
77. Hasegawa M, et al. Alumina ceramic-on-ceramic total hip replacement with a layered acetabular component. J Bone Joint Surg Br. 2006;88:877–82.
78. Park YS, et al. Ceramic failure after total hip arthroplasty with an alumina-on-alumina bearing. J Bone Joint Surg Am. 2006;88:780–7.
79. Keurentjes JC, et al. High incidence of squeaking in THAs with alumina ceramic-on-ceramic bearings. Clin Orthop Relat Res. 2008;466:1438–43.
80. Sugano N, et al. Mid-term results of cementless total hip replacement using a ceramic-on-ceramic bearing with and without computer navigation. Bone Joint Surg Br. 2007;89:455–60.
81. Müller ME. Total hip prostheses. Clin Orthop Relat Res. 1970;72:46–68.
82. Light TR, et al. Anterior approach to hip arthroplasty. Clin Orthop Relat Res. 1980;152:255–60.
83. Hardinge K. The direct lateral approach to the hip. J Bone Joint Surg Br. 1982;64:17–9.
84. Dall D. Exposure of the hip by anterior osteotomy of the greater trochanter. A modified anterolateral approach. J Bone Joint Surg Br. 1986;68:382–6.
85. Berry DJ, et al. Effect of femoral head diameter and operative approach on risk of dislocation after primary total hip arthroplasty. J Bone Joint Surg Am. 2005;87:2456–63.
86. Kwon MS, et al. Does surgical approach affect total hip arthroplasty dislocation rates? Clin Orthop Relat Res. 2006;447:34–8.
87. Berger RA. Total hip arthroplasty using the minimally invasive two-incision approach. Clin Orthop Relat Res. 2003;417:232–41.
88. Woolson ST, et al. Comparison of primary total hip replacements performed with a standard incision or a mini-incision. J Bone Joint Surg Am. 2004;86:1353–8.
89. Sugano N, et al. Does CT-based navigation improve the long-term survival in ceramic-on-ceramic THA? Clin Orthop Relat Res. 2012;470:3054–9.
90. Nakamura N, et al. Does robotic milling for stem implantation in cementless THA result in improved outcomes scores or survivorship compared with hand results of a randomized trial at 10 years. Clin Orthop Relat Res. 2018;476:2169–73.
91. Ando W, et al. Comparison of the accuracy of the cup position and orientation in total hip arthroplasty for osteoarthritis secondary to developmental dysplasia of the hip between the Mako robotic arm-assisted system and computed tomography-based navigation. Int Orthop. 2021;45:1719–25.

THA Development in Japan

2

Yasuharu Nakashima

1 Introduction

Total hip arthroplasty (THA) was first widely performed in Japan in the late 1960s and early 1970s, and since then more than half a century has passed. The reports from those days show that imported implants were unsuitable for the Japanese skeletal structure and that even the tools for osteotomy and excavation were insufficient. In this situation, our predecessors had successfully performed surgeries with their ingenuity and joined forces with domestic companies to challenge the development of new hip implants [1, 2]. This section summarizes the THA developed in Japan. The clinical outcomes of THA developed in the early days were not as good as those of today, but considering that the current THA is based on them, they are all valuable. Looking back at the history of THA, similar mistakes have frequently been repeated. Our mission is to identify the cause of poor outcomes and avoid making the same mistakes. The current treatment results may be the best for us; however, in the eyes of future generations, they are destined to become outdated. The importance of "learning the new from the old" applies well to implant development.

1.1 Jikei Medical School Style THA

The pioneering development of THA in Japan was cementless unipolar hip arthroplasty at Jikei University School of Medicine. In the 1950s, Katayama, Hirakawa et al. fabricated a hand-made acrylic resin unipolar hip arthroplasty for tuberculosis of the hip joint and applied it clinically [3, 4] (See a Jikei Medical School type artificial hip joint and a Maesawa type axial rotation type artificial hip joint [Nemoto Shokai]). Various innovations were subsequently made, and Jikei Medical School-style cementless THA (Nemoto Shokai) was developed by Itami et al. in 1972. The acetabular cup has three characteristic spikes, and both the cup and stem are surface-processed to achieve macroanchoring [4, 5]. Subsequently, improvements were made in metal material, stem shape, and surface processing. Jikei Medical School-type THA has succeeded in the fourth generation and plays an important role in the development of THA in Japan.

1.2 Bearing Surface

1.2.1 Cross-Linked Polyethylene

The wear of bearing surfaces has always been a major concern in THA. The Charnley experienced rapid wear with teflon cups and obtained stable results using a combination of polyethylene cups and 22-mm diameter metal heads [6]. The pioneering work in Japan was the improvement of the wear resistance of cross-linked polyethylene by γ-ray irradiation performed by Shikida et al. in the 1970s [7]. Onishi et al. reported good clinical results [8–10]. They worked on improving polyethylene due to the severe polyethylene wear of Muller-type THA and found that 100 Mrad γ-ray irradiation increased the wear resistance. SOM-type THA (Mizuho) using this type of polyethylene was used in clinical practice until approximately 1978 when its low wear resistance was demonstrated. In the 1990s, these results attracted considerable attention and led to the redevelopment of cross-linked polyethylene. This is a significant achievement because cross-linked polyethylene is now the mainstream bearing surface. In recent years, Moro, Ishihara et al. developed polyethylene with synthetic phospholipid polymers grafted onto the surface of cross-linked polyethylene at the nanometer level. This polyethylene showed improved wear resistance compared to conventional cross-linked polyethylene in a hip simulator comparison and attracted considerable attention and was commercialized as Aquala® Liner (Kyocera) in 2011 [11, 12] (see 4-(4) Polyethylene [MPC polymer]).

Y. Nakashima (✉)
Department of Orthopedic Surgery, Kyushu University, Fukuoka, Japan
e-mail: nakashima.yasuharu.453@m.kyushu-u.ac.jp

Y. Nakashima (ed.), *Advances in Total Hip Arthroplasty*, https://doi.org/10.1007/978-981-97-2461-1_2

1.2.2 Ceramic THA

One of the characteristics of THA in Japan is the use of ceramics. Ceramics are used as the head material and interface with the bone. Alumina is a bioinert material that is harder than metal, scratch-resistant, highly lubricious, and has the advantage of excellent wear resistance. In Europe, THA with ceramic-on-ceramic (COC) sliding surfaces has been used since the early 1970s. The history of alumina heads in Japan dates back to the 1970s when Kyocera Corporation (Kyoto Ceramics Co., Ltd., which entered the THA development field) produced cemented THA Bioceram type 1 and type 2 stems in 1976. Alumina heads were used from the beginning [13–15] (d Bioceram types 1, 2, 4, and 5) (d types 1, 2, 4, and 5 [Kyocera]). Subsequently, improvements have been made to the alumina heads, such as HIP treatment, higher purity, and finer particles. Currently, zirconia-toughened alumina with approximately 20% zirconia is mainly used to increase strength. Zirconia bone heads with high strength were used in the 1990s to compensate for the easy breakability, which is a disadvantage of alumina. However, the phase transformation of zirconia causes increased polyethylene wear, and the use of zirconia bone heads has decreased worldwide [16, 17].

The alumina-bearing surface cup (Kyocera) is a COC THA developed in the 1990s as a polyethylene sandwich of alumina with a metal back cup. Although it was expected to be a solution to polyethylene wear–induced osteolysis, its manufacture and sale were discontinued owing to a series of alumina breakage and liner dislodgement [18] (see 4-(6) COC).

1.2.3 Other Bearing Surfaces

In the 1970s, various models were developed based on different ideas. One is a sliding surface that receives a polyester or polyethylene head on a concave metal surface, and its origin can be seen in Weber-Huggler-type THA. In Japan, Maesawa et al. developed an axial rotation type of THA with a 38 mm diameter polyethylene head and a double-bearing type that moves between a normal bearing part, stem neck, and polyethylene (Nemoto Shokai) [19] (see a Ji Dai-type artificial hip joint and Maesawa-type axial rotation type artificial hip joint).

Early cup loosening and migration are associated with cup arthroplasty. Several cup-and-socket THAs were developed in the 1970s, incorporating THA functionality as an evolution of cup arthroplasty. Owing to the large head diameter, it was necessary to choose either very thin polyethylene or metal-on-metal. Tanaka et al. developed a cemented polyethylene cup, and the femoral side was covered with a cup [20]. Nishio et al. developed a cementless socket with three spikes on the acetabular side and a cementless cup (on the femoral side) with a metal-on-metal bearing surface (Mizuho, 1972) [21] (See c Nishio, Surgical Resurfacing Hip Arthroplasty, 1972 [Mizuho]). However, the problem of fixation, especially on the femoral side, has not yet been solved, and long-term results cannot be expected.

1.3 Surface Treatment of Implants

The success of biological fixation by bone ingrowth on porous or rough surfaces has enabled the long-term durability of cementless THA. Ceramics that bond directly to bones, such as hydroxyapatite (HA), have been investigated worldwide to promote early bone ingrowth [22, 23]. THA with hydroxyapatite coating (HA-THA) began to be applied clinically in the late 1980s. Still, most of the early HA-THA implants used a relatively smooth base material, which made the bone anchorage of the implant dependent solely on HA, resulting in poor clinical results, especially on the cup side [24]. In Japan, the early results of HA-THA were similar. Nonetheless, the AMS HA cup/Kyocera PerFix® stem (Kyocera) commercialized in 1996 achieved stable fixation by coating a rough titanium matrix with HA [25] (see e Kyocera AMS HA CUP+PerFix HA STEM). AW glass ceramic (AWGC), developed by Kokubo et al., is highly biocompatible [26]. The K-MAX ABC HIP system (Kobe Steel, Ltd.), in which AWGC was bottom-coated on the deep layer of a roughened surface, was commercialized in 1996. The K-MAX QHIP system with a modified stem shape was commercialized in 1999, and good results were reported [27] (see f K-MAX ABC [AW-GC glass ceramic bottom coating]). The alkaline heat treatment technology developed later showed a higher bone-bonding ability than the AWGC bottom coating. It was approved in 2007 as AHFIX® Q Stem (Kyocera), in which the rough surface of the K-MAX Q HIP system was treated with alkaline heat, and the implant survival rate was an average of 10 years after surgery was 98% [28]. In contrast, Elance AHFIX® Q Stem (Kyocera), which is an alkaline heat-treated metal matrix with a smooth surface, causes considerable postoperative loosening and has problems with bone fixation [29]. This suggests mechanical microanchoring by bone in/on growth to the porous surface or through the surface is necessary for long-term fixation.

In the 2000s, three-dimensional (3D) porous cups with more than 50% porosity became popular. Geodesic structure cups (Teijin Nakashima) and SQRUM® TT Cup (Kyocera), developed in Japan, are surfaces fabricated using 3D printing technology using the additive manufacturing method, with porosities of 50 and 60%, respectively. However, stable short-term results have also been reported [30].

Stem developed from developmental dysplasia of the hip (DDH) morphology research.

In Japan, most hip osteoarthritis cases are caused by DDH, and the femur exhibits a high neck-shaft angle, excessive anteversion, and a narrow medullary cavity, making it difficult to fit the normal stem shapes in many cases. Therefore, several stems have been developed to fit Japanese patients with DDH [31, 32]. The Centpilar stem (Stryker) and APS Natural Hip (Zimmer) were created by analyzing the proximal femoral morphology based on computed tomography data from DDH cases and healthy Japanese people. These were anatomic-type stems with relatively short stem lengths intended to be fit and filled at the proximal portion. (See h CentPillar Grit Blast HA stem 2003/CentPillar TMZF Plasma Spray HA stem 2007 [Stryker], Z immer APS 2008 [Zimer]). APS is also characterized by its ability to select 0°or 12°according to the anterior tortuosity variation. Good clinical results have been reported for both stems.

1.4 Cement Stem

In 1971, the SOM (acronym of Shikida, Onishi and Mizuho, Mizuho) was developed and applied clinically. COP, a new alloy developed by Okayama University, Daido Steel, and Mizuho, was used as the material [33]. Kyocera developed Bioceram hip prostheses (type 1 and type 2 stems) using COP alloy and alumina ceramic heads in 1976, changed the stem material to Ti-6Al-4 V, and developed various types of stems (1986) [14, 15] (see Bioceram Types 1, 2, 4, and 5 [Kyocera Corporation]). In contrast, Kobe Steel, Ltd., developed a Charnley type H-3 stem in 1995 using a beta-titanium alloy (Ti-15Mo-5Zr-3Al) without vanadium elements, which was improved to a double taper design in 2004 and has been used to this day [34].

The YU Stem (Kyocera) in 1988 was a smooth surface stem made of Ti-6Al-4 V featuring large grooves on the inner and outer sides and the front and rear surfaces; however, in 2001, it was changed to a polished surface made of Ti-6Al-4 V and in 2005, to a smooth surface made of cobalt-chromium-molybdenum alloy [35]. The taper slip theory also spread in Japan in approximately 2000, introducing taper polish stems, such as the Exeter and C-Stem. In 2005, Japan Medical Materials (now Kyocera Corporation) developed an SC stem with a curved triple taper design (See i SC Stem [Kyocera]).

1.5 Development of THA with Antimicrobial Activity

Excellent research and development of implants for antibacterial agents have been reported. Iodine-loaded implants are expected to have an antibacterial effect when iodine is sealed in the micropores of a special oxide film formed on the surface of the implant. In addition, the iodine-loaded implants have been shown to have a high affinity with bone when bone enters the micropores of the oxide film and are used clinically [36, 37] (See 4-(8) Iodine coating). A silver-containing HA coating technology that applies silver, which is known to have antimicrobial activity, to the surface of THA has also been developed [38] (see 4–(9) Cementless silver-containing hydroxyapatite-coated artificial hip joints). HA powder-containing silver oxide was sprayed to form a film, and silver was eluted from this film to provide antimicrobial activity and bone fixation by HA. This technology was commercialized as AG-PROTEX (Kyocera) in 2016.

The history of THA development in Japan has been summarized from several view-points. Many models have been developed over the past 50 years but with limited space, and not all of them can be introduced. In addition, industry-academia collaboration with companies is essential for implant development. Looking back on the development of THA in Japan, many implant manufacturers, such as Kyocera, Mizuho, Nemoto Shokai, Teijin-Nakashima, MDM, Stryker, and Zimmer-Biomet, have cooperated with us. We would like to take this opportunity to thank them for their cooperation.

References

1. Onishi H, editor. General remarks. The results and analysis of prostheses used in Japan for more than 20 years. Shiga: CellTesco Medical Engineering Corporation; 2014. p. 20–7. Japanese
2. Kinoshita M, et al. Experience and material mechanics of artificial joint replacement in the hip joint. JPRS. 1969;12:150–4. Japanese
3. Hiraki H, et al. Methods for the preparation of acrylic resin artificial bone heads in our laboratory and experience in their application to the resection of osteoarticular tuberculosis. J Jpn Orthop Assoc. 1955;29:186–97. Japanese
4. Ohtani Y, et al. The Jidai-style cementless hip prosthesis: its evolution and long-term results of early models. Seikei Saigai Geka. 2007;50:1253–60. Japanese
5. Itami Y, et al. The direct fixation system of total hip prosthesis. Research of the possibilities of fixing total hip prosthesis directly to bone. Arch Orthop Trauma Surg. 1978;1982(100):11–7.
6. Charnley J. Hip sockets – theory. Low friction arthroplasty of the hip: theory and practice. Berlin: Springer; 1979. p. 91–106.
7. Shikida T, et al. Studies on the development of wear-resistant plastic acetabular caps for total hip arthroplasty. J Jpn Orthop Assoc. 1975;49:673–4. Japanese
8. Oonishi H, et al. The optimum dose of gamma radiation-heavy doses to low wear polyethylene in total hip prostheses. J Mater Sci Mater Med. 1997;8:11–8.
9. Oonishi H, et al. Gamma-irradiated cross-linked polyethylene in total hip replacements – analysis of retrieved sockets after long-term. J Biomed Mater Res. 2001;58:167–71.
10. Sugano N, et al. Analysis of a retrieved UHMWPE acetabular cup crosslinked in air with 1000 kGy of gamma radiation. J Orthop Res. 2004;22:828–31.
11. Moro T, et al. Surface grafting of artificial joints with a biocompatible polymer for preventing periprosthetic osteolysis. Nat Mater. 2004;3:829–36.
12. Moro T, et al. Clinical safety and wear resistance of a phospholipid polymer-grafted highly cross-linked polyethylene liner. J Orthop Res. 2017;35:2007–16.

13. Onishi H, editor. Ceramic hip joint prostheses. The results and analysis of artificial joints used in Japan for more than 20 years. Shiga: Celtesco Medical Engineering Corporation; 2014. p. 187–96. Japanese
14. Okawa T, et al. Long-term results of Bioceram stem by first generation cementing technique for more than 20 years. Osteoarthr. Disability. 2007;50:1291–8. Japanese
15. Sugano N, et al. Polyethylene sockets and alumina ceramic heads in cemented total hip arthroplasty. A ten-year study. J Bone Joint Surg Br. 1995;77:548–56.
16. Sato T, et al. Wear-resistant performance of highly cross-linked and annealed ultra-high molecular weight polyethylene against ceramic heads in total hip arthroplasty. J Orthop Res. 2012;30:2031–7.
17. Haraguchi K, et al. Phase transformation of a zirconia ceramic head after total hip arthroplasty. J Bone Joint Surg Br. 2001;83:996–1000.
18. Hasegawa M, et al. Alumina ceramic-on-ceramic total hip replacement with a layered acetabular component. J Bone Joint Surg Br. 2006;88:877–82.
19. Maesawa H. Dis Med. 1972;15(Maesawa type of artificial hip joint):724–32. Japanese
20. Tanaka S. Surface replacement of the hip joint. Clin. Orthop. Relat. Res. (1976–2007). 1978;134:75–9.
21. Nishio A, et al. Socket and cup surface replacement of the hip. Clin Orthop Relat Res. 1978;134:53–8.
22. Ohnishi H, et al. The effect of hydroxyapatite coating on bone growth into porous titanium alloy implant. J Bone Joint Surg Br. 1989;71:213–6.
23. Hayashi K, et al. Effect of surface roughness of hydroxyapatite-coated titanium on the bone-implant interface shear strength. Biomaterials. 1994;15:1187–91.
24. Manley MT, et al. Fixation of acetabular cups without cement in total hip arthroplasty. A comparison of three different implant surfaces at a minimum. J Bone Joint Surg Am. 1998;80:1175–85.
25. Nakashima Y, et al. The results at a minimum of 10 years of follow-up for AMS and PerFix HA-coated cementless total hip arthroplasty: impact of. J Orthop Sci. 2013;18:962–8.
26. Ido K, et al. Cementless total hip replacement. Bioactive glass ceramic coating studied in dogs. Acta Orthop Scand. 1993;64:607–12.
27. So K, et al. Good short-term outcome of primary total hip arthroplasty with cementless bioactive glass ceramic bottom-coated implants: 109 hips. Acta Orthop. 2012;83:599–603.
28. So K, et al. Is the bone-bonding ability of a cementless total hip prosthesis enhanced by alkaline and heat treatments? Clin Orthop Relat Res. 2013;471:3847–55.
29. Kamizono J, et al. Experience with the Zweymuller-type stem Elance. ++Hip Joint. 2016;42:693–6. Japanese
30. Konishi T, et al. Short-term results of SQRUM TT in patients with rheumatoid arthritis. J. Orthop. Sci. 2020;50:772–4. Japanese
31. Noble PC, et al. Three-dimensional shape of the dysplastic femur: implications for THR. Clin Orthop Relat Res. 2003;417:27–40.
32. Kaneuji A, Matsumoto T, Nishino M, Miura T, Sugimori T, Tomita K. Three-dimensional morphological analysis of the proximal femoral canal using a computer-aided design system in Japanese patients with osteoarthrosis of the hip. J Orthop Sci. 2000;5:361–8.
33. Onishi H, et al., editors. SOM-type hip prosthesis. The results and analysis of artificial joints used in Japan for more than 20 years. Shiga: CellTesco Medical Engineering Corporation; 2014. p. 61–8. Japanese
34. Takaoka Y, et al. The long-term results of total hip arthroplasty with a cemented β-titanium stem. J Arthroplast. 2020;35:2167–72.
35. Ishii M, et al. Long-term results of Kyocera-type cemented hip prosthesis: YU-type hip prosthesis. Seikei Saigai Geka. 2007;50:1281–9. Japanese
36. Shirai T, et al. Antibacterial iodine-supported titanium implants. Acta Biomater. 2011;7:1928–33.
37. Kabata T, et al. Iodine-supported hip implants: Short term clinical results. Biomed Res Int. 2015;2015:368124.
38. Noda I, et al. Development of novel thermal sprayed antibacterial coating and evaluation of release properties of silver ions. Appl. Biomater. 2009;89:456–65.

3 Jikei-Type Artificial Hip Joint and Maesawa-Type Axial-Rotation Type Artificial Hip Joint (Nemoto Shokai)

Hajime Sugiyama

The history of the development of THA can be seen in the cup arthroplasty by Smith-Petersen in 1939 [1], vitallium prosthesis by Moore, and acrylic prosthesis by Judet, which were clinically applied. Additionally, In 1956, Mckee and Farrar developed a THA using a combination of a metal socket and a metal head [2]. In 1961, Charnley developed low friction arthroplasty (THA) using a combination of a polyethylene socket and a metal head [3], and Weber-Hugller's THA using a metal socket and a polyethylene head [4]. At the Jikei University School of Medicine, the Katayama-Hirakawa head prosthesis (initially hand-made and later ready-made) using acrylic resin (methyl methacrylate) was developed for the treatment of hip tuberculosis during the tenure of Professor Ryosuke Katayama (1945–1966) [5] and was also used for osteoarthritis of the hip joint (hip joint disease) [6]. Furthermore, The Jikei-type cup arthroplasty was also developed and clinically applied [7]. In the 1960s, the Charnley-type THA was introduced and attracted much attention, and the research and development of THA began in our department. The Jikei-type THA and Maesawa-type axial THA were put into practical use based on the Charnley and Weber-Hugller types. Both THAs were manufactured by NEMOTO SHOKAI Co. and clinical applications were reported at the first meeting of the Japanese Society for Artificial Joint Research (1972). Both THAs have the same concept of cementless fixation with three spikes on the acetabular component, but the sliding surfaces of both THAs are quite different. In this chapter, both THAs are introduced and the process of their development and changes are described.

H. Sugiyama (✉)
Kanagawa Rehabilitation Hospital, Atsugi, Japan
e-mail: sugiyama4460@kanagawa-rehab.or.jp

1 Jikei-Type THA (Manufactured by Nemoto Shokai)

During the tenure of Professor Yasuto Itami (1966–1981), the Jikei -type THA was developed, drawing inspiration from the Charnley-type THA [8]. The core concept was to position the joint prosthesis in a more biologically and biomechanically natural manner. The most important concept was to use a cementless type that fixes the prosthesis to the bone without using cement. To achieve this concept, a three-spike THA that can fix the socket stably to the acetabulum was developed as the second concept. This was an important concept for the stable fixation of the socket because 80% of hip arthrosis in Japan is secondary to acetabular dysplasia, whereas 80% of hip arthrosis in the United States and Europe is primary. This has resulted in the current metal-back system, which uses metal to reinforce the deformational weakness of plastic artificial cartilage while creating a plastic-to-metal combination. The outer diameter of the socket was 42 mm except for the first few cases, which were made of pure titanium, then changed to a titanium alloy (STA-130) in 1972, and then to stainless steel (SUS 316L) in 1977 [9]. The stem side also has a self-locking hole based on the concept of stable fixation in bone without the use of cement, and a collar that fits the osteotomy surface is set at 45 to the stem so that the load is applied to the thick bone cortex of the lesser trochanter. Notches were made on the anterior and posterior surfaces of the collar to prevent rotation, and the stem surface had a groove where macro-anchoring could be expected [10].

The development of the THA began in 1968, but clinical application began in 1970. Figure 3.1 shows the first Jikei-type hip joint prosthesis. The socket is a metal-back type made of pure titanium with three spikes, which is similar to the current THA. However, in the first 11 cases, Teflon was used for the acetabular joint instead of polyeth-

Y. Nakashima (ed.), *Advances in Total Hip Arthroplasty*, https://doi.org/10.1007/978-981-97-2461-1_3

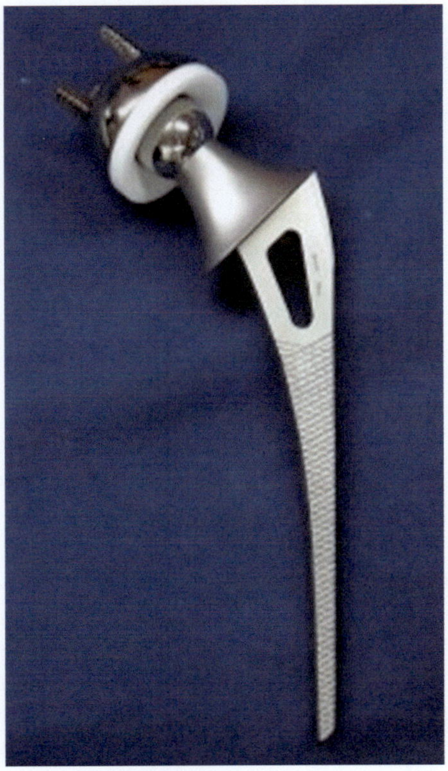

Fig. 3.1 The first Jikei-type THA. The metal socket and stem were made of titanium, and Teflon was used for the liner in the first 11 cases

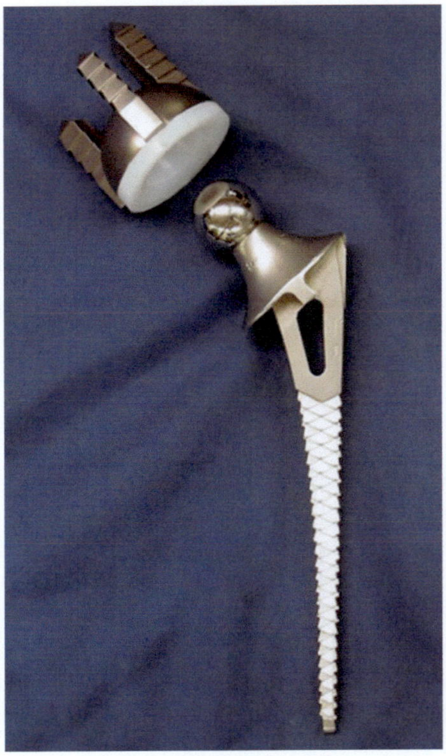

Fig. 3.2 Jikei-type THA. This is a typical Jikei type used since 1972. The stem was one size, but in 1983, four sizes were added, and the indentation became larger, and it has been used in about 800 cases

ylene, as was the case with the Charnley type, resulting in significant wear and laxity, so the Teflon was replaced by polyethylene in late 1971. The diameter of the head was rounded up from 22.225 mm of the Charnley type to 23 mm, aiming for a low-wear THA. In 1972, the spike of the artificial acetabular cup underwent enlargement to accommodate prevalent acetabular dysplasia in Japan, coupled with a shape tailored for macro-anchoring (Fig. 3.2). By 1974, the transition from fragile pure titanium to stainless steel was completed. Furthermore, 1977 witnessed the adoption of four varying sizes for the stem, alongside an enlarged stem notch aimed at enhancing fixation (Fig. 3.3) [11].

This type of THA had been used in more than 2000 cases by 1995, and positive initial clinical outcomes were reported [12, 13]. The JOA score (Japanese Orthopedic Association criteria) stood at 75.1 after 10 years and 73.6 after 15 years, with revision rates relatively stable at about 7% [14]. The author attributes these results to surgical technique involving full hip joint deployment via Ollier variant keratotomy and greater trochanter dissection. Additionally a prolonged postoperative therapy of 3 to 4 months, during which macro-anchoring was gradually accomplished, contributed to these outcomes. Despite minimal revision cases, migration of the prosthesis stem was observed [15], often linked to the shortened postoperative therapy duration. Akamatsu made a Y2-type (University of Yamanashi type) THA [16], lengthened the stem length and changed the angle of the collar from 45° to 35°, and Itami changed the angle of the collar from 45° to 23° to make the third-generation Jikei- type THA [17], and later developed the Kyocera JOHP (Japanese Itami changed the angle of the collar from 45° to 23° and developed the third-generation Jikei-type THA [17]. The change to JOHP was due to the fact that some long-term cases of stem fracture were found to be caused by macro-anchoring, and Jikei University School of Medicine also changed to the fourth generation (new Jikei-type) THA [18]. However, the results were lower than those of THA with a micro-anchoring system, as typified by the AML THA system, which became the mainstream afterwards.

In collaboration with Tomita, the author improved the THA Jikei-type by incorporating a micro-anchoring surface structure, achieved by eliminating the collar [19]. The same approach was applied to Y2 in collaboration with Hamada, resulting in the creation of the Y3 type (Fig. 3.4) [20]. Although the Jikei-type THA underwent many changes since its clinical application in 1970, the core concept of cementless fixation and initial three-spike stabilization f has remained unchanged since its development, playing a pivotal role in the history of THA development.

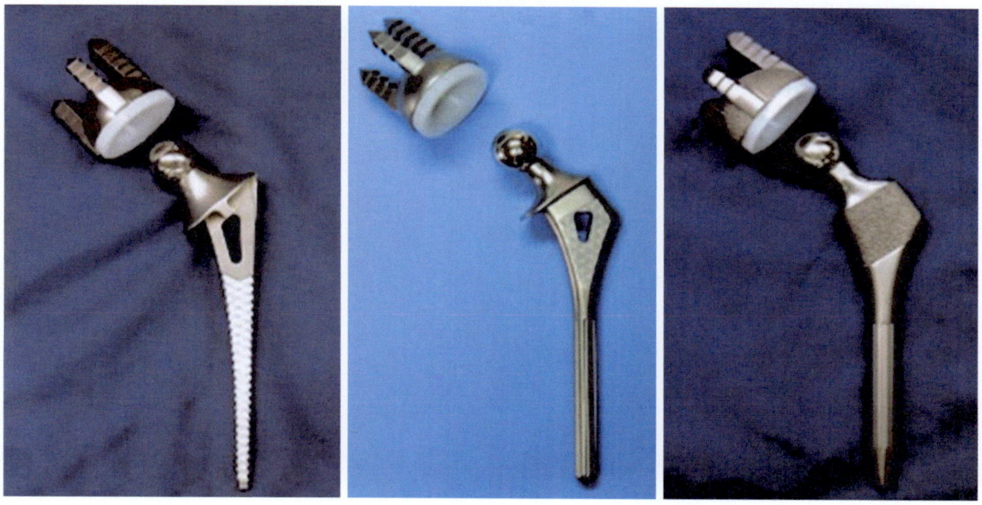

Fig. 3.3 Transition of Jikei-type THA. New Jikei-type THA with improved rotational fixation of the stem to avoid stem breakage (used in about 800 cases) and JT-type THA with Kyocera porous coating (used in about 300 cases)

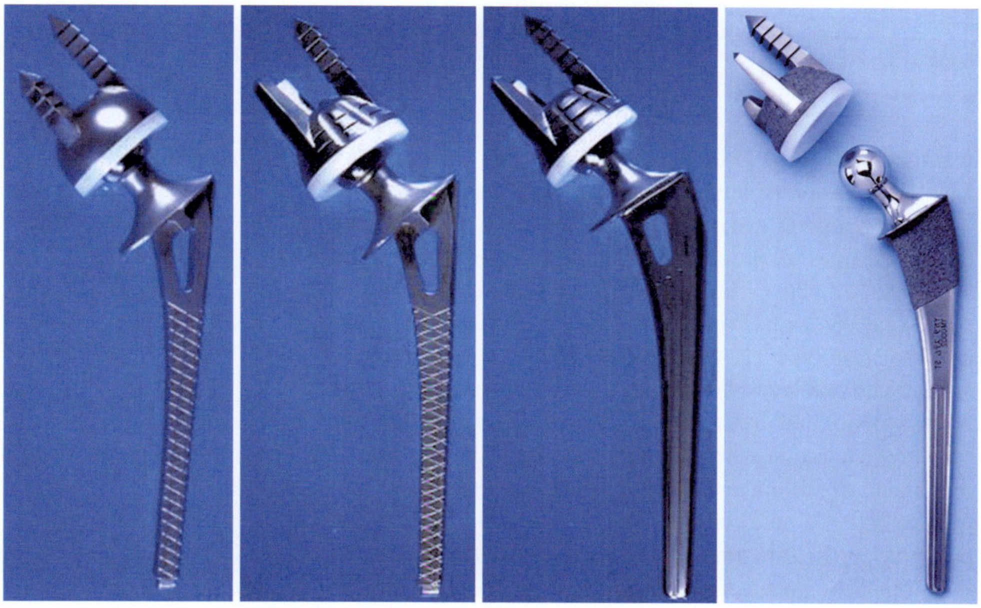

Fig. 3.4 Transition of the Yamanashi University THA. From the left, the Y-type with a slimmer acetabular prosthesis, the Y2-type with a longer stem without incisions, and the Y3-type with a porous coating, which has been used in approximately 700 cases

2 Maesawa Axial-Rotation Type THA (Manufactured by Nemoto Shokai)

Under the guidance of Professor Emeritus Katayama from Jikei University School of Medicine, Maesawa developed a three-spike THA combining a metal socket and a metal head based on the Mckee-Farrar THA and has used it in 40 joints in 36 cases since 1964 [21]. The results were not always good, but Maezawa thought that the cause was the sliding surface between metals and developed an axial-rotation type THA using a polyethylene head (Fig. 3.5), which has been used since 1970 [22]. This THA is a combination of a metallic artificial acetabulum and a polyethylene head, and the central part of the polyethylene head and stem rotates in accordance with the axis of the neck. The polyethylene head is similar to the Weber-Huggler type developed in Europe at that time and has a feature of a bearing-type joint called a trunnion-bearing between the head and the femoral stem. The movable collar is placed at the center of the stem to prevent rotation and to disperse the stress on the stem. The diameter of the head was 38 mm, which was larger than that of the Charnley type and the Jikei- type and was less prone to dislocation. In addition to the Weber-Huggler type, the trunnion-bearing type includes the Christiansen type (metallic head) [23], and the Gifu University type THA has been clinically applied in Japan. This THA was developed as a successor to the first THA with a combination of a metal socket and a metal head developed in Japan, and according to the data of Nemoto Shokai, there were more than 580 cases

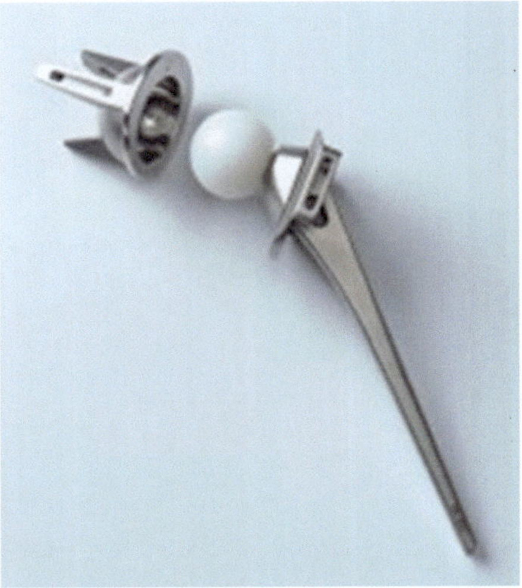

Fig. 3.5 Maesawa axial-rotation type THA. The axial-rotation type involves rotation between the polyethylene bone head and the shaft of the stem. 580 cases of this type have been manufactured by Nemoto Shokai

manufactured from 1970 to 2007. According to Katayama Memorial Hospital, 85% of patients who underwent axial THA are satisfied with the results. The large polyethylene head is designed to prevent dislocation, and the polyethylene head rotates on the cervical axis of the artificial head to minimize wear and avoid complications due to wear debris. This axial-rotation type THA is designed to have a structure and functionality closer to the natural hip joint, leveraging a large polyethylene head to discourage dislocation.. In addition, the polyethylene head rotates axially on the neck axis of the artificial head to minimize wear and prevent complications due to wear debris. Some consider the mechanical strength of the polyethylene head is a problem, but Maezawa and Sasada of the Tokyo Institute of Technology have shown that the Maezawa THA is sufficiently durable in a durability test using a THA simulator, and a study of extracted polyethylene heads by Katayama et al. has shown excellent results regarding its durability [24]. This model had an advanced aspect in that it attempted to deal with dislocation by using a large 38-mm head diameter, and it is also the origin of the double-bearing THA that is currently used mainly in older adults.

The development and evolution of Nemoto Shokai's THA at Jikei University School of Medicine have been described, but other than the development with Nemoto Shokai, the late Prof. Katsuyuki Fujii developed Tri-fix J manufactured by MDM In this prosthesis, polyethylene is directly fixed to a metal socket to reduce the backside wear. Dr. Itami developed the JOHP, manufactured by Kyocera, by modifying the collar angle to 23°, as described earlier. These two THAs have a comparable head diameter of 22.225 mm with other THAs. Moreover, sizes like 28 mm can be used for revision purposes, and they are mindful of the requirement for interchangeability of the prosthesis and stem among different manufacturers.

References

1. Smith-Petersen MN. Evolution of mould arthroplasty of the hip joint. J Bone Joint Surg Br. 1948;30:59–75.
2. Mckee GK, et al. Replacement of arthritic hips by the McKee-Farrar prosthesis. J Bone Joint Surg Br. 1966;48:245–59.
3. Charnley J. Arthroplasty of the hip: a new operation. Lancet. 1961;1(7187):1129–32.
4. Weber BG. Total hip replacement with rotational-endoprosthesis. Clin Orthop Relat Res. 1970;72:79–84.
5. Hirakawa H, et al. Preparation of acrylic resin artificial bone heads in our laboratory and experience in their application to osteoarticular tuberculectomy. J Jpn Orthop Ass. 1955;29:186–97. In Japanese
6. Takemura S. Remote results of the acrylic resin hip prosthesis (Katayama-Hirakawa type) and the new Jikei large (Itami type) hip prosthesis made of titanium. J Jpn Orthop Ass. 1974;48:419–45. In Japanese
7. Hirakawa H, et al. Experience with cup arthroplasty. Orthop Surg. 1963;14:11–7. In Japanese
8. Itami Y, et al. Artificial heads and joint prostheses. Rin to Kenkyu. 1970;47:358–66. In Japanese
9. Itami Y, et al. Preliminary report of our total hip prosthesis. Z Orthop Ihre Grenzgeb. 1976;114:719–21.
10. Akamatsu N, et al. Total hip replacement through self-locking fixation. Int Surg. 1978;63:17–21.
11. Itami Y, et al. The direct fixation system of total hip prosthesis. Research of the possibilities of fixing total hip prosthesis directly to bone. Orthop Trauma Surg. 1982;100:11–7.
12. Tomita Y, et al. Clinical results of hip and knee joint prostheses in Japan. Artif Organs. 1976;5:18–23. In Japanese
13. Itami Y, et al. A clinical study of the results of cementless total hip replacement. Arch Orthop Trauma Surg. 1983;102:1–10.
14. Sugiyama H, et al. Long-term results of Jikei hip arthroplasty. J Jpn Surg Fed. 1991;23:69–71. In Japanese
15. Sugiyama H, et al. Stem fixation and design of cementless hip arthroplasty. In: Intensive discussion on artificial hip joints. Japan Hip Joint Reconstruction Foundation; 1991. p. 151–60. In Japanese.
16. Hamada R, et al. Long-term results of smooth surface cementless hip prosthesis. Chubu Rheumatol. 2005;36:87–93. In Japanese
17. Kutsuna T, et al. Development and clinical application of the three-spike JOHP hip prosthesis. J Jpn Artif Joint Soc. 1998;28:183–4. In Japanese
18. Sugiyama H, et al. Torsional fixation of the femoral component in cementless total hip arthroplasty: newly designed funnel-shaped femoral component. In: Hip biomechanics. Tokyo: Springer; 1993. p. 313–20.
19. Roppongi T, et al. Development of the J-T type hip prosthesis and its radiological study. In Japanese.
20. Horiuchi T, et al. Mid-term results of Y3-type hip prosthesis. In Japanese.
21. Maezawa H, et al. Artificial hip joints. J Jpn Orthop Ass. 1970;44:727–30. In Japanese
22. Maezawa H. The Maesawa type of artificial hip joint. Jpn J Dis Med. 1972;15:724–32. In Japanese
23. Christiansen T. A new hip prosthesis with trunnion-bearing. Acta Chir Scand. 1969;135:43–6.
24. Katayama K, et al. Physicochemical and biomechanical examination of surfaces of retrieved polyethylene heads from total hip prostheses with rotating. J Orthop Sci. 2001;6:503–9.

SOM and KYM Hip Prostheses (Mizuho)

4

Nobuhiko Sugano

1 Development Process

In the 1970s, the introduction of hip prostheses from Europe spread to Japan, but European hip prostheses designed by McKee-Farrar, Charnley, Weber-Huggler, Müller, and others had problems with fitting the hip joints of Japanese patients, most of which were affected by osteoarthritis secondary to hip dysplasia. Shikita was the first to clinically use the Charnley-type hip prosthesis with a metal head versus a polyethylene cup in Japan in February 1968; he switched to the Müller and Weber-Huggler types because of the poor design of the Charnley-type hip prosthesis, which resulted in wear due to the increased surface pressure of the small 22 mm diameter head and dislocation due to the limited range of motion [1]. Concurrently, because polyethylene wear is related to the survival of the artificial hip joint due to wear or breakage and foreign body reactions caused by wear debris, he conducted research on improving wear resistance by cross-linking polyethylene using an electron beam or gamma-ray irradiation and investigated the introduction of high-density polyethylene (HDP), ultra-high molecular weight polyethylene (UHMWPE), and other polymeric materials, such as carbide [2]. Shikita developed the SOM (Shikita-Onishi-Mizuho) hip prosthesis, a domestically produced prosthesis using COP stainless steel (CoNiCr alloy), which had already been reported to be an excellent material for bone fracture fixation devices by Sunami et al. [3]. COP stainless steel is an excellent material that can be polished to achieve an appropriate level of sphericity and surface roughness of the femoral head for sliding parts. The diameters of the femoral head were 28 mm and 32 mm, and a 36 mm head was also fabricated for the large acetabulum to make it like the anatomical form of the hip joint. The stem design was a straight type, and the femoral neck had three different lengths, according to Müller, to reduce the dead space and prevent dislocation by adjusting the leg length. Three types of stem lengths (large, medium, and small) and a mobile polyethylene femoral head around the neck trunnion in combination with a metal articulation cup (KYM: Kumamoto-Yao-Mizuho), like the Weber-Huggler type, were also developed by Hirota to enhance compatibility. The SOM type has been used clinically since 1971 [3]. In the early stage, HDP of HI-ZEX5000B was used as the cup material, RCH1000 (GUR412) of UHMWP was used to improve the wear resistance property [4], and a high dose of gamma-ray irradiated UHMWPE in air was used in 1972. KYM has been clinically used with a 37-mm polyethylene head and a COP cup from 1973 [5].

2 Features of SOM and KYM

The SOM stems were made of COP steel and the cup material was changed from the initial HDP to UHMWP. The head diameter is available in 28 mm, 32 mm, and 36 mm sizes. There were three stem lengths (from the medial collar to the stem tip) of 95 mm, 105 mm, and 115 mm, with a neck shaft angle of 135°. The neck lengths were 10 mm, 15 mm, 20 mm, and 25 mm (Fig. 4.1). The stem has a dent on the anterior and posterior surfaces to improve cementation. The standard cup has an outer diameter of 46 mm for a 28 mm head and 50 mm for a 32 mm head, and a metal wire marker was attached to the periphery of the cup rim to measure the socket angle and wear on postoperative X-rays (Fig. 4.2). The socket with a 36 mm diameter was eccentric (with an outer diameter of 48.5 mm), the polyethylene thickness at the apex of the dome

N. Sugano (✉)
Department of Orthopaedic Medical Engineering, Osaka University Graduate School of Medicine, Osaka, Japan
e-mail: n-sugano@umin.net

Y. Nakashima (ed.), *Advances in Total Hip Arthroplasty*, https://doi.org/10.1007/973-981-97-2461-1_4

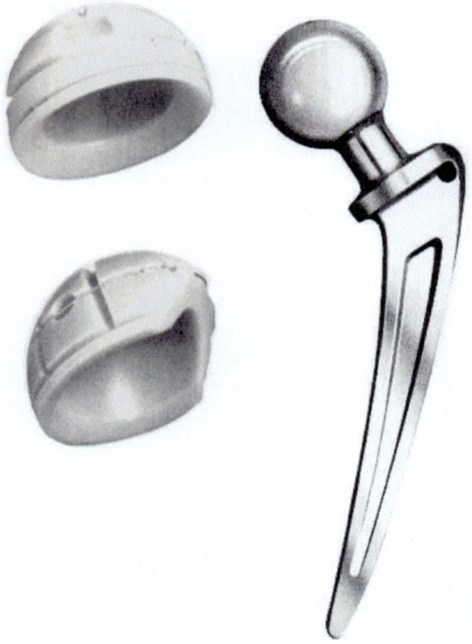

SOM hip stem dimension

Type	A	B	C	D	E	F	G	H	I	J
Head diam. (mm)	28	28	28	28	32	32	32	32	36	36
Neck length (mm)	10	15	20	25	10	15	20	25	20	25
Stem length (mm)	95	95	95	105	105	105	105	115	115	115

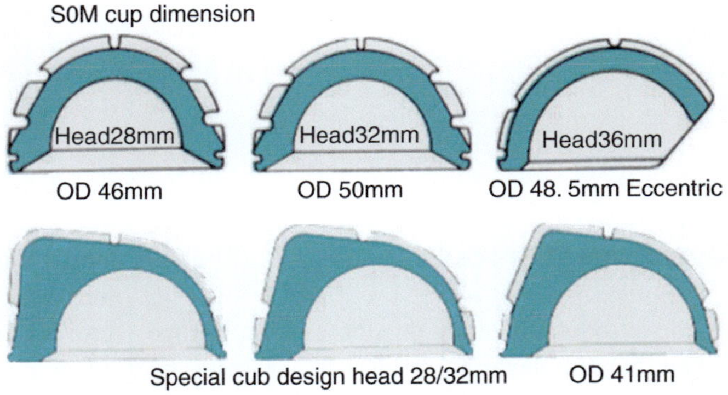

Fig. 4.1 Shikita-Onishi-Mizuho (SOM) total hip prosthesis

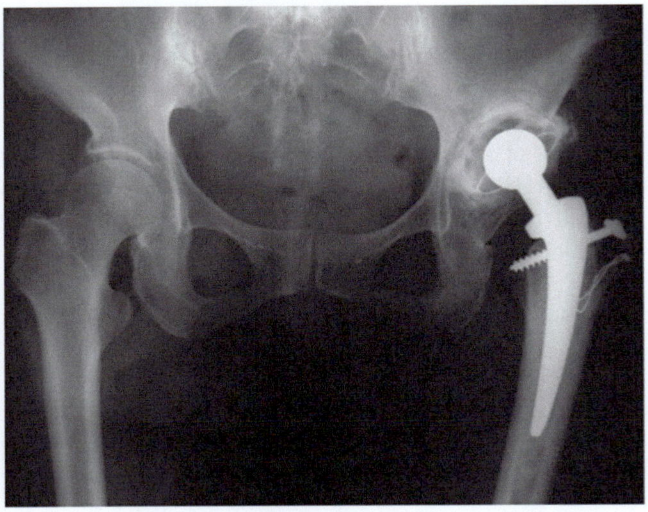

Fig. 4.2 Seven-year postoperative radiograph of a patient who underwent total hip arthroplasty using Shikita-Onishi-Mizuho (SOM) hip prosthesis in 1975

was increased, and the antero-superior portion of the rim was cut off to prevent postoperative groin pain (Fig. 4.1). A deformed socket with an outer diameter of 41 mm and a head diameter of 28 mm or 32 mm was also made for the hypoplastic acetabulum.

While the SOM type was a metal-on-poly joint, the KYM type was a poly on-metal joint like the Weber-Huggler type, and the polyethylene head was designed to rotate around the polished stem neck, which was tapered to reduce the surface friction of the head (Fig. 4.3). The COP stem has the same medullary geometry as the SOM and is available at three different lengths. The stem neck is 16 mm in diameter, polished, and mirror-finished, and a 37 mm diameter polyethylene head with three different neck lengths is modularly attached to the stem neck in a loose fit. The cup was made of COP, 47 mm O.D., and 37 mm I.D. The inner surface was polished and mirror finished. The outer surface of the socket exhibited 30 spikes for cementation. The socket was 3 mm thick for easy placement in the acetabulum.

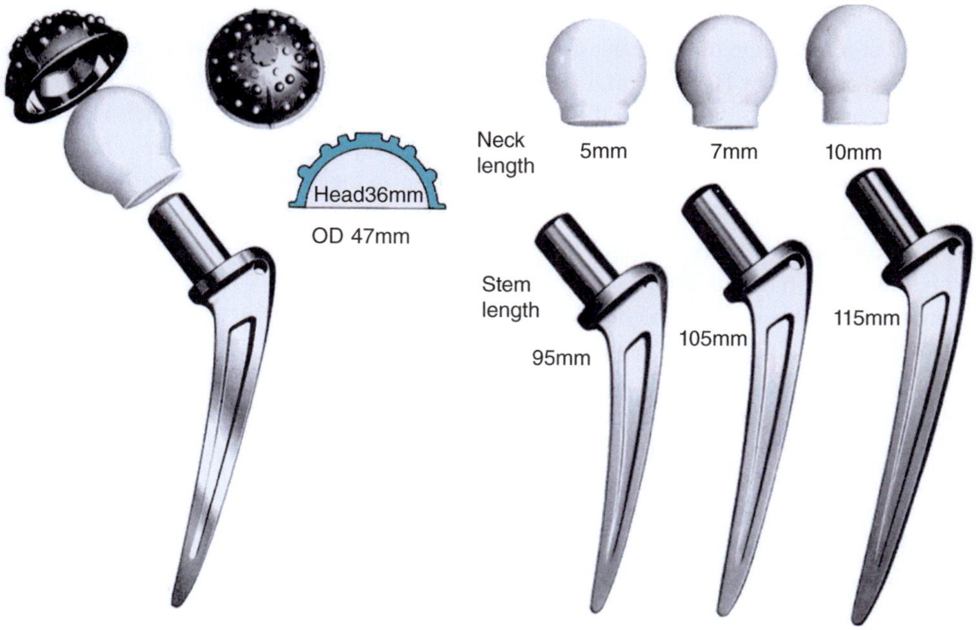

Fig. 4.3 Kumamoto-Yao-Mizuho (KYM) total hip prosthesis

3 Clinical Results

The report of the initial results of SOM in 27 hips for femoral neck fracture and hip arthrosis in elderly patients by Shikita et al. was good and stated that it was easier to place than Weber-Huggler and Müller [6]. Shikita et al. performed SOM-type THA in 43 hips of 39 patients between March 1973 and January 1979. Sixteen hips of 15 patients were directly examined at an average of 11 years and 5 months, and the clinical results were very good [7]. In a report of 146 SOM-type hip arthroplasties performed at National Osaka Minami Hospital, 89 hips in 80 patients were followed up for 5 to 16 years (average, 10 years), and radiological cup loosening was observed in 19 hips, of which revision hip arthroplasty was performed in 6 hips. Stem revision was performed in 7 hips, with 4 hips showing loosening and 3 hips showing stem fracture [8]. Cup loosening was more common in patients with a cup inclination over 60°. In a subsequent report, 37 of the 60 hips that were followed up for more than 9 years, 19 of which were confirmed to have gamma-irradiated cups, were compared with those with trapezoidal and Bioceram cups with the same 28-mm head. The annual wear rate of the gamma-irradiated SOM-type cup was 0.076 mm, which clinically confirmed its superiority [9]. In a study comparing the oxidation indices of gamma-ray 100 Mrad irradiated cups removed 15 and 16 years after surgery with the removed cups of other models, there was no difference in the oxidation indices on the sliding surface regardless of the model, and the gamma-ray 100 Mrad irradiated cups showed a very high oxidation index in the non-wearing area [10]. The results of a comparison between a 24-year postoperative sample of a 100 Mrad-irradiated SOM-type cup and a 20-year postoperative retrieved cup of the same type of normally sterilized cup showed that the oxidation index of the backside of the socket was very high in all samples, with the lowest oxidation index at a depth of 1 mm to 3 mm from the sliding surface, and in this area, the cups 100 Mrad-irradiated in air also showed a very high oxidation index [11]. This suggests that the improvement in wear resistance due to cross-linking far exceeded the oxidative degradation effect of the high level of gamma-ray irradiation.

Shikita et al. performed SOM-type THA in 29 hips at Osaka University from 1979 to 1980. The age at surgery ranged from 35 to 80 years (mean, 55 years), with eight males and 21 females. The cup was UHMWP gamma-sterilized, and COP 28-mm heads were mainly used; however, 32-mm heads were used in five male hips. During the follow-up 12 years postoperatively, both components had been revised in four cases due to cup and stem loosening, and the cup alone had been revised in three cases due to cup loosening. In cases in which only the cup was revised, stem fixation was good afterward, and the joint functioned in some patients for up to 28 years after the initial surgery. Stem loosening was recorded as being due to insufficient cementation of the stem tip.

THA with KY was performed on 44 hips between July 1973 and September 1977 at the Kumamoto Municipal Hospital, and the mean age at surgery ranged from 23 to 81 years (mean, 63 years). Twelve hips in 10 patients had primary osteoarthritis at a mean age of 66 years; 7 hips in 6 patients had osteoarthritis secondary to hip dysplasia at a mean age of 57 years (47 to 64); 8 hips in 6 patients had rheumatoid arthritis at a mean age of 56 years (23 to 72); 3 hips in 2 patients had osteonecrosis of the femoral head at a mean age of 38 years; 8 hips in 7 patients had femoral neck fracture at a mean age of 73 years; and 6 hips in 6 patients were converted from HDP femoral head prosthesis due to wear at a mean age of 70 years. Of the 18 hips followed up for more than 3 years, revision was performed in three hips, but there was no evidence of polyethylene head wear at the sliding area with the metal cup [12]. The three cases of revision were as follows: a case in which the cup was placed with a high inclination due to insufficient acetabular reaming for primary osteoarthritis, and the cup loosened at 1 year and 2 months; a case in which the stem broke due to a fall during farm work at 2 years and 5 months postoperatively, which was revised with a long stem; and a case in which the femoral head was revised 2 years and 9 months after surgery because of frictional wear between the bone at the neck of the head and the neck of the polyethylene head, resulting in osteolysis of the femur [13]. In addition, the number of surgical cases increased to 79 hips, and in the analysis of 28 hips followed up for more than 5 years, the number of revisions was 4. Only one case of femoral fracture due to a fall from stairs was recorded from the previous report, and the remaining hips showed generally good mid-term results [14].

4 Summary

The development of SOM and KYM hip prostheses with a large head diameter and improved wear resistance by gamma-ray irradiation of highly cross-linked UHMWP was a novel idea at the time and marked the beginning of epoch-making Japanese hip prostheses. The COP steel-collared cemented stem is well-suited to the Japanese medullary cavity and has shown good long-term stability as a composite beam-type cemented stem. After examining the problems of cement fixation, the application of a ceramic coating to the metal to provide cementless fixation for young patients was also remarkable [3]. The polyethylene described by Charnley as HDP is actually UHMWP, and Shikita has been researching the improvement in wear resistance by gamma-ray irradiation of various polymer materials since 1970. He was aware of the differences between HDP and UHMWP and arrived at an excellent combination of ceramic femoral heads and gamma-irradiated UHMWP cups [15]. It is impressive that the prototype of the large-diameter ceramic head versus the gamma-irradiated highly cross-linked UHMWP cup hip prosthesis, which has been increasingly used over the past 10 years, was created 40 years ago. It is interesting to observe the long-term results of the KYM-type highly cross-linked UHMWP head and the condition of the extracted polyethylene head, which is also like the dual mobility type of the large-diameter highly cross-linked UHMWP head. Unfortunately, domestic hip prostheses with these innovative designs did not become the global standard in the 1980s.

References

1. Shikida T, et al. Clinical and material mechanical problems in total hip arthroplasty. Central Jpn Ass Orthop Surg Traumatol. 1970;13:172–4. In Japanese
2. Shikita T. Wear and lubrication in total hip arthroplasty. Wear and lubrication in total hip prostheses. Saigai Igaku. 1972;15:683–96. In Japanese
3. Shikida T, et al. Problems in total hip arthroplasty, prosthesis affiliation, dust-free operating room, and autologous anabolic adhesion. Central Jpn Ass Orthop Surg Traumatol. 1972;15:612–4. In Japanese
4. Shikida T, et al. On the SOM and KYM types of total hip arthroplasty. Iryo Kikigaku. 1973;47:368–70. In Japanese
5. Hirota K, et al. Experience with weber-Huggler type hip arthroplasty. J Orthop Dis. 1975;24:100–5. In Japanese
6. Shikita T, et al. Experience of total hip arthroplasty in the elderly. Central Jpn Ass Orthop Surg Traumatol. 1976;19:265–6. In Japanese
7. Sasaki K, et al. Long-term results of SOM-type hip arthroplasty. Nihon Riumachi Kansetsu Geka Gakkaizasshi. 1988;7:365–70. In Japanese
8. Aono M, et al. Postoperative results of SOM-type hip prosthesis for osteoarthritis of the hip. Central Jpn Ass Orthop Surg Traumatol. 1989;32:1463–5. In Japanese
9. Takayama M. A study on the amount of socket wear after total hip arthroplasty. Nihon Riumachi Kansetsu Geka Gakkaizasshi. 1991;10:469–80. In Japanese
10. Oonishi H, et al. Total hip arthroplasty with a modified cementing technique using hydroxyapatite granules. J Arthroplast. 2001;16:784–9.
11. Sugano N, et al. Analysis of a retrieved UHMWPE acetabular cup crosslinked in air with 1000 kGy of gamma radiation. J Orthop Res. 2004;22:828–31.
12. Hirota K, et al. Experience with rotational hip prostheses and short-term prognostic results. Seikei Geka to Saigai Geka. 1978;27:22–5. In Japanese
13. Hirota K, et al. Our experience with K. Y. M. Complications of rotational total hip arthroplasty. Seikei Geka to Saigai Geka. 1978;27:26–8. In Japanese
14. Hirota K, et al. Mid-term outcome of rotational hip prosthesis (KYM type). Seikei Geka to Saigai Geka. 1982;31:210–3. In Japanese
15. Shikita T, et al. Conversion of prosthetic joint materials from metallic to nonmetallic materials. Iryou. 1978;32:449–56. In Japanese

Nishio's Resurfacing Hip Arthroplasty (Mizuho)

5

Goro Motomura

1 Background

In Japan, around 1970, total hip arthroplasty (THA) was already being performed for patients aged 65 years or older, but it was not suitable for adolescent patients due to poor durability. Therefore, arthrodesis or cup arthroplasty was performed for end-stage hip arthritis in adolescent patients who would not benefit from osteotomy or acetabuloplasty. Cup arthroplasty was a "hard-to abandon" treatment according to Nishio's description [1], but its inconsistent results and complicated postoperative treatment led to the development of socket-and-cup resurfacing hip arthroplasty for the improvement of cup arthroplasty. In a special issue on the resurfacing hip arthroplasty of *Clinical Orthopedics and Related Research* published in 1978, in addition to reports from Europe and the United States, Nishio [2], Furuya [3], affiliated with Tokyo Medical and Dental University, and Tanaka [4], affiliated with Kinki University, gave reports from Japan, and each of them had developed and improved their own implants.

2 Features of Nishio's Resurfacing Hip Arthroplasty

In Nishio's resurfacing hip arthroplasty, a metallic socket with three spikes and a concentric metallic cup were fixed to the acetabulum and head, respectively, and mobility was achieved between the socket and cup [2]. This arthroplasty has been performed since 1972, and since October 1975, a high-density polyethylene with a thickness of approximately 4 mm has been inserted (Figs. 5.1 and 5.2) [5]. While most resurfacing THAs developed around the same time used bone cement for implant fixation [6], the Nishio-type arthroplasty was developed based on the concept of not using bone cement [1]. In the first report of this surgery, published in *The Journal of the Japanese Orthopaedic Association* in 1973, Nishio stated that "even if failure occurs in the future, arthrodesis or THA can be performed by removing the socket and cup, so it can be used for younger patients" [7]. This suggests that bone cement was not used in future revision surgeries. According to a report that reviewed cases of revision surgery after this procedure, most revision surgeries were conversion THAs, and that technique was relatively easy [8]. One of the reasons for this is that bone cement was not used during the initial surgery [8].

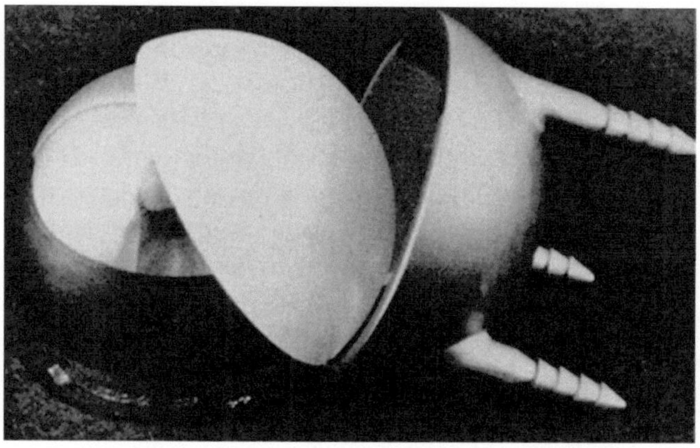

Fig. 5.1 Photographs of Nishio's resurfacing hip arthroplasty implants [5]. (Reproduced from Fig. 5.2 in Eguchi et al. [3])

G. Motomura (✉)
Department of Artificial Joints and Biomaterials, Graduate School of Medical Sciences, Kyushu University, Fukuoka, Japan
e-mail: motomura.goro.014@m.kyushu-u.ac.jp

Y. Nakashima (ed.), *Advances in Total Hip Arthroplasty*, https://doi.org/10.1007/978-981-97-2461-1_5

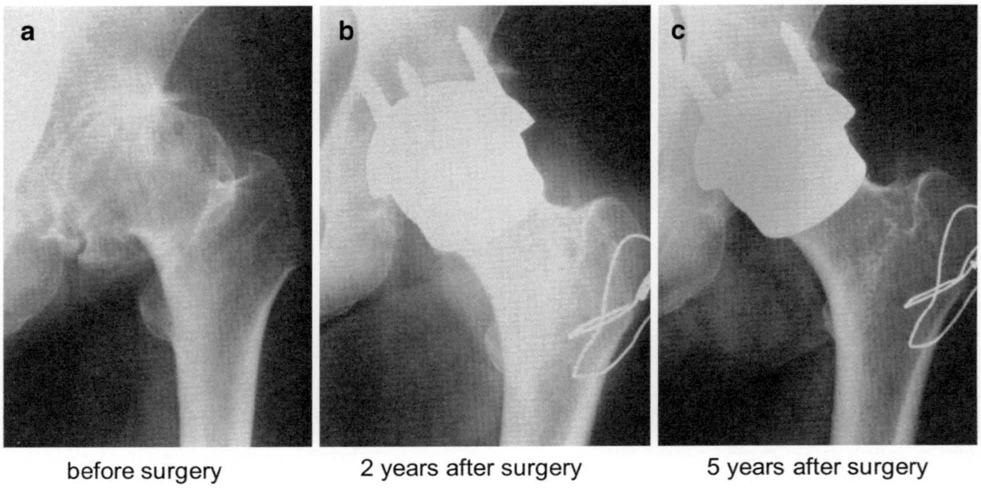

Fig. 5.2 Pre- and postoperative X-ray images of the Nishio's resurfacing hip arthroplasty [5]. Left to right: Preoperative (**a**), 2-year postoperative (**b**), and 5-year postoperative (**c**) anteroposterior radiographs of the left hip joint. The patient was pain-free for 5 years after surgery. (Reproduced from Fig. 8 in Eguchi et al. [3])

3 Clinical Results of Nishio's Resurfacing Hip Arthroplasty

This procedure was performed on 138 joints of 127 patients at Kyushu University Hospital between 1972 and 1982. The mean age at surgery was 44.5 years (range: 20–66 years).

The initial results were reported in *The Journal of the Japanese Orthopaedic Association* in 1976 [1]. Of the 41 joints that had been operated on for more than 6 months (mean follow-up: 2 years), eight had unsatisfactory results, and four of these joints were reoperated on. The goal was to place the socket as close to 40°as possible, but the socket fixation angle was significantly steeper in the poor-performing cases, suggesting that poor socket placement was a factor in the poor results. Since October 1975, high-density polyethylene inserts have been used, as described above, and the authors reported relatively good short-term results for 67 joints in 1978, including the improved type with high-density polyethylene inserts [2]. The authors stated that "Although it is not regarded as a radical operation like THA, it is a good adjuvant operation for young patients."

There have been two reports of mid- to long-term outcomes. In a follow-up study conducted in 1984 (4–12 years postoperatively), reoperation was performed on 23 (18%) of the 130 joints [8]. The average time to reoperation was 5 years and 4 months after surgery, and the reason for reoperation was recurrent pain, most of which was caused by implant loosening [8]. In 1994, a prognostic survey was conducted for 101 joints, and 57% of which underwent reoperation at an average of 9.0 years (range: 5–15 years) after surgery [9]. In contrast, the remaining 43% had no more surgeries over an average of 15.7 years. Although some loosening of the implants, such as cup subsidence and socket migration, was observed, we consider this a time-saving procedure that has achieved a certain degree of success.

References

1. Nishio A. Arthroplasty with cup and socket for osteoarthritis of the hip joint. J Jpn Orthop Assoc. 1976;50:34–6. In Japanese
2. Nishio A, et al. Socket and cup surface replacement of the hip. Clin Orthop Relat Res. 1978;134:53–8.
3. Furuya K, et al. Socket-cup arthroplasty. Clin Orthop Relat Res. 1978;134:41–4.
4. Tanaka S. Surface replacement of the hip joint. Clin Orthop Relat Res. 1978;134:75–9.
5. Eguchi M, et al. Hip arthroplasty with cup and socket. Orthop Surg. 1983;3:245–50. In Japanese
6. Freeman MA. Total surface replacement hip arthroplasty. Clin Orthop Relat Res. 1978;134:2–4.
7. Nishio A, et al. Transition and treatment of osteoarthritis of the hip. J Jpn Orthop Assoc. 1973;47:276–7. In Japanese
8. Eguchi M, et al. A study of revision surgery after socket and cup surface replacement of the hip. Clin Orthop Surg. 1985;20:623–30. In Japanese
9. Inadome T, et al. Postoperative radiological changes in the Nishio-type resurfacing hip arthroplasty. J West-Jpn Soc Orthop Traumatol. 1994;43:923–6. In Japanese

Bioceram Types 1, 2, 4, 5 (Kyocera)

6

Nobuhiko Sugano

1 Development Process

In the 1970s, the use of hip prostheses spread from Europe to Japan, and two major factions were formed, the Charnley and Müller factions. Both sides advocated the excellence of their concepts and awaited the long-term results. Shikita, who was the first to use Charnley's metallic head-on-polyethylene cup hip prosthesis in Japan in 1968, became the leader of the Müller faction. He emphasized the wear problem in Charnley's prosthesis owing to the increased surface pressure of the small 22 mm-diameter head. He researched crosslinked polyethylene using γ-ray irradiation, as polyethylene wear affects the life span of artificial hip joints [1]. In 1971, Shikita was the first to use a crosslinked polyethylene artificial hip joint with a stem fabricated from COP steel (Mizuho Medical Industry) [2]. In contrast, ceramic sliding parts gained attention in Europe for wear resistance [3]. Shikita and Kyocera held an orthopedic colloquium in 1975 and worked on the development of a ceramic hip prosthesis in Japan [4]. To reduce polyethylene cup wear by alumina ceramic heads, 32 mm and 28 mm heads were fabricated with mono- and poly-crystalline alumina. Two types of stem designs were fabricated: KC-1 (Bioceram type 1) and KC-2 (Bioceram type 2) [4]. Subsequently, based on experience with KC-2, modified Type 4 and Type 5 designs were developed to facilitate stem insertion, even in dysplastic hips [5, 6].

2 Characteristics of Bioceram Types 1, 2, 4 and 5

Type 1 was developed for highly deformed femurs and narrow medullary cavities in hip dysplasia (Fig. 6.1). The metal part was made of stainless-steel COP, which had a tensile strength of approximately twice that of previously used SUS316L. The neck-shaft angle was 120°, and the design allowed for fixing of the greater trochanter to the stem during reattachment through a transtrochanteric approach. This screw and wire combination eliminated the problem of non-union of the greater trochanter [7]. There were two stem sizes: "S" and "L." The vertical distance from the head to the collar, stem length, and internal/external stem width were 32.5, 100, and 11 mm for size S and 36, 110, and 15 mm for size L, respectively. The head was composed of polycrystalline alumina, with a diameter of 28 or 32 mm.

Type 2 was designed as a standard (Fig. 6.1). To avoid the risk of stem fracture, the stem was inserted in a valgus position, with three fixation points (proximal medial, mediolateral, and distal medial), and adequate cement injection. The anteroposterior and posterior surfaces of the stem were grooved in an oblique grid pattern to improve cement and stem fixation. The metal part was composed of COP, the neck-shaft angle was 145°, and the head was composed of monocrystalline alumina of 28 or 32 mm-diameter.

The Type 2 stem was difficult to insert in a valgus position, tended to be in a varus position, and could not be used in patients with a narrow medullary cavity. Therefore, Kyocera developed and commercialized Types 4 and 5 stems composed of COP, which are stronger than the Charnley stem and more suitable for the narrow medullary cavities in Japanese patients. The 28 mm femoral head was composed of polycrystalline alumina (AY7).

Type 4 was designed such that the endpoint of the lateral soulder of the stem was at the stem's central axis, such that the stem and femoral medullary axes aligned to pass through the trochanteric fossa. The medial curve radius was 80 mm, neck-shaft angle was 135°, and femoral neck osteotomy angle and collar were 45°. The distal stem tapered 2.5° medially, 2.2° laterally, and 2.8° anteroposteriorly. The distal stem's cross-sectional anteroposterior aspects tapered 10° with a thicker lateral dimension, creating a three-dimensional tapered stem shape. The cross-sectional corners were rounded to a radius of 6.5 mm, and the stem length was

N. Sugano (✉)
Department of Orthopaedic Medical Engineering, Osaka University Graduate School of Medicine, Osaka, Japan
e-mail: n-sugano@umin.net

Y. Nakashima (ed.), *Advances in Total Hip Arthroplasty*, https://doi.org/10.1007/978-981-97-2461-1_6

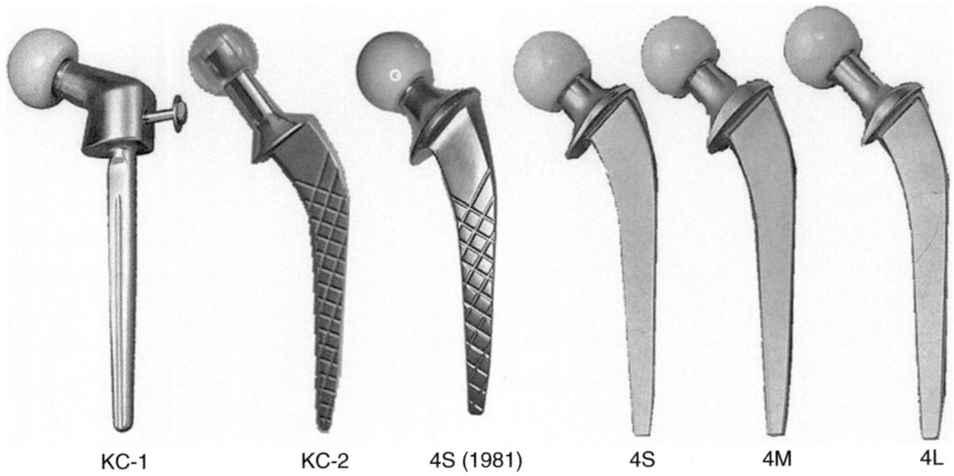

Fig. 6.1 Stems of Bioceram types 1, 2 and 4

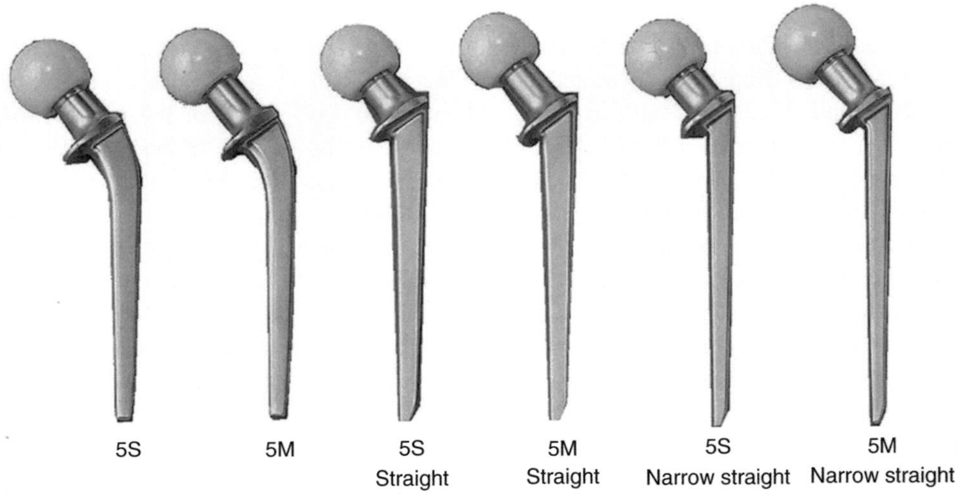

Fig. 6.2 Bioceram Type 5 stems. Bioceram Type 5 has 3 stem designs such as Standard, Straight, and Narroe Straight with 2 head offset lengths of S and M

110 mm. The initial model which was introduced in 1982 had lattice grooves similar to those of Type 2 (Fig. 6.1); however, the surface was subsequently changed to a satin-finish without grooves because of the difficulty of stem removal from the cement mantle. Three different neck lengths with a 6-mm pitch were available, with offset lengths of 28.5 mm for 4S, 32.7 mm for 4 M, and 37 mm for 4 L.

Type 5 was designed for a narrow medullary cavity, whereas type 4 was not (Fig. 6.2). Type 5 was also made of COP, with the stem axis passing through the trochanteric fossa and the endpoint of the lateral stem curve at the central stem axis. The medial curve radius was 40 mm, neck-shaft angle was 140°, and the femoral neck osteotomy angle and collar were 50°. The distal stem was tapered 1° medially, 2° laterally, and 2.7° anteroposteriorly. The anteroposterior aspects of the distal stem section were tapered by 5° with a thicker lateral dimension, creating a three-dimensional tapered stem shape. The stem and neck lengths were 105 mm and 6 mm, respectively. The offset was 26 mm for 5S and 29.3 mm for 5 M (Fig. 6.1). In addition, to accommodate an extremely narrow medullary cavity, "straight" stem designs were also manufactured with 5S and 5 M straight, and 5S and 5 M narrow straight, in which the neck-shaft and femoral neck osteotomy angles were the same. The cross-section of the straight stem was trapezoidal, whereas, that of the narrow straight stem was rectangular with no tapering.

3 Clinical Results

Type 2 prosthesis was first used in a 1978 clinical trial [4]. A 52-year-old woman with end-stage hip dysplasia, after surgery, did not need a cane, could sit upright, had no limp or radiographic laxity for 38 years, and showed no osteolysis [8]. The cup was composed of non-crosslinked, gas-sterilized GUR412 polyethylene, and the 28-mm monocrystalline alumina head was presumed to have contributed to the low wear. Type 1 prosthesis which was combined with a polycrystalline alumina head, has been used clinically since 1979; it was indicated in cases of highly dislocated hips (Fig. 6.3) and post-

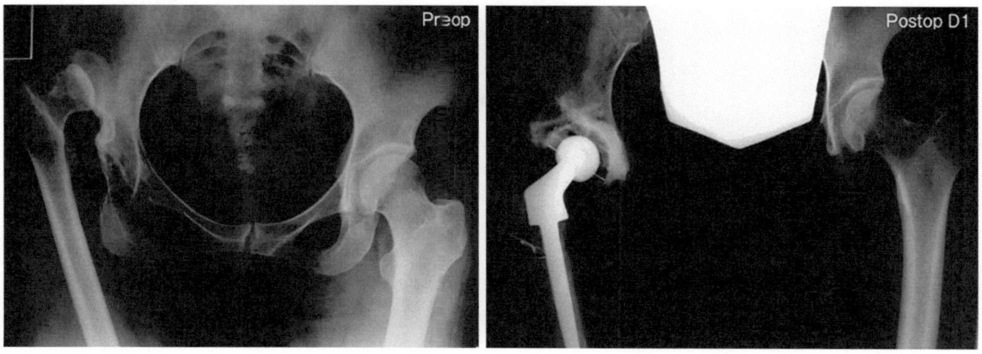

Fig. 6.3 Pre- and post-operative radiographic images of a 41-year-old woman with a Crowe type 4 dislocated hip in whom Bioceram type 1 prosthesis was used in 1980

femoral osteotomy. The distal stem was designed for high rotational stability in the cemented mantle so that long-term fixation of the stem remained good, even if the cup needed revised. The clinical results of Types 4 and 5 were good for up to 10 years, with an annual linear wear rate of approximately 0.1 mm for the non-crosslinked gas-sterilized GUR412 polyethylene cup and 28 mm polycrystalline alumina head [6]. However, as the wear progressed, the number of hips with cup loosening increased due to osteolysis [9]. The survival rates of the cup and stem, with radiological loosening as the end-point at 20 years, were 51 and 94%, respectively [10]. Cup loosening was associated with polyethylene wear; the gas-sterilized non-crosslinked GUR412 cup and 28 mm polycrystalline alumina head combination showed a higher wear rate than the combination of a γ-sterilized crosslinked polyethylene cup and metal head, such as Charnley or Müller prosthesis [11]. In contrast, composite beam types 4 and 5 with a relatively short 3D-tapered stem and satin-finish collar showed good rotational stability and fixation over a long period of >30 years.

4 Summary

Ceramic heads were developed to reduce polyethylene cup wear and increase the long-term outcome of total hip arthroplasty. Monocrystalline alumina head in combination with a non-crosslinked GUR412 polyethylene cup reduced the wear; however, polycrystalline alumina heads did not sufficiently reduce the polyethylene cup wear. Thus, polyethylene crosslinking seems more important for wear reduction. However, it should be noted that the cemented fixation of the composite beam-type short stems with a satin surface and collar, such as Types 1, 2, 4, and 5, showed excellent long-term fixation.

References

1. Shikita T. Wear and lubrication in total hip prostheses. Dis Med. 1972;15:683–96. In Japanese
2. Sugano N, et al. Analysis of a retrieved UHMWPE acetabular cup crosslinked in air with 1000 kGy of gamma radiation. J Orthop Res. 2004;22:828–31.
3. Semlitsch M, et al. New prospects for a prolonged functional lifespan of artificial hip joints by using the material combination. J Biomed Mater Res. 1977;11:537–52.
4. Shikita T. Ceramic artificial hip joint. In: Shikita T, et al., editors. Proceedings of the Orthopedic ceramic implant colloquium. Kyoto: Kyoto Ceramics; 1979. p. 1–8. In Japanese.
5. Shikita T. Artificial materials in orthopedics. Artificial materials in orthopedic surgery. Surgery. 1982;36:1473–86. In Japanese
6. Sugano N, et al. Polyethylene sockets and alumina ceramic heads in cemented total hip arthroplasty: a ten-year study. J Bone Joint Surg Br. 1995;77:548–56.
7. Anwar MM, et al. Total hip arthroplasty in the neglected congenital dislocation of the hip: a five- to 14-year follow-up study. Clin Orthop Relat Res. 1993;295:127–34.
8. Sugano N, et al. History of artificial hip joints. J Jpn Orthop Soc. 2018;92:987–95.
9. Haraguchi K, et al. Analysis of survivorship after total hip arthroplasty using a ceramic head. Clin Orthop Relat Res. 2001;391:198–209.
10. Sugano N, et al. Long-term results of Bioceram cemented and Luebeck cementless hip prostheses. Seikei Saigai Geka. 2007;50:1299–304. In Japanese
11. Haraguchi K, et al. Influence of polyethylene and femoral head surface quality on wear: a retrieval study. Int Orthop. 2001;25:29–34.

AMS HA CUP and PerFix HA STEM (Kyocera)

7

Yasuharu Nakashima

1 Development of HA and Its Application to Artificial Hip Joints

Hydroxyapatite ($Ca_{10}(PO_4)_6(OH)_2$: HA) is a ceramic material with a composition similar to that of inorganic bone which was developed by Aoki et al., in the 1970s [1]. HA is characterized by its direct chemical bonding to the bone and was initially studied as filler for bone defects. The early research on HA for joint prosthesis was conducted in the 1980s as a surface coating material on the base metal, owing to its low mechanical strength. Onishi et al., reported the high bone-conducting ability of HA, in Japan [2–4].

In the late 1980s, clinical application of HA-coated THA (HA-THA) was initiated. However, many of the early HA-THAs relied solely on the HA coating for bone anchorage of the implant, resulting in poor clinical outcomes, especially on the cup side [5–8].

2 Clinical Trial of Moore 7 HA and Lessons Learned

The first Japanese HA-THA was probably Moore 7 HA (Fig. 7.1), which had a relatively smooth surface (average surface roughness 1.4 μm) base metal after the sand blast treatment. HA was coated on the entire cup and on the proximal 1/3rd of the stem. Although HA alone was already in clinical use at that time, HA coating was considered a new material; its clinical trials in Japan began in 1989. Despite the high expectations for this implant, implant survival rate of the cup with revision as the endpoint was 65.7% at 10 years after surgery and 50.2% at 15 years after surgery; showing a high rate of cup loosening [7, 8]. Contrarily, few problems on the stem side were present; the results were good. This trend was similar in other models, with a high rate of cup laxity in early HA-THAs, many of which were HA-coated cups on a smooth surface base metal [6]. These results suggest that, at least on the acetabular side, HA-bone bonding alone is insufficient, and microanchoring with bone tissue, such as porous coating, is necessary.

Fig. 7.1 Moore 7 HA. Implant made of titanium alloy base material with sand blast treatment (average surface roughness 1.4 μm), with HA coating on the entire cup and on the proximal 1/3rd of the stem

3 Development of AMS HA CUP + PerFix HA STEM

The clinical results of HA-THA and the low mechanical strength of the HA-metal interface in basic research [9] necessitates the development of a surface treatment that would promote bone ingrowth. As HA coating is applied by spraying HA at high temperatures in a thermal spraying method, problems regarding partial coating of HA on traditional porous structures existed, such as beads, by clogging of HA in the

Y. Nakashima (✉)
Department of Orthopedic Surgery, Kyushu University, Fukuoka, Japan
e-mail: nakashima.yasuharu.453@m.kyushu-u.ac.jp

Y. Nakashima (ed.), *Advances in Total Hip Arthroplasty*, https://doi.org/10.1007/978-981-97-2461-1_7

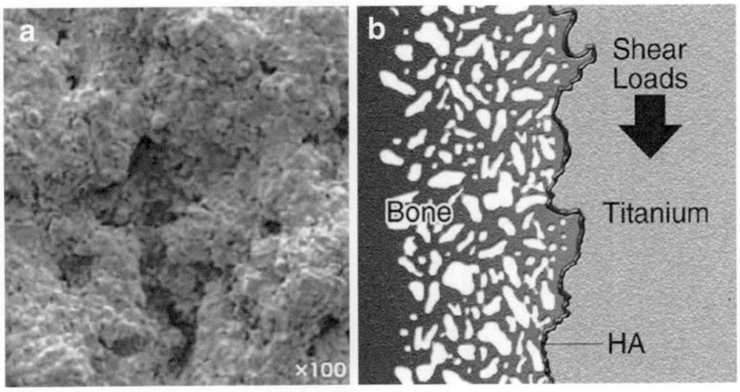

Fig. 7.2 Surface processing of AMS + PerFix HA. (**a**) Surface SEM image: Irregular irregularities with an average surface roughness of 40 μm were formed by the titanium arc spraying method. (**b**) Cross-sectional schema: concept of HA coating to promote early osteoconduction, followed by microanchoring of the formed bone to rough surface irregularities to maintain fixation. (From Kyocera Technical Document, ProArcHA, revised July 2005)

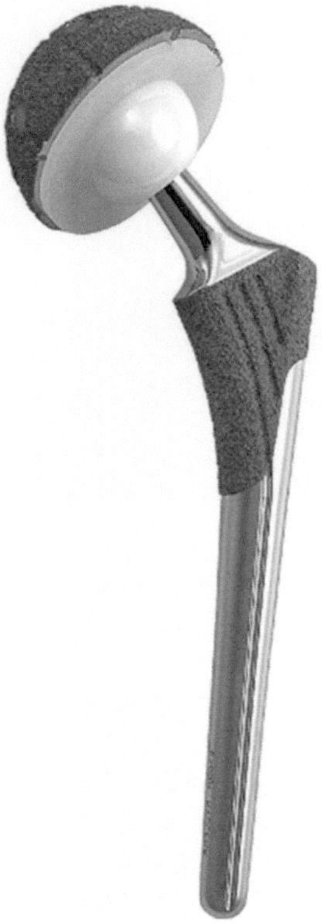

Fig. 7.3 AMS + PerFix HA

porous structures. The titanium arc spray method was developed to obtain a uniform HA coating by spraying pure titanium onto the base metal and roughening the surface as desired. This roughening process made it possible to coat HA uniformly and without loss of surface roughness [9]. The actual implants were roughened on the titanium alloy cup and stem (proximal 1/3rd) by titanium arc spraying (Fig. 7.2) with an average surface roughness of 40 μm, and subsequently coated with HA at a thickness of 20 μm (Fig. 7.3). The cup and stem have been marketed under the names of AMS HA CUP and PerFix HA STEM, respectively, since 1995. The head was made of alumina, zirconia, or metal. Non-cross-linked polyethylene used until 1999; after 2000, cross-linked polyethylene (Aeonian) was used. The stem was a standard fit-and-fill type; according to the recent stem classification, it was a double-wedge metaphyseal filling, with a design that contacted the proximal cortical bone both antero-posteriorly and laterally. The neck shape was a PHS taper (neck diameter 11–13 mm), and the stem neck angle was 135° when it was introduced in 1995. A 910 taper (neck diameter 9–10 mm) was adopted in 2003, and later a high-offset stem with a 130° neck angle was also introduced.

4 Clinical Results of AMS HA CUP + PerFix HA STEM

Clinical results were reported using the AMS HA CUP + PerFix HA STEM from 1998 to 2002, with more than 10 years follow-up [10]. The subjects were 120 consecutive patients with 131 joints. The mean age at surgery was 61.9 years, and the mean observation period was 12.2 years. In all cases, 22 mm zirconia heads were used; 62 joints used conventional polyethylene without cross-linking (CPE group), and 69 joints used cross-linked PE (XLPE group). In the CPE group, 10 (16.1%) had revision of the PE and head only because of wear, and 2 (3.2%) had revision for cup loosening owing to osteolysis. In the XLPE group, cases of obvious osteolysis were not observed, and revision was performed for frequent dislocation, cup laxity, and stem fracture in a total of 3 cases each.

The steady-state wear rate was 0.12 mm/y in the CPE group, whereas it was significantly lower in the XLPE group (0.007 mm/y). The 12-year postoperative implant survival rates at the end of revision were 97.7% for the cup, 93.8% for the PE, and 99.2% for the stem, with revision as the end-point. The only significant risk factor for revision on multivariate analysis was the type of PE, and XLPE reduced the risk of revision by 4.3-fold compared to CPE. In other words, although PE wear reduced implant survival in early cases of CPE use, the adoption of XLPE in 2000 solved the problem of wear and stable long-term clinical results were obtained. From the viewpoint of the bone-implant interface, AMS & PerFix HA showed excellent bone fixation.

The KyoceraPerFix® series is still in use at many institutions, and has been used in approximately 98,000 cases from 1995 to 2020.

References

1. Aoki H, et al. Calcium phosphate formation and effect of insoluble collagen. Med J Jpn. 1974;8:101–4. In Japanese
2. Oonishi H, et al. The effect of hydroxyapatite coating on bone growth into porous titanium alloy implant. J Bone Joint Surg Br. 1989;71:213–6.

3. Hayashi K, et al. Evaluation of metal implants coated with several types of ceramics as biomaterials. J Biomed Mater Res. 1989;23:1247–59.
4. Hayashi K, et al. Effect of surface roughness of hydroxyapatite-coated titanium on the bone-implant interface shear-strength. Biomaterials. 1994;15:1187–91.
5. Geesink RG. Hydroxyapatite-coated total hip prostheses. Two-year clinical and roentgenographic results of 100 cases. Clin Orthop Relat Res. 1990;261:39–58.
6. Chung YY. A comparison of three different implant surfaces at a minimum. J Bone Joint Surg Am. 1998;80:1175–85.
7. Nakajima Y, et al. Clinical results of Kyocera Perfix Ha for more than 10 years. In progress of research and development, clinical application and prospects for the 21st century of artificial joints in Japan, edited by Keiyasu Onishi; 2008, p. 177–184, In Japanese.
8. Kawamura H, et al. Long-term results of Kyocera HA-coated hip prosthesis (investigational type). J Jpn Soc Artif Joints. 2004;34:51–2. In Japanese
9. Nakashima Y, et al. Hydroxyapatite-coating on titanium arc sprayed titanium implants. J Biomed Mater Res. 1997;35:287–98.
10. Nakashima Y, et al. The results at a minimum of 10 years of follow-up for AMS and PerFix HA-coated cementless total hip arthroplasty: impact of. J Orthop Sci. 2013;18:962–8.

K-MAX ABC HIP System 1992 and AHFIX Q HIP System 2007 (Kyocera)

8

Koji Goto

1 Clinical Application of Apatite- and Wollastonite-Containing Glass-Ceramics (AW-GC) for THA

In Japan, many calcium phosphate bone substitute materials were developed and commercialized during the 1980s and 1990s. Among them, AW-GC showed higher mechanical strength and stronger chemical bonding to bone than sintered hydroxyapatite and was launched for sale in the early 1990s. Initially, AW-GC was used for iliac spacers, intervertebral spacers, and artificial vertebrae in clinical settings. In animal experiments, it was proven that titanium implants with AW-GC coating of the deep layer (bottom coating) of the porous coated layer by plasma spraying had high bone-bonding ability under load-bearing conditions [1, 2]. The development of cementless THA implants using this technology was promoted in collaboration with Kobe Steel Ltd., and clinical trials were conducted from 1992 to 1994. The K-MAX ABC HIP System (Fig. 8.1a) was approved for manufacturing and marketing, and sales began in 1997.

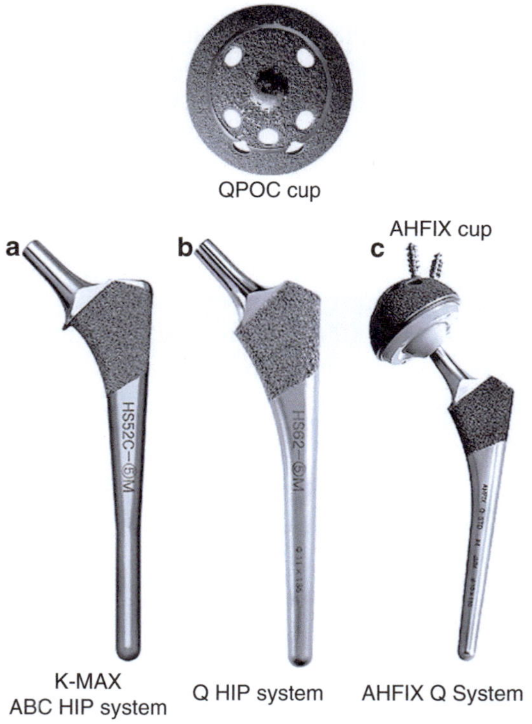

Fig. 8.1 (**a**) ABC HIP System H5 stem. (**b**) Q HIP System QPOC cup and H6 stem. (**c**) Q AHFIX Q System

2 Features of K-MAX ABC HIP System

The implants were made of a vanadium-free titanium alloy (Ti-6Al-2Nb-1Ta-0.8Mo), which has a fatigue strength superior to that of the titanium alloy Ti-6Al-4 V used in other cementless stems. The neck diameter was designed to be as thin as 9 mm, which is believed to contribute to the improvement of the range of motion and prevention of dislocation. Zirconia ceramic heads of 22.225 mm or 26 mm were predominantly used. The outer surface of the cup and the proxi-

K. Goto (✉)
Department of Orthopaedic Surgery, Kindai University, Osaka, Japan
e-mail: kgoto@med.kindai.ac.jp

Y. Nakashima (ed.), *Advances in Total Hip Arthroplasty*, https://doi.org/10.1007/978-981-97-2461-1_8

mal one-third of the stem were coated with a porous coating (pore size: 350–450 μm) by plasma spraying, and the inside of the pore was coated with an AW-GC bottom coating. The stem was a fit-and-fill type proximal fixed stem with a small collar and cylindrical structure in the distal one-third. Long-term results showed bone ingrowth of the implants in all patients; however, osteolysis around the implants due to wear of the non-crosslinked polyethylene was observed in some cases.

3 Development and Clinical Results of the K-MAX Q HIP System

The Q HIP System (Fig. 8.1b) was launched in 1999, with an improvement on the stem shape of the ABC HIP System. The Q HIP System, like the ABC HIP System, was composed of the same vanadium-free titanium alloy implant. The stem collar used in the ABC HIP System was eliminated, and the distal portion was changed from a cylindrical structure to a smooth single-tapered shaped structure. The clinical outcome was excellent, with no revision surgery required for any of the 121 surgeries performed at Kyoto University between 2002 and 2007 after an average follow-up of 7 years (3–9 years) [3]. Bone ingrowth fixation was achieved in all the cases, and there were no cases of osteolysis. However, there was a high incidence of reactive lines around the stem tip (82%), and 5 out of the 109 patients complained of thigh pain when ascending or descending stairs at the last follow-up [3].

4 Clinical Application of Alkali- and Heat-Treatment Technology to THA

In the 1990s, Kokubo et al. developed an alkali- and heat-treatment technique at Kyoto University [4, 5] in which titanium alloys were immersed in sodium hydroxide solution for 24 h, rinsed, and heat-treated at 600 °C for 1 h. Animal experiments have demonstrated higher bone-bonding ability with this procedure than with the AW-GC bottom coating described above [6]. The surface-treated layer was 1 μm thin, and the treated layer was uniformly formed inside the pore regardless of the pore size (Fig. 8.2). Furthermore, compared with coating techniques using bioactive ceramics such as hydroxyapatite or AW-GC, there was no concern about the titanium surface treatment layer falling off, and the process was considered simple and cost-effective. Clinical trials were conducted from 2000 to 2002, and the implant using this technology (AHFIX Q System [Fig. 8.1c]) was approved for clinical use in 2007.

Fig. 8.2 Electron microscopic images of the alkali- and heat-treated surface of the titanium alloy

5 Features and Clinical Results of the AHFIX Q System

For the AHFIX Q System, the method of adding bone-bonding ability to the porous coated layer of titanium was changed from AW-GC bottom coating to alkali- and heat-treatment, and the implant shape was the same as that of the Q HIP System. In a report [7] of a clinical trial conducted with 70 patients at Kyoto University and Kanazawa Medical University between 2000 and 2002 with an average follow-up period of 10 years (8–12 years), there were no cases of radiological loosening, and the 10-year survival rate, with revision for any reason as the endpoint, was 98%. The bone-implant gap around the cup, which was observed in 69% of patients on immediate postoperative radiographs, disappeared by 1 year postoperatively, and bone ingrowth fixation was achieved in all patients. No osteolysis was observed in any of the patients. The AHFIX Q System is still commonly used at many institutions and was used for 18,000 patients between 2007 and 2020.

References

1. Ido K, et al. Cementless total hip replacement: bio-active glass ceramic coating studied in dogs. Acta Orthop Scand. 1993;64:607–12.
2. Yamamuro T. Bone bonding behavior of biomaterials with different surface characteristics under load-bearing conditions. In: Davies JE, editor. Bone biomaterial interface. Toronto: University of Toronto Press; 1991.
3. So K, et al. Good short-term outcome of primary total hip arthroplasty with cementless bioactive glass ceramic bottom-coated implants: 109 hips followed for 3–9 years. Acta Orthop. 2012;83:599–603.
4. Kokubo T, et al. Spontaneous formation of bonelike apatite layers on chemically treated titanium metals. J Am Ceram Soc. 1996;79:1127–9.
5. Kim HM, et al. Bonding strength of bonelike apatite layer to Ti metal substrate. J Biomed Mater Res. 1997;38:121–7.
6. Nishiguchi S, et al. Alkali- and heat-treated porous titanium for orthopedic implants. J Biomed Mater Res. 2001;54:198–208.
7. So K, et al. Is the bone-bonding ability of a cementless total hip prosthesis enhanced by alkaline and heat treatments? Clin Orthop Relat Res. 2013;471:3847–55.

9 S-LOCK Stem 2003 (Kyocera)

Naohiko Mashima

1 Cementless Long Stem with an Interlocking Screw Mechanism

The cementless long stem, which has interlocking screws at the distal and proximal portions of the stem, is a useful implant for revision surgery in cases of severe bone loss and periprosthetic femoral fractures [1–3]. It enables free placement in the long axis and rotation direction and easy leg-length compensation with the healthy side. It has been reported that the distal transverse fixation screw increases torsional stability by 320% and axial stability by 230% [4]. However, there are concerns that stress shielding may occur and that the risk of screw fracture may increase in the long term because of the distal fixation of the stem [1, 2, 6].

Clinical use of the stem started in Europe and Australia in the late 1980s [1, 5], and increased laxity over time and poor results in young individuals were reported in the case of non-porous coated stems [2]. However, it has also been reported that the surface treatment properties of the stem are important, such as a wider range of hydroxyapatite coatings and better clinical results with fully porous stems [1, 3].

2 Results and Problems Reported from Japan

In 1993, the Huckstep hip system (B. Braun-Aesculap, Germany) with an interlocking screw mechanism became available for clinical use in Japan [5]. The Huckstep hip system is a titanium alloy cementless straight stem with a stem diameter of only 12.5 mm and four stem lengths of 160, 210, 260, and 320 mm. The proximal 5 cm of the stem is porous coated, and 2 screws can be fixed in the proximal anteroposterior (AP) and 6 screws in the distal medial-lateral (ML). A good clinical outcome was reported with a cumulative survival rate of 90.4% at 15 years after revision surgery using laxity or screw breakage as the endpoint; however, the absence of spot welds in the proximal porous part suggested the possibility of distal fixation in the long term [6].

Instead of relying solely on the screw mechanism for fixation, developing cementless stems with ingrowth or ongrowth at the porous coating was ultimately needed in clinical practice.

3 Development of S-LOCK Stem

The development of this device was started in 2000, primarily by Okumura, based on the proximal shape of K-MAX Superio and Q-HIP (Kyocera, Kyoto, Japan), which are already commercially available fit-and-fill type devices. The distal diameter of the stem was 12 mm and 13 mm straight, and the stem length consisted of five variations: standard, semilong, and three long types (Fig. 9.1). The surface of the porous coating was treated by titanium plasma spraying and A-W glass-ceramic bottom coating, which is a bioactivation treatment. The area of the porous coating was expanded to half of the stem length for the standard and semilong stems and a full-porous stem for the semilong stem (the full-porous stem was used only in limited facilities). Two screws were used in the proximal AP direction, and two to eight screws were used in the distal ML direction, depending on the stem length. The screw holes were designed to be evenly spaced in the body of the stem to avoid distal fixation.

The S-LOCK stem (Kobelco, Kobe; now Kyocera Corporation, Kyoto) has been marketed since 2002. The S-LOCK stem has a thin neck with a diameter of 9 mm, which is characteristic of K-MAX and was initially limited to use with Co-Cr heads only; however, since 2018, it has become possible to use ceramic heads by attaching a titanium sleeve.

N. Mashima (✉)
Ehime University, Ehime, Japan

Y. Nakashima (ed.), *Advances in Total Hip Arthroplasty*, https://doi.org/10.1007/978-981-97-2461-1_9

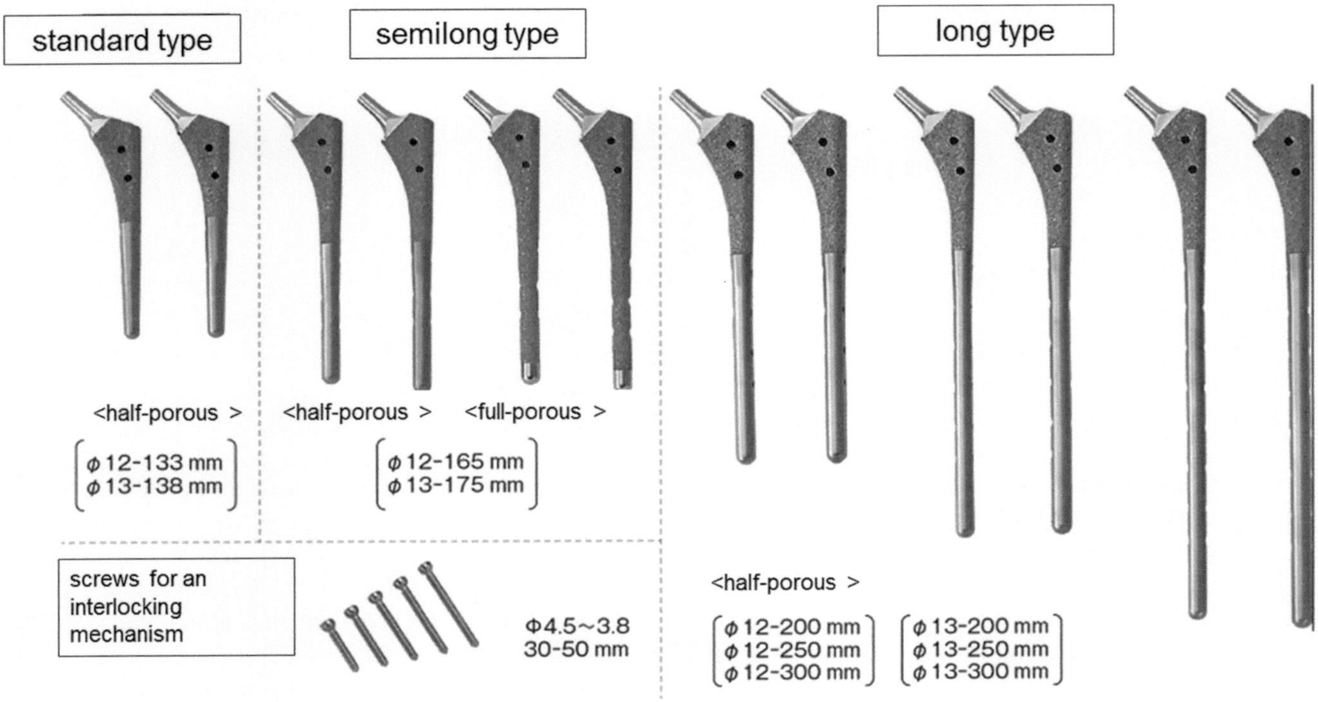

Fig. 9.1 Variations in the S-LOCK stem. Titanium alloy cementless straight stem, only two stem diameters (12 mm and 13 mm), five stem lengths (standard, semilong, and three long stems (200 mm, 250 mm, and 300 mm)), porous range up to half the stem length for standard and semilong stems, and up to half the stem length for long stems. The porous range is half-porous up to half the stem length for the standard and semilong stems and half-porous for the long stem, which is the same length as the semilong stem. The long stem is half-porous, that is, the same length as the semilong stem. The semilong stem is fully porous that can be used at limited facilities. The distal screw holes can be standard 2-hole, semilong 3-hole, 200 mm 4-hole, 250 mm 6-hole, or 300 mm 8-hole

4 Clinical Results of S-LOCK Stem

In the initial THA and artificial head for patients with severe osteoporosis, the S-LOCK stem can help prevent intraoperative fractures and strong initial fixation, and good short-term results have been reported [7]. The S-LOCK stem is also reportedly useful in revision THA for Vancouver Classification Type B2 periprosthetic femoral fractures, which is one of the recent clinical challenges [8].

The clinical outcomes of 56 joints (mean age, 73 years; mean follow-up, 6.7 years) performed between 2003 and 2016 were reported for the mid-term results of revision surgery, with a marked reduction in thigh pain and the appearance of spot welds (32%), confirming that the fixation of the S-LOCK stem has improved steadily. Furthermore, there were no cases of implant fracture or laxity, and the cumulative survival rate with additional surgery as the endpoint was 91.1% at 10 years (infection in 1 joint, fracture of the greater trochanter in 2 joints, graft resorption in 1 joint, and prolonged bone healing in 1 joint), with a good clinical outcome (Fig. 9.2) [9].

Implant fracture was reported, and stem fracture rates of 1.8% and 0.9% were confirmed for 12 mm and 13 mm stem diameters, respectively (Kyocera, in-house data). Both cases occurred in long stem cases applied to periprosthetic femoral fractures and osteotomies. In the case of the S-LOCK stem with a screw hole in the stem body, a case study using Finite Element Analysis (FEM) analysis showed that it is necessary to devise a placement position where the screw hole does not coincide with the fracture or osteotomy line and that additional fixation procedures such as wiring or combined plate fixation are necessary for the fracture or osteotomy area [9].

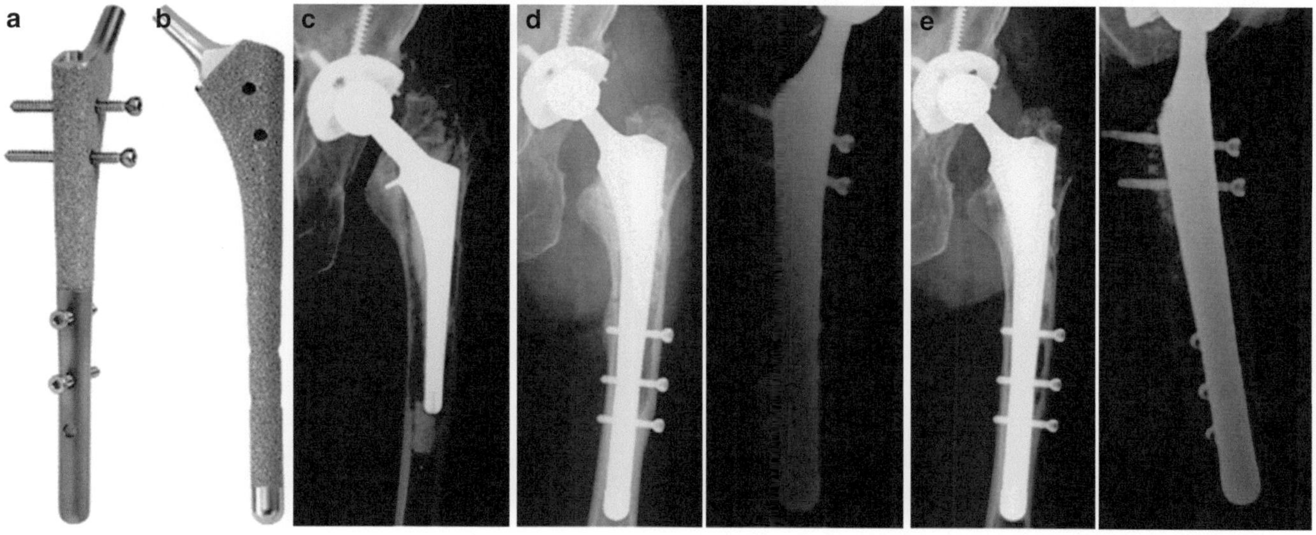

Fig. 9.2 Revision case with S-LOCK semilong stem. (**a**) S-LOCK semilong half-porous, (**b**) S-LOCK semilong full-porous, (**c**) preoperative (cemented stem laxity), (**d**) 1 month after revision surgery with S-LOCK semilong full-porous (only artificial bone graft was used), (**e**) 10 years after revision surgery

References

1. Mertl P, et al. Distal locking stem for revision femoral loosening and peri-prosthetic fractures. Int Orthop. 2011;35:275–82.
2. Sexton SA, et al. The Kent hip prosthesis: an evaluation of 145 prostheses after a mean of 5.1 years. J Bone Joint Surg Br. 2006;88:310–4.
3. Fink B, et al. Distal interlocking screws with a modular revision stem for revision total hip arthroplasty in severe bone defects. J Arthroplast. 2010;25:759–65.
4. Mahomed N, et al. Biomechanical analysis of a distally interlocked press-fit femoral total hip prosthesis. J Arthroplast. 1993;8:129–32.
5. Huckstep RL. Stabilization and prosthetic replacement in difficult fractures and bone tumors. Clin Orthop Relat Res. 1987;224:12–25.
6. Kamada T, et al. Long-term results with an interlocking uncemented long stem in revision hip arthroplasty: a mean 15-year follow-up. J Arthroplast. 2015;30:835–9.
7. Okumura H, et al. Clinical experience with a newly developed cementless long stem with interlocking screw mechanism (S-Lock stem). J Rinsho Gai Gakkaishi. 2007;32:54–6. In Japanese
8. Mashima N, et al. Revision hip arthroplasty for Vancouver classification type B2 and B3 periprosthetic femoral fractures. Hip Joint. 2018;44:562–6. In Japanese
9. Mashima N. Revision hip arthroplasty using cementless femoral implants. J Jpn Orthop Soc. 2019;93:S598. In Japanese

10 CentPillar Grit Blast HA Stem 2003 and TMZF Plasma Spray HA Stem 2007 (Stryker)

Masaaki Matsubara

1 Concept of the CentPillar System

These concepts have been used in the development of artificial hip stems:

1. Intramedullary shape: The stem must conform to the shape of the hip joint in the Japanese hip joint or the normal voluntary femoral cavity, which is different from that of Westerners.
2. Medullary outline: Neck strength that does not require load limitation, correction of excessive anteversion, and sufficient range of motion for daily Japanese living, even with a head measuring 22 mm in diameter, while still reproducing the anatomical head center.
3. To achieve bone preservation of the proximal femur by achieving as much load transfer in the proximal femur as possible. Therefore, the proximal part of the stem should be macrostructured (grit blast on fish scale: GBHA) to achieve biological fixation at the proximal femur, or a hydroxyapatite coating (HA) over a microstructure (plasma spray + TMZF).
4. The anteversion angle can be adjusted (not only reduced but also increased) if necessary to accommodate hyperanteversion, which is common in Japanese patients.
5. Cementless structures, such as bone cement removal, are often difficult during revision surgery.
6. Fit and fill proximally without distal reaming.
7. Attempt to increase proximal contact regardless of torsion of the medullary cavity. In addition, stress shielding was expected to be inevitable because of lengthening of the stem, even if the stem was designed to transmit load at the proximal portion and a stem body that was as short as possible (without affecting stem alignment) was adopted [1] (Fig. 10.1). This made it possible to accommodate small skin incisions.

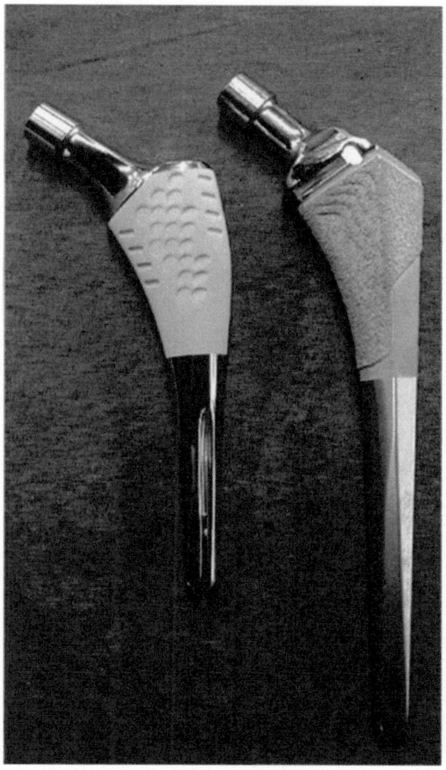

Fig. 10.1 Conventional Stem (right) and GBHA (left)

M. Matsubara (✉)
Department of Orthopedic Surgery, Nissan Tamagawa Hospital, Tokyo, Japan

Y. Nakashima (ed.), *Advances in Total Hip Arthroplasty*, https://doi.org/10.1007/978-981-97-2461-1_10

2 Clinical Results of the CentPillar Stem

The stem was designed to reflect the average Japanese femur [2–11], with an inverted triangular neck and a three-dimensional neck-shaft angle of 127° to minimize implant impingement, an oval distal cross-sectional shape with 35° anterior rotation, and a polished slot distally to provide flexibility at the distal portion (Fig. 10.2). The CentPillar stem has been marketed as the GBHA stem since 2003, following clinical trials at two centers in Japan since 2002 (Fig. 10.2).

From September 2004 to April 2005, 95 THAs using the CentPillar GBHA stem were performed at Nissan Tamagawa Hospital with an average follow-up duration of 17 years (Table 10.1). The progress of these patients (Table 10.1) and radiographic evaluation of stem subsidence, fixation with bone, cortical hypertrophy, stress shielding, cancellous condensation, and reactive lines (all according to Gruen's zone classification) were investigated. The results are shown in Table 10.1 and Fig. 10.3. Other reports showed that the bone reaction was confined to the proximal femur 17 years after surgery, with no postoperative dislocation and good results; the intended bone preservation in the proximal femur was also achieved (Fig. 10.4) [12]. However, because this stem was indicated for use in osteoarthritis in young patients with relatively good bone quality, intraoperative fractures could occur when used in femurs with poor bone quality (osteoporosis). The TMZF HA was subsequently launched in 2007, but as of 2018, Stryker Japan's CentPillar was discontinued in 2018.

Table 10.1 Cases

Cases
Period of surgery: September, 2004 ~ April, 2005
Disease: DDH-OA:92 hips, AVN: 3Hips
No. of THA:95 hips (bilateral simultaneous: 15)
Gender: 74 females (89 hips)/6 males (6 hips)
Age: 59 y. o. (42 ~ 78)
Height: 156 cm, weight: 57 kg (BMI: 23.4)
Stem: CentPillar GBHA (grid blasted hydroxyapatite coated)
Ave. FU.: 17.0 years (16.5 ~ 17.8)
Clinical course
3 patients (3 hips) died 5 ~ 7 years postoperatively (causes unrelated to THA).
1 case (1 hip): Stroke 2 years after surgery, unable to go to the hospital
→final follow-up was possible in 76 cases (91 hips) out of 80 cases (95 hips)*
*No drop-out cases
Results
Stem subsidence*: 4 mm: 1 hip
1 mm: 5 hips
*All cases showed subsidence only in early cases.
*All patients achieved bone ingrown fixation by 1 year postoperatively.
Bone reaction
Cortical hypertrophy:9% (8 hips) occurred in Gruen zone 2, 6, 7
Clinical course:
Postoperative dislocation: 0
Thigh pain: 0
Revision cases: 0
JOA postoperative pain: 38 (out of 40) (Ave. preop. Pain: 17)

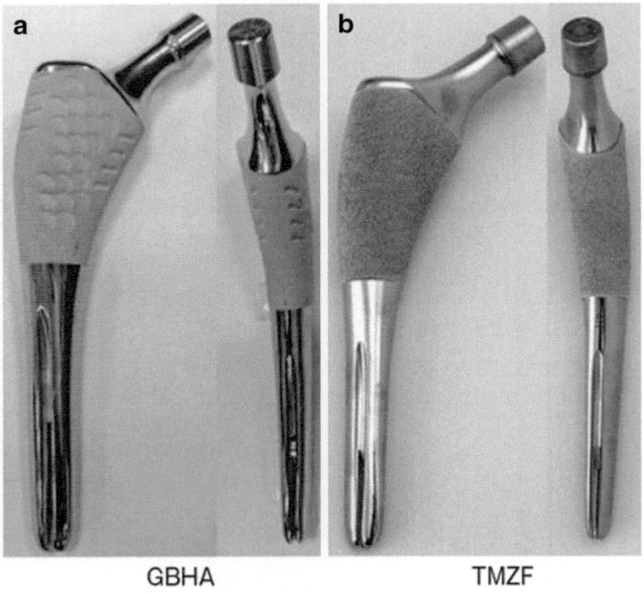

Fig. 10.2 Anatomic Stems. CentPillar GBHA (**a**), and CentPillar TMZF (**b**)

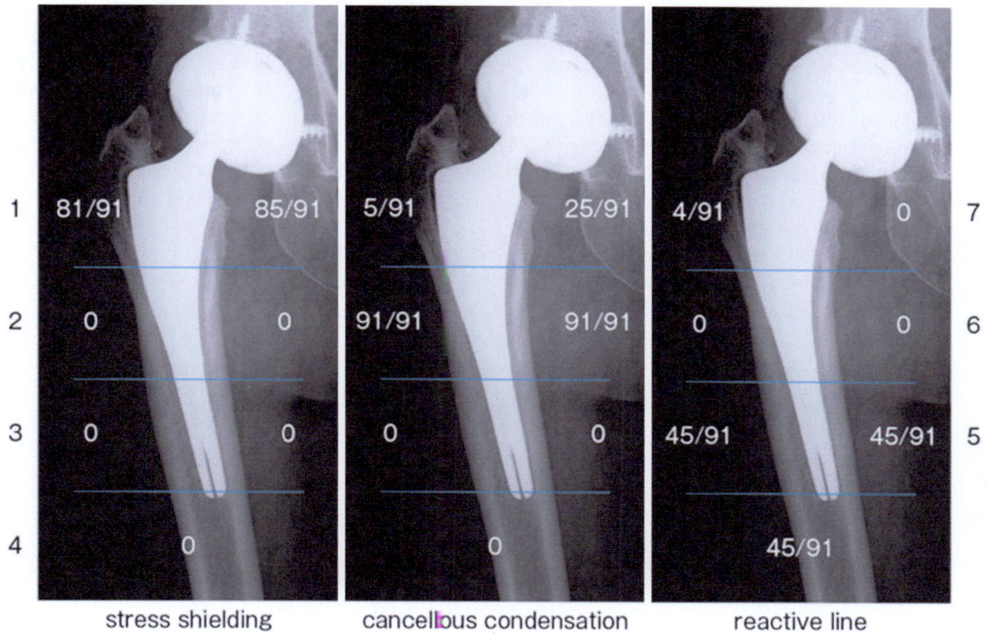

Fig. 10.3 Radiographic evaluation (Gruen's zone classification) at 17 years after THA

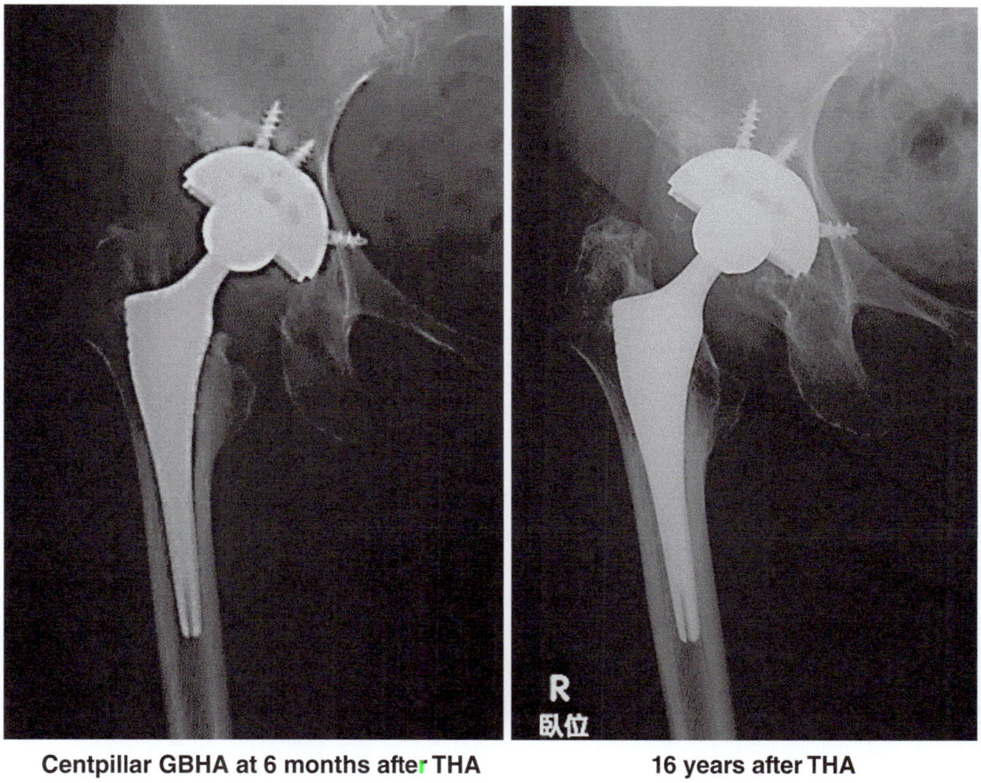

Fig. 10.4 Postoperative Radiographs

References

1. Noble PC, et al. Three-dimensional shape of the dysplastic femur: implications for THR. Clin Orthop Relat Res. 2003;417:27–40.
2. Ikeda K. Anteversion of the femur. Rinshou Seikei Geka. 1977;12:1033. In Japanese
3. Komatsu T. Morphological study of the proximal femur. Morphological studies on the proximal femur. I. Measurement and analysis of the femoral head and neck system. J Jpn Orthop Assoc. 1986;60:755–62.
4. Yanagimoto S, et al. Basic study on shape determination of cementless hip prosthesis: analysis of proximal femoral medullary cavity morphology in Japanese osteoarthritic hip joint cases. J Jpn Orthop Assoc. 1991;65:731–44. In Japanese
5. Okumura Y, et al. A study on the design of cementless hip prosthesis stem compatible with secondary osteoarthritis of the hip. Nihon Seikei Geka Gakkai Zasshi. 1993;67:897–910. In Japanese

6. Fuku S. Basic study on shape determination of cementless hip joint prosthesis: morphological analysis of normal Japanese proximal femur. Nihon Seikei Geka Gakkai Zasshi. 1994;68:763–73. In Japanese
7. Robertson DD, et al. Femoral deformity in adults with developmental hip dysplasia. Clin Orthop Relat Res. 1996;327:196–206.
8. Hagiwara M, et al. A CT study of the morphology of the Japanese proximal femur. Nihon Seikei Geka Gakkai Zasshi. 1995;69:1147–57. In Japanese
9. Akihiko B, et al. Fit and fill analysis of a newly designed femoral stem in cementless total hip arthroplasty for patients with secondary osteoarthritis. J Orthop Sci. 1997;2:301–12.
10. Sugano N, et al. The morphology of the femur in the developmental dysplasia of the hip. J Bone Joint Surg Br. 1998;80:711–9.
11. Kaneuji A, Matsumoto T, Nishino M, Miura T, Sugimori T, Tomita K. Three-dimensional morphological analysis of the proximal femoral canal using a computer-aided design system in Japanese patients with osteoarthrosis of the hip. J Orthop Sci. 2000;5:361–8.
12. Koyano G, et al. Comparison of bone remodeling between an anatomic short stem and a straight stem in 1-stage bilateral total hip arthroplasty. J Arthroplast. 2017;32:594–600.

SC Stem (Kyocera)

11

Kenichi Oe and Hirokazu Iida

1 Development of the SC Stem

Favorable long-term results with cemented polished tapered stems have been clinically proven and theoretically supported [1]. Although the advantages of cemented stems include excellent long-term results, shape compatibility, and flexible applicability to all types of femurs, some changes, such as proximal femoral bone atrophy, cortical thinning, and trabecular ossification of the bone-cement interface, have been observed in many cases due to mechanical environment changes after long-term use. In addition, we were dissatisfied with existing polished taper stems owing to the lack of size variations and instrumentations, leading to the development of a new type of implant in April, 2002.

The SC stem (Kyocera, Kyoto, Japan; Fig. 11.1) was fabricated with a polished triple-taper design to ensure "uniform cement stress" and "proximal load transfer" via a curved tapered stem. The main features of this design include a polished, curved tapered stem made of cobalt-chromium-molybdenum (Co-Cr-Mo) alloy, trapezoidal shape with an additional third taper in the cross-section to promote stress distribution to the proximal femur and improve rotational stability, and neck diameter of 9 mm (910 taper) (Fig. 11.2). Furthermore, there is a guide system to ensure accurate stem placement. The stem design was completed in November, 2003, but the material had a history of change. At first, the SC stem using existing titanium alloy with a smooth surface has been used clinically since May, 2004, because our knowledge of the forging process of Co-Cr-Mo alloy was insufficient at the time of stem development. Since August, 2005, the SC stem has been produced using Co-Cr-Mo alloy with a polished surface.

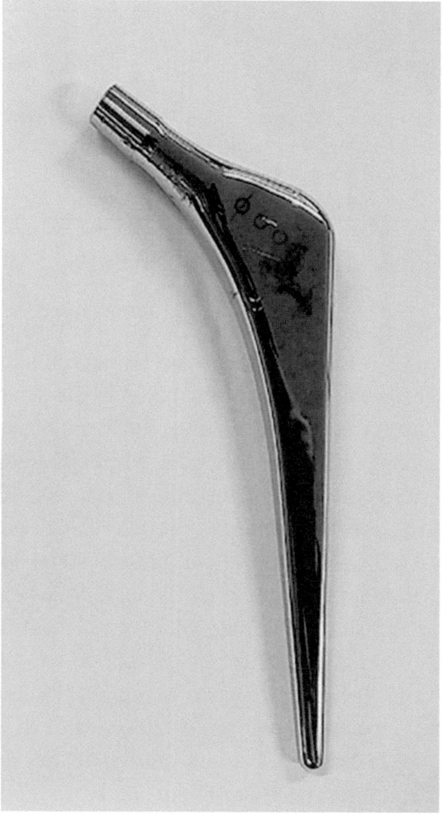

Fig. 11.1 SC stem (cobalt-chromium-molybdenum alloy, #STD-2)

K. Oe (✉) · H. Iida (✉)
Department of Orthopaedic Surgery, Kansai Medical University, Hirakata, Japan
e-mail: oeken@hirakata.kmu.ac.jp; iida@takii.kmu.ac.jp

Y. Nakashima (ed.), *Advances in Total Hip Arthroplasty*, https://doi.org/10.1007/978-981-97-2461-1_11

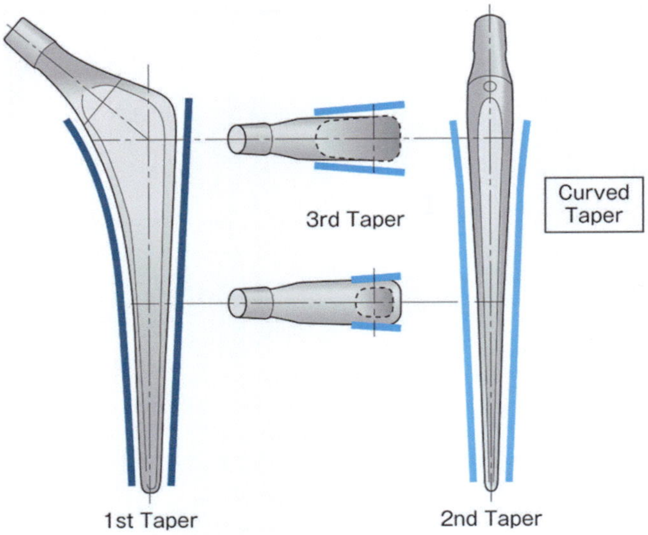

Fig. 11.2 Features of SC stems. The SC stem has a curved triple-taper design, which is expected to reduce stress shielding owing to the uniform distribution of hoop stress

2 Clinical Results of SC Stems

Between 2004 and 2020, 2259 SC stem series, made of both titanium and Co-Cr-Mo alloy, were implanted at the Kansai Medical University. As mentioned above, SC stems made of titanium alloy were initially introduced and are currently used in patients with metal allergies, while SC stems made of Co-Cr-Mo alloy are used in general. All operations were performed using the transgluteal approach (Dall or Hardinge approach), and a bone plug was placed distal to the stem using a third-generation cementation technique [2]. The head size of the SC stem was mainly 22 mm. Endurance bone cement (DePuy CMW, Blackpool, UK) was used from 2004 to 2019, and Palacos R bone cement (Heraeus Kulzer, Wehrheim, Germany) was used 2020 onward. In our data, 0 hips experienced stem loosening and 19 experienced periprosthetic fractures (0.9%). There were no noninfectious radiolucent lines, and good proximal load transmission was observed. However, the rate of periprosthetic fractures was relatively high. In Japan, the SC stem series have been used in many institutions for more than 20,000 joints to date [2–9].

3 Future Tasks

Although the incidence of periprosthetic fractures with the use of cemented stems is approximately one-tenth the incidence of fractures with cementless stems, taper-slip cemented stems have been reported to be associated with a high risk of periprosthetic fracture in recent years [10, 11]. SC stems can cause nontraumatic fractures, namely "atypical periprosthetic fractures," due to excessive taper-slip [12]. The causes can be multifactorial, including implant material, surface properties, compatibility with bone cement, cement mantle, tapered shape, and surgical technique. According to our basic study, the combination of "excessively polished Co-Cr-Mo alloy" and "medium viscosity bone cement" tends to cause excessive taper-slip. The incidence of atypical periprosthetic fractures among patients with SC stems is reported to be 0.4% (9/2, 021 joints) [12, 13]. Therefore, we are currently redeveloping SC stems that are safe and can achieve long-term durability by further clarifying the behavior of cemented stems.

References

1. Ling RSM, et al. The long-term results of the original Exeter polished cemented femoral component: a follow-up report. J Arthroplast. 2009;24:511–7.
2. Oe K, et al. Reattachment of an osteotomized greater trochanter in total hip arthroplasty using an ultra-high molecular weight polyethylene fiber cable. J Orthop Sci. 2018;23:992–9.
3. Usui M, et al. Experience with total hip arthroplasty using the Dall's approach for rheumatoid arthritis. Central Jpn Orthop Surg Traumatol. 2010;53:159–60. [In Japanese]
4. Kato M, et al. Alignment of SC-stem in total hip arthroplasty: comparison of extramedullary guide and centralizer. JSRA. 2015;45:155–6. [In Japanese]
5. Yang Y, et al. Evaluation of subsidence of cemented polished triple tapered stem: influence of stem design and bone cement. JSRA. 2015;45:615–6. [In Japanese]
6. Otani M, et al. Mid-term results of cemented THA in our hospital. JSRA. 2017;47:687–8. [In Japanese]
7. Nishino S, et al. Short-term results of hybrid total hip arthroplasty using SC stem for poor bone quality patients. Hip Joint. 2018;44:105–8. [In Japanese]
8. Otsuka H, et al. Alignment of cemented stem using posterior approach. Gifu Arthroplasty Forum. 2018;22:51–4. [In Japanese]
9. Watanabe R, et al. Bipolar hemiarthroplasty using SC stem. Hip Joint. 2020;46:102–5. [In Japanese]
10. Thien TM, et al. Periprosthetic femoral fracture within two years after total hip replacement: analysis of 437,629 operations in the Nordic arthroplasty register association database. J Bone Joint Surg Am. 2014;96:e167.
11. Palan J, et al. The influence of cemented femoral stem choice on the incidence of revision for periprosthetic fracture after primary total hip arthroplasty: an analysis of national joint registry data. Bone Joint J. 2016;98-B:1347–54.
12. Oe K, et al. An atypical periprosthetic fracture in collarless, polished, tapered, cemented stems of total hip arthroplasty: a report of five SC-stem. J Orthop Sci. 2021; ahead of print
13. Hirata M, et al. Relationship between the surface roughness of material and bone cement: an increased "polished" stem may result in the excessive taper-slip. Materials. 2021;14:3702.

12 DCM-J Natural-Hip System (Zimmer)

Nobuhiko Sugano

1 Development Process

In total hip arthroplasty (THA), cemented stems with various designs and concepts, were used in the early twenty-first century. Among these, some reports on early loosening of polymethylmethacrylate (PMMA) Precoat stems with a collar were published [1–3]. A PMMA precoat was applied to the proximal portion of the stem to prevent debonding of the stem from the cement, and a collar was used to resist subsidence of the stem. Zimmer, the manufacturer of Precoat stems, started a new stem development plan for the U.S., Europe, and Japan in 2002 and established a new stem design policy based on a thorough literature review on (1) collars, (2) stem surface treatment, (3) cement mantle, and (4) stem cross-sectional shape.

The collar was expected to reduce bone resorption by loading on the medial neck cortex, preventing stem subsidence, and providing positioning during stem insertion. However, there are no strong evidence of these benefits. The decision to use the collar was made because of its ease of acceptance by surgeons.

Few surgeons are aware of the definitions of "rough," "satin," "smooth," and "polished" in terms of surface finish. Precoat and other processes that adhere the cement to the stem are "rough," while Exeter and CPT are "polished" and most cemented stems in the United States are "satin" or "smooth". "Polished" stems slide at the interface with the cement; hence, a colorless design matches the concept. The "rough" design keeps the cement and the stem adherent to each other, and good results are obtained when the stem does not subside over the cement. However, when the adhesion between the cement and the stem is broken in cases with Precoat stems, excessive wear occurs between the cement and the "rough" stem surface, leading to the breakdown of stem fixation. On the other hand, the long-term results of Omnifit, Spectron, and Lubinus, which are collared stems with "satin" or "smooth" surface, are good [4–6] and this combination was selected as a candidate.

The most common opinion regarding the cement mantle thickness is 2 mm as a suitable cement mantle. Exeter and Charnley also show good long-term results with a minimum cement mantle of 2 mm. On the other hand, there is no evaluation of results based on a single factor of cement mantle thickness, although US stems, such as Omnifit, Endurance, and VerSys have a 2 mm cement mantle listed in the procedure manual. Surgeons may choose a stem that is one size larger at their own discretion, in which case the VerSys has a 0.5 mm cement mantle. On the other hand, stems such as Charnley-Kerboull remove as much cancellous bone as possible and emphasize canal filling with a thin cement mantle, which is accepted in French-speaking countries as a 'French paradox' [7]. However, a thin cement mantle is a risk factor for osteolysis [8], and 75% of surgeons in the design team favored a 2 mm cement mantle.

In the discussion of stem cross-sectional shape, the European stem has historically been flatter proximally than the US filling type; however, in the literature, rotational stability is important "be it wide and flat or filling".

In 2003, while this project was in progress, Zimmer acquired CenterPulse, and the new stem development plan was taken over by the DCM-J Natural-Hip system development, leaving only the plan for the Japanese market.

2 Features of the DCM-J Natural-Hip System

The stem design is based on CT data from Japanese patients and has a variety of designs and sizes to accommodate various medullary configurations, from developmental hip dislocation to osteoporotic femur with an

N. Sugano (✉)
Department of Orthopaedic Medical Engineering, Osaka University Graduate School of Medicine, Suita, Japan
e-mail: n-sugano@umin.net

Y. Nakashima (ed.), *Advances in Total Hip Arthroplasty*, https://doi.org/10.1007/978-981-97-2461-1_12

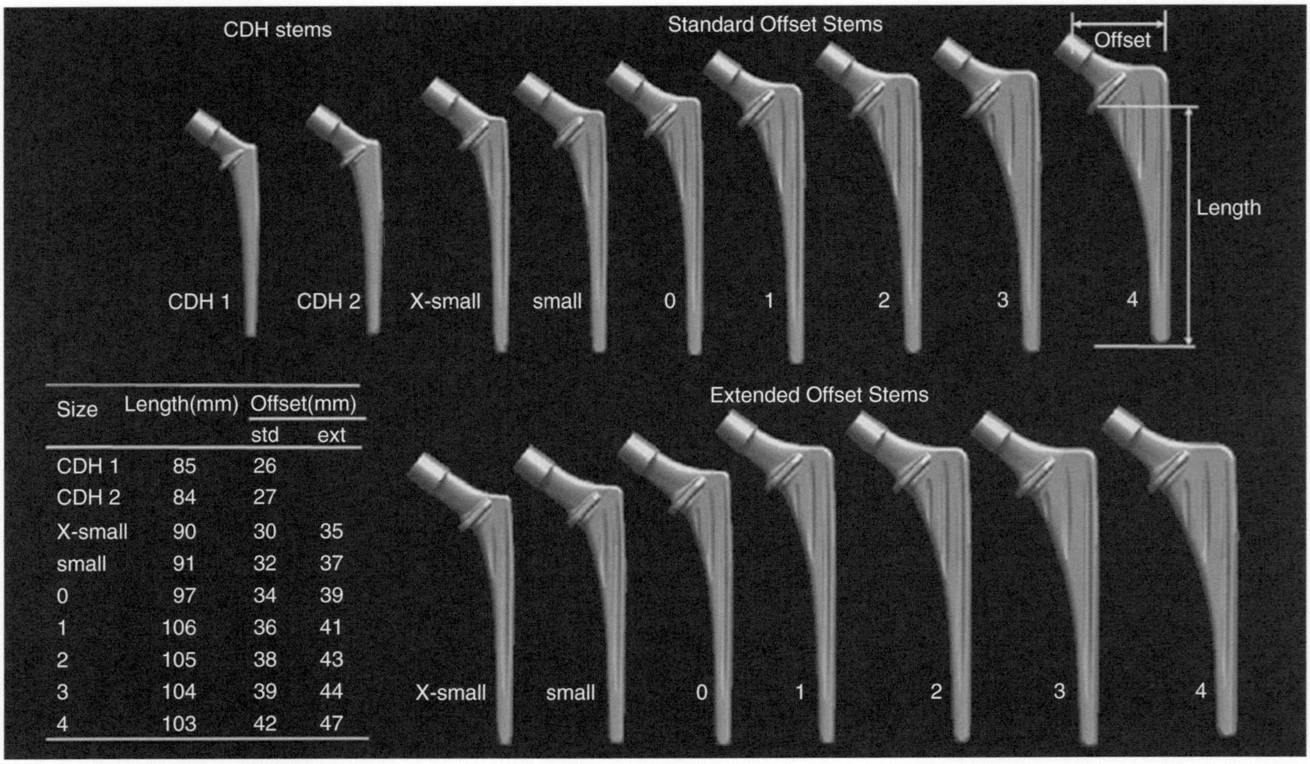

Size	Length(mm)	Offset(mm) std	Offset(mm) ext
CDH 1	85	26	
CDH 2	84	27	
X-small	90	30	35
small	91	32	37
0	97	34	39
1	106	36	41
2	105	38	43
3	104	39	44
4	103	42	47

Fig. 12.1 Size variations of the DCM-J Natural-Hip system

enlarged medullary cavity (Fig. 12.1). A common RASP system allows insertion of either cemented stem with two different offsets, facilitating adjustment of the medullary cavity geometry and leg length offset. Rasps affecting the cement mantle can also be selected from two types: 0.5 mm and 1.5 mm oversize for the entire circumference of the stem. The DCM-J stem has a rectangular cross-sectional shape for high rotational stability, and the tapered design at the distal end and grooves on the surface are expected to reduce strain on the distal cement mantle. The satin-finish surface of the stem does not adhere to the cement, which allows the stem to be removed while leaving the cement mantle.

The neck-shaft angle is anatomical at 127°, and the design provides a wide range of flexion motion, partly owing to the unique reduced thin neck design made of high-strength forged cobalt-chrome alloy.

To ensure that the stem fits centrally within the cement mantle, a polyethylene centralizer with four blades was selectively mounted distally to match the diameter of the medullary cavity, while proximal to the stem, the centralizer was built under the medial collar, and the instrumentation attached to the stem inserter performed the centralizer function on the lateral side.

3 Clinical Results

Clinical use began in 2006, and DCM-J was used in 105 joints in 2010, with no mechanical loosening of the stem [9]. There was no difference in clinical outcomes between the two types of rasps, but the 0.5-mm oversized minimal rasp led to more proximal bone atrophy due to stress shielding than did the 1.5-mm oversized rasp. Distal cortical hypertrophy increased in frequency over time and was observed in 25% of cases at 10 years. However, because the stem length was shorter than that of the U.S. stem, only mild hypertrophy was observed from the distal portion of the lesser trochanter, and the stress concentration on the apical cement of the stem was thought to be reduced as intended by the design (Fig. 12.2).

The DCM-J was also used for five developmentally dislocated hips that required subtrochanteric shortening osteotomy without the need for a CDH stem. All patients were able to walk with full weight bearing immediately after surgery. The design of the subtrochanteric shortening osteotomy was transverse, and the resected bone was cut in half along the axis, which was used as on-lay double bone plates bridging the osteotomized parts. The osteotomy site was fused, which resulted in good long-term outcomes.

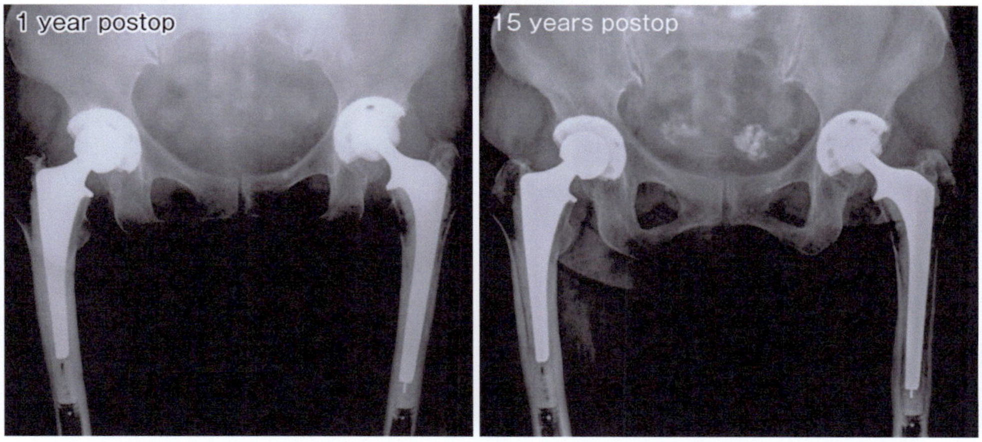

Fig. 12.2 Radiographic images of a 63-year-old woman with bilateral hip osteoarthritis treated with right DCM-J (3 std) and left Omnifit cemented stems at 1 year and 15 years postoperatively. The left hip has more cortical hypertrophy distally than the right hip, suggesting that the distal cement stress concentration was lower with the DCM-J

4 Summary

The DCM-J Natural-Hip system provides good cement fixation and an adjustable leg length offset for dysplastic femora to Dorr C-type osteoporotic femora, with limited stress shielding and excellent clinical results for 15 years. A composite beam with a collar is a design that should be reintroduced because it is thought to have a lower risk of stem subsidence and associated peri-prosthetic fractures than the taper-slip type.

References

1. Collis DK, et al. Loosening rates and bone lysis with rough finished and polished stems. Clin Orthop Relat Res. 1998;355:113–22.
2. Sylvain GM, et al. Early failure of a roughened surface, precoated femoral component in total hip arthroplasty. J Arthroplast. 2001;16:141–8.
3. Ong A, et al. Early failure of precoated femoral components in primary total hip arthroplasty. J Bone Joint Surg Am. 2002;84:786–92.
4. Jaffe WL, et al. Normalized and proportionalized cemented femoral stem survivorship at 15 years. J Arthroplast. 1999;14:708–13.
5. Garellick G, et al. The Charnley versus the Spectron hip prosthesis: clinical evaluation of a randomized, prospective study of two different hip implants. J Arthroplast. 1999;14:407–13.
6. Alho A, et al. Cemented Lubinus and Furlog total hip endoprosthesis: a 12-year follow-up study of 175 hips comparing the cementing technique. Arch Orthop Trauma Surg. 2000;120:276–80.
7. Langlais F, et al. The 'French paradox'. J Bone Joint Surg Br. 2003;85:17–20.
8. Kawate K, et al. Thin cement mantle and osteolysis with a precoated stem. Clin Orthop Relat Res. 1999;365:124–9.
9. Hashimoto Y, et al. The effects of rasp oversize on the clinical and radiographic outcomes of total hip arthroplasty with a collared satin-finished composite beam cemented stem. J Arthroplast. 2021;36:2055–61.

APS 2008 (Zimmer)

13

Ayumi Kaneuji

1 Concept of the Developed Femoral Implant

The APS Natural-Hip™ (APS) system, in which APS stands for anatomic plasma spray, is a stem that can be inserted naturally according to bone morphology and was marketed by Zimmer (Warsaw, IN, USA). The main development concepts were as follows:

1. It is possible to treat cases of excessive neck-shaft angle, excessive femoral anteversion, and narrow femoral canal diameter.
2. Anatomical proximal fit and fill fixation in the proximal femoral canal of Japanese people.
3. Achievement of a short porous coating by proximal morphological adaptation.
4. Short porous coating to minimize stress shielding.
5. Prevention of distal diffusion of osteolysis by full circumferential porous coating.
6. Bone preservation by short stem length.
7. Easy removal of the implant in case of revision surgery.

In the 1990s, anatomic stems were marketed to ensure secure fixation proximal to the medullary cavity and prevent excessive stress shielding. However, the anatomic stem (Zimmer, Inc.) at that time was made based on the anatomical characteristics of Caucasian individuals, and it was difficult to adapt the stem to all Japanese patients. Therefore, the need for a stem that could be easily adapted to Japanese patients was the impetus for development. First, the proximal bone morphology of patients with secondary osteoarthritis (OA), which proceeds from developmental dysplasia of the hip (DDH) and is common in Japan, was analyzed to see if it differed from normal femoral bone morphology. A computer-aided design system was used to examine the angles of inclination of the proximal medullary cortex in coronal and sagittal planes, medial and lateral, and anterior and posterior from CT data of 36 control hips and 113 hips of patients with OA secondary to DDH. It was found that 96% of secondary OA could be classified into three femoral canal morphologies [1]: Type A, which accounted for 89% of the normal morphology, was present in only 42% of the secondary OA group; type B, which had a large anterior tilt at the proximal-posterior region, was present in 26%; and type C, which had a steep proximal-medial tilt, was present in 30%. These results suggest that three types of stems with different shapes are necessary to fit each type of bone morphology in Japanese patients. Additionally, the femoral canal diameter and femoral neck length of Japanese people were used as references for stem size determination [2]. The APS standard anteverted stem, which is suitable for type A, was marketed overseas as an anatomical stem with a 12° stem anteversion. The standard non-anteverted stem, which is more suitable for type B, has a corrected anteversion, and the body is the same as that

A. Kaneuji (✉)
Kanazawa Medical University, Kanazawa, Japan
e-mail: kaneuji@kanazawa-med.ac.jp

Y. Nakashima (ed.), *Advances in Total Hip Arthroplasty*, https://doi.org/10.1007/978-981-97-2461-1_13

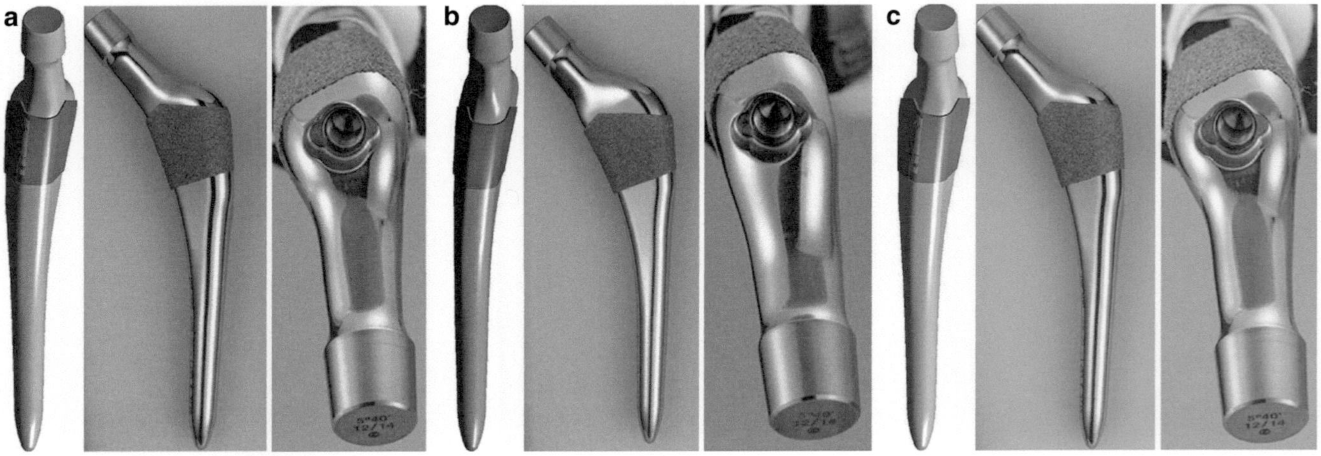

Fig. 13.1 (**a**) APS standard anteverted stem. (**b**) APS standard non-anteverted stem. (**c**) APS reduced metaphysis non-anteverted stem

of the standard stem; however, the anteversion is 0°. The reduced metaphysis type for type C also had 0° anteversion, a steep proximal medial tilt, and a proximal transverse section that was rounder in shape on the lateral side. This was because type C had a significantly greater number of cases with a circular shape cross-section. The latter two types were limited to the domestic market (Fig. 13.1a–c) and were sold from October 2008 to June 2018.

2 Clinical Results

Clinical outcomes and bone mineral density changes in APS stems have been reported. Stress shielding inhibition and bone mineral density loss were less than those of conventional cementless stems (Fig. 13.2); the short-term results were reasonable with the concept about stress shielding [3, 4]. All total hip arthroplasties using APS at the Department of Orthopedic Surgery, Kanazawa Medical University included 740 joints. The mean age was 60.0 years, and the mean follow-up period was 8.0 years; 6 revision cases were present. Three cases underwent revision owing to peri-prosthetic fracture (PPF) within 1 month after surgery, 2 cases of revision were because of PPF at 5 years after surgery, and 1 case of revision at 7 years after surgery was owing to frequent dislocation. The APS had a short porous coating up to the lesser trochanter, but the bone ingrowth and subsequent fixation were excellent; cases of revision owing to loosening were absent. However, some cases of thigh pain, which is a characteristic of cementless stems, could be observed. The frequency of PPF within 1 month after surgery was 3 joints (0.4%); however, this may be because this stem was initially used for bone fragility cases. The frequency did not increase after October 2009 because it was not used for bone fragility cases.

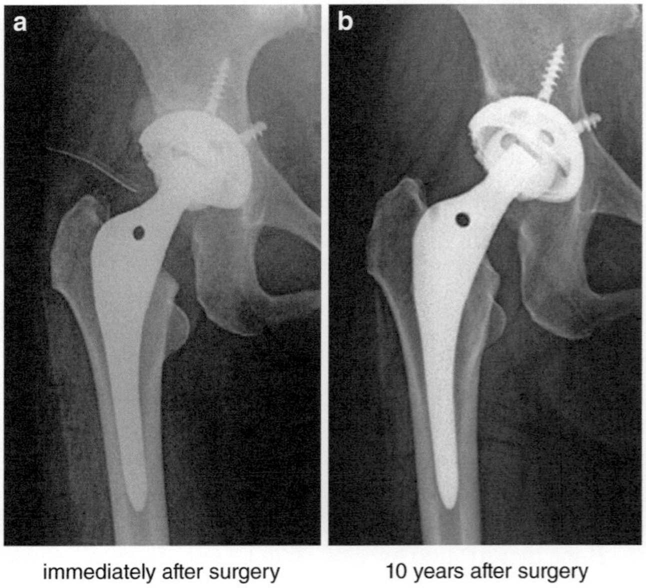

Fig. 13.2 APS use case: A 46-year-old woman with osteoarthritis of the hip joint. A 10-year postoperative case with minimal stress shielding. APS, APS Natural-Hip™ system. (**a**) Immediately after surgery. (**b**) 10 years after surgery

References

1. Kaneuji A, et al. Three-dimensional morphological analysis of the proximal femoral canal using a computer-aided design system in Japanese patients with osteoarthrosis of the hip. J Orthop Sci. 2000;5:361–8.
2. Kaneuji A, et al. Medullary cavity morphology classification and proximal femoral characteristics in secondary osteoarthritis of the hip joint. Hip Joint. 2003;29:74–81. In Japanese
3. Kaneuji A, et al. APS natural-hip™ stem complication study and bone mineral density changes in three different designs. Chubu Nihon Seikei Geka Saigai Geka Gakkai Zasshi. 2013;56:19–20. In Japanese
4. Takahashi E, et al. Femoral bone mineral density changes at 5 years postoperatively in patients with a new anatomical stem. Nihon Jinkou Kansetsu Gakkaishi. 2015;45:147–8. In Japanese

Part II

Anatomy of the Hip Joint

14 Functional Anatomy of the Hip Joint Specific to THA

Hideki Fujii

1 Bone Morphology and Nomenclature Specific to Total Hip Arthroplasty

Understanding the origin and function of the periprosthetic muscles and neurovascular system is important in total hip arthroplasty (THA) from the viewpoint of the use of retractors and implant placement techniques and prevention of complications, such as dislocation and neurovascular injury. The hip joint's anatomy can be easily understood by dividing it into anatomical parts based on its deep structures. The location and names of intraoperative landmarks in bone morphology and the structure and function of intra-articular joints and joint capsules, deep and shallow muscles, tendons, and neurovascular routes must be understood.

1.1 Anatomical Names of the Acetabulum and Proximal Femur

1.1.1 Calcar Femorale

The calcar femorale is an important anatomical structure of the proximal femur. It is a spiculated intramedullary structure posteriorly medial to the femoral neck but deep to the lesser trochanter that mechanically reinforces the medullary cavity of the neck. It is an indicator of anteversion when placing the stem in cases without an advanced deformity of the proximal femur [1–4].

1.1.2 Trochanteric Fossa

The conjoint tendon is attached to the trochanteric fossa, and the ischiofemoral ligament is attached to it. By focusing on the attachment of the trochanteric fossa, the short external rotator muscles can be preserved during THA using the anterior approach.

H. Fujii (✉)
Department of Orthopedic Surgery, The Jikei University School of Medicine, Minato City, Japan
e-mail: h.fujii@jikei.ac.jp

1.1.3 Intertrochanteric Line

The intertrochanteric line is the anterior prominence of the trochanteric section of femur. The lateral ridge of the iliofemoral ligament is attached to the transverse fibrous bundle of the iliofemoral ligament, and the medial ridge extends anteriorly to the lesser trochanter, where the longitudinal fibrous bundle of the iliofemoral ligament is attached to the medial ligament.

1.1.4 Bald Spot

The bald spot is a smooth bony surface without tendon attachment that exists on the lateral aspect of the greater trochanter. Retractor placement in this area can reduce soft tissue damage [5].

1.2 Function and Characteristics of the Hip Capsular Ligament

1.2.1 Attachment of the Hip Capsular Ligament

The hip joint capsule is composed of synovial tissue in the inner layer and ligaments in the outer layer. The ligaments attach to and surround the synovial tissue. The capsular ligament originates at the acetabular rim and terminates distally at the femur. Three ligaments contribute to joint stability: the iliofemoral, pubofemoral, and ischiofemoral ligaments (Fig. 14.1) [6–9].

1.2.2 Iliofemoral Ligament

The iliofemoral ligament is divided into superior and inferior fiber bundles. The superior fiber bundle originates from the acetabular margin below the anterior inferior iliac spine and attaches to the lateral prominence of the intertrochanteric line on the anterior portion of the proximal femoral epiphysis. The inferior fibers run wide from above the neck posteriorly to the ischiofemoral ligament and attach to it. Inferior fiber bundle arises from the acetabular rim and terminates broadly on the anterior prominence of the lesser trochanter, medial to the intertrochanteric line. It is a strong ligament

Y. Nakashima (ed.), *Advances in Total Hip Arthroplasty*, https://doi.org/10.1007/978-981-97-2461-1_14

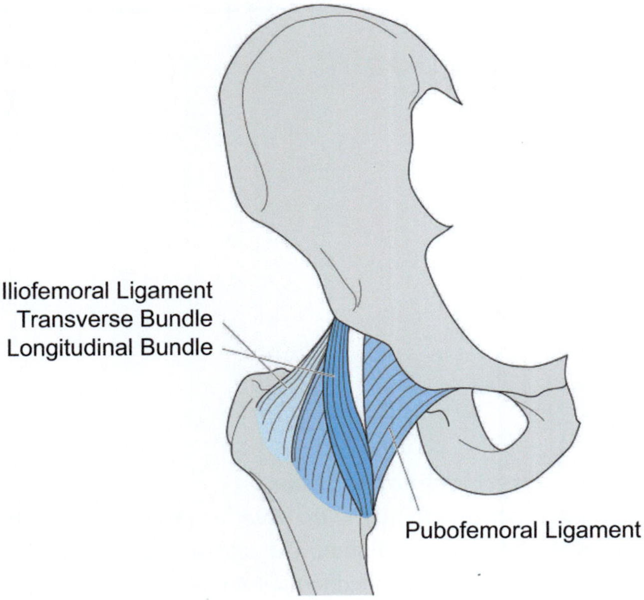

Fig. 14.1 Anterior hip joint and glenohumeral ligaments

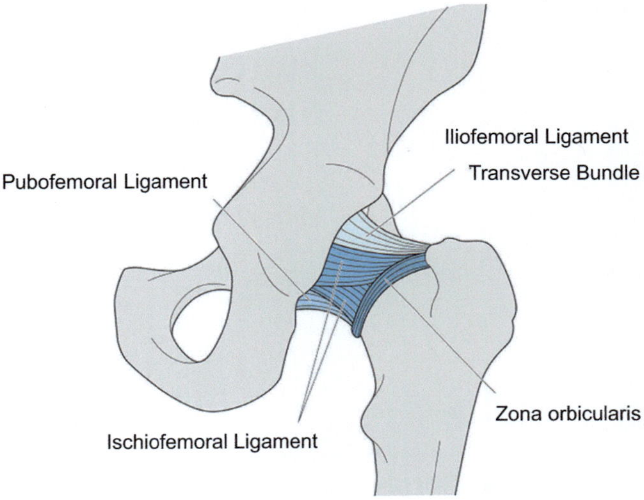

Fig. 14.2 Articular capsule and articular ligament of the posterior hip joint

that limits the hyperextension of the hip joint and also controls adduction and external rotation. It relaxes with a mild flexion of the hip joint.

1.2.3 Pubofemoral Ligament

The pubofemoral ligament originates from the pubic region of the acetabulum pubis and attaches anteriorly to the lesser trochanter. The hip joint is medially reinforced.

1.2.4 Ischiofemoral Ligament

The ischiofemoral ligament originates from the posterior margin of the acetabulum and attaches to the transition from the trochanteric fossa to the concavity of the trochanteric fossa. The conjoined tendon attached to the concavity is often partially fused with the ischiofemoral ligament, but it cannot be separated from the ischiofemoral ligament. Understanding the attachment area of the ischiofemoral ligament allows an anterior approach to be performed without damaging the external rotator muscles, including the joint tendons.

1.2.5 Zona Orbicularis (Fig. 14.2)

A band-like bundle of fibers in the posterior aspect of the joint capsule where the ischiofemoral ligament attaches is the zona orbicularis. It is a free, ring-shaped band that wraps around the central posterior aspect of the neck and is observed intraoperatively as a white band-like prominence 1 cm above the tear scar. It can also be described as a ring-shaped ligament surrounding the narrowest part of the femoral neck and constricting the femoral head in drawstring fashion [10].

1.3 Intra-Articular Structures

1.3.1 Ligamentum Teres of the Femur (Fig. 14.3)

The ligamentum teres connects the acetabulum and femoral head. The ligament is tense when the hip joint is flexed and abducted and relaxes when the hip joint is abducted and flexed. It is covered by the synovial membrane.

1.3.2 Transverse Acetabular Ligament

The transverse acetabular ligament is attached to the acetabular fossa in an anteroposterior direction and forms the articular fossa with the articular lip. The running of the transverse acetabular ligament is reportedly an indicator of cup placement. However, reportedly, the unique morphology of the acetabulum should be considered in cases of strong posterior pelvic tilt and acetabular dysplasia [11–13].

1.3.3 Articular Labrum

The articular labrum is attached to the acetabular rim. It stabilizes the head and plays a role in shock absorption. The sealing effect traps joint fluid, and the suction effect enhances the centripetal afferents of the head. A wide labrum may be observed in acetabular dysplasia. The inferior acetabulum labrum transitions to the transverse acetabular ligament.

1.3.4 Acetabular Fossa (Fig. 14.4)

The center of the acetabular fossa is anterior to the center of the acetabulum, and its position is unaffected by dysplasia or osteoarthritis. The preconception that the center of the acetabulum coincides with the center of the acetabular fossa leads to the risk of anterior wall injury with reaming [14].

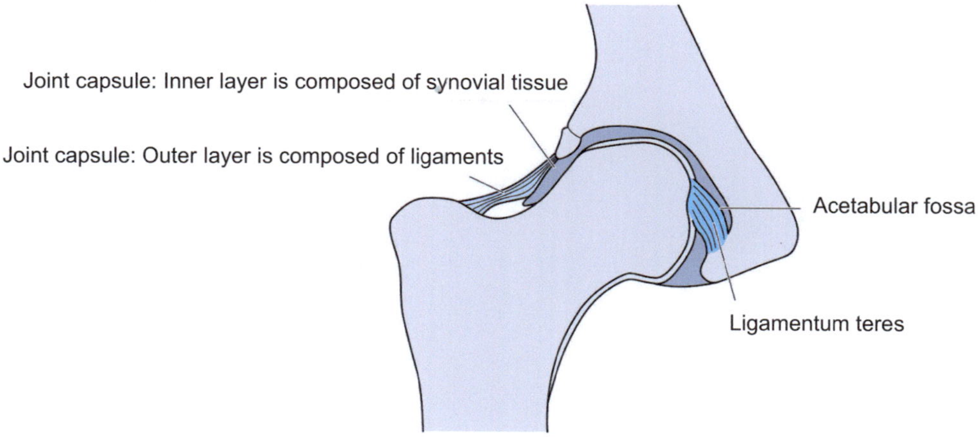

Fig. 14.3 Articular capsule and osseous ligament. The inner layer of the joint capsule is composed of synovial tissue, and the outer layer is composed of ligaments

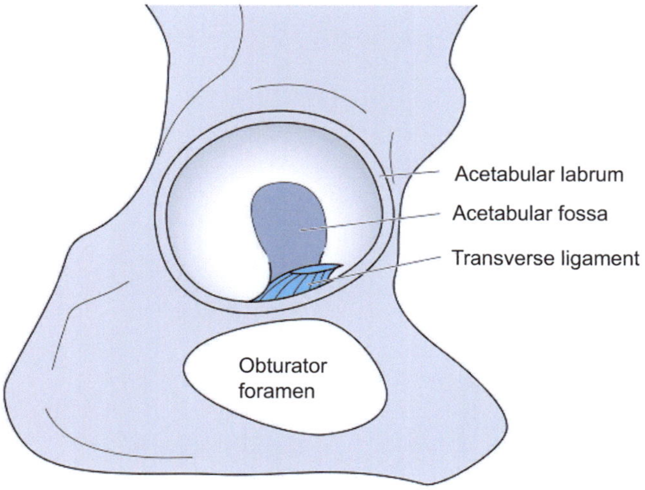

Fig. 14.4 Hip labrum and transverse ligament, with the lacrimal scar

1.4 Posterior Hip Joint Supporting Tissues and Short External Rotators (Figs. 14.5, 14.6, and 14.7)

1.4.1 Conjoined Tendon

Three muscles are collectively referred to as the conjoined tendons: gemellus superior, obturator internus, and gemellus inferior. The conjoined tendon is not clearly delineated from the ischiofemoral ligament at the attachment of the trochanteric fossa. Its position should be assessed during the anterior approach to avoid unintentional dissection [15, 16]. The function of the obturator internus is to control pelvic rotation in the standing position. The obturator internus originates from the medial obturator foramen and obturator membrane and stops at the trochanteric fossa while inverting below the sciatic spine. The medial obturator foramen is a muscle that controls hip joint rotation. External rotation occurs at 90° of flexion, and internal rotation occurs at 90° and above of flexion.

1.4.2 Piriformis Muscle

The pisiform muscle begins at the lateral aspect of the anterior foramen of the sacrum and the margin of the greater sciatic notch of the sciatic bone and attaches to the trochanteric fossa. It also acts as a lengthening and external rotator for the hip joint.

1.4.3 External Obturator Muscle

The external obturator muscle performs an external rotatory action during flexion. It runs posteriorly upward from the foramen obturatum and the foramen obturatum membrane on the lateral surface of the os pubis/ischii to the femoral trochanteric fossa. The external obturator muscle is connected to the lower part of the joint capsule as it travels along the inferior aspect of the femoral neck, and the posterior fibers of the external obturator muscle are connected to the quadratus femoris muscle. The external obturator muscles extend the hip joint as the angle of flexion increases, producing adduction and external rotation in the anterior and horizontal planes. The external rotator muscles of the piriformis muscles do not change their external rotation action, even if the hip flexion angle changes. The posterior approach in THA has been reported to increase internal rotation braking by reattaching the external obturator muscle, piriformis muscle, and common tendon to the greater trochanter [17, 18].

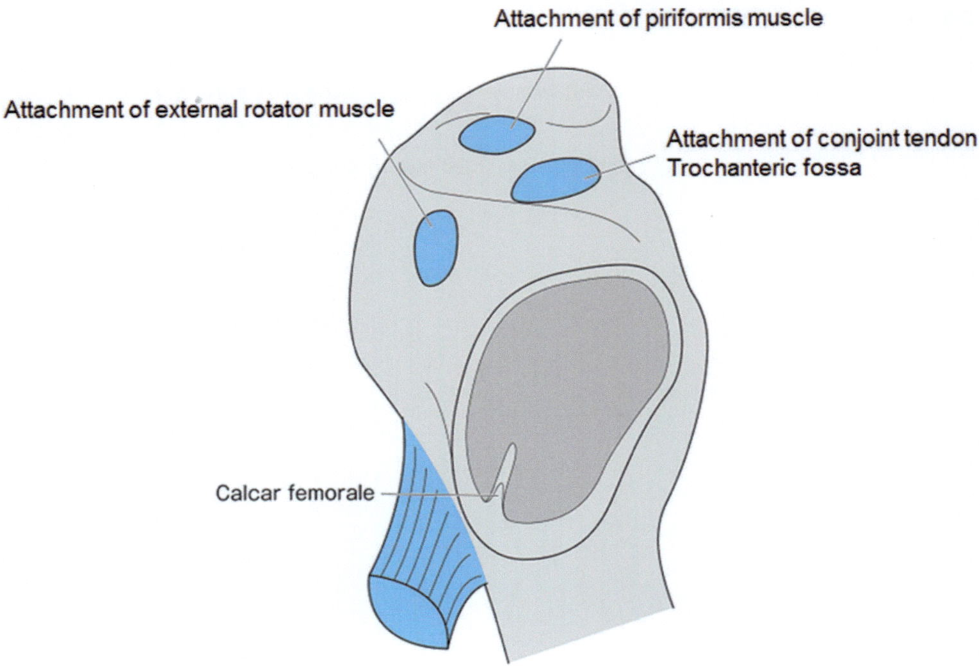

Fig. 14.5 Bone attachments of the short external rotators

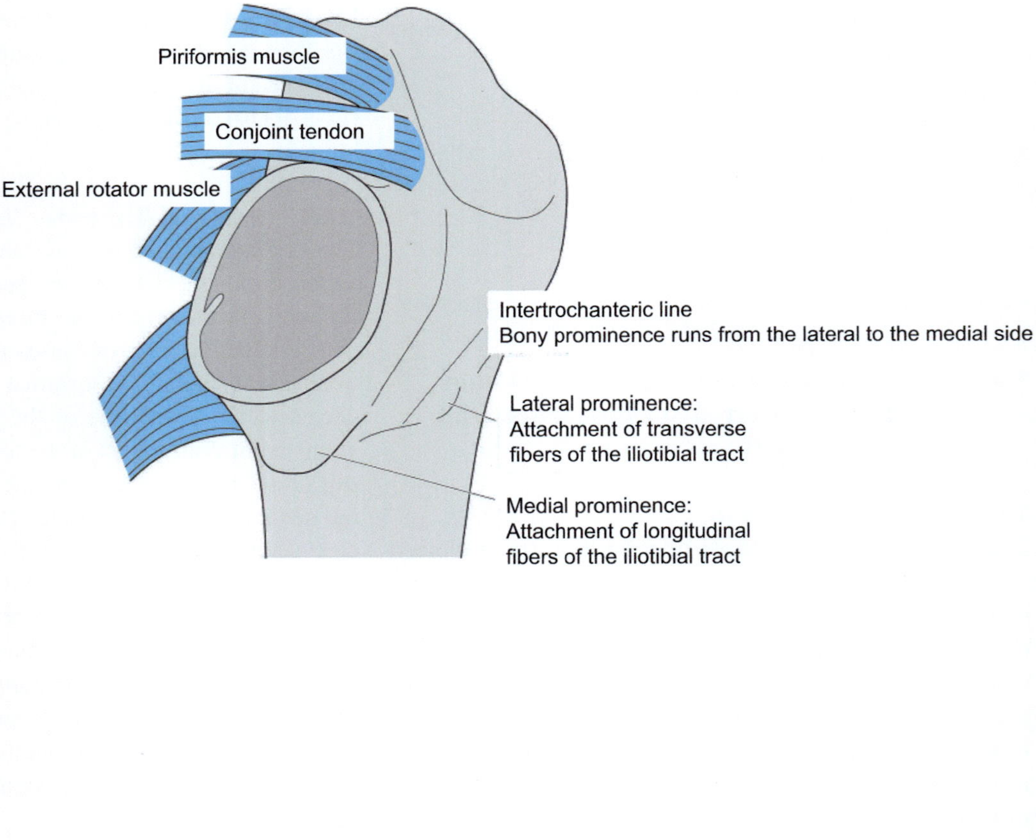

Fig. 14.6 Attachment area of the short external rotator muscle group and tendon running

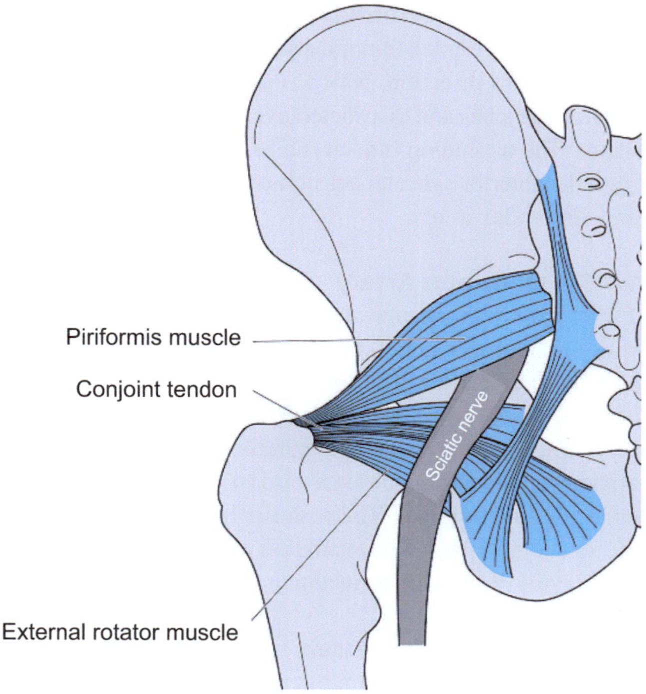

Fig. 14.7 Short external rotators viewed from behind the hip joint

1.5 Muscles of the Anterior Surface of the Hip Joint Connecting the Pelvis and Femur

1.5.1 Gluteus Maximus Muscle

Superficial fibers of the gluteus maximus muscle connect to the shallow part of the iliotibial ligament. The deep fibers of the iliotibial tubercle migrate into the deep fibers of the gluteus maximus muscle, penetrating the vastus lateralis muscle, passing through a 3- to 5-cm-long hole separated by the fascia femoris, and stopping at a location approximately 2 cm caudal to the greater trochanter and 2 cm anterior to the cranial surface of the gluteal tuberosity.

1.5.2 Gluteus Medius Muscle

The proximal portion consists of three parts: gluteal fossa, gluteal posterior surface, and posterior inferior end of the iliac crest. The distal tendon is divided into lateral and posterior tendons. The function of the anterior fibers of the middle gluteus medius muscle is abduction, and those of the posterior fibers are stabilization and external rotation of the hip joint [19, 20].

1.5.3 Gluteus Minimus Muscle

The gluteus minimus muscle presses the femoral head against the acetabulum and contributes to hip stability by acting as a barrier to the upward movement of the femoral head. The gluteus minimus covers a wide area of the outer surface of the articular capsule and is partially connected to it. The tendinous attachments of the gluteus minimus are concentrated anterior to the greater trochanter. Care must be taken not to detach the gluteus minimus attachment when placing the stem in the anterior approach during THA [10, 21].

1.5.4 Iliocapsularis Muscle

The iliocapsularis muscle covers the anterior and medial surfaces of the joint capsule. However, its true function remains unknown. The possibility of its involvement in preventing impingement of the anterior surface of the synovium is supported. Hypertrophy of the iliocapsularis due to hip dysplasia has been speculated to allow the iliocapsularis to stabilize the anterior part of the hip joint [22, 23].

1.5.5 Rectus Femoris and Reflected Head

The straight and reflected heads of the rectus femoris muscle originate over a wide area of the inferior side of the anterior inferior iliac spine. Owing to their proximity to femoral neurovascular structures, care must be taken to avoid injury [24].

1.5.6 Iliopsoas Muscle

The iliopsoas muscle arises from the iliac fossa and anterior inferior iliac spine and merges with the psoas major muscle to form the iliopsoas muscle, with the iliopsoas muscle fibers ending inferiorly beyond the obturator teres major muscle fibers anteriorly. The fibers of the anterior inferior iliopsoas muscle act during the initial flexion of the hip joint and function as a starter for hip flexion. The psoas tendon may cause pain due to impingement by the edge of acetabular cup.

1.6 Nerves around the Hip Joint

1.6.1 Femoral Nerve (Femoral N) (Fig. 14.8)

The femoral nerve is closest to the acetabular rim at the anterior 3 o'clock position of the labrum, and the risk of femoral nerve injury during retractor placement in this region is high [25] This risk has been reported to be greater with supine hip replacement [25, 26].

1.6.2 Sciatic Nerve (Fig. 14.7)

The sciatic nerve is the longest and thickest nerve in the body and originates from the sacral plexus (L4-S3). It leaves the lumbosacral plexus, passes over the anterior surface of the piriformis muscle, passes through the foramen ischiadicum majus with the inferior pudendal nerve, exits the pelvis, descends into the posterior surface of the thigh, and branches to the femoral flexor muscles and the adductor magnus muscle.

1.6.3 Lateral Femorocutaneous Nerve

Injury has been reported in THA using the direct anterior approach (DAA) in patients with a fan-type branching pattern of

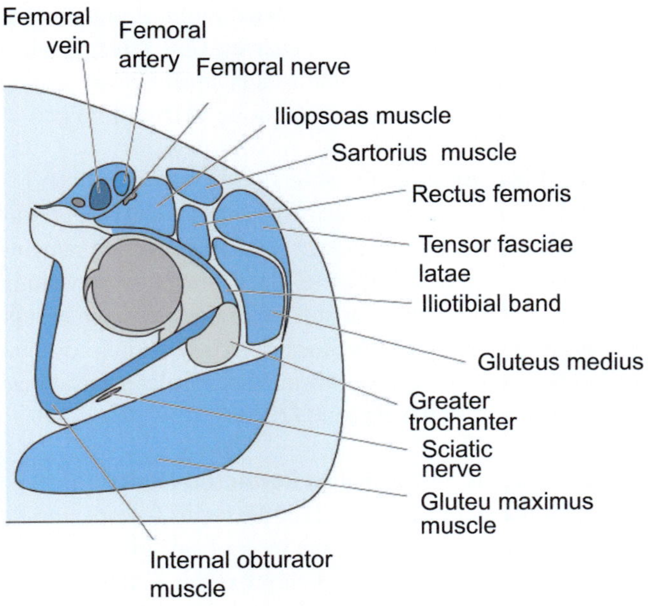

Fig. 14.8 Position of the hip labrum and neurovasculature

the lateral femorocutaneous nerve (LFCN) [27, 28] Forty-two percent of LFCNs were reported to cross the DAA skin incision along the midline of the tensor fascia femoris muscle [29].

1.6.4 Superior Gluteus Nerve

In severely dysplastic hips, the superior gluteus nerve may travel within 3 cm of the tip of the greater trochanter [30].

1.7 Blood Vessels around the Hip Joint

1.7.1 Femoral Artery (Fig. 14.8)

In Crowe III and IV hips, the femoral vessels pass close to the anterior wall of the acetabulum, and vascular injury should be noted [31].

1.7.2 Superior Gluteal Artery

The superior gluteal artery is the largest branch of the internal iliac artery. It passes through the foramen magnum (foramen piriformis), together with the superior gluteal vein and superior gluteal nerve, exiting the gluteal region, and dividing into shallow and deep branches. The shallow branch runs between the gluteus maximus and middle gluteus medius and is distributed to the upper part of the gluteus maximus and middle gluteus medius. The deep branch runs between the gluteus medius and gluteus minimus and is distributed to these muscles before reaching the tensor fascia femoris.

1.7.3 Medial Circumflex Femoral Artery

The medial circumflex femoral artery runs posteriorly and medially above the adductor muscle group, rounds the femoral head, and reaches the upper posterior aspect of the femur.

1.7.4 Lateral Circumflex Femoral Artery

The lateral circumflex femoral artery is a large artery running in an outward direction, which is divided into superior and inferior branches and distributed over the entire length of the femur. The ascending branch runs around the femoral head from the anterior side and anastomoses with the medial femoral circumflex artery.

1.7.5 Obturator Artery

The obturator artery runs along the lateral wall of the pelvis with the obturator nerve, through the obturator canal to the anterior pelvic wall. The anterior branch of the obturator nerve divides into anterior and posterior branches after branching into the external obturator muscle. The anterior branch runs inward and downward to the upper part of the adductor pollicis brevis. The posterior branch passes between the sciatic tubercle and acetabulum and is distributed in the deep layers of the gluteus medius.

1.7.6 Pelvic Blood Vessels

In the posterior acetabulum, the pelvic vessels are absent or located 31 mm from the acetabular surface, allowing for safe acetabular screw fixation [32, 33].

References

1. Hammer A. The calcar femorale-a new perspective. J Orthop Surg (Hong Kong). 2019;27:2309499019848778.
2. Mei J, et al. Strategies for managing the destruction of calcar femorale. BMC Musculoskelet Disord. 2021;22:460.
3. Tetsunaga T, et al. Calcar femorale in patients with osteoarthritis of the hip secondary to developmental dysplasia. Clin Orthop Surg. 2017;9:413–9.
4. Tsukeoka T, et al. The T-line as an intraoperative landmark for reproducing native femoral anteversion during hip arthroplasty. Arch Orthop Trauma Surg. 2014;134:873–9.
5. Gardner MJ, et al. Anatomy of the greater trochanteric 'bald spot': a potential portal for abductor sparing femoral nailing? Clin Orthop Relat Res. 2008;466:2196–200.
6. Hidaka E, et al. Ligament strain on the iliofemoral, pubofemoral, and ischiofemoral ligaments in cadaver specimens: biomechanical measurement and anatomical observation. Clin Anat. 2014;27:1068–75.
7. Walters BL, et al. New findings in hip capsular anatomy: dimensions of capsular thickness and pericapsular contributions. Arthroscopy. 2014;30:1235–45.
8. Wagner FV, et al. Capsular ligaments of the hip: an anatomic, histologic, and positional study in cadaveric specimens with MR arthrography. Radiology. 2012;263:189–98.
9. Walters J, et al. Gluteus minimus: observations on its insertion. J Anat. 2001;198:239–42.
10. Ito H, et al. The proximal hip joint capsule and the zona orbicularis contribute to hip joint stability in distraction. J Orthop Res. 2009;27:989–95.
11. Fujita K, et al. The use of the transverse acetabular ligament in total hip replacement: an analysis of the orientation of the trial acetabular component using a navigation system. Bone Joint J. 2014;96-B:306–11.

12. Miyoshi H, et al. Anteversion of the acetabular component aligned with the transverse acetabular ligament in total hip arthroplasty. J Arthroplast. 2012;27:916–22.
13. Abe H, et al. Is the transverse acetabular ligament a reliable cup orientation guide. Acta Orthop. 2012;83:474–80.
14. Shimodaira H, et al. The acetabular fossa may not be located at the true center of the acetabulum: a detailed analysis using preoperative CT images. J Bone Joint Surg Am. 2018;100:e27.
15. Tamaki T, et al. An anatomic study of the impressions on the greater trochanter: bony geometry indicates the alignment of the short external rotator. J Arthroplast. 2014;29:2473–7.
16. Ito Y, et al. Anatomic mapping of short external rotators shows the limit of their preservation during total hip arthroplasty. Clin Orthop Relat Res. 2012;470:1690–5.
17. Fujii H, et al. Preventing postoperative prosthetic joint dislocation by repairing obturator externus in total hip arthroplasty performed via the posterior approach. Arthroplasty. 2020;2:33.
18. Kawaguchi Y, et al. Functional and clinical anatomy of the obturator externus muscle: cadaveric studies and clinical findings for total hip arthroplasty in the posterior approach. J Orthop. 2021;25:93–7.
19. Gottschalk F, et al. The functional anatomy of the tensor fasciae late and gluteus medius and minimus. J Anat. 1989;166:179–89.
20. Flack NAMS, et al. The anatomy of the hip abductor muscles. Clin Anat. 2014;27:241–53.
21. Beck M, et al. The anatomy and function of the gluteus minimus muscle. J Bone Joint Surg Br. 2000;82B(3):358–63.
22. Mashriqi F, et al. The functional and surgical relevance of the iliocapsularis muscle: an anatomical review. Anatomy. 2017;11:146–8.
23. Saito M, et al. The role of iliocapsularis in hip pathology: a scoping review. J Hip Preserv Surg. 2021;8:145–55.
24. Ryan JM, et al. Origin of the direct and reflected head of the rectus femoris: an anatomic study. Arthroscopy. 2014;30:796–802.
25. Yoshino K, et al. Anatomical implications regarding femoral nerve palsy during a direct anterior approach to total hip arthroplasty: a cadaveric study. J Bone Joint Surg Am. 2020;102:137–42.
26. Takada R, et al. Does surgical body position influence the risk for neurovascular injury in total hip arthroplasty? A Magnetic resonance imaging study. Orthop Traumatol Surg Res. 2021;107:102817.
27. Ozaki Y, et al. Preoperative ultrasound to identify the distribution of the lateral femoral cutaneous nerve in total hip arthroplasty using a direct anterior approach. SICOT J. 2018;4:42.
28. Rudin D, et al. The anatomical course of the lateral femoral cutaneous nerve with special attention to the anterior approach to the hip joint. J Bone Joint Surg Am. 2016;98:561–7.
29. Sugano M, et al. Anatomical course of the lateral femoral cutaneous nerve with special reference to the direct anterior approach to total hip arthroplasty. Mod Rheumatol. 2020;30:752–7.
30. Ikeuchi M, et al. Safe zone for the superior gluteal nerve in the transgluteal approach to the dysplastic hip: intraoperative evaluation using a nerve stimulator. Acta Orthop. 2006;77:603–6.
31. Maeda Y, et al. Risk of injury to the femoral blood vessels based on the extent of acetabular dysplasia in total hip arthroplasty. J Artif Organs. 2019;22:324–9.
32. Ohashi H, et al. Surgical anatomy of the pelvic vasculature, with particular reference to acetabular screw fixation in cementless total hip arthroplasty in Asian population: a cadaveric study. J Orthop Surg (Hong Kong). 2017;25:2309499016685520.
33. Kawasaki Y, et al. Location of intrapelvic vessels around the acetabulum assessed by three-dimensional computed tomographic angiography: prevention of vascular-related complications in total hip arthroplasty. J Orthop Sci. 2012;17:397–406.

Pelvic and Femoral Morphology in Developmental Dysplasia of the Hip

15

Masanori Fujii

1 Bone Morphology in the Pelvis and Acetabulum

A recent epidemiological study showed that the prevalence of hip osteoarthritis (OA), defined as grade ≥ 3 based on the Kellgren and Lawrence classification system, was 1.34% in men and 2.54% in women in Japan [1]. The epidemiological feature of OA in Japan is characterized by the high prevalence of OA secondary to developmental dysplasia of the hip (DDH), 81% of which are reported to be secondary OA caused by DDH [2]. DDH is currently used as a generic term for a broad spectrum of conditions, including complete dislocation of the hip, subluxation, and acetabular dysplasia. This section will focus primarily on the morphological abnormality in adult acetabular dysplasia, which should be considered when reconstructing the adult hip.

Typical morphological features of DDH includes a shallow and steep articular concavity, excessive femoral anteversion, coxa valga, and insufficient acetabular coverage of the femoral head, especially in the anterosuperior area [3]. These deformities can lead to structural hip instability, a decreased load-transferring area, abnormal distribution of stress on the articular cartilage, and concentration of joint contact force. The contact area of the articular bearing surface in DDH patients is reportedly reduced by approximately 50% (484 mm^2 vs 954 mm^2) compared to the normal hip [4]. Therefore, patients with DDH likely develop anterosuperior acetabular cartilage degeneration and labral injury in the overload area, resulting in premature OA [5]. However, recent three-dimensional analysis using CT images has revealed that morphological abnormalities in DDH are not uniform and that each individual case presents a variety of bone morphologies.

M. Fujii (✉)
Faculty of Medicine, Department of Orthopaedic Surgery, Saga University, Saga, Japan
e-mail: mfujii@cc.saga-u.ac.jp

1.1 Acetabular Morphology

Various morphological parameters have been reported for the evaluation of hip joint using plain radiographs, the most common of which is the lateral center-edge angle (LCEA) on the anteroposterior pelvic radiograph, an indicator of the coronal acetabular coverage of the femoral head. The normal LCEA in Japanese adults is approximately 30° [6], and it is generally accepted that a LCEA <20° is classified as DDH, 20° ≤ LCEA <25° as borderline DDH, and a LCEA ≥25° as normal, as in the western countries [7]. Although the acetabular rim is used to measure LCEA, it should be noted that 83% of patients with DDH present with upsloping lateral sourcil (ULS), a finding suggestive of hip instability. It has been reported that measuring the LCEA at the lateral margin of the acetabulum may overestimate the acetabular coverage in patients with ULS, and that measuring it at the medial margin of the USL better reflects the acetabular coverage (Fig. 15.1) [8].

Computed tomography images enable to evaluate the three-dimensional acetabular coverage of the femoral head. Using the multiplanar reconstruction images, the anterior and posterior acetabular coverages are often assessed using the anterior and posterior center-edge angles on the sagittal plane through the femoral head center. However, these angles are not measurable when the LCEA is less than 0 degrees, and the acetabular sector angles are also useful for evaluating three-dimensional acetabular coverage (Fig. 15.2) [9]. In a study on 50 patients (82 hips) with DDH (LCEA <20°), the anterior and posterior ASA on the axial plane were measured with reference to the anterior pelvic plane (APP). The patients were classified into four subtypes: the anterior deficiency group (anterior ASA <50° + posterior ASA ≥ 90°), the global deficiency group (anterior ASA < 50° + posterior ASA < 90°), the posterior deficiency group (anterior ASA ≥ 50° + posterior ASA < 90°), and mild deficiency group (anterior ASA ≥ 50° + posterior ASA ≥ 90°). The

Y. Nakashima (ed.), *Advances in Total Hip Arthroplasty*, https://doi.org/10.1007/978-981-97-2461-1_15

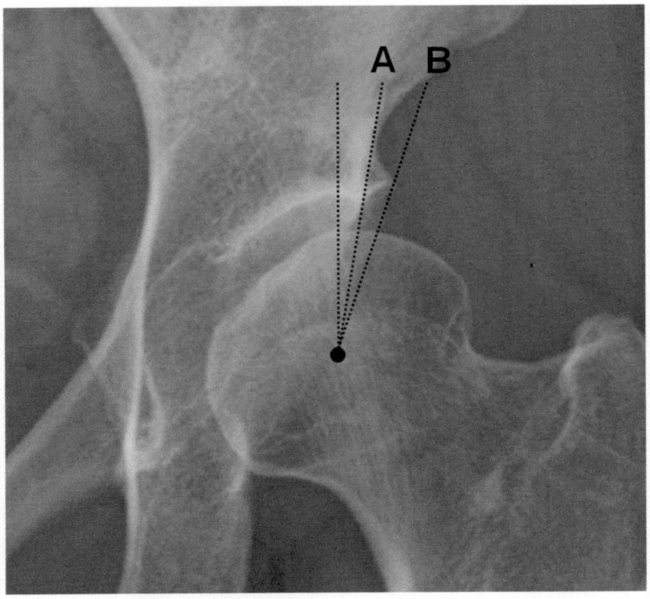

Fig. 15.1 Most patients with DDH present with upsloping lateral sourcil (ULS), a finding suggestive of hip instability. The LCEA should be measured at the medial margin of the USL (**a**) because measuring the LCEA at the lateral margin of the acetabulum (**b**) overestimates the effective acetabular coverage of the femoral head

results showed that 57% of the patients were in the anterior deficiency group, 28% in the global deficiency group, 11% in the posterior deficiency group, and 4% in the mild deficiency group [10]. On the other hand, although a simple comparison cannot be made because of the different evaluation methods, it has been reported that among American patients with DDH, 30% were in the anterosuperior deficiency group, 36% in the global deficiency group, and 34% in the posterosuperior deficiency group [11]. These reports suggest that the acetabular morphology of DDH is not uniformly deficient in anterosuperior coverage but that a variety of acetabular morphological abnormalities exist, ranging from cases with poor anterior coverage to those with nearly normal anterior coverage but poor posterior coverage and that racial differences exist in the acetabular morphology in DDH.

Periacetabular osteotomies (PAO), including rotational acetabular osteotomy and transposition osteotomy of the acetabulum, are well-established procedures for treating adolescent and young adult patients with DDH, and good clinical results have been reported even in the long-term. In PAO, morphological parameters in the coronal plane, i.e., LCEA and Tönnis angle, are commonly used to evaluate the correction of the acetabular fragment, with normalization of LCEA and Tönnis angle to 0° being considered an adequate correction. Although improvement of the lateral coverage is most important, considering substantial individual variations in the acetabular version and deficiency as described above, acetabular correction must be customized for each patient rather than applying a uniform correction. Recent studies have shown that the anterior center-edge angle has a more significant impact on improving the natural history of the disease after PAO than coronal parameters. Hip dysplasia typically presents as an anterolateral acetabular deficiency, in which shearing stress and contact stress are concentrated on the anterolateral acetabular rim. Therefore, adequate sagittal plane correction of the anterior undercoverage is crucial for successful hip preservation. However, care should be taken not to overcorrect the sagittal plane because excessive anterior coverage can result in iatrogenic femoroacetabular impingement. Therefore, three-dimensional morphological correction, including sagittal and axial planes, is essential to achieve favourable outcomes after PAO [12].

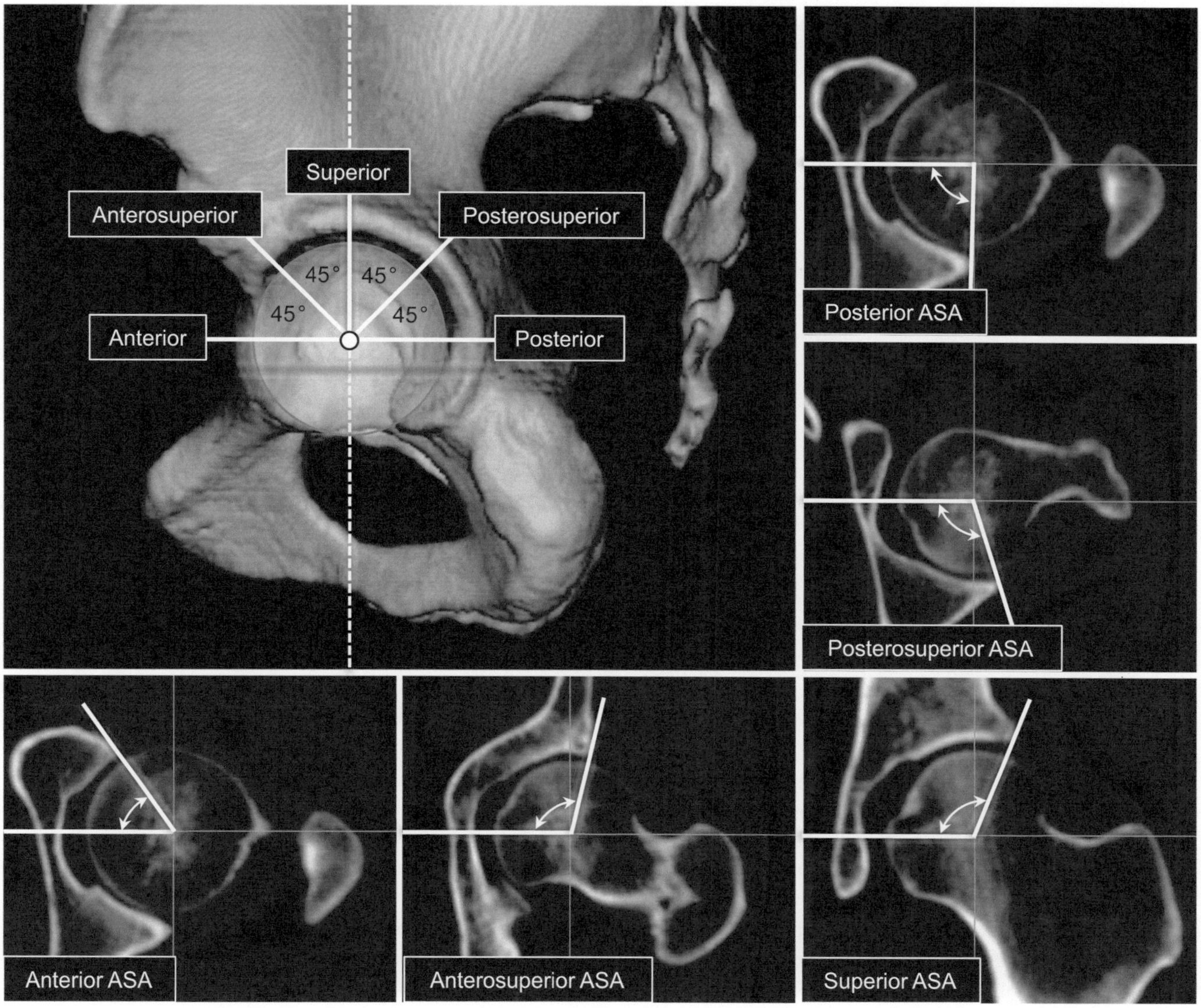

Fig. 15.2 Acetabular sector angles (ASA). (Cited from Tachibana et al. [9])

1.2 Pelvic Morphology

The acetabular morphological abnormalities in DDH have been reported to be associated with overall pelvic morphology. A previous study evaluated the rotational alignment of the innominate bone at three levels in the axial plane: the superior iliac angle (SIA), inferior iliac angle (IIA), and ischiopubic angle (IPA). The results showed that the mean values of the innominate rotation angles showed greater internal rotation from the iliac wing through the ischiopubic level in DDH patients compared with the normal controls, i.e., most DDH cases presented with a so-called "inward wing ilium" [10]. This study also showed a correlation between the rotational alignment of the innominate bone and the acetabular version and coverage. In subjects with an internally rotated innominate bone, the acetabula tended to rotate anterosuperiorly, resulting in decreased anterosuperior coverage and increased posterior coverage. When the acetabular morphology was classified into four subgroups based on dysplastic portion, 85% were classified as having an anterior or global deficiency. However, 11% were classified as having a posterior deficiency, and the innominate rotation of this group was more externally rotated than other dysplasia types. These findings suggest that structural abnormalities exist throughout the pelvis in DDH, and the morphologic abnormalities of the acetabulum are not caused solely by local dysplasia around the hip but are influenced by the morphologic features of the entire pelvis.

1.3 Pelvic Tilt

When evaluating acetabular morphology, it is necessary to establish a reference coordinate axis to minimize measure-

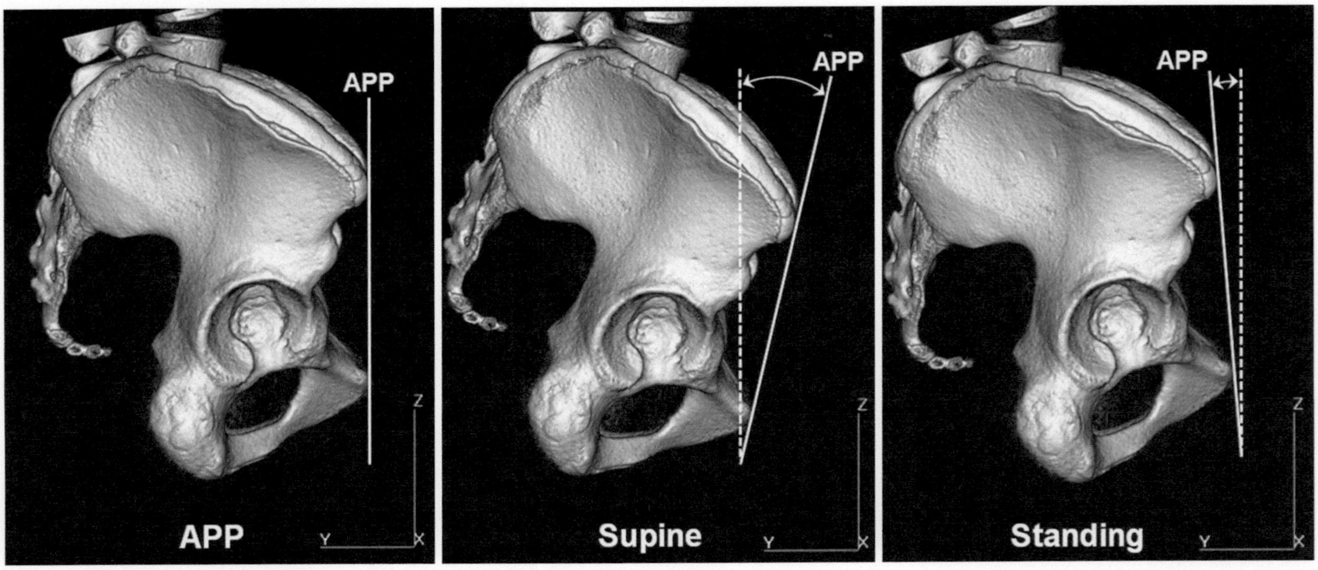

Fig. 15.3 Three-dimensional acetabular coverage varies depending on the reference pelvic coordinate axis (anterior pelvic plane [APP], supine, and standing positions). (Cited from Kitamura et al. [4])

ment error, and previous studies performed 3D evaluations of the morphology of the hip with reference to standardized pelvic positions such as the anterior pelvic plane (APP) and the supine functional pelvic plane (FPP). However, given the individual and postural variation in physiologic pelvic tilt and its effect on acetabular orientation, the reference position of the pelvis for morphologic analysis remains controversial. Standardizing the pelvic position may be advantageous for anatomic and epidemiologic studies. However, an abnormal mechanical environment in dysplastic hips, such as shearing stress forces and overload on the anterosuperior acetabulum, becomes a problem when the patient's hip is in a weight-bearing position rather than in the supine position. Thus, assessment with a standardized pelvic position may overlook changes in the acetabular version and coverage in weight-bearing positions. Tachibana et al. evaluated postural change in sagittal pelvic tilt between the supine and standing position in 65 patients with DDH [9]. They showed that the pelvis tilted posteriorly by 5° to 6° from the supine to the standing position in patients with hip dysplasia, resulting in increased acetabular version and decreased anterosuperior acetabular coverage in the standing position [9] (Fig. 15.3). Kitamura et al. reported that the average APP angle in the standing position in DDH patients was 2°, but the range of APP angles varied widely among individuals, from 11° anteversion to 14° posterior tilt. The assessment in a supine or standard pelvic position may overlook alterations in the hip's contact mechanics in the weight-bearing positions [4]. Based on these findings, it is recommended that postural change in sagittal pelvic tilt and functional pelvic tilt in the weight-bearing position should also be considered when evaluating the acetabular coverage in DDH and planning hip preservation surgeries such as PAO. Further studies are needed to elucidate the effect of patient-specific functional pelvic tilt on the degeneration process of dysplastic hips, the acetabular reorientation maneuver, and the clinical result of joint preservation surgery.

2 Bone Morphology of the Femur

Sugano et al. compared the femoral morphology of 31 DDH cases and 15 normal controls using three-dimensional computer reconstruction models created from CT images. They reported that the femoral anteversion angle was 10–14° larger in DDH cases than in the normal controls, regardless of the degree of hip subluxation (Crowe's classification) [13]. They also showed that even in the most mildly dysplastic joints, the femur had a smaller and more anteverted canal than the normal control and that with increased subluxation, additional abnormalities were observed in the size and position of the femoral head. Therefore, when performing total hip arthroplasty (THA) on patients with DDH, it is recommended that the femoral prosthesis be selected based on the severity of the subluxation and the degree of femoral anteversion of each individual case. Akiyama et al. also compared the femoral morphology of 49 patients with DDH and 49 normal controls using CT images. They reported that the mean femoral anteversion angle of DDH was 22.1°, significantly larger than that of the normal group (14.3°) [14]. Notably, the standard deviation (range) of the femoral anteversion angle was 12.0° (−13.2° to 58.2°) in DDH cases and 6.9° (−0.6° to 37.3°) in the normal controls, indicating that the femoral anteversion angle in DDH was more widely dis-

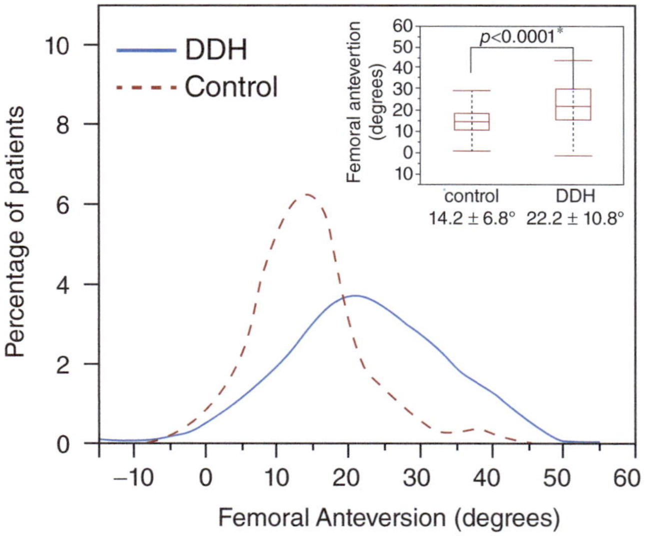

Fig. 15.4 Distribution of the femoral anteversion angle in the DDH and normal groups. (Cited from Akiyama et al. [14])

tributed in DDH cases (Fig. 15.4). Therefore, it is crucial to evaluate the femoral anteversion angle before THA to prevent dislocation because the anteversion angle of the femur in DDH is not uniformly large, and there are not a few cases with small anteversion.

In addition, it has been reported that there exists a left-right difference in femur length in patients with DDH. Tamura et al. used CT images to compare the distance from the top of the greater trochanter to the center of the knee (femoral length GT) and the distance from the medial prominence of the lesser trochanter to the center of the knee (femoral length LT) in 90 cases of unilateral secondary OA due to DDH with 43 normal control subjects [15]. The results showed that the percentages of hips with an absolute difference of 5 mm or more in the femoral GT and LT lengths were significantly greater in the DDH group (24% for both) than in the control group (2 and 7%, respectively). The authors also reported that the femoral length GT of the affected femur was significantly shorter in Crowe classification I and longer in Crowe classification IV compared to the unaffected side. Therefore, when performing THA for unilateral secondary OA due to DDH, it is recommended to evaluate the actual leg-length difference by whole-leg radiographs or CT measurements because of the possible existence of femoral length asymmetry.

References

1. Iidaka T, et al. Prevalence of radiographic hip osteoarthritis and its association with hip pain in Japanese men and women: the ROAD study. Osteoarthr Cartil. 2016;24:117–23.
2. Jingushi S, et al. Multiinstitutional epidemiological study regarding osteoarthritis of the hip in Japan. J Orthop Sci. 2010;15:626–31.
3. Nakahara I, et al. Three-dimensional morphology and bony range of motion in hip joints in patients with hip dysplasia. Bone Joint J. 2014;96:580–9.
4. Kitamura K, et al. Does patient-specific functional pelvic tilt affect joint contact pressure in hip dysplasia? A finite-element analysis study. Clin Orthop Relat Res. 2021;479:1712–24.
5. Kitamura K, et al. Effect of sagittal pelvic tilt on joint stress distribution in hip dysplasia: a finite element analysis. Clin Biomech(Bristol, Avon). 2020;74:34–41.
6. Iidaka T, et al. Radiographic measurements of the hip joint and their associations with hip pain in Japanese men and women: the research on osteoarthritis/osteoporosis against disability (ROAD) study. Osteoarthr Cartil. 2017;25:2072–9.
7. Gala L, et al. Hip dysplasia in the young adult. J Bone Joint Surg Am. 2016;98:63–73.
8. Wong TY, et al. Upsloping lateral sourcil: a radiographic finding of hip instability. J Hip Preserv Surg. 2018;5:435–42.
9. Tachibana T, et al. Does acetabular coverage vary between the supine and standing positions in patients with hip dysplasia? Clin Orthop Relat Res. 2019;477:2455–66.
10. Fujii M, et al. Pelvic deformity influences acetabular version and coverage in hip dysplasia. Clin Orthop Relat Res. 2011;469:1735–42.
11. Nepple JJ, et al. Three patterns of acetabular deficiency are common in young adult patients with acetabular dysplasia. Clin Orthop Relat Res. 2017;475:1037–44.
12. Wyles CC, et al. Hitting the target: natural history of the hip based on achieving an acetabular zone following periacetabular osteotomy. J Bone Joint Surg Am. 2020;102:1734–40.
13. Sugano N, et al. The morphology of the femur in the developmental dysplasia of the hip. J Bone Joint Surg Br. 1998;80:711–9.
14. Akiyama M, et al. Femoral anteversion is correlated with acetabular version and coverage in Asian women with anterior and global deficient subgroups of hip dysplasia: a CT study. Skeletal Radiol. 2012;41:1411–8.
15. Tamura K, et al. Femoral morphology asymmetry in hip dysplasia makes radiological leg length measurement inaccurate. Bone Joint J. 2019;101:297–302.

Part III

Biomechanics of the Hip Joint

Gait and Movement Analysis

16

Satoshi Hamai

1 Gait Analysis

The basic parameters include the gait cycle (initial ground contact, loading response, mid-stance, terminal-stance, pre-swing, initial-swing, mid-swing, terminal-swing; double-support phase, single-support phase), walking speed, step length, step rate, step width, foot angle, joint angle, angular acceleration, joint moment, and the projected point of body center of gravity. These can be measured using a marker-based motion capture system and force platforms (Fig. 16.1). A gait analysis using a markerless motion capture system or sheet-type pressure sensors has also been conducted, although the measurement items and accuracy are different. Generally, it has been reported that (1) the time and ratio of the stance phase and the swing phase change with the walking speed, and the double support phase decreases or disappears as the walking speed increases, and (2) the walking speed, stride length, walking cadence, and the hip flexion-extension range of motion (RoM) decreases in elderly individuals because of the decrease in the dynamic balance and the increase in the step width.

Gait analysis is also used as an objective evaluation method for postoperative recovery in total hip arthroplasty (THA). In a comparison of patients with unilateral THA at 1 month postoperatively and normal subjects, the walking speed, stride length, and walking cadence were all lower in the former group (unilateral THA at 1 month: 49 m/min, 0.99 m, 103 steps/min; normal subjects: 69 m/min, 1.20 m, 115 steps/min), the lateral trunk inclination was greater (7.9 cm vs. 5.0 cm), and the walking efficiency was reported to be approximately 21% lower [1]. Lateral trunk tilt is a factor that decreases the gait efficiency and has been shown to increase the energy expenditure due to the compensatory trunk tilt caused by reduced hip abductor muscle strength [2]. In a study investigating the relationship between the gait parameters and patient-reported outcomes at 1 month postoperatively, the motion items of the Japanese Orthopedic Association Hip Disease Evaluation Questionnaire (JHEQ) were positively correlated with the knee joint extension moment at the first peak, and the JHEQ total score and mental status were positively correlated with the hip joint extension moment at the second peak and the forward acceleration phase caused by hip extension [3]. No improvement in gait parameters was observed at 3 weeks after THA; however, there were clear improvements in the gait speed, stride length, and muscle strength at 3 months postoperatively, indicating that age and the maximum lower limb load were factors affecting the walking status at postoperative 6 months [4]. The cutoff value was 8.2 N/kg, which was more predictive of the autonomous walking ability than pain or hip RoM.

The walking speed in unilateral THA correlated with the hip RoM on the operative side at 1 month postoperatively and with bilateral hip extension moments at 3–6 months postoperatively and reached the same level as that of normal subjects at 1 year postoperatively [5]. Preoperatively, asymmetry was observed in the hip RoM, maximum flexion and extension angles, maximum knee flexion angle, and pelvic tilt and rotation. The preoperative hip flexion angles on the affected side were 21° (47° on the healthy side) and 35° (44° on the healthy side) at 1 year postoperatively. The remaining asymmetry in hip RoM after surgery was primarily due to the decreased maximum extension RoM on the affected side at the end of stance. The anterior-posterior pelvic tilt and RoM of the healthy hip were inversely correlated with the RoM of the operated hip, indicating compensatory motion. The walking speed and stride length did not differ from those of normal subjects in either bilateral THA or unilateral THA after 6 months of the postoperative period [6]. In bilateral THA, claudication was more pronounced on the side with a 47% or greater decrease in hip RoM during walking, and the ipsilateral ankle plantar flexion moment, which functions as a push-off, was increased. The decrease in the hip power at the end of stance was observed to correlate

S. Hamai (✉)
Department of Medical-Engineering Collaboration for Healthy Longevity, Kyushu University, Fukuoka, Japan
e-mail: hamai.satoshi.075@m.kyushu-u.ac.jp

Y. Nakashima (ed.), *Advances in Total Hip Arthroplasty*, https://doi.org/10.1007/978-981-97-2461-1_16

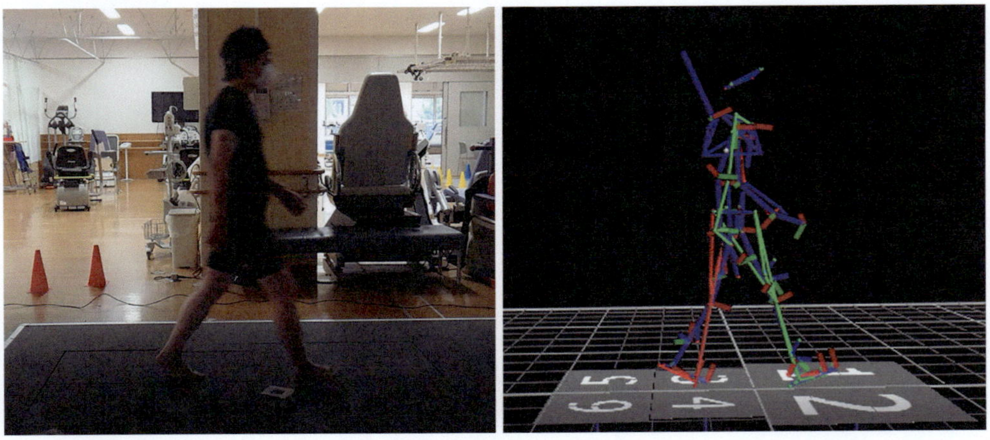

Fig. 16.1 Motion analysis using a surface marker motion capture system and a floor reaction force meter

with the increase in contralateral knee power at the beginning of stance. In unilateral THA, the decrease in hip power at the end of stance was correlated with the increase in plantar flexion moment of the healthy ankle joint. Furthermore, gait training to decrease the ankle plantar flexion after THA was found to increase the hip power and may help improve the muscle power distribution during walking [7].

The gait parameters in unilateral THA (stride length: 62% stature, percentage of single-support phase on the affected side: 40% cycle, walking cadence,: 104 steps/min, velocity: 0.83 m/sec) were inferior to those of healthy subjects (stride length: 71% stature, percentage of single-support phase on the affected side: 42% cycle, walking cadence,: 117 steps/min, velocity: 1.02 m/sec); however, significant improvements (stride length: 52% stature, single-leg support time on the affected side: 35% cycle, walking rate: 96 steps/min, velocity: 0.63 m/sec) were observed compared to preoperative subjects [8]. The change in the preoperative disease stage and leg length difference correlated with the percentage of single-support phase throughout the follow-up period up to 1 year postoperatively and was described as an important factor in the improvement of walking function. The predictive factor for early postoperative recovery of ambulatory function was preoperative maximal walking speed [9]. In addition, the preoperative contralateral single-support phase was a predictor of the postoperative comfortable walking speed, preoperative contralateral knee extension muscle strength was a predictor of the postoperative maximal walking speed, and leg length difference and offset were not influential factors in postoperative walking speed [10].

The center of gravity trajectory was asymmetric preoperatively, and the abnormal trajectory remained despite an improvement in pain at 3 months postoperatively [11]. The asymmetry of the trajectory of gravity shift in the sagittal plane was correlated with a decrease in the isometric hip flexor strength, and the trajectory of gravity shift in the vertical and anteroposterior directions showed a large change with increasing walking speed at 1 year postoperatively compared with preoperative values; however, no change was observed in the internal and external directions [12]. The coefficient of variation of the trajectory of gravity shift in the medial and lateral directions before surgery was larger than that of normal subjects but improved to the same level as that of normal subjects at 1 year postoperatively. The coefficient of variation of the overlap walking time decreased with an improvement in pain at 1 year postoperatively and was comparable to that of normal subjects [13]. Additionally, the trunk harmonic ratio improved at 1 year postoperatively, although it was inferior to that of healthy subjects and was correlated with improvements in the hip abductor strength, range of extension motion, and leg length.

2 Motion Analysis

The motion capture, 4D motion analysis systems, shape matching, 2D/3D registration, image-matching methods, and open MRI were used for the analysis. In the analysis of standing up from bed using the motion capture system, it was reported that the maximum internal/external rotation angle was smaller when the motion was made from the affected side [14]. The maximum angle of flexion and internal rotation, time required, and degree of difficulty were not different among the six different types of movements, including movements from the healthy side. The ground reaction force during stair climbing correlated with the hip extensor muscle strength at 6 months postoperatively and with the knee extensor muscle strength at 1 year postoperatively, suggesting that an improvement of not only hip extensor muscle strength, but also knee extensor muscle strength is important for rehabilitation [15]. In the analysis of getting into and out of a car using the motion capture method, it was reported that even high-functioning unilateral THA patients have limited RoM and power production in the surgical hip; they compensate for this loss by producing power in other joints, especially when the surgical side is the pivot limb, which may increase risk of osteoarthritis of those joints [16]. The authors stated that the restoration of RoM and muscle strength and adopting the normal side as the pivot limb may allow for a more appropriate balance in the motion of getting into and out of a car, which will lead to safe mobility, assist in social participation, and improve the quality of life.

Postoperative evaluation using a 4D motion analysis system combined with CT and motion capture system was reported to be useful in motion guidance [17–20] for patient-specific prevention of dislocation. In a study that used an *in vivo* dynamic analysis to search for the cause of dislocation in patients with a history of dislocation, it was shown that impingement between the anterior liner and stem neck occurred in cases of posterior dislocation, and impingement between the posterior liner and stem neck occurred in cases of anterior dislocation [20]. In addition, an analysis using intraoperative fluoroscopic images with the shape matching method reported that implant impingement was the main factor of dislocation in the extended external rotation position and impingement between the pelvis and femur or soft tissue contracture in the flexion and internal rotation positions [21].

The squat motion analysis in unilateral THA using the 2D/3D registration method showed that the maximum flexion angle was 86° and pelvic retroversion was 26°. Additionally, the theoretical minimum angle to implant impingement was 26° [22]. The ratio of posterior pelvic tilt to femoral flexion in the maximum flexion position was 23%. An analysis using the image-matching method showed that the maximum flexion angle was 81° and the posterior pelvic tilt was 10° during the squatting motion (Fig. 16.2) [23], which was consistent with the previous study [22]. The implant placement angle was 39° cup inclination, 19° cup anteversion, 32° stem anteversion, and 50° combined anteversion (CA), and the closest distances between the liner and neck were 11 mm anteriorly (35° angle to impingement) and 8 mm posteriorly, neither anterior nor posterior implant impingement was observed. The factors influencing the anterior and posterior distances between the liner and neck were the hip flexion angle, cup and stem anteversions, and use of elevated liner [24, 25]. In the analysis of chair standing, the maximum flexion after THA (72°) was improved compared with that before THA (63°); however, the flexion angle was smaller than that of the normal hip (81°) [26]. In the half-rising posture, the hip was 12° more flexed, and the pelvis was 26° more anteriorly tilted than in the relaxed sitting posture. For the first time, an *in vivo* analysis showed that the factors influencing the anterior distance between the liner and neck during the movement were the hip flexion angle, cup inclination, stem anteversion, and head diameter; additionally, the factors influencing the posterior distance were the hip extension angle, cup anteversion, neck length, and liner shape. Furthermore, based on the hip dynamics during maximum flexion/extension (72°/−4°) of chair standing, we simulated only the angle of internal rotation at maximum flexion and the angle of external rotation at maximum extension from 0° to 60° (in 15° increments), measured the closest distance between the liner and neck, and investigated the optimal cup anteversion and CA to avoid impingement [27]. The optimal cup anteversion and CA to reduce the risk of contact both anteriorly and posteriorly were 12–25° and 38–62°, respectively. In cases with a stem anteversion of less than 13° and more than 50°, the stem anteversion should be adjusted because the adjustment based on the cup anteversion alone is out of the optimum range. Microseparation between the liner and head of up to 3.3 mm has been reported during THA side leg raising and just before initial ground contact of walking [28]. There have been reports that offset reconstruction, leg length difference correction, and minimally invasive surgery [29] have reduced the separation and that THA as usual did not result in a measurable separation of more than 1 mm during gait [30].

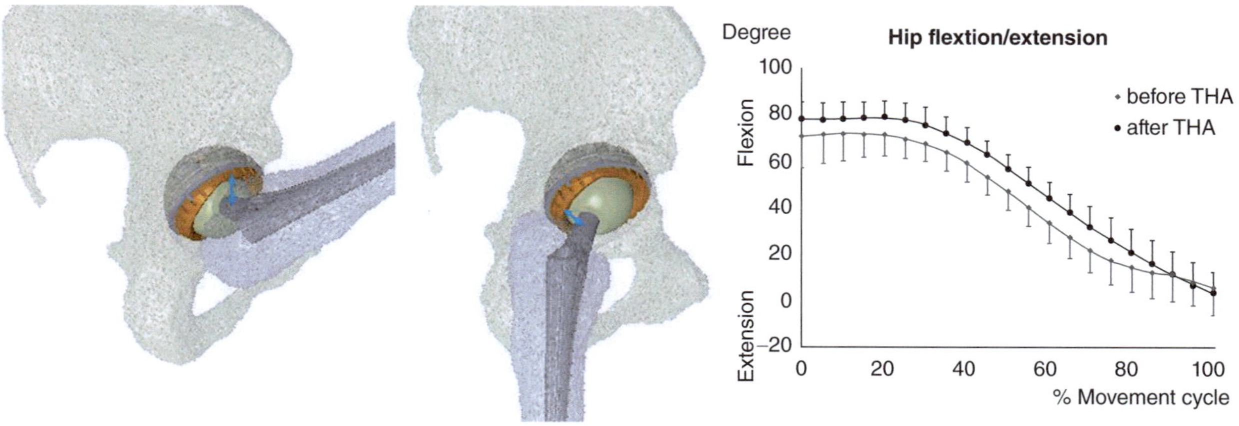

Fig. 16.2 Analysis of squat standing dynamics before and after THA (left: liner-neck distance measurement, right: hip flexion/extension angle). (Cited from Komiyama et al. [23])

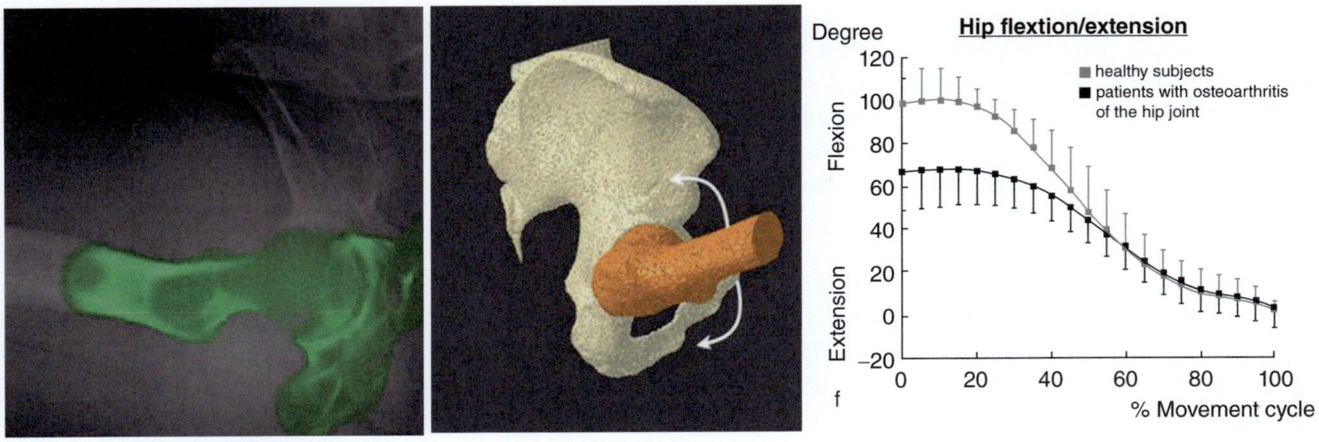

Fig. 16.3 Analysis of squat standing dynamics in healthy subjects and patients with osteoarthritis of the hip (left: image-matching, center: 3D dynamics, right: hip flexion/extension angle). (Cited from Hara et al. [31])

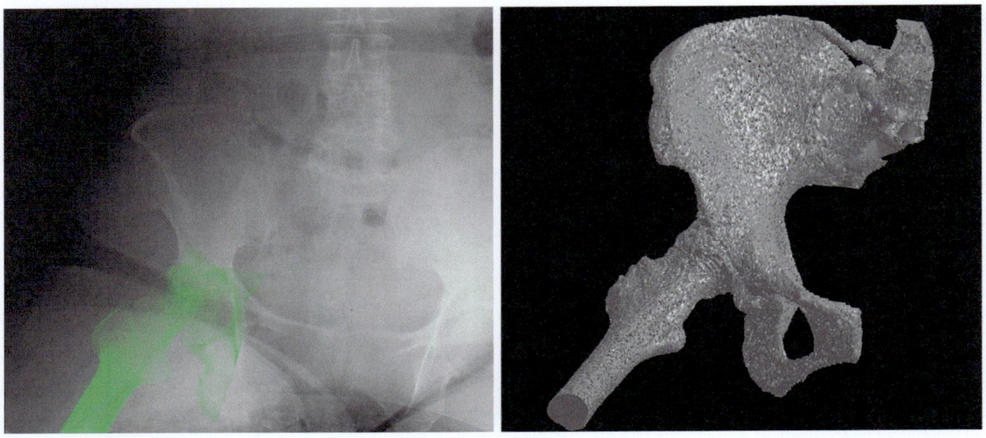

Fig. 16.4 Analysis of chair standing dynamics in healthy subjects and patients with osteoarthritis of the hip (left: image-matching, right: 3D dynamics)

There have been several reports from Japan on the motion analysis of normal and osteoarthritic hip joints. In the analysis of the normal hip using a 4D motion analysis system, the maximum hip flexion angle was 80° (9° abduction, 9° external rotation) in squatting, 76° (3° abduction, 11° external rotation) in chair standing, 86° (6° abduction, 12° external rotation) in picking up an object in a chair, 84° (2° abduction, 12° external rotation) in bowing while sitting straight, and 61° (1° abduction, 15° external rotation) in the seiza posture [31]. In the analysis of healthy hips using the image-matching method, the maximum hip flexion angle was 81° (12° anterior pelvic tilt) during the movement in chair standing and 102° (11° posterior pelvic tilt) during the movement in squatting. The maximum hip internal and external rotation angles were 29° and 31° (0–18° flexion, 0–6° abduction) during trunk twisting (Fig. 16.3) [32]. The closest distance between the acetabular rim and femoral neck during squatting was 11.3 mm, with a standard deviation of 5.2 mm [33]. In osteoarthritis of the hip joint, the maximum hip flexion angle was 64° (0° anterior pelvic tilt) in chair standing, 68° (16° posterior pelvic tilt) in the squatting position, and 3° and 13° (3–8° flexion, −1° to 2° abduction) in trunk twisting (Fig. 16.4) [34]. Additionally, an analysis of healthy hips using open MRI reported 55° flexion, 1° external rotation, and 5° internal rotation in upright sitting, 107° flexion, 42° external rotation, and 25° external rotation in sitting on the floor, 110° flexion, 8° internal rotation, and 4° internal rotation in sitting bow, and 111° flexion, 10° internal rotation, and 2° external rotation in squatting [35].

References

1. Nankaku M, et al. Gait analysis of patients in the early stages after total hip arthroplasty: effect of lateral trunk displacement on walking efficiency. J Orthop Sci. 2007;12:550–4.
2. Nankaku M, et al. Effects of vertical motion of the center of mass on walking efficiency in the early stages after total hip arthroplasty. Hip Int. 2012;22:521–6.

3. Kurihara Y, et al. Early postoperative relationship between patient-reported outcome measures and gait biomechanical factors after total hip. Gait Posture. 2022;91:14–8.
4. Nankaku M, et al. Factors associated with ambulatory status 6 months after total hip arthroplasty. Physiotherapy. 2014;100:263–7.
5. Miki H, et al. Recovery of walking speed and symmetrical movement of the pelvis and lower extremity joints after unilateral THA. J Biomech. 2004;37:443–55.
6. Tateuchi H, et al. Dynamic hip joint stiffness in individuals with total hip arthroplasty: relationships between hip impairments and dynamics of the other joints. Clin Biomech (Bristol, Avon). 2011;26:598–604.
7. Tateuchi H, et al. Immediate effects of different ankle pushoff instructions during walking exercise on hip kinematics and kinetics in individuals with total hip arthroplasty. Gait Posture. 2011;33:609–14.
8. Tanaka R, et al. Factors influencing the improvement of gait ability after total hip arthroplasty. J Arthroplast. 2010;25:982–5.
9. Shibuya M, et al. The maximal gait speed is a simple and useful prognostic indicator for functional recovery after total hip arthroplasty. Musculoskelet Disord. 2020;21:84.
10. Ohmori T, et al. Contralateral lower-limb functional status before total hip arthroplasty: an important indicator for postoperative gait speed. Joint Surg Am. 2021;103:1093–103.
11. Gomi M, et al. Early clinical evaluation of total hip arthroplasty by three-dimensional gait analysis and muscle strength testing. Gait Posture. 2018;66:214–20.
12. Wada O, et al. Root mean square of lower trunk acceleration during walking in patients with unilateral total hip replacement. Gait Posture. 2017;58:19–22.
13. Wada O, et al. Gait variability in women with hip osteoarthritis before and after total hip replacement: a prospective cohort study. Am J Phys Med Rehabil. 2019;98:866–71.
14. Yamahara J, et al. Analyzing the characteristics of rising from the bed in patients having undergone total hip arthroplasty immediately after surgery. Gait Posture. 2018;64:50–4.
15. Arima Y, et al. Relationship between vertical ground reaction force and muscle strength while climbing stairs after total hip arthroplasty. Clin Biomech (Bristol, Avon). 2020;78:105088.
16. Harada T, et al. Three-dimensional kinematics and kinetics of getting into and out a car in patients after total hip arthroplasty. Gait Posture. 2022;98:305–12.
17. Otake Y, et al. 4-dimensional computer-based motion simulation after total hip arthroplasty. Stud Health Technol Inform. 2003;94:251–7.
18. Hagio K, et al. A novel system of four-dimensional motion analysis after total hip arthroplasty. J Orthop Res. 2004;22:665–70.
19. Otake Y, et al. Real-time motion analysis for patients after total hip arthroplasty by using a 4-dimensional patient-specific model. Int Congr Ser. 2005;1281:696–701.
20. Miki H, et al. Detecting the cause of dislocation after total hip arthroplasty by patient-specific four-dimensional motion analysis. Clin Biomech (Bristol, Avon). 2013;28:182–6.
21. Tanino H, et al. An in vivo model for intraoperative assessment of impingement and dislocation in total hip arthroplasty. J Arthroplast. 2008;23:714–20.
22. Koyanagi J, et al. In vivo kinematic analysis of squatting after total hip arthroplasty. Clin Biomech (Bristol, Avon). 2011;26:477–83.
23. Komiyama K, et al. Dynamic hip kinematics during squatting before and after total hip arthroplasty. J Orthop Surg Res. 2018;13:162.
24. Harada S, et al. Squatting after total hip arthroplasty: patient-reported outcomes and in vivo three-dimensional kinematic study. J Arthroplast. 2022;37:734–41.
25. Harada S, et al. Elucidation of target implant orientations with safety range of hip rotation with adduction or abduction during squatting: simulation based on in vivo replaced hip kinematics. Front Bioeng Biotechnol. 2022;10:1023721.
26. Shiomoto K, et al. In vivo kinematics, component alignment and hardware variables influence on the liner-to-neck clearance during chair-rising after total hip arthroplasty. J Orthop Sci. 2020;25:452–9.
27. Shiomoto K, et al. Computer simulation based on in vivo kinematics of a replaced hip during chair-rising for elucidating target cup and stem positioning. Clin Biomech (Bristol, Avon). 2022;91:105537.
28. Dennis DA, et al. "In vivo" determination of hip joint separation and the forces generated due to impact loading conditions. J Biomech. 2001;34:623–9.
29. Glaser D, et al. In vivo comparison of hip mechanics for minimally invasive versus traditional total hip arthroplasty. Clin Biomech (Bristol, Avon). 2008;23:127–34.
30. Kiyohara M, et al. Do component position and muscle strength affect cup-head translation during gait after total hip arthroplasty? Eur J Orthop Surg Traumatol. 2019;29:1263–9.
31. Sugano N, et al. Dynamic measurements of hip movement in deep bending activities after total hip arthroplasty using a 4-dimensional motion analysis. J Arthroplast. 2012;27:1562–8.
32. Hara D, et al. Kinematic analysis of healthy hips during weight-bearing activities by 3D-to-2D model-to-image registration technique. Biomed Res Int. 2014;2014:457573.
33. Yoshimoto K, et al. Dynamic hip kinematics before and after periacetabular osteotomy in patients with dysplasia. J Orthop Sci. 2020;25:247–54.
34. Hara D, et al. Dynamic hip kinematics in patients with hip osteoarthritis during weight-bearing activities. Clin Biomech (Bristol, Avon). 2016;32:150–6.
35. Yamamura M, et al. Open-configuration MRI study of femoroacetabular impingement. J Orthop Res. 2007;25:1582–8.

17 Computer Simulation

Takeshi Shoji and Nobuo Adachi

The number of total hip arthroplasty (THA) procedures for hip osteoarthritis (OA) has been increasing annually. However, some patients still require revision surgery because of complications, such as postoperative dislocation and implant loosening, and avoiding these complications is an urgent issue. In particular, the frequency of postoperative hip dislocation is still not small, and it is currently the main revision factor [1]. These complications are largely due to the position/angle of the implant, which is closely related to the range of motion (ROM) after THA. Understanding the relationship between the two is necessary to avoid postoperative hip dislocation and to achieve a high level of activity of daily life (ADL) in patients receiving the surgery.

The factors associated with postoperative hip dislocation and postoperative ROM include (1) patient factors, such as age, sex, and previous surgery; (2) surgery-related factors, such as implant position/angle, surgical approach, and soft tissue tension; and (3) implant factors, which are thought to be involved in a complex manner. In 2000, Bartz [2] et al. classified the mechanisms of postoperative hip dislocation as (1) implant impingement (II), (2) bony impingement (BI), and (3) spontaneous dislocation, and they also reported that impingement was deeply involved in postoperative hip dislocation and ROM. Shon [3] et al. also reported that the majority (approximately 94%) of cases of repeated dislocation after THA surgery had findings suggestive of impingement based on the analysis of removed implants of revision cases. Understanding the mechanism of impingement and developing a treatment strategy based on the aforementioned factors are important to improve postoperative ROM and avoid dislocation in THA. Furthermore, understanding the effect of the position and angle of implant placement on impingement is also important because it is more closely related to the occurrence of impingement.

Recent advances in computer technology have made it possible to perform detailed three-dimensional preoperative planning, dynamic analysis, and postoperative evaluation using various types of imaging information. These techniques enable us to understand the relative positions of bone and other tissues in three dimensions and to recognize the relative positions more objectively, whereas previously physicians relied on their experience and sensory perception. In our previous studies, we analyzed the relationship between implant placement and ROM, as well as dislocation factors, by conducting three-dimensional dynamic analysis of THA using computer simulation and applied these analyses to clinical settings. In this section, we review the relationship between implant alignment/placement and ROM in THA.

1 Impingement in THA

Impingement is reportedly associated with ROM, dislocation, increased wear, implant fracture, and postoperative pain [4]. Impingement is mainly classified into implant to implant (II), implant to bone (implant-bone impingement), bone to bone (BI), and soft tissue impingement, whereas most of the impingements associated with implant placement are II and BI. The factors associated with II and BI are bony morphological factors, such as pelvis and femur morphology, implant-related factors, such as implant position/angle/design, and spinal alignment, which act independently of each other.

1.1 Implant Impingement (II)

II is the impingement of the cup and stem, which mainly depends on the implant alignment. To avoid II, Widmer et al.

T. Shoji (✉)
Artificial Joints and Biomaterials, Graduate School of Biomedical and Health Sciences, Hiroshima University, Higashihiroshima, Japan
e-mail: readyma@hiroshima-u.ac.jp

N. Adachi
Department of Orthopaedic Surgery, Graduate School of Biomedical and Health Sciences, Hiroshima University, Higashihiroshima, Japan

Y. Nakashima (ed.), *Advances in Total Hip Arthroplasty*, https://doi.org/10.1007/978-981-97-2461-1_17

[5] proposed a safe zone based on the cup alignment, and Dorr et al. [6] reported implant placement using a combined anteversion (CA) technique.

1.2 Bony Impingement (BI)

BI occurs mainly in the anterior hip joint at the inferior anterior iliac spine (AIIS) and proximal femur, and in the posterior hip joint at the inferior margin of the acetabulum, proximal femur, sciatic bone, and lesser trochanter (Fig. 17.1). They are associated with multiple factors, including the pelvis, femur bone morphology, and implant placement position/alignment. The authors found that an anterolateral overhang of the AIIS [7], an anteversion of the femur, anterior and posterior bone morphology of the proximal femur, and postoperative ROM were associated with BI [8]. In addition, in the narrow pelvis morphology characteristic of developmental dysplasia of the hip (DDH), the incidence of BI increased posteriorly at the hip joint [9], suggesting an association with sex and racial differences related to bone morphology.

2 Three-Dimensional (3D) Dynamic Analysis

In our past study, we used ZedHip® (LEXI, Tokyo) as 3D dynamic motion analysis software. This software enables the quantitative measurement of the position and ROM of impingement by creating a bone model from computed tomography (CT) data, segmenting the pelvis and femur from multiplanar reconstruction images, and performing hip movement simulation with the implants in place.

The choice of the reference coordinate system is important for the analysis. There are anatomical, functional and International Society of Biomechanics (ISB) coordinate systems for the pelvic coordinate system and posterior condylar plane and ISB coordinate systems for the femoral coordinate system. The interpretation of the results differs depending on which coordinate system is used. In our past study, the evaluation was performed using the anatomical coordinate system in the supine position and the ISB coordinate system on the femoral side to maintain the pelvic reference as a constant pelvic coordinate system.

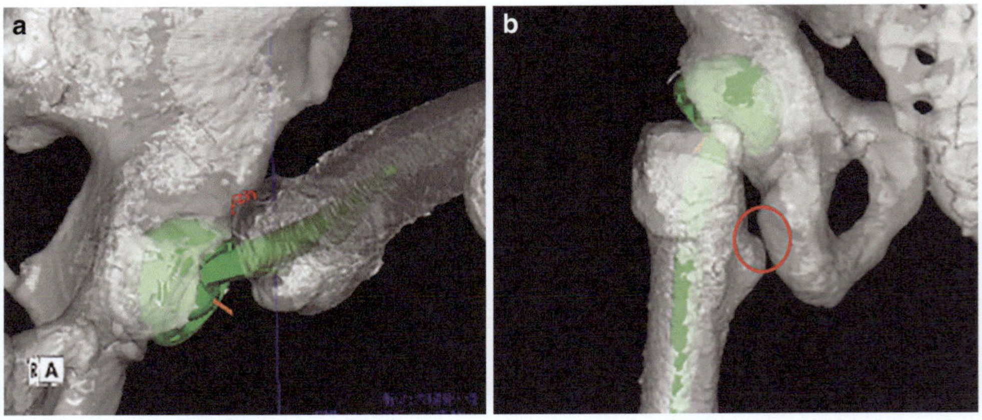

Fig. 17.1 Evaluation of impingement by 3D dynamic analysis. (**a**) Anterior impingement of the hip joint. Bony impingement at the inferior anterior iliac spine and proximal femur. (**b**) Impingement at the posterior hip joint. Bony impingement at the sciatic and popliteal sites

3 Relationship Between Implant Selection, Implant Position/Angle, and ROM

To obtain satisfactory outcomes in THA, it is necessary to select an appropriate implant design and size for each patients and place the implant at an appropriate position and alignment. The safe zone (external opening angle 30–50°, anterior opening angle 5–25°) proposed by Lewinnek et al. [10] has been widely used as the optimal implant alignment; however, many reports showed that an insignificant correlation exists between dislocation and the safe zone [11], and the CA theory [6] is widely used for implant placement angles.

3.1 Implant Selection

Implants, such as the stem, neck, head, liner, and cup, affect the postoperative ROM and dislocation. To prevent impingement, selecting an implant with a high oscillation angle is necessary, and the stem design is particularly important [12]. When selecting a stem, fully understanding its concept is important because the fixation site varies depending on the design of the stem. For stem implants, the neck shaft angle (NSA) and neck length are important to avoid postoperative dislocation. NSA and neck length are closely related to femoral offset (FO) and leg length, which are directly related to the occurrence of impingement. FO is especially important to avoid hip dislocation because it is involved not only in impingement but also in soft tissue tension. Our previous study showed that the NSA of the stem and frequency of BI in the anterior hip joint were positively correlated, particularly with a decrease in the range of flexion and internal rotation [12]. Conversely, the negative correlation between the NSA and frequency of BI in the posterior hip joint increases the range of external rotation, suggesting that BI in the anterior/posterior region and ROM have a trade-off relationship. FO has also shown that overextension of the vertical offset attenuates the effect of ROM improvement [13]. The selection of the head diameter is also important to avoid impingement and dislocation. Previous reports of dynamic analysis have shown that the ROM to the onset of II increases with an increase in the diameter of the head and jumping distance, resulting in an increase in ROM and resistance to dislocation. Clinically, the dislocation rate decreases with the use of a femoral head diameter ≥ 32 mm. Conversely, the revision rate increased with the use of a large-diameter head of ≥40 mm [1], which has to be focused on in clinical practice.

3.2 Cups; Relationship Between Implant Placement and ROM

The cup abduction/anteversion angle and position affect the postoperative ROM, where the abduction/anteversion angle is related to II onset and the cup position to BI onset (under the condition of the same FO).

As the cup abduction angle increased, the ROM up to the occurrence of II in the anterior/posterior direction also increased, resulting in an increase in ROM (Fig. 17.2). However, once BI occurs, an increase in the cup abduction angle leads to a decrease in the jump distance of the implant itself, resulting in a decrease in the ROM expansion range. In addition, an increase in the cup abduction angle may lead to

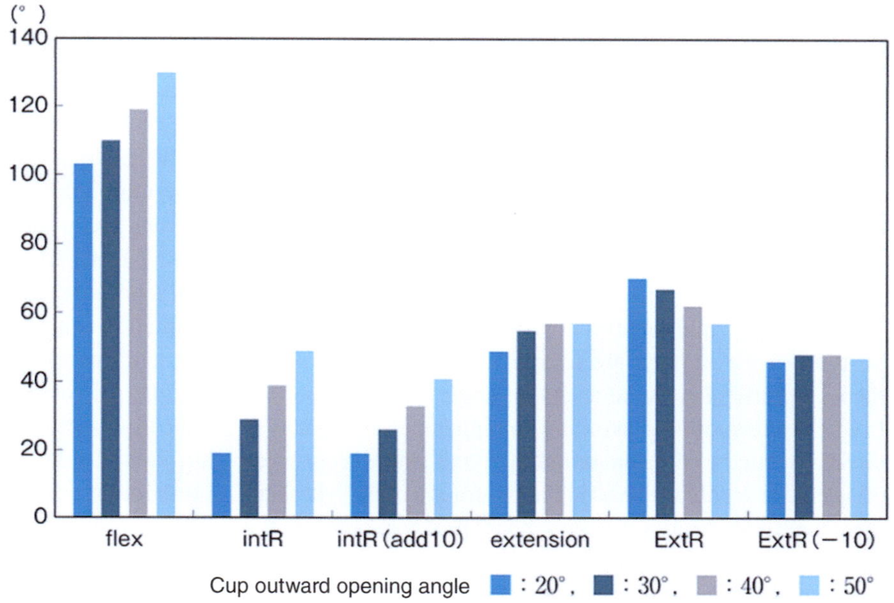

Fig. 17.2 Change in hip joint ROM with change in cup external opening angle. Flex: flexion angle, intR: internal rotation angle (flexion 90°, internal rotation 0°), intR(add10): internal rotation angle (flexion 90°, internal rotation 10°), Ext: extension angle, ExtR: external rotation angle (flexion 0°, internal rotation 0°), ExtR(−10): external rotation angle (flexion 0°, internal rotation 10°). Implant used: kinectiv stem (Zimmer Biomet®), combined anteversion; set at 45°

wear and damage of the implant, and an abduction angle of 45° or less is considered desirable based on previous reports.

The cup anteversion angle is also important for II generation. Sugano et al. [14] proposed that the cup anteversion angle should be adjusted according to the stem anteversion based on the cup external opening angle of 40° according to CA theory. Many previous reports using simulations and cadavers suggest that implant placement in accordance with CA theory is useful in avoiding II [5, 15]. However, once BI occurs, the influence of the cup anteversion angle is greatly weakened, and it is difficult to avoid a decrease in ROM due to BI by adjusting the cup installation angle, so other countermeasures are necessary.

The cup position (medialization, elevation) is closely related to the onset of BI. Our study showed that the ROM until the onset of II (ROMCI) and the ROM until the onset of BI (ROMBI) did not change regardless of the horizontal position of the cup under the condition of constant global offset; however, under the condition of decreasing global offset (FO: difinite), ROMBI decreased with cup medialization, and the ROM of flexion and internal rotation decreased [13]. On the other hand, both ROMCI and ROMBI of the ROM in external rotation decrease with cup medialization under the condition of constant global offset, and no change in ROM under the condition of decreasing global offset, indicating a trade-off relationship between the ROM of flexion/internal rotation and external rotation. Therefore, both global offset and FO reduction lead to a decrease in postoperative ROM, and offset selection should be based on careful preoperative planning.

With respect to the cup position in the sagittal plane, it is generally considered that the cup should be placed at the original position, there are cases in which the cup must be allowed to be placed at a higher level. However, it has been shown that the frequency of BI increases and the ROM of flexion/internal rotation decreases as the cup is proxilalized [7]. Komiyama et al. [16] reported from the results of a simulation analysis that elevated placement up to 24 mm from the intertearal line is acceptable.

3.3 Stem; Implant Placement and its Relationship to ROM

The insertion angle (anteversion/flexion angle) and inserted position of the implant are related to ROM in stem placement. In particular, the anteversion angle is very important from the viewpoint of avoiding impingement because it is directly related to the generation of II and anterior/posterior offset and is deeply involved in the generation of BI. Although there have been many studies on the relationship between the stem anteversion angle and ROM, our study showed that the stem anteversion angle was positively related to the ROM of flexion/internal rotation, especially in patients with a small stem anteversion angle, even under conditions in which the implant was placed with the CA theory, because the anterior hip offset was reduced, BI occurred at a higher rate, and postoperative ROM decreased because the anterior hip offset was reduced [17]. The insertion angle of the stem in flexion/extension and the insertion position in the anteroposterior direction in the sagittal view were also related to the anterior/posterior offset, leading to BI occurrence and a decrease in postoperative ROM with a decrease in ROMBI.

4 Discussion

Hip joint requires mobility and stability at the same time, and in THA, ensuring sufficient hip ROM for the required ROM and achieving hip stability by preserving/reconstructing the soft tissues are important. Especially in the posterior approach where the posterior supporting tissues are separated, an understanding of impingement is essential to improve hip ROM and especially, to avoid postoperative dislocation. Although the anterior approach has attracted attention as a muscle/tendon-sparing technique in recent years, understanding of the mechanism of impingement generation is also essential because it may lead to decreased ROM, deterioration of implant durability, and even dislocation.

In recent years, the CA technique, large diameter heads, 3D preoperative planning, and navigation systems have been recommended to avoid postoperative dislocation, and the number of cases of postoperative THA dislocation has been decreasing. However, even now when stable implant placement and postoperative ROM can be expected, the number of dislocation cases has not been reduced to "zero". One of the reasons for this may be that the frequency of BI has increased, although the ROM defined by II has expanded due to surgery using CA theory, navigation system etc.; however, the frequency of BI has relatively increased. Several studies have been conducted to search for a safe zone that can avoid II and BI [20], however, whether a single CA theory can provide the necessary ROM is not clear. Factors related to BI include bone morphology and implant factors, and taking measures for each factor independently is necessary.

Regarding the frequency of impingement, Kessler et al. [18] reported that 44% of all impingements were BI based on simulation analysis using a 3D bone model. In this study, the frequency of BI in the anterior and posterior hip during flexion/internal rotation was 66%, 77%, and 34%, respectively, indicating that the incidence of BI in the anterior/posterior hip joint is relatively high and causes ROM restriction [5]. Furthermore, the rate of decrease in the ROMBI of flexion/internal rotation was significantly higher than that of external rotation, suggesting that BI is particularly likely to be a cause of ROM limitation in the anterior hip. Vigdorchik et al. [19]

similarly reported that in a cadaver study of the frequency of anterior/posterior impingement, more than half of the patients with anterior impingement were BI, and all patients with posterior impingement were II, a result that supports the studies. We further investigated the factors contributing to dislocation using postoperative CT in 53 patients of postoperative THA dislocation to examine the validity of these simulation analyses [20]. The results showed that anterior dislocation was associated with a high rate of II posterior to the hip joint, whereas posterior dislocation was associated with a significantly higher rate of BI anterior to the hip joint, indicating that BI leads to posterior dislocation after THA surgery. In addition, implant placement in patients with dislocation is characterized by a small stem anteversion angle and a small FO, which are associated with the occurrence of BI in the anterior hip joint, resulting in posterior hip dislocation. However, dislocation exists on the cup side both inside and outside the safe zone proposed by Lewinnek, and the position of the cup side was not directly related to dislocation, suggesting that the safe zone on the cup side is wider than that on the stem side. These results suggest that stem anteversion is important to avoid II and BI, cup placement according to the CA technique can avoid the occurrence of II, and impingement correction of BI is necessary to improve ROM, particularly in the anterior part of the hip joint in necessary cases. In addition, impingement correction of the BI in the anterior part of the hip joint is necessary to improve ROM in necessary cases.

In THA, fully understanding the theory related to impingement, considering the bone morphology, spinal alignment, and lifestyle of each patient, performing 3D dynamic analysis and preoperative planning before surgery, selecting and placing the implant, and selecting the approach according to the patient to achieve hip stability and a high level of ADL in the patient are necessary.

References

1. AAOS American Joint Replacement Registry. 2019 Annual Report.
2. Bartz RL, et al. The effect of femoral component head size on posterior dislocation of the artificial hip joint. J Bone Joint Surg Am. 2000;82:1300–7.
3. Shon WY, et al. Impingement in total hip arthroplasty a study of retrieved acetabular components. J Arthroplast. 2005;20:427–35.
4. Jolles BM, et al. Factors predisposing to dislocation after primary total hip arthroplasty: a multivariate analysis. J Arthroplast. 2002;17:282–8.
5. Widmer KH, et al. Compliant positioning of total hip components for optimal range of motion. J Orthop Res. 2004;22:815–21.
6. Dorr LD, et al. Combined anteversion technique for total hip arthroplasty. Clin Orthop Relat Res. 2009;467:119–27.
7. Shoji T, et al. Anterior inferior iliac spine bone morphology in hip dysplasia and its effect on hip range of motion in Total hip arthroplasty. J Arthroplast. 2016;31:2058–63.
8. Shoji T, et al. Bony impingement depends on the bone morphology of the hip after total hip arthroplasty. Int Orthop. 2013;37:1897–903.
9. Shoji T, et al. Factors affecting the potential for posterior bony impingement after total hip arthroplasty. Bone Joint J. 2017;99-B:1140–6.
10. Lewinnek GE, et al. Dislocations after total hip-replacement arthroplasties. J Bone Joint Surg Am. 1978;60:217–20.
11. Biedermann R, et al. Reducing the risk of dislocation after total hip arthroplasty: the effect of orientation of the acetabular component. Joint Surg Br. 2005;87:762–9.
12. Shoji T, et al. The influence of stem offset and neck shaft angles on the range of motion in total hip arthroplasty. Int Orthop. 2016;40:245–53.
13. Shoji T, et al. The effect of cup medialization and lateralization on hip range of motion in total hip arthroplasty. Clin Biomech (Bristol, Avon). 2018;57:121–8.
14. Sugano N, et al. Dynamic measurements of hip movement in deep bending activities after total hip arthroplasty using a 4-dimensional motion analysis. J Arthroplast. 2012;27:1562–8.
15. Nakashima Y, et al. Combined anteversion technique reduced the dislocation in cementless total hip arthroplasty. Int Orthop. 2014;38:27–32.
16. Komiyama K, et al. Does high hip Center decrease range of motion in Total hip arthroplasty? Computer simulation study. J Arthroplast. 2016;31:2342–7.
17. Shoji T, et al. Low femoral antetorsion and total hip arthroplasty: a risk factor. Int Orthop. 2015;39:7–12.
18. Kessler O, et al. Bony impingement affects range of motion after total hip arthroplasty: a subject-specific approach. J Orthop Res. 2008;26:443–52.
19. Vigdorchik JM, et al. Does prosthetic or bony impingement occur more often in Total hip arthroplasty: a dynamic preoperative analysis. Arthroplasty. 2020;35:2501–6.
20. Shoji T, et al. Factors affecting impingement and dislocation after total hip arthroplasty – computer simulation analysis. Clin Biomech (Bristol, Avon). 2020;80:105151.

Bone Remodeling and Stress Shielding

18

Junya Shimizu and Satoshi Nagoya

1 Bone Remodeling

Stable long-term results after THA using cementless stems have been reported. They depend on how the skeleton adapts mechanically to load transfer following implant insertion [1]. Irregular load transfer patterns are detrimental to bone remodeling and are referred to as stress shielding (SS), which can cause bone resorption around the implant and lead to implant failure, making revision difficult [2]. However, it is not clear whether changes in bone resorption are associated with clinical symptoms and complications.

2 Stress Shielding

When a cementless stem is used, initial fixation is obtained by the stem being press-fitted to the cortical bone. Once the stem is fixed to the femur, loading stresses are transferred through the stem to the bone distal to the stem, resulting in SS in the proximal femur [3, 4]. The degree of SS varies depending on stem geometry; a four-step qualitative evaluation method by Engh et al. [5] is commonly used to assess SS. Cementless stems are generally classified using the stem geometry categories reported by Khanuja et al. [6]. This section summarizes the occurrence of SS for each stem geometry (Fig. 18.1).

2.1 Type 1: Tapered Proximal Fixation (Single Wedge Type)

The coronal surface of the stem is tapered, and the stem is intended to be fixed at the end of the diaphysis. The proximal part of the stem is coated, and the anteroposterior part is thin and flat.

Tri-Lock (Depuy-Synthes) and Taperloc (Zimmer-Biomet), which are representative of this type of stem, have reported rates of SS occurrence of 96% [7] and 98% [8], respectively, though most occurrences were grade 1 or 2. However, SS rates as low as 4% have been reported for Accolade TMZF (Stryker) [9]. Differences in the stiffness of the titanium stem metal composition, the length of the follow-up period, and the subjectivity of the examiner's evaluation of SS may be factors. On the other hand, the incidence of thigh pain is low at only 2–3% [7, 8] and does not always correspond to clinical symptoms.

2.2 Type 2: Tapered Proximal Fixation (Double Wedge Metaphyseal Filling Type)

Similar to type 1, this type of stem is intended for fixation at the distal end of the diaphysis, but is tapered in two planes, anteroposterior and mediolateral, and is wider than type 1.

Kawaji et al. [10] reported that 71 of 106 patients (67%) with VerSys HA/TCP Fiber Metal Taper (Zimmer-Biomet) had SS, and an SS of grade 3 or 4 using Engh's classification was more common in varus alignment. Capello et al. [11] reported SS in 61 (49%) of 124 patients using the Omnift HA (Stryker). Thigh pain occurred in only two patients (1%). Koyano et al. [12] reported SS in the proximal femur in all 36 patients with Super Secur-Fit HA (Stryker). Type 2 is often reported to have a relatively high frequency of SS.

2.3 Type 3A: Tapered Proximal Fixation (Tapered Round Type)

This type is intended for fixation at the boundary between the metaphyseal and diaphyseal ends. The proximal two-

J. Shimizu
Department of Orthopedic Surgery, Sapporo Medical University, Sapporo, Japan

S. Nagoya (✉)
Department of Orthopaedic Surgery, Sapporo Kojinkai Memorial Hospital, Sapporo, Japan
e-mail: nagoya@sapmed.ac.jp

Y. Nakashima (ed.), *Advances in Total Hip Arthroplasty*, https://doi.org/10.1007/978-981-97-2461-1_18

Types	Author (year)	Stem Type	Cases	Mean Follow -up (years)	Frequency of SS
type1	Teloken (2002)[7]	TRI–LOCK	49	15	96% (grade 1, 2 : 84%/ grade 3, 4 : 12%)
	McLaughlin (2008)[8]	Taperloc	57	20	98% (grade 1, 2 : 93%/ grade 3, 4 : 5%)
	Cho (2016)[9]	Accolade TMZF	437	5	4%
type2	Kawaji (2016)[10]	VerSys HA/TCP Fibermetal Taper	106	7	67% (grade 1, 2 : 28%/ grade 3, 4 : 39%)
	Capello (2016)[11]	Omnifit HA	124	15	49%
	Koyano (2017)[12]	Super Secur–Fit HA	36	9	100%
type3A	Kanto (2020)[13]	Bicontact	240	7	17%
	Bourne (2001)[14]	Mallory Head	307	10	50% (grade 1, 2 : 50%/ grade 3, 4 : 0%)
	Park (2003)[15]		76	10	89% (grade 1, 2 : 88%/ grade 3, 4 : 0%)
	Müller (2010)[17]	CLS–Spotorno	74	17	84% (grade 4 : 14%)
	Aldinger (2009)[18]		101	15	100% (grade 1, 2 : 100%/grade 3, 4 : 0%)
type3B	Pak (2021)[20]	Wagner	28	7	25% (grade 1, 2 : 25%/grade 3, 4 : 0%)
type3C	Suckel (2009)[22]	Alloclassic–SL	97	15	34%
	Delaunay (2014)[23]		193	10	1%
	Tanaka (2020)[24]	SL–PLUS MIA	98	1	100% (grade 1, 2 : 94%/ grade 3, 4 : 6%)
		SL–PLUS MIA HA	52	1	100% (grade 1, 2 : 100%)
type4	Engh (1988)[25]	AML	411	2	72% (grade 1, 2 : 68%/ grade 3 : 4%)
	Hennessy (2009)[26]	Prodigy	76	11	80% (grade 1, 2 : 79%/ grade 3, 4 : 1%)
	Chen (2006)[27]		145	6	83% (grade 1, 2 : 79%/ grade 3 : 4%)
type5	Bawale (2021)[29]	S–ROM	167	28	20%
	Tanzer (2001)[29]		59	8	78%
type6	Matsuyama (2016)[30]	Spongiosa	16	10	94% (grade 1 : 18, 2 : 63% grade 3, 4 : 13%)
short stem	Kutzner (2017)[31]	Optimys	201	2	8%

Fig. 18.1 Frequency of SS with stem type

thirds of the stem is porous-coated, round and tapered. Fins and ribs are placed for rotational stability.

Kanto et al. [13] reported that SS occurred in 41 (17%) of 240 patients using the Bicontact (B/Braun-Aesculap). However, Bourne et al. [14] and Park et al. [15] reported SS in 50% and 89% of 307 and 76 patients, respectively, using the Mallory Head stem (Zimmer-Biomet). Dhillon et al. [16] reviewed 12 reports including 1348 patients with CLS-Spotorno (Zimmer-Biomet) and reported findings of SS in 358 patients (27%). Of these, 72% were grade 1 or 2 on the Engh classification [3] and showed little clinically significant bone atrophy or loosening. However, the incidence of SS in CLS-Spotorno has been reported to be as high as 84–100% [17, 18].

2.4 Type 3B: Tapered Distal Fixation (Tapered Spline/Cone Type)

The Wagner stem (Zimmer-Biomet) is a type of stem intended for fixation at the diaphysis or the boundary between the metaphysis and diaphysis. It is a tapered stem with a circular cross-section and a groove for rotational stability. It is often used in revision surgery and following trochanteric osteotomy; it is effective in revision surgery but not often used in typical cases [19]. Park et al. [20] found grade 1 and 2 SS in 17.9 and 7.1%, respectively, of 28 patients who underwent revision surgery using the Wagner stem during a follow-up period of 7.6 years after revision surgery, but grade 3 or higher SS was not seen in any of these patients.

2.5 Type 3C: Tapered Distal Fixation (Tapered Rectangle Type)

This type is intended for fixation at the diaphysis and at the boundary between the metaphysis and diaphysis. It is a tapered stem with a square cross-section that achieves rotational stability. The Zweymüller [21] stem, which is widely used in Europe, is a representative example. Suckel et al. [22] used the Alloclassic-SL stem (Zimmer-Biomet) in 97 patients and reported that SS was observed in approximately one-third of them; Delaunay et al. [23] reported SS in 3 (1%) of 193 patients with Alloclassic (Zimmer-Biomet), of which one had hydroxyapatite (HA) coating and two did not. Tanaka et al. [24] reported that all patients had SS at 1 year. They found that 94% of patients without HA coating had grade 1–2 SS, and 6% had grade 3–4 SS, whereas all patients with HA coating had grade 1–2 SS. In type 3C stems, while a radiolucent line (RL) may appear at the proximal end of the stem, few patients develop SS, of whom most are grade 1–2. HA coating may further reduce the incidence of SS.

2.6 Type 4: Distal Fixation (Cylindrical Fully Coated Type)

This type is intended for press-fit fixation at the diaphysis and has a broad porous coating for distal fixation. The AML stem (Depuy-Synthes) is a typical example. Engh et al. [25] reported that SS occurred in 72% (grade 1: 53%, grade 2: 14%, grade 3: 4%) of AML stems. After that report, the Prodigy stem (Depuy-Synthes), which has an expanded porous coating intended to reduce thigh pain and SS, came into use. However, Hennessy et al. [26] reported that SS occurred in 80% of 76 patients who used the Prodigy stem for over 10 years after surgery. Most of these were grade 1 or 2. Chen et al. [27] reported that SS was present in 83% of patients with Prodigy stems. The frequency of SS is high, and some patients have severe SS.

2.7 Type 5: Modular Type

This stem is intended for fixation at the metaphysis and diaphysis. The proximal and distal parts can be used in combination. It is most often indicated in cases of complex femoral morphology, such as after osteotomy, and is rarely used in normal cases. Bawale et al. [28] and Tanzer et al. [29] reported that SS was observed in 20% and 78% of 167 and 59 patients, respectively, using S-ROM (Depuy-Synthes) for initial THA at 28 years and 8 years of follow-up.

2.8 Type 6: Curved Anatomic Type

This type is intended for fixation at the metaphysis. The curve is convex posteriorly at the metaphyseal end and convex anteriorly at the diaphyseal end, differentiating it from the straight stems (types 1–4). The stem differs from left to right and the Spongiosa stem (Zimmer-Biomet) is a representative example. Matsuyama et al. [30] investigated SS at 10 years after surgery in 16 patients using the Spongiosa stem and reported grade 1 SS in 3 patients (18%), grade 2 in 10 patients (63%), and grade 3 in 2 patients (13%).

2.9 Short Stem

Short stems have come into use in recent years for bone and muscle tendon preservation, but long-term results have not yet been reported, and there are few reports on SS. Kutzner et al. [31] reported 16 cases (8%) of SS in 201 patients treated with Optimys (Mathys). Further reports are needed.

References

1. Sumner DR. Long-term implant fixation and stress-shielding in total hip replacement. J Biomech. 2015;48:797–800.
2. Kobayashi S, et al. Poor bone quality or hip structure as risk factors affecting survival of total-hip arthroplasty. Lancet. 2000;355:1499–504.
3. Skinner HB, et al. Correlation of computed finite element stresses to bone density after remodeling around cementless femoral implants. Relat Res. 1994;305:178–89.
4. Kerner J, et al. Correlation between preoperative periprosthetic bone density and postoperative bone loss in THA can be explained by strain-adaptive remodeling. J Biomech. 1999;32:695–703.
5. Engh CA, et al. Porus-coated hip replacement. The factors governing bone ingrowth, stress shielding, and clinical results. J Bone Joint Surg Br. 1987;69:45–55.
6. Khanuja HS, et al. Cementless femoral fixation in total hip arthroplasty. J Bone Joint Surg Am. 2011;93:500–9.
7. Teloken MA, et al. Ten to fifteen-year follow-up after total hip arthroplasty with a tapered cobalt-chronium femoral component (tri-lock) inserted. J Bone Joint Surg Am. 2002;84:2140–4.
8. McLaughlin JR, et al. Total hip arthroplasty with an uncemented tapered femoral component. J Bone Joint Surg Am. 2008;90:1290–6.
9. Cho YJ, et al. Distal femoral cortical hypertrophy after hip arthroplasty using a cementless double-tapered femoral stem. J Orthop Surg. 2016;24:317–22.
10. Kawaji H, et al. Influence of femoral implant alignment in uncemented total hip replacement arthroplasty: varus insertion and stress shielding. J Nippon Med Sch. 2016;83:223–7.
11. Capello WN, et al. Hydroxyapatite-coated femoral components: 15-year minimum follow up. Clin Ortho Relat Res. 2006;453:75–80.
12. Koyano G, et al. Comparison of bone remodeling between an anatomic short stem and a straight stem in 1-stage bilateral total hip arthroplasty. J Arthroplaty. 2017;32:594–600.
13. Kanto M, et al. Radiological evaluation of the relationship between cortical hypertrophy and stress shielding after total hip arthroplasty using a cementless stem. Arthroplasty Today. 2020;6:894–900.

14. Bourne RB, et al. Tapered titanium cementless total hip replacements: a 10- to 13-year followup study. Clin Orthop Relat Res. 2001;393:112–20.
15. Park MS, et al. Plasma spray-coated Ti femoral component for cementless total hip arthroplasty. J Arthroplast. 2003;18:626–30.
16. Dhillon MS, et al. Long-term survival of CLS Spotorno femoral stem: a systematic review of the literature. Arch Orthop Trauma Surg. 2021; https://doi.org/10.1007/s00402-021-03975-0.
17. Müller LA, et al. Seventeen-year survival of the cementless CLS Spotorno stem. Arch Orthop Trauma Surg. 2010;130:269–75.
18. Aldinger PR, et al. Uncemented grit-blasted straight tapered titanium stems in patients younger than fifty-five years of age. J Bone Joint Surg Am. 2009;91:1432–9.
19. Böhm P, et al. Femoral revision with the Wagner SL revision stem: evaluation of one hundred and twenty-nine revisions followed for a mean of 4.8 years. Bone Joint Surg Am. 2001;83:1023–31.
20. Park KS, et al. Long-term outcomes of cementless femoral stem revision with the Wagner cone prosthesis. J Orthop Surg Res. 2021;16:375.
21. Zweymüller K, et al. Concept and material properties of a cementless hip prosthesis system with Al_2 O_3 ceramic ball heads and wrought Ti-6Al-$_4$ V stems. Arch Orthop Trauma Surg. 1982;100:229–36.
22. Suckel A, et al. Long-term results for the uncemented Zweymüller/Alloclassic hip endoprosthesis. A 15-year minimum follow-up of 320 hip. J Arthroplast. 2009;24:846–53.
23. Delaunay C. Effect of hydroxyapatite coating on the radio-clinical results of a grit-blasted titanium alloy femoral taper: a case–control study of 198 cementless primary total hip arthroplasty with the Alloclassis™ system. Orthop Traumatol Surg Res. 2014;100:739–44.
24. Tanaka A, et al. Comparison of early femoral bone remodeling and functional outcome after total hip arthroplasty using the SL-PLUS MIA stem with and without hydroxyapatite coating. Musculoskelet Surg. 2020;104:313–20.
25. Engh CA, et al. The influence of stem size and extent of porous coating on femoral bone resorption after primary cementless hip arthroplasty. Orthop Relat Res. 1988;231:7–28.
26. Hennessy DW, et al. Second-generation extensively porous-coated THA stems at a minimum 10-year followup. Clin Orthop Relat Res. 2009;467:2290–6.
27. Chen CJ, et al. Second-generation porous-coated cementless total hip arthroplasties have high survival. Clin Orthop Relat Res. 2006;451:121–7.
28. Bawale R, et al. Long-term survivorship of modular cementless femoral stem in complex primary total hip arthroplasty: a concise minimum 15-year follow-up report. J Arthroplast. 2021;36:3221–5.
29. Tanzer M, et al. Primary cementless total hip arthroplasty using a modular femoral component. J Arthroplast. 2001;16(Suppl 1):64–70.
30. Matsuyama K, et al. Finite element analysis of cementless femoral stems based on mid- and long-term radiological evaluation. BMC Musculoskelet Disord. 2016;17:397.
31. Kutzner KP, et al. Outcome of extensive varus and valgus stem alignment in short-stem THA: clinical and radiological analysis using EBRA-FCA. Arch Orthop Trauma Surg. 2017;137:431–9.

Part IV

Biomaterials in Arthroplasty

Polyethylene Wear and Osteolysis

19

Yuya Takakubo and Michiaki Takagi

Particles formed by wear on the sliding surfaces of artificial hip joints are known as wear debris or wear particles. Apart from the sliding surfaces, wear and corrosion may also occur in the contact surfaces between implants. Wear debris derived from ultra-high molecular weight polyethylene (UHMWPE) and metals of sizes varying from submicron to several microns are phagocytosed by monocytes and macrophages, forming foreign body granulomas and causing osteolysis around the implants. When osteolysis is extensive, implant fixation is reduced, which leads to loosening of the artificial hip joints [1–3] (Fig. 19.1). Dealing with osteolysis associated with polyethylene wear remains a clinical challenge. The biological reactions associated with metal wear debris, fine particles associated with corrosion, and ion generation are known as adverse local tissue reactions (ALTRs) to metals [4] and adverse reactions to metallic debris [5]. They can form pseudotumors around the prosthesis or cause necrotic destruction of soft tissues [4–7]. The biological reactions involved in wear are diverse. Here, we focused on polyethylene wear and osteolysis, which are the starting points of osteolysis research.

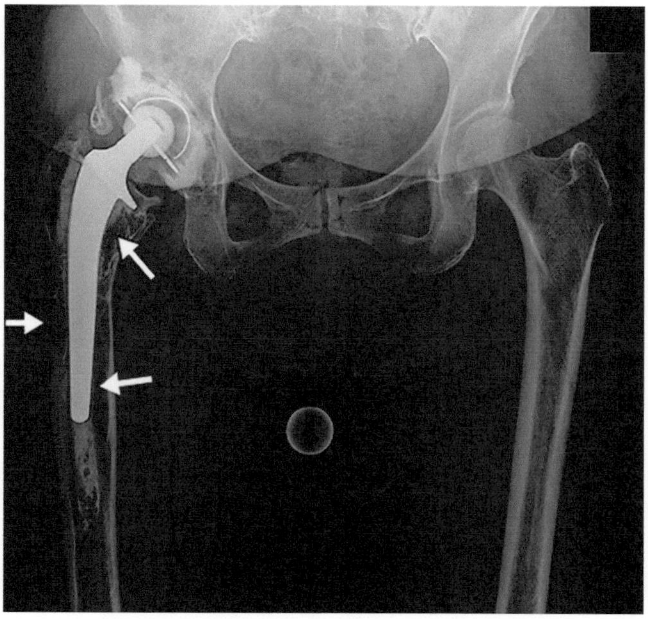

Fig. 19.1 Osteolysis around the artificial hip joint associated with polyethylene wear

Y. Takakubo · M. Takagi (✉)
Faculty of Medicine, Department of Orthopedic Surgery, Yamagata University, Yamagata, Japan
e-mail: mtakagi@med.id.yamagata-u.ac.jp

1 Wear Products and Biological Reactions

When solids come into contact with each other and relative motion occurs in a tangential direction, friction is generated. Wear is a phenomenon in which the contact surface is damaged by friction, the surface of the solid peels off, and the weight is reduced. This phenomenon is inevitable for solid biomaterials that are used in hip joint prostheses. However, lubrication by the joint fluid and intercellular fluid components can reduce friction and wear. Although joint fluid plays a lubricating role in a normally functioning hip prosthesis, the possibility of wear cannot be completely eliminated. An interface without lubrication is formed at the sliding surface of metal on metal combination and at the bone–cement–implant interface. Wear and corrosion also occur in tapered metal implants, screw fixation areas, and bone–cement implants. Wear is significantly enhanced when particles are interposed between the opposing implants. This phenomenon is called third body wear [8, 9]. The wear of polyethylene becomes more pronounced when fragments of metal, ceramics, or bone cement are embedded in its sliding surfaces.

The biological reactivity differs depending on the size of the wear and corrosion products. According to size, wear products can be classified into bulk (solids), large (particles), moderate (fine particles), small (microparticles), and very

Y. Nakashima (ed.), *Advances in Total Hip Arthroplasty*, https://doi.org/10.1007/978-981-97-2461-1_19

small (metal ion/metal ion complexes). This classification is useful for understanding their properties [2].

Implant products are mainly classified as "bulk." Titanium alloys and hydroxyapatite, with good bone compatibility, bind firmly to bone tissue. Other materials with low bone compatibility bind to the surrounding bone tissue by thin fibrous capsules. Under mechanically stable conditions and without excessive inflammatory reaction of the surrounding tissues, the prostheses remain stable and function well, with good compatibility with the surroundings, including the bone–cement interface without loosening and the bone–metal implant interface without bone induction. Fragments and fractures resulting from implants are not clinically problematic if they are covered by a fibrous capsule. However, when the fracture sites come into contact with each other or with other sites, biological reactions associated with wear and corrosion are induced.

Implant-derived particles (associated with wear) of size ≥10 μm are classified as "large". They are formed by wear due to micromotion or impingement between the implants. As the particle size increases, the phagocytes are unable to phagocyte the particles, and the wear particles are eventually separated by a fibrous capsule. Small particles are embedded in the fibrous matrix of regenerating arthrodesis and do not elicit prolonged inflammatory responses. However, particles classified as "large" can contribute to wear when they enter articular surfaces. When particles, such as bone cement, metal, ceramics, and hydroxyapatite, enter the sliding surface, they cause wear between the polyethylene, metal, and ceramics used as sliding surface materials (three-body wear). Despite using polyethylene with high cross-linking and wear resistance for the sliding surface, a large amount of wear powder can be generated owing to the three-body wear.

Wear particles that are classified as "moderate" vary in size from submicron to several microns. Particles in this size range are mainly processed by phagocytosis by monocytes and macrophages. If the amount of wear is small, phagocytes undergo apoptosis after phagocytosis, and the wear debris is eventually buried in the fibrous tissue and becomes unproblematic. However, when particles are continuously produced and their quantity is large, the biological reaction triggered by phagocytosis persists and foreign body granulomas are formed, as described below [10, 11].

Wear particles that are classified as "moderate" in size, together with particles classified as "small," serve as scaffolds to which bacterial endotoxins and bioconstituents adhere, thus eliciting biological responses via innate immunity [12]. The cellular reactivity to the particles is reported to differ depending on the shape of the particles, but the surface properties of the particles are not sufficiently comparable. Further studies are required to confirm this hypothesis.

Particles that are classified as "small" are also formed by wear. Wear particles, a few nanometers in size, reside within the extracellular matrix (ECM) or float in the intercellular fluid. Non-aggregated particles are internalized by cells via pinocytosis [2]. The biological effects of particles of this size are unknown and require further analysis.

Metal ions and their complexes are classified as "very small." They are formed by wear and corrosion of metal-derived materials. Metal ions and their complexes have the potential to elicit an immune response and cause delayed hypersensitivity, which is classified as a type IV allergy. However, the details of the mechanism, including which substances are haptens, remain unknown.

2 Polyethylene Wear and Osteolysis

2.1 Phagocytosis of Wear Particles

The polyethylene wear particles are phagocytosed by monocytes and macrophages, and foreign body granulomas are formed (Figs. 19.2 and 19.3). When the number of particles associated with submicron- to several micron-sized wear is large, the foreign body granuloma reaction is prolonged and persistent. This occurs mainly at the implant–bone interface and in the regenerating arthrodesis around the implant [13]. Chronic inflammation induces osteolysis.

The foreign body granuloma reaction, which is mainly triggered by polyethylene abrasion powder, primarily involves monocytes, macrophages, and fibroblasts; however, multinucleated giant cells, epithelial-like cells, vascular endothelial cells, a few lymphocytes, mast cells, and very few multinucleated leukocytes have also been observed [3, 11]. Macrophages and monocytes are the major components of foreign body granulomas and are CD68-positive. CD68-positive macrophages are classified as naïve macrophages

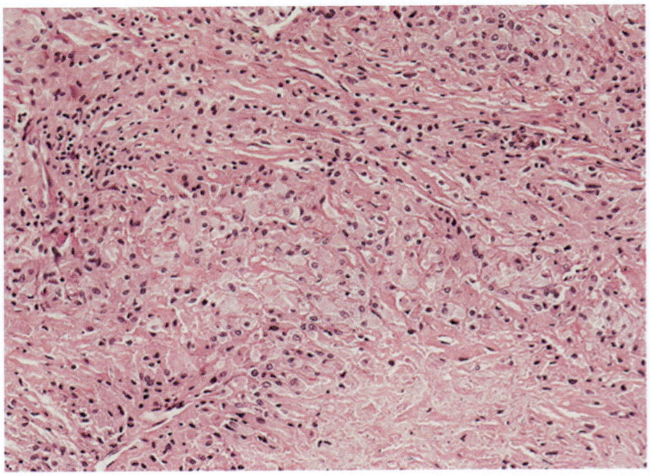

Fig. 19.2 Granulomatous tissue observed in an artificial joint with polyethylene wear. A foreign body reaction mainly comprising foamy macrophages was observed

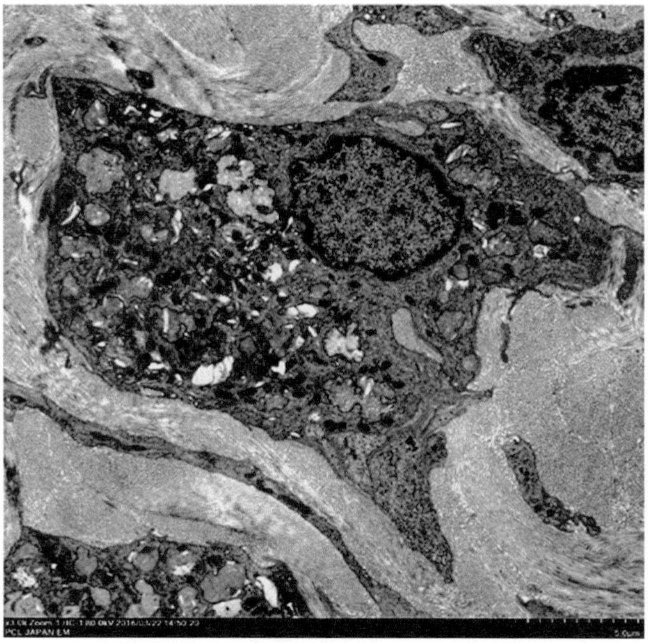

Fig. 19.3 Phagocytosis image of polyethylene wear particle observed by transmission electron microscopy

(M0), inflammatory macrophages (M1), and anti-inflammatory macrophages (M2). M1 macrophages are considered to be the major macrophages involved in the foreign body granuloma response [14]. Various biologically active substances are produced by various cell types. Activated monocytes, macrophages, and fibroblasts produce and secrete cytokines and ECM proteolytic enzymes. Bone-resorptive cytokines, such as TNF-α, IL-1β, and M-CSF; chemokines, such as CCL2, 3, and CXCL8, 9, and 10 [3]; and proteolytic enzymes, such as cathepsin K, G, MMP-1, −2, −3, −9, and −13, are known to be involved in the granuloma reaction [15, 16]. This leads to pathological osteoclastic and non-osteoclastic bone resorption [17] in the tissues surrounding the implant and induces osteolysis. Previous studies have reported that osteoclast activation is mediated by RANKL and MCS-F expressed in osteoblasts and stromal cells [18, 19]. In addition to osteoclast and osteoblast activation, osteocyte hyperfunction is also observed in bone tissues adjacent to granulomas [17]. This phenomenon suggests the involvement of sclerostin, which has a bone resorption activity. Prolonged and sustained activation of ECM proteolytic enzymes produced by granulomas and imbalance with endogenous inhibitors weakens the fibrous connective tissue around the implants [11, 15]. This pathology also plays a significant role in aseptic hip joint laxity owing to wear debris.

Recently, the roles of Toll-like receptors (TLRs) and nucleotide-binding oligomerization domain (NOD)-like receptors (NLRs), which function as innate immune systems, in these biological responses have attracted considerable attention [20]. In granulomatous reactions, the expression of NALP3 and ASC is upregulated in macrophages in association with TLRs 1, 2, 5, 6, and 9, and NLRs. TLRs are activated by exogenous microbial ligands, as well as endogenous ligands derived from inflammatory components, such as heart shock protein, high mobility group box-1, fibronectin, and hyaluronan, while NLRP3 is activated by ATP, radical oxygen species, and uric acid. Therefore, in chronic inflammation with granuloma formation, these endogenous substances activate the innate immune system and enhance the inflammatory response by producing cytoreductive contents and substances associated with inflammation [3, 20]. These substances may adhere to wear particles and enhance the inflammatory response upon phagocytosis, similar to exogenous substances, such as lipopolysaccharide, which have already been shown to enhance the inflammatory response upon attachment to wear particles.

When the sustained overproduction of submicron- to several micron-sized wear particles exceeds the response limit of the periprosthetic tissue, periprosthetic bone tissue fragilization with osteolysis occurs [21]. Despite a foreign body granuloma reaction occurring with extensive bone resorption around the implant, the patient will often experience mild or almost asymptomatic pain around the hip joint if there are no surrounding microfractures, bleeding, or significant implant instability. Pathological elevations in leukocytes, C-reactive protein (CRP), and erythrocyte sedimentation rate (ESR) are rarely observed in blood biochemistry immunological examination. When elevated CRP or ESR is observed, the possibility of a biological reaction due to latent infection or its coexistence, immune response to metals, or inflammatory disease other than localized inflammation should be considered [3].

2.2 Access Theory

When excessive wear debris is produced near the sliding surface, the foreign body granulomatous reaction, mainly by monocytes and macrophages, is induced and persists for a long time, resulting in the increased production and accumulation of joint fluid due to inflammation. The pressure loading caused by joint motion widens the gap between the bone and implant, and the wear debris contained in the fluid reaches a site distant from its source, inducing a new resorptive foreign body granulomatous reaction. The synergistic effect of pressure loading and the foreign body granulomatous reaction widens the access space between the bone tissue and implant, thus weakening the supporting tissue around the implant and causing extensive osteolysis (Access Theory) [22]. This results in loss of implant stability and can lead to periprosthetic fractures, dislocations, and implant loosening. Polyethylene wear particles also cause local biological reactions and are transported to the surrounding area, causing the expansion of the osteolysis lesion, also known as small-particle disease [23].

2.3 Improved Clinical Outcomes Associated with Upgrading UHMWPE

(1) The first polyethylene used by Charnley for sliding surfaces was a resin with a molecular weight of 1–2 million containing calcium stearate [24], and later, with the aim of achieving increased strength, even larger molecular weight resins (2–4 million and 4–6 million) without calcium stearate were used. In addition to the improvement in wear resistance by crosslinking treatment by irradiation with gamma rays or electron beams, the remelt method of heating above the melting point of UHMWPE and the annealing method of heating below the melting point were devised to remove free radicals by heat treatment or the method of applying free radicals [25]. Clinical results using cross-linked UHMWPE have shown improved wear resistance with reduced osteolysis, reduced sagging, and good long-term results [26]. However, while the remelt method can completely eliminate free radicals, the strength is reduced, and in the annealing method, the residual free radicals remain [25]. To overcome these limitations, several methods have been devised, such as annealing multiple times [27], adding vitamin E to polyethylene to suppress the internal oxidation of UHMWPE [28], and coating the surface of UHMWPE with a special phospholipid polymer (MPC polymer) to reduce the coefficient of friction on sliding surfaces [29]. Other technologies have also been devised furthermore to improve the performance.

3 Imaging Findings of Periprosthetic Biological Reactions

3.1 Osteolysis and Implant Laxity

Simple radiographs are the basis of evaluation. The degree of implant loosening is evaluated according to the fixation method (cemented or cementless) and site (acetabular or femoral side).

The decreased fixation of a cemented acetabular cup usually appears as radiographic transparency of the bone–cement interface inferiorly within the cup. In a simple frontal radiograph of the hip joint, the boundary between the cemented hemispherical socket and bone is classified into three sections: Zones I, II, and III, from medial to posterior, and loosening is defined as a permeability of 2 mm or more [30]. The presence of circumferential loosening suggests acetabular implant instability, but in cemented cups, intraoperative cup instability can be confirmed, and the cup can easily be removed even without the presence of circumferential loosening. If revision is considered, a three-dimensional evaluation using computed tomography (CT) should be performed. Upward or medial cup migration and breakage of the cup and cement mantle are also findings of loosening. If there is significant polyethylene wear, the center of the femoral head may shift upward.

In cementless acetabular cups, shell migration, breakage of the shell and screws, and loss of the porous surface finish suggest the presence of loosening. Through screw holes, wear debris may induce cystic osteolysis, and subclinical osteolysis may be present without evident radiographic permeability. In contrast, increased radiolucency around implants may reflect bone atrophy; therefore, it is important to investigate bone defects in detail on CT.

The characteristic features associated with cemented femoral stem loosening include the appearance of radiolucency at the bone–cement interface, thinning of the cortical bone, stem migration or breakage, and separation between the bone cement and stem. Periprosthetic radiographic images may show a continuous linear shadow or a focal oval-shaped radiographic image (endosteal scalloping sign) that locally erodes the bone cortex from the medullary cavity. Both may be present in some cases. The separation of the implant from the cement is observed in the early stages of loosening. However, slight stem subsidence in colorless polished tapered stems is not necessarily an early sign of loosening. The Harris classification [31] is often used to evaluate the loosening of cemented stems and is classified into three levels: definite loosening, which is defined as stem movement and radiographic evidence of continuity among stem cements, stem fracture, and cement mantle failure; probable loosening, which is defined as the appearance of a continuous radiolucency between the cement and bone or separation between the stem and cement in either frontal or lateral view; and possible loosening, which is when there is more than 50% and less than 100% transmission between the cement and bone in the front or side.

In cementless stems, movement and widening of the medullary cavity at the distal tip of the stem suggest unstable fixation. In the case of a foreign body granulomatous reaction caused by the intrusion of wear particles between the bone and stem, irregular cortical bone atrophy and resorption are observed. If osteosclerosis is observed in the medullary cavity distal to the stem tip, the stem is considered inadequately fixed (bony pedestal). In a well-fixed stem, an osteosclerotic region known as "spot weld," which is derived from the metallurgical term spot weld, is formed between the cortical bone and the stem where the implant surface with good bone affinity is adjacent to the cortical bone. Proximal to the spot weld, there may be increased radiolucency owing to cortical bone atrophy and stress shielding. The proximal portion of the stem may show irregular thinning of the cortical bone and increased radiolucency, which should be carefully evaluated using CT (Fig. 19.1).

3.2 Soft-Tissue Evaluation

Pathological soft tissue lesions may also be detected on plain radiographs. The shadow is enhanced by a soft-tissue mass. The shadow is relatively more apparent when there is significant metal wear. MRI, which can be performed in a mode that reduces metallic artifacts, is useful for identifying the lesion. CT and ultrasound can also detect the lesion.

4 Treatment

Osteolysis associated with granulomas of the wear particles is often asymptomatic. Pain is often reported after bone loss becomes severe and implant instability or periprosthetic fractures occur. Revision surgery is indicated when there is pain associated with osteolysis and dislocation due to implant instability. Revision should also be considered before the procedure becomes difficult in the future, even if the patient is asymptomatic. When a revision is performed for loosening, the point is how to restore the site of the bone defect and stabilize newly inserted implants. Reconstruction should be considered depending on the extent of the bone defect, fixation technique used for the initial implant, and tissue damage expected at the time of removal. Extended transtrochanteric osteotomy may be used for femoral implants that are difficult to remove. The cement mantle, which adheres firmly to the bone in the medullary cavity and is difficult to remove, can be removed using a device with ultrasonic vibration.

If the acetabular side has a large bone defect, allogeneic bone grafting is performed. If a cementless cup is used, a massive bone graft is placed over the bone defect and fixed with screws. If the defect is large, the graft is stabilized using a plate. The gap after molar base or massive bone grafting can be filled with impaction bone grafting (IBG) [32]. The use of large hemispherical or eccentric components and porous metallic fillers has also been attempted to restore such bone defects. When the bone defect is restored with a cemented cup, allogeneic bone grafting using the IBG technique can be useful to facilitate remodeling of the grafted bone and promote the restoration of bone stock. Depending on the degree of bone loss, allogeneic or plate-shaped bone grafts can be used in combination with a mesh, cross-plate, or cage to reconstruct the acetabulum in the anatomical position as much as possible (Fig. 19.4). The most difficult case is an extensive bone defect with pelvic ring instability classified as pelvic discontinuity (AAOS type IV) in which a cemented cup is inserted using the IBG technique in combination with screws and plates to stabilize the pelvic ring or using massive strut allogeneic bone with a cross plate, mesh, or cage. A cementless cup is also used after stabilizing the grafted bone in the same manner.

If the bone defect is small on the femoral side, a cementless stem can be used alone. If the bone defect is massive on the proximal side, a long stem or megaprosthesis can be used to restore the defect with a modular metal part. A strut allograft in which a homologous femur is used as a post to replace the defect or the entire proximal portion is replaced with homologous bone can also be used. In Japan, allografts are uncommon because of their limited availability in limited facilities and regions. Cemented femoral stems may be used without bone grafting if the patient is relatively old and inactive and the proximal bone is preserved. If the bone defect is large, the IBG technique using morselized allogeneic bone is effective, according to the method of Gie et al. [33] (Fig. 19.4). If the proximal bone defect is large, IBG can be performed with a metal mesh. If the risk of periprosthetic fracture is high, the femur is reinforced with a plate or wire

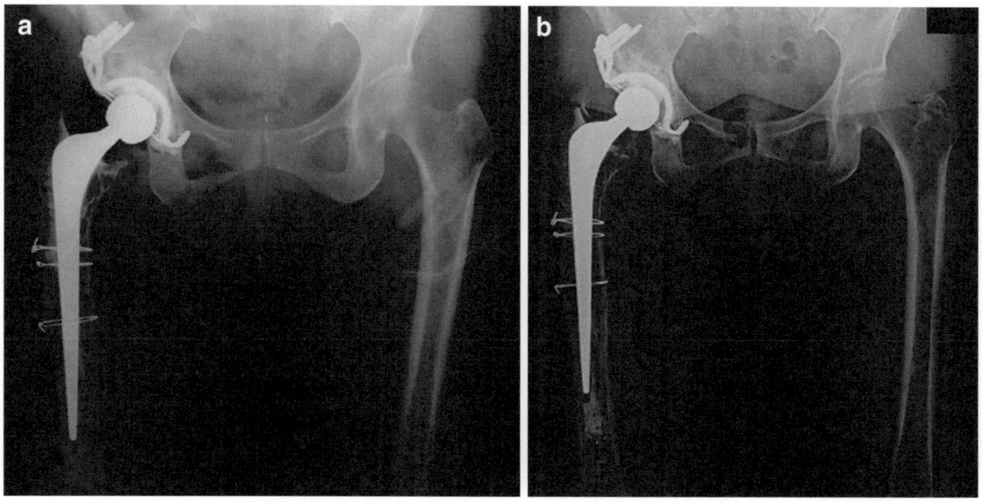

Fig. 19.4 Immediately after the functional reconstruction of the case shown in Fig. 19.1 (**a**), the acetabular side was reconstructed with a Kerboul-type plate and a strut allogeneic bone graft, and the femoral side was reconstructed with the impaction bone grafting technique after 2 years of surgery (**b**)

prior to IBG. If the bone defect is large, an allogeneic femur is a useful replacement method, but it is difficult to achieve, as mentioned above. Cemented megaprostheses can also be used for this purpose.

The same is true for bone defects caused by ALTRs to metals, but if cobalt-chromium-related metal allergy is suspected, other materials, including sliding surfaces and joints, should be used. In case the removal of the stem is difficult, a ceramic head and titanium sleeve adaptor may be used if the fitting part permits, considering the risk-benefit ratio. If necrotic destruction of the surrounding soft tissues is extensive, soft tissue reconstruction should also be performed, considering muscle and fascia migration.

Wear debris production can be reduced with improved polyethylene, resulting in reduced osteolysis and implant sagging. However, when three-body wear occurs on sliding surfaces, the risk of osteolysis increases even with improved polyethylene. Although stable long-term results have been reported, the long-term outcomes should be carefully monitored in the future. Sufficient knowledge of the effects of wear and corrosion on joint prostheses is crucial for effective diagnosis and treatment.

References

1. Konttinen YT, et al. The microenvironment around total hip replacement prostheses. Clin Orthop Relat Res. 2005;430:28–38.
2. Goodman SB, et al. Biocompatibility of total hip replacement: a review. J Biomed Mater Res, Part A. 2009;90:603–18.
3. Goodman SB, et al. Diagnosis and management of implant debris associated-inflammation. Expert Rev Med Devices. 2020;17:41–56.
4. Cooper HJ, et al. Corrosion at the head neck taper as a cause for adverse local tissue reactions in total hip arthroplasty. J Bone Joint Surg Am. 2012;94:1655–61.
5. Macnair RD, et al. Metal ion levels not sufficient as a screening measure for adverse reactions in metal-on-metal hip arthroplasty. J Arthroplast. 2013;28:78–83.
6. Boardman DR, et al. A benign psoas mass following metal-on-metal resurfacing of the hip. J Bone Joint Surg Br. 2006;88:402–4.
7. Pandit H, et al. Pseudotumors associated with metal-on-metal hip resurfacings. J Bone Joint Surg Br. 2008;90:847–51.
8. Tribology: Definition, Terminology, Testing. Moers, Germany. German Society of Tribology; 2002, GfT Arbeitsblatt 7.
9. Wimmer MA, et al. Tribology. In: Callaghan JJ, et al., editors. The adult hip, vol. 1. 2nd ed. Philadelphia: Lippincott Williams & Wilkins; 2006. p. 215–26.
10. Schmarzried TP. Mechanical failure – loosening and wear. In: Pellicci PM, et al., editors. Orthopedic knowledge update, hip and knee reconstruction 2. 2nd ed. Rosemont: American Academy of Orthopedic Surgeons; 2000. p. 167–74.
11. Takagi M. Bone-implant interface biology: foreign body reaction and periprosthetic osteolysis in artificial hip joints. J Clin Exp Hematol. 2001;41:81–7.
12. Takagi M, et al. Toll-like receptors in the interface membrane around loosening total hip replacement implants. J Biomed Mater Res A. 2007;81:1017–26.
13. Takagi M. Neutral proteinases and their inhibitors in the loosening of total hip prostheses. Acta Orthop Scand Suppl. 1996;271:3–29.
14. Rao AJ, et al. Revision joint replacement, wear particles, and macrophage polarization. Acta Biomater. 2012;8:2815–23.
15. Takagi M, et al. Matrix metalloproteinases and tissue inhibitors of metalloproteinases in loose artificial hip joints (review). Clin Orthop Relat Res. 1998;352:35–45.
16. Konttinen YT, et al. Acid attack and cathepsin K in bone resorption around total hip replacement prosthesis. J Bone Miner Res. 2001;16:1780–6.
17. Takagi M, et al. High turnover periprosthetic bone remodeling and immature bone formation around loose cemented total hip joints. J Bone Miner Res. 2001;16:79–88.
18. Takei I, Takagi M, Ida H, Ogino T, Santavirta S, Konttinen YT. High macrophage-colony stimulating factor levels in synovial fluid of loose artificial hip joints. J Rheumatol. 2000;27(4):894–9.
19. Wang CT, et al. Overexpression of receptor activator of nuclear factor-kappaB ligand (RANKL), inflammatory cytokines, and chemokines in periprosthetic osteolysis of loosened total hip arthroplasty. Biomaterials. 2010;31:77–82.
20. Takagi M, et al. Danger of frustrated sensors: role of toll-like receptors and NOD-like receptors in aseptic and septic inflammations around total. J Orthop Translat. 2017;10:68–85.
21. Dumbleton JH, et al. A literature review of the association between wear rate and osteolysis in total hip arthroplasty. J Arthroplast. 2002;17:649–61.
22. Manley MT, et al. Osteolysis: a disease of access to fixation interfaces. Clin Orthop Rel Res. 2002;405:129–37.
23. Schmalzried TP, et al. Wear in total hip and knee replacements. J Bone Joint Surg Am. 1999;81:115–36.
24. Charnley J. The long term results of low friction arthroplasty of the hip performed as a primary intervention. J Bone Joint Surg Br. 1972;54:61–76.
25. Bradford L, et al. Wear and surface cracking in early retrieve highly cross-linked polyethylene acetabular liners. J Bone Joint Surg Am. 2004;86:1271–82.
26. de Steiger R, et al. Cross-linked polyethylene for total hip arthroplasty markedly reduces revision surgery at 16 years. J Bone Joint Surg Am. 2018;100:1281–8.
27. Dumbleton JH, et al. The basis for a second-generation highly cross-linked UHMWPE. Clin Orthop Relat Res. 2006;453:265–71.
28. Oral E, et al. Migration stability of alpha-tocopherol in irradiated UHMWPE. Biomaterials. 2006;27:2434–9.
29. Moro T, et al. Surface grafting of artificial joints with a biocompatible polymer for preventing periprosthetic osteolysis. Nat Mater. 2004;3:829–36.
30. DeLee JG, et al. Radiological demarcation of cemented socket in total hip replacement. Clin Orthop Relat Res. 1976;121:20–32.
31. Harris WH, et al. Femoral component loosening using contemporary techniques of femoral cement fixation. J Bone Joint Surg Am. 1982;64:1063–7.
32. Slooff TJ, et al. Bone grafting in total hip replacement for acetabular protrusio. Acta Orthop Scand. 1984;55:593–6.
33. Gie GA, et al. Impacted cancellous allografts and cement for revision total hip arthroplasty. J Bone Joint Surg Br. 1993;75:14–21.

Ultra-High Molecular Weight Polyethylene

20

Kengo Yamamoto, Yasuhito Takahashi, and Toshiyuki Tateiwa

1 History of the Development of Cross-Linked Ultra-High Molecular Weight Polyethylene (XLPE)

1.1 Early History: 1950s–1980s

In 1953, Ziegler in West Germany succeeded in polymerizing linear polyethylene with a high molecular weight from ethylene gas, and the industrialization of high-density polyethylene (HDPE) was initiated. Later, Ruhrchemie (currently known as Celanese, Germany) started to produce ultra-high molecular weight polyethylene (UHMWPE, see Sect. 2.1) powder. In 1962, a British orthopedic surgeon, Charnley developed the world's first acetabular cup made of UHMWPE (RCH 1000, Hoechst [formerly Ruhrchemie], average molecular weight 1 million g/mol) and put it into clinical use.

In the 1960s, the phenomenon of radiation-induced cross-linking of polymers (see Sect. 2.2) was already well known in the field of materials science. In 1968, Charnley began irradiating his RCH 1000 cups with γ-rays (25 kGy) in air; however, at that time, the main focus was on the high sterilization effect rather than the modification of physical properties. The first example of radiation cross-linking of UHMWPE with the explicit intention of improving wear resistance was reported by Oonishi et al. in 1970 [1]. They irradiated RCH 1000 with 1000 kGy of γ-rays in air and confirmed by wear tests that the wear of UHMWPE was dramatically reduced. As a result of its clinical use since 1971, the wear was successfully reduced to 1/6 of that of non-cross-linked UHMWPE. In 1978, Grobbelaar et al. conducted a similar study in South Africa and reported that γ-ray irradiation (up to 800 kGy) in an acetylene atmosphere improved the mechanical properties with increasing dose [2]. In the early 1980s, chemical cross-linking with silane was used instead of radiation, and Thackray, Wrightington Hospital, and Leeds University in the U.K. jointly developed silane-cross-linked HDPE; in 1986, Wroblewski et al. reported that the wear was reduced to 1/3 of that of the Charnley hip prosthesis when it was used clinically [3].

1.2 Development of First-Generation XLPE: 1990s

In the early history of the development of UHMWPE hip prostheses, the improvement of wear resistance was the primary objective, but subsequently, research and development were also promoted from the viewpoint of chemical durability. In the 1980s, increases in the density and crystallinity of UHMWPE cups were observed during the shelf-life period after manufacturing and several years after implantation, indicating changes in their physicochemical properties [4, 5]. Since no similar changes were observed in the unsterilized or ethylene oxide (EtO)-sterilized products, it was considered that this was an aging degradation characteristic of the γ-sterilized products in air. This was eventually interpreted as the consequence of molecular chain scission associated with oxidation reactions, but at the time, it was unclear how oxidation would affect wear and clinical outcomes. In the early 1990s, several implant retrieval studies showed the formation of a distinct oxidized layer called a white band at a depth of 0.5 to1.5 mm below the surface of γ-sterilized cups in air. This was reported to facilitate high wear, cup rim fracture, delamination, and oxidative degradation, thus emerging as a new clinical challenge [6]. In this period, the understanding of radiation chemistry and oxidation mechanisms was deepened significantly, and it was concluded that

K. Yamamoto (✉) · T. Tateiwa
Department of Orthopedic Surgery, Tokyo Medical University, Tokyo, Japan
e-mail: kyamamoto@pf6.so-net.ne.jp; tateiwa@tokyo-med.ac.jp

Y. Takahashi
Department of Orthopedic Surgery, Tokyo Medical University, Tokyo, Japan

Department of Bone and Joint Biomaterial Research, Tokyo Medical University, Tokyo, Japan
e-mail: yasuhito@tokyo-med.ac.jp

Y. Nakashima (ed.), *Advances in Total Hip Arthroplasty*, https://doi.org/10.1007/978-981-97-2461-1_20

the oxidation reaction was triggered by free radical formation after γ-irradiation. Therefore, quenching residual free radicals in the manufacturing process is central to improving both the wear and oxidation resistance.

Post-irradiation heat treatment (remelting and annealing) has been developed as a technology for eliminating free radicals. Heat treatment was performed around the melting point of UHMWPE (135 °C) to induce cross-linking reactions between unreacted radicals by promoting molecular mobility with the aim of reducing the residual free radical concentration.

In the 1990s, with the development of hip joint simulators, the relationship between radiation dose and wear resistance was verified in a more clinically relevant manner [7]. In addition to γ-rays, electron beams were also used as a radiation source;γ-rays were irradiated at room temperature and electron beams at elevated temperatures, and the effects of the dose and temperature on the microstructure and mechanical properties were studied in detail. Based on these findings, the balance between wear and fatigue resistance of hip prostheses was examined, and 50–100 kGy was considered the optimum dose range from the viewpoint of mechanical durability. GUR1050 (Ticona [formerly Hoechst], with an average molecular weight of 5.5 million to 6 million g/mol), which has excellent wear resistance, has been used as a raw resin material.

In the late 1990s, conventional air-permeable packaging was abolished, and barrier packaging filled with inert gases such as nitrogen and argon was adopted to improve the oxygen barrier performance during γ-ray sterilization [8]. On the other hand, EtO sterilization did not cause radical formation or changes in material properties, so it continued to be enforced as before. However, in the 1990s, low-temperature gas plasma sterilization was introduced. Plasma sterilization has many advantages over EtO, such as shorter processing time, lower cost, and the absence of hazardous residues.

Irradiated and heat-treated (remelted or annealed) UHMWPE is called first-generation XLPE and has been used clinically worldwide since it was launched in 1998 (Fig. 20.1). In Japan, Ishihara et al. succeeded in further reducing wear by grafting a highly hydrophilic polymer (2-methacryloyloxyethyl phosphorylcholine [MPC]) onto the articulating surface of annealed XLPE to improve lubrication properties. MPC polymer-grafted XLPE (Aquala) was launched in 2011.

1.3 Development of Second-Generation XLPE: 2000s

While many good clinical results have been reported for first-generation XLPE, which outperformed non-cross-linked UHMWPE, some minor improvements were identified in the 2000s. First, in the remelting process, XLPE is heated above the melting point, which enables the complete removal of residual free radicals; however, it is known to cause a decrease in mechanical properties such as fatigue resistance. On the other hand, the annealing process heats XLPE below the melting point, so the structure and properties are preserved, but the material still contains unstable free radicals. Especially in remelted products, there were many cases of liner fracture due to impingement between the femoral neck and acetabular rim [9]. However, these failures were not attributed to the remelting process alone but to a combination of material, design, surgery, and patient factors, such as high cross-linking, use of thin liners, use of elevated rim liners, stress concentration at the locking groove, vertical cup placement, poor liner fitting, and recurrent dislocation. However, wear and fracture of the liner rims were also observed in the annealed products, which were mainly attributed to a high degree of oxidation [10]. Although the aforementioned failures were not frequent, the consideration of the application of XLPE to large-diameter head and knee joint components provided motivation for further material improvement.

The concept of second-generation XLPE was to maintain the wear resistance of the previous generation while improving oxidation and fatigue resistance. This is not easy from a materials science point of view, because optimization of existing manufacturing methods alone will inevitably lead to compromising fatigue properties as either wear or oxidation resistance increases [11]. To overcome this trade-off, it is necessary to create new manufacturing methods.

The annealing process was modified in 2002. To maximize the removal efficiency of residual free radicals, Dumbleton, Wang, and Yau proposed a sequential annealing process in which the conventional manufacturing process of high-dose irradiation followed by annealing was divided into separate cycles. In this method, low-dose γ-ray irradiation (30 kGy) and annealing (130 °C, 8 h) are performed in one cycle, and a total of three consecutive cycles were performed to improve the cross-linking efficiency. As a result, the residual free radical concentration was reduced to 1/8 of the conventional single-cycle annealing process [12]. This is the level at which only one free radical exists in the 160 UHMWPE molecules. The raw material resin was also changed to improve the fatigue resistance, and highly crystalline GUR1020 (Ticona [now Celanese], average molecular weight 3.5 million g/mol) was used. The sequentially annealed XLPE (X3) was launched in the United States in 2005 (2009 in Japan).

Next, as an alternative to post-irradiation heat treatments, the addition of antioxidants (α-tocopherol, vitamin E) to UHMWPE (blending and infusing) was devised to neutralize free radicals. Vitamin E has been used as an antioxidant in propylene resins for food packaging in Japan since the 1980s. In the late 1990s, Tomita et al. developed a vitamin E-blended UHMWPE in Japan and reported the results of fatigue test

Trade name (Manufacturer)	Generation of XLPE	Year of clinical approval (Japan/ U.S.)	Processing				Total dose
			Starting material	Radiation crosslinking	Post-irradiation stabilization	Sterilization	
Crossfire (Stryker)	1st	1998 / 1998	GUR1050 (Ram extruded rod)	γ-ray (75 kGy, room temperature)	Annealed (130°C, 8 h)	γ-ray (30 kGy, in N_2 gas)	105 kGy
Aeonian (Kyocera Medical)	1st	1999 / -	GUR1050 (Compression molded sheet)	γ-ray (35 kGy, room temperature)	Annealed (110°C, 10 h)	γ-ray (25 kGy, in N_2 gas)	60 kGy
Longevity (Zimmer Biomet)	1st	2000 / 1998	GUR1050 (Compression molded sheet)	e-beam (100 kGy, ~40 °C)	Remelted (150°C, 6 h)	Gas plasma or ethylene oxide	100 kGy
Durasul (Zimmer Biomet)	1st	2000 / 1998	GUR1050 (Compression molded sheet)	e-beam (95 kGy, ~125 °C)	Remelted (150°C, 2 h)	Ethylene oxide	95 kGy
XLPE (Smith & Nephew)	1st	2002 / 2001	GUR1050 (Ram extruded rod)	γ-ray (100 kGy, room temperature)	Remelted (150°C, 2 h)	Ethylene oxide	100 kGy
Marathon (Depuy Synthes)	1st	2009 / 1998	GUR1050 (Ram extruded rod)	γ-ray (50 kGy, room temperature)	Remelted (155°C, 24h) and then annealed (120°C, 24h), both in reduced O_2 atmospheres	Ethylene oxide	50 kGy
Aquala (Kyocera Medical)	1st	2011 / -	GUR1020 (Compression molded sheet)	γ-ray (50 kGy, in N_2 gas)	Annealed (20°C, 7.5h, in N_2 gas) and then UV-induced poly-MPC grafted (60 °C, 1.5 h)	γ-ray (25 kGy, in N_2 gas)	75 kGy
X3 (Stryker)	2nd	2009 / 2005	GUR1020 (Compression molded sheet)	γ-ray (30 kGy, room temperature) in three steps	Annealed (130°C, 8 h) after each γ-irradiation step	Gas plasma	90 kGy
E1 (Zimmer Biomet)	2nd	2010 / 2007	GUR1050 GUR1020 (Isostatic compression molded rod)	γ-ray (100 kGy, room temperature)	Infused with liquid vitamin E and then homogenization annealed (130°C, 24 h)	γ-ray (30 kGy, in Ar gas)	130 kGy
Vivacit-E (Zimmer Biomet)	2nd	2013 / 2012	Vitamin E-blended GUR 1020 (Compression molded rod)	e-beam (100 kGy, elevated temperature)	-	Ethylene oxide	100 kGy
Blend-E XL (Teijin Nakashima Medical)	2nd	2013 / -	Vitamin E-blended GUR 1050 (Direct compression molded)	e-beam (300 kGy, room temperature)	Annealed (110°C, 72 h)	Ethylene oxide	300 kGy

Fig. 20.1 Comparison of processing conditions of main commercial XLPE acetabular liners approved in Japan

ing for the first time, which showed that the presence of vitamin E prevented crack propagation partly due to reduced hardness at grain boundaries [13]. In 2002, Parth et al. in Austria pointed out the problem of reduced radiation cross-linking efficiency of UHMWPE by a vitamin E blend [14]. In the same year, Wolf et al. in Austria tested the chemical safety of vitamin E-blended UHMWPE and reported high antioxidant activity and biocompatibility [15, 16]. Subsequently, Costa et al. held the first international meeting in Italy to discuss vitamin E stabilization in 2003 and 2005, taking a step toward international standardization for the clinical application of vitamin E-stabilized prostheses. In 2007, the American Society for Testing and Materials [currently, ASTM International] standardized UHMWPE blended with vitamin E for medical applications. In the same year, Muratoglu et al. in the U.S. developed XLPE (E1) infused with vitamin E after radiation cross-linking to avoid a decrease in cross-linking efficiency (launched in Japan in 2010). In 2012, a vitamin E-blended XLPE (Vivacit-E) was launched in the U.S. (launched in Japan in 2013), and the following year, Blend-E XL was launched in Japan.

2 Fundamentals of XLPE

2.1 Material Structure

Polyethylene is a generic term for polymers obtained by polymerizing ethylene ($CH_2 = CH_2$, molecular weight 28) and is represented by the general formula ($-C_2H_4-)_n$. A polyethylene molecule consists of repeating units of methylene ($-CH_2-$) in a planar zigzag form (Fig. 20.2a). There are various types of polyethylene, such as low-density polyethylene (LDPE, LLDPE), HDPE, and UHMWPE, depending on the molecular weight and number of branches. The average molecular weight of UHMWPE used for joint prostheses is currently defined as at least 1 million g/mol (ISO 11542) or 3.1 million g/mol (ASTM D4020). UHMWPE is a linear molecule with no branches protruding from the polymer backbone, allowing the molecules to closely approach each other and form a highly ordered crystalline phase, resulting in a high physical strength. The impact strength is highest in the molecular weight range of 2.4 million to 3.3 million g/mol, and the abrasion resistance is almost constant above ~3.3 million g/mol [17]. In the first-generation XLPE acetabular cups, UHMWPE (GUR1050) with a molecular weight of 5.5 million to 6 million g/mol was used to enhance wear resistance, but in some second-generation XLPEs, the resin was changed to UHMWPE (GUR 1020) with a molecular weight of 3.5 million g/mol to improve fatigue resistance.

The crystalline structure of polyethylene has been elucidated by transmission electron microscopy, and Keller (1957) proposed that polyethylene has a folded chain structure in which molecular chains are aligned vertically on the crystal lamella surface (Fig. 20.2a) [18]. Polyethylene usually forms orthorhombic crystals, and the thickness and width of the crystal lamellae are approximately 10–50 nm and 10–50 μm, respectively. The gaps of approximately 50 nm between the lamellar structures are filled with amorphous polyethylene molecules that could not form lamellae. The crystalline region shows elastic behavior owing to the highly ordered and dense arrangement of molecules, whereas the disordered and liquid-like amorphous region shows viscous behavior. The tie molecules in the amorphous region connect the crystalline lamellae, through which stress propagates to the crystalline phase. There is also an anisotropic intermediate phase (third phase) located at the transition zone between the crystalline and amorphous layers. The UHMWPE joint prosthesis can be considered a type of composite material because of the mixture of three distinct phase structures and properties, and thus it exhibits viscoelastic behavior.

2.2 Radiation Cross-Linking

Cross-linking is defined as the formation of covalent linkages between two or more molecular chains, resulting in a branched structure and a significant increase in molecular weight, which leads to changes in mechanical properties. Cross-linking can be performed by chemical or radiation cross-linking, but since the chemical cross-linking is not commonly used in orthopedics today, only the radiation cross-linking is discussed in this section. When UHMWPE is irradiated, radicals are formed by the cleavage of molecular chains, and the recombination reaction causes the amorphous phase molecules to link together in a mesh-like network structure, thereby improving wear resistance. However, embrittlement of the material due to the formation of unsaturated bonds or the reaction with oxygen by free radicals that cannot recombine may increase wear.

There are two main types of cross-linked structures in UHMWPE: "Y-crosslink" and "H-crosslink" [19, 20]. Although UHMWPE is known for the chemical formula ($-C_2H_4-)_n$, it actually contains not only simple methylene repeats but also vinyl groups ($-CH=CH_2$), methyl groups ($-CH_3$), and tertiary carbons (branching structure) at the ends of molecular chains. The Y-crosslink is formed when the double bond of the terminal vinyl group reacts with a free radical (secondary alkyl macroradical) of the adjacent UHMWPE molecule, as shown in Fig. 20.2b. Y-crosslink formation is generally observed under γ-ray and electron-beam irradiation. Cross-linking is less likely to progress rapidly when the radiation dose exceeds 100 kGy, which is thought to be caused by the fact that most of the initial vinyl groups are consumed by Y-crosslinking. In contrast, H-crosslinking is a termination reaction process in which radicals are deacti-

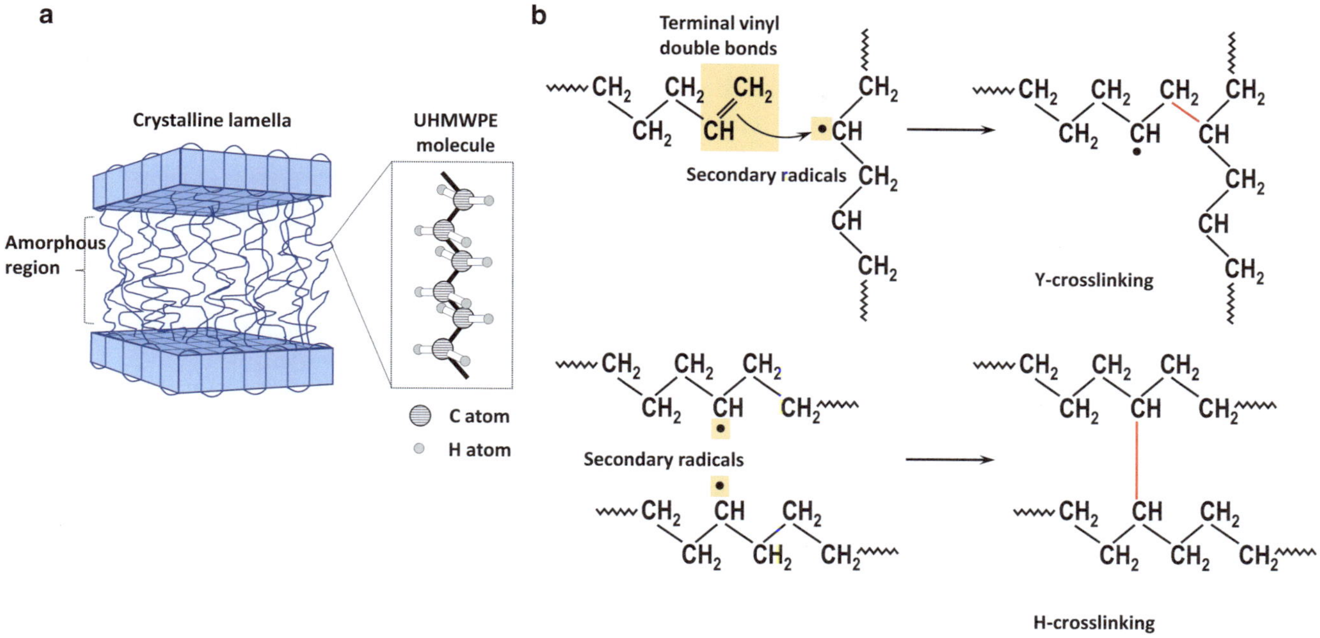

Fig. 20.2 Morphological features of UHMWPE (**a**) and radiation cross-linking in non-crystalline regions (**b**)

vated, and it is caused by the recombination of neighboring secondary alkyl macroradicals (Fig. 20.2b). In the solid state, this reaction is limited by intermolecular steric hindrance but is more likely to occur when molecular motion is promoted by annealing or remelting. The average distance between C-C atoms in UHMWPE is 1.54 Å, but the C-C distance in H-crosslinks is more than 4 Å. No cross-links are formed in the crystalline phase, which maintains an ordered structure. Radicals are randomly formed in the crystalline, intermediate, and amorphous phases, but cross-links are selectively formed only in the non-crystalline phases; thus, free radicals tend to remain in the crystalline phase.

The cross-linked structure limits the molecular mobility of the amorphous region in response to external stress, making it difficult to relax the applied energy through deformation and plastic flow, thereby increasing the brittleness of the material [21]. Therefore, excessive radiation cross-linking for the sake of wear resistance must be avoided because it reduces several important mechanical properties, such as strength, ductility, fracture toughness, and crack propagation resistance, leading to an increased risk of implant failure [22, 23]. In the orthopedic field, products irradiated at 0 kGy and 25–40 kGy (sterile level) are defined as conventional UHMWPE (CPE), and those cross-linked to doses above this level are called XLPE. XLPE is commonly classified into moderately cross-linked and highly cross-linked, with approximately 50 kGy being moderately cross-linked and approximately 100 kGy or more being highly cross-linked. However, this definition is not unambiguous; therefore, care should be taken when reading or using texts containing these terms.

3 Clinical Performance of XLPE

Although many studies have clinically demonstrated the benefits of cross-linking, in many cases, the indirect effect of the reduced dislocation rate due to the use of a large-diameter head seems to be more significant than the direct effect of cross-linking in short- to medium-term postoperative results. Therefore, long-term validation is necessary to determine the influence of cross-linking. The superiority of XLPE over CPE has not always been proven when the effects of postoperative years, head diameter, head material, XLPE type, cup design, and fixation method are considered.

In a systematic review of short- to mid-term post-THA, Kurtz et al. reported an 87% reduction in the incidence of osteolysis in XLPE compared with that in CPE [24]. The wear rate was 0.137 mm/year for CPE (18 studies, n = 695 hips) and 0.042 mm/year for XLPE (28 studies, n = 1503 hips), indicating a 69% reduction. In contrast, a systematic review by Shen et al. [25] found no significant difference in the revision risk between CPE and XLPE. Furthermore, a systematic review by Wyles et al. [26] found no reduction in the revision risk with XLPE when ceramic heads were used.

Six long-term clinical studies, including four randomized controlled trials (RCTs), were reported between 2012 and 2015 [27–32]. The minimum 10-year survivorship of XLPE was excellent, ranging from 93.5% to 100%, but only one study by Engh et al. [27] showed superiority over CPE. Considering that these RCTs were initiated in the early 2000s, the use of large-diameter heads was uncommon at that time, and small-diameter heads of 28 mm were predominantly used in XLPE, which may have influenced the outcomes.

In recent years, several excellent analyses based on national arthroplasty register data have been reported. Johanson et al. [33] analyzed data from the Nordic Arthroplasty Register Association (NARA) at 5–11 years postoperatively and performed a comparative analysis of 163,368 joints treated with CPE and XLPE. The results showed a significant difference in the overall survival rate with aseptic loosening as the endpoint between the two groups with cementless fixation, but not with cement fixation. Furthermore, when analyzed by cup design, one cementless cup (Trilogy) and two cemented cups (ZCA and Charnley Elite) did not show a significant survival benefit with XLPE use. The Australian Orthopaedic Association National Joint Replacement Registry (AOANJRR) reported in 2020 [34] that a significant reduction in the revision rate with XLPE was seen at 6 months, with a cumulative revision rate of 14.9% for CPE and 8.2% for XLPE at 19 years postoperatively. In the AOANJRR study, 80.5% of the XLPEs used a large-diameter head (≥32 mm). With respect to head diameter, CPE showed a trend toward higher revision rates with larger diameter heads, whereas XLPE showed the lowest revision rate when coupled with a 32 mm head. In a comparison by cup design up to 16 years postoperatively, there was no significant difference in the revision rates between CPE and XLPE in one of the six models (Vitalock). Davis et al. [35] utilized data from the National Joint Registry for England, Wales, Northern Ireland, and the Isle of Man (NJR) to analyze the risk of revision in cementless and hybrid THA by radiation dose and post-cross-linking stabilization. The NJR data up to 12 years postoperatively showed that XLPE with a total dose of ≥50 kGy markedly reduced revision due to aseptic loosening; however, irradiation of ≥100 kGy did not further reduce the risk of revision. The risk of revision due to aseptic loosening at 8.5 years after surgery was lower for vitamin E-stabilized and remelted XLPE than for annealed XLPE. It has been pointed out that annealed XLPE might have been more affected by wear because the oxidation reaction cleaves the cross-linked structure. As long-term results of second-generation XLPE are obtained, the effect of XLPE type on clinical outcomes will be further clarified.

References

1. Oonishi H, et al. Wear of high-dose gamma-irradiated polyethylene in total hip replacements. J Orthop Sci. 2000;5:223–8.
2. Grobbelaar CJ, et al. The radiation improvement of polyethylene prostheses. A preliminary study. J Bone Joint Surg Br. 1978;60-B:370–4.
3. Wroblewski BM, et al. Prospective clinical and joint simulator studies of a new total hip arthroplasty using alumina ceramic heads and cross-linked polyethylene cups. J Bone Joint Surg Br. 1996;78:280–5.
4. Grood ES, et al. Analysis of retrieved implants: crystallinity changes in ultrahigh molecular weight polyethylene. J Biomed Mater Res. 1982;16:399–405.
5. Eyerer P, et al. Property changes of UHMW polyethylene hip cup endoprostheses during implantation. J Biomed Mater Res. 1984;18:1137–51.
6. Sutula LC, et al. The Otto Aufranc award. Impact of gamma sterilization on clinical performance of polyethylene in the hip. Clin Orthop Relat Res. 1995;319:28–40.
7. McKellop H, et al. Development of an extremely wear-resistant ultra high molecular weight polyethylene for total hip replacements. J Orthop Res. 1999;17:157–67.
8. Costa L, et al. Oxidation and oxidation potential in contemporary packaging for polyethylene total joint replacement components. J Biomed Mater Res B Appl Biomater. 2006;78:20–6.
9. Duffy GP, et al. Fracture of a cross-linked polyethylene liner due to impingement. J Arthroplast. 2009;24(158):e15–9.
10. Kurtz SM, et al. 2006 Otto Aufranc award paper: significance of in vivo degradation for polyethylene in total hip arthroplasty. Clin Orthop Relat Res. 2006;453:47–57.
11. Atwood SA, et al. Tradeoffs amongst fatigue, wear, and oxidation resistance of cross-linked ultra-high molecular weight polyethylene. J Mech Behav Biomed Mater. 2011;4:1033–45.
12. Dumbleton JH, et al. The basis for a second-generation highly cross-linked UHMWPE. Clin Orthop Relat Res. 2006;453:265–71.
13. Tomita N, et al. Prevention of fatigue cracks in ultrahigh molecular weight polyethylene joint components by the addition of vitamin E. J Biomed Mater Res. 1999;48:474–8.
14. Parth M, et al. Studies on the effect of electron beam radiation on the molecular structure of ultra-high molecular weight polyethylene under the influence of alpha-tocopherol with respect to its application in medical implants. J Mater Sci Mater Med. 2002;13:917–21.
15. Wolf C, et al. Examination of the suitability of alpha-tocopherol as a stabilizer for ultra-high molecular weight polyethylene used for articulating surfaces in joint endoprostheses. J Mater Sci Mater Med. 2002;13:185–9.
16. Wolf C, et al. Tests of biocompatibility of alpha-tocopherol with respect to the use as a stabilizer in ultrahigh molecular weight polyethylene for articulating surfaces in joint endoprostheses. J Mater Sci Mater Med. 2002;13:701–5.
17. Kurtz SM, et al. Advances in the processing, sterilization, and crosslinking of ultra-high molecular weight polyethylene for total joint arthroplasty. Biomaterials. 1999;20:1659–88.
18. Keller A. A note on single crystals in polymers: evidence for a folded chain configuration. Philos Mag. 1957;2:1171–5.
19. Bracco P, et al. Radiation-induced crosslinking of UHMWPE in the presence of co-agents: chemical and mechanical characterisation. Polymer. 2005;46:10648–57.
20. Pérez E, et al. A ^{13}C CP-MAS NMR study of irradiated polyethylene. J Polym Sci B Polym Phys. 1988;26:1979–93.
21. Takahashi Y, et al. Effects of vitamin E blending on plastic deformation mechanisms of highly crosslinked ultrahigh molecular weight polyethylene (HXL-UHMWPE) in total hip arthroplasty. Acta Biomater. 2015;15:227–36.
22. Pruitt LA. Deformation, yielding, fracture and fatigue behavior of conventional and highly cross-linked ultra high molecular weight polyethylene. Biomaterials. 2005;26:905–15.
23. Takahashi Y, et al. Improved resistance to neck-liner impingement in second-generation highly crosslinked polyethylene-the role of vitamin E and crosslinks. J Arthroplast. 2016;31:2926–32.
24. Kurtz SM, et al. History and systematic review of wear and osteolysis outcomes for first-generation highly crosslinked polyethylene. Clin Orthop Relat Res. 2011;469:2262–77.

25. Shen C, et al. Does cross-linked polyethylene decrease the revision rate of total hip arthroplasty compared with conventional polyethylene? A meta-analysis. Orthop Traumatol Surg Res. 2014;100:745–50.
26. Wyles CC, et al. There are no differences in short- to mid-term survivorship among total hip-bearing surface options: a network meta-analysis. Clin Orthop Relat Res. 2015;473:2031–41.
27. Engh CA Jr, et al. A prospective, randomized study of cross-linked and non-cross-linked polyethylene for total hip arthroplasty at 10-year follow-up. J Arthroplast. 2012;27(Suppl):2–7.e1.
28. Johanson PE, et al. Highly crosslinked polyethylene does not reduce aseptic loosening in cemented THA 10-year findings of a randomized study. Clin Orthop Relat Res. 2012;470:3083–93.
29. Babovic N, et al. Total hip arthroplasty using highly cross-linked polyethylene in patients younger than 50 years with a minimum 10-year follow-up. J Arthroplast. 2013;28:815–7.
30. García-Rey E, et al. New polyethylenes in total hip replacement: a ten- to 12-year follow-up study. Bone Joint J. 2013;95:326–32.
31. Bedard NA, et al. Fixation and wear with a contemporary acetabular component and cross-linked polyethylene at a minimum 10-year follow-up. J Arthroplast. 2014;29:1961–9.
32. Glyn-Jones S, et al. The John Charnley award: highly cross-linked polyethylene in total hip arthroplasty decreases long-term wear: a double-blind randomized trial. Clin Orthop Relat Res. 2015;473:432–8.
33. Johanson PE, et al. Outcome in design-specific comparisons between highly crosslinked and conventional polyethylene in total hip arthroplasty. Acta Orthop. 2017;88:363–9.
34. Australian Orthopaedic Association National Joint Replacement Registry (AOANJRR). Hip, Knee & Shoulder Arthroplasty: Annual Report 2020, Adelaide; AOA, 2020: 1–474. Accessed from: https://aoanjrr.sahmri.com/annual-reports-2020
35. Davis ET, et al. Polyethylene manufacturing characteristics have a major effect on the risk of revision surgery in cementless and hybrid total hip arthroplasties. Bone Joint J. 2020;102:90–101.

Polyethylene (Vitamin E Mixed)

21

Naohide Tomita

1 Characteristics of Artificial Hip Joints

Figure 21.1 shows a schematic and symbolic structures of hip and knee joint prostheses. The upper panel shows the hip prosthesis upside down. Since the hip prosthesis has a ball-socket structure with multidirectional degrees of freedom of motion, the contact area between the femur (black) and acetabulum (gray) is relatively large. In contrast, extension and internal/external rotation of the knee joint are strongly restricted. The knee joint prosthesis shown in the lower panel of Fig. 21.1 requires a degree of freedom of motion in the anterior-posterior direction to enable deep flexion of the joint. Thus, the contact at the sliding surface is relatively narrow. Besides, as shown in the frontal view, the contact surface tends to tilt, so the tension applied to the ligament as a limiting force during motion is added to the load on the sliding contact area, resulting in excessive stress on the sliding contact area [1].

In knee joint prostheses, when the strength of the material is increased, the modulus of elasticity generally increases and the stress concentration in the convex contact area may become more pronounced due to the increased modulus of elasticity. In contrast, ball-socket structures, which do not have a convex-contact interface, allow for a combination of materials with a relatively high elastic modulus, such as hard-on-hard sliding parts. Reinforcement methods that increase the elastic modulus, such as cross-linking with polyethylene, are also effective in improving resistance to wear.

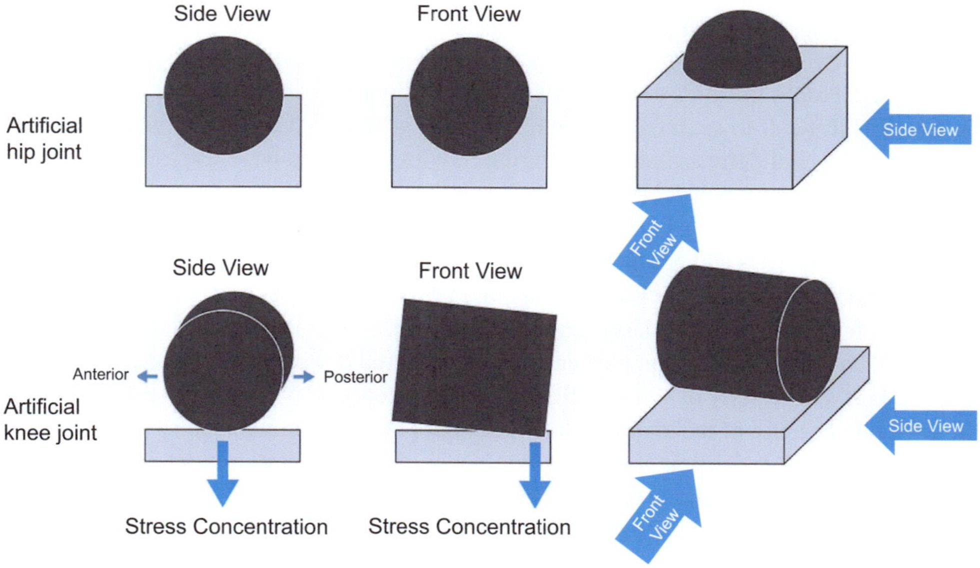

Fig. 21.1 Schematic structure of artificial hip and knee joints

N. Tomita (✉)
Kyoto City University of Arts, Kyoto, Japan
e-mail: n_tomita@kcua.ac.jp

Y. Nakashima (ed.), *Advances in Total Hip Arthroplasty*, https://doi.org/10.1007/978-981-97-2461-1_21

2 Basic Characteristics of Polyethylene

Polymeric materials are composed of chain-like covalent bonds. Various properties can be imparted by changing the higher-order structures of the chain-like molecules. In crystalline polymers such as polyethylene, the basic mechanical properties depend on the proportion of crystalline portions (crystallinity) with a regular folded arrangement. The crystalline portion is divided into solid and liquid phases at the melting point, whereas the amorphous portion is divided into solid and liquid phases at the glass transition temperature. The glass transition temperature of polyethylene is approximately −120 °C. Thus, the amorphous part of polyethylene is in a liquid state at room temperature. Ultra-high-molecular-weight polyethylene (UHMWPE), which has a large molecular weight, has an extremely high viscosity. Even if UHMWPE is a liquid, the time required for its flow is extremely long. Flow deformation, similar to the creep of metals, occurs when the material is under load for an extended time. The crystallinity, fluidity of the amorphous portion, cross-links between radicals in the amorphous portion, and molecule chain connecting the crystalline and crystalline parts are involved in this flow.

The molecular weight of polyethylene varies depending on the measurement method. Polyethylene with a molecular weight of approximately 2 million or more is called UHMWPE. This molecule is characterized by high mechanical properties, such as abrasion resistance. However, UHMWPE is difficult to mold and process. As mentioned above, the amorphous portion of polyethylene is in a liquid state at room temperature, but when its molecular weight is high, it is difficult to melt and perform injection molding because of its high viscosity, even if the temperature is increased. Because the thermal conductivity of polyethylene is low, it is difficult to set molding conditions such as temperature, time, and deformation speed during molding. The degree of fusion at the powder interface differs depending on the molding conditions, which significantly changes the mechanical properties and permeability of the powder.

UHMWPE has a simple molecular structure in which carbon and hydrogen are regularly arranged. However, as described above, it is a very complex and difficult material to handle. Further, property of UHMWPE changed in the order of 1–10 mm because of the distribution of heat and pressure during molding and deformation speed. Moreover, there is property distribution in the order of 10–100 μm related to powder fusion and property distribution on the order of nanometers that is derived from the structure of the molecular chain and movement in the crystalline and amorphous parts. Property distribution on the nanometer scale is derived from the structure of molecular chains and is the subject of research by polymer scientists. In contrast, property distribution in the 1–10 mm and 10–100 μm order is the subject of controll by implicit intellectual experience. This implicit-intellectual technology significantly affects the actual function of products. Thus, the complexity of powder interface fusion and moldability is one factor that makes the study of UHMWPE difficult.

3 History of Polyethylene Material Development and Bone Dissolution

Onishi et al. reported excellent clinical results for hip prostheses over a 30-year period using polyethylene irradiated with doses of 100 Mrad γ-rays in air [2–4]. These results suggest that cross-linking in the amorphous portion of polyethylene from γ-ray irradiation is useful for hip prosthesis function. Based on these results, various types of cross-linked UHMWPE have been developed and are now used in most hip prostheses.

Onishi used high-dose irradiated cross-linked polyethylene in the clinic, focusing on results from experiments that were close to clinical situations. As mentioned above, cross-linking in the amorphous portion of polyethylene is thought to be catalyzed by irradiation. However, it is possible that radiation treatment (100 Mrad irradiation in air) has other effects on polyethylene.

The cross-linked polyethylene that was put into practical use was irradiated with a relatively low dose. One reason is that other effects of high-dose irradiation, such as oxidation, were of concern. Another reason is a lack of engineering experience with materials irradiated with doses as high as 100 Mrad.

One cause of reoperation is osteolysis, which starts with phagocytosis or encrustation of UHMWPE wear debris by macrophages, resulting in loosening of prostheses. A decrease in the number of osteolysis cases in the prosthetic joint registry has been reported, suggesting that cross-linking may suppress osteolysis. However, microinfections in the body and adsorption of microscopic amounts of endotoxin produced by bacteria onto the surface of polyethylene particles may be involved in osteolysis [5]. Therefore, it is possible that the decrease in osteolysis might be caused by the removal of small amounts of endotoxin from bone cement [6] or by the use of drugs such as antibacterial agents, anti-inflammatory drugs, osteoporosis treatment agents, and proton-pump inhibitors.

4 Development Polyethylene Mixed with Vitamin E for Artificial Hip Joints

Some vitamin E polyethylenes developed in Europe and the United States are technologies that impregnate polyethylene with vitamin E after molding and cross-linking, mainly because the addition of vitamin E inhibits cross-linking. In contrast, vitamin E polyethylene developed in Japan is mixed with vitamin E before molding and cross-linking. We developed a technique to mix D,L-α-tocopherol, a type of vitamin E, with polyethylene to prevent delamination, which is a characteristic feature of knee joint prostheses [7]. The advantage of mixing D,L-α-tocopherol before molding polyethylene is that D,L-α-tocopherol acts as a lubricant, thereby densifying the grain boundary created by powder molding. This improvement in the powder interface and formability decreases the permeability of oxygen and lipids at the boundary, which is thought to reduce the formation of cracks [8, 9]. Yamamoto and Takahashi et al. reported that a mixture of vitamin E increases the mobility of polyethylene molecules [10, 11], particularly by allowing deformation of polyethylene at the nanometer level. There is a possibility that the molecular mobility affects on molecular entanglement at the powder fusion area.

Delamination fracture, which is a problem with artificial knee joints, has not been an obvious problem with artificial hip joints [12]. Application of vitamin E mixed with polyethylene in artificial hips was considered after the report that including d,l-α-tocopherol reduces the biological reactivity of polyethylene wear powder [13]. However, as mentioned above, vitamin E inhibits cross-linking. Therefore, there is an urgent need to optimize cross-linking in polyethylene containing d,l-α-tocopherol. Uetsuki et al. reported that polyethylene containing 0.3% d,l-α-tocopherol irradiated at 300 kGy electron irradiation showed wear resistance equivalent to that of cross-linked polyethylene irradiated with gamma irradiation (95 kGy) [14, 15]. Uetsuki et al. also defined the oxidative induction time (OIT) as the time required for the oxidation reaction to occur in an oxygen atmosphere, measured the offset time of the oxidation reaction at various temperatures, and predicted the long-term antioxidant capacity at 37 °C. Their results showed that the antioxidant period of polyethylene containing 0.3% d,l-α-tocopherol treated with 300 kGy electron irradiation was sufficiently longer than the period of prosthesis implantation [16].

5 Next-Generation Materials for Artificial Hip Joints

In addition to vitamin E-mixed polyethylene, a graft polymerization technique for 2-methacryloyloxyethyl phosphorylcholine (MPC) immobilization on the sliding surface of polyethylene is currently being commercialized in Japan [17]. This is an application of the hydration-lubrication effect, which has been proposed as a mechanism of ultralow friction of cartilage. These technologies may improve artificial joint function. In recent years, the term "human augmentation" has been used for technologies that assist or enhance human functions. "Human augmentation" is commonly used to describe expanding the functions of the body and senses by combining robotics, augmented reality, and virtual reality. Artificial joints are a human augmentation technology in the broad sense of the word and are required to extend the range of life activities.

In contrast, in the industrial world, viewpoints such as "attachment" and "ANSHIN," which are different from "function," are beginning to be regarded as important. For example, it is becoming important to consider not only functionality but also how these augmentations affect the real life of people. For example, it is important to determine how much an artificial joint allows for the diversity of human activities. The next generation of joint prostheses will be required to have high durability to cope with dynamic movements such as work and sports. Thus, the "durability" of joint prostheses cannot be evaluated by simple measurements such as wear resistance and fatigue resistance. Historically, analysis and measurement methods in engineering have guaranteed high reproducibility and accuracy. However, tests that attempt to reproduce diverse clinical situations have low reproducibility and are difficult to interpret, owing to the complex involvement of multiple factors. It is important to always consider real-life situations in which various factors are intricately intertwined during product development. However, it is also true that various arbitrary interpretations can be easily introduced. To achieve both safety and quality of life, it is important for clinicians to acquire basic engineering knowledge and for engineers to learn about the diversity of clinical situations and to have in-depth discussions with patients.

References

1. Tomita, N. Biomaterials. Hiza Kansetsu Gekagaku, edited by H. Tsumura et al. (in Japanese).
2. Onishi, H, et al. Wear of high-dose gamma irradiated polyethylene in total joint replacement: long term radiological evaluation. Orthop Res Soc. 1998;23:97.
3. Onishi H, et al. Wear resistance of gamma-ray irradiated UHMW polyethylene sockets in total hip prostheses-Wear test and long term clinical results. J Biomech. 1990;23:382.
4. Onishi K, et al. Development of ultralow wear cross-link polyethylene over 30 years and its evolution. Nihon Riumachi Kansetsu Geka Gakkai Zasshi. 2003;21:205–13. (in Japanese)
5. Xing Z, et al. Accumulation of LPS by polyethylene particles decreases bone attachment to implants. J Orthop Res. 2006;24:959–66.
6. Brooks RA, et al. Accumulation of LPS by polyethylene particles decreases bone attachment to implants. J Bone Joint Surg Br. 2002;84:295–9.
7. Tomita N, et al. Prevention of fatigue cracks in UHMWPE joint components by addition of vitamin E. J Biomed Mater Res. 1999 48:474–8.
8. Shibata N, et al. Numerical simulations on fatigue destruction of ultrahigh molecular weight polyethylene using discrete element analyses. Biomed Mater Res A. 2003;64:570–82.

9. Shibata N, et al. The antioxidative properties of α-tocopherol in γ-irradiated UHMWPE with respect to fatigue and oxidation resistance. Biomaterials. 2005;26:5755–62.
10. Takahashi Y, et al. Effects of vitamin E blending on plastic deformation mechanisms of highly crosslinked ultrahigh molecular weight polyethylene (HXL-UHMWPE) in total hip arthroplasty. Acta Biomater. 2015;15:227–36.
11. Yamamoto K, et al. Vitamin E-stabilized highly crosslinked polyethylenes: the role and effectiveness in total hip arthroplasty. J Orthop Sci. 2017;22:384–90.
12. Koizumi M, et al. Detection of cracks in polyethylene components of retrieved knee joint prostheses. J Orthop Sci. 1998;3:330–5.
13. Bladen CL, et al. Analysis of wear, wear particles, and reduced inflammatory potential of vitamin E ultrahigh-molecular-weight polyethylene for use in total joint replacement. J Biomed Mater Res B Appl Biomater. 2013;101:458–66.
14. Uetsuki K, et al. Controversial effects of blending vitamin E with UHMWPE on the wear resistance of hip and knee prostheses. ORS 2012 Annual Meeting, Poster No 1074; 2012.
15. Turner AC, et al. Mechanical and oxidative performance of high-dose electron-beam irradiated, dl-α-tocopherol (vitamin E) blended UHMWPE. J Biomech Sci Eng. 2015;10:14–00238.
16. Uetsuki K, et al. Prediction of long-term stability against oxidation of UHMWPE by thermal analysis. ORS. 2016; Poster No. 1099
17. Moro T, et al. Surface grafting of artificial joints with a biocompatible polymer for preventing periprosthetic osteolysis. Nat Mater. 2004;3:829–36.

Polyethylene (MPC Polymer)

22

Toru Moro, Kazuhiko Ishihara, and Yoshio Takatori

Total hip replacement (THR) is an established treatment modality for patients with end-stage hip disorders such as osteoarthritis and rheumatoid arthritis. Recent surveys have revealed that the number of surgical procedures is growing faster than ever due to the increase in elderly population. However, aseptic loosening caused by periprosthetic osteolysis is a serious concern that limits patient survival and clinical success, and extensive research has indicated that wear particles from polyethylene (PE) liners are responsible for osteolysis. Hence, many approaches have been adopted to reduce the production of wear particles, thereby improving the survival of THR. Aiming at the reduction of wear particles and the elimination of periprosthetic osteolysis, we successfully developed a novel highly cross-linked PE (HXLPE) liner with the surface graft of a biocompatible phospholipid polymer, poly(2-methacryloyloxyethyl phosphorylcholine) (PMPC). In this section, we introduce the characteristics and clinical outcomes of THR using PMPC-grafted HXLPE liners.

1 Properties of MPC Polymers Mimicking the Neutral Phospholipids of Cell Membranes

The methacrylate monomer 2-methacryloyloxyethyl phosphorylcholine (MPC) has a phospholipid polar group in the side chain, and MPC polymers are cell membrane-inspired biomaterials, designed and developed by Ishihara et al., that mimics the neutral phospholipids of cell membranes [1, 2]. The MPC polymers are one of the most common biocompatible hydrophilic polymers studied thus far and have potential applications in a variety of fields, such as biology, biomedical science, and surface chemistry because they possess unique properties such as biocompatibility, high lubricity, low friction, anti-protein adsorption ability, and cell membrane-like surface. Further, MPC polymers have been used on the surfaces of several medical devices such as guide wires, intravascular stents, contact lenses, artificial lungs, and artificial hearts. These implants were introduced into clinical practice in 1997 under the authorization of the Food and Drug Administration (FDA) in the United States and Ministry of Health, Labor, and Welfare (MHLW) in Japan. Since then, no adverse reactions to MPC polymers have been reported; therefore, the efficacy and safety of MPC polymers as biomaterials has been well established.

2 Basic Studies on the PMPC-Grafted HXLPE Liners

To reduce the production of wear particles and subsequent aseptic loosening, it is important to understand the tribology of the natural synovial joints and the role of lubrication. In addition, the bearing surfaces of artificial joints should mimic the surface conditions of healthy human articular cartilage to provide ideal surfaces for smooth motion of the joint with daily activity.

Articular cartilage covers the surface of natural synovial joints and is composed of a dense extracellular matrix, comprising proteoglycans, collagen, and water, with sparsely distributed chondrocytes and surface-active phospholipids, such as phosphatidylcholine derivatives, which protect the joint interface from mechanical wear and facilitate the smooth motion of joints during daily activities [2]. The charge carried by the extracellular matrix, either zwitterionic or alternating charge, helps trap water to maintain water-fluid and electrolyte balance in the articular cartilage, which provides hydrophilicity and works as an effective boundary lubricant. The hydrated polyelectrolyte layer of the articular cartilage provides fluid thin-film lubrication, known as hydration lubrication, that is essential for the smooth motion of natural synovial joints [3, 4]. Hence, grafting a

T. Moro (✉) · K. Ishihara · Y. Takatori
The University of Tokyo, Tokyo, Japan
e-mail: moro-ort@h.u-tokyo.ac.jp

Y. Nakashima (ed.), *Advances in Total Hip Arthroplasty*, https://doi.org/10.1007/978-981-97-2461-1_22

phospholipid-like layer on the surface can help realize ideal hydrophilicity and lubricity, resembling the surface conditions of healthy articular cartilage. Based on this hypothesis, we successfully produced a biocompatible and highly hydrophilic surface by grafting PMPC on HXLPE at a nanometer-scale.

Grafting PMPC on HXLPE liner surfaces (100–200 nm in thickness) was performed using a photo-initiated "grafting-from" polymerization process (Fig. 22.1) [5]. This technique has several important benefits: direct grafting of PMPC on HXLPE, thereby forming stable C–C covalent bonds between PMPC and HXLPE substrate; high mobility of the chains of PMPC; high density; and control of the length of the introduced PMPC grafting layer. Notably, PMPC-grafted HXLPE is prepared using a simple procedure. The MPC monomer was dissolved in degassed pure water to a concentration of 0.5 mol/L. Subsequently, the HXLPE liners coated with benzophenone were immersed in an aqueous MPC solution. Photo-induced graft polymerization on the HXLPE surface was performed with ultraviolet (UV) light irradiation. After polymerization, the PMPC-grafted liners were removed, washed with pure water and ethanol, and then dried. The PMPC-grafted liners were gamma-sterilized at a dose of 25 kGy under N_2 gas. As this procedure, which is used in the photoinduced graft polymerization of MPC, involves nanometer-scale surface modifications, it does not affect the physical or mechanical properties of the HXLPE substrate [6]. Clinically, these findings are important.

The PMPC grafting significantly increased the hydrophilicity of the hydrophobic HXLPE surfaces. The molecular structure of the PMPC chains is relatively rigid because of the bulky phosphorylcholine groups in the side chains; particularly in water, the PMPC chains have a polymer brush-like structure that seems to rise from the HXLPE surface. In other words, the PMPC grafting layer on the surface cooperatively formed a hydration layer with water molecules and produced a good lubrication mechanism (surface hydration gel lubrication mechanism) on the articulating surface (Fig. 22.1). Further, PMPC grafting on the HXLPE surface decreased the friction torque by 80–90% compared with the untreated HXLPE surface and increased lubrication to levels that resembled articular cartilage under physiological conditions [6, 7]. In addition, PMPC-grafted polymer particles were not recognized by macrophages as foreign bodies and suppressed bone resorptive responses in a murine osteolysis model [4].

Our previous hip simulator study (ISO 14242-3) revealed that the PMPC grafting layer markedly reduced wearing of HXLPE liners and lowered the total number and volume of wear particles by 99.9% during 20 million loading cycles compared with that of untreated HXLPE liners (Fig. 22.2) [8]. We extended the simulator test to 70 million cycles for a single liner, and the results confirmed that the PMPC layer was maintained after the test [8]. In addition, we confirmed that the improvement caused by PMPC grafting surpassed that which is attributable to linear cross-linking alone or changes in femoral head material or diameter. Furthermore, although the nanometer-scaled PMPC grafting layer (approximately 200 nm in thickness) does not affect the physical or mechanical properties of the HXLPE substrate, [6] limited knowledge is available regarding the wear conditions after the PMPC-grafted layer is removed from the liner surface due to long-term usage and/or under severe wear conditions. Hence, we evaluated the effect of PMPC grafting on the wear resistance of the HXLPE substrate by removing the PMPC-grafted layer using a severely roughened femoral head [9]. The wear rate of the PMPC-grafted HXLPE liners did not exceed that of the untreated HXLPE liners. This result confirms the clinical safety and advantages of grafting PMPC for wear resistance of the HXLPE substrate, suggesting that if the PMPC-grafted layer is removed from the surface of the liner under severe wear conditions, the HXLPE substrate indicates the same wear resistance as that of the untreated HXLPE liner.

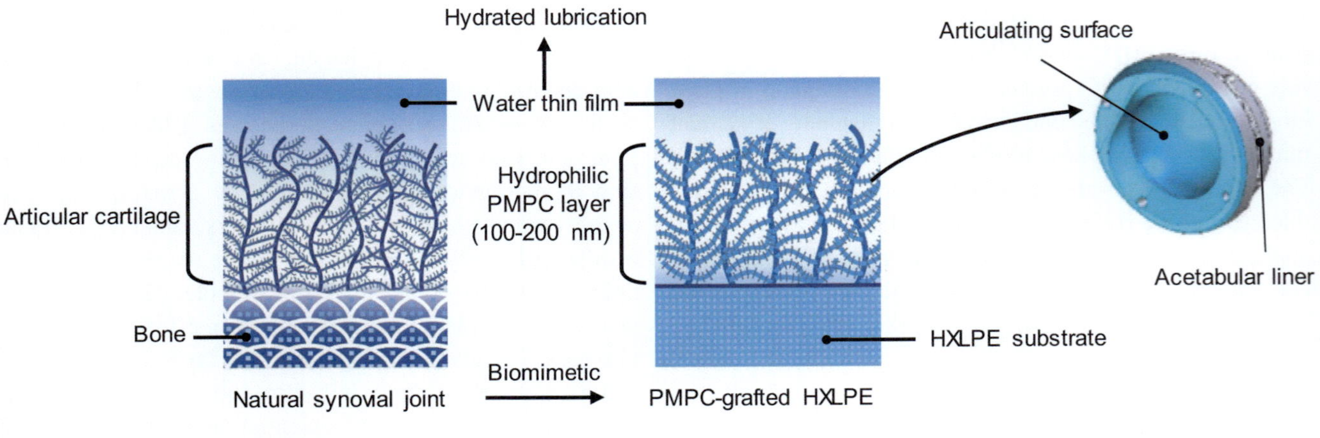

Fig. 22.1 A scheme of the PMPC-grafted HXLPE surface

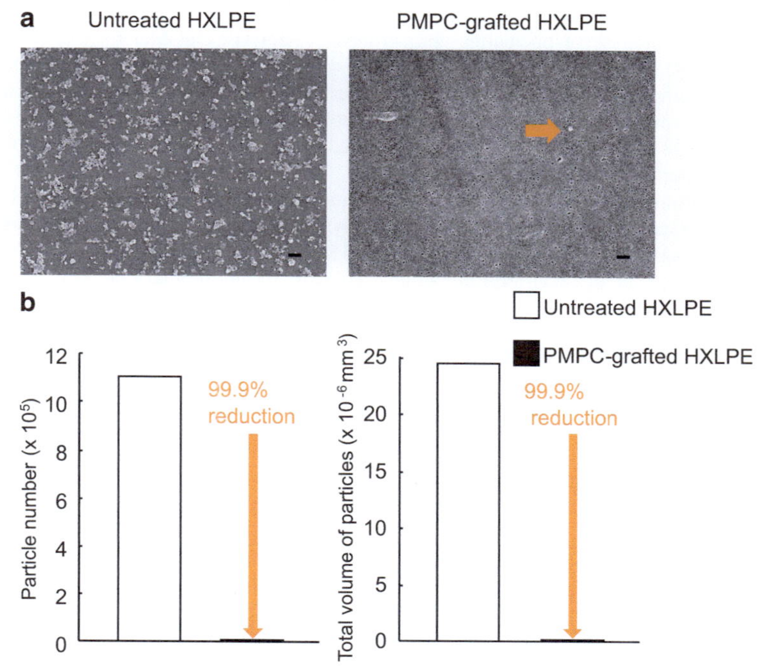

Fig. 22.2 Analyses of wear particles. (**a**) SEM images of wear particles. Scale bar 1.0 μm. (**b**) Total number and volume of wear particles

3 Clinical and Radiographic Outcomes of the PMPC-Grafted HXLPE Liners

Although the results of the basic study mentioned above suggest that the PMPC-grafted HXLPE liner is a promising approach to extend the longevity of artificial hip joints, in vitro findings do not always translate into clinical success. Therefore, we performed a clinical trial of a prospective cohort of 80 patients who underwent primary cementless THR with a PMPC-grafted HXLPE liner between April 2007 and September 2008. In addition, we are still conducting follow-up studies on this cohort [10]. Before the end of the 5-year postoperative follow-up period, 12 patients were lost to follow-up. Therefore, the final analysis included 68 patients (M/F, 13/55) with a mean age of 64.1 (range, 48–74) years and mean body mass index (BMI) of 22.5 kg/m^2 at the time of entry. The preoperative diagnosis was osteoarthritis in the hips of 66 patients and osteonecrosis in 2 patients. No reoperations were performed for any reason. No adverse events were found to be correlated with the implanted liners. The average Japanese Orthopaedic Association (JOA) score significantly improved from a preoperative value of 43.0 to a postoperative value of 94.1 at 5 years, corresponding to Harris hip scores of 46.7 and 95.6, respectively. The index of health-related quality of life (HRQoL), SF-36, improved 5 years after operation. Neither periprosthetic osteolysis nor implant migration was detected in radiographic analysis. The mean amount of bedding-in (penetration during the first year) was 0.225 mm. Between 1 and 5 years, the mean steady wear rate was 0.002 mm/year, representing a marked reduction compared with 0.042 mm/year mentioned in a review of 28 articles (1503 joints) of first-generation HXLPE by Kurtz et al. [11]. The PMPC-grafted HXLPE liners indicated similar steady wear rates at intervals of 1–3 (0.001 mm/year) and 3–5 (0.002 mm/year). The mean steady wear rate did not correlate with patient age, BMI, sex, Charnley category, preoperative diagnosis, liner thickness, or component orientation. This cohort was also postoperatively evaluated at 7 and 10 years, and the results were similar to those found at 5 years after operation, confirming the stable clinical results of the PMPC-grafted HXLPE liners. A multicenter parallel group study (19 centers, 760 joints) coupling a 32/36-mm diameter cobalt-chrome alloy/ceramic fenoral head and a PMPC-grafted HXLPE liner was initiated from October 2012 to September 2021, with favorable results.

Although studies involving longer follow-up periods are required to elucidate the true clinical benefit of PMPC-grafted HXLPE liners in THR, the results of the basic and clinical studies described in this section suggested that the PMPC-grafted HXLPE liner will improve the longevity of artificial hip joints and, therefore, substantially benefit public health.

References

1. Ishihara K, et al. Preparation of phospholipid polymers and their properties as polymer hydrogel membranes. Polym J. 1990;22:355–60.
2. Ishikawa Y, et al. Role of water in the lubrication of hydrogel. Wear. 2006;261:500–4.
3. Hills BA, et al. Surfactants identified in synovial fluid and their ability to act as boundary lubricants. Ann Rheum Dis. 1984;43:641–8.
4. Kyomoto M, et al. Polymers for artificial joints. In: Dumitriu S, editor. Polymeric biomaterials. Structure and function. CRC Press; 2013. p. 851–83.

5. Moro T, et al. Surface grafting of artificial joints with a biocompatible polymer for preventing periprosthetic osteolysis. Nat Mater. 2004;3:829–36.
6. Kyomoto M, et al. Effects of photo-induced graft polymerization of 2-methacryloyloxyethyl phosphorylcholine on physical properties of cross-linked polyethylene in artificial hip joints. J Mater Sci Mater Med. 2007;18:1809–15.
7. Yarimitsu S, et al. Influences of dehydration and rehydration on the lubrication properties of phospholipid polymer-grafted cross-linked polyethylene. Proc Inst Mech Eng H. 2015;229:506–14.
8. Moro T, et al. Long-term hip simulator testing of an artificial hip joint bearing surface grafted with biocompatible phospholipid polymer. J Orthop Res. 2014;32:36–376.
9. Moro T, et al. Effects of a roughened femoral head and the locus of grafting on the wear resistance of a phospholipid polymer-grafted acetabular liner. Acta Biomater. 2019;86:338–49.
10. Moro T, et al. Clinical safety and wear resistance of a phospholipid polymer-grafted highly cross-linked polyethylene liner. J Orthop Res. 2017;35:2007–16.
11. Kurtz SM, et al. History and systematic review of wear and osteolysis outcomes for first-generation highly crosslinked polyethylene. Relat Res. 2011;469:2262–77.

Metal-on-Metal THA

23

Wataru Ando

1 The World's First Total Hip Arthroplasty

Philip Wiles (1899–1966) performed the world's first total hip arthroplasty (THA) in 1938. The implant used in this case was made of stainless steel, including the bearing; therefore, this case was the world's first metal-on-metal THA (MoM THA). The metal acetabular component was fixed with several screws, and the metal femoral head prosthesis was fixed with a screw inserted from the distal site of the greater trochanter through the femoral neck. THA was performed on six patients with Still's disease who were almost bedridden, and all patients were able to walk after the procedure. One patient was followed up for 13 years postoperatively without any incidence of pain. Although many radiographs of their hips were destroyed during World War II, the remaining plain radiographs showed that the lateral acetabular screw had broken. Consequently, it is important to select a material with sufficient strength and low friction and to fix the acetabular component to the pelvis for a successful THA [1].

2 The First-Generation MoM THA

George K McKee (1906–1991) had focused on cobalt-chromium alloy as a biocompatible material since 1940; however, its bone fixation was a concern associated with its use. In 1948, he performed a MoM THA in three patients using screws that were similar to those used in Wiles' method. MoM THA using stainless steel implants was performed in two of the patients, and all implants loosened within the first postoperative year. Conversely, MoM THA with cobalt-chromium implants in one of the three patients showed good results up to 3 years postoperatively. In 1956, MoM THA was performed using a Thompson-type stem that was inserted intramedullary and an acetabular component with screw fixation. The results improved; however, implant loosening remained a concern [2].

In 1960, McKee and John W. Farrar (1926–1999) developed the McKee–Farrar prosthesis, in which the acetabular and femoral components were fixed using acrylic cement. Although short-term results were good, clinical failure increased over time and implant loosening occurred in 50% of the femoral stem and 51% of the acetabular component at 8 and 14 years postoperatively, respectively [3]. Since the thickness of the stem neck design caused impingement at the edge of the acetabular component and stem neck, the design was further improved. However, a relationship between hypersensitivity to metal debris and implant loosening was not observed [4].

MoM THA with McKee–Farrar hip prosthesis performed between 1965 and 1973 had an 81% survival rate at a mean follow-up of 13.9 (10–22) years postoperatively [5]. When comparing MoM THA with McKee-Farrar hip prosthesis and metal on polyethylene (MoP) THA with Charnley prosthesis performed in patients with hip osteoarthritis from 1975–1976, the survival rates at 15 years postoperatively were 58% and 90% for MoM THA McKee-Farrar hip prosthesis and MoP THA with Charnley prosthesis, respectively. However, the survival rates at 20 years postoperatively reversed to 81 and 73%, respectively, suggesting the possibility of using MoM THA for long-term durability [6]. The results of MoM THA with McKee–Farrar hip prosthesis in Japan showed that 10 of 38 hips survived up to an average of 21.8 years postoperatively. Patients aged 60 years or younger had significantly better outcomes than those aged 60 years or older [7].

Retrieval analysis of implants obtained at the time of revision showed that the volumetric wear rate of MoM bearings was much lower than that of MoP bearings [8]. The MoM bearing with a cobalt-chromium-molybdenum alloy produced considerably less wear debris than the standard MoP bearing [9].

W. Ando (✉)
Department of Orthopaedic Surgery, Kansai Rosai Hospital, Amagasaki, Japan
e-mail: w-ando@umin.ac.jp

Y. Nakashima (ed.), *Advances in Total Hip Arthroplasty*, https://doi.org/10.1007/978-981-97-2461-1_23

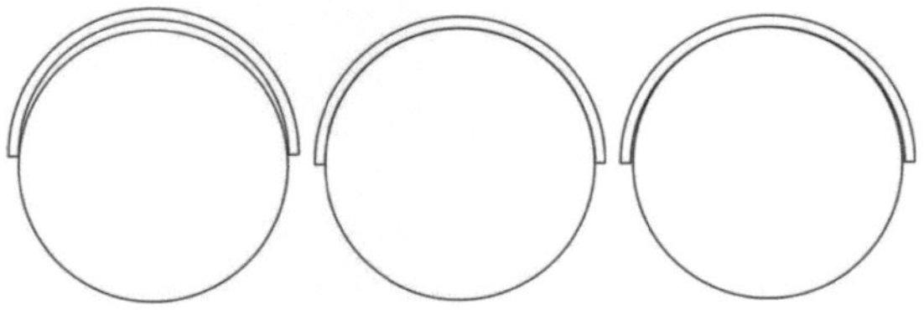

Fig. 23.1 Clearance between acetabular component and femoral head

However, quality control of the product was insufficient. When the diameter of the femoral head was larger than the inner diameter of the acetabular component, the friction was higher in "coronal contact," in which the femoral head contacted only the acetabular edge, or in "total contact," in which the inner diameter of the acetabular component and the femoral head diameter were the same. However, if the diameter of the femoral head was smaller than that of the acetabular component, a "loose fit" was appropriate (Fig. 23.1). The clearance for low friction was not consistent among products, and McKee discontinued the use of the McKee–Farrar prosthesis in 1982 [10].

3 The Second-Generation MoM THA

In the late 1980s and early 1990s, Bernhard G Weber (1927–2002) developed a new MoM THA. Weber noted that some McKee–Farrar-type prostheses used in MoM THA functioned properly for more than 25 years. Experimentally, the annual linear wear was 0.002 mm/year for MoM THA compared to 0.1–0.2 mm/year for MoP THA. Low wear and friction were further improved by adding a small amount of carbon to the cobalt-chromium-molybdenum alloy. Using this metal for the inlay of a polyethylene acetabular component, they developed the Metasul prosthesis, for use in MoM THA, with a metal bearing designed to fit the loose clearance, as previously described [10]. A 28 mm or 32 mm diameter of the femoral head was available depending on the size of the acetabular component.

The revision rate of the Metasul prosthesis was 4.7% at 13 years, and the cause of revision was dislocation with no adverse biological reactions [11]. A case report from Japan described metallosis caused by an impingement between the stem neck and acetabular component [12]. The survival rate at 15 years with revision as the endpoint was 71.2%, and it was thought that the problem was failure of the locking mechanism and impingement of the implant due to the small diameter of the femoral head [13].

Metal ion concentrations in patients who underwent MoM THA using the Metasul prosthesis were analyzed. Blood chromium concentration in patients who underwent MoM THA at 5 years postoperatively was reported not to be significantly different from that in patients who underwent MoP THA [14], whereas serum chromium ions were mildly elevated up to 3 years postoperatively [15]. Serum chromium ions did not change for up to 7 years postoperatively, but some patients showed high metal ion concentrations. When the MoM bearing was replaced with an MoP bearing, the elevated serum chromium ion concentration decreased [16]. Furthermore, while metal concentrations remained low and unchanged in approximately half of the patients (16/30) up to 12 years postoperatively, metal ion concentrations were elevated in approximately 30% (8/30) of patients despite the absence of implant malposition and good clinical outcomes [17].

4 Hip Resurfacing Arthroplasty

Derek JW McMinn developed a resurfacing arthroplasty with an MoM bearing in the late 1980s and performed the first MoM resurfacing arthroplasty in 1991 [18]. Further refinements of this model led to the Birmingham hip resurfacing procedure, with good results reported, especially in young, active patients and patients with osteonecrosis of the femoral head [19, 20]. For more information on resurfacing prostheses, see a separate chapter (Part 6, Chap. 39).

5 MoM THA with a Large-Diameter Femoral Head

Compared with conventional polyethylene liners, the MoM bearing was expected to reduce the wear of the bearing and enable the use of large-diameter femoral heads, which is expected to reduce dislocation. Therefore, the use of MoM bearings spread rapidly at the beginning of 2000. MoM bearings require high manufacturing accuracy in terms of sphericity, polishing, diameter, and clearance, which can be achieved through technological progress and a stable product supply.

When comparing the use of 28 mm and 38 mm heads in MoM THA, no dislocation was observed in the 38 mm heads regardless of approach within 3 months postoperatively, and the dislocation rate of the 38 mm heads was significantly lower than that of the 28 mm heads [21]. The hip simulator test with 5 million cycles for the wear of the MoM bearing (Conserve) of 28 mm, 36 mm, and 54 mm revealed that the 28 mm configuration had almost twice the wear of the 54 mm configuration; however, there was no significant difference between the 36 mm and 54 mm configurations [22]. MoM THA with a 28 mm/32 mm head (M2a taper) was compared with MoM THA with a 38–52 mm large-diameter head

(Magnum). Five years postoperatively, MoM THA with a large-diameter head had significantly lower blood cobalt ion levels [23]. Thus, MoM THA with ≥36 mm head was performed frequently until 2008 [24].

However, many MoM THA cases with poor outcomes due to metal debris have been reported, and an adverse reaction to metal debris (ARMD) has received focus as a complication of MoM THA. For more information on ARMD, refer to a separate chapter (Part 15, Chap. 85). The US Food and Drug Administration recommended a recall for the Zimmer Durom Acetabular Component in 2008 [25] and for the DePuy ASR-XL Acetabular System in 2013 [26]. In ASR-XL MoM THAs performed between 2005 and 2007, 54.9% of patients underwent revision at an average of 11.1 years postoperatively [27].

In addition, the Medicines and Healthcare products Regulatory Agency in the UK recommended annual careful follow-up, including blood and imaging tests, as screening for ARMD for all MoM THAs with a large-diameter femoral head in 2012 [28] and clearly proposed a follow-up procedure for MoM THA in 2017 [29]. In fact, the use of MoM THA with a large-diameter femoral head increased steadily in the UK but peaked in 2008 and started to decline thereafter [24].

In Japan, 62 of 239 patients (26%) were reported to have pseudotumors on a mean 2-year postoperative MRI after MoM THA [30]. A case was reported in which a pseudotumor compressed the iliac artery and vein after MoM THA with the Conserve Acetabular System, resulting in edema of the ipsilateral lower extremity [31]. In a survey conducted by the Japanese Society for Replacement Arthroplasty, the incidence of ARMD in large surgical-volume hospitals in Japan from 2000 to 2011 was estimated to be 1.2% with stem MoM THA. The incidence of ARMD was higher in Cormet, Conserve, and Ultamet components [32].

MoM THA was rapidly discontinued in Japan and overseas, and all companies stopped selling MoM THA in 2015.

6 Trunnionosis

The wear of the MoM bearing was initially thought to be the cause of ARMD. Impact forces during press-fit fixation of the cementless cup induced a slight deformation of the acetabular component, and an average deformation of 160 μm was reported during the insertion of the Pinnacle acetabular component [33]. Considering the possibility of a deformation occurring during cementless cup insertion, the diameter clearance, which is the difference between the inner diameter of the acetabular component and acetabular head diameter, should be greater than or equal to this deformation to achieve a loose fit. One of the reasons for the ASR MoM THA performing worse than other MoM THAs was that the diameter clearance was smaller than the diameter of the acetabular component [34], indicating that the cause of ARMD depended on the design of the MoM bearing.

ARMD was also observed in the MoP bearings. Black deposits were found not only at the bearing, but also at the head-neck junction. ARMD was observed in 23 (10.9%) of the 211 MoM THAs with Asian-type AML stems at 87.5 months of follow-up, and black deposits were found between the head and neck in 13 of the 14 revised hips [35]. Ultrasonography is useful for examining pseudotumor in MoM THA using S-ROM A and Pinnacle, and the occurrence of pseudotumor was regardless of the cup placement angle, suggesting a head-neck junction corrosion-related complications was distinguished from the bearing-related complication [36].

The head-neck junction is called a trunnion, which was originally based on a cylindrical projection on a cannon forming an axis on which it pivots. Symptoms of trunnion problems are called trunnionosis. Subsequently, more attention was given to the design of the trunnion, diameter of the head, and the combination of stem and head materials to avoid trunnionosis (Fig. 23.2).

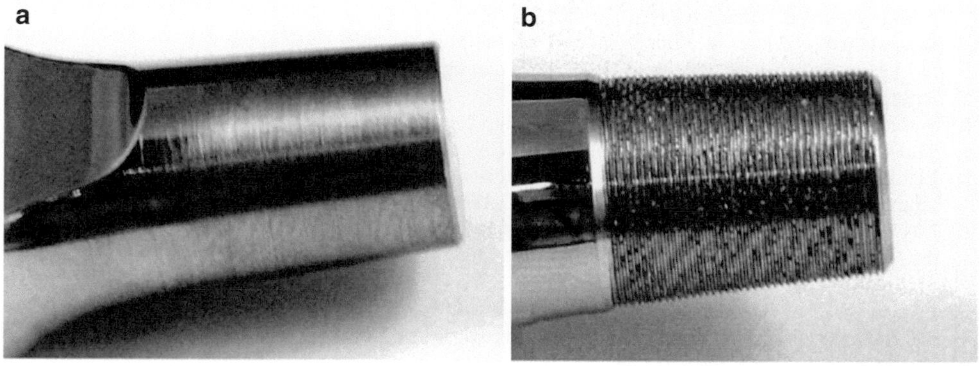

Fig. 23.2 Design of trunnion. (**a**) S-ROM-A (Deputy Synthes) (**b**) AML Plus (Deputy Synthes)

7 The Future of MoM THA

MoM THA has undergone developments since it was first performed and has a recurrent history of problems following clinical application, resulting in discontinuation. MoM THAs are currently not used due to complications such as ARMD; however, if the implant design, implant position, and patient selection are appropriate, large-diameter MoM THAs are excellent in terms of the reduction of bearing wear and fewer complications, including dislocation. Dual mobility, which consists of a small femoral head held captive yet mobile within the large polyethylene liner and articulated with a highly polished metallic acetabular component, is sometimes used as an alternative bearing to MoM THA, with the expectation that the large-diameter head will prevent dislocation. Long-term results of dual mobility remain unclear, and MoM THA may yet receive attention. If a new MoM THA procedure is introduced in the future, close monitoring and long-term follow-up will be necessary based on the experience with large-diameter MoM THAs.

References

1. Wiles P. The surgery of the osteoarthritic hip. Br J Surg. 1958;45:488–97.
2. McKee GK. Total hip replacement – past, present and future. Biomaterials. 1982;3:130–5.
3. McKee GK, et al. Replacement of arthritic hips by the McKee-Farrar prosthesis. J Bone Joint Surg Br. 1966;48:245–59.
4. Brown GC, et al. Sensitivity to metal as a possible cause of sterile loosening after cobalt-chromium total hip-replacement arthroplasty. Surg Am. 1977;59:164–8.
5. August AC, et al. The McKee-Farrar hip arthroplasty. A long-term study. J Bone Joint Surg Br. 1986;68:520–7.
6. Jacobsson SA, et al. Twenty-year results of the McKee-Farrar versus Charnley prosthesis. Clin Orthop Relat Res. 1996;329(Suppl):S60–8.
7. Higuchi F, et al. Metal-on-metal CoCrMo McKee-Farrar total hip arthroplasty: characteristics from a long-term follow-up study. Arch Orthop Trauma Surg. 1997;116:121–4.
8. Walker PS, et al. The wear on removed McKee-Farrar total hip prostheses. J Bone Joint Surg Am. 1974;56:92–100.
9. Schmalzried TP, et al. Long-duration metal-on-metal total hip arthroplasties with low wear of the articulating surfaces. J Arthroplast. 1996;11:322–31.
10. Weber BG. Experience with the Metasul total hip bearing system. Clin Orthop Relat Res. 1996;329(Suppl):S69–77.
11. Randelli F, et al. Cementless metasul metal-on-metal total hip arthroplasties at 13 years. J Arthroplast. 2012;27:186–92.
12. Iida H, et al. Metallosis due to impingement between the socket and the femoral neck in a metal-on-metal bearing total hip prosthesis. A case report. Bone Joint Surg Am. 1999;81:400–3.
13. Haraguchi K, et al. Results of a long-term follow-up study after metasul metal-on-metal cementless total hip arthroplasty. In: Knahr K, editor. Tribology in Total hip and knee arthroplasty. Heidelberg: Springer; 2014. p. 145–51.
14. Saito S, et al. Midterm results of Metasul metal-on-metal total hip arthroplasty. J Arthroplast. 2006;21:1105–10.
15. Maezawa K, et al. Seven years of chronological changes of serum chromium levels after Metasul metal-on-metal total hip arthroplasty. J Arthroplast. 2010;25:1196–200.
16. Maezawa K, et al. Serum chromium levels before and after revision surgery for loosened metal-on-metal total hip arthroplasty. J Arthroplast. 2009;24:549–53.
17. Maezawa K, et al. Changes in serum chromium levels over 12 years after Metasul metal-on-metal total hip arthroplasty. J Orthop. 2018;15:201–4.
18. McMinn D, et al. Metal on metal surface replacement of the hip. Experience of the McMinn prothesis. Clin Orthop Relat Res. 1996;329(Suppl):S89–98.
19. Daniel J, et al. Metal-on-metal resurfacing of the hip in patients under the age of 55 years with osteoarthritis. J Bone Joint Surg Br. 2004;86:177–84.
20. Nakasone S, et al. Does the extent of osteonecrosis affect the survival of hip resurfacing? Clin Orthop Relat Res. 2013;471:1926–34.
21. Cuckler JM, et al. Large versus small femoral heads in metal-on-metal total hip arthroplasty. J Arthroplast. 2004;19(8 Suppl 3):41–4.
22. Affatato S, et al. Larger diameter bearings reduce wear in metal-on-metal hip implants. Clin Orthop Relat Res. 2007;456:153–8.
23. Ando W, et al. A comparison of the effect of large and small metal-on-metal bearings in total hip arthroplasty on metal ion levels and the incidence of pseudotumour: a five-year follow-up of a previous report. Bone Joint J. 2018;100-B:1018–124.
24. Smith AJ, et al. Failure rates of stemmed metal-on-metal hip replacements: analysis of data from the National Joint Registry of England and Wales. Lancet. 2012;379:1199–204.
25. U.S. Food and Drug Administration. Class 2 Device Recall Zimmer Durom. June 4, 2008. https://www.accessdata.fda.gov/scripts/cdrh/cfdocs/cfRes/res.cfm?ID=65270. Accessed 11 Oct 2021.
26. U.S. Food and Drug Administration. Class 2 Device Recall Depuy ASR XL Acetabular cup system. August 29, 2013. https://www.accessdata.fda.gov/scripts/cdrh/cfdocs/cfRes/res.cfm?id=96135. Accessed 11 Oct 2021.
27. Kelly GA, et al. Ten-year outcomes following a cohort of ASR XL total hip arthroplasties. Hip Int. 2021;31(6):759–65.
28. Medical Device Alert (MDA/2012/036). All metal-on-metal (MoM) hip replacement. Medicines and Healthcare products Regulatory Agency (MHRA). June 25, 2012. https://assets.publishing.service.gov.uk/media/5485abf6ed915d4c10000273/con155767.pdf. Accessed 11 Oct 2021.
29. Medical Device Alert (MDA/2017/018). All metal-on-metal (MoM) hip replacements. Medicines and Healthcare products Regulatory Agency (MHRA). June 29, 2017. https://assets.publishing.service.gov.uk/media/5954ca1ded915d0baa00009b/MDA-2017-018_Final.pdf. Accessed Oct 11 2021.
30. Hasegawa M, et al. Factors associated with symptomatic pseudotumors following metal-on-metal total hip arthroplasty. BMC Musculoskelet Disord. 2016;17:456.
31. Kawakita K, et al. Leg edema due to a mass in the pelvis after a large-diameter metal-on-metal total hip arthroplasty. J Arthroplast. 2013;28(197):e1–4.
32. Sugano N, et al. Nationwide investigation into adverse tissue reactions to metal debris after metal-on-metal total hip arthroplasty in Japan. Orthop Sci. 2014;19:85–9.
33. Squire M, et al. Acetabular component deformation with press-fit fixation. J Arthroplast. 2006;21(6 Suppl 2):72–7.
34. Jameson SS, et al. Articular surface replacement of the hip: a prospective single-surgeon series. J Bone Joint Surg Br. 2010;92:28–37.
35. Nakamura M, et al. Failure rates of Asian-type anatomic medullary locking stemmed metal-on-metal total hip replacement: a cause for adverse tissue reactions to metal debris (ARMD). J Orthop Sci. 2016;21:779–85.
36. Muraoka K, et al. Usefulness of ultrasonography for detection of pseudotumors after metal-on-metal total hip arthroplasty. J Arthroplast. 2015;30:879–84.

Ceramic-on-Ceramic THA

24

Taisuke Seki

1 History of Development

Ceramics is a general term that refers to nonmetallic inorganic materials produced by ceramic manufacturing procedures, such as molding and firing. Engineering and scientific high-performance ceramics manufactured from refined nonmetallic raw materials are used in the medical field for prosthetic joints. Ceramics are durable, nonmagnetic, and resistant to heat, corrosion, and scratch.

Because of these characteristics, total hip arthroplasty (THA) with a ceramic-on-ceramic (COC) bearing surface was developed to solve the problem of wear of the polyethylene liner that serves as the THA's bearing surface. The advantages of using COCs for the bearing surfaces of THAs include minimal wear, a moist surface, and low biological reactivity (bioinert) [1, 2]. Alumina (Al_2O_3) is the most commonly used material for COCs.

In 1970, Boutin et al. were the first to clinically perform THA with COCs using alumina in humans in France, and clinical results have been reported since then [3–5]. Early implants used ceramic-integrated (monoblock) cups. However, there were many problems, such as fractures due to a lack of clinical knowledge and poor alumina quality [4, 6]. The high rigidity of the cement-fixed type, in particular, causes early loosening and detachment [7]. Cement-free COC-THA had a high fracture rate of 21.5% but no incidence of osteolysis [8]. Mechanical safety has increased substantially since the 1980s due to machining accuracy and alumina material. COC-THA with an alumina liner fitted inside a metal cup was developed in the 1990s, and good clinical results have since been reported [9].

Two types of COC bearing surfaces with a ceramic liner mated to a metal cup include a direct-mating type and an interlaced type with polyethylene or metal interlaced at the back of the liner. (Fig. 24.1). The polyethylene-mediated type was used to prevent ceramic liner edge and neck impingement. However, it is no longer recommended because of the

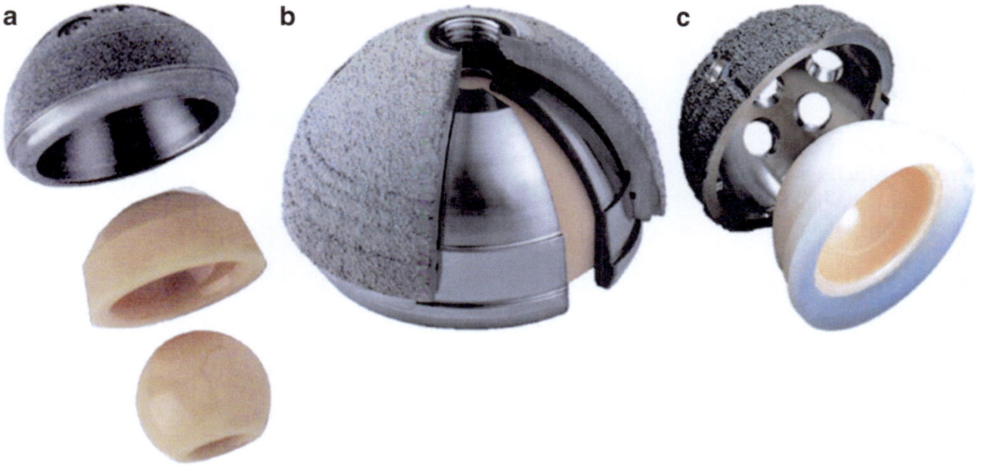

Fig. 24.1 Ceramic Liner. (**a**) Direct-fit Stryker SECUR-FIT AD ceramic on ceramic shell, alumina insert, alumina head (BIOLOX forte). (**b**) Metallic intervention type Stryker Trident hip arthroplasty cup trident ceramic liner. (**c**) Kyocera ABS Cup with polyethylene inclusion, alumina ceramic inlay

T. Seki (✉)
Department of Orthopaedic Surgery, Aichi Medical University Medical Center, Aichi, Japan
e-mail: taiseki@aichi-med-u.ac.jp

Y. Nakashima (ed.), *Advances in Total Hip Arthroplasty*, https://doi.org/10.1007/978-981-97-2461-1_24

early deformation and fracture of the ceramic liner. However, the metal-inserted type showed good clinical results.

2 Basis

Because of its big grain size and low purity, first-generation COC-THA using alumina failed due to insufficient strength because of the large grain size and low purity. Second-generation alumina, introduced in the 1980s, has a failure rate of <5%, owing to the higher densities achieved by finer alumina particles. Third-generation alumina, introduced in the 1990s, has dramatically improved strength by increasing purity through hot isostatic pressing [10]. Since 2000, fourth-generation COC bearing surfaces have been introduced, in which zirconia (ZrO_2) is blended into the alumina matrix to increase its strength compared to that of alumina. Because of the homogeneous dispersion of small crystal grains and zirconia particles, zirconia-toughened alumina offers great strength and toughness. (Fig. 24.2). At temperatures between 200 and 300 °C, zirconia undergoes a remarkable phase transition to the monoclinic phase, which is accelerated by the presence of moisture (low- temperature degradation). The addition of a small amount of yttria (Y_2O_3) as a stabilizer prevents the phase transition during cooling and results in a partially stabilized zirconia with a tetragonal crystal structure. This zirconia-toughened alumina minimizes stress concentration at the crack tip because of the approximately 4% volume expansion caused by the phase transition of zirconia and overcomes fractures through aluminous platelet crystals (Fig. 24.3), which are the reasons for its high strength and toughness. Kyocera AZ209 and CeramTec BIOLOX delta were used as fourth-generation ceramic heads. Furthermore, ceramic heads with cobalt replacing some of the aluminum and strontium in the platelet crystal are now available, providing both excellent mechanical properties and in vivo stability (no crystalline phase transition of zirconia in water) (Kyocera AZUL).

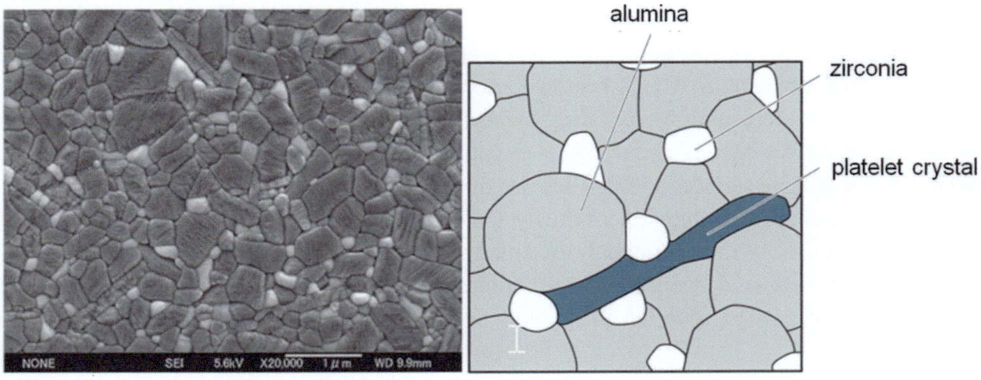

Fig. 24.2 Zirconia-toughened alumina enlarged image from Kyocera AZ209 Technical document (with permission)

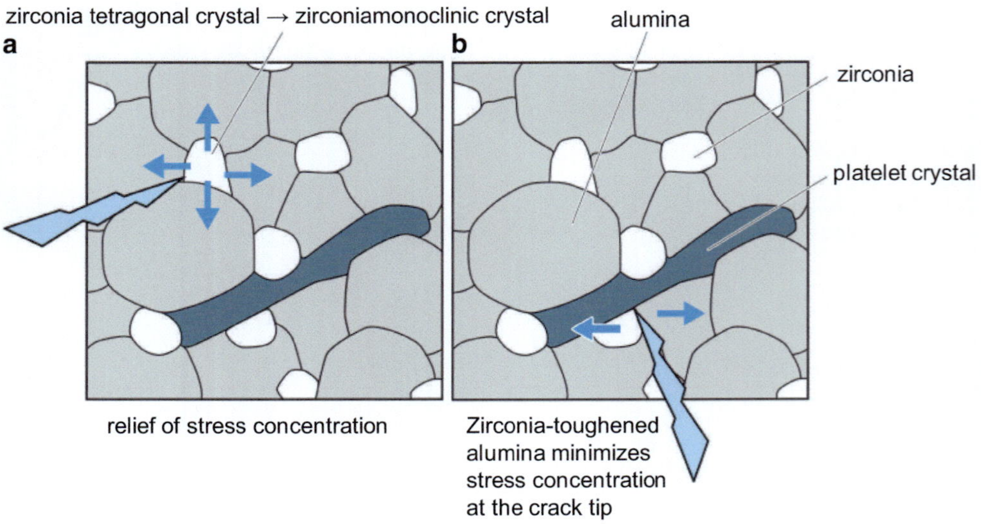

Fig. 24.3 Mechanism of high strength and high toughness. (**a**) Homogeneously dispersed zirconia particles reduce stress concentration at the crack tips by inhibiting the development of fracture-causing cracks by a phase transformation from tetragonal to monoclinic. (**b**) Crack bypassing effect by aluminous platelet crystals

3 Clinical Outcome, Complications

3.1 Clinical Outcome

For third-generation COC-THA, long-term results of more than 10 years have been reported [11–13]. The appearance of osteolysis was low when compared to conventional polyethylene, and mechanical laxity was 0% (Fig. 24.4). The 5-year survival rates for BIOLOX delta, a fourth-generation COC, and BIOLOX forte, a third-generation COC, were 98.6 and 98.4%, respectively, indicating good results. However, even with the new COC glide-bearing surface, the squeaking sound was reported to be 5% in forte and 2% in delta, and this problem has not been resolved [14]. However, the polyethylene-mediated type resulted in early alumina liner fracture and dislocation and poor clinical outcome (Fig. 24.5). At an average follow- up of 5.8 years for 35 joints, the survival rate was reported to be 83% [15]. The fracture was caused by the deformation of the polyethylene lining and excessive stress concentration on the margins [16].

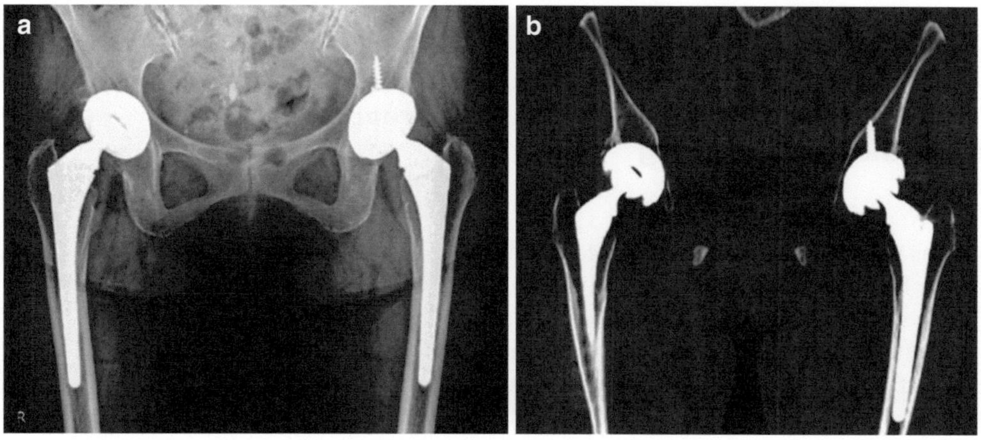

Fig. 24.4 Left: 15 years after ceramic-on-ceramic total hip arthroplasty (COC-THA) surgery. A 64-year-old woman with bilateral end-stage hip osteoarthritis. A 64-year-old woman underwent COC-THA for bilateral end-stage hip osteoarthritis. She had no pain or limitation of activity of the daily living. (**a**) Model using Stryker. SECUR-FIT AD ceramic on a ceramic shell 46 mm, alumina insert SUPER SECUR-FIT HA 132°#7, V40 Alumina femoral head 28 mm + 0 mm. (**b**) Plain computed tomography multiplanar reformation images (coronal section) of both the hip joints: no osteolysis or breakage, good fixation

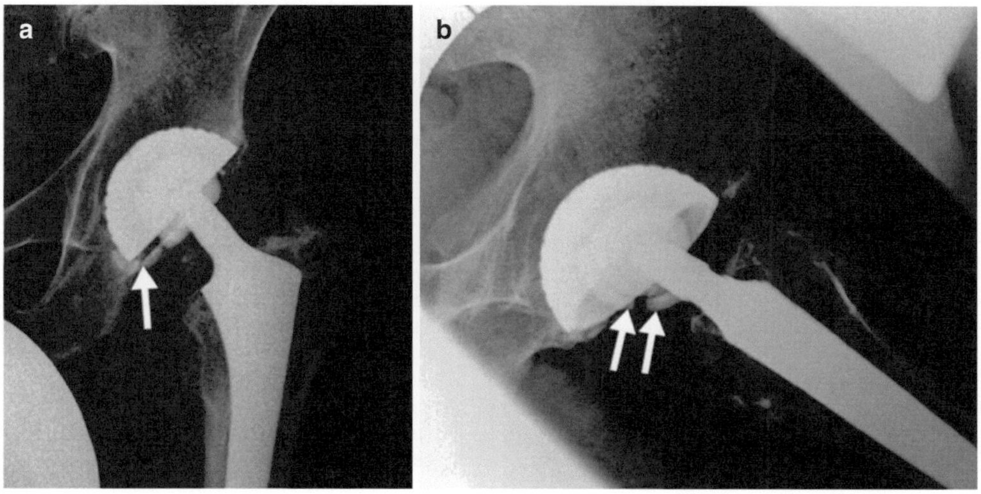

Fig. 24.5 Case of ceramic liner breakage. 61-year-old man, total hip arthroplasty after transtrochanteric rotational osteotomy of the left femoral head for left idiopathic osteonecrosis of the femoral head; the bearing surface of the ceramic-on-ceramic was made of a Kyocera ABS liner and alumina head. 8 years and 10 months postoperatively, the ceramic liner broke when the patient stood. (**a**) Plain radiographic frontal view of the hip joint showing a broken alumina ceramic inlay and small fragments. (**b**) Lauenstein image of hip joint; the inlay is derotated

3.2 Poor Installation of Ceramic Liners

Impingement and chipping are risk factors for fractures caused by the tapered fit of the liner at an angle. During surgery, titanium cups are deformed by 0.6 mm [17], which may interfere with the accurate placement of the liner. A ceramic liner with a metal-backed titanium sleeve or an integrated metal cup can be used as a countermeasure. However, with metal- backed ceramic liners, 20% of the patients had poor placement, and osteolysis was significantly more common; however, there was no fracture, squeaking, or revision [18]. Caution should be exercised because a small cup thickness, high bone rigidity, and soft tissue or bone inclusions may interfere with the proper fit of the ceramic liner.

3.3 Damage to the Ceramic Bearing Surfaces

The head and liner on the COC glide-bearing surface failed. In the early days, the failure rate of the mushroom-type alumina head was 0.4%, while the ball-type without neck had a low failure rate of 0.07% [19]. Since the late 1980s, improvements in properties of materials, such as particle size and purity, and improvements in implant designs have reduced the failure rate of ceramic heads to <0.1%. Poor placement, chipping, and impingement are all causes of ceramic liner failure. In the case of revision, a metal sleeve can be attached to the original neck, and a new ceramic head can be placed instead of a metal head if there are no visual problems with the neck. However, if the stem neck is severely damaged, in that case, it may be necessary to remove the stem, and it is recommended that the bearing surface should be replaced with a new ceramic-on- polyethylene.

3.4 Wear

The amount of wear on the acetabular side was <0.04 mm^3/year and < 0.1 mm^3/year on the ceramic head, which was less than that of conventional polyethylene, according to the study of the extracted implants [20]. On radiograph images, the annual linear wear of the fitted ceramic liner without a metal back and the ceramic liner with a metal-backed titanium sleeve was 0.0049 mm/year and 0.0046 mm/year, respectively, indicating very low wear [21]. The powder's foreign body granulation reaction triggers osteolysis; however, the bioreactivity of alumina ceramics is small, and the incidence of osteolysis is as low as 1.4% [22].

3.5 Squeaking Noise

COCs have low wear characteristics, although their hardness has been reported to generate noise concerns. The prevalence of noise ranges from 0% to 21.9% [12, 22–24]. The incidence of squeaking is 53% in recent years, including transitory noises [25], and squeaking is common. Lateral cup placement, short stem neck, poor liner placement, specific combinations (Accolade TNZF stem), failure of surface lubrication by a third body, metal transfer, and microseparation are the factors that lead to squeaking. No deterioration in clinical or imaging results in the natural history of the disease in the absence of obvious damage on imaging or clinical findings; however, careful follow-up is needed.

References

1. Hulbert SF, et al. Potential of ceramic materials as permanently implantable skeletal prostheses. J Biomed Mater Res. 1970;4:433–56.
2. Jeffers JR, et al. Ceramic-on-ceramic bearings in hip arthroplasty: state of the art and the future. J Bone Joint Surg Br. 2012;94:735–45.
3. Boutin P, et al. The use of dense alumina-alumina ceramic combination in total hip replacement. J Biomed Mater Res. 1988;22:1203–32.
4. Griss P, et al. Five-year experience with ceramic-metal-composite hip endoprostheses: I. Clinical evaluation. Arch Orthop Trauma Surg. 1978;1981(98):157–64.
5. Hamadouche M, et al. Alumina-on-alumina total hip arthroplasty: a minimum 18.5-year follow-up study. J Bone Joint Surg Am. 2002;84:69–77.
6. Shishido T, et al. A study for a retrieved implant of ceramic-on-ceramic total hip arthroplasty. J Arthroplast. 2006;21:294–8.
7. Nizard RS, et al. Ten-year survivorship of cemented ceramic-ceramic total hip prosthesis. Clin Orthop Relat Res. 1992;282:53–63.
8. Huo MH, et al. Total hip replacements using the ceramic Mittelmeier prosthesis. Clin Orthop Relat Res. 1996;332:143–50.
9. Sugano N, et al. Mid-term results of cementless total hip replacement using a ceramic-on- ceramic bearing with and without computer navigation. Joint Surg Br. 2007;89:455–60.
10. Willmann G. Ceramic femoral head retrieval data. Clin Orthop Relat Res. 2000;379:22–8.
11. Higuchi Y, et al. 32-mm ceramic-on-ceramic total hip arthroplasty versus 28-mm ceramic bearings: a 5- to 15-year follow-up study. Hip Int. 2019;29:65–71.
12. Sugano N, et al. Eleven- to 14-year follow-up results of cementless total hip arthroplasty using a third-generation alumina ceramic-on-ceramic. J Arthroplast. 2012;27:736–41.
13. Tsukada S, et al. Uncemented third-generation ceramic-on-ceramic total hip arthroplasty using metal acetabular shell with direct taper locking. Arch Orthop Trauma Surg. 2013;133:861–8.
14. Kim SC, et al. Fourth-generation ceramic-on-ceramic THA results in improvements in midterm outcomes compared to third-generation THA but does not resolve noise problems: a cohort study of a single-hip system. BMC Musculoskelet Disord. 2019;20:263.
15. Hasegawa M, et al. Alumina ceramic-on-ceramic total hip replacement with a layered acetabular component. J Bone Joint Surg Br. 2006;88:877–82.

16. Hasegawa M, et al. Ceramic acetabular liner fracture in total hip arthroplasty with a ceramic sandwich cup. J Arthroplasty. 2003;18(5):658–61.
17. Squire M, et al. Acetabular component deformation with press-fit fixation. J Arthroplast. 2006;21(6 Suppl 2):72–7.
18. Higuchi Y, et al. Incidence of ceramic liner malseating after ceramic-on-ceramic total hip arthroplasty associated with osteolysis: a 5- to 15-year follow-up study. J Arthroplast. 2017;32:1641–6.
19. Fritsch EW, et al. Ceramic femoral head fractures in total hip arthroplasty. Clin Orthop Relat Res. 1996;328:129–36.
20. Lusty PJ, et al. Wear and acetabular component orientation in third generation alumina- on-alumina ceramic bearings: an analysis of 33 retrievals. Bone Joint Surg Br. 2007;89:1158–64.
21. Higuchi Y, et al. Survivorship between 2 different ceramic-on-ceramic total hip arthroplasty with or without a metal-backed titanium sleeve bearing: a 5- to 14-year follow-up study. J Arthroplast. 2017;32:155–60.
22. Capello WN, et al. Ceramic-on-ceramic total hip arthroplasty: update. J Arthroplast. 2008;23(7 Suppl):39–43.
23. Inagaki K, et al. Natural history of noise and squeaking in cementless ceramic-on-ceramic total hip arthroplasty. J Orthop. 2020;21:544–9.
24. Kiyama T, et al. Can squeaking with ceramic-on-ceramic hip articulations in total hip arthroplasty be avoided? J Arthroplast. 2013;28:1015–20.
25. Taniguchi K, et al. Squeaking is common and increases over time among patients with long-term follow-up after ceramic-on-ceramic THA. Clin Orthop Relat Res. 2021;479:736–44.

Surface Treatment of Cementless Cups and Stems

25

Hiroshi Ito

1 Cementless Hip Arthroplasty

In recent years, cementless total hip arthroplasty (THA) has become mainstream, and from the viewpoint of long-term durability of cementless cups and stems, it is important to achieve biologically firm fixation with direct contact between bone tissue and implant surface without fibrous soft tissue intervention. This requires good initial implant fixation. Factors that influence long-term durability and initial fixation include implant material, design, and surface finish [1]. For cementless cup stems to adhere to bone, the implant must be made of a material with a high affinity to bone, and a scaffold for osteogenesis must be formed by roughening the surface or forming a porous structure. Metals such as pure titanium (Ti), titanium alloys, and cobalt-chromium (Co-Cr) alloys have been used. It is also known that the surface coating of ceramic biomaterials with high biological activity, such as hydroxyapatite (HA), promotes new bone formation.

2 Osteogenesis

The porous surface of metallic materials means porosity, which is often used to describe the shape of materials used for artificial joints. Osteogenesis is defined as a state in which bone is formed directly on the implant surface without the intervention of fibrous tissue, and the implant functions as part of the bone. The porous structure has the characteristics of being an excellent scaffold for mesenchymal cells and blood vessels to penetrate and proliferate and is useful as a scaffold to generate osteogenesis. In cementless THA, porous coatings, which add a porous finish to the implant surface to achieve osseointegration by bone ingrowth, have become common, as first described by Austin-Moore in the early 1950s [2, 3]. The body of the stem had an opening to accommodate the grafted bone, and the concept was that the grafted bone and surrounding bone would be joined at the macro level. In the late 1960s and the 1970s, a number of fundamental experiments on surface processing and tissue ingrowth were conducted. In metal implants with smooth surfaces, fibrous tissue intervenes and fixes the implant to the bone. In the 1970s, Judet [4] and Lord [5] were the first to use porous hip implants, and the demand for cementless fixation increased rapidly. In recent years, the demand for cementless fixation has increased with an increasing proportion of hip arthroplasty registry systems in different countries, although there are regional variations.

Osteogenesis involves the formation of bone tissue in the gap between the implant and the bone or the porous structure on the implant surface after implant insertion. When the implant is inserted into the bone, local tissue is damaged, bleeding occurs, hematoma is formed, and an inflammatory reaction is induced. Subsequently, undifferentiated mesenchymal cells and fibroblasts infiltrate the hematoma, forming granulation tissue, and capillary proliferation occurs within the granulation tissue. The hematoma is replaced by woven bone by osteoblasts in approximately 3 weeks, and fibrous bone is formed around the implant. Although fibrous bone is fragile in terms of strength, resorption and formation of bone occur repeatedly when fibrous bone is mechanically stimulated. This remodeling fixes the implant to the laminar bone for effective load transfer in approximately 6 weeks. If osteogenesis is not achieved, fibrous tissue remains at the implant-bone interface, leading to complaints such as pain and the appearance of implant laxity. The factors on the implant side that affect osteogenesis include the biocompatibility of the material, pore size, and initial fixation. The biological factors include local blood flow, susceptibility to infections, and allergic reactions [3].

H. Ito (✉)
Department of Orthopaedic Surgery, Asahikawa Medical University, Asahikawa, Japan
e-mail: itohiro@asahikawa-med.ac.jp

Y. Nakashima (ed.), *Advances in Total Hip Arthroplasty*, https://doi.org/10.1007/978-981-97-2461-1_25

3 Problems of Porous Materials

The types of porous coatings include wire mesh made of pure Ti and porous bead made of cobaltactom alloy, and many good long-term results of hip prostheses with acetabular components (cups) with these surface finishes have been reported [6–8].

Implant biocompatibility is an important factor in osteogenesis. Biocompatibility is classified into biotorelant, bioinert, and bioactive, according to the osteogenic response [9]. Titanium alloys (mainly Ti-6Al-4 V), cobalt-chromium alloys, and tantalum alloys are used for cementless hip prostheses because of their mechanical strength. The porosity of the porous structure, which is advantageous for osseointegration, has been reported to be 50–400 μm based on the results of animal experiments [10]. Porosity should be 30–50% to maintain mechanical strength [11, 12], but this limits the amount of bone penetration. The stability of the implant-bone interface remains a concern in terms of long-term durability. To obtain strong biological fixation in cementless cups, it is important to minimize micromotion between the implant and bone as much as possible until fixation is achieved, because fixation between the implant and bone depends on mechanical fixation, and in the early stages until osteogenic union. It has been reported that micromotion should be less than 100 μm [13]. The gap between the implant and bone interface in the early stages is also important for osteogenesis, and should be kept as small as possible. In animal studies, the larger the gap between the implant and bone, the less favorable it is for osteogenesis and the more likely it is to be affected by fibrous tissue. The gap should not exceed 0.5 mm [14].

4 Metal Implant Surface Processing Method

Osteogenesis is classified into bone ingrowth and bone ongrowth according to the surface treatment method. Bone ingrowth is the formation of bone within the porous structure of the implant surface, whereas bone ongrowth is the formation of bone on a roughened implant surface.

The current main surface finishing methods for metal implants include sintered beads (bead coating), fiber mesh (fiber mesh coating), and plasma-spray coating [3] (Fig. 25.1). Sintered beads and fiber mesh are classified as bone ingrowth surfaces, and plasma-spray is classified as bone ongrowth surface [15].

Sintered beads are made of cobalt-chromium alloy or titanium alloy spheres 100–250 μm in diameter, sintered at high temperature [15]. Fiber mesh is made by diffusion bonding titanium alloy fibers [15]. The porosity of both types of mesh is approximately 50–400 μm. Fiber mesh has a higher porosity and is said to maintain a higher mechanical strength. Titanium plasma spray is a titanium alloy powder melted at a high temperature and flame-emitted onto the implant surface. Compared with sintered beads and fiber mesh, it can maintain the strength of the implant [15]. Good clinical results and long-term survival rates have been reported for implants using either surface treatment method [7, 8, 16].

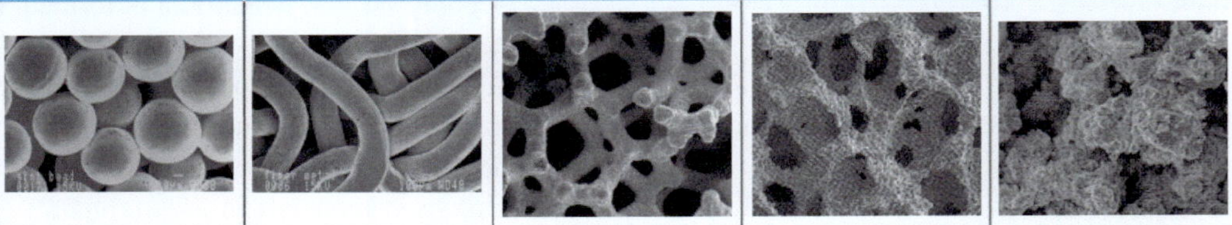

	Co-Cr beads	Fiber mesh	Trabecular Metal	Tritanium	Plasma spray
Material	Co–Cr alloy	pure titanium	tantalum	pure titanium	pure titanium
Pore Size (μm)	100~400	100~400	550	546	various
Porosity (μm)	30~50	40~450	80	72	N/A

Fig. 25.1 Porous surfaces

5 Materials to Solve Problems

As materials that can solve the problems of existing porous structures, metallic materials have a porosity of 60–80% and pore diameter of 300–500 μm. Metallic materials with structures similar to cancellous bone structures have been developed and are now being applied clinically [10]. These are called highly porous metals and are characterized by a higher degree of porosity and permeability than existing porous materials. Are three-dimensional and are also called 3-dimensional porous. Good clinical results of primary and revision THA using highly porous metal cups have already been reported [17, 18]. One of the characteristics of the 3D porous structure is its low stiffness due to high porosity, which is expected to suppress bone loss due to stress shielding. Tantalum (Ta) 3D porous structures developed in the 1990s (Trabecular Metal®: Zimmer Biomet) have the longest history and have been used for implant surface coating and bone defect replacement [18].

6 Model Selection by the Difference in the Cementless Cup Surface Processing Method

It is important to understand the differences in implant surface processing methods and to select a model. Typical porous structures of various companies are listed in Fig. 25.2. Recently, dual mobility cups (G7 or Mexera cup; both from Zimmer Biomet) have been used in cases where postoperative dislocation is a concern or where the pelvis is tilted backward significantly in the standing position. The Trabecular Metal® cup (Zimmer Biomet) is often used for revision cups because of the poor condition of the metaphyseal floor.

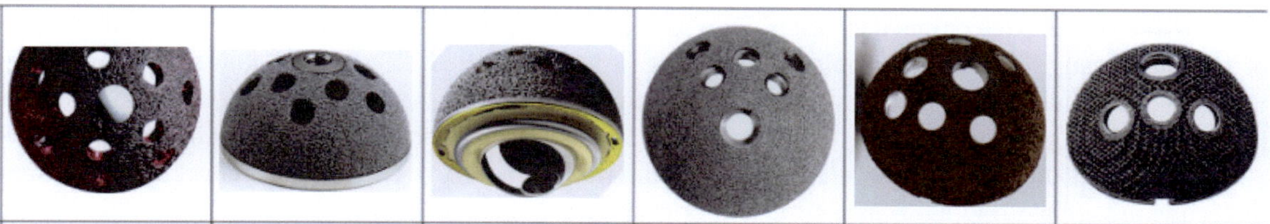

Company Name	Zimmer Biomet	Zimmer Biomet	Zimmer Biomet	Stryket	Kyocera	Kyocera
Product Name	Trilogy	Continuum	G7	Tritanium	AG-PROFIX	SQRUM TT
Material	titanium alloy	tantalum	titanium alloy	pure titanium	pure titanium	titanium alloy
Pore Size (μm)	100~400	550	475	546	350~450	640
Porosity (μm)	40~50	80	70	72	N/A	60
3D Porous		○	○	○		○

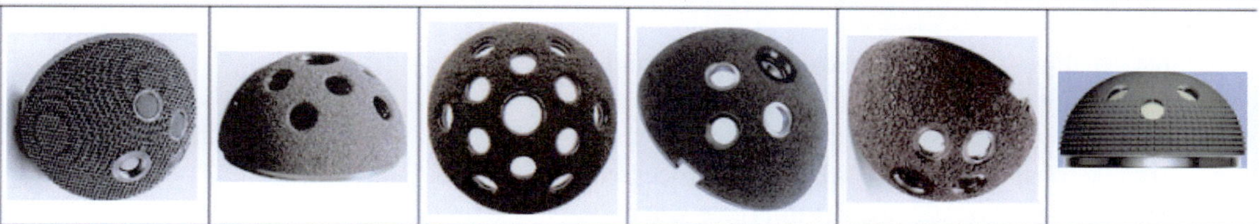

Company Name	Lima	DePuy Synthes	DePuy Synthes	Smith & Nephew	Smith & Nephew	AECECULAP
Product Name	Delta TT	POROCOAT	GRIPTION	REFLECTION	STIKTITE	Plasmafit
Material	titanium alloy	pure titanium	pure titanium	titanium alloy	titanium alloy	pure titanium
Pore Size (μm)	640	250	300	100~400	550	Teeth
Porosity (μm)	64	45	63	40~450	62	50
3D Porous	○	○	○		○	

Fig. 25.2 Porous acetabular components

7 Surface Treatment of Cementless Stems

Ingrowth surfaces are processed by sintering beads of cobalt-chromium or titanium alloy particles under high temperatures (bead coating), surface treatment of fiber mesh by diffusion bonding, macrotexturing, and three-dimensional modeling (porous metal) [19]. Macrotexturing includes casting and surface cutting methods. The lost-wax method, which is a casting method, is a method in which a precise model of the same shape as the implant is made of wax, the surrounding area is hardened with a refractory material, the wax model is heated to make the wax model disappear, and metal is poured into the space afterwards and cooled to produce a stem. Porous metals have a higher porosity (75–85%) than beads and fiber meshes and are interconnected three-dimensionally and uniformly [10].

Ongrowth surfaces are processed using grit blasting or plasma spraying. Grit blasting is a method of forming a rough surface by rubbing the implant surface with fine particles of aluminum oxide or the like. Grit blasting is a method of forming a rough surface by rubbing the implant surface with fine aluminum oxide particles, resulting in a surface roughness of 3–5 μm [20].

Plasma is an ultrahigh-temperature gas state, in which molecules dissociate one atom at a time and release electrons. Plasma spraying is a coating method in which super energy is used as a heat source to melt the coating material, and the expanding gas is accelerated by a converged plasma jet to strike the base metal. Although the effective porosity is lower than that of the ingrowth surface, plasma spraying has the advantage of maintaining 90% of the fatigue strength of the implant, whereas diffusion bonding and sintering methods maintain only approximately 50% of the fatigue strength [21, 22].

Hydroxyapatite is a calcium phosphate compound processed directly on the implant surface by plasma spraying, which has osteoconductive ability and promotes the formation of calcified bone [23, 24]. When processed on porous surfaces, deterioration of the interface where bond strength with the base material is concerned can lead to implant sagging, but the optimal coating thickness that does not degrade bond strength is approximately 50 μm [24, 25]. However, no significant difference has been reported in clinical and radiological outcomes between implants with and without hydroxyapatite coating [26, 27]. Recently, it has been reported that pure titanium and titanium alloys can form apatite by forming a thin layer containing sodium titanate on the surface when they are heat-treated after being eroded in aqueous sodium hydroxide solution [28].

Sugino et al. reported the GRAPE technology, which can impart apatite formation ability to the surface of Ti-15Zr-4Nb-4Ta alloy in a pseudobody fluid by forming an oxide film in an appropriate space (groove depth and spacing) and by appropriate heat treatment [30]. The deposition of an apatite layer was confirmed in the groove under in vivo conditions, and vigorous conduction of new bone cells was also observed in animal experiments. The advantage of this technology is that there is no bonding or coating of the base material and the surface-treated part does not peel off.

Surface treatment of cementless joint prostheses often involves joining dissimilar materials, which causes joint detachment. The solution to the problem of joint separation is the molding of implants using an additive manufacturing method, which has been commercialized [31]. The layered fabrication method can be used to fabricate base materials and porous structures as a single piece by irradiating metal powder with an electron beam based on three-dimensional computer-aided design (CAD) model data and fabricating a lattice-like object by piling up layers. This manufacturing method can mold complex-shaped products created by three-dimensional CAD in a single process, and when applied to artificial joints, it is possible to manufacture implants with a porous shape in which the base material and surface porous material are molded as a single piece, which is expected to reduce the occurrence of exfoliation defects [31].

A continuous all-perimeter surface finish on the implant promotes osteogenesis and good stress transfer at the stem end, reduces stress shielding at the proximal portion, and results in better outcomes [31–34]. Full-perimeter coating also reduces the migration of wear debris to the distal portion of the stem and decreases distal osteolysis [29]. Emerson et al. reported that full-perimeter coating on implants significantly reduced the incidence of osteolysis in the distal stem [31].

8 Problems with Cementless Stems

Various types of cementless stems are available depending on the length of the implant and the extent of surface treatment for osteogenesis, but they cause stress concentrations in the distal part of the stem, resulting in femoral pain and bone atrophy in the proximal femur. Once the stem is fixed to the femur, bone remodeling occurs due to changes in the mechanical environment around the stem. Load stresses transmitted to the proximal femur are transferred via the stem to the bone around the distal stem, thereby causing stress shielding of the proximal femur [30]. Because the difference in implant stiffness is closely related to stress shielding, titanium alloys, which have lower stiffness than stainless steel or cobalt-chromium alloys, have become mainstream cementless stem materials in recent years.

References

1. Khanuja HS, et al. Cementless femoral fixation in total hip arthroplasty. J Bone Joint Surg Am. 2011;93:500–50.
2. Moore AT. Metal hip joint: a new self-locking vitallium prosthesis. South Med J. 1952;45:1015–9.
3. Nakahara, I, et al. Metal surface treatment (cementless fixation). Kokansetsu Gaku, Shun-ichi Kubo, editors. Kyoto: Kinpo-Do; 2014. pp 125–130. In Japanese.
4. Judet R, et al. A noncemented total hip prosthesis. Clin Orthop Relat Res. 1978;137:76–84.
5. Lord GA, et al. An uncemented total hip replacement: experimental study and review of 300 madreporique arthroplasties. Clin Orthop Relat Res. 1979;141:2–16.
6. Kaneuji A, et al. Porous processing ② shell. Jinko Kokansetsu no Biomaterials—Zairyo Sentaku kara Design made, Kengo Yamamoto, editors, Tokyo: Medical View; 2017, pp. 101–107. In Japanese.
7. Anseth SD. Fifteen- to twenty-year results of cementless Harris-Galante porous femoral and Harris-Galante porous I and II acetabular components. J Arthroplast. 2010;25:687–91.
8. McLaughlin JR, et al. Cementless total hip replacement using second-generation components: a 12- to 16-year follow-up. J Bone Joint Surg Br. 2010;92:1636–41.
9. Osborn JF. Biomaterials and their application to implantation. SSO Schwaiz Monstsechr Zahnheikd. 1979;89:1138–9.
10. Bobyn JD, et al. Characteristics of bone ingrowth and interface mechanics of a new porous tantalum biomaterial. J Bone Joint Surg Br. 1999;81:907–14.
11. Albrektsson T, et al. Osseointegrated titanium implants: requirements for ensuring a long-lasting, direct bone-to-implant anchorage in man. Acta Orthop Scand. 1981;52:155–70.
12. Jasty M, et al. In vivo skeletal responses to porous-surfaced implants subjected to small induced motions. J Bone Joint Surg Am. 1997;79:707–14.
13. Hadded RJ, et al. Biological fixation of porous-coated implants. J Bone Joint Surg Am. 1987;69:1459–66.
14. Dalton JE, et al. The effect of operative fit and hydroxyapatite coating on the mechanical and biological response to porous implants. Surg Am. 1995;77:97–110.
15. Manning DW, et al. The proximally ingrown stem. In: Callaghan JJ, et al., editors. The adult hip. 2nd ed. Philadelphia: Lippincott Williams & Wilkins; 2006. p. 1014–24.
16. Bourne RB, et al. Ingrowth surfaces. Plasma spray coating to titanium alloy hip replacements. Clin Orthop Relat Res. 1994;298:37–46.
17. Schmalzried TP, et al. The Harris-Galante porous acetabular component press-fit without screw fixation. Five-year radiographic analysis of primary cases. J Arthroplast. 1994;9:235–42.
18. Banerjee S, et al. Highly-porous metal option for primary cementless acetabular fixation: what is the evidence? Hip Int. 2013;23:509–21.
19. Malahias MA, et al. Highly porous titanium acetabular components in primary and revision total hip arthroplasty: a systematic review. Arthroplasty. 2020;35:1737–49.
20. Sakai K, et al. Cementless fixation of stem. In: Kubo S-i, editor. Hip arthroplasty. Kyoto: Kinpo-Do; 2014. p. 934–45.
21. Zweymüller KA, et al. Biologic fixation of a press-fit titanium hip joint endoprosthesis. Clin Orthop Relat Res. 1988;235:195–206.
22. Callaghan JJ. The clinical results and basic science of total hip arthroplasty with porous-coated prostheses. J Bone Joint Surg Am. 1993;75:299–310.
23. Collier JP, et al. Symposium: porous-coating methods: the pros and cons. Contemp Orthop. 1993;27:269–96.
24. Søballe K, et al. Histologic analysis of a retrieved hydroxyapatite-coated femoral prosthesis. Clin Orthop Relat Res. 1991;272:255–8.
25. Nakashima Y, et al. Hydroxyapatite-coating on titanium arc sprayed titanium implants. J Biomed Mater Res. 1997;35:287–98.
26. Søballe K, et al. The current status of hydroxyapatite coating of prostheses. J Bone Joint Surg Br. 1996;78:689–91.
27. Camazzola D, et al. A randomized trial of hydroxyapatite-coated femoral stems in total hip arthroplasty: a 13-year follow-up. J Arthroplast. 2009;24:33–7.
28. Incavo SJ, et al. Total hip arthroplasty with the Secur-fit and Secur-fit plus femoral stem design a brief follow-up report at 5 to 10 years. Arthroplasty. 2008;23:670–6.
29. Kim HM, et al. Preparation of bioactive Ti and its alloys via simple chemical surface treatment. J Biomed Mater Res. 1996;32:409–17.
30. Sugino A, et al. Effect of spatial design and thermal oxidation on apatite formation on Ti-15Zr-4Nb-4Ta alloy. Acta Biomater. 2009;5:298–304.
31. Emerson RH Jr, et al. Effect of circumferential plasma-spray porous coating on the rate of femoral osteolysis after total hip arthroplasty. Joint Surg Am. 1999;81:1291–8.
32. Dorr LD, et al. Failure mechanisms of anatomic porous replacement. I cementless total hip replacement. Clin Orthop Relat Res. 1997;334:157–67.
33. Marshall AD, et al. Cementless titanium tapered-wedge femoral stem: 10- to 15-year follow-up. J Arthroplast. 2004;19:546–52.
34. Clohisy JC, et al. The Harris-Galante uncemented femoral component in primary total hip replacement at 10 years. J Arthroplast. 1999;14:915–7.

26 Iodine Coating Artificial Hip Joints

Tamon Kabata and Hiroyuki Tsuchiya

1 Background, Rationale, and History of Development

Although total hip arthroplasty (THA) has developed remarkably in the twentieth century and provided great benefits, complications such as infection, loosening, wear, and dislocation have not yet been completely overcome. Postoperative infection is one of the most stressful complications for surgeons, and the number of revision surgeries owing to periprosthetic joint infection (PJI) has been increasing with the increase in the number of joint arthroplasties in recent years [1]. This problem needs to be solved urgently.

The first step in infection development in implants, such as artificial joints, is the adhesion of bacteria to the implant surface [2]. Once a biofilm is formed, the effectiveness of antibiotics is significantly reduced, which makes the treatment significantly more difficult. Therefore, antimicrobial treatment of implant surfaces to inhibit bacterial adhesion and growth is promising. One of the innovative solutions is coating of implant surfaces with antimicrobial substances. In Japan, a technique for coating titanium implants with iodine compounds has been developed and is currently being prepared for practical use.

Iodine compounds have been used as disinfectants since the 1800s [3]. Iodine compounds have a broad antimicrobial spectrum against not only general bacteria, but also *Mycobacterium tuberculosis*, viruses, and fungi.

Excessive use of antibiotics can cause bacterial resistance, but iodine-based disinfectants are effective against the resistant bacteria [4]. The major advantages of iodine compounds are that they are essential elements of thyroid hormones; possess a broad antibacterial spectrum and do not cause drug resistance; and have excellent antibacterial activity, biocompatibility, and low cytotoxicity.

The processing technology to coat iodine compounds on metals was originally developed by Professor Matsufumi Takaya of Chiba Institute of Technology as an antimicrobial surface treatment for aluminum [5]. Iodine compound-loaded processing technology (iodine coating) has been applied to titanium and titanium alloys, which are widely used in orthopedic implants [6]. Since the late 2000s, basic research and exploratory clinical studies have been conducted.

2 Surface Structure

In the iodine-loading process, a thin oxide film with microscopic pores is deposited on the surface of a titanium implant using anodization, and an iodine compound is loaded (impregnated) around the pores using anodic oxidation technology to impart metal antimicrobial activity. The number of micropores in the oxide film deposited using anodic oxidation technology was approximately 50,000 per 1 mm^2 and their thickness was approximately 5–10 μm. The formed micropores were impregnated with 10–15 μg of iodine compound per cm^2 using electrical force (Fig. 26.1) [6, 7]. The iodine compound is placed in the micropores by electrochemical treatment so that the structure of the iodine compound cannot be easily altered. The thickness of the oxide film, the number of micropores, and the amount of iodine compound loaded can be adjusted using the treatment method. It has also been found to be applicable to all surface treatments of implants, such as polishing, blasting, and plasma spraying (Fig. 26.2) [7].

T. Kabata (✉) · H. Tsuchiya
Department of Orthopaedic Surgery, Graduate School of Medical Sciences, Kanazawa University, Kanazawa, Ishikawa, Japan

Y. Nakashima (ed.), *Advances in Total Hip Arthroplasty*, https://doi.org/10.1007/978-981-97-2461-1_26

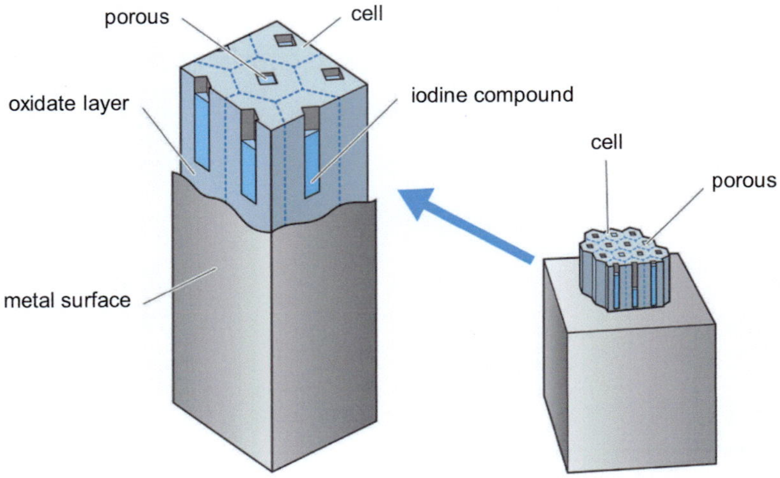

Fig. 26.1 Schema of iodine compound-loaded titanium (iodine coating). (Adapted from Tsuchiya et al. [6])

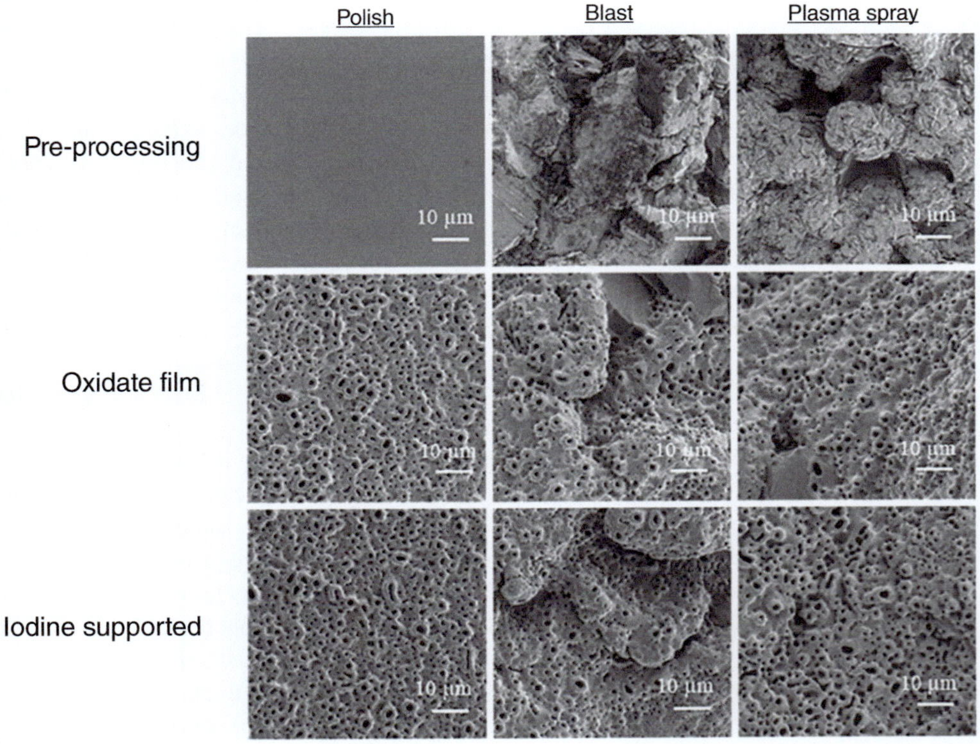

Fig. 26.2 The oxide film has approximately 50,000 micropores per mm^2, and iodine loading can be performed on any surface shape of polish, blast, or plasma spray. (Reprinted with permission from Ueoka et al. [7])

3 Basic Research

(a) Antibacterial activity

The antimicrobial activity of iodine compound-loaded processing has been demonstrated in in vitro and in vivo antibacterial tests [8, 9]. In vitro antimicrobial susceptibility testing was conducted in accordance with JIS Z 2801, and significant bacterial adhesion inhibition was observed against these bacteria, which are the typical causative agents of PJI, including methicillin-sensitive *Staphylococcus aureus* (MSSA), methicillin-sensitive *Staphylococcus epidermidis* (MSSE), *Escherichia coli*, Methicillin-resistant *Staphylococcus aureus* (MRSA), *Pseudomonas aeruginosa*, and *Candida albicans* (Fig. 26.3) [8, 9].

In vivo antibacterial studies were conducted using animal models. Various types of half pins were inserted into the femur of rabbits and their efficacy in preventing infection at 14 days postoperatively was verified. Iodine compound-loaded half pins showed significantly lower inflammatory and infection scores in histological evaluation around the puncture site and in the evaluation of infection findings [8]. In a rat femoral nail infection model using MSSA, MSSA was administered into the femoral bone marrow cavity, a Kirschner wire (K-wire) was inserted as an implant, and bacterial counts were

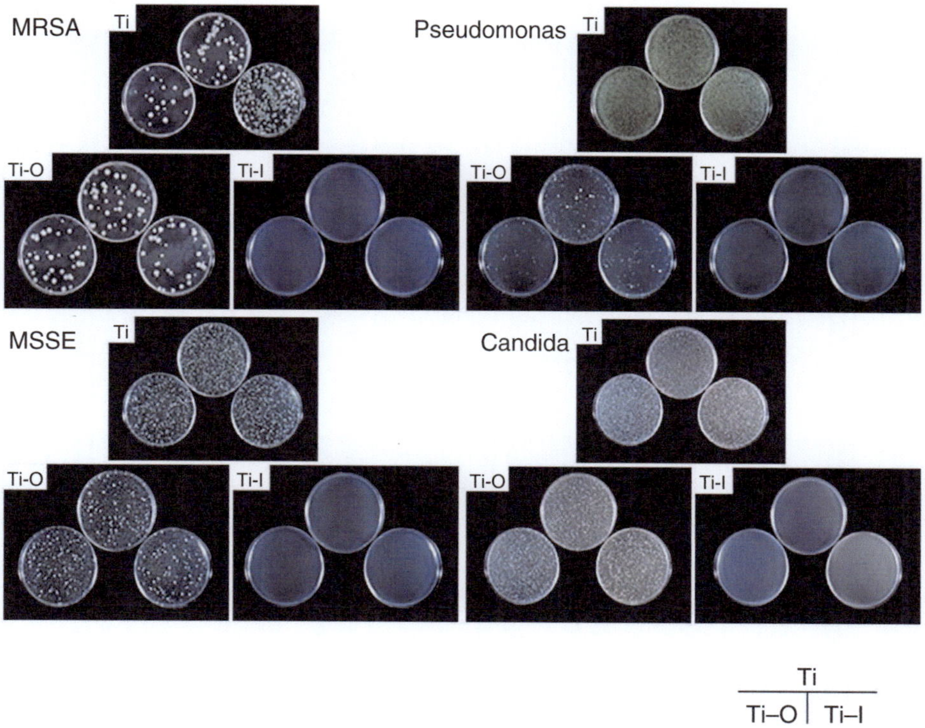

Fig. 26.3 The bacterial adhesion inhibitory effects of titanium (Ti), oxidized titanium (Ti-O), and iodine compound-loaded titanium (Ti-I) were compared. Significant inhibition of bacterial adhesion was observed against *Staphylococcus epidermidis*, methicillin-resistant *Staphylococcus aureus*, *Pseudomonas aeruginosa*, and Candida. (Reprinted with permission from Inoue et al. [9])

measured at 24, 48, and 72 h [9]. The number of bacteria adhering to the iodine-loaded K-wire was significantly reduced at all time points.

(b) Effect on biofilm formation
The inhibitory effect of iodine compound-loaded titanium on the deposited biofilm was demonstrated in vitro using MSSA, a biofilm-forming strain. The biofilms on the specimens were analyzed after MSSA biofilms were formed, and it was observed using scanning electron microscopy or fluorescence microscopy that the iodine compound-loaded titanium significantly inhibited biofilm deposition. The number of bacteria in the biofilm was also evaluated over time and significant reduction was observed (Fig. 26.4) [10].

(c) In vivo attenuation and antimicrobial effects of iodine compounds
Iodine compounds loaded on metal surfaces decay over time, and in vitro and in vivo studies have been conducted to evaluate the duration of in vivo decay and the antimicrobial effect associated with this decay [11]. The amount of iodine compounds on the surface of the specimens gradually decreased after immersion in PBS. However, approximately 30% of the iodine compounds remained on the surface of the implants after 365 days compared to the amount before immersion. In vivo, iodine compound-loaded metal specimens were implanted under the skin of the back, knee joint, and femur of rabbits, and the amount of iodine remaining on the surface was evaluated over time. The results showed that approximately 30% of iodine compounds remained on the surface of the specimens after 365 days compared to the amount before implantation [11]. In addition, inductively coupled plasma–mass spectrometry (ICP-MS) was performed on metal specimens implanted in rats to verify the amount of iodine compounds loaded on the surface as well as the mass spectrometry of all iodine compounds loaded using an alkaline dissolution of the entire oxide film. The results showed that 72% and 65% of iodine compounds remained at 4 and 8 weeks after implantation, respectively [12].
The antimicrobial activity of iodine compounds has also been verified under the condition that they decay over time. The antimicrobial activity of iodine compounds was evaluated by preparing test specimens in which the iodine compound loading was attenuated to 20% of the industrial standard value. The results showed that the iodine compounds significantly reduced the number of bacteria such as *S. aureus* and *E. coli* even after the loading was reduced [11]. In addition, antibacterial activity was observed against MSSA, MRSA, and *P. aeruginosa* in implants removed from rats after 8 weeks of in vivo implantation and attenuation of iodine compound loading [12].

(d) In vivo safety and bone affinity
The safety of titanium loaded with iodine compounds at the cellular level was verified using the colony formation method, which is a typical method for cytotoxicity testing. When Chinese hamster lung-derived fibroblast cells (V79) were seeded and cultured on stainless steel, pure titanium, and iodine compound-loaded titanium, the

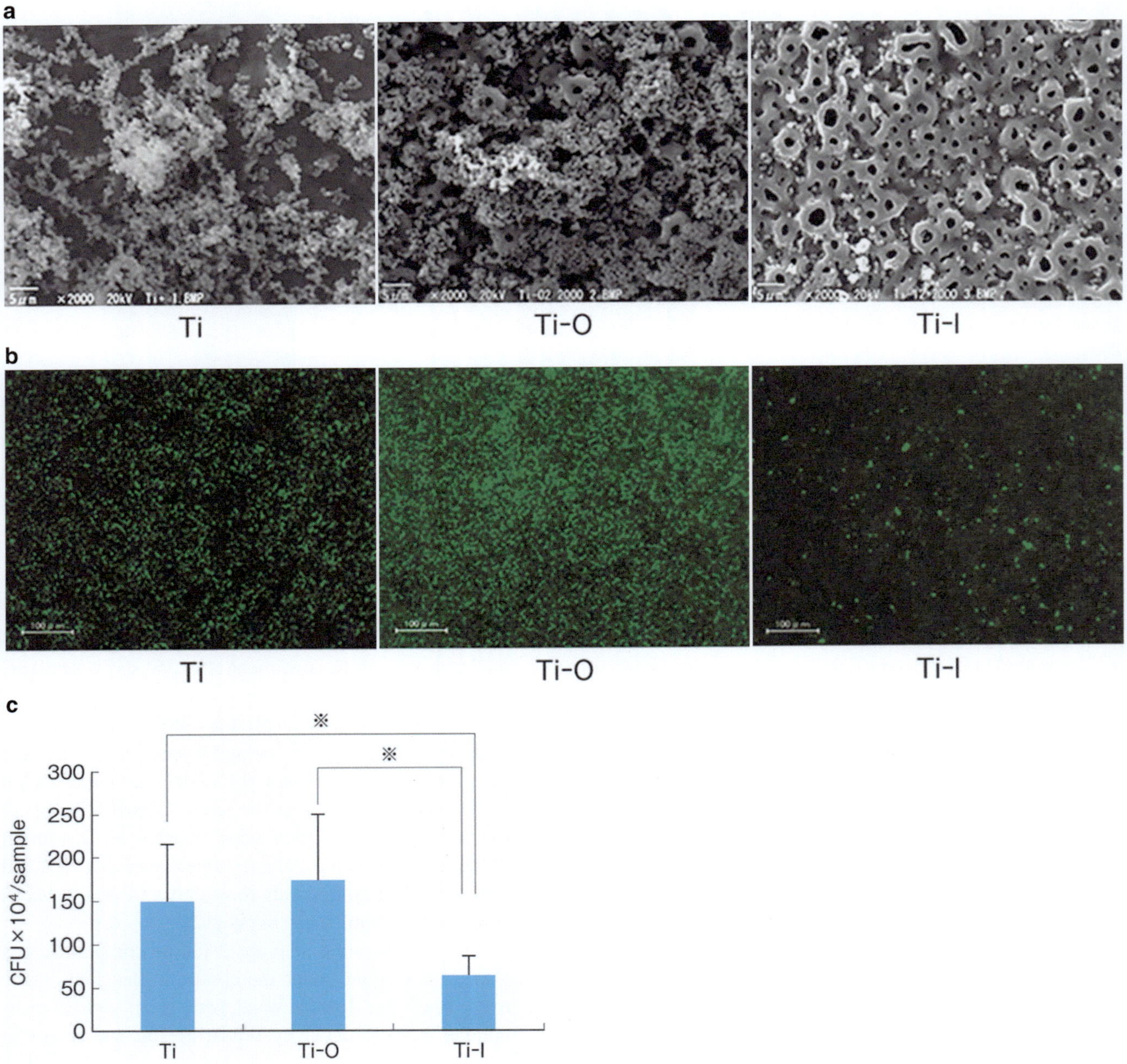

Fig. 26.4 (**a**) Biofilm formation observed using scanning electron microscopy. (**b**) Biofilm formation observed using fluorescence microscopy. Compared to titanium (Ti) and titanium oxide (Ti-O), iodine-compound-loaded titanium (Ti-I) formed fewer biofilms. (**c**) Number of bacteria in biofilm. Bacteria in the biofilm were cultured for 72 h, and the number of bacteria in the biofilm was significantly reduced by the iodine compound-loaded titanium (Ti-I). (Reprinted with permission from Inoue et al. [10])

same level of V79 cell colony formation was observed in all metal surfaces [8]. This means that there is no increase in cytotoxicity compared to existing implant materials such as titanium and stainless steel.

In the bone affinity study, titanium implants were inserted into the femur of a rabbit, and bone formation capacity around the implants was evaluated. As a result, iodine-loaded implants showed higher bone formation capacity around the implants than conventional titanium implants (Fig. 26.5) [13]. This suggests that loading iodine compounds onto conventional titanium implants does not reduce the bone affinity of titanium.

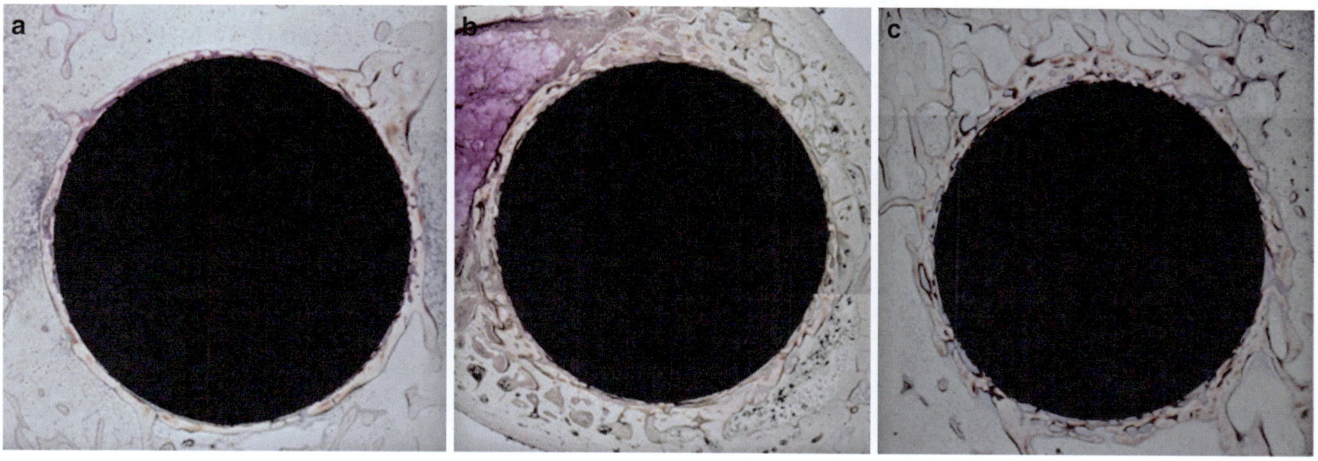

Fig. 26.5 Vigorous bone formation around iodine-loaded titanium implants. (**a**) Titanium implant (Ti). (**b**) Titanium implant with oxide film (Ti-O). (**c**) Iodine compound-loaded titanium implant (Ti-I). (Reprinted with permission from Taga et al. [13])

4 Clinical Research

Based on the results of the basic research described above, an exploratory clinical study of iodine compound-loaded titanium implants was conducted at the development center with the approval of a medical ethics committee in 2008. In this study, THA was prophylactically used in patients with severe immunosuppression and susceptibility to infection, as represented by patients with cancer, and therapeutically used in patients with PJI or pyogenic arthritis, and PJI and suppurative arthritis.

(a) Antibacterial effects
Iodine-loaded THA implants were used in 30 joints (primary revision: 1 joint, secondary revision: 13 joints, prophylactic use: 16 joints), and short-term clinical outcomes (mean postoperative follow-up: 33 months) have been reported [14]. In 14 joints used for treatment after PJI, the causative organisms were methicillin-resistant coagulase-negative *Staphylococcus aureus* (MRCNS) (7), MRSA (1), and culture-negative (6). In all cases, there was no recurrence of PJI and the infection was successfully stopped. No cases of SSI/PJI were observed in any of the 16 prophylactic joints during the follow-up period.

(b) Biological safety
It was hypothesized that the local toxicity of high concentrations of iodine in the body, especially on thyroid function, would be a concern when this implant is used in vivo.

Therefore, blood thyroid hormone (FT3 and FT4) and thyroid-stimulating hormone (TSH) levels at 1 month, 6 months, and 1 year postoperatively in patients who underwent iodine-loaded THA were measured. The results showed that none of the patients had abnormal thyroid hormone levels [14]. These results suggest that iodine-loaded implants are non-toxic to the body at high local iodine concentrations.

(c) Implant fixation
We investigated bone ingrowth/on growth in patients who underwent iodine-compound-loaded THA by monitoring the postoperative radiographic images over time. None of the patients required revision due to implant laxity during the follow-up period [14]. In THA, fine radiolucent lines were observed only in two cases in the acetabular side, but the alignment of the acetabular implants and femoral stems remained unchanged. All other acetabular implants and femoral stems showed good bone ingrowth or out growth [14].

References

1. Schwartz AM, et al. Projections and epidemiology of revision hip and knee arthroplasty in the United States to 2030. J Arthroplasty. 2020;35:S79–85.
2. Romanò CL, et al. Antibacterial coating of implants in orthopaedics and trauma: a classification proposal in an evolving panorama. J Orthop Surg Res. 2015;10:157.
3. Naito S. Utilization technology of iodine disinfectant for microbiological control in food factories. Shokuhin BUnseki Kaihatsu Center Mail Magazine 2012; 5. http://www.mac.or.jp/mail/120501/03.shtml, in Japanese.
4. Manual of clinical microbiology, 11th ed. Jorgensen JH, et al, editors. ASM Press; 2015, pp. 184–186.
5. Takaya M. Oxide film structure of iodine or iodine compound impregnated metal material base material, its formation method and applied articles having such film structure. Patent No. 2932437, 1999. https://www.j-platpat.inpit.go.jp/p0200, in Japanese.
6. Tsuchiya H, et al. Innovative antimicrobial coating of titanium implants with iodine. J Orthop Sci. 2012;17:595–604.
7. Ueoka K, et al. The feasibility of iodine-supported processing for titanium with different surfaces. J Orthop Sci. 2020;25:1095–100.

8. Shirai T, et al. Antibacterial iodine-supported titanium implants. Acta Biomater. 2011;7:1928–33.
9. Inoue D, et al. Iodine-supported titanium implants have good antimicrobial attachment effects. J Orthop Sci. 2019;24:548–51.
10. Inoue D, et al. Inhibition of biofilm formation on iodine-supported titanium implants. Int Orthop. 2017;41:1093–9.
11. Kato T, et al. Temporal attenuation of iodine content and its effect on the antibacterial activity of iodine-supported titanium implants. J Microb Biochem Technol. 2016;8:285–9.
12. Ueoka K, et al. Antibacterial activity in iodine-coated implants under conditions of iodine loss: study in a rat model plus in vitro analysis. Orthop Relat Res. 2021;479:1613–23.
13. Taga T, et al. Comparison with the osteoconductivity and bone-bonding ability of iodine supported titanium, titanium with porous oxide layer and the titanium alloy in the rabbit model. J Orthop Sci. 2018;23:585–91.
14. Kabata T, et al. Iodine-supported hip implants: short term clinical results. Biomed Res Int. 2015;2015:368124.

Silver-Containing Hydroxyapatite-Coated Artificial Hip Joints

27

Masaya Ueno and Masaaki Mawatari

1 Development of Silver-Containing Hydroxyapatite-Coated Artificial Hip Implants

All medical devices used in human body, including artificial joints, are associated with the risk of infection because immunosuppression can occur locally around the inserted implant and bacteria can adhere to the implant surface [1, 2]. Implant-related infections are characterized by biofilm formation on the implant surface. Once a biofilm has formed, it is considerably more difficult to remove the bacteria from the implant [3].

Although temporary implants, such as catheters, can be removed and the infection can be cured, prostheses are often difficult to remove because they are intended to be permanent and can easily become intractable, resulting in serious complications that require multiple surgical procedures. According to the "Survey of postoperative infections after joint replacement and spinal instrumentation," an academic research project of the Japanese Orthopaedic Association, the incidence of postoperative infection was reported as 134/9882 (1.36%) in joint replacement [4]. Infection, often due to staphylococci, after initial total hip arthroplasty (THA) occurs in approximately 0.2–2.5% of patients, not only as a surgical-site infection but also as a delayed infection [5]. Additionally, the infection rate is higher in revision THA than that in initial THA [6]. The risk of infection is increased in immunocompromised patients, and those undergoing prolonged or multiple surgeries. To reduce this risk, the surgical environment should be improved, patients' preexisting conditions should be controlled, preoperative procedures should be performed, intraoperative washing should be performed, appropriate antimicrobial agents should be used, and antimicrobial agents should be added to the cement.

In recent years, research on and the development of antibacterial materials for implants have been actively pursued as a preventive measure against prosthetic joint infection (PJI) to inhibit bacterial adhesion and biofilm formation on the surfaces of joint prostheses [7–11]. The only cementless antibacterial hip prosthesis available for clinical use in Japan is the silver-containing hydroxyapatite-coated hip prosthesis, developed by Saga University and Kyocera Corporation. This section summarizes the research results on implant development and describes the clinical results to date.

2 Development of an Antibacterial Joint Prosthesis

The key to the development of an antimicrobial artificial hip joint is to impart antimicrobial properties to the implant itself, thus, we focused on silver, an inorganic antimicrobial agent. Silver has been known for its antimicrobial activity since ancient times, and Hippocrates has been considered to have used it to treat wounds [12]. Among inorganic antimicrobial agents, silver has a broad antimicrobial spectrum and excellent antimicrobial performance; moreover, it is relatively safe and less to cause bacterial resistance. We focused on the application of hydroxyapatite (HA), which is a major component of vertebrate bone and has good biocompatibility and excellent osteoconductivity with respect to metal surfaces. We chose the flame spraying method for the coating, which has a proven performance in HA coating for orthopaedic implants, and developed the silver-containing hydroxyapatite (silver HA) thermal spraying technology (AG-PROTEX®) [13]. This technology has been patented based on four cases in Japan, three in the United States, one in Europe, and one in China. We were the first to commercialize an antimicrobial artificial hip joint with a silver HA coating on the bone contact surface of the artificial hip joint [14–16].

M. Ueno · M. Mawatari (✉)
Saga University, Saga, Japan

Y. Nakashima (ed.), *Advances in Total Hip Arthroplasty*, https://doi.org/10.1007/978-981-97-2461-1_27

Silver HA coating is a film which formed by flame-spraying HA powder containing silver oxide. Silver ions are eluted from the silver HA coating to produce an antimicrobial effect. We evaluated the in vivo antimicrobial ability, biofilm formation-inhibitory effect, bone conduction ability, and synostosis ability, as well as the safety of serum silver concentrations and acute and subacute toxicity. After clinical trials, the device was approved as an antimicrobial hip joint prosthesis by the Ministry of Health, Labor and Welfare in September 2015 and has been used clinically since April 2016.

The most remarkable feature of the silver HA coating is that it combines the antibacterial properties of silver ions with the osteoconductive properties of HA. Biofilms are involved in the onset and severity of PJI related to artificial joint surgery [17]. Bacterial adhesion to the substrate surface is the first step in the biofilm formation process, and reducing this adhesion is considered to inhibit the subsequent steps involved in colony and biofilm formation. The silver HA coatings have osteoconductive properties due to HA [18] and the antimicrobial ability of silver ions promotes aseptic bonding between the silver HA coating and bone during the early postoperative period [19], and it is considered to inhibit bacterial penetration into the bone–implant interface. Thus, silver HA coatings are expected to reduce the risk of infection by inhibiting biofilm formation and bacterial invasion into the bone–implant interface during the early postoperative period.

3 Performance and Safety of Silver HA Coating

Prior to clinical evaluation, the properties and safety of the silver HA coating in basic experiments are described.

(a) Silver ion elution characteristics
Silver HA coating is a technology in which HA powder containing silver oxide is flame-sprayed at 2700 °C to form a film. Silver ions are eluted from this film and exhibit antimicrobial properties [13, 20]. Silver HA is slightly yellowish, as opposed to the color of HA. No obvious difference in appearance or in the structure of the surface observed by electron microscopy (Fig. 27.1). The results based on a quasi-in vivo environment showed that the dissolution of silver ions is slow after an initial burst, but that the slow release continues even after 6 months of immersion [14].

(b) Antibacterial activity
In a film adhesion test to evaluate the antibacterial activity of the silver HA coating in a quasi-in vivo environment based on the JIS Z 2801 standard, six bacteria namely methicillin-resistant *Staphylococcus aureus* (MRSA), *Staphylococcus aureus*, *Staphylococcus epidermidis*, *Pseudomonas aeruginosa*, methicillin-resistant *Staphylococcus epidermidis* (MRSE), and Escherichia coli were tested, and antimicrobial activity was observed against all bacteria (Fig. 27.2). In an in vivo antimicrobial efficacy study using rat tibiae, the number of viable MRSA in silver HA-coated implants was significantly lower than that in HA-coated implants [19]. Further, in a model in which infection was established via the blood on the implants in rats, the number of bacteria was reduced [21].

The barrier function against bacteria at surgical wound sites is re-established in approximately 48 h by the body's own repair mechanism by the physiological closure of the surgical wound. Practical guidelines for the appropriate use of antimicrobial agents for the prevention of postoperative infection, proposed by the Japanese Society of Chemotherapy and the Japanese Society for Surgical Infection, recommend that antimicrobial agents be administered for ≤48 h after total joint replacement surgery [22]. In basic experiments, antimicrobial agents have been shown to have antimicrobial efficacy in vivo for ≤3 days. During the high-risk period for postoperative infection (≤48 h of surgery), antimicrobial agents can be used to kill bacteria that are introduced during surgery and invade the surgical wound, which can cause an early infection, and these compounds are expected to be effective in this scenario.

(c) Inhibition of biofilm formation
In vitro and in vivo experiments using MRSA showed that the biofilm area on the specimens after 14 days of incubation was significantly reduced on silver HA compared to that on HA [23]. The biofilm properties of the specimens indicate that only immature biofilms can form on the silver HA-coated specimens, indicating significant enhancement of the antibacterial effect of vancomycin [24, 25].

(d) Bone conductivity and synostosis
Cementless hip joint prostheses work for a long period, and it is very important to ensure synostosis ability. Therefore, in the development of silver HA-coated hip prostheses, it is necessary to obtain the same synostosis ability as with a conventional HA coating. An in vitro evaluation of cell adhesion and differentiation using an osteoblast progenitor cell line (MC3T3-E1) showed that silver inhibits osteoblast adhesion in a concentration-dependent manner [26]. In vivo, a 3% silver HA coating in the rat tibiae showed same level bone formation as a HA coating, but the coating containing higher concentrations of silver was showed significantly lower bone formation [18]. In a pull-out strength experiment using

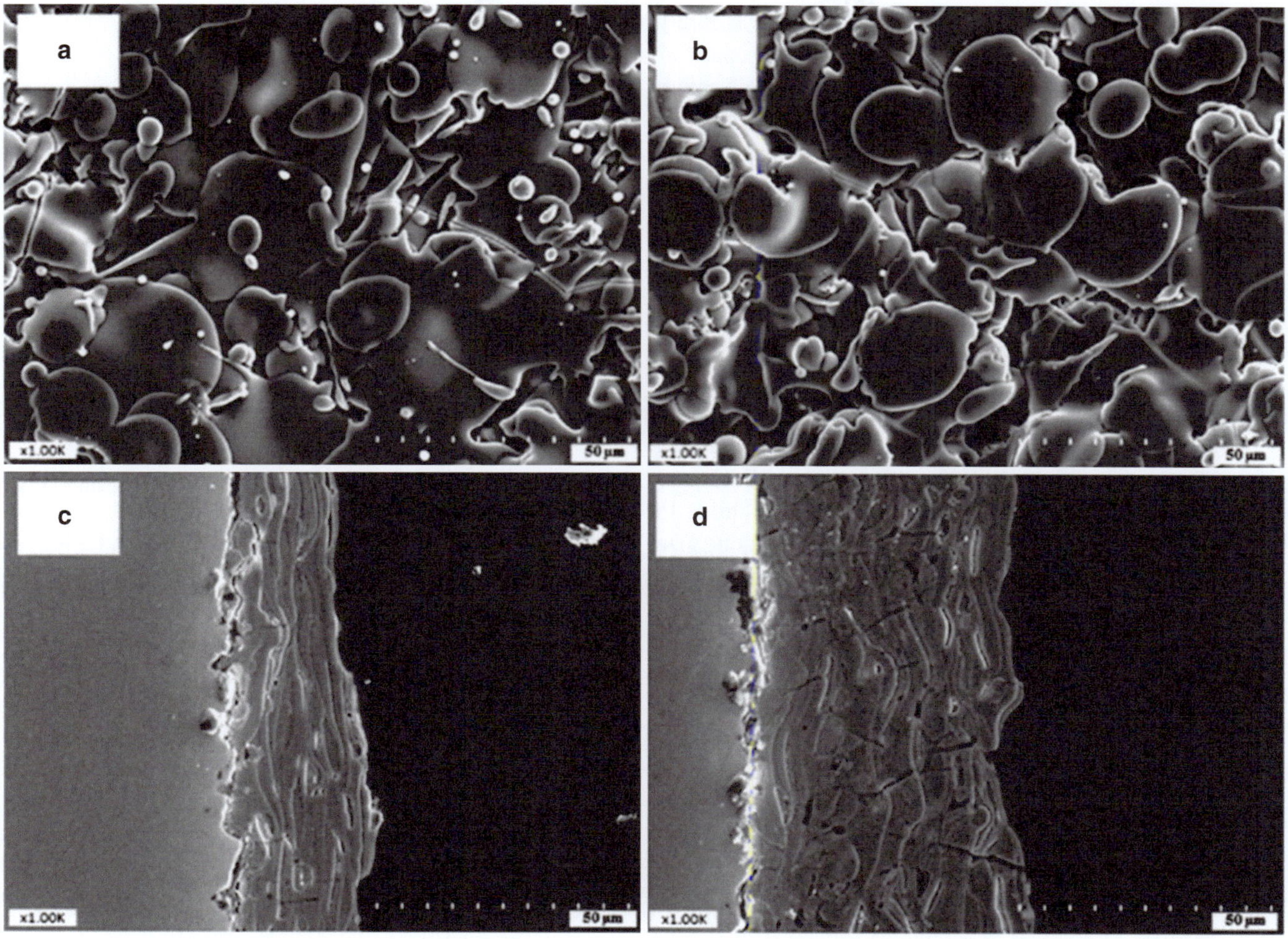

Fig. 27.1 Electron micrographs of hydroxyapatite (HA) coating and silver HA coating. (**a**, **c**) HA coating; (**b**, **d**) silver HA-coated surfaces. (Reprinted from Akiyama et al. [19])

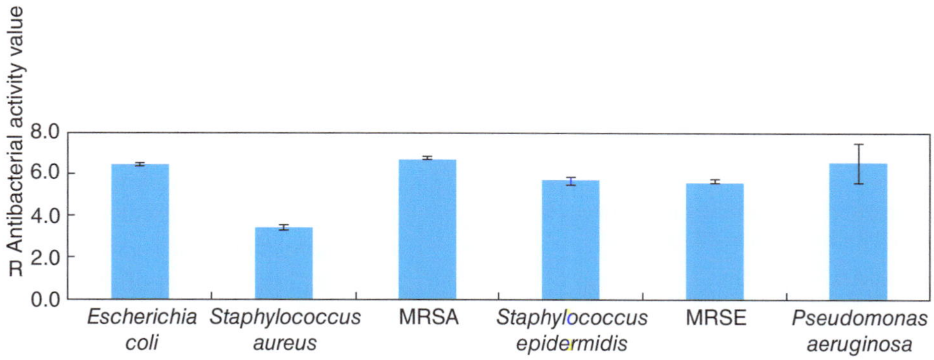

Fig. 27.2 In vitro antibacterial evaluation of silver hydroxyapatite (HA) coating. The antibacterial activity of the silver HA coating was measured using the ISO 22196 antibacterial testing method. The antibacterial activity value was the logarithm of the number of viable bacteria. Number 2 indicates that the growth of bacteria was inhibited by ≥99% (≤1%). High antibacterial activity was observed against *Escherichia coli*, *Staphylococcus aureus*, methicillin-resistant *Staphylococcus aureus*, *Staphylococcus epidermidis*, methicillin-resistant *Staphylococcus epidermidis*, and *Pseudomonas aeruginosa*

the rat femurs, 3% silver HA remained as rigid as HA, [26] and it was confirmed that the silver HA coating has synostosis ability equivalent to those of the HA coating when the concentration is <3%.

(e) Biological safety

Implants were placed in the bone marrow cavity of the rat tibiae to evaluate the accumulation of silver in the organs. The accumulation of silver in the brain, liver, kidney, and spleen at concentrations of 2% was rarely observed even 12 weeks after surgery, and no hepatic or renal damages were observed [27]. Biological safety studies based on ISO 10993-1 were also conducted and no toxicity was observed in any of these studies.

4 Clinical Evaluation of Antimicrobial Hip Prostheses

(a) Characteristics of antimicrobial hip prostheses
The silver HA-coated cementless antibacterial hip prosthesis consists of a hemispherical titanium alloy cup with 0.5 mm porous silver HA coating on the external surface, and a fit-and-fill-shaped titanium alloy stem with a porous coating on the proximal one-third. In 2016, only silver-HA full-coated stems with a one size larger distal diameter were available. From April 2019, partial coat stems with silver HA coating only on the proximal one-third of the porous process have also been available.

(b) Results of clinical trials
After informed consent was obtained, initial hip arthroplasty was performed in 20 patients with 20 hips, and the patients were followed up for 1 year after surgery (Saga University Ethics Committee, approval number: 0344-CT021). The patient background, primary disease, and pre-existing medical conditions are shown in Table 27.1. The silver content of the implants ranged from 1.9 to 2.9 mg (mean 2.28 mg).

Clinical evaluation using the Japanese Orthopeadic Association hip function criteria (JOA hip score) showed that the preoperative hip score improved from 47.1 to 85.8 at 1 year after surgery ($p < 0.01$). A radiographic evaluation showed that there was no problem with fixation of the cup and stem, and spot welds were observed

Table 27.1 Demographic descriptions of the patient cohort, the indication for THA, and the total silver mass of the prosthesis

Number of cases	20 cases (male: 6, female: 14)		
Age	77.1 (range: 66~86)		
Medical history	Hepatitis: 5	Indication for THA	Osteoarthritis with subluxation: 13
	Malignant tumor: 4		Primary osteoarthritis: 4
	Chronic renal failure: 1		Avascular necrosis femoral head: 1
	Diabetes mellitus: 2		Osteoarthritis after tuberculosis: 1
	Psoriasis vulgaris: 1		Osteoarthritis after pyogenic arthritis: 1
	Tuberculosis of hip 1例		

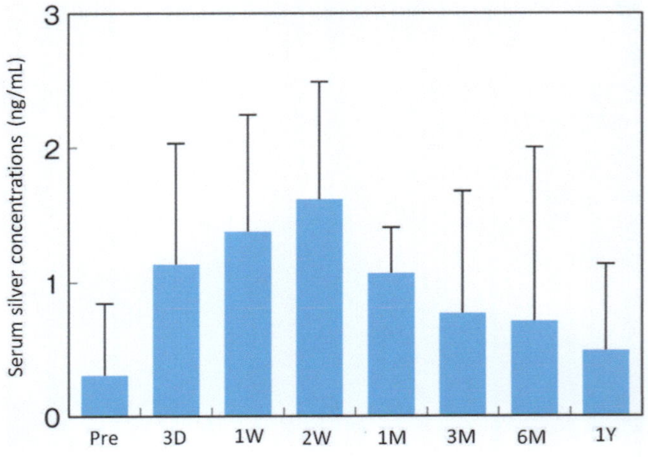

Fig. 27.3 Blood silver concentrations in clinical trials. Blood silver concentrations in 20 patients who underwent total hip arthroplasty (THA) in a clinical trial. The median silver concentration at each time point peaked 2 weeks after surgery and gradually decreased thereafter. The median silver concentration did not increase to the upper limit of the normal range (14 ng/mL)

on the stem side in 19 of 20 patients. Blood silver levels peaked at 1–2 weeks after surgery in most cases, increasing from 0.6 (0.2–2.3) ng/mL, before surgery, to 1.4 (0.6–4.3) ng/mL and 1.6 (0.7–3.8) ng/mL, 1 and 2 weeks after surgery, respectively. These values were much lower than the upper normal limit of 14 ng/mL and decreased to preoperative values 1 year after surgery (Fig. 27.3). Blood biochemical tests showed no elevation in the hepatobiliary enzymes (AST, ALT, γ-GTP), except in one patient with acute cholecystitis on the 9th postoperative day, as well as no deterioration in the renal function (Cr, BUN), including in those patients who had been on dialysis before surgery. There was one case of delayed wound healing due to subcutaneous hematoma and one case of postoperative dislocation, but no cases of deep vein thrombosis, implant migration, fracture, or peripheral neuropathy. A typical side effect of silver is argyria, a blue-gray discoloration of the skin caused by silver deposition, which can sometimes be severely disfiguring [28]; no cases of argyria were observed in the study. Patients with a history or comorbidities that would have increased the risk of infection were included, no case of postoperative infection were observed [29].

(c) Cementless silver-containing hydroxyapatite-coated hip prosthesis.
In our department, we initially started to use this implant in patients at risk of infection (history of hip arthritis, hip trauma, or surgery or rheumatoid arthritis, older age, diabetes, liver disorder, etc.) and in patients who wanted to use it. After the partial coat stem (Fig. 27.4) became available, we began its use in all patients with primary THA. Based on a comparison of the HA-coated implants

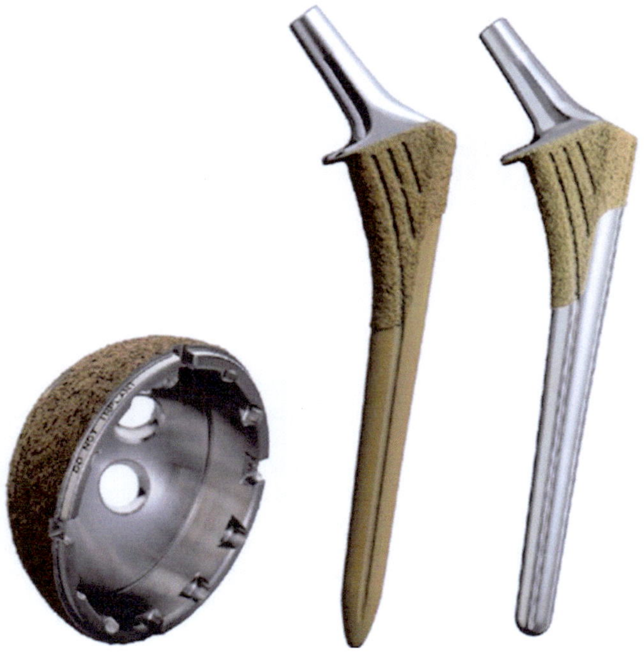

Fig. 27.4 Silver-containing hydroxyapatite-coated artificial hip joint AG-PROTEX®. From left to right: cup, fully coated stem, and partially coated stem that were coated with silver hydroxyapatite on the outer surface of the cup, proximal third of the stem, and all surfaces of the stem, respectively

with the same shape used before the introduction of the silver HA-coated cementless antibacterial THA, the incidence of PJI decreased from 40 in 6921 (0.6%) before introduction to 5 in 1704 (0.3%) after introduction. It was also reported that the incidence of infection does not increase even when PJI-risk cases are included [30], and this is considered to help with PJI preventative measures.

Silver HA-coated implants have potential applications not only for the prevention of PJI but also for implant preservation when it occurs and for the treatment of infection with adult-onset pyogenic hip arthritis. The Infectious Diseases Society of America treatment guidelines [31] recommend the use of antimicrobial-loaded acryl cement for primary revision surgery. The use of silver HA-coated implants as an alternative to antimicrobial-containing cement is anticipated to shorten the waiting period before second-stage surgery and enable one-stage cementless joint replacement.

Although the number of cases was small, all three patients who underwent one-stage reconstruction of the acetabulum silver HA-coated implants and eight patients who underwent two-stage revision hip arthroplasty had controlled infection and no laxity of the implants. In addition, no recurrence of infection was observed in adult patients with pyogenic hip arthritis who underwent curettage, cleaning, and insertion of an antimicrobial cement mold, followed by two-stage surgery using silver HA-coated implants [32]. However, a large number of patients and long-term studies are required to verify the clinical efficacy against PJI, then we continue to evaluate.

Currently, basic research is being conducted to evaluate the antimicrobial resistance of implants to late-stage infection and to determine whether HA remains antimicrobial even after a certain amount of time has elapsed since the initial burst. An analysis of extracted implants coated with HA, via a flame spraying method similar to the one used for silver HA coating, showed that HA remains for more than 10 years, and silver HA coating can be expected to have an antimicrobial effect for more than 10 years, leading to long-term antimicrobial ability. In addition, a silver HA-coated spinal cage (Resitage Box Gauge®) has been available for clinical use since April 2020, and we plan to continue the evaluation of this cage as well.

References

1. Zimmerli W, et al. Pathogenesis of implant-associated infection: the role of the host. Semin Immunopathol. 2011;33:295–306.
2. Higgins DM, et al. Localized immunosuppressive environment in response to implanted biomaterials. Am J Pathol. 2009;175:161–70.
3. von Eiff C, et al. Pathogenesis of infections due to coagulasenegative staphylococci. Lancet Infect Dis. 2002;2:677–85.
4. Yamamoto K, et al. Epidemiology of surgical site infection in implant surgery. Seikei Saigai Geka. 2010;53:419–25. in Japanese
5. Kono S, et al. Infection prevention of artificial head replacement using silver-containing hydroxyapatite-coated cementless antimicrobial stem – a multicenter safety evaluation study. Seikei Saigai Geka. 2019;68:34–6. in Japanese
6. Leekha S, et al. Should national standards for reporting surgical site infections distinguish between primary and revision orthopedic surgeries? Infect Control Hosp Epidemiol. 2010;31:503–8.
7. Norowski PA Jr, et al. Biomaterial and antibiotic strategies for peri-implantitis: a review. J Biomed Mater Res B Appl Biomater. 2009;88:530–43.
8. Ando Y, et al. Development of antimicrobial biomaterials. Seikei Saigai Geka. 2010;53:467–75. in Japanese
9. Neut D, et al. A gentamicin-releasing coating for cementless hip prostheses – longitudinal evaluation of efficacy using in vitro. J Biomed Mater Res A. 2012;100:3220–6.
10. Nablo BJ, et al. Inhibition of implant-associated infections via nitric oxide release. Biomaterials. 2005;26:6984–90.
11. Shirai T, et al. Antibacterial iodine-supported titanium implants. Acta Biomater. 2011;7:1928–33.
12. Alexander JW. History of the medical use of silver. Surg Infect (Larchmt). 2009;10:289–92.
13. Noda I, et al. Development of novel thermal sprayed antibacterial coating and evaluation of release properties of silver ions. Appl Biomater. 2009;89:456–65.
14. Noda I, et al. Development of an artificial hip joint with antimicrobial performance that is expected to reduce the risk of infection. J Soc Inorg Mater Japan. 2021;28:159–63. in Japanese
15. Ueno M, et al. Development of silver HA-coated joint prosthesis. Kansetsu Geka. 2017;36:458–62. in Japanese

16. Noda I, et al. Silver-containing apatite coating. Biomaterials Seitai Zairyou. 2011;29:266–70. in Japanese
17. Dougherty SH. Pathobiology of infection in prosthetic devices. Rev Infect Dis. 1988;10:1102–17.
18. Yonekura Y, et al. Osteoconductivity of thermal-sprayed silver-containing hydroxyapatite coating in the rat tibia. J Bone Joint Surg Br. 2011;93:644–9.
19. Akiyama T, et al. Silver oxide-containing hydroxyapatite coating has in vivo antibacterial activity in the rat tibia. J Orthop Res. 2013;31:1195–200.
20. Ando Y, et al. Effect of bacterial media on the evaluation of the antibacterial activity of a biomaterial containing inorganic antibacterial reagents or antibiotics. Biocontrol Sci. 2010;15:15–9.
21. Kobatake T, et al. Antibacterial activity of Ag-hydroxyapatite coating against hematogenous infection by methicillin-resistant Staphylococcus aureus in the rat femur. J Orthop Res. 2019;37:2655–60.
22. The Japanese Society for Chemotherapy/Japanese Society for Surgical Infectious Diseases, Committee for the Development of Guidelines for the Appropriate. Use of Antimicrobial Agents for Prevention of Postoperative Infection. Practical guidelines for the appropriate use of antimicrobial agents for prevention of postoperative infection; 2016. http://www.chemotherapy.or.jp/guideline/jyutsugo_shiyou_jissen.pdf, in Japanese.
23. Ueno M, et al. Silver-containing hydroxyapatite coating reduces biofilm formation by methicillin-resistant Staphylococcus aureus in vitro and in vivo. Biomed Res Int. 2016;2016:80670597.
24. Hashimoto A, et al. The combination of silver-containing hydroxyapatite coating and vancomycin has a synergistic antibacterial effect on. Bone Joint Res. 2020;9:211–8.
25. Hashimoto A, et al. Time-dependent efficacy of combination of silver-containing hydroxyapatite coating and vancomycin on methicillin-resistant. BMC Res Notes. 2021;14:81.
26. Eto S, et al. Silver oxide-containing hydroxyapatite coating supports osteoblast function and enhances implant anchorage strength in rat femur. Orthop Res. 2015;33:1391–7.
27. Tsukamoto M, et al. Acute and subacute toxicity in vivo of thermal-sprayed silver containing hydroxyapatite coating in rat tibia. Biomed Res Int. 2014;2014:902343.
28. Bouts BA. Images in clinical medicine. Argyria. N Engl J Med. 1999;340:1554.
29. Eto S, et al. First clinical experience with thermal-sprayed silver oxide-containing hydroxyapatite coating implant. J Arthroplasty. 2016;31:1498–503.
30. Shunsuke K, et al. Infection prevention using silver-containing hydroxyapatite-coated antimicrobial cementless hip prosthesis. Hip Joint. 2019;45:351–4. in Japanese
31. Osmon DR, et al. Diagnosis and management of prosthetic joint infection: clinical practice guidelines by the Infectious Diseases Society of Clinical Infectious Diseases. Clin Infect Dis. 2013;56:e1–e25.
32. Hashimoto A, et al. First experience with a thermal-sprayed silver oxide-containing hydroxyapatite coating implant in two-stage total hip. Int J Surg Case Rep. 2020;77:434–7.

Shape of Cemented Cups and Shape & Surface Roughness of Cemented Stems

28

Koji Goto

1 Prototype of the Cemented Cup [1]

The history of successful cemented total hip arthroplasty (THA) began with the pioneering development of the low-friction arthroplasty in England by Sir Dr. Charnley (1911–1982), who combined an ultrahigh molecular weight polyethylene (UHMWPE) cup with a 22.2 mm nonmodular metal head. This UHMWPE cup is the prototype of the present cemented cups. Charnley, born in 1911, moved to Wrightington Hospital, West Lancashire, in 1958, to set up a hip surgery center where he conducted extensive research on hip biomechanics. He continued performing basic research on THA in a hospital laboratory. To mimic the very low coefficient of friction (0.001) of the cartilage-to-cartilage sliding surfaces of biological joints, he thought that a material with a smooth surface (low coefficient of friction) that did not rely on joint fluid lubrication would be most suitable to construct the sliding surface of the joint prosthesis; he thus introduced polytetrafluoroethylene (PTFE) cups. However, PTFE wear debris was so large and plentiful, and the tissue reaction was so strong that most cases required early revision surgery. This led Charnley to recognize the importance of wear debris, and he discontinued the use of PTFE in 1961. While searching for other materials, he discovered UHMWPE, which, despite having a coefficient of friction more than five times higher than that of PTFE, had 500–1000 times higher wear resistance. In addition, boundary lubrication by the joint fluid promoted reduced friction for UHMWPE. In 1962, Charnley began to perform cemented THA with UHMWPE cups and 22.2 mm stainless steel heads. He termed THA with this combination as a low-friction arthroplasty, and excellent clinical results were reported in sequence in the 1970s [2–4]. Subsequently, low-friction arthroplasty was adopted worldwide. The 22.2-mm-inner diameter cup was the most commonly used cup for cemented THA until 1998, when highly crosslinked polyethylene (HXLPE) began to be applied clinically, and its excellent abrasion resistance was confirmed in the 2000s.

K. Goto (✉)
Department of Orthopaedic Surgery, Kindai University, Osaka, Japan
e-mail: kgoto@med.kindai.ac.jp

2 Why Were Cemented Cups Used in Combination with 22.2 mm Heads? [1]

In the 1960s, McKee-Farrar's design of metal-on-metal THA, which had achieved a certain level of clinical success, used a large-diameter 41-mm cobalt-chrome head [5]. However, Charnley's experiments confirmed that the frictional torque generated on the sliding surface was lower with a smaller-diameter head, and he thought that the smaller-diameter head was advantageous for the sliding surface of metal-on-UHMWPE, in which the wear was more problematic. Furthermore, basic experiments showed that using a smaller-diameter head allowed the use of a cup with a larger polyethylene (PE) thickness, which distributed the stress at the cement-bone interface more uniformly. A 22.2 mm diameter head was subsequently chosen as the smallest acceptable diameter to reduce the risk of dislocation, and Charnley believed that a smaller-diameter head would allow THA to be performed on smaller, more pathological acetabular bones.

3 Evolution of the Shape of Cemented Cups (Fig. 28.1)

In 1963, a year after introducing low-friction arthroplasty to medical practice, Charnley began using cups with wire markers to facilitate radiographic follow-up. After 1972, a cup with a posterior wall was used to increase the resistance against posterior dislocation, and after 1977, a flanged cup was made available to improve the effect of pressurization for bone cement intrusion. The flange could be cut with scissors to fit the shape of the acetabulum, allowing cup place-

Y. Nakashima (ed.), *Advances in Total Hip Arthroplasty*, https://doi.org/10.1007/978-981-97-2461-1_28

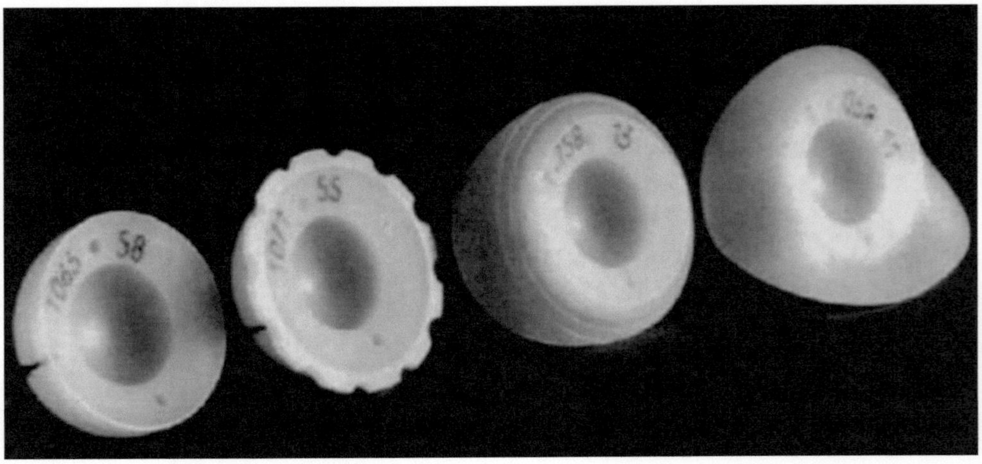

Fig. 28.1 The cups developed by Charnley. (Right: OGEE cup)

ment in the original acetabular position and stable cup retention until cementation. In 1982, the OGEE cup (Fig. 28.1) was introduced, which had a curved flange. It increased the pressure on the cement at the back of the cup. Basic experiments have shown that a flanged cup provides more sustained cement pressurization than a cemented cup without a flange [6], and clinical evaluations have also shown a decreased incidence of radiolucent lines around the cup on postoperative radiographs of the hip joint [7, 8]. However, research showed that cement pressurization with the flange became relatively less effective as the double-cementing method using a pressurizer [9] had become common as cementing techniques progressed [10]. Since the 1980s, it has become clear that an insufficient cement mantle at the loading zone could cause cement failure and acceleration of PE wear [11, 12] Therefore, some cemented cups have been modified with PMMA spacers on the outer surface of the cup to ensure a uniform and sufficient cement mantle [13]. However, although PMMA spacers provide a uniform cement mantle, the clinical results are not as good as those of cemented cups without PMMA spacers, which may lead to insufficient cementation [14]. HXLPE cemented cups have proven to have excellent long-term results, demonstrating the long-term wear resistance of PE and the reduction of osteolysis around the cups [15–17]. In accordance with this trend, the use of small-diameter heads has become less significant from the viewpoint of reducing frictional torque, and thinner PE-cemented cups have begun to be commonly used with larger-diameter heads to reduce the risk of dislocation. However, ISO 21535, the international standard for THA implants, specifies that the PE thickness of the PE liner placed in a cementless cup at the loading portion must be at least 5 mm, and the PE thickness of cemented cups must be at least 6 mm [18]. Conversely, metal-backed cemented cups, in which the outer surface of the PE was covered with metal, were widely used in the United States and other countries before the 1980s to prevent polyethylene deformation, but most of the reports indicated that their clinical results were not as good as those of polyethylene cups. Recently, metal-backed cemented cups have begun to be used in a limited number of cases, such as acetabular tumors requiring pelvic reconstruction, while metal-backed dual-mobility cemented cups are applied for cases with a high risk of dislocation [19, 20]. The cemented cups commonly used in recent years have not changed much in shape from those developed by Charnley, although the presence or absence of posterior walls, PMMA spacers, and flanges varies among the products (Fig. 28.2).

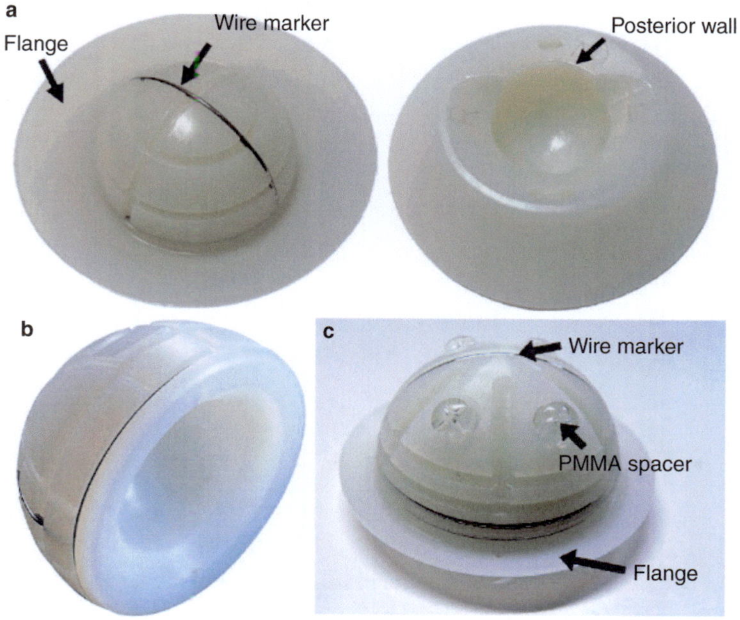

Fig. 28.2 Structure of the cemented cup. (**a**) Flanged cup (Kyocera), (**b**) Non-flanged cup (Kyocera), (**c**) Flanged cup with PMMA spacer (Stryker)

4 Shape & Surface Roughness of Cemented Stem (Fig. 28.3)

(a) Classification of the cemented stem (composite beam or taper-slip)

Since cemented stems were introduced by Charnley for low-friction arthroplasty in 1962, cemented stems of various shapes and sizes have been clinically applied worldwide; those with proven good long-term clinical results continue to be used while others got eliminated. Reports from the National Joint Registry in various countries have clarified the superiority or inferiority of different cemented stems in terms of clinical outcomes, and orthopedic surgeons now prefer to use cemented stems that have been proven to have good outcomes [21–23]. Cemented stems can be divided by shape into a composite-beam type (shape-closed design) and a taper-slip type (force-closed design). In the femoral bone marrow cavity, the stem sustains the load together with the cement mantle that surrounds it, with both cement-bone and stem-cement interfaces. In the composite-beam type, which includes stems with a collar or with a rough or uneven surface, the cement mantle and stem are integrated to sustain the load. Theoretically, the stem-cement interface around the stems of the composite-beam type is adherent, and the bone cement is susceptible to tensile and shear stresses. It is believed that stress relaxation by creep deformation of the bone cement does not occur around the stems of the composite-beam type. However, as described below, even in the case of composite-beam type stems, when the surface roughness is polished or smooth, a certain percentage of stem sinking due to creep deformation of the cement is observed, even when the stem is not loosened [24, 25]. Taper-slip type stems, which include polished tapered stems with no collar, allow the stem to sink by slipping in the cement mantle. Theoretically, the loading stress is converted into outward radial stress and sustained hoop stress on the cement-bone interface, which is repeated by an increase in hoop stress with the sinking of the stem and stress relaxation with creep deformation of the cement, resulting in a strong anchorage of the cement-bone interface. Studies using radiostereometric analysis (RSA) have shown that sinking of the Exeter stem, a typical taper-slip type stem, progresses rapidly within the first postoperative year, while progressing more slowly as decades pass [26].

(b) Cemented stem materials

The materials of cemented stems currently in clinical use can be broadly classified into stainless steel, cobalt chromium, and titanium alloy, each of which has specifications for composition and mechanical properties (Table 28.1). Although stainless steel has inferior biocompatibility and corrosion resistance than other materials, the original Charnley and Exeter stems which are historically successful, are made of stainless steel, as are many currently used cemented stems.

Cobalt-chromium alloy is widely used to construct the sliding surface of hip-joint prostheses because of its high elastic modulus, hardness, strength, corrosion resistance, and wear resistance. It is also widely used as a material for stems. Titanium alloys are widely used to produce cemented stems in Japan, but there is a concern that their relatively low modulus of elasticity could cause debonding of the stem-cement interface at the proximal portion of the stem. As many reports worldwide have indicated

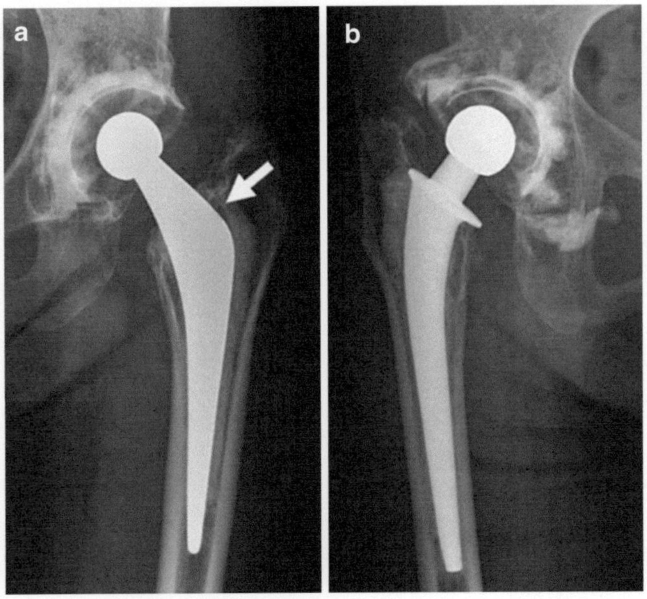

Fig. 28.3 Anteroposterior radiographs of the cemented implants in THA. (**a**) Taper-slip type C-stem (DePuy) at 21-years after cemented THA. The stem was not loosened, although a gap was observed at the stem-cement interface in the arrowhead area owing to sinking of the stem. (**b**) A composite-beam type PHS (type-6) stem (Kyocera) 21 years after cemented THA. No evidence of stem sinking or loosening was observed. *THA* total hip arthroplasty

Table 28.1 Mechanical properties of metallic biomaterials according to the ASTM[a] standards

	Co-Cr alloy		Ti-6Al-4 V	
	Casting material	Machining material		SUS316 L
Tensile strength (MPa)	> 655	> 855	> 853	> 482
0.2% proof stress (MPa)	> 448	> 310	> 770	> 172
Elongation (%)	> 8	–	10	> 40

[a]*ASTM* American Society for Testing and Materials

poor clinical results on cemented stems made of titanium alloys, the use of titanium alloy for cemented stems has been limited in countries outside of Japan. In Japan, the Ti-6Al-4 V alloy is the most commonly used, but the Ti-15Mo-5Zr-3Al alloy, which does not contain cytotoxic vanadium, is also used. Excellent long-term results with titanium alloy stems have been reported in Japan, possibly due to the low body weight and relatively high proportion of female Japanese THA patients [24, 27].

(c) Surface roughness of the cemented stem

The surface properties of cemented stems differ according to the surface roughness, and the processing methods also differ, as shown in Table 28.2.

Table 28.2 Definition of metal implant surface properties and manufacturing methods [31]

Surface properties	Processing method	Surface roughness (μm)
Polished (shiny)	Polishing	0–0.1
Smooth	Machining, grinding	0.1–0.4
Satin	Beads blasting	0.4–1.0
Matte	Grid blasting	1.0–2.5
Rough	Plasma spray	2.5–12.5
Textured	Machining, casting	12.5 <

In the taper-slip type, the stem-cement interface is subject to micromotion and sinking owing to debonding; but wear debris is usually not produced from polished surfaces made of stainless steel or cobalt chromium, which have excellent wear resistance. However, titanium alloys are less wear resistant and are therefore more prone to micromotion-induced wear (fretting wear), and when comparing polished (shiny) and smooth surfaces, wear is more likely produced from polished surfaces, as has been revealed by friction tests using PMMA pins and titanium plates (Fig. 28.4) [28]. Therefore, titanium alloys are not used as materials for taper-slip type stems. Conversely, in composite-beam type stems, the stem-cement interface is usually adhered, but if micromotion occurs owing to debonding of the interface, abrasive wear or fretting wear may be produced, resulting in a high risk of rapid loosening from osteolysis. In the history of cemented stem development, the original Charnley stem and CMK stem, which were composite-beam types, were originally modified from polished surfaces (Ra: 0.01–0.03 μm) to satin surfaces (Ra: 0.01–0.03 μm) to improve adhesion at the stem-cement interface. However, the satin-surface stems exhibited a higher incidence of loosening, and the polished surface was applied again. It has subsequently been proven that debonding and slight sinking of the stem-cement interface occur with creep deformation of the cement in the original Charnley stems, even if loosening does not occur [24, 25, 29]. Therefore, many of the currently used composite-beam type collared stems have polished surfaces (smooth surfaces for titanium alloy stems). However, Lubinus SP II (Link), which has a matte surface with a relatively large collar and an anatomical design, has been widely used in Scandinavia, and has proven to have excellent long-term results, indicating that even a stem that has strong bonding at the stem-cement interface can achieve excellent long-term results [30].

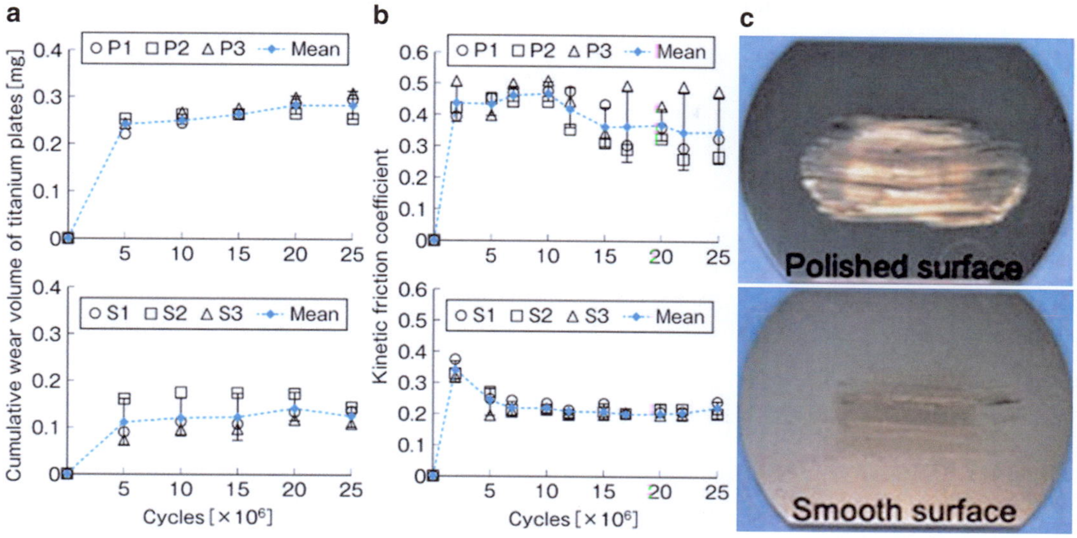

Fig. 28.4 Evaluation of titanium wear on polished and smooth surfaces in pin-on-disk friction tests [28]. Both the amount of wear (**a**) and the coefficient of friction (**b**) were larger on the polished surface, and friction marks could be clearly observed on the polished surface of the plate

5 Differences in the Clinical Results of Cemented Stem Depending on Taper Design

The currently used cemented stems (Table 28.3, Fig. 28.5) are generally double-tapered, with the anteroposterior and mediolateral widths decreasing distally, or triple-tapered, with cross-sectional anteroposterior width tapering in the medial direction, in addition to the double-tapered design. However, the pioneering cemented stem, the Charnley stem, had a constant anteroposterior width and a single tapered shape that tapered distally only in the mediolateral direction. The Charnley stem, which has a polished surface and the KC stem (Kyocera, Japan), made of titanium alloy with a single-taper shape similar to the Charnley stem, both have good long-term results [27, 32]. However, the double-taper cemented stem has better long-term results because the stress transmission to the cement mantle around the stem is more evenly dispersed, and the stress concentration on the cement mantle at the tip of the stem is reduced, where stress is more likely to be concentrated [24, 33, 34].

As shown in the results of finite element (FEM) analysis (Fig. 28.6), the distal portion of the stem with a cylindrical or a quite small taper angle tends to cause a greater stress concentration on the cement mantle at the distal portion of the stem. Therefore, the stem is more likely to loosen due to the breakage of the cement mantle at the distal portion of the stem, and its clinical results are therefore inferior to those of single-taper stems [35, 36]. FEM analysis (Fig. 28.6) also showed that the stress on the cement mantle and cortical bone at the distal end of the stem is greater in the medial direction, regardless of the taper geometry. Distal femoral cortical hypertrophy, a finding that indicates stress concentration at the distal portion, is more likely to occur in the medial direction and occurred more often when the stem was inserted in the valgus position [24, 37, 38]. Conversely, a study using RSA to compare the migration of Exeter stems (double-taper) and CPCS stems (triple-taper stem with a wider and more rectangular design proximally than the Exeter stem) showed no significant difference in sinking, whereas the Exeter stem tended to move to the retroversion and valgus position as compared to the CPCS stem [39]. Therefore, the triple-taper design is logically designed in that it relieves medial stress concentration at the distal portion of the stem; however, its clinical superiority over the double-

Table 28.3 Stem materials, surface roughness, tapered shape, and presence/absence of collar used in Japan and overseas

Implant name	Manufacturer	Material surface	Roughness	Taper shape	Collar
Exeter V40	Stryker	SUS[a]	Polished	Double	–
CPT	Zimmer	oCr alloy[b]	Polished	Double	–
C-stem	DePuy	SUS	Polished	Triple	–
CPCS	Smith & Nephew	CoCr alloy	Polished	Triple	–
MS-30	Zimmer	SUS	Polished	Triple	+
Lubinus SPII	Link	CoCr alloy	Matte	Double	+
CMK	Zimeer	SUS	Polished	Double	+
SC	Kyocera	CoCr alloy	Polished	Triple	–
SS hip taper	Kyocera	Ti alloy[A]	Smooth	Double	+
PHS type-6	Kyocera	Ti alloy[B]	Smooth	Triple	+

A&B Ti alloy, titanium alloy (A: Ti-15Mo-5Zr-3Al, B: Ti-6Al-4 V)
[a]*SUS* stainless steel
[b]*Co-Cr alloy* cobalt-chromium alloy

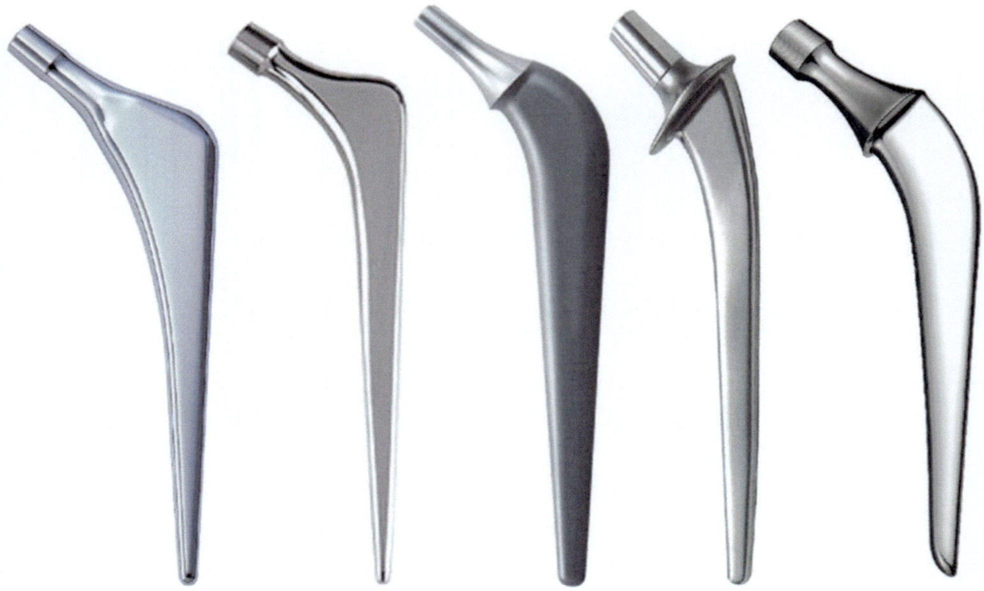

Fig. 28.5 Stems used in Japan and overseas. From left to right: SC (Kyocera), Exeter (Stryker), SS hip taper (Kyocera), PHS type 6 (Kyocera), and CMK (Zimmer). SC and Exeter are taper-slip types, SS hip taper, PHS type 6, and CMK are composite-beam types

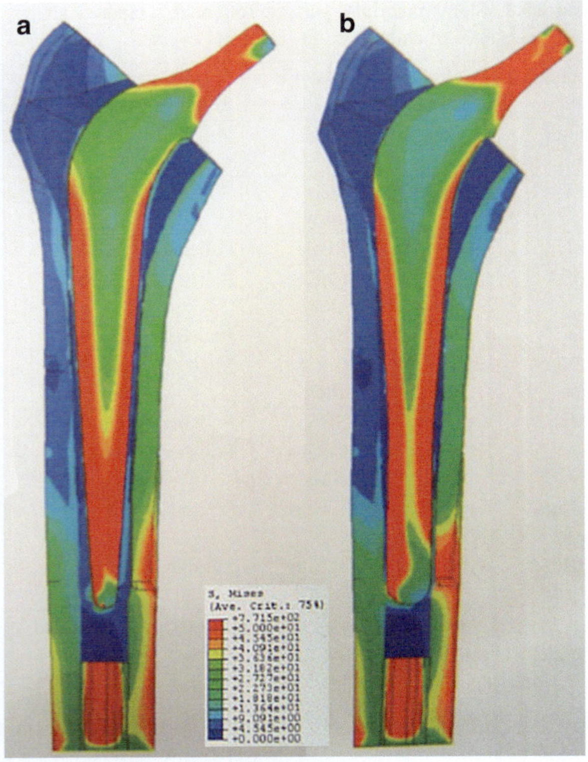

Fig. 28.6 Stress analysis using FEM on a tapered stem (**a**) and a distally cylindrical stem (**b**). The stress increased as the color of the sample changed from blue to red

taper design is currently not clear, with no conclusive results in the registries in various countries [21–23, 40]. The larger cross-sectional width and more rectangular shape of the proximal portion of the stem are considered to contribute to the rotational stability of the stem. However, it is possible that they may also increase the risk of cement fracture [39].

References

1. Charnley J. Low friction arthroplasty of the hip: theory and practice. Berlin/Heidelberg: Springer; 1979.
2. Charnley J. Total hip replacement by low-friction arthroplasty. Clin Orthop Relat Res. 1970;72:7–21.
3. Charnley EN. "Low friction torque" arthroplasty. A study of long-term results. Clin Orthop Relat Res. 1971;81:93–104.
4. Cupic Z. Long-term follow-up of Charnley arthroplasty of the hip. Clin Orthop Relat Res. 1979;141:28–43.
5. McKee GK, et al. Replacement of arthritic hips by the McKee-Farrar prosthesis. J Bone Joint Surg Br. 1966;48:245–59.
6. Shelley P, et al. Socket design and cement pressurisation in the Charnley low-friction arthroplasty. J Bone Joint Surg Br. 1988;70:358–63.
7. Hodgkinson JP, et al. Flanged acetabular components in cemented Charnley hip arthroplasty. Ten-year follow-up of 350 patients. J Bone Joint Surg Br. 1993;75:464–7.
8. Kobayashi S, et al. Radiology of low-friction arthroplasty of the hip. A comparison of socket fixation techniques. J Bone Joint Surg Br. 1990;72:439–43.
9. Hirose S, et al. Outcomes of Charnley total hip arthroplasty using improved cementing with so-called second- and third-generation techniques. J Orthop Sci. 2012;17:118–23.
10. Parsch D, et al. Acetabular cementing technique in THA—flanged versus unflanged cups, cadaver experiments. Acta Orthop Scand. 2004;75:269–75.
11. Wroblewski BM, et al. External wear of the polyethylene socket in cemented total hip arthroplasty. J Bone Joint Surg Br. 1987;69:61–3.
12. Jasty M, et al. The initiation of failure in cemented femoral components of hip arthroplasties. J Bone Joint Surg Br. 1991;73:551–8.
13. Lichtinger TK, et al. Improvement of the cement mantle of the acetabular component with bone cement spacers. A retrospective analysis of 200 cemented cups. Arch Orthop Trauma Surg. 1998;118:75–7.
14. Faris PM, et al. 14. The cemented all-polyethylene acetabular cup: factors affecting survival with emphasis on the integrated polyethylene spacer: an analysis of the effect of cement spacers, cement mantle thickness, and acetabular angle on the survival of total hip arthroplasty. J Arthroplast. 2006;21:191–8.

15. Kawata T, et al. Polyethylene and highly cross-linked polyethylene for cemented total hip arthroplasty: a comparison of over ten-year clinical and radiographic results. J Orthop. 2017;14:520–4.
16. Takaoka Y, et al. Radiolucent lines do not affect the longevity of highly cross-linked polyethylene cemented components in total hip arthroplasty. Bone Joint J. 2021;103–B:1604–10.
17. Johanson PE, et al. Outcome in design-specific comparisons between highly crosslinked and conventional polyethylene in total hip arthroplasty. Acta Orthop. 2017;88:363–9.
18. ISO21535 Non-active surgical implants-Joint replacement implants-specific requirements for hip-joint replacement implants.
19. Ritter MA, et al. Metal-backed acetabular cups in total hip arthroplasty. J Bone Joint Surg Am. 1990;72:672–7.
20. Morscher EW. Current status of acetabular fixation in primary total hip arthroplasty. Clin Orthop Relat Res. 1992;274:172–93.
21. Australian Orthopaedic Association Natrional Joint Replacement Registry (AOANJRR). Hip, knee and shoulder arthroplasty: annual report 2020. https://aoanjrr.sahmri.com/documents/10180/689619/Hip%2C+Knee+%26+Shoulder+Arthroplasty+New/6a07a3b8-8767-06cf-9069-d165dc9baca7
22. National Joint Registry: 17th Annual Report 2020. https://reports.njrcentre.org.uk/Portals/0/PDFdownloads/NJR%2017th%20Annual%20Report%2020.pdf
23. Karrholm JRC, et al. Swedish hip arthroplasty register: annual report 2019. https://registercentrum.blob.core.windows.net/shpr/r/VGR_Annual-report_SHAR_2019_EN_Digital-pages_FINAL-ryxaMBUWZ.pdf
24. Takaoka Y, et al. The long-term results of total hip arthroplasty with a cemented β-titanium stem. J Arthroplast. 2020;35:2167–72.
25. Weber FA, et al. A radiological study of fractures of acrylic cement in relation to the stem of a femoral head prosthesis. J Bone Joint Surg Br. 1975;57:297–301.
26. Murray DW, et al. Ten-year RSA-measured migration of the Exeter femoral stem. Bone Joint J. 2013;95–B:605–8.
27. Okutani Y, et al. Long-term outcome of cemented total hip arthroplasty with the Charnley-type femoral stem made of titanium alloy. J Orthop Sci. 2019;24:1047–52.
28. Akiyama H, et al. In-vitro characteristics of cemented titanium femoral stems with a smooth surface finish. J Orthop Sci. 2013;18:29–37.
29. Numata Y, et al. Biomechanical behaviour of a French femoral component with thin cement mantle: the "French paradox" may not be a paradox after all. Bone Joint Res. 2018;7:485–93.
30. Prins W, et al. Excellent results with the cemented Lubinus SP II 130-mm femoral stem at 10 years of follow-up: 932 hips followed for 5–15 years. Acta Orthop. 2014;85:276–9.
31. Crowninshield RD, et al. Cemented femoral component surface finish mechanics. Clin Orthop Relat Res. 1998;355:90–102.
32. Callaghan JJ, et al. Charnley total hip arthroplasty with cement. Minimum twenty-five-year follow-up. J Bone Joint Surg Am. 2000;82:487–97.
33. El Masri F, et al. Is the so-called "French paradox" a reality?: long-term survival and migration of the Charnley-Kerboull stem cemented line-to-line. J Bone Joint Surg Br. 2010;92:342–8.
34. Ling RS, et al. The long-term results of the original Exeter polished cemented femoral component: a follow-up report. J Arthroplast. 2009;24:511–7.
35. Rowsell M, et al. Survivorship of the Charnley elite plus cemented femoral stem. Int Orthop. 2005;29:214–8.
36. Kimura G, et al. Characteristic loosening of the distally straight cylindrical femoral stem. J Orthop Sci. 2014;19:437–42.
37. Fujita H, et al. Multi-center study of use of the Exeter stem in Japan: evaluation of 1000 primary THA. J Orthop Sci. 2012;17:370–6.
38. Goto K, et al. Long-term results of total hip arthroplasty using Charnley elite-plus stem and the effect of stem geometry on radiographic distal femoral cortical hypertrophy. J Orthop Sci. 2018;23:365–70.
39. Glyn-Jones S, et al. Influence of stem geometry on the stability of polished tapered cemented femoral stems. J Bone Joint Surg Br. 2005;87:921–7.
40. Fujita H, et al. Prospective, comparative study of cemented, smooth-surfaced titanium stems and polish-surfaced, stainless steel stems at a minimum follow-up of 10 years. Eur J Orthop Surg Traumatol. 2020;30:501–12.

Part V

Surgical Techniques

29 Basic Surgical Techniques for Cementless THA

Kazuhiro Oinuma

1 Advantages and Disadvantages of Cementless Cups

The advantages of the cementless cup are as follows: (1) the technique is simpler, which shortens the operative time and the surgeon's learning curve; (2) there are many options for the weight-bearing surface, including ceramic liners, large diameter heads, and dual mobility; and (3) even after the cup is placed, it can be easily repositioned if malaligned or hip instability is detected intraoperatively. The disadvantages are (1) early postoperative dislodgment and migration risk if the initial fixation is inadequate and (2) intraoperative fractures may occur during implantation, causing early postoperative cup migration and loosening. Therefore, it is necessary to become proficient in this simple technique to overcome these drawbacks.

2 Indication of Cementless Cup

The cementless cup can be used in the majority of total hip arthroplasty (THA), but some cases are difficult to handle, such as (1) a severely dislocated hip without a thickened acetabular floor, (2) cases with a large bone defect in the acetabular wall after acetabular osteotomy, and (3) posttraumatic osteosynthesis, such as cases with exposed screws or plates that are difficult to remove from the acetabular surface in contact with the cup. Although cementless reconstruction is possible for (1) and (2), excessive medialization and high hip center can be avoided using a cemented cup with a cross plate and a massive bone graft. In case (3), a cemented cup is a good option to prevent interference between metals.

K. Oinuma (✉)
Funabashi Orthopedic Hospital, Funabashi, Japan
e-mail: oinuma@fff.or.jp

3 Surgical Technique of the Cementless Cup

The basic technique for cementless cups is to insert a hemispheric metal shell into the acetabulum with additional screw fixation if necessary. Since the method differs depending on the presence or absence of acetabular dysplasia, we will first describe the basic technique for cases without acetabular dysplasia and then describe additional techniques for patients with acetabular dysplasia.

A. Cases without acetabular dysplasia
   1. Reaming size and depth
      Reaming is completed by starting with a small reamer and gradually increasing the size until the cortical bone of the acetabular fossa is exposed. We begin with a 43-mm diameter reamer and increase the size by 2 mm, gradually increasing the depth. One method is to ream to the acetabular fossa with the first reamer and then ream up to the size of the acetabular fossa. However, since the reaming depth increases slightly as the size increases, there is a tendency toward over-excavation in the end, so care should be taken. The appropriate size for final reaming is the size at which the acetabular cartilage has been sufficiently removed visually, and petechial hemorrhages can be seen on the acetabular surface. When approaching the final size, the surgeons can feel the resistance of the reamer against the acetabular bone through the hand holding the reamer. This is indicated by the reamer gripping the bone and the surgeon's hand twisting. The cup size is usually 5 to 6 mm larger than the head diameter, so the diameter of the extracted head is also helpful. Excessive reaming increases blood loss, weakens the initial fixation force, and increases the risk of acetabular fracture at the time of implantation [1]. The low-wear properties of crosslinked polyethylene have been demonstrated, so it is not necessary to risk fracture or bleeding to use a larger cup to maintain the

Y. Nakashima (ed.), *Advances in Total Hip Arthroplasty*, https://doi.org/10.1007/978-981-97-2461-1_29

thickness of the polyethylene. Therefore, it is reasonable to use the required minimum size. For example, we used to use a 50 mm cup but now use a 48 mm cup frequently. Especially in fragile cases such as femoral neck fractures, the safe reaming size range is narrow, so the reaming depth should be checked often to avoid excessive reaming.

2. Reaming direction
   First, reaming is performed inward, and as the size increases, reaming is gradually performed posteriorly and upward, taking the anterior opening angle of the acetabulum into consideration. The reamer follows the direction of the shaft, so reaming should be started with the shaft facing medially and then gradually reamed posteriorly and upward as the size increases. The reaming operation tends to be left to the surgeon alone, but the direction of the reamer shaft indicates the direction in which the surgeon is reaming, allowing the assistant and surgeon to confirm the direction.
3. Cup placement
   Before inserting the cup, one should ensure that the labrum and bursa do not get stuck between the cup and the acetabulum, and then insert the cup into the acetabulum. There are two methods of cup fixation: the press-fit method, in which the cup is reamed 1 to 2 mm smaller than the cup size and does not rely on screw fixation, and the line-to-line method, in which the cup is reamed to the same size as the cup and screw fixation is added. The disadvantage of the press-fit technique is that it results in a rim-fit, which may cause a gap between the acetabulum and cup, and there is a risk of acetabular fracture if this gap is forced [1]. Since good results have been obtained with the line-to-line technique [2], under-reaming by 2 mm or more may not be necessary. If a gap develops between the cup and the acetabular floor using the press-fit technique, the reaming should be increased.
4. Addition of a screw
   Screw fixation should be added when the bone is fragile in elderly patients, when the press-fit is weak, or when the line-to-line technique is used. The drill tip may cause vascular injury, so the site and depth of insertion should be carefully selected. The most effective and safest site is the posterior-superior aspect of the cup, and screw insertion anterior portion of the cup should be avoided due to the risk of vascular injury [3]. In cases without acetabular dysplasia, sufficient screw fixation can be achieved within 30 mm, so drills longer than this should not be used from the start.

B. Cases with acetabular dysplasia
   1. Complete exposure of the acetabular rim
      In cases of acetabular dysplasia, it is important to identify the position of the cup, so the acetabular margin should be exposed completely. Osteophytes developed on the acetabular margin are good indicators of the position of the cup. They should be preserved as much as possible, except for those where their removal is necessary for the dislocation or removal of the femoral head. For good acetabular exposure, retracting the surrounding soft tissues with a retractor or pin is essential. Retracting the iliopsoas tendon by applying a retractor to the teardrop's inferior margin is useful for identifying the original acetabulum and protecting the blood vessels around the iliopsoas tendon. It is a fundamental THA surgical technique for hip dysplasia.
   2. Reaming up the size and correcting the position of the reaming
      Using the distance from the outer edge of the acetabulum and the inferior border of the teardrop as a reference, imagine the original acetabulum and draw a line of the outer edge of the final reaming on the acetabular surface (Fig. 29.1) and start reaming with the smallest size reamer while assuming the center of the original acetabulum. The acetabular base usually forms a double floor (the newly formed acetabular base due to subluxation and the acetabular base of the original acetabulum), and once the bony part of the superficial new acetabular base has been excavated with the smallest size reamer, insert the tip of a Hohmann hook into the cancellous bone to measure the distance to the acetabular base of the original acetabulum and check the depth to final reaming. The

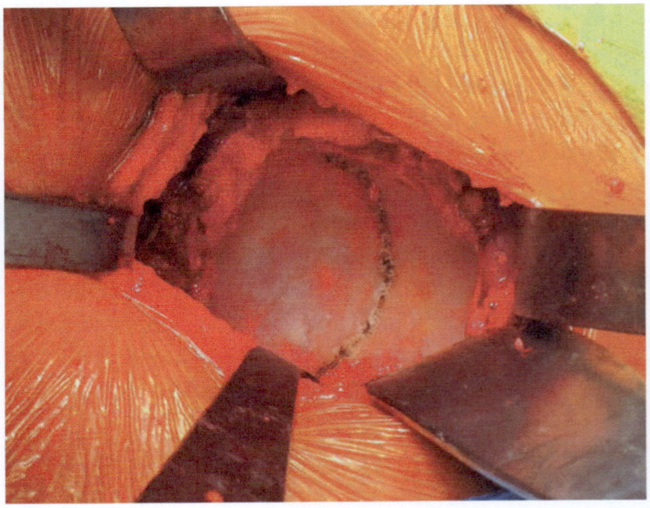

Fig. 29.1 Virtual line of the lateral border of the final reaming drawn by electrocautery on the acetabular surface of the dysplastic hip. Acetabulum of the left hip viewed from the direct anterior approach (anterior above, cephalad on the right)

distance between the outer edge of the reaming and the anterior and posterior edges of the acetabulum is visually observed to check the anterior-posterior deviation of the reamed area. The reaming is continued to the bottom of the acetabulum, correcting the anteroposterior deviation as the reaming size increases (Fig. 29.2). The smaller the diameter of the starting reamer is, the easier it is to correct. In cases of severe acetabular dysplasia, the author starts with a reamer diameter of 38 to 40 mm instead of 43 mm.

3. Confirmation of the strength of the acetabular matrix and bone grafting
   It is important to check the strength of the acetabular surface after the final reaming, as it is the sill for the initial fixation of the cup. In elderly patients with severe acetabular dysplasia or well-developed osteophytes at the margins of the acetabulum, the cancellous bone and acetabular base near the original acetabulum are so fragile that it may be possible to dent by pressing with a finger alone. If a cup is driven into the acetabulum, a fracture of the acetabular base may occur, or even if no fracture occurs, a press-fit may not be obtained, and the bone at the screw insertion site may be too weak to allow screw insertion or fixation. In such cases, bone grafting from the femoral head is necessary. In many cases, a good metaphyseal bone graft is obtained by filling the acetabulum with a soybean-sized fragment of the femoral head using a Riehl and then crimping it by counter-rotational reaming. If the acetabular base is fractured or over-reamed, a plate of grafted bone is made from a slice of the loading surface of the femoral head, with the articular side of the femoral head facing the acetabular base, and is formed to fit into the cup by counter-rotational reaming. In such cases, increasing the cup size by 2–4 mm is advisable to avoid the risk of central cup migration.

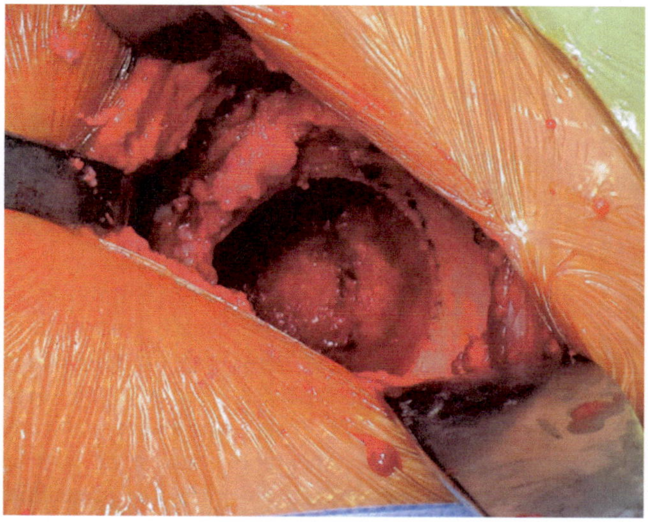

Fig. 29.2 Acetabulum after final reaming

4. Cup placement
   In cases of acetabular dysplasia, the acetabular bone at the upper edge of the reaming is often osteosclerotic, which increases the risk of acetabular fracture during cup insertion, so special attention should be given to the under-reaming press-fit technique. Since the cup is fixed at the anterior and posterior acetabular edges, a press-fit can often be obtained even if the upper acetabular coverage is insufficient. Therefore, it is not recommended to use a higher reaming position from the beginning to obtain upper acetabular coverage, since this will deprive the last possibility of original acetabular placement. As mentioned earlier, the advantage of cementless cups is that they can be reinstalled repeatedly. Although stable results can be obtained without screw fixation if the cup CE angle is 10° or more, [4] it has been reported that stable results can be obtained even if the cup CE angle is less than 0° if screw fixation is added [5]. Therefore, reaming should be started with the goal of original acetabular placement, especially in younger patients with good bone quality. If press fit cannot be obtained due to insufficient acetabular coverage, reaming may be performed several millimeters upward, or the medial protrusion technique may be used to ream further inward [6]. We use the former method to avoid the risk of central migration of the cup, and a high hip center (hip center more than 35 mm above the teardrop line), which is feared to cause poor outcomes due to high cup placement, is rarely the result.
5. Screw fixation
   Screw fixation is necessary in cases of severe acetabular dysplasia or bone fragility in elderly patients. Screw fixation is often performed on the superior-posterior side of the cup, with little risk of vascular damage. Because this area is a loading site, the bone quality of the acetabulum is good, and the screws are firmly fixed and are very effective in reinforcing the initial fixation of the cup. If the cup does not move even slightly during screw tightening, it has good fixation to withstand full-load walking from the first day postoperatively. However, if the cup moves even slightly as the screw is tightened, the reaming operation should be redone again, as the initial fixation is poor. In most cases, reaming the cup upward by a few millimeters will significantly improve the fixation of the cup.
6. Bone grafting
   Unlike cemented cups, cementless cups do not require the creation of a bone cement containment,

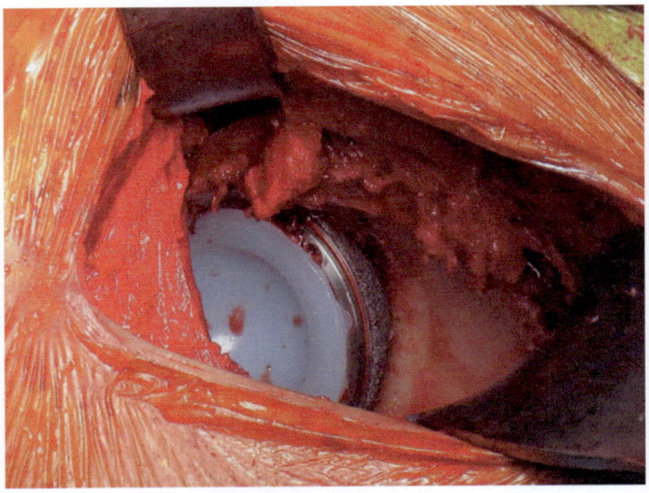

Fig. 29.3 Acetabulum after cup placement. The acetabulum does not cover the outer surface of the cup

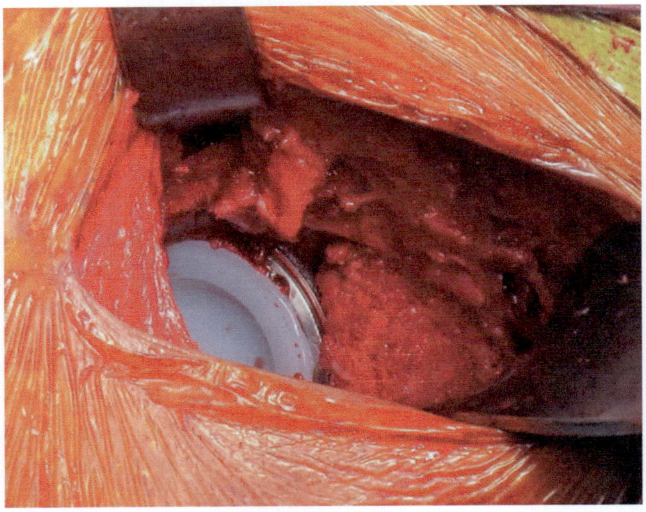

Fig. 29.4 The gap between the outside of the cup and the acetabulum filled with morselized bone

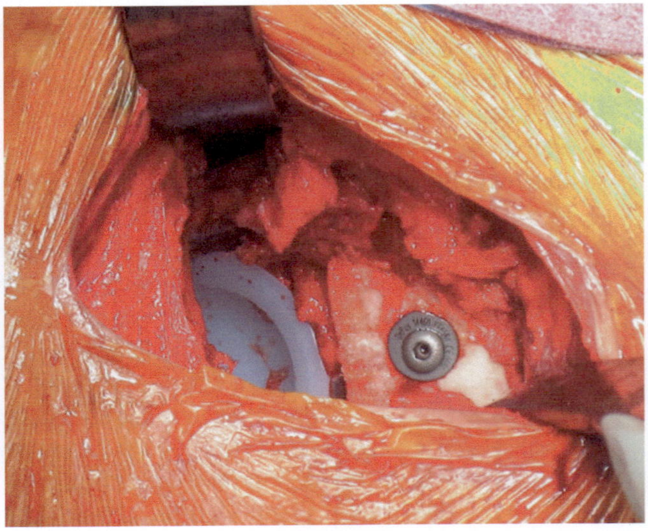

Fig. 29.5 Crescent-shaped tubercle bone taken from the femoral head fixed with a screw; the morselized bone is crimped to the acetabulum, bridging the lateral margin of the acetabulum and the outer side of the cup

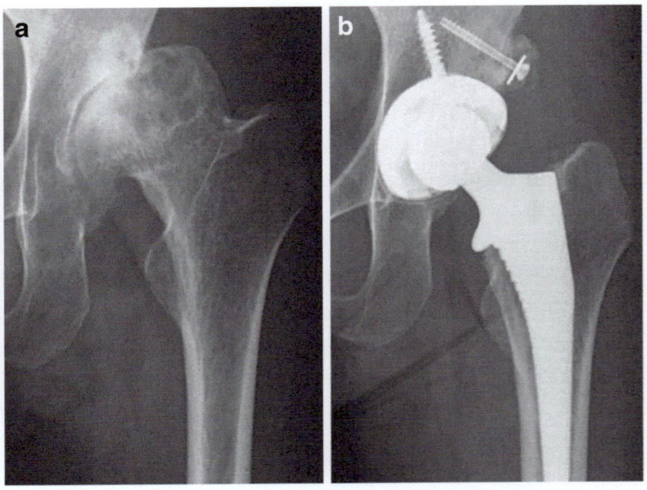

Fig. 29.6 Figs. 1–6: Preoperative (left) and 5-year postoperative (right) frontal X-ray views of the patient (52-year-old female). The boundary between the grafted bone and the matrix bone disappeared

so bone grafting requires only the filling of the gap between the top of the cup and the acetabulum with morselized bone after cup placement to achieve sufficient osteogenesis. This is an advantage of the cementless cup technique, simplifying the procedure and shortening the operative time. Some surgeons place a massive bone graft before acetabular reaming like cemented cup technique when the gap is large, but there is no evidence for how large the size of the gap requires the massive bone graft. However, it is undeniable that the larger the gap, the weaker the initial fixation force of the cup by press-fit, so we decided to use a massive bone graft if the defect was large (cup overhang of ≥20 mm) (Fig. 29.3), the morselized bone was filled (Fig. 29.4), and then a crescent-shaped tubercle of bone taken from the femoral head was screwed in place to cover the defect (Fig. 29.5). Because of the screw fixation to the hardened bone above the outer surface of the acetabulum, the massive bone graft was fixed very firmly with a single screw, there was no risk of early postoperative crushing of the grafted bone, and the remodeling above the cup was excellent in the mid- to long-term (Figs. 29.6 and 29.7). This is a simple and good method to improve the initial fixation of the cup without changing the consistent flow of cementless cup technique, which is bone grafting after cup placement.

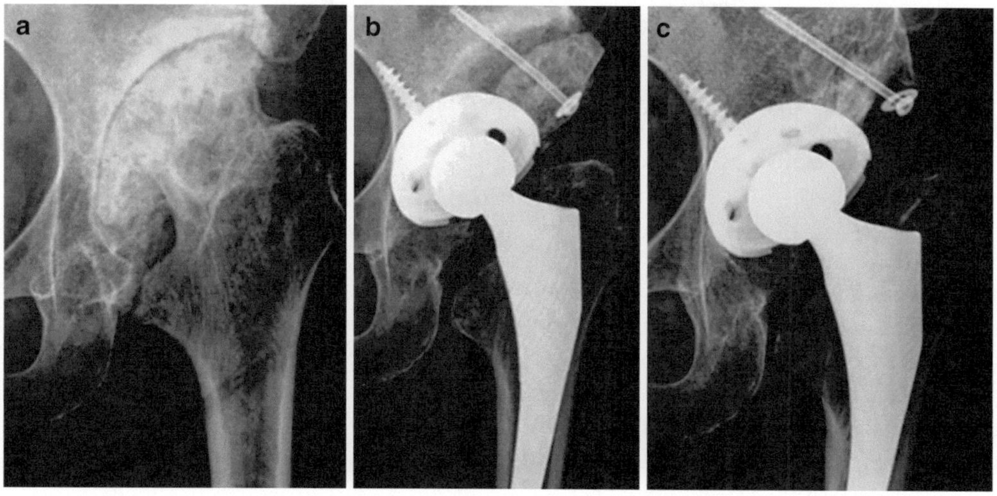

Fig. 29.7 Grafted bone before and after left THA surgery in a 61-year-old woman. Preoperatively (**a**), the patient had a highly dislocated hip with a little thickening of the acetabular floor, and the postoperative cup CE angle was −24° (**b**), but the cup was press-fit at this position. The gap between the cup's outer surface and the acetabular base bone was filled with morselized bone, and a crescent-shaped tubercle bone taken from the head was screw-fixed. Ten years after surgery (**c**), the boundary between the cup and the acetabulum had disappeared, and the resorption of the tubercle bone was mild

4 Advantages and Disadvantages of Cementless Stems

The advantages of cementless stems are (1) simpler technique compared to cemented stems, which is advantageous in reducing the operative time and the operator's learning curve, and (2) stable long-term results can be expected once osteointegration has been achieved. On the other hand, the disadvantages are (1) the risk of early postoperative sinking and fracture and (2) the wide variety of models used. Early postoperative sinking and fracture are devastating complications of cementless stems because the fixation of the stem depends on the surgeon's broaching sensation, resulting in sinking if the stem is loose and fracture if it is tight. Intraoperative fractures are less problematic if appropriate measures are taken, but postoperative fractures often require reoperation and are a serious complication. Another disadvantage is the wide variety of stems. Cementless stems come in a wide range of shapes and lengths, and each has its own surgical technique, but even models of the same design require slightly different techniques depending on the thickness of the coating and surface finish. Model selection is an important topic to reduce the risk of early postoperative sinking and fracture. The model that is the least likely to sink or cause fractures, or even if intraoperative fractures do occur, the model that allows easy detection of the fracture site should be selected, and the technique should be familiar to the surgeon. The fewer models selected, the better it is so that the surgeons can become familiar with the technique in a limited time. It is important to emphasize that choosing a safe and widely applicable model is an extremely important topic that needs to be addressed before discussing cementless techniques.

5 Indication of Cementless Stem

The indications vary from institution to institution and from surgeon to surgeon. Some surgeons use cementation in all cases, while others use a cementless technique. This is mainly because one of the major complications of cementless stems in early postoperative subsidence and fracture, as described above. The risk of such complications is also lower with cemented stems, and this is also because the cementing technique on the stem side is easier than on the cup side. Therefore, the indication for cementless stems is relative and can be used in most cases, but in some cases, cemented stems are technically easier and avoid the risk of major complications, such as postoperative fractures. We recommend cemented stems for patients with fragile bones, such as older patients, patients with a low osteotomy level due to a flat head or highly dislocated hip, and patients requiring a reduction of the rotation angle by more than 15°.

6 Surgical Technique of Cementless Stem

Unlike cups, cementless stems come in a wide variety of designs and different stem lengths and fixation styles, making it difficult to discuss surgical techniques for the stem side in a limited number of words. Therefore, we describe the basic surgical techniques for the stem side, assuming a flat wedge-shaped stem, which is currently the most popular type of stem in the registry in each country. We used a collared, fully hydroxyapatite-coated stem mainly because of its advantage in reducing postoperative subsidence and fractures [7].

A. Osteotomy of the femoral neck

Accurate osteotomy of the femoral neck is an important maneuver for cementless stems to reduce the risk of fractures. First, the osteotomy of the femoral neck for head resection should be approximately 5 mm higher than the preoperative planning level. This is because after resection of the head, the entire proximal femur is unfolded, and the distance from the osteotomy level indicators (superior margin of the lesser trochanter, neck base, and intertrochanteric ridge) to the osteotomy plane can be measured, which allows the correct osteotomy level and angle to be corrected if a higher level is used. In addition, the lower the osteotomy level, the closer it is to the intertrochanteric ridge, the more blood flow is available in this area, and the more blood loss occurs. Thus, the initial osteotomy level should always be high because low osteotomies can lead to complications.

B. Checking the anteversion angle of the stem

Once the osteotomy of the femoral neck is completed, the anteversion angle of the stem is determined. With the knee joint flexed at 90°, the axis of the lower leg is used as an indicator of the anteversion angle of the stem (Fig. 29.8). The line of the anteversion angle of the stem is imagined and marked on the osteotomy surface of the femoral neck using an electrocautery scalpel (Fig. 29.9). Usually, the line of the anteversion angle is parallel to the direction of the posterior cortical bone of the femoral neck.

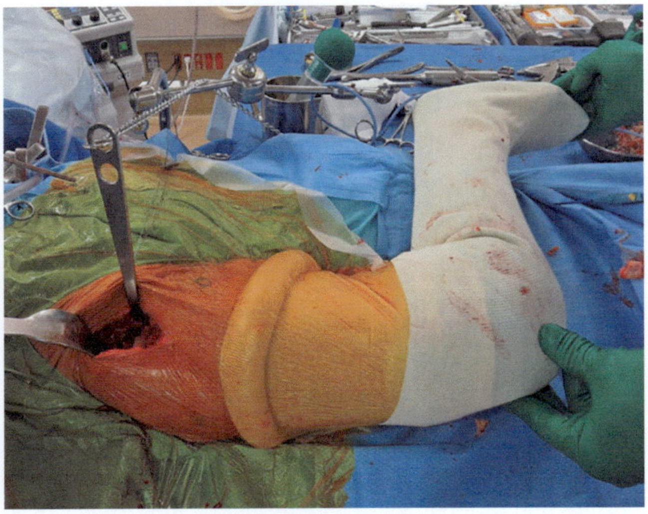

Fig. 29.8 The right knee joint is flexed 90° in the direct anterior approach

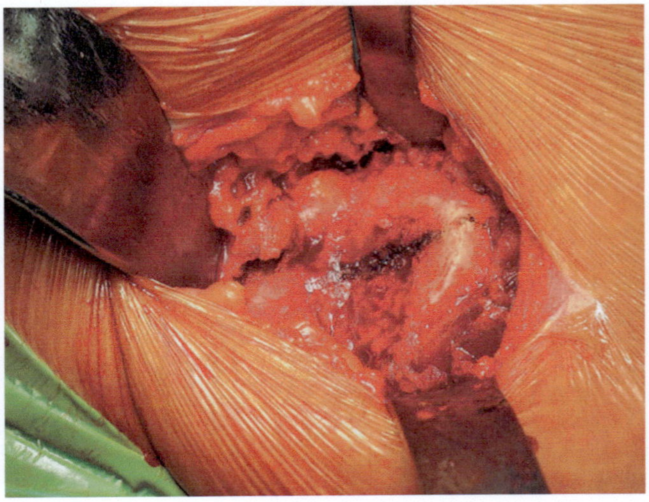

Fig. 29.9 Virtual line of the anterior torsion angle of the stem drawn by electrocautery on the osteotomy surface of the right femoral neck

C. Confirmation of the femoral axis

The femoral axis is identified by inserting a canal finder through the neck osteotomy. The direction of the femoral axis is the direction of broach insertion. The intersection of the femoral axis and the surface of the neck osteotomy is the entry point for the broach, usually near the trochanteric fossa.

D. Broaching

The broach's entry point and insertion direction are determined, and broaching is initiated. Since the femoral axis is usually located posteriorly on the lateral side of the femoral neck osteotomy surface and the anteversion angle of the stem is parallel to the posterior femoral cortical bone, the broaching should be inserted to make contact with the posterior femoral cortical bone. The size of the broach is gradually increased, and the final size is the size at which the broach can no longer be inserted.

The surgeon's feel of the broach determines the final size, which is the most important technique to master. As mentioned above, a loose broach results in postoperative sinking, and a tight broach results in a fracture. Bone strength varies from case to case, so the surgeon's factors should be kept as constant as possible. As some surgeons recommend using a woodpecker, always strike the broach with a constant force, speed, rhythm, and the same hammer, paying attention to changes in the speed of broach insertion and the sound of the hammer strike [8] as you approach the final size. It is important to note that the strength of the trabecular bone becomes weaker in older individuals, rapidly approaching the final size, and the strength of the cortical bone is also weaker, so if these changes are noticed too late, fractures may occur.

E. Collared stem

In the case of a collared stem, the final rasp should ideally be inserted one to several millimeters under the osteotomy level, and the osteotomy site is finished with a calcar reamer. Contact of the collar with the calcar, the strongest part of the proximal femur, adds to the collar's extramedullary fixation force and increases the stem's initial fixation force of the stem [9]. This is probably the main reason for the reduced postoperative stem sinking and femoral fractures. The additional extramedullary fixation force of the collar is more reasonable than relying on the intramedullary fixation force of the stem itself because the contact of the collar to the calcar can be visually confirmed.

F. When the final brooch is smaller than the planned size

The insertion is likely mal-aligned, and the alignment should be confirmed by X-ray fluoroscopy or other means. In many cases, the size can be increased by downsizing the broach by one size and repeatedly broaching it again. This is thought to be because the alignment of the stem is corrected when a broach of reduced size is inserted intramedullary.

G. When a fracture occurs

If there is no resistance to the broach after it reaches the expected size, or if the broach suddenly loses resistance, there is a possibility that a fracture has occurred, and intraoperative radiography or visual observation of the proximal femur should be performed to confirm the presence of fracture lines. Careful observation of the proximal femur is necessary, as there is a possibility of a fissure fracture that cannot be confirmed by radiography. The neck osteotomy should be checked circumferentially, but if the direct anterior approach is used, the vastus intermedius should be split, and the medial, anterior, and lateral cortical bones should be observed. If the fracture is only at the calcar proximal to the lesser trochanter, no additional treatment is necessary, but if the fracture extends below the lesser trochanter, a cerclage wire is necessary. If the fracture extends to the tip of the stem, the stem must be removed, and plate fixation is needed. However, as mentioned above, we should not select a model with a risk of such a fracture. If the stem is cementless, the risk of fracture is always present, and if a fracture does occur, it should be in an area that is easy to detect, i.e., on the femoral neck osteotomy surface.

H. Stem insertion

In general, cementless stems are manufactured larger than broaches (several tens to several hundred micrometers), so it is not always possible to insert the stem to the same height as the broach. The difference in stem insertion depends on the surface finish and coating, so it is necessary to be familiar with the characteristics of the model used. The difference in stem insertion also relies on the strength of the cancellous bone, and this difference tends to be larger in males and younger patients with stronger cancellous bone.

References

1. Hasegawa K, et al. Periprosthetic occult fractures of the acetabulum occur frequently during primary THA. Clin Orthop Relat Res. 2017;475:484–94.
2. Fukui K, et al. A radiological study of the true anatomical position of the acetabulum in Japanese women. Hip Int. 2011;21:311–6.
3. Ohashi H, et al. Surgical anatomy of the pelvic vasculature, with particular reference to acetabular screw fixation in cementless total hip arthroplasty in Asian population: a cadaveric study. J Orthop Surg (Hong Kong). 2017;25:2309499016685520.
4. Takao M, et al. The results of a press-fit-only technique for acetabular fixation in hip dysplasia. J Arthroplast. 2011;26:562–8.
5. Kim M, et al. High long-term survival of bulk femoral head autograft for acetabular reconstruction in cementless THA for developmental hip dysplasia. Clin Orthop Relat Res. 2010;468:1611–20.
6. Dorr LD, et al. Medial protrusion technique for placement of a porous-coated, hemispherical acetabular component without cement in a total hip arthroplasty in patients who have acetabular dysplasia. J Bone Joint Surg Am. 1999;81:83–92.
7. Lamb JN, et al. A calcar collar is protective against early periprosthetic femoral fracture around cementless femoral components in primary total hip arthroplasty: a registry study with biomechanical validation. Bone Joint J. 2019;101-B:779–86.
8. Morohashi I, et al. Acoustic pattern evaluation during cementless hip arthroplasty surgery may be a new method for predicting complications. SICOT J. 2017;3:13.
9. Demey G, et al. Does a collar improve the immediate stability of uncemented femoral hip stems in total hip arthroplasty? A bilateral comparative. J Arthroplasty. 2011;26:1549–55.

Basic Surgical Techniques for Cement THA

30

Hirotsugu Ohashi

1 Basics of Bone Cement

(a) Material properties of bone cement
Bone cement is a biomaterial that is used to fix implants to bone. The main raw materials of bone cement are a powder made of polymethylmethacrylate (PMMA), which is a polymer, and a liquid made of methyl methacrylate (MMA), which is a monomer (Fig. 30.1). When the powder and liquid are mixed, polymerization reaction initiates, and the bone cement hardens in approximately 10 min. Polymerized bone cement is a viscoelastic material with the property of gradual deformation (creep) after hardening. As shown in Table 30.1, the mechanical properties of bone cement are high in compressive strength but weak in tension, shear, and bending. The mechanical strength of bone cement is slightly weaker than that of cortical bone and much weaker than that of metals. Therefore, bone cement should be used under compressive stress, and usage under shear and bending stresses should be avoided.

(b) Mixing
When the powder and liquid are mixed, the viscosity gradually increases. The curing process is divided into four stages, depending on the viscosity of the bone cement. The bone cement is mixed when it is in a liquid state. The mixture is left to harden when its viscosity is low and adheres to the gloves. When the viscosity becomes high, and the bone cement does not adhere to gloves, it is in the right condition for cement injection and implant insertion. The bone cement is then left to harden after implantation. The curing time is affected by various conditions. It becomes faster when the temperature and humidity in the operating room are high and slower if the bone cement is cooled. Additionally, the faster the agitation speed, the faster the curing time. Therefore, it is essential to store bone cement under stable conditions at each facility, so that it can always be used under the same conditions.

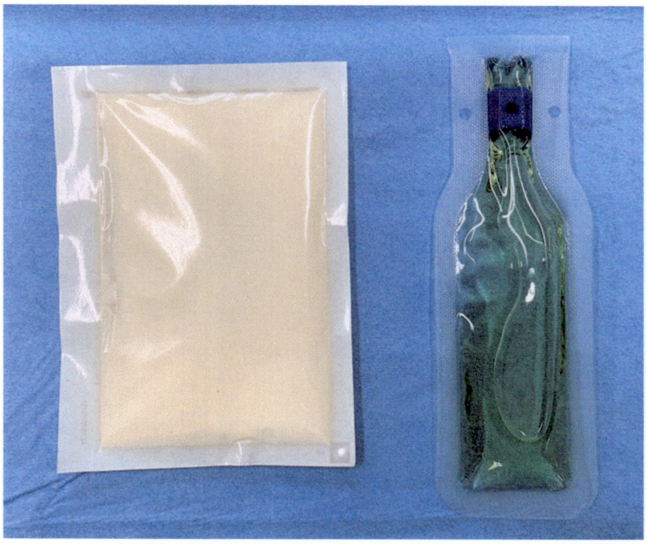

Fig. 30.1 Bone cement. Powder, polymethyl methacrylate (PMMA); liquid, monomeric methyl methacrylate (MMA)

H. Ohashi (✉)
Osaka Saiseikai Nakatsu Hospital, Osaka, Japan
e-mail: hohashi@grace.ocn.ne.jp

Y. Nakashima (ed.), *Advances in Total Hip Arthroplasty*, https://doi.org/10.1007/978-981-97-2461-1_30

Table 30.1 Mechanical properties of various biomaterials and bone

	bone cement	cortical bone	trabecular bone	titanium alloy	CoCr Alloy
Tensile strength (MPa)	25.0–49.2	143	1.2	860	655–1172
Shear strength (MPa)	37.0–69.0	82–84	–	–	–
Compressive strength (MPa)	72.6–114.3	100–230	2–12	–	–
Bending strength (MPa)	12.1–74.0	50–150	10–20	–	–
Young's modulus (GPa)	2–3	7–30	0.05–0.5	115	234

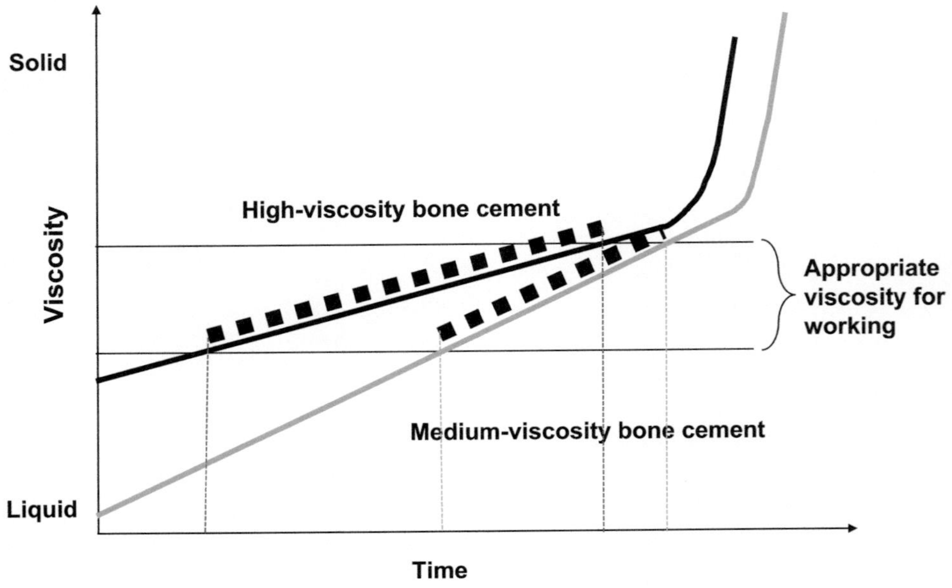

Fig. 30.2 Curing process in high- and medium-viscosity bone cement. High-viscosity cement can be used without waiting time due to its high viscosity from the early stage after mixing, resulting in a longer working time

Bone cements are classified into high, medium, and low viscosities after mixing. High-viscosity cements, such as Paracos R and BBC R (Biomet Bone Cement R), are used globally in many countries, while medium-viscosity cements, such as Simplex P and Endurance, have been generally used in Japan. However, high-viscosity cement has recently become available in Japan. The viscosity of high-viscosity cements is already high after mixing; thus, they can be used without waiting time, enabling longer working time and easier of use once the surgeon becomes accustomed to them (Fig. 30.2).

Mixing in air causes air bubble contamination in the bone cement, and vaporized monomers during mixing are toxic to the human body. Therefore, it is preferable to perform the mixing under a vacuum. Vacuum mixing reduces the porosity of bone cement which enhances its strength. Recently, a mixing system in which powder and liquid are already set has become available, which enables mixing without exposure to vaporized monomers (Fig. 30.3), making it safer for medical staff.

Fig. 30.3 Vacuum mixing system with powder and liquid already set up. (Courtesy of Zimmer-Biomet)

2 Femoral Side Cementing Technique

(a) Preparation of the bony bed
Strong fixation of bone cement to the bone is considered to be attainable as the bone cement penetrates into the trabecular bone structures. This theory is reasonable; however, if the trabecular bone is weak, strong fixation cannot be obtained. Therefore, it is necessary to fix the bone cement to a strong trabecular bone. Since cancellous bone is found near cortical bone, the weak cancellous bone should be removed during rasping and the strong cancellous bone near cortical bone should be maintained. For this purpose, it is important to rasp to the largest rasp possible, such as the techniques for a cementless stem. Like impaction grafting, a compression type rasp involves the compression of the trabecular bone in the femoral canal, and the use of a woodpecker is effective. However, reaming is undesirable because it removes cancellous bone and flattens the inner surface of the cortical bone, which loses mechanical fixation to the bone cement.

After rasping, the cement plug is fixed approximately 1.5–2 cm distal to the stem tip. The purpose of the cement plug is to prevent bone cement implantation syndrome, which is caused by excessive cement injection into the distal femoral canal, and to facilitate the penetration of bone cement into the cancellous bone by pressurization of the bone cement during bone cement injection and stem insertion. Cement plugs of various shapes and sizes are commercially available. Bone chips from the femoral head can be pressed into the femoral canal as a substitute for the cement plug.

Finally, the femoral canal is cleaned with pulse lavage to remove blood and bone fragments, so that the bone cement can bond directly to the bony bed. Cleaning also removes intramedullary fat and reduces the risk of fat embolization. After lavage, the medullary cavity is filled with gauze to achieve hemostasis before cement injection, and the bony bed is kept as dry as possible.

(b) Pressurization of bone cement
The cementing technique for the stem has progressed progressively, and the long-term results have improved. In the first-generation technique, cancellous bone was removed up to the planned size, and the bone cement was made into a cement ball that was pressed into the femoral canal with the fingers. It was also known as the two-thumb technique because it was pushed in with the thumbs of both hands.

In the second-generation technique, the usage of a cement plug, pulse lavage, and retrograde cement injection with a cement gun was introduced. Cement injection is initiated when the bone cement extruded from the cement gun does not adhere to the gloves and has a high viscosity (working state). The nozzle of the cement gun is inserted distally into the femoral canal and bone cement is injected retrogradely from the distal end. The nozzle is pushed back to the proximal part of the femur by pressurized cement. When the bone cement overflows, the proximal femur is sealed with a pressurizer and the bone cement is further injected slightly to apply continuous pressure to the bone cement. Then, the stem is slowly inserted manually or with a stem insertion device. The overflowing bone cement is then removed. Once the stem is inserted to the planned depth and angle, the position is maintained until the bone cement hardens. As bone cement implantation syndrome is most likely to occur between cement injection and stem insertion, it is important to collaborate with an anesthesiologist. In elderly patients and patients with femoral neck fractures who are prone to this complication, care should be taken to not apply excessive pressure.

In the third-generation technique, a vacuum mixing system is used to mix bone cement under negative pressure. Vacuum mixing mixes the bone cement more uniformly and improves its strength by reducing the amount of air bubble contamination during mixing.

Recently, the concept of line-to-line technique has been advocated, and the combination of stem and cement techniques has been emphasized [1]. In the line-

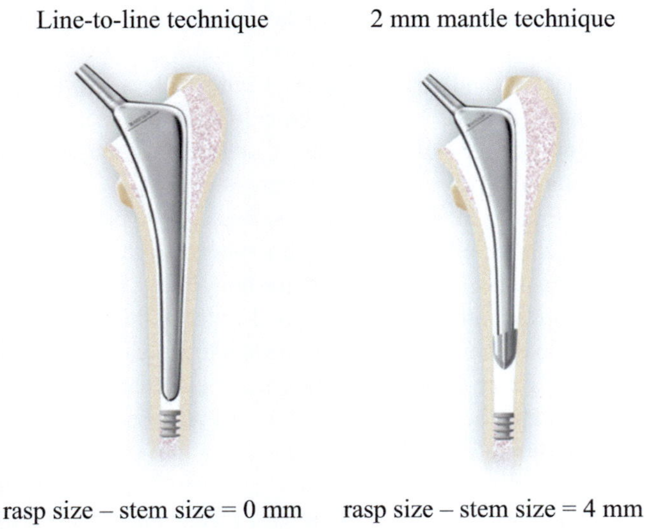

Fig. 30.4 Line-to-Line Technique. In the line-to-line technique, the rasp and stem have the same size; therefore, the cement mantle created by the rasp and stem is zero In contrast, in the 2-mm mantle technique, the stem is 4 mm smaller than the rasp, and the cement mantle is secured in the gap between the rasp and stem. (Courtesy of B-Braun, Aesculap)

to-line technique, the size of the rasp is the same as that of the stem; thus, the cement mantle around the stem becomes thinner (Fig. 30.4). Clinically, bone cement penetrates into the remaining cancellous bone and fills in the gaps, owing to the difference in the medullary cavity and stem geometry. As a result, although there are a few areas where the cement mantle is very thin, an average of approximately 2 mm of cement mantle is present [2]. The purpose of this technique is not to make the cement mantle thin; instead, it is to enhance bone cement pressurization during stem insertion. Therefore, this technique does not require pressurization during cement injection with a cement gun, and it is sufficient to fill the femoral canal with bone cement. Furthermore, this technique has the advantage that the stem can only be inserted at the same depth and in the same angle as the rasping; therefore, the stem cannot be inserted too deeply or in the incorrect orientation, with respect to that of the rasping. Although there is concern that the mechanical strength of the thin bone cement mantle may decrease, the use of a polished tapered stem allows the bone cement to be withstand mainly compressive stress, and good long-term results have been reported.

A comparison of cementing techniques is presented in Table 30.2. Although it is important to follow the concept of the stem, it is conceivable that the easiest and most reliable cementing technique is the line-to-line technique using high-viscosity cement.

Table 30.2 Comparison of cementing techniques

	First generation	Second generation	Third generation	Line-to-line
Cement plug	Not used	Used	Used	Used
Intramedullary washing	Not enough	Pulse lavage	Pulse lavage	Pulse lavage
Cement mixing	Cement bowl	Cement bowl	Vacuum Mixing	Vacuum Mixing
Cement gun	Not used	Used	Used	Used
Pressurization	With fingers	Pressurizer	Pressurizer	During stem insertion
Cement mantle	Secured to some extent	Secured to some extent	Secured to some extent	Thin
Stem	All sorts	All sorts	All sorts	Polished Tapered stem

3 Acetabular Side Cementing Technique

(a) Preparation of the bony bed

Unlike the femoral canal, the acetabulum is not a deep enclosed space; thus, it is difficult to press the bone cement into the acetabulum, and skills in bony bed preparation and bone cement pressurization are required. First, it is necessary to understand the differences in fixation modes between cemented and cementless cups. In cementless fixation, initial fixation is achieved by frictional forces due to compressive stresses between the implant and the bone. On the acetabular side, the acetabular rim is the fixation site (rim fit); therefore, it is important to keep a strong bone at the acetabular rim. Conversely, the medial floor is not involved in fixation; therefore, the center of the metal shell does not need to be in contact with the bone, and bone defects in the medial floor can be accepted. On the other hand, the main cement fixation site is the area where compressive stress is applied between the bone cement and the bone. On the acetabular side, the acetabulum from the superior margin to the medial floor is the main area that is subjected to compressive stress; therefore, it is necessary to preserve the bone in this area. The purpose of preserving the bone at the superior margin is to create an enclosed shape as much as possible and to reduce the shear forces applied to the bone-cement interface by leveling the acetabular roof. Bone grafting is often used for this purpose, especially in cases of acetabular dysplasia.

Reaming is initiated at the planned position. The bone is reamed to the largest extent possible preserving the anterior and posterior acetabular walls because a thicker cemented cup is prone to polyethylene wear. Medialization should be performed while maintaining approximately 5 mm of the cancellous bone on the medial floor. Care should be taken to avoid excessive medialization. After reaming, anchor holes are drilled in the remaining eburnation bone or strong cancellous bone in the acetabulum to increase the bone-cement contact area.

Bone grafting is often performed when defects exist at the superior margin of the acetabulum. A graft is made from the resected femoral head. In most cases, the mechanical strength of the femoral head near the articular surface is weak because of the formation of bone cysts and osteophytes; in contrast, the femoral head near the femoral neck is slightly smaller but consists of hard bone. The femoral head is shaped to fit the defect using a small reamer or a Surgairtome. The grafted bone is temporarily fixed with two Kirschner steel wires (K-wire) and then fixed with two metal screws or absorbable screws.

Once the bony bed is prepared, blood and bone fragments are washed away with pulse lavage, gauze is packed, and pressure hemostasis is applied. Hypotensive anesthesia is effective for hemostasis. If bleeding is still observed, the reamed bone chips can be pressed into the bleeding spot to achieve hemostasis. The bony bed is kept as dry as possible until cement injection.

(b) Pressurization of bone cement

The cup is set in the holder and the flange of the cup is trimmed to cover the acetabular defect. If an excessive flange remains, the redundant flange may catch on the soft tissue around the acetabulum during insertion, causing insufficient medialization. Therefore, smooth insertion should be confirmed before cementation.

After mixing the bone cement with a vacuum mixing system, the cement is injected into the acetabulum using a cement gun. Since pressurization cannot be achieved by the cement gun on the acetabular side, bone cement may be filled into the acetabulum by hand. If a medium-viscosity cement is used, it should be used after its viscosity becomes slightly higher than that of the femoral side. The remaining bone cement can be applied to the posterior surface of the flange and back of the cup. After cement injection, a pressurizer was used to interdigitate the bone cement into the cancellous bone. High pressure is applied to the bone cement by pressing the pressurizer vigorously, rather than gently. However, it is not recommended to push too deeply into the acetabulum, as the bone cement in the acetabulum would be lost. The acetabulum was refilled with the overflowing bone cement, and this procedure is repeated two or three times.

The cup was initially inserted into the acetabulum at an abduction angle of less than 45°. After sufficient medialization, the cup is adjusted to the planned angle, which facilitated the pressurization of the bone cement at the superior part of the acetabulum. The overlying bone cement is removed, and the periacetabular osteophytes are resected.

4 Interface Bioactive Bone Cement Technique (IBBC)

In conventional cementing techniques, bone and bone cement are only mechanically fixed, but the interface bioactive bone cement technique (IBBC technique) is a method that increases the fixation force between bone and bone cement by adding biological fixation through the interposition of hydroxyapatite granules at the interface [3]. This technique was first performed by Oonishi in 1982 and can be used on the femoral or acetabular side. By spreading hydroxyapatite granules on the dry bony bed immediately before cement injection, the interface becomes a graded functional material of bone, hydroxyapatite granules, and bone cement, in that order. Since the hydroxyapatite granules used are hardly absorbed, the bone formed around the hydroxyapatite granules persists, and the effect is expected to last for a long period of time. Compared to the hydroxyapatite-coated cementless implants, this technique can be recognized as a hydroxyapatite-coated bone cement. As the hydroxyapatite coating of the cementless implant is thin and absorbable, its long-term effect is not anticipated. The procedure is relatively easy to perform on the acetabular side. After preparation of the bony bed, gauze packing is applied, and the bony bed is kept dry. Once the bone cement is mixed and ready for use, the gauze is removed. After washing with hydrogen peroxide to achieve hemostasis, 2–3 g of hydroxyapatite granules of 0.3–0.6 mm diameter are quickly spread over the bone surface with a spoon and wiped with gauze, resulting in an entire bone surface covered with several layers of hydroxyapatite granules. Bone cement is then injected to fix the cups.

Hemostasis on the femoral side is not always easy, and it is difficult to apply hydroxyapatite granules over the entire surface of the femoral canal. Hemostasis is achieved by hypotensive anesthesia and washing with hydrogen peroxide before the application of hydroxyapatite granules. For the application of hydroxyapatite granules, a half-cut silicone tube with a diameter of 1 cm and length of 3–5 cm is held using long forceps to form a spatula. As in the acetabular side, 2–3 g of hydroxyapatite granules with a diameter of 0.3–0.6 mm are placed in the proximal part of the femoral canal, and then, hydroxyapatite granules are spread from proximal to distal over the entire surface of the femoral canal with a prepared spatula. The spatula is moved in and out several times to spread granules. After spreading, the bone cement is injected with a cement gun to fix the stem.

5 Antibiotic-Loaded Bone Cement (ALBC)

One of the advantages of bone cement is that it can function as a drug delivery system by mixing antibiotics with bone cement. Bone cement premixed with gentamicin or tobramycin at concentrations of 2.1–4.22% is commercially available and is used in most cases in the United States and Europe. However, in Japan, its use is limited because it is only approved for use in the second stage of revision arthroplasty, after periprosthetic infection. Although the elution of antibiotics rapidly decreases within 2–3 days and long-term efficacy is not expected, it is considered to be effective in the prevention of early infection. It has been reported that the use of antibiotic-loaded bone cement not only reduces the incidence of periprosthetic infections but is also effective against aseptic loosening [4].

6 Clinical Results

The outcome of cemented THA (total hip arthroplasty) is more influenced by the implant than by the cementing technique. The femoral side is mainly influenced by the design of the stem, and the acetabular side is mainly influenced by the cup material, which is described in another chapter. Clinical results of the cementing techniques are presented below.

In a report comparing cementing techniques, the 20-year survival rates of the second-generation technique were 89% and 94% for the acetabular and femoral sides, respectively, and 97% and 100% for the third-generation technique 10 years after surgery, respectively [5] The results of the third-generation technique were excellent, with a reported 100% survival rate of the stem at 15 years postoperatively with an endpoint of revision due to loosening [6].

The line-to-line technique on the femoral side has been used for a long time, but it is not generally accepted because a thick cement mantle is used as the mainstay of the technique. In 2003, a proposal called the 'French paradox' was published, in which the significance of the thin cement mantle was stated [7]. Instead of making a thin cement mantle, the intended aim is to enhance cement pressurization during stem insertion. It has been reported that cement cracks are less likely to occur with a thin cement mantle after mechanical tests [8], and clinically, the 17-year postoperative survival rate was reported to be 96.8% for the CMK stem when radiographic loosening was used as the endpoint, [9] and the 10-year postoperative survival rate for the Trilliance stem was 100% [10]. The line-to-line technique makes cementing easy, especially for surgeons who are not accustomed to using bone cement.

On the acetabular side, bone grafting is an essential procedure in case of acetabular dysplasia. It has been reported that there were no cases of radiographic loosening at an average follow-up of 11 years when bone grafting was performed for acetabular dysplasia [11]. The acetabulum is reconstructed using L-shaped or D-shaped bone grafts depending on the size and shape of the bone defect, and the technique needs to be mastered.

The IBBC technique aims to improve clinical outcomes in the long term. Histological examination of the interface taken 14 and 18 years postoperatively still showed bone formation around the interposed hydroxyapatite granules, proving the long-term effect of hydroxyapatite granules against osteoporosis [12]. Clinically, good outcomes have been reported, with a low incidence of osteolysis and radiolucent lines even more than 25 years after surgery [3]. In a report focusing on the acetabular side, the survival rate with revision as the endpoint was 97.0% at an average of 8.3 years after surgery [13]. Unfortunately, it is not a commonly performed procedure, it is a good option for young patients who need long-term improvement and for elderly patients with poor bone formation capacity.

7 Bone Cement Implantation Syndrome (BCIS)

Hypoxemia, hypotension, shock, and other respiratory and circulatory dysfunctions may occur with the use of bone cement. This is known as bone cement implantation syndrome (BCIS). It has been reported to be fatal in severe cases; thus, the risk should be emphasized. Although the cause of BCIS has not been identified, systemic effects, such as vasodilatation caused by bone cement monomers and fat embolization after bone cement injection, have been considered. In Japan, the Ministry of Health, Labor, and Welfare (MHLW) issued an alert in 1992, but there have been reports of outbreaks and fatalities since then. Recently, the incidence of BCIS was reported in 24% of total hip arthroplasty patients, 31% of bipolar hemiarthroplasty patients, and 28% of total knee arthroplasty patients, and the mortality rate within 30 days after surgery was high in severe cases (BCIS grade 2 or higher) [14]. It is important that not only orthopedic surgeons, but also anesthesiologists and nurses, cooperate to prevent the occurrence of BCIS by understanding the risk factors for its occurrence and by mastering appropriate cementing techniques to prevent it.

Patient risk factors include age > 75 years, poor general condition (ASA class III and IV), impaired renal function, osteoporosis, and femoral neck fractures. Pressurization of low-viscosity bone cement into the femoral canal with fat tissue is the most dangerous practice. It is important for orthopedic surgeons to maintain a constant temperature in the operating room, so that bone cement can be used under constant conditions, to use a cement plug to prevent cement injection into the femoral canal too distally, to lavage the medullary cavity to remove fat and blood, to inform the anesthesiologist before bone cement injection, and to avoid excessive pressurization of bone cement. The anesthesiologist should check the patient's general condition frequently before the use of bone cement and prepare for hypotension caused by vasodilatation by providing adequate infusion and hypertensive drugs.

It has been reported that only 2% of BCIS cases causes severe cardiorespiratory complications and 0.5% of BCIS cases result in cardiopulmonary arrest [15]. Although the frequency of severe complications is low, orthopedic surgeons, anesthesiologists, and nurses should coordinate efforts to prevent BCIS.

References

1. Cassar-Gheiti AJ, et al. Current concepts and outcomes in cemented femoral stem design and cementation techniques: the argument for a new EFORT. Open Rev. 2020;5:241–52.
2. Scheerlinck T, et al. CT analysis of defects of the cement mantle and alignment of the stem: in vitro comparison of Charnley-Kerboul femoral hip. J Bone Joint Surg Br. 2006;88:19–25.
3. Oonishi H, et al. Total hip arthroplasty around the inception of the Interface bioactive bone cement technique. Clin Orthop Surg. 2016;8:237–42.
4. Leong JW, et al. Is the use of antibiotic-loaded bone cement associated with a lower risk of revision after primary total hip arthroplasty? Bone Joint J. 2020;102-B:997–1002.
5. Hirose S, et al. Outcomes of Charnley total hip arthroplasty using improved cementing with so-called second- and third-generation techniques. Orthop Sci. 2012;17:118–23.
6. Rasquinha VJ, et al. Fifteen-year survivorship of a collarless, cemented, normalized femoral stem in primary hybrid total hip arthroplasty with a modified third-generation cement technique. J Arthroplast. 2003;18(7 Suppl 1):86–94.
7. Langlais F, et al. The 'French paradox'. J Bone Joint Surg Br. 2003;85:17–20.
8. Cristofolini L, et al. Effect of undersizing on the long-term stability of the Exeter hip stem: a comparative in vitro study. Clin Biomech (Bristol, Avon). 2010;25:899–908.
9. El Masri F, et al. Is the so-called 'French paradox' a reality?: long-term survival and migration of the Charnley-Kerboull stem cemented. J Bone Joint Surg Br. 2010;92:342–8.
10. Ohashi H, et al. Minimum ten-year outcome of a triple-tapered femoral stem implanted with a line-to-line cementing technique. BMC Musculoskelet Musculoskeletal Disord. 2021;22:601.
11. Oe K, et al. Long-term results of acetabular reconstruction using three bulk bone graft techniques in cemented total hip arthroplasty for developmental dysplasia. Int Orthop. 2016;40:1949–54.
12. Kim SC, et al. Histologic findings at 14 and 18 years after cemented total hip arthroplasty with interface bioactive bone cement technique. J Arthroplast. 2007;22:1067–9.
13. Miyamoto S, et al. Minimum 5 year follow-up of clinical and radiographic results of cemented acetabular components with an interface bioactive bone cement technique in primary cemented total hip arthroplasty. Arch Orthop Trauma Surg. 2021;141:139–47.
14. Rassir R, et al. What are the frequency, related mortality, and factors associated with bone cement implantation syndrome in arthroplasty surgery? Clin Orthop Relat Res. 2021;479:755–63.
15. Griffiths R, et al. Guideline for the management of hip fractures 2020: guideline by the Association of Anesthetists. Anesthesia. 2021;76:225–37.

Part VI

Implant Design

Anatomical Stem

31

Takashi Sakai

1 Concept

An anatomical stem has a large canal fill occupied proximal to both medially and laterally as well as anteroposteriorly. It is intended to conform to the femoral canal and acquire osseointegration [1]. It is classified as an anatomical type according to Kuhanuja's classification [2].

The curved stem has a left-right distinction with a lateral curvature. A custom-made stem is the ultimate curved anatomic type, created based on the femoral canal cavity in each case. Typical stem types include the PCA stem, anatomic stem, S + G (Lübeck) stem, ANCA-fit stem, CentPillar stem, and mainstay stem, which are designed based on the femoral canal cavity morphology of Japanese people [3] (Figs. 31.1 and 31.2).

Recently, short anatomical stems have been introduced from the viewpoint of bone preservation [4] and are classified as type 3 in Kuhanuja's short stem design classification [5].

Fig. 31.1 Anatomical stem. (**a**) CentPillar stem; (**b**) Ancafit stem; (**c**) Profemur Am stem; (**d**) Mainstay stem

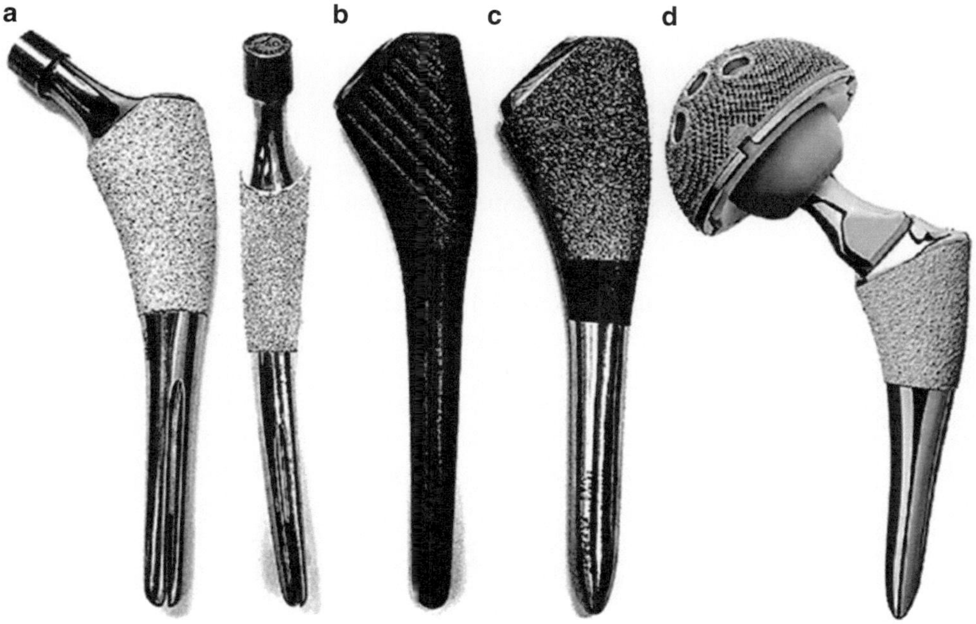

T. Sakai (✉)
Yamaguchi University, Ube, Yamaguchi, Japan
e-mail: cozy@yamaguchi-u.ac.jp

Y. Nakashima (ed.), *Advances in Total Hip Arthroplasty*, https://doi.org/10.1007/978-981-97-2461-1_31

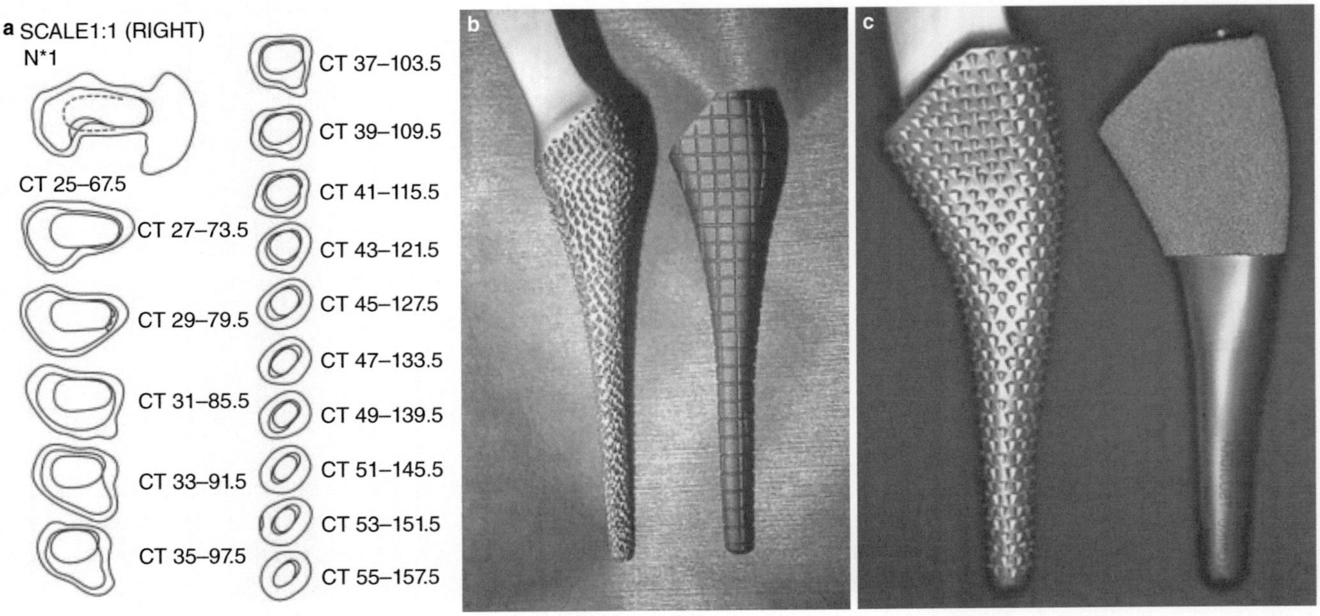

Fig. 31.2 Custom-made stem. (**a**) Determination of the stem cross-sectional shape in each CT section of the femur. (**b**) Custom-made stem with a special rasp and fully blasted surface. (**c**) Custom-made stem with a special rasp and proximal plasma spray processing

2 Theoretical Advantages

The anatomical stem has a large proximal canal fill, is easily adapted to the femoral canal, and is bonded to the bone. The anatomical stem is inserted and placed along the shape of the femoral medullary cavity so that both femoral anteversion and insertion level can be placed according to the native femoral anatomy and preoperative planning [6]. The alignment in the internal/external and extension-flexion directions can be easily achieved in the mid-position, regardless of the anterior or posterior surgical approach [7, 8].

The advantages of a short anatomical stem are that it promotes load stress transfer at the proximal portion, reduces stress shielding, reduces the difficulty of implant insertion through a small skin incision, limits the area of bone treatment while removal of implant, and facilitates implant fixation during revision surgery. However, obtaining initial stability in axial and rotational directions with limited stem length is necessary.

3 Surgical Techniques

Preoperative 3D templating is important for anatomical stems to reduce intraoperative stem size mismatches and poor stem alignment [9]. The femoral center should be elevated 5 mm proximal to the neck saddle position (Figs. 31.3 and 31.4). The stem size should be selected to maximize the fit and to fill the proximal medullary cavity from the level of the neck osteotomy to approximately 1 cm below the lesser trochanter.

Both anterior and posterior surgical approaches can be used. Recently, the stem length has been 100–125 mm, which is relatively short and easy to operate. It is difficult to intentionally increase or decrease stem anteversion excessively because of the shape of the femoral medullary cavity and its placement in accordance with the anteversion. The large canal fill occupied proximally makes it difficult to insert the stem in the neutral position without excavating the medial side of the greater trochanter, which may result in a varus position.

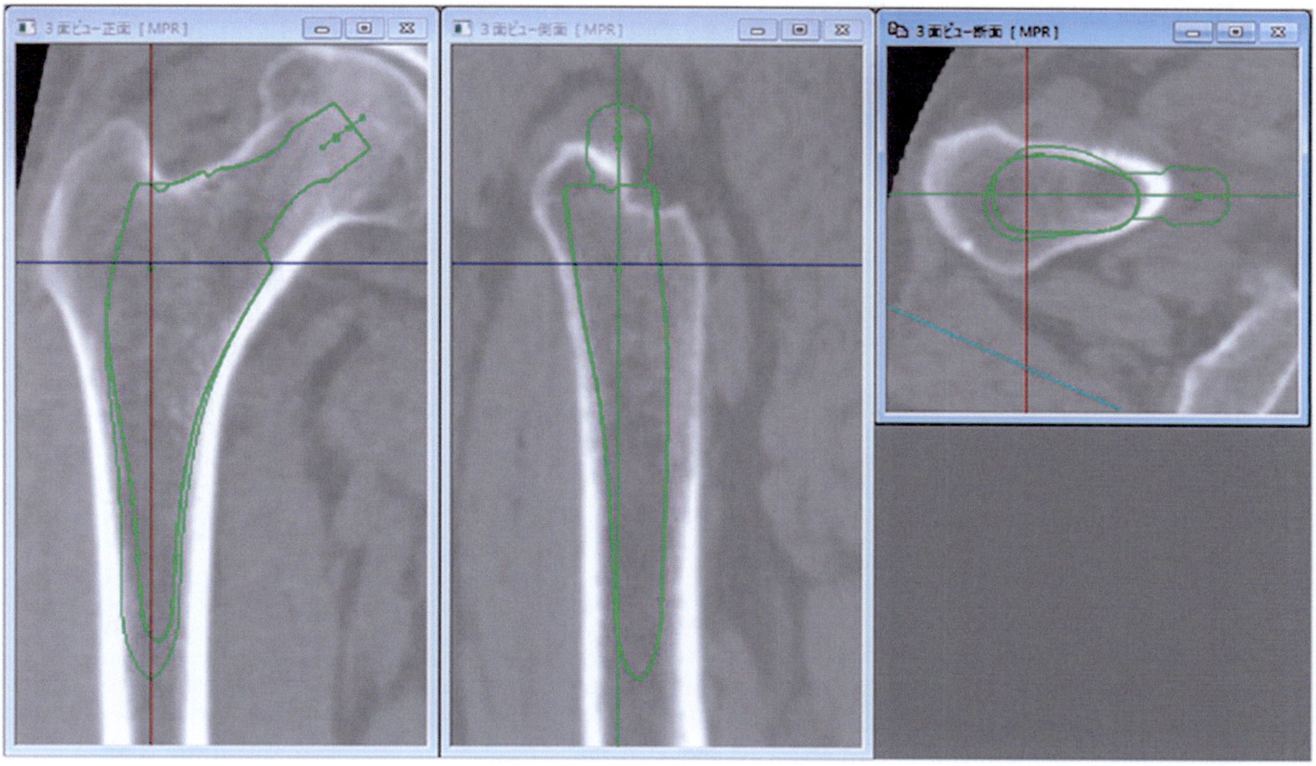

Fig. 31.3 Preoperative 3D template of the anatomical stem

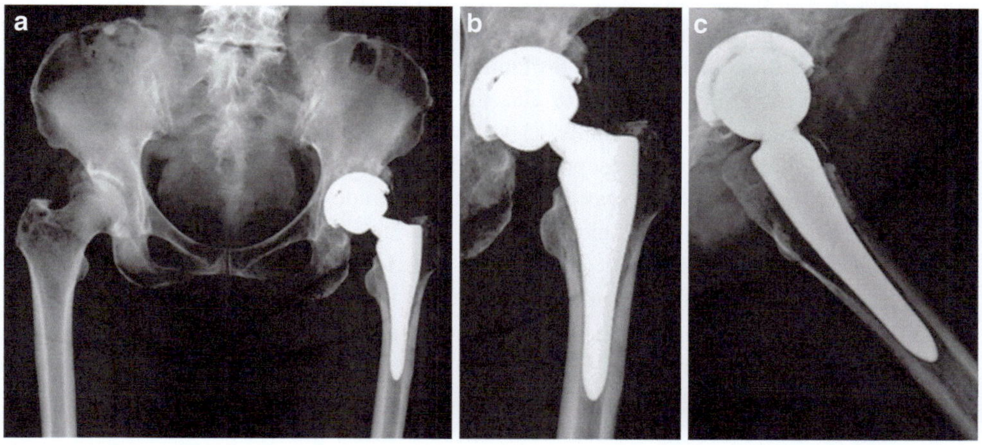

Fig. 31.4 Postoperative simple radiograph of anatomical stem. (**a**) Bilateral anteroposterior view of the hip joint; (**b**) anteroposterior view of the left femur; (**c**) Lateral view of the left femur

4 Indication

Anatomical stems can be used in almost all deformed femurs in various hip diseases. However, anatomical stems are not suitable in two situations: 1. In cases of a narrow femoral canal where both the proximal and distal medullary spaces are extremely narrow, even the smallest stem size may be difficult to insert. 2. In case of design incompatibilities, such as severe champagne-flute cases with an extremely narrow distal medullary cavity and large proximal cavity, distal fixation is not suitable because the proximal medullary cavity occupancy rate is low.

5 Clinical Results

The results are dependent on the stem design and surface finish, and anatomical stems with a design that provides a high proximal canal fill and an appropriate surface finish that allows osteogenesis to have good results [10–19]. Porous surface treatments such as blasting [12], spongiosis metal [10], fiber metal [15], plasma spray [14], and hydroxyapatite-coated stems [4, 11, 13, 15, 17–19] have shown favorable outcomes. PCA made of cobalt chromium with bead coating on the proximal one-third showed inferior cumulative survival [16] (Table 31.1).

Table 31.1 Medium- and long-term results of anatomical stems

Reporter	Implant	Material	Surface processing	Number of joints	Average postoperative period (years)	Cumulative survival rate (%)
Al Muderis et al. [10]	Lübeck (S + G)	CoCr	Spongiosa	209	21.8	86.0
Kim et al. [11]	IPS	Ti	HA	630	15	99.4
Sakai et al. [12]	Custom-made	Ti	Blast	74	14.5	98.6
Pakos et al. [13]	Custom-made	Ti	HA	86	10.5	95.4
Sugano et al. [14]	Axcel	Ti	Plasma spray	56	14	98.0
Kaneuji et al. [15]	Anatomic hip	Ti	Fiber metal/HA	135	13.5	100
Tezuka et al. [16]	PCA	CoCr	Beads	60	15.2	82
Uemura et al. [17]	CentPillar	Ti	HA	222	13	99.0
Herrera et al. [18]	ABG II	Ti	HA	196	11.3	98.4
Sariali et al. [19]	SPS	Ti	HA	131	10	100
Kim et al. [4]	Ultrashort	Ti	HA	858	16.5	97.6

Regarding the CentPillar stem, a Ti-6Al-4 V alloy stem with grid blasting and hydroxyapatite treatment on the proximal 1/3 of the stem showed good results than those by a conventional fit and fill stem [20]. In addition, a comparison of a beta-titanium alloy stem made of a Ti-12Mo-6Zr-2Fe alloy and an $\alpha + \beta$-titanium alloy stem made of a Ti-6Al-4 V alloy with different materials using the same stem design showed good clinical results for both stem types. In contrast, the width of the radiolucent line that appeared in Gruen zone 1 was reported to be wider in the $\alpha + \beta$-titanium alloy stem [21]. In a recent report on long-term results, the cumulative survivorship of implants in 222 joints 13 years postoperatively was 99% and 99.5% when aseptic loosening was used as the endpoint. Stress shielding was grade 1 (39.5%), grade 2 (52.9%), grade 3 (0.6%), and grade 4 [none] [17].

In the custom-made, proximal hydroxyapatite-coded stem intended to fit in the flare of the metaphyseal cavity, it was 95.4% at 10.5 years postoperatively [13]. In the case of a stem with fully blasted surface, it was reported to be 98% 14.5 years postoperatively [12].

The long-term results of anatomical stems with extremely short stem lengths showed that the cumulative survivorship was as good as 97.6% and 96.6% at 16.5 and 17.5 years postoperatively compared to anatomical stems with normal stem lengths. However, femoral pain and stress shielding were significantly more frequent and severe in the anatomical stem with normal stem length [4].

References

1. Noble PC, et al. The anatomic basis of femoral component design. Clin Orthop Relat Res. 1988;235:148–65.
2. Khanuja HS, et al. Cementless femoral fixation in total hip arthroplasty. J Bone Joint Surg Am. 2011;93:500–9.
3. Noble PC, et al. Three-dimensional shape of the dysplastic femur: implications for THR. Clin Orthop Relat Res. 2003;417:27–40.
4. Kim YH, et al. A prospective, randomized comparison of the long-term clinical and radiographic results of an ultra-short versus a conventional length cementless anatomic femoral stem. J Arthroplast. 2021;36:1707–13.
5. Khanuja HS, et al. Short bone-conserving stems in cementless hip arthroplasty. J Bone Joint Surg Am. 2014;96:1742–52.
6. Yoshitani J, et al. Anatomic stem inserted according to native anteversion could reproduce the native anterior distance of the femoral head and decrease bony impingement in total hip arthroplasty. Int Orthop. 2020;44(2):245–51.
7. Abe H, et al. Difference in stem alignment between the direct anterior approach and the posterolateral approach in total hip arthroplasty. J Arthroplast. 2015;30:1761–6.
8. Nakahara I, et al. Does difference in stem design affect accuracy of stem alignment in total hip arthroplasty with a CT-based navigation system? J Artif Organs. 2021;24:74–81.
9. Ogawa T, et al. Factors related to disagreement in implant size between preoperative CT-based planning and the actual implants used intraoperatively for total hip arthroplasty. Int J Comput Assist Radiol Surg. 2018;13:551–62.
10. Muderis M, et al. Cementless total hip arthroplasty using the Spongiosa-I fully coated cancellous metal surface: a minimum twenty-year follow-up. Bone Joint Surg Am. 2011;93:1039–44.
11. Kim YH, et al. Long-term results and bone remodeling after THA with a short, metaphyseal-fitting anatomic cementless stem. Clin Orthop Relat Res. 2014;472:943–50.
12. Sakai T, et al. A modular femoral neck and head system works well in cementless total hip replacement for patients with developmental dysplasia of the hip. J Bone Joint Surg Br. 2010;92:770–6.
13. Pakos EE, et al. Long term outcomes of total hip arthroplasty with custom made femoral implants in patients with congenital disease of the hip. J Arthroplast. 2015;30:2242–7.
14. Sugano N, et al. Eleven- to 14-year follow-up results of cementless total hip arthroplasty using a third-generation alumina ceramic-on-ceramic. J Arthroplast. 2012;27:736–41.
15. Kaneuji A, et al. Cementless anatomic total hip femoral component with circumferential porous coating for hips with developmental dysplasia: a. J Arthroplast. 2013;28:1746–50.
16. Tezuka T, et al. Long-term results of porous-coated anatomic total hip arthroplasty for patients with osteoarthritis of the hip. J Arthroplast. 2014;29:2251–5.

17. Uemura K, et al. Minimum 10 years clinical results of an anatomical short stem with a proximal hydroxyapatite coating. Mod Rheumatol. 2021;31:1066–72.
18. Herrera A, et al. Long-term outcomes of a new model of anatomical hydroxyapatite-coated hip prosthesis. J Arthroplast. 2013;28:1160–6.
19. Sariali E, et al. High 10-year survival rate with an anatomic cementless stem (SPS). Clin Orthop Relat Res. 2012;470:1941–9.
20. Koyano G, et al. Comparison of bone remodeling between an anatomic short stem and a straight stem in 1-stage bilateral total hip arthroplasty. J Arthroplast. 2017;32:594–600.
21. Miyatake K, et al. Comparison of different materials and proximal coatings used for femoral components in one-stage bilateral total hip. J Arthroplast. 2015;30:2237–41.

Fit and Fill Stem

32

Tetsuya Jinno

1 Concept and Development History

The concept of "fit-and-fill" is to obtain a strong and secure initial fixation by filling the femoral canal and fitting a large area with the inner surface of the cortical bone. Even though anatomical and custom-made stems also aim to fill the femoral canal, fit-and-fill stems in the narrow sense usually refer to the straight-type stem described in this chapter. A straight fit-and-fill stem is often referred to as a "dual taper stem" or "double-wedge stem" because it is tapered not only in the coronal plane but also in the sagittal plane [1, 2]. The distal portion of the stem is cylindrical or slightly tapered along the shape of the intramedullary diaphyseal canal. However, a fin-spline or longitudinal groove structure may be added to provide rotational fixation by diaphyseal engagement and a slot structure to reduce stiffness. Therefore, the primary site for medullary filling is mainly the metaphysis, and the stem is also called the metaphyseal-filling stem [1, 2]. Although the femoral canal filling rate is theoretically not as high as that of the anatomic or custom-made type if inserted properly, the straight type may have a higher fill rate than the anatomic type [3].

Since cemented stems fit the femoral canal with cement as a filling material, it was natural to fill the canal with the stem itself in the case of cementless stems. First-generation cementless stems such as AML (the anatomical medullary locking stem made of CoCr, DePuy), PCA (the porous-coated anatomic stem made of CoCr, Howmedica), and the Harris-Galante (made of Ti6Al4V, Zimmer) were designed to fill the femoral canal with the stem. However, thigh pain was the main problem in PCA stem. Additionally, while surface treatment for osteoconductivity is essential for cementless stems [4], in the case of AML with a fully porous surface finish, severe bone atrophy due to stress shielding was a problem. In the case of the Harris-Galante stem, although the surface finish was limited to the proximal region, extensive osteolysis due to the distal migration of wear particles from the articulating surface was observed because the proximal surface finish was not circumferential. Insufficient canal filling by the stem also caused long-term fixation failures [5]. The basic form of the fit-and-fill straight stem, a typical cementless stem for primary total hip arthroplasty (THA) with a proximal surface finish, has been developed on the basis of these results [1, 2]. Titanium alloys are used as the base material, and titanium fiber mesh, bead porous, and calcium phosphate coatings such as hydroxyapatite, are sometimes added to the proximal surface finish for bone ingrowth (or ongrowth), all of which have reported good long-term results [1, 2, 6–11].

2 Theoretical Advantages

The stem occupies the femoral canal and makes extensive contact with the cortical bone, resulting in a firm and stable initial fixation, leading to a long-lasting biological fixation (Fig. 32.1). Some anatomical variations and uncertainties in surgical techniques are compensated for by a stem design that can conform to the femoral canal over a wide area, thus reducing the initial fixation failure risks. Distal reaming can also be used to confirm the optimal size and stem axis, which has the advantage of facilitating the placement of the planned stem size in the planned neutral position. Even though the medullary canal diameter defines the implant size to some extent, there are many systems with multiple options for the size and shape of the proximal metaphysis and neck offset [1, 3, 11], which can be adapted to most cases of initial THA.

T. Jinno (✉)
Department of Orthopaedic Surgery, Dokkyo Medical University Saitama Medical Center, Koshigaya, Saitama, Japan
e-mail: jinnot@dokkyomed.ac.jp

Y. Nakashima (ed.), *Advances in Total Hip Arthroplasty*, https://doi.org/10.1007/978-981-97-2461-1_32

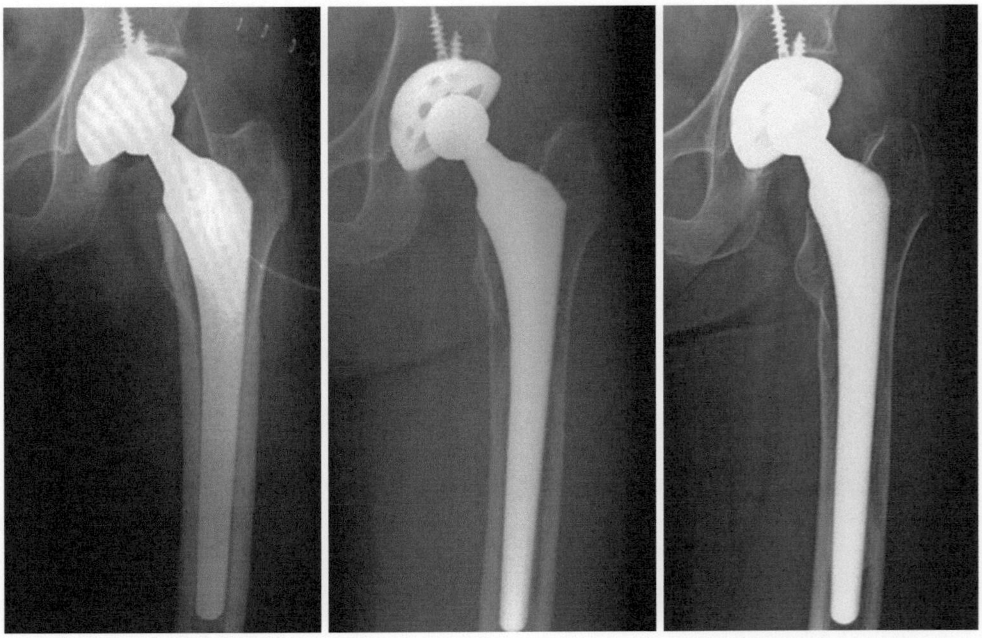

Fig. 32.1 Long-term radiographic change of fit-and-fill stem. A fit-and-fill stem (Secur-Fit AD-HA stem, Stryker) maintains bony fixation for a long period of time. However, in this case, the bone atrophy due to stress shielding becomes more severe over time

Owing to the easy-to-understand concept of filling the femoral canal and the simple insertion procedure for a straight stem, which is common to both the left and right sides, the fit-and-fill stem was the most widely used stem for primary THA in Japan until the 2000s, when tapered wedge stems came into use [3–30].

3 Surgical Techniques

(a) Stem insertion point and partial release of the posterior capsule and short external rotators

It is important to understand that the stem insertion point is the extension of the proximal diaphyseal axis, i.e., the posterior medial portion of the apex of the greater trochanter, because the stem is straight and medullary canal filling (Fig. 32.2). Inevitably, the posterior neck cortex will be partially resected, and the higher the neck osteotomy, the greater the amount of cortex that will be resected. Muscle tendons that insert the trochanteric fossa and medial portion of the greater trochanter also often require release. The posterior edge of the gluteus medius muscle only needs to be partially released, but the short external rotators need to be extensively released. So, it is difficult to take advantage of preservation of the posterior capsule and short external rotators, which is an advantage of the anterior approaches [13]. However, in recent years, a fit-and-fill stem with a shorter length and reduced shoulder has been developed and used. It can be inserted with a slightly flexed alignment relative to the diaphyseal axis (without aiming for a complete filling at the diaphysis) to take advantage of the advantages of anterior approaches [31].

(b) Reaming and rasping of the femoral canal

The medullary canal is reamed with a cylindrical or conical reamer, and the proximal metaphyseal cavity is rasped with a double taper. Although there are fit-and-fill stems that only require rasping, meticulous rasping is necessary to avoid undersizing and malalignment in cases with good bone quality. Rasping is complete when the planned size of the rasp does not progress near the planned height, even after multiple careful taps. Note that the stem is placed at the same or slightly higher position than the final rasp because of press-fit fixation. Intraoperative fractures should always be considered with cementless stems, particularly with a fit-and-fill stem [2, 14]. Occult fractures should also be suspected when the rasp or stem fixation is insufficient or when the stem is easily inserted deeper than the rasp.

(c) Anteversion adjustment

In the fit-and-fill stem, stem alignment is defined by the proximal diaphyseal axis, and the anteversion angle is defined by the principle of inserting the largest stem. Typically, the anteversion angle is 10–20° greater than the anatomical anteversion angle (Fig. 32.2) [15–17]. Intentional adjustment of stem alignment, including anteversion, is difficult and not recommended. However, this can be regarded as an advantage of the fit-and-fill stem, since the anteversion angle of the stem can be easily predicted [17, 18].

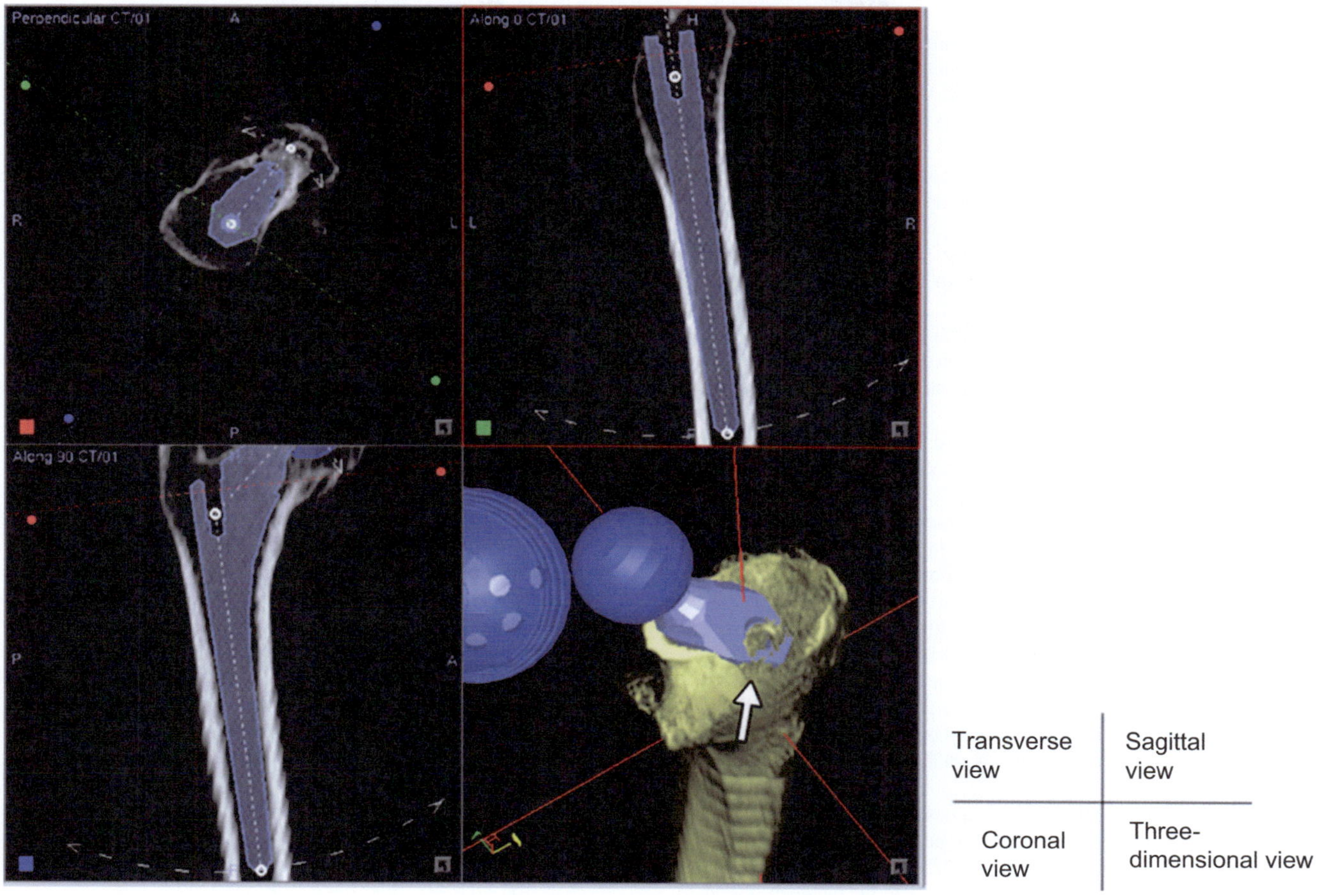

Fig. 32.2 Preoperative planning using a fit-and-fill stem. Three-dimensional planning is performed using a fit and fill stem (Secur-Fit AD-HA stem, Stryker). The stem is inserted along the proximal diaphyseal axis. Sagittal and coronal views show high femoral canal filling. The transverse section shows that the anteversion of the stem is larger than that of the neck. In the three-dimensional proximal posterior view, a portion of the posterior cortical bone at the proximal end of the neck cut is resected (arrow)

4 Indication

Fit-and-fill straight stems are widely indicated for usual primary THA cases, but in cases of abnormal (excessive or too small) anteversion requiring stem anteversion adjustment, the straight fit-and-fill straight stem without options such as a changeable neck is not indicated [17, 18]. Although fit-and-fill straight stems are expected to achieve long-term durability, even for Dorr type C bones, they are not a good indication because of concerns about intraoperative and postoperative fractures and bone atrophy due to stress shielding [19–22], in addition to the need to avoid excessive offset due to the unproportionally widened intramedullary canal.

5 Clinical Results

Survival rates with endpoints of revision or loosening are extremely good; 95–100% have been reported at 10–20 years, including Dorr type C cases and younger cases, indicating good long-term durability [1, 2, 6–11].

However, maintenance of bone quality around the stem is an important issue for stems that are expected to last for a long time. In the case of the fit-and-fill stem, the amount of cancellous bone removed by surgery is naturally larger because the femoral canal is intended to be filled. Additionally, bone atrophy due to stress shielding may progress over a long period around the fit-and-fill stem because of the relatively high stiffness of the stem with a large volume (Fig. 32.1) [8, 9, 19–22], which may lead to periprosthetic fractures and difficulties during revision [22, 32]. Many studies on stress shielding for fit-and-fill stems have been conducted in Japan, including implant factors such as stem shape, length, size, and surface finish [23–26, 31]; surgical technique [27, 28]; and patient factors such as femoral canal shape and activity [19–21, 29]. The efficacy of bisphosphonates as a countermeasure to stress shielding has also been suggested [30].

References

1. Khanuja HS, et al. Cementless femoral fixation in total hip arthroplasty. J Bone Joint Surg Am. 2011;93:500–9.

2. Kheir MM, et al. An update on cementless femoral fixation in total hip arthroplasty. J Bone Joint Surg Am. 2020;102:1646–61.
3. Nishihara S, et al. Comparison of the fit and fill between the anatomic hip femoral component and the VerSys taper femoral component using virtual implantation on the ORTHODOC workstation. J Orthop Sci. 2003;8:352–60.
4. Kitamura S, et al. Catastrophic failure of cementless total hip arthroplasty using a femoral component without surface coating. J Arthroplast. 1999;14:918–24.
5. Inoue S, et al. A 10- to 13-year follow-up study of Harris-Galante type prosthesis in total hip arthroplasty. J Orthop Sci. 2000;5:561–6.
6. Nakashima Y, et al. Results at a minimum of 10 years of follow-up for AMS and PerFix HA-coated cementless total hip arthroplasty: impact of cross-linked polyethylene on implant longevity. J Orthop Sci. 2013;18:962–8.
7. Miyakawa S, et al. Grit-blasted and hydroxyapatite-coated total hip arthroplasty: an 11- to 14-year follow-up study. J Orthop Sci. 2004;9:462–7.
8. Saito S, et al. The Harris-Galante cementless THA: a 19- to 25-year follow-up study. Orthopedics. 2011;34:12.
9. Naito Y, et al. Minimum 10-year follow-up of cementless total hip arthroplasty with a 32-mm cobalt-chromium head on highly cross-linked polyethylene and a tapered, fiber metal proximally coated femoral stem. J Arthroplast. 2021;36:647–52.
10. Kawamura H, et al. The 21- to 27-year results of the Harris-Galante cementless total hip arthroplasty. J Orthop Sci. 2016;21:342–7.
11. Takatori Y, et al. Ten-year follow-up of a proximal circumferential porous-coated femoral prosthesis: radiographic evaluation and stability. J Orthop Sci. 2002;7:68–73.
12. Bo A, et al. Fit and fill analysis of a newly designed femoral stem in cementless total hip arthroplasty for patients with secondary osteoarthritis. J Orthop Sci. 1997;2:301–12.
13. Xu D, et al. Comparison of hip muscle volume between fit-and-fill stem and tapered-wedge stem after total hip arthroplasty using the anterolateral approach. J Orthop Sci. 2017;22:1102–6.
14. Kawaji H, et al. Mid-term clinical results of Versys hip system (Zimmer) uncemented total hip replacement arthroplasty. J Nippon Med Sch. 2016;83:184–7.
15. Nakashima Y, et al. Combined anteversion technique reduced the dislocation in cementless total hip arthroplasty. Int Orthop. 2014;38:27.
16. Hirata M, et al. Influencing factors for the increased stem version compared to the native femur in cementless total hip arthroplasty. Int Orthop. 2014;38:1341–6.
17. Taniguchi N, et al. Cementless hip stem anteversion in the dysplastic hip: a comparison of tapered wedge vs metaphyseal filling. J Arthroplast. 2017;32:1547–52.
18. Imai H, et al. Preoperative planning and operative techniques of the shorter tapered stem compared to the metaphyseal fit stem in cementless total hip arthroplasty. J Arthroplast. 2017;32:1192–9.
19. Tsubosaka M, et al. Patients with a Dorr type C femoral bone require attention for using a summit cementless stem: results of total hip arthroplasty. J Orthop Sci. 2018;23:671–5.
20. Kusano T, et al. Preoperative canal bone ratio is related to high-degree stress shielding: a minimum 5-year follow-up study of a proximally hydroxyapatite-coated straight tapered titanium femoral component. J Arthroplast. 2018;33:1764–9.
21. Nishino T, et al. Follow-up results of 10-12 years after total hip arthroplasty using cementless tapered stem – frequency of severe stress shielding. J Arthroplast. 2013;28:1736–40.
22. Hirao M, et al. Periprosthetic fragility fracture of the femur after primary cementless total hip arthroplasty. Mod Rheumatol. 2022;32:626–33.
23. Sano K, et al. Changes of bone mineral density after cementless total hip arthroplasty with two different stems. Int Orthop. 2008;32:167–72.
24. Inaba Y, et al. Difference in postoperative periprosthetic bone mineral density changes between 3 major designs of uncemented stems: a 3-year follow-up study. J Arthroplast. 2016;31:1836–41.
25. Yamauchi Y, et al. Comparison of different distal designs of femoral components and their effects on bone remodeling in 1-stage bilateral total hip. J Arthroplast. 2012;27:1538–43.
26. Koyano G, et al. Comparison of bone remodeling between an anatomic short stem and a straight stem in 1-stage bilateral total hip arthroplasty. J Arthroplast. 2017;32:594–600.
27. Sanki T, et al. Ten-year outcomes of total hip arthroplasty using fit-and-fill type cementless collared straight stem implants: relationship between the initial contact status and stress shielding. Acta Med Okayama. 2020;74:7–15.
28. Hananouchi T, et al. Effect of robotic milling on periprosthetic bone remodeling. J Orthop Res. 2007;25:1062–9.
29. Hayashi S, et al. Periprosthetic bone mineral density with a cementless triple tapered stem is dependent on daily activity. Int Orthop. 2012;36:1137–42.
30. Yukizawa Y, et al. Efficacy of alendronate for the prevention of bone loss in the calcar region following total hip arthroplasty. J Arthroplast. 2017;32:2176–80.
31. Hayashi S, et al. Hydroxyapatite-coated compaction short stem represents a characteristic pattern of peri-prosthetic bone remodelling after total hip arthroplasty. Arch Orthop Trauma Surg. 2022;142:2903–10.
32. Carli AV, et al. Periprosthetic femoral fractures and trying to avoid them: what is the contribution of femoral component design to the increased risk of periprosthetic femoral fracture? Bone Joint J. 2017;99-B(1 Supple A):50–9.

Taper Wedge Stem

33

Toru Nishiwaki

1 Concept

The history of tapered wedge (TW)-type stems is long; the representative stems of the first generation are the TRI-LOCK stem (DePuy, Warsaw, IN, USA; Fig. 33.1a) and the Taperloc stem (Zimmer Biomet, Warsaw, IN, USA; Fig. 33.1b), which were developed in the early 1980s. The standard TW-type stems initially introduced in Japan were the Linear Hip System (approved and marketed in 1999; Fig. 33.1c) by SENKO MEDICAL INSTRUMENT Mfg. CO., LTD. and the CTiII™ (Corin, Cirencester, UK) (approved in 1996 and marketed in 2003; Fig. 33.1d). Later, in 2012, the short TW-type stem was introduced. The fixation concept is designed to fit the tapered medial and lateral cortical bone at the proximal metaphyseal end with the tapered coronal cross-section of the stem and to obtain rotational fixation with a flat shape with a thin anteroposterior width without a collar.

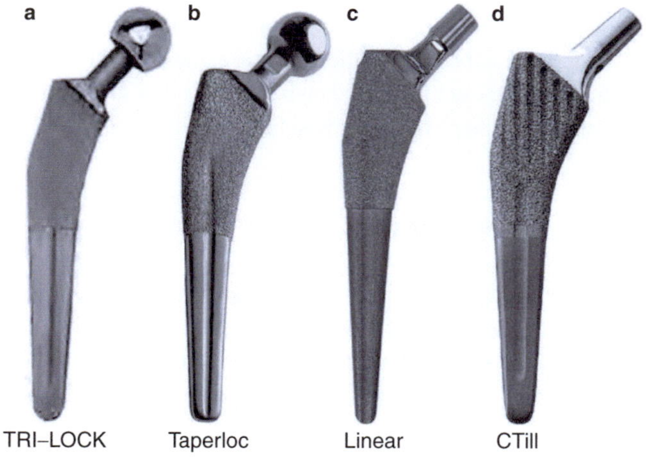

Fig. 33.1 Taper wedge stems

T. Nishiwaki (✉)
Shizuoka Red Cross Hospital, Shizuoka, Japan

2 Theoretical Advantages

Minimally invasive surgery (MIS) has been advocated for since the early 2000s, and its frequency has increased yearly due to improvements in surgical instruments. The TW-type stem is one of the most used implants in MIS because the outer side of the stem is cut and does not protrude, making insertion easy. Additionally, its thin anteroposterior width allows bone preservation in the greater trochanter and anteroposterior region. Since the load is transmitted to the proximal femur, proximal femoral atrophy due to stress shielding is considered less severe [1, 2]. Furthermore, it has a high degree of freedom and ease of control in the medullary cavity [3, 4]. The short TW-type stem is less painful in the femur, allows bone preservation, and is especially suitable for young patients.

3 Surgical Techniques

TW-type stems do not require distal reaming, and broaching provides sufficient contact between the bone and implant to achieve primary stability, resulting in biological fixation. There are three primary broach surface shapes: angled pattern, diamond pattern, and compression (Fig. 33.2). The broaches used in the TW-type stem are the first two patterns. The angled pattern is used in Accolade II (Stryker, Mahwah, MJ, USA), Trilock BPS (DePuy, Warsaw, IN, USA), and Taper Kinectiv (Zimmer Biomet, Warsaw, IN, USA); meanwhile, the diamond pattern is used in J-taper (Kyocera, Kyoto, Japan) and Anthology (Smith & Nephew, Memphis, TN, USA). The sharp teeth of the angled pattern, the friction coefficient during broaching, and the broaching efficiency are high [5]. The diamond pattern and the compression broach have the same friction coefficient. However, the internal pressure on the medullary cavity during broaching and the fracture risk are high. Therefore, angled pattern broaches are recommended in patients with poor bone quality [5]. In our study, however,

Y. Nakashima (ed.), *Advances in Total Hip Arthroplasty*, https://doi.org/10.1007/978-981-97-2461-1_33

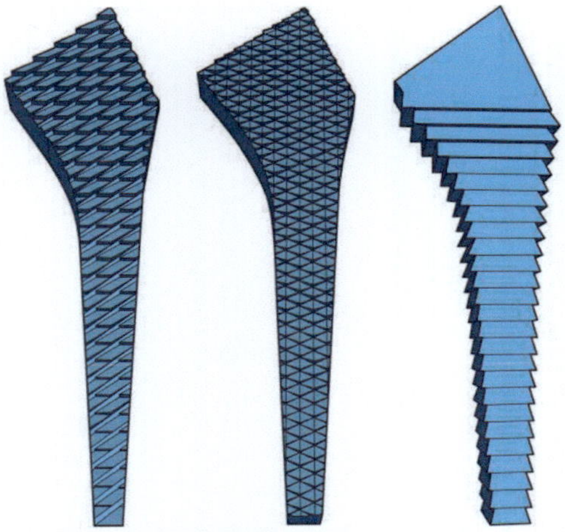

Fig. 33.2 Broach surface shape

using a simulated bone, the rotational stability of the diamond pattern broach was higher than that of the angled pattern broach. The groove in the angled pattern broach is perpendicular to its axis, which may result in lower rotational resistance in the medullary cavity. Angled pattern broaches should be used with caution when performing a trial reduction or dislocation, as these are susceptible to rotational forces.

Stem alignment is affected by the femoral marrow cavity morphology and stem shape. TW-type stems can be placed in various alignments because they are thinner both anteriorly and posteriorly, and they are more flexible than fit- and fill-type stems [3, 4]. This can be both an advantage and a disadvantage for the surgeon. It is reported that the anterior torsion angle of the femoral neck, which has varied to various angles due to developmental dysplasia, can be surgically controlled to reduce the variation of the anterior torsion [6]; this is because fixation of the tapered portion can be achieved, even if the hip is rotated to some extent and the anterior-posterior torsion can be easily adjusted. Conversely, there are reports that variations in anterior torsion and sagittal and coronal plane alignment can be increased [3].

The degree of freedom of placement alignment of the TW-type stem is affected by the position and entry route at the time of surgery and the patient's physique [4, 7, 8]. In an anterior approach, the TW-type stem is inserted with more flexion than in a posterior approach [7]. In a study examining the effect of body position (supine or lateral recumbency) on the same entry path (anterolateral intermuscular entry), the anteversion angle of the stem and the difference in anteversion angle before and after surgery were more significant in the lateral position, while the flexion insertion was greater in the supine position [4]. It has been suggested that a decrease in intraoperative femoral maneuverability may lead to increased flexion insertion, which may be resolved by using a leg positioner [4]. In addition, it has been reported that patients with a high body mass index (BMI) are more likely to have flexion insertion [8]. Insertion in more than 3° of flexion increases the frequency of intraoperative fractures [9]. Therefore, especially in patients with a high BMI, care should be taken to avoid excessive flexion when using a TW-type stem in the supine anterior approach.

In muscle-sparing MIS, muscle preservation is essential for postoperative stability and activity. For these cases, a TW-type stem is reported to preserve the obturator internus muscle better than a fit & fill type stem [10]. The short TW-type stem is considered the best solution for preserving the obturator internus muscle because of its stem length and its cut-off outer portion, allowing bone preservation at the medial part of the greater trochanter. However, there is a report [11] that the short TW-type stem is more likely to cause peri-prosthetic fracture intraoperatively or postoperatively when inserted with muscle-sparing anterior approaches.

4 Criteria

TW-stems can be applied to a wide range of cases. However, radiographic results are poor in patients with high-stem occupancy in the diaphysis and low-stem occupancy in the proximal femur [12]. Therefore, for cases with a disproportionate medullary cavity diameter between the proximal and distal femur, stem selection should consider stem design and the extent of surface treatment [12].

In Dorr classification type C, there are concerns about insufficient initial fixation, such as rotational fixation and subsidence. However, the TW-type stem can be applied to Dorr classification type C because it is initially fixed by the femoral neck and metaphyseal end fixation without depending on hoop stress in the medullary canal [13]. Dorr classification type C is seen in older women and young adults with inflammatory joint disease. Especially in the latter case, the advantage of the short TW-type stem is significant because it allows reconstruction using the diaphysis if a revision is considered [14].

5 Clinical Results

The 25-year survival rate of the standard TW-type stem with revision as the endpoint is 99% [15], and the 5-year survival rate of the short TW-type stem is reported to be 99.2% [16], both of which are excellent. Prospective comparative studies of standard and short TW-type stems have shown no difference in clinical outcomes or longevity [17]. However, some reports have pointed out an increased risk of malposition, persistent thigh pain, laxity, subsidence, and proximal stress shielding with short TW-type stems [18, 19].

The number of flexion insertions in the sagittal plane increased due to the increase in MIS. There is no difference in clinical and radiographic outcomes compared with intermediate insertion in the sagittal plane [9]. The change in bone mineral density around the stem was negatively correlated with the anteversion of the stem but not with flexion [20]. Accordingly, flexion alignment will not be considered to affect load transfer.

Dorr classification type C is an unfavorable environment for implant fixation due to the wide femoral marrow cavity and thinning of cortical bone, which may result in impaired biological activity. Intraoperative fracture, osteolysis with stem migration, periprosthetic fracture, and poor long-term outcomes may be associated with various adverse events. However, rotational stability can be achieved with a short TW-type stem immediately after surgery [21]. The 20-year postoperative survival rate of a standard TW-type stem for Dorr classification type C is 98% [22].

Stem subsidence, one of the complications, affects clinical outcomes and is related to periprosthetic fractures. In the conventional fit-and-fill type stem, subsidence correlates with BMI; however, in the TW type stem, it does not, and the amount of subsidence is small [23]. In TW-type stems, it is important to fit the stem firmly in the coronal plane to prevent subsidence [23].

Thigh pain occurs more frequently with cementless stems than with cemented ones. This is associated with various factors, such as patient activity, surgical approach, prosthesis design, size, modulus of elasticity, the extent of porous coating, alignment, osteoporosis, and bony ingrowth. The degree of pain is less than VAS 3 in approximately 75% of patients with TW-type femoral pain and often resolves within 1 year [24]. Furthermore, risk factors for thigh pain in TW-type stems include the UCLA activity score [24], Dorr classification type C [24], stem tip contact with cortical bone [24], and the area of contact between the stem and cortical bone in the distal portion of the porous process of the stem in zone two of the Gruen classification [25]. To prevent postoperative thigh pain, it is necessary to pay attention to patient activity, Dorr classification type C, and intermediate insertion [24], as well as to confirm the contact area between the stem and cortical bone in the zone two of the Gruen classification using density mapping in three-dimensional template software during preoperative planning [25].

References

1. Yamaguchi K, et al. Evaluation of periprosthetic bone-remodeling after cementless total hip arthroplasty. The influence of the extent of porous coating. J Bone Joint Surg Am. 2000;82:1426–31.
2. Gallart X, et al. Early clinical and radiological outcomes for the Taperloc complete microplasty stem. Eur J Orthop Surg Traumatol. 2019;29:619–24.
3. Taniguchi N, et al. Cementless hip stem anteversion in the dysplastic hip: a comparison of tapered wedge vs metaphyseal filling. J Arthroplasty. 2017;32:1547–52.
4. Takada R, et al. Difference in tapered wedge stem alignment between supine and lateral position in cementless total hip arthroplasty via modified Watson-Jones anterolateral approach. Eur J Orthop Surg Traumatol. 2022;32:497–503.
5. Bätz J et al. The influence of broach design on bone friction and osseodensification in total hip arthroplasty. Clin Biomech (Bristol, Avon) 2020;73:234–40.
6. Hayama T, et al. Clinical results of a short stem with flat tapered wedge design in primary total hip arthroplasty for hip dysplasia in Asians. J Orthop Surg (Hong Kong). 2020;28:2309499020956742.
7. Abe H, et al. Difference in stem alignment between the direct anterior approach and the posterolateral approach in total hip arthroplasty. J Arthroplasty. 2015;30:1761–6.
8. Hayashi S, et al. The contributing factors of tapered wedge stem alignment during mini-invasive total hip arthroplasty. J Orthop Surg Res. 2015;10:52.
9. Yoshitani J, et al. The effect of flexion alignment in total hip arthroplasty with a cementless tapered-wedge femoral stem. Eur J Orthop Surg Traumatol. 2018;28:1625–32.
10. Xu D, et al. Comparison of hip muscle volume between fit-and-fill stem and tapered-wedge stem after total hip arthroplasty using the anterolateral approach. J Orthop Sci. 2017;22:1102–6.
11. Tamaki T, et al. Cementless tapered-wedge stem length affects the risk of periprosthetic femoral fractures in direct anterior total hip arthroplasty. J Arthroplasty. 2018;33:805–9.
12. Ishii S, et al. Does the canal fill ratio and femoral morphology of Asian females influence early radiographic outcomes of total hip arthroplasty with an uncemented proximally coated, tapered-wedge stem? J Arthroplasty. 2016;31:1524–8.
13. Zhen P, et al. Primary total hip arthroplasty using a short bone-conserving stem in young adult osteoporotic patients with Dorr type C femoral bone. Orthop Surg Res. 2021;16:17.
14. Kim YH, et al. Is diaphyseal stem fixation necessary for primary total hip arthroplasty in patients with osteoporotic bone (Class C bone)? J Arthroplasty. 2013;28:139–46.
15. McLaughlin JR, et al. Second-generation uncemented total hip arthroplasty: a minimum 20-year follow-up. Bone Jt Open. 2021;2:33–9.
16. Albers A, et al. Favorable results of a short, tapered, highly porous, proximally coated cementless femoral stem at a minimum 4-year follow-up. J Arthroplasty. 2016;31:824–9.
17. Barrington JW, et al. The short and "shorter" of it: >1750 tapered titanium stems at 6- to 88-month follow-up. J Arthroplasty. 2013;28:38–40.
18. Khanuja HS, et al. Short bone-conserving stems in cementless hip arthroplasty. J Bone Joint Surg Am. 2014;96:1742–52.
19. McLaughlin JR, et al. The outcome of total hip replacement in obese and nonobese patients at 10- to 18-years. J Bone Joint Surg Br. 2006;88:1286–92.
20. Hayashi S, et al. Stem anteversion affects periprosthetic bone mineral density after total hip arthroplasty. Hip Int. 2016;26:260–4.
21. Iwamoto Y, et al. Initial rotational instability of the tapered wedge-shaped type cementless stem. Adv Orthop. 2020;2020:2180260.
22. McLaughlin JR, et al. Long-term results of uncemented total hip arthroplasty with a Taperloc femoral component in patients with Dorr type C proximal femoral morphology. Bone Joint J. 2016 98-B:595–600.
23. Grant TW, et al. Cementless tapered wedge femoral stems decrease subsidence in obese patients compared to traditional fit-and-fill stems. J Arthroplasty. 2017;32:891–7.
24. Hayashi S, et al. Risk factors for thigh pain following total hip arthroplasty with short, tapered-wedge stem. Int Orthop. 2020;44:2553–8.
25. Yoshitani J, et al. The use of density mapping in the analysis of thigh pain after total hip arthroplasty in patients with well-fixed tapered wedge stems. J Orthop Surg (Hong Kong). 2020;28:2309499020930306.

34 Short Stem

Katsuya Nakata

1 Concept

In view of the evolution of THA (Total Hip Arthroplasty), classic short stems without femoral bone marrow cavity excavation, such as Smith–Petersen's vitalium cup (1938), Phillip Wiles' stainless steel metal on metal THA (1938) and Judet's acrylic femoral head, can be considered to be the origin of the modern short stems. The McMinn surface-replacing metal-on-metal THA (1996) was reintroduced for the purpose of bone preservation and avoidance of dislocation, but problems such as adverse reactions to metal debris (ARMD) and pseudotumors [1] have become impossible to be ignored, and its application has been gradually reduced. The first step in understanding short stems is to familiarize oneself with their taxonomy [2].

Khanuja et al. [3] proposed a short stem classification method based on the fixation site and stem shape. The four types of short stems are type 1 (neck fixation type): trapezoidal cross section, rounded stem geometry, and threaded geometry; type 2 (calcar load type): trapezoidal cross section, rounded cross section, threaded geometry, and thrust plate design; type 3 (lateral flare type); and type 4 (short tapered conventional type).

Feyen and Shimmin [4] also classified cementless stems according to osteotomy level, stem length, and fixation site. They defined short stem as the total stem length less than twice the distance from the tip of the greater trochanter to the lower end of the lesser trochanter, and standard stem as more than twice the distance. Types I, II, and III are classified as short stems.

The type 2A, 2B and type 4 short stems in the short stem classification of Khanuja et al. [3] have already been introduced to Japan, which can be considered as curved short stems (CSSs). Conventional tapered stems have medial curve and lateral flare taper with inflection point. On the other hands, CSSs have medial and lateral curves without inflection point. Namely, CSSs can be defined short stems with double curves without inflection point at medial and lateral sides of stem.

CSSs were developed in the late 1980s, with Mayo [5] and Pipino Biodynamic/CFP [6] as their origins. These CSSs, which were proposed at an early stage, have been eliminated due to problems such as poor initial fixation, malalignment, poor design of porous surfaces, early loosening, and revision risk. Against this background, many CSSs have been proposed with improved shapes and concepts from the late 1990s to the present, including Metha, Nanos, Proxima, Silent, Fitmore, and MiniHip in approximately 2000 and optimys, GTS, Preserve, SMF, Minima, etc., have been developed one after another since 2010.

Short tapered wedged stems [7] are classified as type 4 in the short stem classification of Khanuja et al. [3], which are not CSSs but conventional short stems. Short tapered wedged stems [7] are explained in other sections of this journal, so in this section, the CSSs are mainly described.

2 Theoretical Advantages

The expected benefits of CSSs are bone and soft tissue preservations.

Bone preservation does not simply refer to a small amount of femoral neck resection but has a comprehensive meaning that includes how to preserve the bone quality of the femur during revision, such as preservation of the diaphysis by omitting the amount of medullary excavation, bone ingrowth at the appropriate site, proximal load transfer, localization of proximal femoral atrophy by stress shielding, reduction of periprosthetic fracture risk, easy removal of the stem at revision, and minimizing bone damage during removal [8].

The reason why bone preservation is required is that the age at which THA is performed is getting younger. In the Japanese Arthroplasty Registry (2017), more than 30% of patients at the time of primary THA were aged 60 years or

K. Nakata (✉)
JCHO Osaka Hospital, Osaka, Japan
e-mail: nakata-katsuya@osaka.jcho.go.jp

Y. Nakashima (ed.), *Advances in Total Hip Arthroplasty*, https://doi.org/10.1007/978-981-97-2461-1_34

younger. However, it should be kept in mind that the risk of revision is higher in younger patients (Australian Orthopedic Association National Joint Replacement Registry 2021). In other words, as the number of patients undergoing THA becomes younger, the choice of implants should be made with an awareness of revision surgery and bone preservation in such cases.

The advantages of CSS include reduction in the amount of head-to-neck resection [4], allocation of femoral marrow cavity excavation due to shorter stem length [9], load transfer to the proximal femur [10, 11], reduction of femoral atrophy due to suppression of stress shielding [12], and avoidance of thigh pain [13]. Liang et al. [12] reported a systematic review and meta-analysis of RCTs on bone mineral density of short and standard stems. They reported that there was no significant difference in bone mineral density loss in Gruen zone 1, but bone mineral density loss in Gruen zone 7 was significantly lower with the short stem.

The CSSs, have shoulderless shape that omits the overhang of the stem shoulder and double curves at medial and lateral sides of stem, making it extremely easy to insert the stem. The anterior MIS-THA rapidly increases to 40.5%, which was reported in the Japanese Arthroplasty Registry System in 2018, especially in anterior muscle sparing MIS-THA [14–16], and CSS is frequently used because it can preserve the external rotator muscles and capsular ligaments at a high rate [17].

3 Surgical Technique (Fitting First Theory)

In general, a higher level of cervical osteotomy results in a larger stem size and more varus and flexion alignment. On the other hand, if the neck osteotomy level is low, the stem size tends to decrease, resulting in valgus and extension alignment. If the stem size is increased at the same osteotomy level, the stem insertion site moves posterolaterally, the stem tends to be in valgus and extension alignment, and the anteversion angle tends to increase [18].

Standard stems such as cemented stems, fit and fill straight stems, and distally fixed stems are inserted in neutral alignment to the proximal femoral axis to obtain the fixation concept of the stem. In other words, the stem alignment and the realization of the stem concept are the same. This stem insertion principle can be referred to as the alignment first theory [7, 8].

On the other hand, since the fixation site and insertion method of CSSs differ slightly depending on the model, it is important to be familiar with the surgical technique and to make a detailed surgical planning. 3D preoperative planning of CSSs often reveals that different alignments can be achieved even if the osteotomy level is the same. This means that while the alignment of the CSS is adjustable, there is a potential risk that the technique may result in a different alignment from that of the preoperative plan.

In CSSs, the cervical osteotomy level is based on the saddle cut (trochanteric sparing cut) and can be varied from mid-neck cut to low neck cut (trochanteric harming cut). The cervical osteotomy level should be determined by considering leg length compensation, offset adjustment, and conformity to the medullary cavity (intramedullary fitting).

The CSSs should not be inserted with an emphasis on alignment but rather with an emphasis on where in the femoral bone marrow cavity it should fit. In other words, achieving an appropriate intramedullary fitting is consistent with the realization of the CSSs concept. This insertion principle of CSSs is called fitting first theory [7, 8]. This insertion principle is based on the principle that the CSSs should achieve intramedullary fitting by contact between the proximal medial portion of the CSSs and the medial calcar and between the distal lateral porous portion of the CSSs and the lateral flare. Therefore, slightly metaphyseal varus alignment of CSSs can be accepted (Fig. 34.1). To realize this concept, it is essential to reproduce the preoperative plan intraoperatively by accurately designing the cervical osteotomy height and angle and confirming that the osteotomy line and the

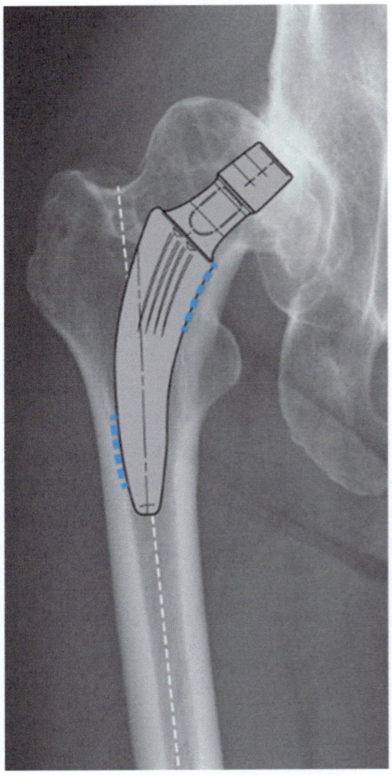

Fig. 34.1 Fitting first theory. In principle, the contact of the proximal medial part of the CSS with the medial calcar and the contact of the distal lateral porous part of the CSS with the lateral flare are used to achieve intramedullary fitting

upper line of the CSS porous line [18–21]. In addition, it is also important to confirm the degree of osteocortical exposure by direct observation of the medial calcar and lateral flare excavation with medullary probing, and intraoperative radiographic imaging should be considered to confirm stem size and fitting in the early surgical stages of CSS introduction.

Sas et al. [22] conducted a biomechanical analysis of the degree of stress shielding by CSS size and alignment. According to their study, the degree of stress shielding in the proximal medial and posterior femur tended to be stronger with lower bone density in the proximal femur, larger CSS size, and more valgus alignment. In other words, to suppress stress shielding, it is necessary to insert a CSS without selecting a large size in a slightly varus alignment in younger patients with less bone mineral density loss. Namely, the importance of obtaining medial calcar and lateral flare fitting of CSS by fitting first theory has been biomechanically recommended and clinically reported [23].

4 Indication

The CSSs have various shapes and concepts [8] and should be taken into consideration when deciding on a case-by-case basis without compromising the initial fixation, age, primary disease, possibility of revision, anatomical specificity, leg length dicrepancy, offset difference, medullary canal shape and bone quality, and intramedullary fitting of the CSS. The indication for Metha, a CSS, should be carefully selected in cases of poor bone quality, severe acetabular dysplasia, excessive anteversion, coxavara, and wide metaphysis [24].

However, in Japan, the majority of THA patients have DDH, and CSSs are currently used even in cases with leg length or offset difference, and femoral anteversion adjustment. As mentioned above, if the height and angle of the cervical osteotomy and the intramedullary fitting of the stem are accurately planned, the range of application of CSS can be expanded to include DDH patients.

5 Clinical Results

Most of the clinical results of CSS have been reported in the short to medium term, and there are only a few CSSs with clear long-term results.

According to a meta-analysis [13] by Hou et al., there is no significant difference in hip function, such as HSS and WOMAC, between the short and standard stems, but thigh pain is observed less frequently in the short stem.

Khanuja et al. [3] reported short stem survival rates of 92.9–99.8% at 2.1–6.3 years postoperatively. van Oldenrijk et al. in a systematic review [25] reported revision rates of 0–3.2% at 0.5–5.1 years postoperatively, varying with the stem. The 5-year survival rate of Mayo was 92.3%, with a high early loosening rate of 7.6% within 5 years [26]. The 10-year results of Metha showed that the revision rate was 1.9%, which was generally good [27].

A mid-term registry comparison study (short vs. standard stem) [28] reported that there was no significant difference in revision rates excluding infection, but the short stem was also superior in revision rate due to infection. In a 15-year registry [29] comparing three groups of short (type 2/4) and standard stems, no significant differences were found in the frequency of intraoperative fracture, periprosthetic fracture, stem loosening, and survival (all >90% at 15 years), but the incidence of pain was higher in type 4 (Fitmore) The revision rate due to pain was also slightly higher in type 4 (Fitmore).

The incidence of proximal femoral bone atrophy due to stress shielding varied from 0% to 24.6%, with a large difference depending on the stem [3]. Mayo: 5.6%, Metha: 2.5%, while CFP: 24.6%, CUT: 12.8%, with a high frequency. A comparative DEXA study of Metha and Nanos [30] also reported differences in the degree of stress shielding. In the study on Fitmore, moderate stress shielding in Gruen zone 1/7 and cortical hypertrophy in zone 3/5 were reported to occur [31]. In the medium term study of optimys [23], stress-shielding was seen in 3.9% (2 years postoperatively) increased to 42.3% (5 years postoperatively). In other words, some CSSs have not achieved the original goal of CSS, i.e., suppression of bone atrophy by load transfer to the proximal femur and the expected bone reaction.

A CT analysis of CFP 7 years postoperatively [32] reported that contact with the bone cortex at the distal portion of the stem induced bone atrophy in the proximal femur due to stress shielding. Brinkmann et al. [33] reported that proximal medial load transfer by CSS medial alignment leads to suppression of stress shielding, suggesting the efficacy of the fitting first theory.

References

1. Sugano N, et al. Nationwide investigation into adverse tissue reactions to metal debris after metal-on-metal total hip arthroplasty in Japan. Orthop Sci. 2014;19:85–9.
2. Nakata K, et al. Short stem. Bone Joint Nerve. 2017;7:93–102. Japanese
3. Khanuja HS, et al. Short bone-conserving stems in cementless hip arthroplasty. J Bone Joint Surg Am. 2014;96:1742–52.
4. Feyen H, et al. Is the length of the femoral component important in primary total hip replacement? Bone Joint J. 2014;96B:442–8.
5. Morrey BF. Short-stemmed uncemented femoral component for primary hip arthroplasty. Clin Orthop Relat Res. 1989;249:169–75.
6. Pipino F, et al. Biodynamic total hip prosthesis. Ital J Orthop Traumatol. 1987;13:289–97.
7. Nakata K, et al. Characteristics of cementless tapered wedge stem and minimally invasive THA. Clin Orthop Surg. 2018;53:205–16. Japanese

8. Nakata K, et al. Current status and challenges of short stem. J Musculoskelet Med. 2017;28:356–66. Japanese
9. Østbyhaug PO, et al. Shortening of an anatomical stem, how short is short enough? An in vitro study of load transfer and primary stability. Proc Inst Mech Eng H. 2013;227:481–9.
10. Boyle C, et al. Comparison of different hip prosthesis shapes considering microlevel bone remodeling and stress-shielding criteria using. J Biomech. 2011;44:1722–8.
11. Tamura S, et al. Intramedullary fitting mode of short stem (MiniHip). Hip Joint. 2015;41:597–601. Japanese
12. Liang HD, et al. Are short-stem prostheses superior to conventional stem prostheses in primary total hip arthroplasty? A systematic review and meta-analysis of randomized controlled trials. BMJ Open. 2018;8:e021649.
13. Huo SC, et al. Short-stem prostheses in primary total hip arthroplasty: a meta-analysis of randomized controlled trials. Medicine (Baltimore). 2016;95:e5215.
14. Siguier T, et al. Mini-incision anterior approach does not increase dislocation rate: a study of 1037 total hip replacements. Clin Orthop Relat Res. 2004;426:164–73.
15. Bertin KC, et al. Anterolateral mini-incision hip replacement surgery: a modified Watson-Jones approach. Clin Orthop Relat Res. 2004;429:248–55.
16. Pflüger G, et al. Minimally invasive total hip replacement via the anterolateral approach in the supine position. Int Orthop. 2007;31(Suppl 1):S7–S11.
17. Nakata K. Pitfalls and future prospects of ALSA. J Orthop Assoc. 2021;95:S34. Japanese
18. Nakata K, et al. Implant selection and surgical techniques in cementless stem arthroplasty. In: Iida H, et al., editors. Hip replacement [THA/BHA]: basic knowledge, basic techniques, and countermeasures for complications. Osaka: MEDICUS SHUPPAN, Publishers Co., Ltd; 2016. p. 81–94. Japanese.
19. Nakata K, et al. Indications for cementless stem based on femoral morphology and bone quality. J Joint Surg. 2019;38:198–207. Japanese
20. Kitada M. 3D printer × tailor-made medicine. For THA. In: Nakata K, et al., editors. Practical hip surgery. Kyoto: KINPODO, INC; 2016. p. 82–105. Japanese.
21. Nakata K, et al. Application of 3D printer to hip arthroplasty: its evolution and practice. Jpn J Orthop Surg Tech. 2018;8:110–5. Japanese
22. Sas A, et al. Effect of anatomical variability on stress-shielding induced by short calcar-guided stems: automated finite element analysis of 90 femora. J Orthop Res. 2019;37:681–8.
23. Kutzner KP, et al. Mid-term results of a new-generation calcar-guided short stem in THA: clinical and radiological 5-year follow-up of 216 cases. J Orthop Traumotol. 2019;20:31.
24. Gruner A, et al. Patient selection for shorter femoral stems. Orthopedics. 2015;38:S27–32.
25. van Oldenrijk J, et al. Revision rate after short-stem total hip arthroplasty: a systematic review of 49 studies. Acta Orthop. 2014;85:250–8.
26. Martins LGG, et al. Aseptic loosening rate of the Mayo femoral stem with medium-term follow up. J Arthroplast. 2014;29:2122–6.
27. von Lewinski G, et al. 10-year experience with short stem total hip arthroplasty. Orthopedics. 2015;38:S51–6.
28. Steinbrück A, et al. Short versus conventional stem in cementless total hip arthroplasty: an evidence-based approach with registrydata of mid-term. Orthopade. 2021;50:296–305.
29. Giardina F, et al. Short stems versus conventional stems in cementless total hip arthroplasty: a long-term registry study. J Arthroplast. 2018;33:1794–9.
30. Brinkmann V, et al. A prospective randomized radiographic and dual-energy X-ray absorptiometric study of migration and bone remodeling after implantation of two modern short-stemmed femoral prostheses. J Orthop Traumatol. 2015;16:237–43.
31. Freitag T, et al. Bone remodeling after femoral short stem implantation in total hip arthroplasty: 1-year results from a randomized DEXA study. Orthop Trauma Surg. 2016;136:125–30.
32. Kress AM, et al. Stress-related femoral cortical and cancellous bone density loss after collum femoris preserving uncemented total hip arthroplasty: a prospective 7-year follow-up with quantitative computed tomography. Arch Orthop Trauma Surg. 2012;132:1111–9.
33. Brinkmann V, et al. Influence of varus/valgus positioning of the Nanos® and Metha® short-stemmed prostheses on stress shielding of metaphyseal bone. Acta Orthop Belg. 2017;83:57–66.

Zweymüller Stem

35

Hiroyuki Ike

1 Concept

The Zweymüller-type stem is a cementless straight stem with a rectangular cross-section and double taper design that was developed by Zweymüller et al. in 1979. The initial design had a collar on the proximal medial side of the stem (Fig. 35.1a). However, the collar was eliminated in 1983 when the design was changed because it caused insufficient bone contact at the distal part of the stem and interfered with the observation of the femoral bone marrow cavity during stem insertion and bone grafting into the marrow cavity (Fig. 35.1b). The distal portion of the stem, which had a groove structure on the medial and lateral sides, was changed to a central ridge structure in 1985 (Fig. 35.1c), and the ridge structure was later eliminated. The material of the stem in the initial design was a Ti6Al4 V alloy with a surface roughness of 1 μm by fine blasting; however, in 1985, the material was changed to a Ti6Al7Nb alloy with a surface roughness of 4–6 μm that was achieved using grit blasting to improve bone growth. Alloclassic (Zimmer Biomet) was improved in 1986 with a thinner neck and an offset stem; however, the basic concept and design were inherited from the initial design and are still currently being sold and used (Fig. 35.1d) [2].

SL-PLUS® stem (Smith & Nephew, Memphis, TN, USA) is a stem modified by Zweymüller, who made the following changes to Alloclassic: (1) The corners of the rectangular cross-section were rounded; (2) the surface area in 1/3 of the proximal stem was increased by 5% to improve bone ongrowth; and (3) the size of the outer wing was slightly reduced [3, 4] (Fig. 35.2a).

The SL-PLUS™ MIA (Smith & Nephew, Memphis, TN, USA) was developed by Pfluger et al. in 2005 based on SL-PLUS. To accommodate minimally invasive surgery (MIS), the lateral wing of the conventional Zweymüller-type stem was redesigned to a rounded back shape to preserve the bone of the greater trochanter and musculotendons attached to the medial side of the greater trochanter (Fig. 35.2b). SL-PLUS MIA HA with hydroxyapatite (HA) coating on the proximal portion was developed to promote osseointegration of the bone to the stem in the early postoperative period and has been available in Japan since 2014 (Fig. 35.2c).

H. Ike (✉)
Yokohama City University, Yokohama, Kanagawa, Japan
e-mail: hike@yokohama-cu.ac.jp

Y. Nakashima (ed.), *Advances in Total Hip Arthroplasty*, https://doi.org/10.1007/978-981-97-2461-1_35

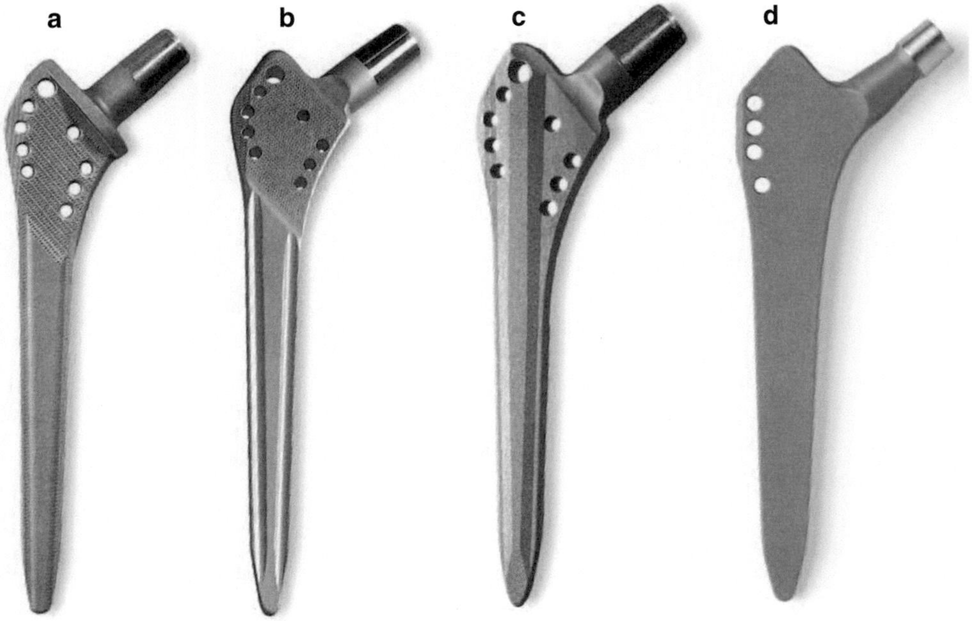

Fig. 35.1 Geometry of Alloclassic stems (Zimmer Biomet, Warsaw, IN). (**a**) In 1979, the early design had a collar inside the proximal stem and a groove structure on the medial and lateral sides of the distal stem. (**b**) In 1983, the collar was discontinued. (**c**): In 1985, the groove structure was discontinued and replaced with a central ridge structure. (**d**) The stem concept of double taper design and rectangular cross-section has been inherited since 1979. (Reproduced from Inaba et al. [1] (Fig. 1))

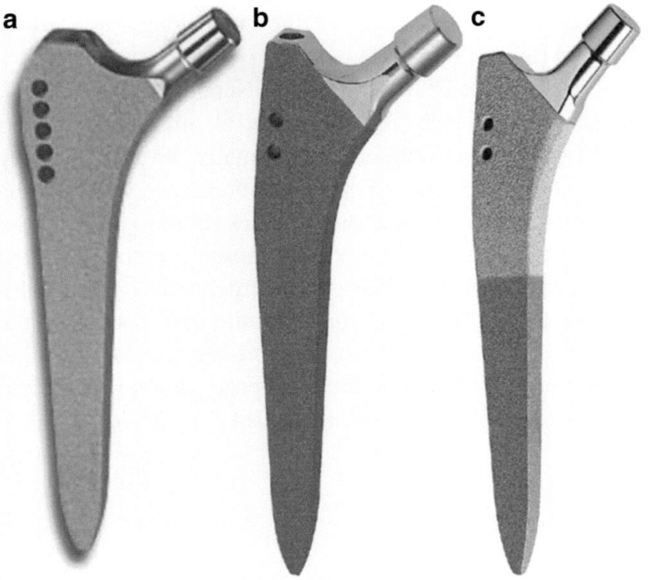

Fig. 35.2 Geometry of SL-PLUS (Smith & Nephew, Memphis, TN, USA). (**a**) The following modifications were made to the SL-PLUS Alloclassic stem. (1) The corners of the rectangular cross-section were rounded, (2) the surface area of the 1/3 of proximal stem was increased by 5%, and (3) the size of the lateral wing was slightly reduced. (**b**) The lateral wing section of the SL-PLUS MIA was modified to a round-back shape. (**c**) SL-PLUS MIA HA has the hydroxyapatite coating on the proximal portion. (Reproduced from Inaba [5], Fig. 2)

2 Theoretical Advantages

According to the classification of cementless stems by Khanuja et al., the Zweymüller-type stem is classified as tapered and rectangular (Type 3C) [6]. It is designed for proximal fixation from the metaphyseal-epiphyseal transition to the metaphysis, which makes it widely applicable and versatile. Osteotomy can be performed distal to the osteotomy site of the stem intended for proximal fixation, making this stem particularly useful in cases where poor proximal fixation is a concern. It can also be used in elderly patients and in those with bone fragility [1].

The Zweymüller stem with a rounded-back design at the lateral wing allows bone preservation of the greater trochanter and preservation of the musculotendons attached to the medial side of the greater trochanter. This design change also facilitated easier insertion of the Zweymüller stem in the midline position using the MIS approach.

3 Surgical Techniques

In the Zweymüller-type stem, the stem and neck may interfere with each other when the femoral neck is left too long, resulting in loss of fixation from the metaphyseal-epiphyseal transition to the proximal metaphysis. In addition, there is a risk of fracture during stem insertion; therefore, it is desirable to perform sufficient osteotomy. The osteotomy line should be tilted at an angle of approximately 45° to the femoral axis and extend from 1 to 2 cm proximal to the superior margin of the lesser trochanter to the anterior part of the trochanteric fossa. If the proximal portion of the femoral neck interferes with the stem owing to a severe proximal deformity, the osteotomy line should be lowered to a level just above the lesser trochanter. A conventional Zweymüller-type stem with an overhanging lateral wing should be used with caution to prevent a varus alignment. The direction of the femoral axis and medullary cavity expansion should be confirmed using straight and curved rasps. A curved rasp can be used to remove the outer cortex of the neck.

The following points should be noted when rasping [2]:

- Maintain a constant anterior torsional angle to achieve rotational stability.
- Be aware of resecting the lateral cortex of the femoral neck to avoid placing the stem in a varus position.
- When rasping with smaller sizes, avoid going deeper than the planned placement depth to prevent fractures.
- The rasp should be pulled out carefully to avoid the shoulder of the rasp being caught on the greater trochanter, which can cause greater trochanter fracture.

4 Adaptation

The Zweymüller-type stem can be used in many cases with no restrictions on femur bone quality or patient age. In general, patients with bone fragility are at a higher risk of intraoperative and postoperative fractures, and cemented stems are often used in elderly patients. The Zweymüller-type stem is cementless, and good clinical results have been reported in elderly patients [7, 8].

In general, there are no cases in which the Zweymüller-type stem is unsuitable for use. However, its slightly distal fixation site causes postoperative atrophy and thinning of the proximal cortical bone. Compared to fit-and-fill and taper-wedge stems, the Zweymüller-type stem significantly decreases bone density in the medial part of the proximal femur [9, 10]. From the viewpoint of preservation of the bone stock, the use of Zweymüller-type stems in young patients should be carefully considered.

In cases in which this stem was used, a radiolucent line (RLL) was often observed in the proximal femur after surgery; however, it did not affect stem stability or progress over time (Figs. 35.3 and 35.4) [11]. Although the mechanism of the occurrence of RLL has not yet been elucidated, Zweymüller himself suggested that it may be a gap caused by the removal of cancellous bone by rasping larger than the inserted implant or a gap due to micromotion between the femur and the stem caused by the traction force of the musculature attached to the proximal femur [11]. The incidence of proximal RLL is high in SL-PLUS MIAs without HA coating; furthermore, the importance of the lateral wing in load transfer to the proximal femur and rotational stability has been reported [12]. In SL-PLUS MIAs with HA coating, the proximal RLL decreased, suggesting that HA coating contributes to the improvement of initial fixation and achievement of early osseointegration [13].

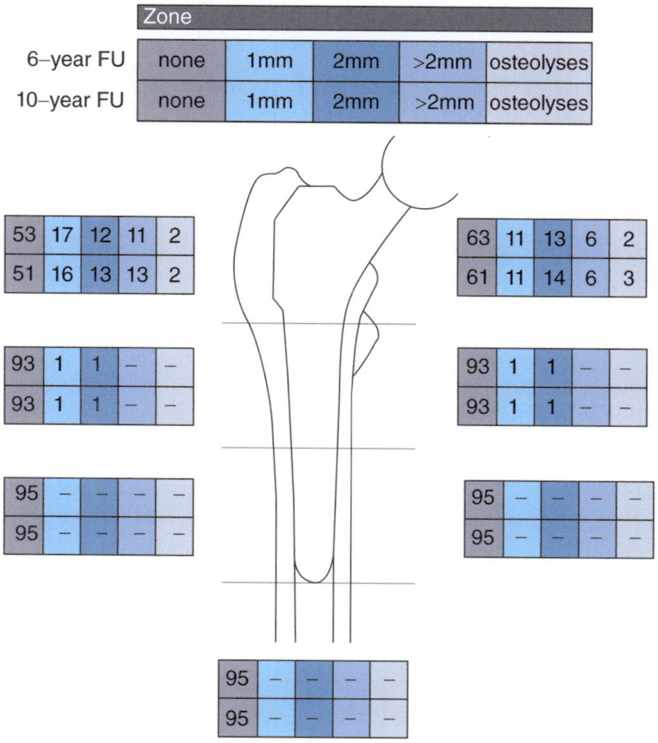

Fig. 35.3 Radiolucent line and osteolysis in the Gruen Zone. The frequencies of radiolucent lines and osteolysis at 6 years (upper panel) and 10 years (lower panel) postoperatively in each Gruen zone are shown. Radiolucent lines were mainly located in the Gruen zones 1 and 7, and most were 1–2 mm wide. The frequency of radiolucent lines at 10 years postoperatively was almost the same as that at 6 years postoperatively, and they rarely progressed over time. (Reproduced from Zweymüller [11], Fig. 2)

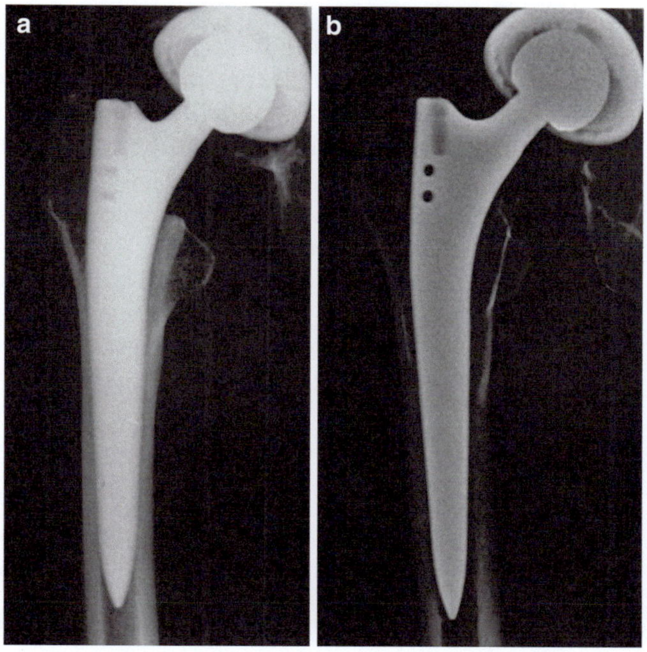

Fig. 35.4 A 50-year-old female patient after THA; AP X-ray of the hip joint. (**a**) Immediately after surgery; right THA with SL-PLUS MIA. (**b**) Six years after surgery with radiolucent lines in Gruen zones 1 and 7

5 Clinical Results

Good long-term results of the Zweymüller-type stem have been reported, with a 20-year postoperative survival rate of 96% with revision as the endpoint, and 98% with aseptic loosening as the endpoint [14]. In a report analyzing registries in various countries, the revision rate of SL-PLUS was significantly lower, with approximately 25% of the overall average, and good results can be expected even among less experienced surgeons [15]. In a study comparing SL-PLUS and SL-PLUS MIA with a mean postoperative observation period of 6.5 years, there were no significant differences in the stem subsidence survival rates and clinical outcomes. The study also reported that good initial and biological fixation could be achieved, even if the lateral wing was removed [16]. In a study comparing SL-PLUS MIA with and without HA coating, there was no significant difference in Harris hip score at 1 year postoperatively but significantly less stem subsidence in the HA-coated group ($p = 0.022$) [13].

Cortical bone atrophy and thinning in the proximal femur, and cortical bone thickening of the femoral diaphysis are often observed in this stem after surgery. Although most reports indicate that cortical bone atrophy in the proximal femur is not associated with stem instability or clinical outcomes, there are concerns regarding an increased risk of greater trochanteric fractures. In cases with a stovepipe shape due to enlargement of the medullary cavity, a larger stem is used, which is fixed more distally and may further increase stress shielding [17]. In cases using SL-PLUS MIA, loss of bone mineral density in the preoperative lumbar spine and small canal flare index (CFI) were reported to be associated with a decrease in bone mineral density in the Gruen zone 7, the medial part of the proximal femur [18]. In addition, administration of denosumab, an anti-receptor activator of nuclear factor-κB ligand (RANKL) monoclonal antibody, can maintain bone mineral density in the Gruen zone 7 [19]. Thus, postoperative bone mineral density loss should be noted in elderly patients with a wide medullary cavity. In addition, osteoporosis drugs can be expected to maintain bone mineral density in the proximal femur; therefore, preoperative or early postoperative treatment should be considered in patients with osteoporosis.

References

1. Inaba Y, et al. Characteristics and minimally invasive THA of Zweymüller-type stem (SL-PLUS®, SL-PLUS® MIA) used in elderly patients. Clin Orthop Surg. 2018;53:191–7. Japanese
2. Ike H, et al. Surgery using the Zweymüller-type stem. In: Inaba Y, et al., editors. BHA and THA hip arthroplasty perfect. 1st ed. Tokyo: YODOSHA CO. 2021. Japanese.
3. Korovessis P, et al. High medium-term survivorship and durability of Zweymüller-Plus total hip arthroplasty. Arch Orthop Trauma Surg. 2011;131:603–11.
4. Böhm G, et al. Morphometric examination of straight, tapered titanium stems: a retrieval study. Clin Orthop Relat Res. 2001;393:13–24.
5. Inaba Y, et al. Characteristics of Zweymüller type stem (SL-PULS® and SL-PLUS® MIA stem) for the elderly and minimally invasive total hip arthroplasty. Rinsho Seikei Geka. 2018;53:191–7. Japanese.
6. Khanuja HS, et al. Cementless femoral fixation in total hip arthroplasty. J Bone Joint Surg Am. 2011;93:500–9.
7. Veen EJ, et al. Outcome of the cementless Zweymüller BICON-PLUS cup and SL-PLUS stem in very elderly individuals. Geriatr Orthop Surg Rehabil. 2016;7:74–80.
8. Pieringer H, et al. Cementless total hip arthroplasty in patients older than 80 years of age. J Bone Joint Surg Br. 2003;85:641–5.
9. Inaba Y, et al. Difference in postoperative periprosthetic bone mineral density changes between 3 major designs of uncemented stems: a 3-year follow-up study. J Arthroplast. 2016;31:1836–41.
10. Hirata Y, et al. Comparison of mechanical stress and change in bone mineral density between two types of femoral implant using finite element analysis. J Arthroplast. 2013;28:1731–5.
11. Zweymüller KA, et al. Radiolucent lines and osteolysis along tapered straight cementless titanium hip stems: a comparison of 6-year and 10-year follow-up results in 95 patient. Acta Orthop. 2006;77:871–6.
12. Henmi N, et al. The shape of the Zweymuller-type stem learned from the SL MIA stem. J Jpn Soc Replacement Arthroplasty. 2015;45:293–4. Japanese
13. Tanaka A, et al. Comparison of early femoral bone remodeling and functional outcome after total hip arthroplasty using the SL-PLUS MIA stem with and without hydroxyapatite coating. Musculoskelet Surg. 2020;104:313–20.
14. Kolb A, et al. Cementless total hip arthroplasty with a rectangular titanium Zweymüller stem: a concise follow-up, at a minimum of twenty years, of previous reports. J Bone Joint Surg Am. 2012;94:1681–4.
15. Labek G, et al. The outcome of the cementless tapered SL-Plus stem: an analysis of arthroplasty register data. Int Orthop. 2012;36:1149–54.
16. Caron E, et al. Prospective randomized study using EBRA-FCA to compare bone fixation between cementless SL-PLUS Zweymüller versus SL-PLUS MIA femoral implants in primary total hip arthroplasty with clinical assessment at a minimum 5years' follow-up. Orthop Traumatol Surg Res. 2020;106:519–25.
17. Kondo S, et al. Zweymuller-type stem. Bone Joint Nerve. 2017;7:103–14. Japanese
18. Morita A, et al. Preoperative factors predicting the severity of BMD loss around the implant after total hip arthroplasty. BMC Musculoskelet Disord. 2021;22:290.
19. Nagoya S, et al. Restoration of proximal periprosthetic bone loss by denosumab in cementless total hip arthroplasty. Eur J Orthop Surg Traumatol. 2018;28:1601–7.

36 Full HA Stem

Makoto Osaki

1 Concept, Theoretical Merits, etc.

Full hydroxyapatite (HA) stems have been widely used recently for hip arthroplasty worldwide. CORAIL is one of the most widely used full HA stems because of its excellent long-term results over 25–30 years [1, 2]. Currently, various types of full HA stems are available, including full HA stems with a similar shape based on CORAIL or different concepts based on CORAIL, such as short stems. The currently used full HA stems in Japan and overseas include CORAIL, POLAERSTEM, ACTIS, Avenir Muller, Universia, QUADRA H, twinSys, MetaFix, optimys, AMIStem, HYDRA, H-MAX, Novation Element, Entrada, HIPFORTRESS-ND, Paragon, Alteon HA, and KORUS.

CORAIL (Fig. 36.1a) was developed by the Association for Research in Traumatology and Orthopaedics (ASTRO) group and Landos in France in 1986, and the HA coating was performed by Bioland. The name is derived from the French word for coral.

CORAIL has a double-tapered geometry with a trapezoidal proximal and square distal cross section. It has a longitudinal and transverse step design to provide initial mechanical fixation. The grid-blasted titanium surface of the nonporous stem is coated with approximately 150 micron-thick HA, which induces bone union and long-term biological fixation. The compaction broach compresses the trabecular bone, intervening between the bone cortex and the implant. The stem is in partial contact with the bone cortex. There are several contact patterns depending on the bone morphology [3]. Collared and uncollared types of stems are available. The collared stem is superior to uncollared stems in terms of initial fixation and prevention of femur fracture [1]. Up to five different neck shapes with different neck lengths and offsets can be selected. According to distributors, the standard offset with a short neck is the most commonly used in Japan.

Full HA stems, such as POLAERSTEM (Fig. 36.1b), are based on CORAIL with various modifications. POLAERSTEM has a similar shape to that of CORAIL, but it is slightly shorter. It has a plasma-sprayed titanium surface coated with approximately 50 microns of HA. The anterior and posterior trabecular bone are compressed by compaction teeth, as in CORAIL, but the inner and outer trabecular bones are removed with extraction teeth, as in many conventional stems. Collared and uncollared types of stems are available. Three different neck shapes with different offsets can be selected.

ACTIS (Fig. 36.1c) has been used internationally since 2016. It has been increasingly used in the United States in recent years [4]. ACTIS has a similar shape to that of the SUMMIT stem with a metaphyseal-filling design. Compared to the SUMMIT stem, the ACTIS stem has a smaller proximal external shoulder overhang and a shorter stem length. The triple-tapered shape and collar reinforce the initial fixation. The proximal half of the stem is coated with porous POROCOAT, and the entire circumference is coated with approximately 35 microns of HA. Similar to POLAERSTEM, the anterior and posterior trabecular bone are compressed with a compaction broach. The inner and outer trabecular bones are removed with extraction teeth. Two neck shapes with different offsets can be selected.

M. Osaki (✉)
Department of Orthopedic Surgery, Nagasaki University Graduate School of Biomedical Sciences, Nagasaki, Japan
e-mail: mosaki@nagasaki-u.ac.jp

Y. Nakashima (ed.), *Advances in Total Hip Arthroplasty*, https://doi.org/10.1007/978-981-97-2461-1_36

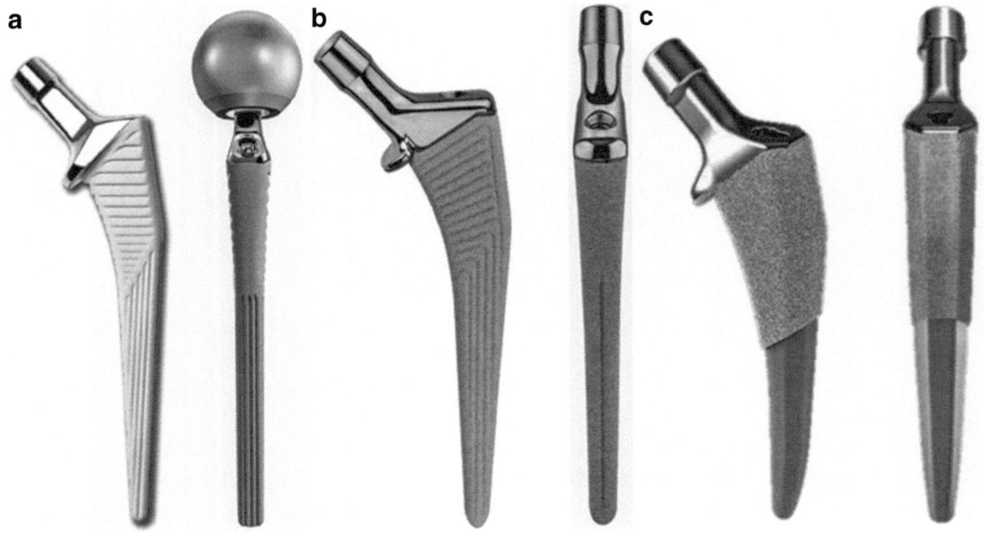

Fig. 36.1 (**a**) CORAIL, (**b**) POLAERSTEM, (**c**) ACTIS

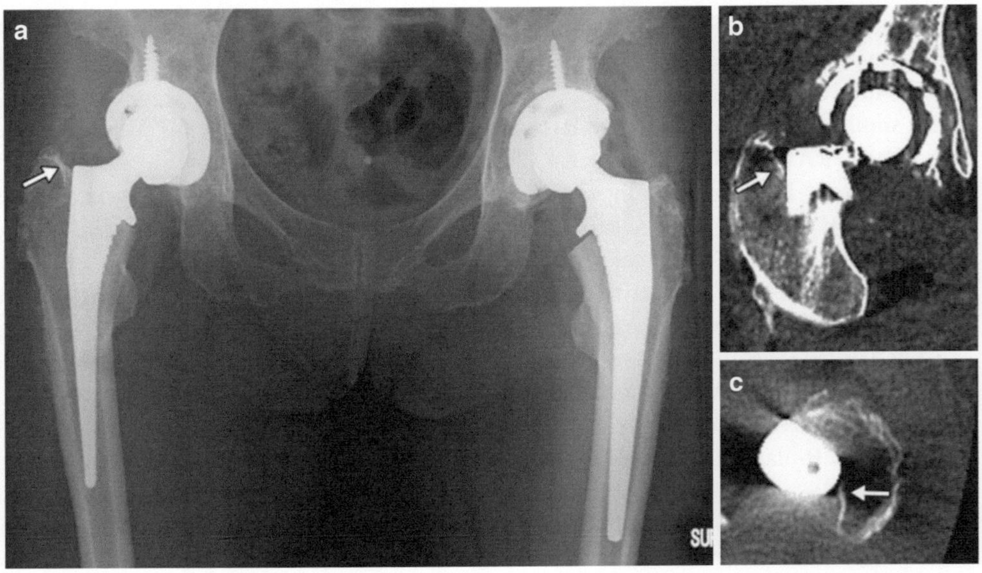

Fig. 36.2 (**a**) Female in her 60 s; right CORAIL #11 and left CORAIL #9 were used. (**b**) Postoperative computed tomography (CT) coronal section. (**c**) Postoperative CT transverse section. Arrowheads indicate that the proximal femoral cortical bone is in contact with the proximal lateral surface of the stem due to inadequate bone resection. An undersized stem is inserted in the varus position into the right femur

2 Surgical Technique and Indication

The basic surgical technique and indications are the same as those for conventional stems. The currently used full HA stems generally have a smaller lateral overhang of the proximal shoulder. Although it requires less bone resection on the lateral side of the proximal femur, sufficient bone resection at the stem insertion site is essential to prevent varus stem placement and insertion of an undersized stem (Fig. 36.2).

As mentioned above, CORAIL uses a compaction broach, which compresses the trabecular bone without removing it. This allows as much trabecular bone as possible to intervene between the bone cortex and the stem. The feeling of inserting the compaction broach into the medullary cavity is different from that of inserting the conventional broach, which requires accustomization. In POLAERSTEM, the trabecular bone is compressed on the anterior and posterior surfaces of the stem by compaction teeth, as in CORAIL, but the trabecular bone is removed on the inner and outer surfaces of the stem with extraction teeth, as in conventional broaches. The feeling of inserting the POLAERSTEM broach into the medullary cavity is similar to that of inserting conventional broaches, which have been widely used.

If the patient has osteoporosis or the compaction of the trabecular bone is weak, test reduction of the implant may result in instability of the broach due to insufficient fixation of the broach. Confirmation of broach stability is necessary before deciding on the stem size. The stability of the broach should be confirmed by applying vertical and rotational forces with the broach handle connected. If the stability of the broach is insufficient, it is necessary to adjust the size of the broach by inserting it deeper or increasing its size to obtain stability.

A collared stem is recommended for initial stem stability and fracture prevention. When a collared stem is used, the femoral neck needs to be treated with a calcar reamer after the depth of the broach is determined by a stability test. Although this extra step is required compared to using a collarless stem, once the stem of the planned size is inserted into the planned position where the collar meets the femoral neck, the stability test can be omitted.

The indications for full HA stems are the same as those for conventional cementless stems. In cases of extreme periprosthetic deformity or after osteotomy, the use of a modular stem, such as the S-ROM or a cemented stem, may be considered.

If the tapered stem is fixed distally in the Dorr A-type medullary cavity geometry, a stem with a smaller volume than that of the proximal femoral medullary cavity geometry is placed. This may result in a mismatch between the smaller proximal stem and the wider medullary cavity geometry. This has been pointed out as a risk for poor stability of the stem in the proximal femur, resulting in no stem-to-bone fixation or stem loosening. The same problem has been reported with the use of CORAIL. In cases with Dorr A-type 1 medullary cavities with an extremely narrow space at the stem tip, the distal medullary cavity should be enlarged with a reamer. It helps to insert an appropriately sized stem and avoid a mismatch between the stem and the medullary cavity.

In general, cementless stems are associated with more fracture problems in the early stage of implantation than cemented stems, and cemented stems tend to be selected in patients with advanced osteoporosis. As shown in the next chapter, CORAIL has a relatively lower short- and long-term fracture risk than conventional cementless stems. It can be used in femurs with osteoporosis, as shown in Fig. 36.3, with prophylactic wiring as needed.

3 Clinical Results

Excellent results have been reported with the use of CORAIL. The Australian Orthopedic Association National Joint Replacement Registry (AOANJRR) reported a 10-year revision rate of 4.9% and a 15-year revision rate of 8.0% for the combination of CORAIL and a PINNACLE cup [5]. The National Joint Registry (NJR) in the United Kingdom also reported a 10-year revision rate of 4.55% and a 15-year revision rate of 7.39% for the combination of CORAIL and a PINNACLE cup [6]. In the NJR, the 15-year revision rate for CORAIL and PINNACLE cups with metal-on-metal articulation was not good (17.76%). However, excellent 15-year revision rates were reported for metal-on-polyethylene bearing surfaces (4.55%) and ceramic-on-polyethylene bearing surfaces (3.51%), [6]. The American Joint Replacement Registry (AJRR) 2012–2019 survey in the United States also reported revision rates of 1.03% at 3 years, 1.23% at 5 years, and 1.28% at 7 years for CORAIL [4].

The 10-year revision rate for the combination of POLAERSTEM and R3 cup, which has a similar stem shape to CORAIL and has relatively long-term results, is 4.9% in the AOANJRR and 1.77% in the NJR [4, 6]. The AJRR also reports that the revision rate for POLAERSTEM is 1.42% at 3 years and 1.42% at 5 years [5].

The 10-year revision rate for QUADRA H in combination with Versafitcup CC is 5.9% in the AOANJRR [5]. In the AJRR, the revision rate for ACTIS is 0.79% at 3 years; that for Avenir Muller is 1.75% at 3 years and 1.75% at 5 years; and that for AMIStem is 1.39% at 3 years, 1.95% at 5 years, and 1.14% at 7 years [6].

In an evaluation by the Orthopaedic Data Evaluation Panel (ODEP), CORAIL along with Exeter cemented stems and the Taperloc Hip System was assigned a 15A star, with a 10-year survival rate of at least 95% and strong evidence for >15 years [7]. POLAERSTEM, QUADRA H, and twinSys were assigned a 10A star; Avenir Muller and Metafix were assigned a 10A star; optimys was assigned a 7A star;

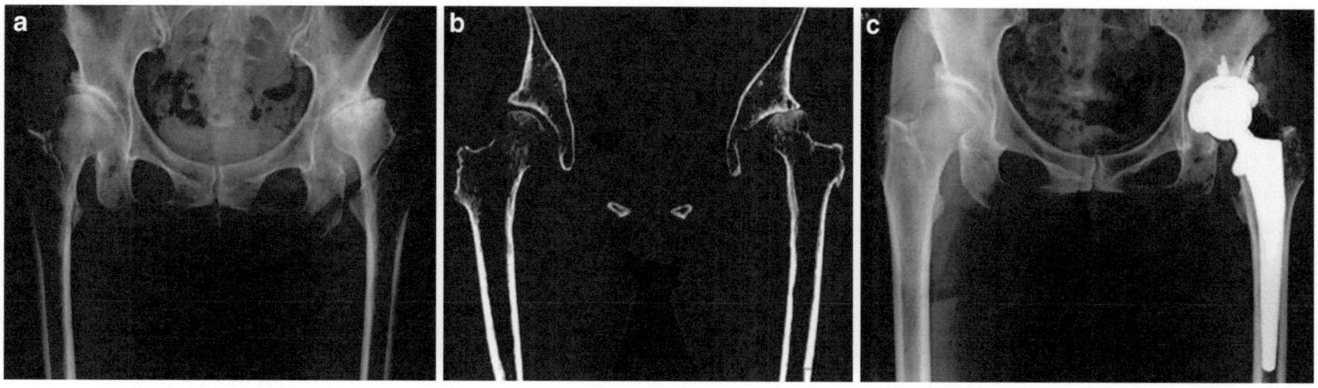

Fig. 36.3 (**a**, **b**) Female in her 70 s. Preoperative radiographs and computed tomography (CT) showing an enlarged medullary cavity and thinning of cortical bone. (**c**) A polyethylene fiber tape is used at the femoral neck to prevent fracture

AMIStem and HYDRA stem were assigned a 7A star; H-MAX was assigned a 5A star; and Novation Element was assigned a 5A star. ACTIS was assigned a PRE-ENTRY A star due to its short period of use. However, other stems need to be evaluated in the future.

Although cementless stems show excellent long-term results, problems such as thigh pain and stress sealing have been reported [8]. Two types of full HA stem models, including CORAIL, were reported to have a lower incidence of thigh pain, stress sealing, and radiolucent lines than the other cementless stems investigated [8]. In addition, an analysis using dual energy X-ray absorptiometry reported that CORAIL had a smaller area of decreased bone mass in the femur and, conversely, a larger area of increased bone mass than in other cementless stems [9].

In recent years, peri-prosthetic fractures have become an increasing problem with the aging of patients. An analysis of the causes of revision total hip arthroplasty (THA) by age in the AOANJRR showed that fracture was the most common cause of revision in patients aged ≥75 years. Even in other age groups, the frequency of fracture increases with time [5]. Moreover, the incidence of fracture increases to 13.2% at 29 years after surgery, even with a relatively low fracture rate of 1.6% at 10 years after surgery [10]. Therefore, it is necessary to consider peri-prosthetic fractures after long-term follow-up, even in younger patients, by selecting stems with a low fracture risk. The frequency of peri-prosthetic fractures in the early postoperative period was 0.42% for cementless stems, while it was 0.04% for cemented stems, indicating a higher risk of fracture with cementless stems [11]. As a cementless stem CORAIL, has a lower fracture risk (0.28%). However, there is no difference between cemented and cementless stems in terms of the number of fractures and frequency of fractures in medium- and long-term periods after THA [12]. CORAIL had 30 fractures out of 5947 primary THA. Exeter, the most commonly used cemented stem had 548 fractures out of 46,808 primary THA. In a study of older adults with proximal femoral fractures, Corail had a lower risk of a post-operative femoral fracture ($p = 0.042$; odds ratio (OR) 0.19, 95% CI 0.01 to 1.42) than cemented Exeter [13]. Corail also had a significantly lower all-cause rate of further surgery ($p = 0.016$; OR 0.18, 95% CI 0.04 to 0.84) and the number of hips undergoing major further surgery ($p = 0.029$; OR 0.13, 95% CI 0.01 to 1.09).

As described above, CORAIL, a typical full HA stem, has excellent long-term results and is highly evaluated. Various types of full HA stems are available, including full HA stems with similar shapes based on CORAIL with various innovations and full HA stems with different concepts based on CORAIL. However, there are a limited number of reports on the long-term results of these implants. Hence it is unclear whether other full HA stems have clinical outcomes comparable to those of CORAIL.

References

1. Vidalain J-P, et al., editors. The CORAIL® hip system: a practical approach based on 25 years of experience. Berlin: Springer; 2011.
2. Jacquot L, et al. Clinical and radiographic outcomes at 25-30 years of a hip stem fully coated with hydroxylapatite. J Arthroplast. 2018;33:482–90.
3. Kobayashi K, et al. Stem-bone contact patterns of a long straight tapered uncemented stem for primary THA. Arch Orthop Trauma Surg. 2022; https://doi.org/10.1007/s00402-021-04273-5.
4. American Joint Replacement Registry (AJRR): 2021 annual report. Rosemont: American Academy of Orthopedic Surgeons (AAOS). Available from: www.aaos.org/ajrr. Accessed 20 Nov 2021.
5. Australian Orthopedic Association National Joint Replacement Registry (AOANJRR). Hip, knee & shoulder arthroplasty: 2021 annual report. Adelaide: AOA; 2021. p. 1–432. Available from: https://aoanjrr.sahmri.com/annual-reports-2021. Accessed 20 Nov 2021
6. The National Joint Registry (NJR). 18th annual report, Hempstead UK; 2021, p. 1–373. Available from: www.njrcenter.org.uk. Accessed 20 Nov 2021.
7. Orthopedic Data Evaluation Panel (ODEP), NEC Software Solutions (UK) Ltd., Hempstead UK. Available from: https://www.odep.org.uk/. Accessed 20 Nov 2021.
8. Fernández-Fernández R, et al. Comparison of an uncemented tapered stem design in cobalt-chrome vs titanium at 15-year follow-up. J Arthroplast. 2018;33:1139–43.
9. Slullitel PA, et al. Influence of femoral component design on proximal femoral bone mass after total hip replacement: a randomized controlled trial. Bone Joint Surg Am. 2021;103:74–83.
10. Peitgen DS, et al. Cumulative long-term incidence of postoperative periprosthetic femoral fractures using an uncemented tapered titanium hip stem: 26- to 32-year results. J Arthroplast. 2019;34:77–81.
11. Thien TM, et al. Periprosthetic femoral fracture within two years after total hip replacement: analysis of 437,629 operations in the Nordic arthroplasty register association database. J Bone Joint Surg Am. 2014;96:e167.
12. Chatziagorou G, et al. Incidence and demographics of 1751 surgically treated periprosthetic femoral fractures around a primary hip prosthesis. Hip Int. 2019;29:282–8.
13. Bell KR, et al. A comparison of the use of uncemented hydroxyapatite-coated bipolar and cemented femoral stems in the treatment of femoral neck fractures: a case–control study. Bone Joint J. 2014;96-B:299–305.

Modular: S-ROM Stem

37

Takuya Otani

1 Concept

1.1 Basic Concept of the S-ROM Stem

Developed in the early 1980s, the S-ROM stem (DePuy, Johnson & Johnson company, Warsaw, IN, USA) is a modular femoral prosthesis made of titanium alloy (Ti6Al4V). It is composed of a proximal component, called the sleeve, and a neck-to-distal component known as the stem. This prosthesis was developed to address the issue of mismatch between the proximal and distal shapes of the femoral medullary canal, which worked against the fit-and-fill concept of conventional cementless stems. The main concept behind its design is that the shape and size for each fit can be independently selected and combined [1]. The sleeve is coated with a single layer of pure titanium beads to achieve biological fixation. In addition, the sleeves come in a variety of shapes and sizes, and the thickness of the coating is press-fitted to the reaming size. The distal stem has a smooth surface to avoid "fixation" with the bone, which plays an important role in obtaining initial stability [2]. The combination of the claw structure, comprising eight longitudinal flutes with an under-reaming mechanism (the flute diameter of the stem is larger than the reaming diameter) provides strong rotational stability. In addition, the distal stem has a tuning fork-like shape (slot) to reduce rigidity, particularly in the direction of the femoral curvature [1, 2].

1.2 S-ROM-A Stem

In 2004, the S-ROM-A stem was developed for Asian patients with high frequency of developmental dysplasia of the hip (DDH) [3]. The basic modular mechanism and sleeve structure of this stem are the same as those of the S-ROM, but the stem components have been modified to have a shorter length, tapered tip shape, and more sophisticated surface polish finish. In addition, the S-ROM-A has a 9/10 taper, which is thinner than the 11/13 neck taper of the S-ROM [4] (Fig. 37.1).

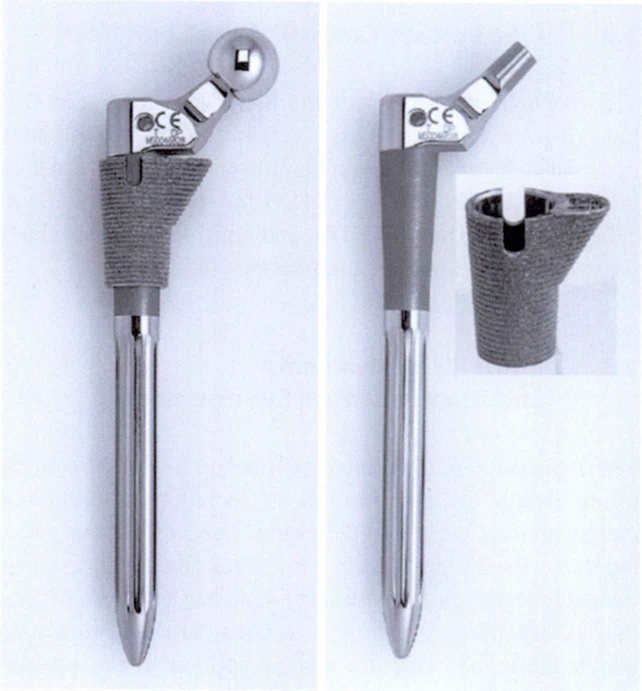

Fig. 37.1 S-ROM-A modular type femoral stem. Left: S-ROM-A stem. Right: Stem and sleeve before bonding

2 Theoretical Advantages

2.1 Compatibility of Proximal and Distal Fixation

The S-ROM has an extremely wide variety of shapes and sizes for both proximal and distal components, with an estimated 8364 possible combinations [5]. In addition, a major

T. Otani (✉)
Jikei University School of Medicine, Minato, Tokyo, Japan

Y. Nakashima (ed.), *Advances in Total Hip Arthroplasty*, https://doi.org/10.1007/978-981-97-2461-1_37

advantage of this stem is that it enables accurate press-fit placement of the proximal sleeve while providing strong rotational stability of the distal stem by an under-reaming mechanism. This implies that even if a mono-block stem of the same design as the S-ROM was available, achieving an accurate press-fit placement proximally would be extremely difficult because once the distal stem begins to press-fit at the diaphysis during stem insertion, further rotation would be impossible. The modular design of the S-ROM allows for a press-fit insertion of the stem after the sleeve has been placed in an optimal rotational position, enabling both proximal and distal fixation.

2.2 Free Version Control of the Stem Neck

The modular mechanism of the S-ROM is a rotation-free tapered joint that allows free version control of the stem neck. This feature confers a significant advantage for both initial total hip arthroplasty (THA) for various morphological abnormalities such as DDH, and revision hip arthroplasty for various deformities of the proximal femur.

2.3 Modular Mechanisms and Biomechanical Environment

The application of a modular mechanism to weight-bearing joints always involves the risk of metallic complications due to fretting, wear, and corrosion. The S-ROM has a biologically fixed thin sleeve with a relatively large taper-diameter at the joint, whereas the stem has a smooth surface that is not fixed, creating a mechanical environment in which the axial loads are always applied to the stem in order to reinforce the taper fixation, resulting in a logical structure.

2.4 Design Considerations in Low Stiffness Stems

The stiffness of the S-ROM femoral prosthesis is reduced by the slot structure of its distal stem. A study examining the relationship between stem stiffness, stress transfer, and micromotion reported that a low-stiffness stem increased the risk of proximal stem micromotion, and that it is crucial to strengthen the proximal design [6, 7]. In this respect, the proximal stability of S-ROM is sufficiently enhanced by a variety of available shape options and a unique reaming system, making it a reasonable design.

3 Indications

Although some reports [8] suggest that the S-ROM stem is widely applicable for standard primary THA, it is often indicated in select cases considering the inherent mechanical weakness of this modular stem. For primary THA, it is indicated in cases with morphological abnormalities, such as DDH, and in cases with a mismatch between proximal and distal medullary canal geometry. The reaming system is useful for osteosclerosis and bone formation around the osteotomy site, as well as morphological adaptability to remaining deformities after femoral osteotomies. In addition, this stem is often indicated for THA combined with subtrochanteric shortening osteotomy in highly displaced Crowe III/IV hips because of the advantages of compatibility of both proximal and distal fixation and the ability to freely set the rotational alignment of its components. Furthermore, the S-ROM stem has several reconstructive advantages in revision arthroplasty, including multiple options for proximal and distal components, a distinct reaming system, and a rotation-free design, allowing its application in a wide range of cases. In cases with poor bone quantity and quality in the proximal femur, the stability of the distal stem can be strengthened to support the stability of the proximal stem. Thus, this stem can be indicated for femoral defects with Paprosky classifications up to Type IIIB [9].

4 Surgical Techniques

The complexity of this model makes it is critical to understand its structural characteristics and techniques for optimal placement to take advantage of these characteristics. The goal is to achieve sufficient stability of the entire stem, successful biological fixation to the sleeve, and a balance between proximal and distal stability.

4.1 Balanced Distribution of Proximal and Distal Stability

Proximal stability is enhanced by extensive cortical bone contact due to abundant sleeve geometry and reaming. However, it is difficult to stabilize the entire stem only proximally, and biomechanical experiments have shown that it is important to obtain both proximal and distal stability [10]. Therefore, it is necessary to ensure constant distal stability; however, excessive reinforcement of distal stability may increase the risk of fracture in the same region and worsen stress shielding of the proximal bone.

4.2 Control of Distal Stability

The basic principle of the S-ROM model is to obtain sufficient distal stability, and therefore the distal reaming system incorporates an under-reaming of 0.75 to 1.25 mm (depending on the size). Distal stability is controlled by grasping the fit during reaming and size selection. An experimental study investigating the distal stability of a cylindrical stem using the under-reaming method, reported a strong distal stability when a fit of 1–4 cm was obtained in the distal medullary canal; however, the risk of fracture increased with an increase in the fit range or stem size [11]. Similarly, in the distal reaming of the S-ROM stem, the size of the stem was selected such that a fit of 1–4 cm was obtained in the medullary canal, after considering proximal stability and stem diameter.

4.3 Precautions to Ensure Proximal Stability

The sleeve has a variety of geometrical options, and the reaming system allows extensive contact with the cortical bone. However, overly aggressive reaming of hard cortical bone may decrease the biological activity of the bone due to thermal damage that may lead to a partial clear zone in the postoperative radiographic findings [4] and may also increase the risk of fracture due to bone thinning [9].

4.4 Proximal and Distal Femur Fractures

While both the proximal and distal initial stabilities of the S-ROM stem are strong, the risk of fracture should always be considered because reaming itself causes a reduction in bone volume and strength. In addition, the implants are press-fitted (proximally by the thickness of the coating and distally by the under-reaming) and axial impaction is required for modular bonding, both of which enhance the risk of fractures. Thus, it is important to perform prophylactic wiring as needed and to carefully check for fractures both intraoperatively and postoperatively [9].

5 Clinical Results

During the development of the S-ROM stem, the performance results of cementless stems were still unstable, and several cases of mechanical loosening had been reported [12, 13]. In addition, orthopedic surgeons had limited experience with modular heads, and there were various concerns about the use of the modular mechanism in a weight-bearing environment. However, contrary to the popular belief, no metallic complications were reported in either primary [1, 2, 8] or revision [14, 15] THA cases in the early reports on the use of S-ROM stem. Additionally, the revision rate due to mechanical loosening was as low as 0.25–0.6% after primary THA [2, 8] and 0.8–0.9% after revision surgery [14, 15]. Recently, a group of researchers, including the developers of the S-ROM studied the long-term outcomes of 167 hip surgeries after more than 15 years of the initial surgery (average: 26.2 years), and found that radiological fixation was seen as stable bony ingrowth in 86.2% of cases and stable fibrous ingrowth in 13.8% of cases. Revision surgery due to mechanical loosening was performed in only 3 (1.8%) cases. The estimated survival rate was 97.6% at 30 years, assuming revision of the stem for all reasons as the endpoint [5].

The short-term [3, 4, 16] as well as medium to long-term [17–19] results of primary THA using S-ROM or S-ROM-A for DDH in Japan have been reported to be excellent. On the other hand, it has been reported [20] that loosening may occur if surgery is performed without sufficient medullary filling of the stem. Good surgical results have also been reported for primary THA combined with corrective osteotomy [21], owing to the compatibility of both proximal and distal fixation. In addition, a recent report on the mid-term results of revision THA using the S-ROM-A stem in Japan [9] yielded good clinical and radiological results in patients with femoral defects up to Paprosky stage IIIB by optimizing the distal and proximal stability and paying attention to the characteristic complications of fracture. Regarding the design changes from the S-ROM to S-ROM-A series, some reports have pointed out that the shortened stem length and tapered tip may lead to ease of use in small Japanese patients and a reduced risk of fractures and thigh pain [4, 9, 18]. On the contrary, there have been reports of stem fractures [19, 22] and cases of ARMD caused by trunnionosis [9, 17, 19]; therefore, caution is paramount during prosthesis insertion. Due to the smaller neck size (9/10 taper) of the S-ROM-A, we recommend the following measures to reduce the risk of metallic complications at the point of taper: selecting patients with low body weight and low activity, using ceramic heads instead of CoCr heads, selecting a smaller-diameter head for direct connection with the neck, and avoiding metal-on-metal articulating surfaces [19].

References

1. Cameron HU. The 3-6-year results of a modular noncemented low-bending stiffness hip implant. A Prelim Study J Arthroplasty. 1993;8:239–43.
2. Cameron HU, et al. The role of modularity in primary total hip arthroplasty. J Arthroplast. 2006;21(4 Suppl 1):89–92.

3. Kido K, et al. Short-term results of the S-ROM-A femoral prosthesis operative strategies for Asian patients with osteoarthritis. J Arthroplast. 2009;24:1193–9.
4. Tamegai H, et al. A modified S-ROM stem in primary total hip arthroplasty for developmental dysplasia of the hip. J Arthroplast. 2013;28:1741–5.
5. Bawale R, et al. Long-term survivorship of modular cementless femoral stem in complex primary total hip arthroplasty: a concise minimum 15-year follow-up report. J Arthroplast. 2021;36:3221–5.
6. Otani T, et al. Strain distribution in the proximal femur with flexible composite and metallic femoral components under axial and torsional loads. J Biomed Mater Res. 1993;27:575–85.
7. Otani T, et al. Effects of femoral component material properties on cementless fixation in total hip arthroplasty. A comparison study between carbon composite, titanium alloy, and stainless steel. J Arthroplast. 1993;8:67–74.
8. Christie MJ, et al. Primary total hip arthroplasty with use of the modular S-ROM prosthesis. Four to seven-year clinical and radiographic results. J Bone Joint Surg Am. 1999;81:1707–16.
9. Abe T, et al. Revision total hip arthroplasty using a modular proximal stem modified for Asians: medium-term clinical results and perioperative complications. J Arthroplast. 2022;37:770–6.
10. Ohl MD, et al. Torsional fixation of a modular femoral hip component. Clin Orthop Relat Res. 1993;287:135–41.
11. Otani T, et al. Reaming technique of the femoral diaphysis in cementless total hip arthroplasty. Clin Orthop Relat Res. 1995;311:210–21.
12. Morscher EW, et al. Cementless fixation of "isoelastic" hip endoprostheses manufactured from plastic materials. Clin Orthop Relat Res. 1983;176:77–87.
13. Kim YH, et al. Early migration of uncemented porous coated anatomic femoral components related to aseptic loosening. Clin Orthop Relat Res. 1993;295:146–55.
14. Christie MJ, et al. Clinical experience with a modular noncemented femoral component in revision total hip arthroplasty: 4-to 7-year results. J Arthroplast. 2000;15:840–8.
15. Cameron HU. The long-term success of modular proximal fixation stems in revision total hip arthroplasty. J Arthroplast. 2002;17(4 Suppl 1):138–41.
16. Hozumi A, et al. Total hip arthroplasty using the S-ROM-A prosthesis for anatomically difficult Asian patients. Biomed Res Int. 2015;2015:690539.
17. Sato A, et al. Midterm outcomes and frequency of osteolysis of total hip arthroplasty using cementless modular stem for Asian patients. J Arthroplast. 2020;35:811–7.
18. Takahashi K, et al. Long-term clinical and radiological outcomes of primary total hip arthroplasty with modular femoral stem in middle-aged Asian patients. J Arthroplast. 2020;35:3650–5.
19. Matsushita Y, et al. A modified S-ROM stem in primary total hip arthroplasty for developmental dysplasia of the hip: average 11-year follow-up in cases with previously reported 3-year clinical results. J Arthroplast. 2022;37:1832–8.
20. Kobayashi K, et al. Higher incidence of aseptic loosening caused by a lower canal filling ratio with a modified modular stem in total hip arthroplasty. J Orthop Surg Res. 2020;15:568.
21. Onodera S, et al. Cementless total hip arthroplasty using a modular S-ROM prosthesis combined with corrective proximal femoral osteotomy. J Arthroplast. 2006;21:664–9.
22. Yoshitani J, et al. Modular stem fracture at the stem-sleeve junction after primary total hip arthroplasty. J Orthop Sci. 2019;24:170–3.

Modular Neck Stem

38

Kazuo Hirakawa

1 Concept

In total hip arthroplasty (THA), proximal modularity plays an important role in primary and revision hip arthroplasty. Furthermore, the new modularity of the neck and stem (distal modularity) has the advantage that leg length, offset, and anteroposterior torsion can be independently corrected. This is especially necessary to manage acetabular dysplasia and the associated complex geometry of the femur in Japan. A fixed neck-type prosthesis alone causes many intraoperative impingement problems for dysplasia deformities, and a changeable neck stem could be used for femoral arthroplasty up to 47° in Japanese patients with severe anteversion. Furthermore, we reported that using a neck-changeable stem prosthesis is useful to obtain a larger range of motion (ROM), appropriate offset, and correct leg length [1, 2].

2 History

Stem modularity can be classified into the following three types:

1. Proximal modularity: general head and neck variants
2. Mid stem modularity: represented by Sivash range of motion (S-ROM)
3. Distal modularity: dual modularity that combines the neck and stem parts (ANCA-Fit, Kinectiv, Profemur Z, and Mainstay)

The history of neck-adjustable stems includes the McBride [3], Lippman [4], and Weber [5] stems, all of which were of the trunnion type, wherein the neck stem junction could be rotated at 30° intervals, like a clock. The S-ROM stem was used in Japan for THA for severe anteversion of the femoral neck. In 1985, Cremascolli in Milan, Italy, developed a stem with an elliptical junction that had modularity of both the neck head and neck stem (ANCA-Fit type) [6, 7]. In the 1990s, Wright Medical's Profemur was used, and Zimmer-Biomet's M/L taper Kinectiv was introduced in 2007. After that, Kyocera Medical released its Mainstay from Japan. All these stems are elliptical in shape.

Furthermore, the head-neck junctions are called "morse tapers," and most have various trunnion shapes from the 9/10 to 14/16 taper. The neck stem junction can be circular, oval, or double-tapered, and a wide variety of materials can be used, including CoCr and titanium alloys [3–8].

In contrast, the Kinectiv stem (Zimmer-Biomet Inc., Warsaw, IN, USA) had a large contact area between the neck and stem portion. The stem and neck were made of titanium alloy, and only a single size, called zero head, was used for the head-neck junction to reduce wear and corrosion. Sixty neck variations were created for each stem. The Kinectiv stem was launched worldwide in 2007, and in Japan, it was covered by insurance in 2009. Developmental dysplasia of the hip is common in Japan, and the Kinectiv stem is used not only for THA but also for bipolar hemiarthroplasty after femoral neck fracture.

3 History of Dual Modularity Stem Development

As osteolysis due to the wear of ultrahigh-molecular-weight polyethylene was solved by reinforcing crosslinks and remelting to prevent oxidation, the development of modular necks and stems progressed to avoid postoperative dislocation, which was the most common complication. In addition, the large metal-on-metal method has been developed to prevent dislocation; however, it is no longer used worldwide because of the high frequency of adverse local tissue reactions (ALTRs) and pseudotumors caused by CoCr [9]. In 2008, the American Association of Hip and Knee Surgeons (AAHKS) published a study on using CoCr in metal-on-metal

K. Hirakawa (✉)
Shonan Kamakura Joint Reconstruction Center, Kamakura, Kanagawa, Japan
e-mail: hirakawa-k1@skjrc.jp

Y. Nakashima (ed.), *Advances in Total Hip Arthroplasty*, https://doi.org/10.1007/978-981-97-2461-1_38

for hip and knee dislocation treatment in the USA. The "Reference Book on Total Hip Arthroplasty" was published, and modular neck and stem development has been gaining momentum. There was also a movement to boost the development of the modular neck and stem (Figs. 38.1 and 38.2).

However, the occurrence of ALTR owing to crevice corrosion in the combination of titanium alloy and CoCr alloy head-neck junction parts has been reported since approximately 1990 [10, 11]. However, the continued use of neck-changeable stems has been greatly criticized as undesirable by the 2012 Stryker (Mahwah, NJ, USA) study, which used ABG2 and cobalt-chromium alloys in the femoral neck with the new titanium-based alloy femoral stem. The major blow to the continued use of rejuvenated stems (titanium-molybdenum-zirconium-iron alloy [TMZF]) was the recall due to its high metal wear powder and revision rate [12]. Fortunately, it was not introduced to Japan.

Fig. 38.1 Presentation of the neck-changeable stem materials discussed at the symposium held by the American Association of Hip and Knee Surgeons (AAHKS) in 2008

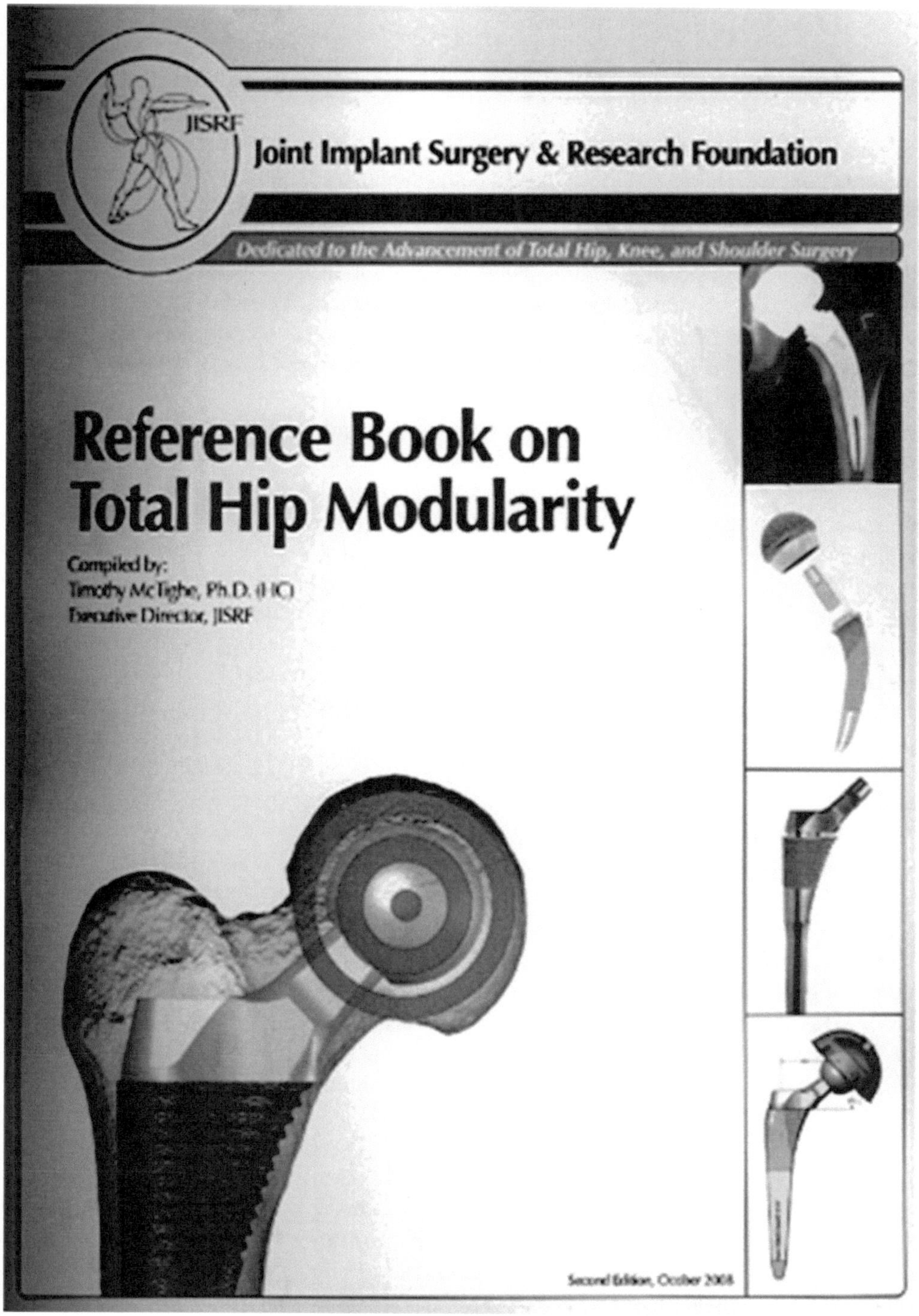

Fig. 38.2 Reference material on several previous neck-changeable stems

A possible cause of the high wear and revision rate is the use of a neck made of CoCr, as opposed to a stem made of TMZF. The Accolade I stem made using the same material by the company was discontinued for the same reason. Additionally, the small junction area of the neck stem junction, which caused sagging and corrosion over time, was considered a likely cause of the severe ALTR [13]. The use of neck-changeable stems was drastically decreased in Western countries, and implant design to avoid dislocation shifted to dual mobility on the acetabular component side (however, the risk of ALTR between the titanium alloy acetabular component and the CoCr shell inside the acetabular component was still high in these countries) [14]. When a type of neck-changeable stem with poor clinical results was recalled, the use of neck-changeable stems as a whole was discouraged, despite different materials and designs being used; this trend has continued to date [15]. However, in some European countries, neck-changeable stems are still useful [16].

4 Theoretical Advantages

In biomechanics, the theoretical advantage of neck-changeable stems is that leg length correction and offset correction can be performed independently, eliminating the weakness of proximal modularity, wherein both can move at the same time. A custom-made stem is possible, but unfortunately, it is costly and inappropriate. The ability to adjust the femoral offset independently with leg length compensation, including the neck version, and reproduce it for each patient, has the advantage of maintaining soft tissue balance and abductor muscle strength and reducing limping. In addition, the ability to adjust offset and anteroposterior torsion is a major advantage in preventing dislocation because it avoids impingement of the neck and acetabular components as well as interference between bone and bone or between the bursa, muscles, and osteophytes. Miki et al. used a CAD model with an ANCA-Fit and found that a femoral anteversion of up to 60° could avoid impingement and achieve an appropriate ROM using a modular neck [17]. Furthermore, Sakai et al. argued that using an extended lip, which causes osteolysis, is no longer necessary [18]. Therefore, stem modularity is an important option for revision surgery or when intraoperative soft tissue imbalance cannot be repaired.

5 Material and the Problem of Modular Junction

Currently, uncemented femoral stems are made of titanium alloys. However, when the neck part is made of CoCr, although it has high strength, it causes corrosion, and using CoCr for a titanium alloy stem is undesirable. This is because of the risk of wear, ALTR, and breakage of the stem side. Thus, the combination of TMZF and CoCr employed in the aforementioned rejuvenated stem is a good example [12].

Using a ceramic head is recommended because ALTR at the neck head junction may occur when a CoCr head is used [10, 11]. Although there is no problem with any neck size or version change, it is important to carefully control the offset and version to prevent intraoperative dislocation, especially for short neck lengths, because neck-to-cup and bone-to-bone impingement may occur. Therefore, carefully controlling the version preventing intraoperative dislocation are important [19].

6 Indications and Complications

The head size must be zero in length when the Kinectiv is used, although it varies depending on the neck-changeable stems. This is because it was adopted as the head-neck junction with the lowest wear in the mechanical tests. There have been reports of breakage using other head sizes (minus or plus); therefore, it is vital to select the appropriate head size. In addition, the use of CoCr heads should be avoided as much as possible to avoid ALTR; hence, ceramic heads are currently used in all cases. Moreover, the femoral neck-changeable stem is useful not only for primary and revision THA, but also for patients with osteoarthritic hip with short neck deformity or a history of femur osteotomy, which are common in Japan.

7 Surgical Techniques and Clinical Outcomes

We have been involved in developing the changeable neck THA stem since 2002, originally designed for the 2-incision approach, and the M/L taper Kinectiv stem has been available in Japan since November 2009. The characteristics of the Kinectiv stem are shown in Figs. 38.3 and 38.4. In addition, the Kinectiv stem is capable of patient-specific offset and leg length compensation compared to the neck-integrated stem.

The Kinectiv stems were used for 12 years (November 2009 to November 2021) in 6446 joints and 805 males (12.5%). Preoperative diagnoses included developmental hip dysplasia 5479 (85%), primary (OA) (525), AVN (155), RDC (109), posttraumatic OA (80), RA (10), deformity after Perthes' disease (76), and unknown causes in 12 joints. The mean age at surgery was 62.5 (34–96) years, with the mean age for men and women being 63.8 years (44–90) and 61.4 (34–96) years, respectively. There were 1,896 (20%), 2965 (60%), and 618 (20%) cases of Crowe I, II, and III hip dysplasia, respectively. The approaches were as follows: mini-anterolateral, 4447 (69%); mini-Watson-Jones, (OCM) 1290 (20%); and Hardinge, 709 (11%). The cups used were Trilogy, Trabecular Metal, Converge, and G2 cups (all Zimmer-Biomet Inc. Warsaw, IN, USA). The mean cup size was 50 (48–63) mm, and the mean stem size was 9 (5–16.25). Neck selection was straight in 5286 joints, anteversion in 308, and retroversion in 852. Version control was required in 18% (1160) of all joints.

The endpoint of revision (86 joints) was a survivorship rate of 98.7% at 12 years, including 45 (0.7%) infections, 18 (0.28%) dislocations, 11 ALTRs (all with CoCr heads, 0.17%), 4 fractures, 1 iliopsoas tendon impingement, 1 iliopsoas hematoma, and 6 avulsion fractures of the greater trochanter (0.12%). The number of patients with dislocation in the Versys stem series had decreased from 0.64% to 0.28%. There were only two cases of postoperative DVT or PE requiring new treatment and no cases of stem fracture. As shown in Fig. 38.5, the ability to independently compensate for the leg length and offset is important for obtaining the proper tension of the hip abductor muscle and tendon.

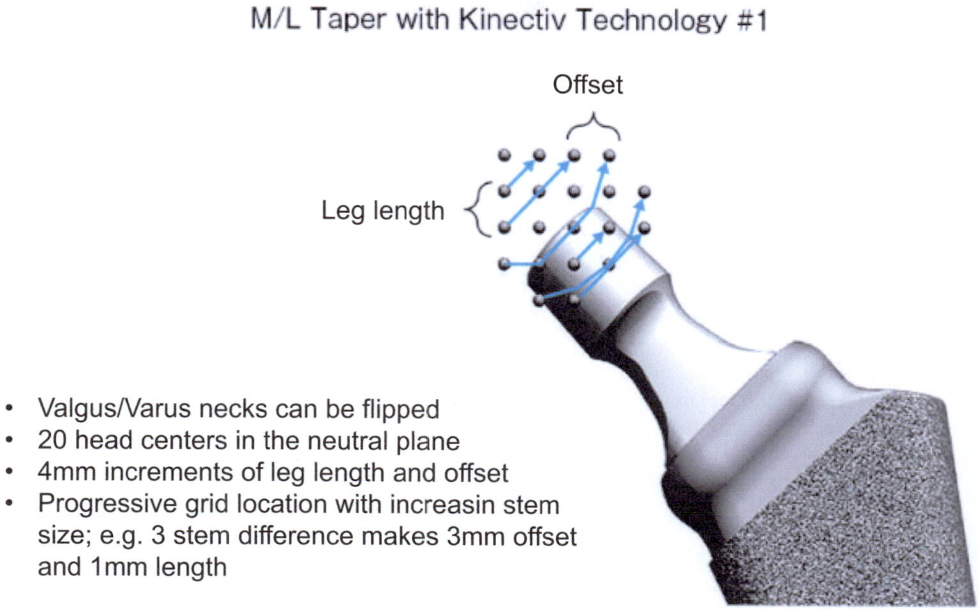

Fig. 38.3 The selection of pitch and internal/external warp of the neck of the Kinectiv stemTwenty different neck sizes can be selected for the front view alone. The leg length and offset can be changed independently by a 4 mm pitch, depending on the stem size. Internal/external rotation can also be selected. This change is beneficial for both the surgeon and the patient to avoid intraoperative impingement

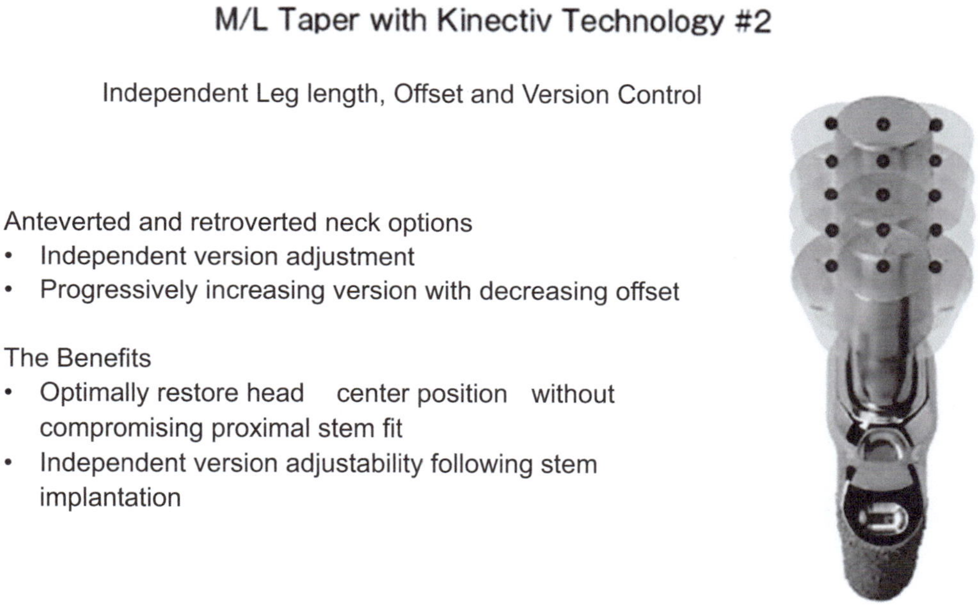

Fig. 38.4 The advantage of the Kinectiv stem is that patients can have one type of straight neck with anterior and posterior twists. Overall, there are 60 types of Kinectiv stems

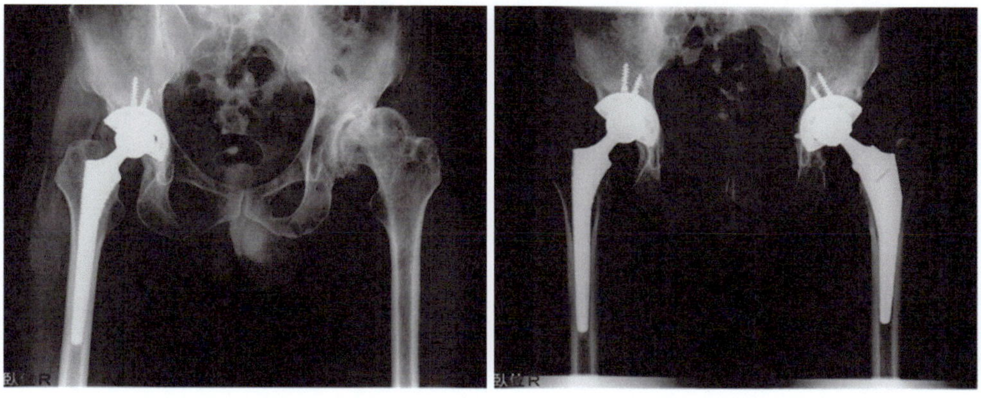

Fig. 38.5 Postoperative radiograph of a patient with bilateral hip dysplasiaThe right hip had an integrated neck stem, and the left hip had a Kinectiv stem with sufficient offset and leg length correction to match the right side. The left hip has an appropriate Shenton line

References

1. Sakai T, et al. Optimizing femoral anteversion and offset after total hip arthroplasty, using a modular femoral neck system: an experimental study. Orthop Sci. 2000;5:489–94.
2. Sakai T, et al. Femoral anteversion, femoral offset, and abductor lever arm after total hip arthroplasty using a modular femoral neck system. J Orthop Sci. 2002;7:62–7.
3. McBride ED. A femoral head prosthesis for the hip joint; four years' experience and the results. J Bone Joint Surg Am. 1952;34-A:989–96.
4. Lippmann RK. Transfixation hip prosthesis. Sequential clinical and roentgenographic observations of sixty-four patients followed for five to fifteen years. J Bone Joint Surg Am. 1967;49:876–90.
5. Weber BG. Total hip replacement with rotational-endoprosthesis. [Trunnion-bearing prosthesis]. Clin Orthop Relat Res. 1970;72:79–84.
6. Toni A, et al. Anatomical ceramic arthroplasty (AN. C. A.): preliminary experience with a new cementless prosthesis. Chir Organi Mov. 1990;75:81–97.
7. Toni A, et al. Cementless hip arthroplasty with a modular neck. Chir Organi Mov. 2001;86:73–85.
8. Sarmiento A, et al. Clinical experiences with a titanium alloy total hip prosthesis: a posterior approach. Clin Orthop Relat Res. 1979;144:166–73.
9. Munro JT, et al. High complication rate after revision of large-head metal-on-metal total hip arthroplasty. Clin Orthop Relat Res. 2014;472:523–8.
10. Jacobs JJ, et al. The effect of porous coating processing on the corrosion behavior of cast Co-Cr-Mo surgical implant alloys. J Orthop Res. 1990;8:874–82.
11. Kawarec JS, et al. Mixed metal fretting corrosion of Ti6Al4 V and wrought cobalt alloy. J Biomed Mater Res. 1995;29:867–73.
12. U.S. Food and Drug Administration. Class 2 Device Recall Rejuvenate Modular Stems. https://www.accessdata.fda.gov/scripts/cdrh/cfdocs/cfres/res.cfm?id=110699. Accessed on 3 Feb 2022.
13. Cooper HJ, et al. Adverse local tissue reaction from corrosion at the femoral neck-body junction in a dual-taper stem with a cobalt-chromium modular neck. J Bone Joint Surg Am. 2013;95:865–72.
14. Sassoon AA, et al. Pseudotumour formation and subsequent resolution in metal-on-metal total hip arthroplasty following revision: an instructional review and an illustrative case report with revision using a dual mobility design. Bone Joint J. 2016;98-B:736–40.
15. Hirakawa K. Letter to the editor on "Titanium neck-titanium stem corrosion in a modular neck stem". Arthroplast Today. 2020;6:129.
16. Maniscalco P, et al. Neck modularity in total hip arthroplasty: a retrospective study of nine hundred twenty-eight titanium neck implants with a maximum follow-up of eighteen years. Int Orthop. 2020;44:2261–6.
17. Miki H, et al. Modular neck for prevention of prosthetic impingement in cases with excessively anteverted femur. Clin Biomech (Bristol, Avon). 2011;26:944–9.
18. Sakai T, et al. A modular femoral neck and head system works well in cementless total hip replacement for patients with developmental dysplasia of the hip. J Bone Joint Surg Br. 2010;92:770–6.
19. Hariri S, et al. Range of motion in a modular femoral stem system with a variety of neck options. J Arthroplast. 2013;28:1625–33.

Surface Replacement

39

Tamon Kabata

1 Concept

Hip resurfacing (HR) is an artificial hip joint with the same physiological shape as that of the normal human hip joint, and only the surface of the joint is replaced with an artificial material.

Most current total hip arthroplasty (THA) procedures have a non-physiological design in which the femoral head is removed from the neck, the head is much smaller in diameter than the physiological head, and a stem connected to the implant is inserted into the femoral marrow cavity to support load bearing. When considering the simple design of an artificial hip joint, it is considered simple and anatomically physiological to replace only the surface of the joint with an artificial material without resecting the femoral head. However, there/ had been no materials or fixation methods that can stabilize the prosthesis over a long period without wear or breakage.

The development of hip joint prostheses that replaced only the joint surface began in the 1950s and the 1960s, but all failed in the short-term [1]. Factors contributing to this failure include material wear due to the use of a physiological bone head diameter and loosening due to bearing torque [1].

In the 1990s, McMinn et al. developed an HR procedure that used a metal-on-metal (MOM) bearing surface, with cementless fixation on the acetabular side and cemented fixation on the femoral side. This design was revolutionary. The difference between McMinn's HR implant and previous HR implants that had failed in the past is that the MOM, which has excellent wear resistance, was selected as the bearing surface, and the cup side was fixed without cement to enable biological fixation, thereby achieving fixation that can overcome the torque generated by the physiological size of the femoral head. As a result, MOM-HR has been performed worldwide since 2000 and has caused great movement in recent years.

T. Kabata (✉)
Department of Orthopaedic Surgery, Graduate School of Medical Sciences, Kanazawa University, Kanazawa, Ishikawa, Japan

2 Implant Design

All MOM-HRs marketed after 1990 were made of a high carbon content cobalt-chromium (CoCr) alloy. The cup was a thin monoblock cementless press-fit fixation type of cup with a porous coating of titanium plasma spray or beads on the cup surface. The femoral head has a stem that is inserted into the center of the neck and was cemented, with a few exceptions (Fig. 39.1).

The jig manufacturing method, heat treatment method, clearance of the bearing surface, properties of the porous surface, profile of the cup, and thickness and length of the stem were different in each model (Table 39.1).

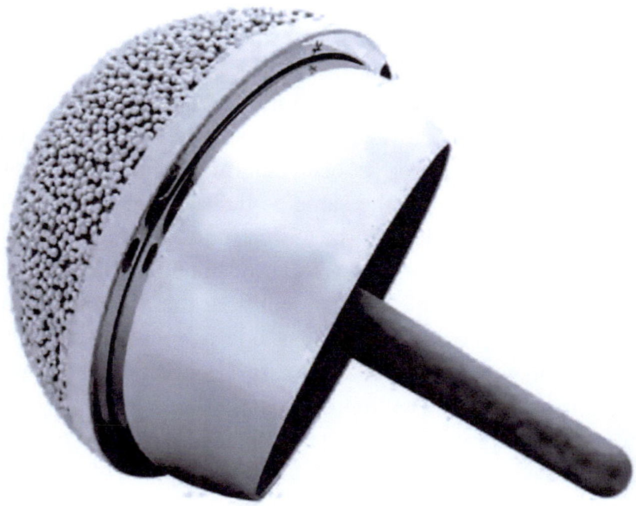

Fig. 39.1 Birmingham Hip Resurfacing with Metal-on-metal Bearing SurfacesBoth the cup and head were made of a high-carbon-content cobalt-chromium (CoCr) alloy. The cup was a monoblock cementless press-fit fixation type with a porous coating on the surface of the pelvic junction. The head has a stem inserted into the center of the neck. (Photograph reproduced with permission from the Birmingham Hip Resurfacing catalog)

Y. Nakashima (ed.), *Advances in Total Hip Arthroplasty*, https://doi.org/10.1007/978-981-97-2461-1_39

Table 39.1 Types and characteristics of MoM-HRA

Model name	Sales company	Year of sale	Sales conditions	Sliding part		Cup
				Metallurgical process	Heat treatment	Porous section
BHR	Smith & Nephew	1997	On sale	Cast	None	Cast CoCr beads, HA
Conserve Plus	MicroPort	1997	Ended	Cast	HIP	Sintered CoCr beads
Cormet	Corin	1997	Ended	Cast	HIP/SHT	Ti VPS, HA
Durom	Former Zimmer	2001	Recalled(not introduced in Japan)	Wrought		Ti VPS
ASR	Former DePuy	2004	Recalled(not introduced in Japan)	Cast	HIP	Sintered CoCr beads
ReCap	Former Biomet	2004	Ended	Cast	None	Ti VPS
Adept	MatOrtho	2006	Ended	Cast	None	Cast CoCr beads, HA

Sugano et al. [1] [Partially reorganized tables from Hip Arthroplasty]
HIP hot isostatically pressed, *SHT* solution heat treated, *Ti VPS* titanium vacuum plasma sprayed, *HA* hydroxyapatite

3 Theoretical Advantages

The theoretical advantages of HR include overwhelming bone preservation, physiological load transfer, low wear, and increased resistance to dislocation due to the large-diameter head.

3.1 Bone Preservation and Physiological Load Transfer

The bone stock around the neck was completely preserved, because the head was not resected at the level of the neck. Therefore, reconstruction with a normal stem is possible for femoral-side revision. The amount of bone resection on the acetabular side has not yet been determined, as some scholars consider that it is the same as that for a normal THA, while others consider it to be a little more. There are differing opinions regarding the amount of bone resection on the acetabular side. Some scholars consider that the amount of bone resection on the acetabular side is approximately the same as that in conventional primary THA, while others suggest that it is somewhat more [2, 3].

Since HR provides almost physiological load transfer to the proximal femur, there is almost no stress shielding of the proximal femur, which is usually observed in normal-stemmed THA, and bone density in the neck is known to increase with HR [4]. The loading stress between the implant and femur in HR is mainly compressive, and there is no shear stress between the implant and the femur, as in normal-stemmed THA. Therefore, patients who were treated using HR can run marathons and participate in contact sports after surgery [5].

3.2 Low Wear of Bearing Surfaces

Because the bearing surface of the HR is an MOM surface with a large femoral head diameter, fluid-film lubrication on the bearing surface occurs, and wear from the bearing surface will be significantly reduced if the implant is placed with proper alignment [6]. However, poor implant alignment may cause edge loading on bearing surfaces, resulting in adverse reactions to metal debris (ARMD) due to metal wear.

In Birmingham Hip Resurfacing (BHR) (Smith & Nephew), which is still on sale and is said to have the best performance [7], the as-cast method was used to sufficiently preserve the surface carbide of the bearing area that exhibits wear resistance. The clearance between the bearing surfaces was set slightly larger to account for the slight deformation of the cup caused by cup press fitting, and it is thought that these factors lead to low wear and good results [1]. On the other hand, the ASR (DePuy-Synthes), which was recalled early, was prone to abnormal wear owing to the low-profile cup design in addition to the small clearance of the bearing surfaces [1].

3.3 Resistance to Dislocation Due to the Large Diameter of the Femoral Head

Since the HR has a physiological head diameter, the dislocation rate is said to be lower than that of conventional THA [8].

4 Indications for HR

HR is best indicated in young active male patients but may or may not be appropriate depending on the patient's background. Therefore, selection of patients for HR is critical.

It has also been reported that female patients treated with HR have significantly more complications, such as ARMD and loosening, with poor outcomes [9, 10]. Although no clear reason has been identified for this causal relationship, it has been pointed out that placement of smaller implants may lead to increased wear, edge loading of the bearing surface, and weakening of the neck [11].

Acetabular dysplasia (DDH) has also been reported to be one of the factors that contribute to poor outcomes [12]. The following factors are thought to be responsible for DDH: patients with DDH are mostly female, the femoral anteversion is large, which affects wear, and the bony coverage of the acetabular side is small, which tends to make cup placement steeper [9]. Patients with highly dislocated or deformed DDHs often have difficulty with press-fit fixation of the cup or placement of a femoral implant due to femoral head deformity, which may also affect them.

Osteonecrosis of the femoral head is also known to have a higher revision rate than osteoarthritis (OA) [13]. The influence of underlying diseases and related factors, such as steroids and alcohol, as well as mechanical problems due to the extent of necrosis are considered to be factors. The extent of necrosis is important in determining the indications. Good clinical results have been reported when all necrotic areas in the femoral head were resected and cemented [14]. However, HR should not be used in cases in which the extent of necrosis exceeds the radius of the femoral head [11].

Renal disease should not be considered as an indication for HR because the metal ions from the bearing surface are eliminated from the kidneys. In addition, osteoporosis is unlikely to be an indication for the risk of neck fractures.

Therefore, the best indication for HR is likely to be a young active male patient with OA with minimal deformity.

5 Surgical Techniques

Regarding the surgical approach, HR can be performed using the same approach as in conventional THA. However, it is difficult to use the so-called muscle-sparing minimally invasive approach, which preserves the musculotendinous units because it is necessary to expand the acetabulum while leaving the femoral head intact. The posterior approach is the most common approach for HR; however, in some cases, the tendinous attachment of the gluteus maximus to the femoral shaft must be partially dissected during the surgical procedure on the femoral head. The femoral head is shaved off the surface using a special reamer. The position and alignment of the femoral implant are important and directly affect the treatment outcomes. It is important to avoid varus alignment and avoid making a notch in the neck as these are the risk factors for postoperative neck fractures [9, 15]. The cup is usually fixed to the acetabulum by press fitting only, which requires a sufficient surgical field of view and reaming of the appropriate size and depth. Cup alignment is also important, and excessive inclination or anteversion may cause edge loading, which may lead to postoperative ARMD or metal ion elevation [9, 16].

In addition, the residual neck can easily cause impingement between the cup rim and neck. Therefore, the ideal cup alignment range is extremely narrow [17]. Amstuts et al. reported that poor cup alignment was associated with implant survival based on the 10-year survival rate of HR and recommended an ideal cup alignment of 42° ± 10° for inclination and 15° ± 10° for anteversion [16].

6 Clinical Outcomes and Complications

Although overall survival rates of 91–97% at 10 years have been reported for BHR [7, 18–24], even BHR is worse than conventional THA [7, 22]. However, a 99% survival rate at 10 years in male patients with OA aged ≤50 years has been reported in two different centers [20, 24]; therefore, appropriate patient selection is also important in this regard. In Japan, a limited number of institutions have reported mid- to long-term results, which are generally good, probably because of the use of navigation for cup placement [25, 26].

Important complications of concern include femoral neck fractures, loosening, elevated serum metal ion levels, and ARMD.

Femoral neck fractures are the most common early complications. As mentioned previously, varus alignment of the femoral implant and making a notch on the femoral neck are the greatest risk factors [9, 15]. Once a neck fracture occurs, internal fixation with screws is possible, and should be attempted at least once [27].

National Joint Registry data have shown that mechanical loosening is the most frequent cause of revision [23], with loosening of the femoral component being the most common. In addition, poor cup alignment is related to loosening of the cup side [9, 16].

The elevated serum metal ion concentrations after HR are said to always occur postoperatively and peak at 9–12 months, followed by a gradual decrease; however, a state of elevated ion concentrations without gradual decrease is considered to indicate wear with abnormal edge loading [28, 29]. The American Hip Society has developed its own algorithm for the follow-up of MOM-THA and has set the monitoring threshold for serum cobalt (Co) and chromium (Cr) ions at 7 μg/L [30]. The mean serum metal ion concentrations of BHR that maintained good function 10 years after surgery were reported to be 1.95 μg/L for Cr and 1.62 μg/L for Co in unilateral cases and 3.46 μg/L for Cr and 2.66 μg/L for Co in bilateral cases [29]. The acceptable concentrations were reported to be 4.6 μg/L for Cr and 4.0 μg/L for Co in unilateral cases and 7.4 μg/L for Cr and 5.0 μg/L for Co in bilateral cases [28], suggesting that 7 μg/L as a monitoring threshold is appropriate. Unfortunately, measurement of serum metal ion concentrations is not currently covered by insurance in

Japan. The revision of asymptomatic elevated serum metal ion concentrations remains controversial.

Although ARMD is a synonymous term for MOM-bearing surfaces, there is no ARMD derived from trunnionosis because the HR does not have a stem neck. According to the results of a nationwide survey by the Committee for the Investigation of Metal-on-Metal Hip Arthroplasty Complications of the Japanese Society for Arthroplasty and a review of overseas literature, the incidence of ARMD (including asymptomatic) after HR in Japan was 0.5%, which is less than half that of MOM-THA [31], with an incidence of approximately 1/4 in overseas reviews [32]. Although individual differences in response as well as tissue response to wear are thought to be involved in the occurrence of ARMD [33], ideal implant placement that does not cause excessive wear by causing edge loading is required for its prevention.

References

1. Sugano N, et al. Surface-replacement hip joint prostheses. In: Kubo T, editor. VII: artificial hip joints and artificial head therapy. Hip Hip Science. 1st ed. Kyoto: KINPODO, INC; 2014. p. 961–5. Japanese.
2. Brown NM, et al. Hip resurfacing and conventional THA: comparison of acetabular bone stock removal, leg length, and offset. Orthopedics. 2013;36:e637–41.
3. Tanzer M, et al. Surface replacement of the hip can result in decreased acetabular bone stock. Clin Orthop Relat Res. 2012;470:541–6.
4. Kishida Y, et al. Preservation of the bone mineral density of the femur after surface replacement of the hip. J Bone Joint Surg Br. 2004;86:185–9.
5. Fouilleron N, et al. Running activity after hip resurfacing arthroplasty: a prospective study. Am J Sports Med. 2012;40:889–94.
6. Rieker CB, et al. Influence of the clearance on in vitro tribology of large diameter metal-on-metal articulations pertaining to resurfacing hip. Orthop Clin North Am. 2005;36:135–42.
7. Seppänen M, et al. Poor 10-year survivorship of hip resurfacing arthroplasty. Acta Orthop. 2016;87:554–9.
8. Marshall DA, et al. Hip resurfacing versus total hip arthroplasty: a systematic review comparing standardized outcomes. Clin Orthop Relat Res. 2014;472:2217–30.
9. Haughom BD, et al. Do complication rates differ by gender after metal-on-metal hip resurfacing arthroplasty? A systematic review. Clin Orthop Relat Res. 2015;473:2521–9.
10. Society CA. The Canadian Arthroplasty Society's experience with hip resurfacing arthroplasty. An analysis of 2773 hips. Bone Joint J. 2013;95-b:1045–51.
11. Sakagoshi D, et al. A mechanical analysis of femoral resurfacing implantation for osteonecrosis of the femoral head. J Arthroplast. 2010;25:1282–9.
12. Gross TP, et al. Prevalence of dysplasia as a source of worse outcome in young female patients after hip resurfacing arthroplasty. Int Orthop. 2012;36:27–34.
13. Gross TP, et al. Comparative study between patients with osteonecrosis and osteoarthritis after hip resurfacing arthroplasty. Acta Orthop Belg. 2012;78:735–44.
14. Nakasone S, et al. Does the extent of osteonecrosis affect the survival of hip resurfacing? Clin Orthop Relat Res. 2013;471:1926–34.
15. Kohan L, et al. Early complications of hip resurfacing. J Arthroplast. 2012;27:997–1002.
16. Amstutz HC, et al. Socket position determines hip resurfacing 10-year survivorship. Clin Orthop Relat Res. 2012;470:3127–33.
17. Kajino Y, et al. Strict component positioning is necessary in hip resurfacing. J Orthop Sci. 2013;18:290–7.
18. Mehra A, et al. Birmingham hip resurfacing: a single surgeon series reported at a minimum of 10 years follow-up. J Arthroplast. 2015;30:1160–6.
19. Azam MQ, et al. Survivorship and clinical outcome of Birmingham hip resurfacing: a minimum ten years' follow-up. Int Orthop. 2016;40:1–7.
20. Daniel J, et al. Results of Birmingham Hip Resurfacing at 12 to 15 years: a single-surgeon series. Bone Joint J. 2014;96-b:1298–306.
21. Matharu GS, et al. The outcome of Birmingham Hip Resurfacing in patients aged < 50 years up to 14 years post-operatively. Bone Joint J. 2013;95-b:1172–7.
22. Holland JP, et al. Ten-year clinical, radiological and metal ion analysis of the Birmingham hip resurfacing: from a single, non-designer surgeon. Bone Joint Surg Br. 2012;94:471–6.
23. Jameson SS, et al. Independent predictors of revision following metal-on-metal hip resurfacing: a retrospective cohort study using National Joint. J Bone Joint Surg Br. 2012;94:746–54.
24. Murray DW, et al. The ten-year survival of the Birmingham hip resurfacing: an independent series. J Bone Joint Surg Br. 2012;94:1180–6.
25. Uemura K, et al. Long-term results of Birmingham Hip Resurfacing arthroplasty in Asian patients. J Artif Organs. 2018;21:117–23.
26. Inoue D, et al. Mid- to long-term results of resurfacing hip arthroplasty in Japanese patients: a comparison of osteoarthritic vs non-osteoarthritic patients. J Artif Organs. 2019;22:77–83.
27. Kutty S, et al. Intracapsular fracture of the proximal femur after hip resurfacing treated by cannulated screws. J Bone Joint Surg Br. 2009;91:1100–2.
28. Van Der Straeten C, et al. The interpretation of metal ion levels in unilateral and bilateral hip resurfacing. Clin Orthop Relat Res. 2013;471:377–85.
29. Van Der Straeten C, et al. Metal ion levels from well-functioning Birmingham hip resurfacings decline significantly at ten years. Bone Joint J. 2013;95-B:1332–8.
30. Lombardi AV Jr, et al. The Hip Society: algorithmic approach to the diagnosis and management of metal-on-metal arthroplasty. J Bone Joint Surg Br. 2012;94(11 Suppl A):14–8.
31. Sugano N, et al. Nationwide investigation into adverse tissue reactions to metal debris after metal-on-metal total hip arthroplasty in Japan. Orthop Sci. 2014;19:85–9.
32. The Japanese Society for Replacement Arthroplasty Metal-on-Metal Sliding Surface Complication Committee. Guideline for the treatment of complications of metal-on-metal total hip arthroplasty (recommendation). Available from: http://jsra.info/pdf/Metal-on-Metal.pdf Japanese.
33. Matthies AK, et al. Pseudotumors are common in well-positioned low-wearing metal-on-metal hips. Clin Orthop Relat Res. 2012;470:1895–906.

Custom Made Stem

40

Kenji Kawate

1 Concept

In the late 1980s, custom-made stems were clinically applied based on the idea that it is important to reduce micromotion to obtain mechanical stability of cementless stems, which depends largely on the medullary cavity occupancy [1–3]. In these reports, partial mesh structures were used in the antero-posterior and posterior surfaces, but the results varied. In some reports, a method was presented in which a truck was placed next to the operating room and a stem-first procedure was performed to shorten the time required to insert the stem into the patient (Identifit, Stewal implants, Belgium, and Orthogenesis, Orlando, Florida), during a mold taken from the medullary cavity was measured by laser, and the metal was shaved and disinfected in approximately 40 min in the truck [4]. However, many cases of mechanical loosening were reported in the short postoperative period because of the smooth surface of the stem without any surface treatment [5, 6]. This series of studies showed that stem stability is also affected by surface characteristics.

At that time, imported off-the-shelf foreign-sized stems were the mainstream in Japan, but with the spread of CT in Japan, research on hip joint morphology in petite Japanese female patients became active, and the development of cementless stems and custom-made stems in sizes more suited to their physique and femoral bone morphology progressed [7, 8]. In 1992, Nara Medical University launched the Kyocera custom-made system.

The Kyocera custom-made system was developed for the treatment of congenital dislocation of the hip, secondary hip arthroplasty due to acetabular dysplasia, and deformed femur after femoral osteotomy for these conditions. Total hip arthroplasty for DDH (developmental dysplasia of the hip) requires careful preoperative planning. The development of custom-made stems based on CT data, which became popular in the late 1980s, has led to the use of three-dimensional templating for preoperative planning, which is now common preoperative planning. In DDH, the acetabular bone stock is problematic, and it is important to place the acetabular implant. Once the position and angle of the acetabular implant are determined, the femoral stem should be adjusted with the combined anteversion theory. It is not only the acetabulum, but the shape of the femur is also characterized by an excessive anterior neck torsion angle (+distortion), a steep proximal medial cortex, a shortened neck, and a narrow medullary cavity. Furthermore, the anteversion angle of the neck after femoral intertrochanteric osteotomy cannot be predicted without CT data, and there are many problems that cannot be simply solved by cementation alone, such as deformation of the medullary cavity, new bone around the osteotomy site and screws, and large granulations that interfere with the interdigitation of cement.

The Kyocera Custom-Made System was initially developed to better match the femoral shape of a Japanese petite woman, using 39 slices from the acetabulum to the mid femur at 5 mm intervals and one slice through the posterior margin of the posterior knee condyle. In total, 40 slices of preoperative CT data were used, with the window level (CT value that distinguishes the cortex from the marrow cavity) set at 700. The axis was set at 3 cm and 8 cm distal to the lesser trochanter. The distal slice was tapered so that it was not wider than the proximal slice in each slice and was designed to follow the medial and lateral medial cortical bone in Ver. 1 and to follow the medial cortical bone circumferentially in Ver. 2 (Fig. 40.1). The surface was initially fabricated using a titanium alloy (Ti-6Al-4 V) with a sandblasted roughness over the entire circumference. Then proximal 1/3 was coated with a 400 μm-thick proarc coating, a 20 μm-thick hydroxyapatite coating was sprayed on top of the proarc coating, and the distal 2/3 was polished [9, 10].

K. Kawate (✉)
Nara Prefecture General Rehabilitation Center, Shiki, Nara, Japan
e-mail: kkawate@naramed-u.ac.jp

Y. Nakashima (ed.), *Advances in Total Hip Arthroplasty*, https://doi.org/10.1007/978-981-97-2461-1_40

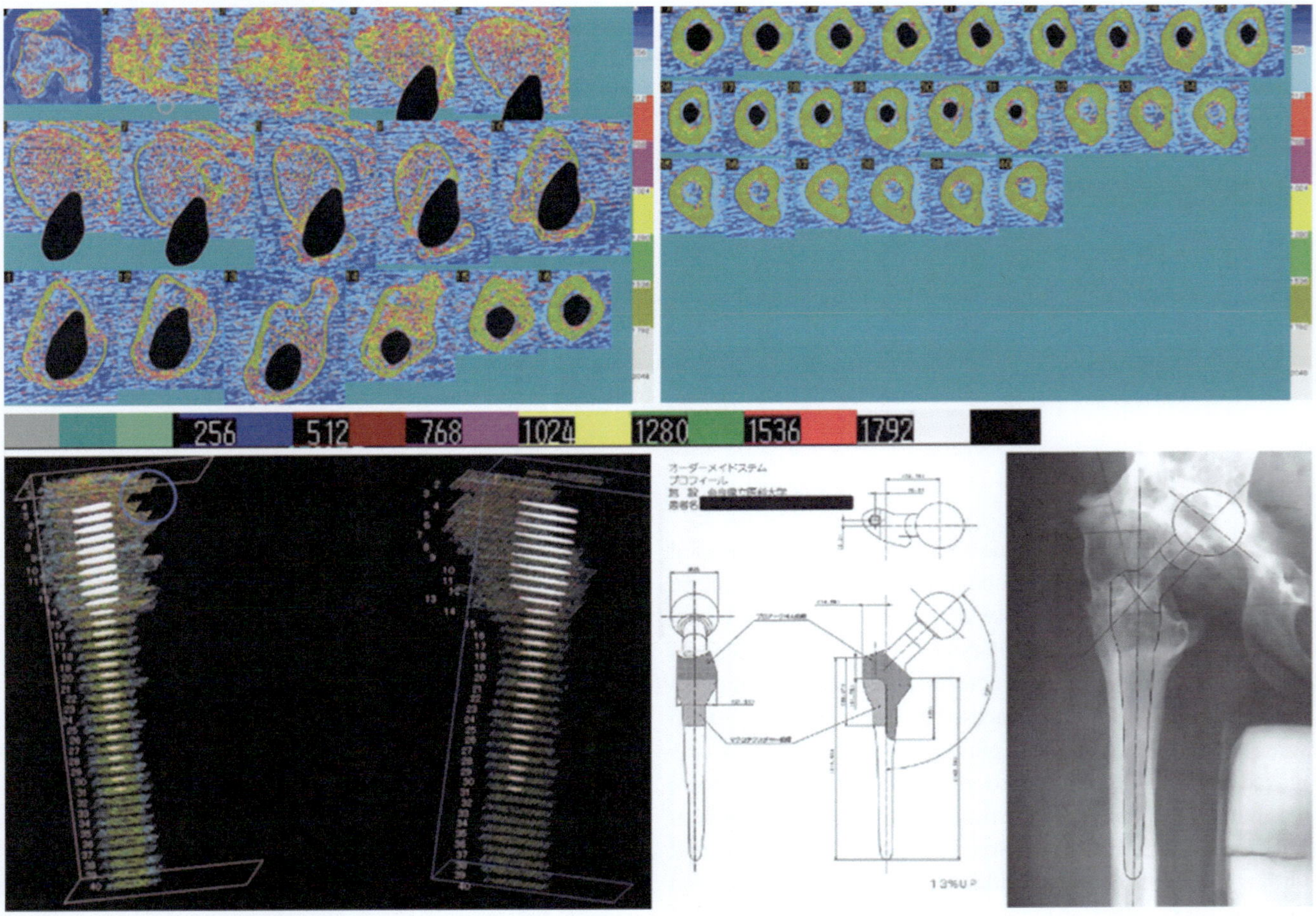

Fig. 40.1 Example of a drawing of a three-dimensional custom-made stem

2 Theoretical Advantages

The advantages of the custom-made stem are good initial fixation by Fit &Fill, short stem length (10 cm on average), combined anteversion theory for cases with large femoral anteversion by correcting the anteversion of the neck in advance, and the ability to set the neck length and offset freely. In cases of excessive anteversion of the femur, correction of the anteversion can avoid postoperative in-toe and can prevent the patient from falling down due to stumbling on the inner forefoot. A special case is the extremely narrow marrow cavity, which may cause stem fracture with existing stems. The three-dimensional custom-made stem is stronger due to its wider cross-sectional area.

3 Surgical Techniques

The initial method was to insert an elevatorium after osteotomy of the femoral neck to confirm the direction of the femoral bone marrow cavity and to place a custom-made stem without reaming to compact the cancellous bone. The initial fixation was excellent even in patients with osteoporosis. However, in cases with very hard bone marrow, such as young osteonecrosis of the femoral head, it is difficult to insert the stem into the bone marrow, and a fracture could occur if the direction of insertion is wrong. For this reason, a rasp of the same type as the stem was later fabricated, and after reaming to some extent, the rasp was inserted after rasping with an off-the-shelf rasp smaller than the custom-made stem (Fig. 40.2). Recently, a navigation system has been used to obtain a more accurate insertion direction and angle. In our experience, the knack is to design the stem offset 5 mm longer than the tip of the greater trochanter to obtain the ideal tension when the rasp is inserted to the target point.

4 Adaptation

The custom-made stem was developed for the treatment of osteoarthritis caused by congenital dislocation of the hip joint and acetabular dysplasia in Japanese patients with large

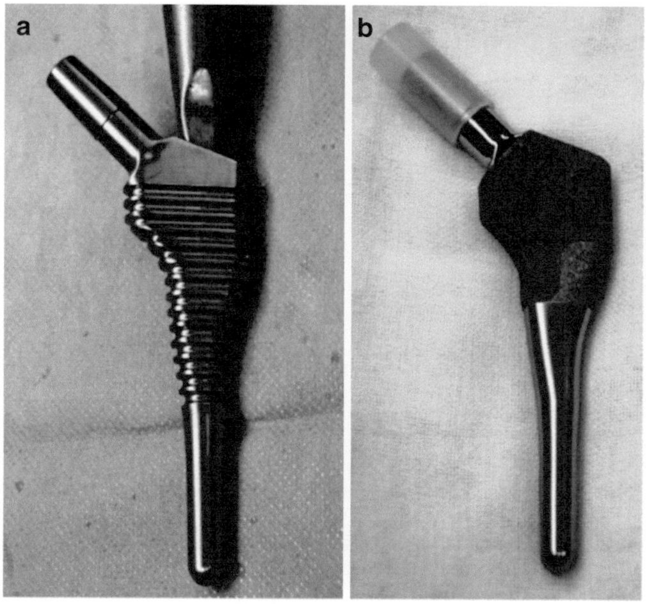

Fig. 40.2 A homomorphic rasp, B Custom-made stem

and steep anteversion of the femoral neck. However, due to health insurance coverage issues, the current version is limited to patients with severe deformities, such as after femoral osteotomy, bone tumors, or traumatic injuries.

5 Clinical Results

We report the results of Kyocera Expert System Ver. 1 with Proarc HA coating on the proximal 1/3 of the stem [10]. The number of cases was 48 females with 50 joints and 5 males with 5 joints, for a total of 53 cases with 55 joints. The mean age at surgery was 60 (40–73) years, and the mean postoperative follow-up period was 7 [5–11] years. The mean JOA score was 45 (15–80) preoperatively, and the average postoperative score was 90 (57–100). Complications included 1 dislocation (1.8%) and 4 fractures (7.2%). No infection or revision surgery was observed. Radiographic evaluation showed bone-ingrown fixation in all 55 joints by Engh's evaluation [11]. Eleven joints (20%) had severe stress shielding.

In addition, we report the results of Kyocera Expert System Ver. 1 and Ver. 2 in 40 joints in 35 patients after femoral intertrochanteric osteotomy [12]. The patients included ~~4~~38 joints in 33 women, 2 joints in 2 men, 20 joints after intertrochanteric varus osteotomy, 19 joints after valgus osteotomy, and 1 joint after Shantz osteotomy. The mean age at surgery was 63 (29–77) years, the mean postoperative ~~course~~ follow-up period was 15 (5–25) years, and the mean time from osteotomy to THA (total hip arthroplasty) was 8.5 [2–17] years. The mean preoperative JOA score was 50 (19–74), and the mean postoperative score was 82 (51–97). Complications included thigh pain in 2 joints (5%), fracture in 2 joints (5%), peroneal nerve palsy in 1 joint (2.5%), and dislocation in 3 joints (7.5%). There was no infection or revision surgery. Thirty-nine joints had bone-ingrown fixation, and 1 joint had stable fibrous fixation as evaluated radiologically by Engh. Five joints (12.5%) had severe stress shielding (Fig. 40.3). Custom-made stems have recently attracted worldwide attention as a solution for ~~femoral~~ patients with strong proximal deformities. Although the manufacturing process is time-consuming and costly, the custom-made stem is of significance to both patients and hip surgeons because it leads the surgeon to deepen his or her anatomical understanding of the patient's case and has evolved into the three-dimensional templating that is common today.

Regarding the surgical method of this custom-made stem system, in which the stem is inserted directly without using a reamer or rasp, Clark et al. [13] experimentally reported that the initial fixation was better than that of the conventional method. On the other hand, careful impact is required because of the risk of fracture if the direction of impact is not correct. In our experience, older patients with bone atrophy can be inserted more easily into the cancellous bone, and the difference in the sensation of impact between the cancellous bone and cortical bone makes the insertion relatively safe. On the other hand, in younger patients with firm cancellous bone, it is difficult to insert stem into the cortical bone because there is no difference in the sensation of impaction between the cancellous bone and the cortical bone. It is recommended that radiographs be taken at a certain point of insertion to confirm the insertion. Currently, we use a navigation system for insertion.

In addition, it has been reported that the mechanical stability of the grit-blasted surface is excellent [14, 15]. However, in a previous study of cases in which only sandblasting was used, stress shielding was observed in many cases, and the surface texture was subsequently changed. However, the technology at that time forced us to apply sandblast coating to the proximal two-third and matte blast coating to the distal one-third. Currently, the proximal 1/3 of the surface is coated with a 400-μm-thick proarc coating and a 20-μm-thick hydroxy coating on it, and the distal 2/3 of the surface is polished.

Recently, ultrashort custom-made stems have been developed, and good postoperative results have been reported, but there is concern that mechanical problems may cause future ~~laxity~~ mechanical loosening [16].

The above custom-made stem is useful for small Japanese women, after osteotomy patients, and deformed femurs, and further innovations in materials and 3D modeling methods are expected in the future [17].

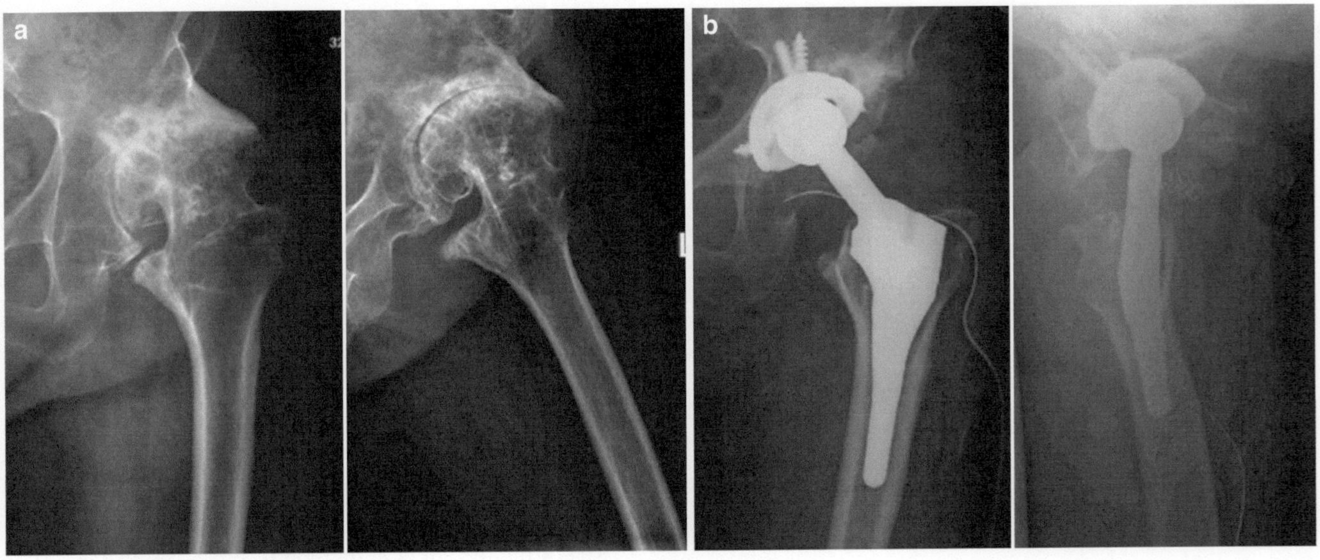

Fig. 40.3 A 73-year-old woman with navigation after intertrochanteric vulgus osteotomy. *A* preoperative, *B* postoperative

References

1. Barger WL. Shape the implant to the patient. A rationale for the use of a custom-fit cementless total hip implant. Clin Orthop Relat Res. 1989;249:73–8.
2. Stulberg SD, et al. The rationale, design characteristics, and preliminary results of a primary custom total hip prosthesis. Clin Orthop Relat Res. 1989;249:79–96.
3. Bert JM. Custom total hip arthroplasty. J Arthroplast. 1996;11:905–15.
4. Mulier JC, et al. A new system to produce intraoperatively custom femoral prosthesis from measurements taken during the surgical procedure. Clin Orthop Relat Res. 1989;249:97–112.
5. Lombardi AV Jr, et al. Failure of intraoperatively customized non-porous femoral components inserted without cement in total hip arthroplasty. J Bone Joint Surg Am. 1995;77:1836–44.
6. Robinson RP, Clark JE. Uncemented press-fit total hip arthroplasty using the identifit custom-molding technique. A prospective minimum 2-year follow-up study. J Arthroplast. 1996;11:247–54.
7. Yanagimoto S. [Basic study on shape determination of cementless hip prosthesis: analysis of proximal femoral medullary cavity morphology in Japanese osteoarthritic hip joint cases.] J Jpn Orthop Assoc. 1991;65:731–44. Japanese.
8. Ooneda Y, et al. [CT and computer simulation of cementless hip prosthesis stem insertion.] Hip Joint. 1993;19:168–71. Japanese.
9. Muirhead-Allwood SK, et al. Uncemented custom computer-assisted design and manufacture of hydroxyapatite-coated femoral components: survival at 10 to 17 years. J Bone Joint Surg Br. 2010;92:1079–84.
10. Kawate K, et al. Computed tomography-based custom-made stem for dysplastic hips in Japanese patients. J Arthroplast. 2009;24:65–70.
11. Engh CA, et al. The case for porous-coated hip implants. The femoral side. Clin Orthop Relat Res. 1990;261:63–81.
12. Kawate K. Custom-made stem for hip after trochanteric osteotomy and dysplastic hip. Korean Hip Soc. 2019;63:100.
13. Clark YL, et al. Improving the initial mechanical fixation of a porous coated femoral stem by a cancellous bone compaction method. Trans ORS. 1999;45:863.
14. Goldberg VM, et al. Biology of grit-blasted titanium alloy implants. Clin Orthop Relat Res. 1995;319:122–9.
15. Sakai T, et al. The custom femoral component is an effective option for congenital hip dysplasia. Clin Orthop Relat Res. 2006;451:146–53.
16. Santori FS, et al. Mid-term results of a custom-made short proximal loading femoral component. J Bone Joint Surg Br. 2010;92:1231–7.
17. Sakai T. [Artificial hip implants by three-dimensional laminate fabrication.] Clin Orthop Surg. 2019;75:222–5. Japanese.

Polished Tapered Stem

41

Toshiki Iwase

1 Concept and History

Cemented stems are classified into taper-slip and composite beams on the basis of their mechanical behavior in the femoral canal [1, 2]. In the taper-slip concept, the polished tapered (wedge-shaped) stem subsides into the cement mantle when loaded. This causes compressive stress on the bone–cement mantle, and the load is transferred to the femur through the bone–cement interface [2]. Bone cement is weak against bending stress but strong against compressive stress, and these properties support the long-term stability of polished tapered stems [3] The polished tapered stem subsides slightly into the bone cement mantle under load, rises slightly when unloaded, and subsides slightly again when reloaded [4]. The polished surface has a very low risk of generating metal or cement wear debris at the stem–cement interface during this cyclic movement as well as of osteolysis derived from wear debris.

There was no understanding regarding the theoretical background of the polished tapered stem prior to its development; therefore, its theoretical construction was serendipitous. The first Exeter stem, developed by Ling and Lee at the Princess Elizabeth Orthopaedic Centre in Exeter, England, was designed with a wedge-shaped geometry in two planes—antero-posteriorly and medially—to provide effective bone–cement penetration when the stem is linearly inserted into the femoral canal via a posterior approach. In the United Kingdom, during the development of the first Exeter stem, the polished finish surfaces were adopted because of the regulation that metallic materials to be placed inside the human body must be polished [3, 5].

In the early stage of development, the neck of the first-generation Exeter stem tended to break, and so the neck shape was changed. However, during this time, the regulation was lifted, and only the base of the neck was reshaped to increase the strength of the stem neck, resulting in a second-generation Exeter stem with a rough metal surface that was inexpensive to manufacture. This change in the metal surface finish caused many cases of early loosening with periprosthetic osteolysis, and detailed observations of the clinical course helped the developers realize the importance of the polished finish, thereby finalizing the use of the polished surface finish for the stems [5]. This led to the development of the taper-slip concept of stem behavior, which is a theory regarding the importance of polished surfaces in wedge-shaped structures.

The polished tapered stem was introduced in Japan in the late 1990s. It was initially introduced for use in impaction bone grafting, a surgical technique for revision hip arthroplasty on the femoral side, by many institutions [6], and its use was expanded to initial total hip arthroplasty (THA), after experiencing good postoperative results in these cases.

As of 2021, many types of "polished tapered stems" (Fig. 41.1) are available in Japan. Stems classified as polished tapered stems include models (1) with different taper structures, such as 2- or 3-sided tapers; (2) with different transverse sections, such as rectangular, trapezoidal, or elliptical transverse sections proximal to the stem; (3) with different concepts for cemented stem placement, such as with a cement mantle thickness of approximately 2 mm, or with a line-to-line technique, in which a stem with approximately the same size as that of the femoral canal to be rasped is placed.

T. Iwase (✉)
Hamamatsu Medical Center, Hamamatsu, Shizuoka, Japan

Y. Nakashima (ed.), *Advances in Total Hip Arthroplasty*, https://doi.org/10.1007/978-981-97-2461-1_41

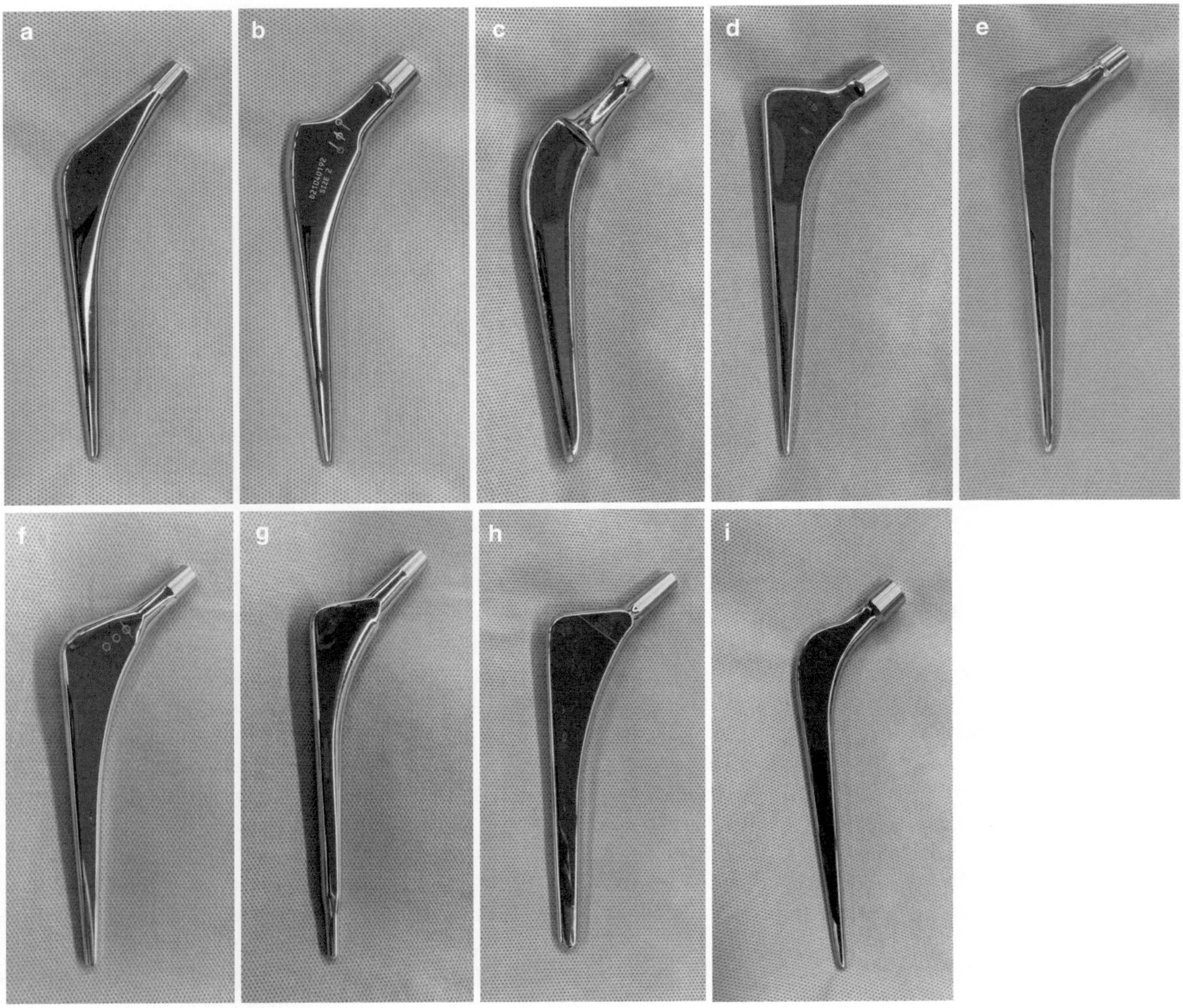

Fig. 41.1 Examples of polished tapered stems marketed in Japan as of October 2021. (**a**) C-Stem (Depuy Syhnthes), (**b**) C-Stem AMT (Depuy Syhnthes), (**c**) CMK (Zimmer-Biomet), (**d**) CPT (Zimmer-Biomet), (**e**) Exeter V40 (Stryker), (**f**) SC stem (Kyocera), (**g**) SN-C HIP stem (Kyocera), (**h**) Trilliance stem (B.BRAUN AESCULAP) and (**i**) VLIANstem (Teijin Nakashima Medical)

2 Theoretical Advantages

Polished tapered stems are characterized by slight subsidence in the cement mantle while loading, and some models are equipped with a hollow centralizer at the distal end of the stem that permits subsidence. According to a systematic review, the average subsidence of a well-functioning polish tapered stem was reported to be 0.97 mm, 1.07 mm, 1.47 mm, and 1.61 mm at 1, 2, 5, and 10 years after surgery, respectively [7]; in general, it is considered to be approximately 2 mm even in long-term cases (Fig. 41.2).

The bone–cement mantle is subjected to hoop stress as the stem subsides, and the load is effectively transferred from the proximal femur through the bone–cement interface, which prevents bone atrophy of the proximal femur due to stress shielding, as is often seen after cementless stem fixation [8].

Some polished tapered stems are available in a wide variety of series configurations, including stem neck offset, stem length, and stem thickness, making it easy to adjust the appropriate stem position according to the individual differences in body size and hip joint construction in surgical cases, which is advantageous for achieving the main objective of THA—to reconstruct the normal biomechanics around the hip joint in any case.

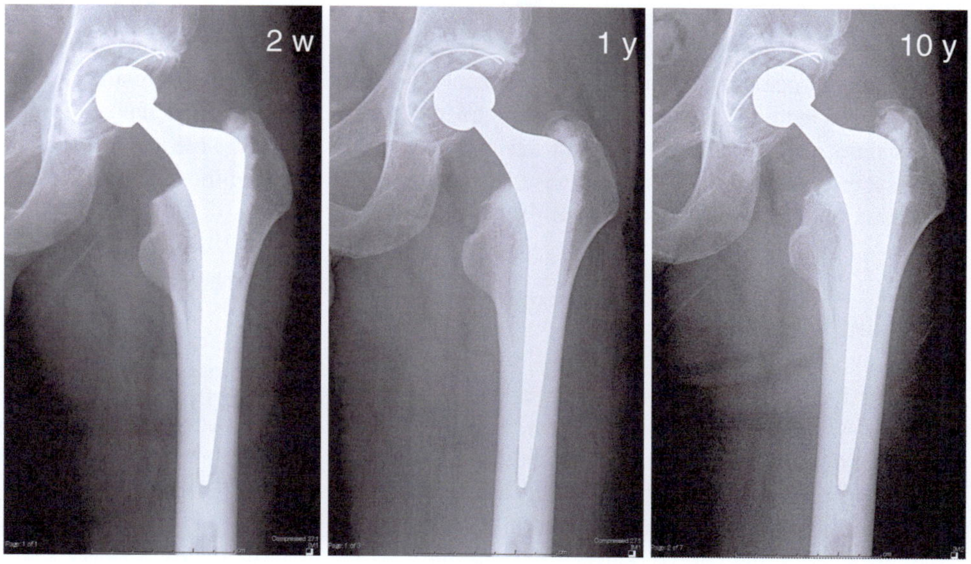

Fig. 41.2 Postoperative subsidence of the polished tapered stem in the cement mantle can be seen as the gap between the stem and cement mantle proximal to the stem shoulder or subsidence of the stem tip in the hollow centralizer

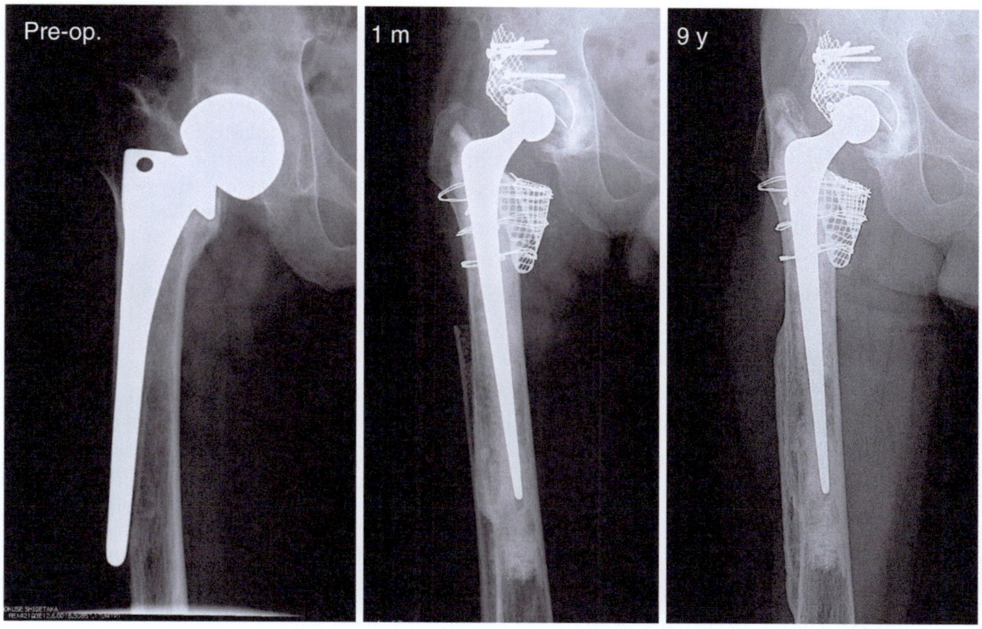

Fig. 41.3 Case of revision femoral impaction bone grafting with a polished tapered stem. (pre-op, one momnth post-operatively, and 9 years post-operatively)

3 Surgical Techniques

Proper exposure of the femoral canal and the use of modern cementing techniques are important for the insertion of a polished tapered stem.

Modern cementing techniques include using an intramedullary plug, cleaning with pulsatile lavage, drying the intramedullary cavity prior to cementing, retrograde injection with a cement gun, and pressurization with a proximal seal.

The surgical field should be exposed for the stem to be inserted straight into the medullary canal, leading to stem placement with good alignment. The so-called "flexion insertion" of the stem insertion axis, in which the stem is inserted from the anterior part of the cut surface of the femoral neck and the stem tip reaches the posterior part of the medullary canal, is likely to cause cement mantle defects and uneven cement mantle thickness in the Gruen Zone 8–9. Therefore, exposing a surgical field that allows good stem alignment is important.

Polished tapered stems are useful for impaction bone grafting [6] in revision hip arthroplasty for patients with femoral bone defects, and special surgical instruments are available. In this procedure, the bone wall defects that often occur in the proximal medial aspect of the femur are contained with metal mesh and wires, and the femoral tube structure is augmented with metal or strut allobone plates, if there are significant bone fragility areas in the femoral diaphysis. Subsequently, a guide wire is used to confirm the center of the medullary canal, allogeneic bone is grafted firmly into the marrow cavity using special instruments to form a "new medullary canal," and a polished tapered stem is inserted (Fig. 41.3).

In addition, polished tapered stems are versatile for other stem exchange methods in THA revision, such as cement-in-cement and in-cement techniques. The cement-in-cement technique is a method in which a new stem is fixed with new cement onto residual old cement, and the in-cement technique is a method in which a new stem is taper-locked to the old cement mantle without new cement [9].

4 Indication

THA, revision hip arthroplasty, and hemiarthroplasty can be performed with a polished tapered stem, regardless of the underlying disease, age, or bone quality of the patient. In the case of revision hip arthroplasty, the ability to use techniques such as impaction bone grafting, cement-in-cement, and in-cement is an advantage of the polished tapered stem.

5 Clinical Results

Among polished tapered stems, the first Exeter stem is associated with the majority of the long-term clinical results, with a reported stem survival rate of 93.5% (95% confidence interval (CI): 90.0–97.0%) at 33 years after surgery, based on the endpoint of aseptic loosening [10]. In Japan, clinical outcomes at an average of 11 years (range 10–13 years) after surgery in 100 patients undergoing primary THA using a polished tapered stem (Exeter stem) showed a stem survival rate of 99.1% (95% CI: 97.5–100%) at an endpoint of ≥5 mm stem subsidence; stem survival at the endpoint of any revision surgery, including cup revision, was 98.3% (95% CI: 96.0–100%). The stem survival rate at the end of revision surgery for aseptic loosening was 100% [11]. These results were reported mainly from a series of primary THA for osteoarthritis cases.

When cemented stems are used in older patients with bone fragility and/or femoral neck fractures, the intraoperative and postoperative risks of periprosthetic femoral fractures are significantly lower than those with cementless stems. If these risks are compared between the fixation concepts within the cemented stem, the use of a polished tapered stem may result in a higher frequency of periprosthetic femoral fractures than that of a composite beam stem [12]. However, even in polished tapered stems, the incidence of periprosthetic femoral fractures differs in terms of taper shape, stem length, stem thickness, and metal composition [13, 14]. Recently, it has become clear that the risk of periprosthetic femoral fracture is higher with a polished tapered stem made using cobalt and chromium than with a stainless-steel owing to a larger amount of subsidence in the cement mantle [14]. Furthermore, within the same brand of polished tapered stems, a longer and/or larger stem is unlikely to cause a periprosthetic femoral fracture [15]. When using a polished tapered stem in patients with high bone fragility, it is advisable to pay attention to the selection of the model and size of the stem, including the choice of metal material.

References

1. Shen G. Femoral stem fixation. An engineering interpretation of the long-term outcome of Charnley and Exeter stems. J Bone Joint Surg Br. 1998;80:754–6.
2. Howell JR, et al. Stem design – the surgeon's perspective. In: Breusch SJ, et al., editors. The well-cemented total hip arthroplasty. Heidelberg: Springer; 2005. p. 180–9.
3. Lee C. The mechanical properties of PMMA bone cement. In: Breusch SJ, et al., editors. The well-cemented total hip arthroplasty. Heidelberg: Springer; 2005. p. 60–7.
4. Takahashi E, et al. The influence of cement thickness on stem subsidence and cement creep in a collarless polished tapered stem: when are thick cement mantles detrimental? Bone Joint Res. 2017;6:351–7.
5. Gie GA. The Exeter Hip: history & development. In: Gie GA, et al., editors. The Exeter Hip: 50 years of innovation in total hip arthroplasty. Exeter: Exeter Hip Publishing; 2020. p. 7–15.
6. Iwase T, et al. Impaction bone grafting for femoral revision hip arthroplasty with Exeter universal stem in Japan. Arch Orthop Trauma Surg. 2012;132:1487–94.
7. Baryeh K, et al. Temporal subsidence patterns of cemented polished taper-slip stems: a systematic review. EFORT Open Rev. 2021;6:331–42.
8. Iwase T, et al. Peri-prosthetic bone remodeling and change in bone mineral density in the femur after cemented polished tapered stem implantation. Eur J Orthop Surg Traumatol. 2019;29:1061–7.
9. Fujita H, et al. Multi-centre study of cement-in-cement and in-cement femoral revision total hip arthroplasty using polished, stainless steel stems. J Orthop Sci. 2022;27:1073–7.
10. Ling RSM, et al. The long-term results of the original Exeter polished cemented femoral component: a follow-up report. J Arthroplast. 2009;24:511–7.
11. Iwase T, et al. The effects of patient characteristics and stem alignment on distal femoral cortical hypertrophy after cemented polished tapered stem implantation. Eur J Orthop Surg Traumatol. 2020;30:559–67.
12. Kristensen TB, et al. More reoperations for periprosthetic fracture after cemented hemiarthroplasty with polished taper-slip stems than after anatomical and straight stems in the treatment of hip fractures: a study from the Norwegian Hip Fracture Register 2005 to 2016. Bone Joint J. 2018;100-B:1565–71.
13. Palan J, et al. The influence of cemented femoral stem choice on the incidence of revision for periprosthetic fracture after primary total hip arthroplasty: an analysis of national joint registry data. Bone Joint J. 2016;98-B:1347–54.
14. Lamb JN, et al. Risk factors for revision of polished taper-slip cemented stems for periprosthetic femoral fracture after primary total hip replacement: a registry-based cohort study from the National Joint Registry for England, Wales, Northern Ireland and the Isle of Man. J Bone Joint Surg Am. 2020;102:1600–8.
15. Morishima T, et al. Periprosthetic fracture torque for short versus standard cemented hip stems: an experimental in vitro study. J Arthroplast. 2014;29:1067–71.

Composite Beam Stem

42

Hiroshi Fujita

1 Concept

Unlike in Western countries, the majority of causative diseases of total hip arthroplasty (THA) in Japan are secondary to acetabular dysplasia, and the skeletal structure of Japanese individuals is small [1]. Hence, we should select a suitable surgical technique for Japanese patients rather than simply introducing a foreign product.

The advantages of cemented stems include easy preoperative planning and a wide range of intraoperative options. Femoral neck hyperanteversion, narrow medullary canal, stovepipe, and osteoporotic bony structure can all be treated with a single system, and good initial fixation can be achieved. Additionally, good long-term results have been demonstrated in several systems [2, 3]. Cemented stems are classified into composite beam and taper slip, according to material, shape, and surface roughness [4].

2 Theoretical Advantages

In Japan, the concept of a composite beam or taper-slip did not emerge until the twenty-first century, and everyone vaguely believed that the stem and cement were fixed. Since the introduction of the Exeter stem to Japan in 2002, the dramatic difference in results due to changes in the surface finish of the Exeter stem has led to it becoming popular. In June 2000, I attended an educational course at the Princess Elizabeth Orthopedic Center of Exeter, and I remember as if it were only yesterday that I first encountered the "taper-slip" stem design and had many scales fall from my eyes.

George Shen stated that in composite beams, the stem and cement must be perfectly bonded, and the cement must have sufficient strength. He stated that the strains in the stem and cement at the stem–cement interface are always the same, and the stress in the cement is released only when the interface fails [4]. The load applied to the head is transmitted through the stiffer stem to the tip, and then further into the cement and distal bone. This leads to stress shielding in the proximal region and atrophy of the proximal femur.

Composite beam-cemented stems, such as Lubinus SP, are designed to adhere to the cement and require a strict cementation technique and thickness of the cement mantle. However, recent radio-stereometric analysis (RSA) studies have demonstrated that all cemented stems migrate within the cement to varying degrees [5, 6]. Even in the case of composite beam-cemented stems, the stem–cement interface is debonded, which may eventually lead to failure.

By contrast, the taper slip system, as typified by Exeter, induces radial compression and hoop tension in the cement as the stem slips in the cement, resulting in good load transfer to the bone–cement interface. This is the theoretical basis for the system to obtain good long-term results, notwithstanding a slight lack of techniques or a thin cement mantle in some parts of the cement. Radial compression and hoop tension in the cement prevents bone atrophy due to bone irritation at the bone–cement interface, which is the theoretical basis for the good long-term results of this system.

"Taper slip" is a direct translation of the term "slip by tapering." George Shen stated that in the taper slip design, the stem slipping in the cement is inevitable, which results in strong fixation of the stem and cement [4]. Centrifugal stress is generated in the cement by the slip, which is transmitted to the bone–cement interface as hoop stress because the stress is transmitted to the bone from the proximal side, and stress shielding of the proximal femur is unlikely to occur.

However, classifying all cement stems into one of these categories is often difficult [7]. In this section, taper slip is defined as "having a polished surface, a taper shape, and no collar," and composite beam is defined as "not conforming to any of the definitions of a taper slip." However, some stems with a "polished surface, taper shape, and collar" demonstrate a similar behavior to that of taper slip stems. Thus, some stems are classified differently depending on the definition.

H. Fujita (✉)
Rakuyo Hospital, Kyoto, Japan

Y. Nakashima (ed.), *Advances in Total Hip Arthroplasty*, https://doi.org/10.1007/978-981-97-2461-1_42

3 Clinical Results

The results for Elite-Plus in Japan are generally favorable [8]. However, distal femoral cortical hypertrophy (DFCH) reportedly occurs in 14.8% of cases [9], suggesting that proximal stress shielding is common (Fig. 42.1).

The long-term performance of cemented stems composed of Ti alloys is generally poor [10, 11]. Oonishi et al. have reported excellent long-term results of cementation of a taper-shaped, smooth-surfaced Kyocera Type 6 stem using an interface bioactive bone cement (IBBC) technique [12, 13]. Okutani et al. have reported that distally tapered Ti alloy stems demonstrated better long-term results than distally cylindrical stems [14]. Takaoka et al. have reported that DFCH occurred in 17.0% of the Ti alloy stems, although the results were good for >10 years [15].

Lubinus SP II is the most commonly used stem in Nordic countries. It is a typical composite beam with an anatomical shape, and according to the Nordic Registry, the 10-year survival rate with it is >95%, which is comparable to that of Exeter [16]. Hashimoto et al. have reported good results for DCM-J stems with vertical grooves, collars, and satin surface characteristics over 10 years [17]. Comparing the two types of rasps, stress shielding occurred more frequently in the 0.5-mm oversized rasp than in the 1.5-mm oversized rasp [17].

Palan et al. analyzed UK registry data of >250,000 cases and compared the frequency of postoperative peri-prosthetic fractures between the Charnley stem and several types of a polished taper stem (PTS) [18]. They have reported that the Charnley stem had a significantly lower incidence of postoperative periprosthetic fractures than the PTS (Exeter stem, C-stem, and CPT). In elderly patients with bone fragility, the composite beam has a slight advantage over the PTS in terms of postoperative periprosthetic fractures. However, the overall revision rate should be carefully evaluated before model selection.

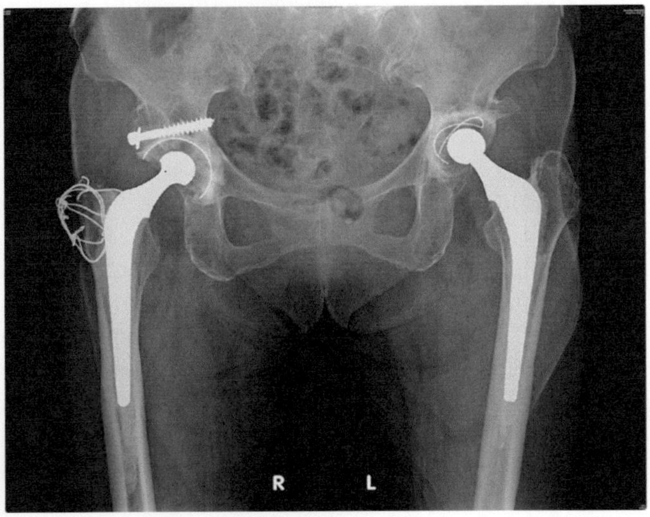

Fig. 42.1 Example of the long-term progress of the composite beam. This figure presents the radiographic images of 32 years (surgeon: Makoto Kobori) and 25 years (surgeon: Kaori Kobori) after bilateral cemented total hip arthroplasty in the right and left images, respectively. A composite beam Charnley Elite stem was used on the femoral side. No loosening or radiolucent lines were observed on either side. Cortical hypertrophy was observed in zones 3, 4 L, 4 M and 5 on both the sides

Taper slips are relatively easy to remove during revision [19]. Among the composite beams, those with macro-uneven surfaces or anatomical shapes are very difficult to remove compared to those with tapered and smooth surfaces. In addition to the frequency of loosening and periprosthetic fractures, ease of removal should be considered in model selection.

4 Comparison of the Composite Beam and Taper Slip in Our Department

We prospectively observed Type 6 (Kyocera) and Exeter (Stryker) stems cemented by IBBC at different times for >10 years [20]. The Type 6 stems were composed of Ti alloy with a smooth surface, triple taper, and collar and belonged to the composite beam group. Meanwhile, the Exeter stem was composed of stainless steel, with a polished surface, double taper, and no collar and belonged to the taper slip group. We compared 46 hips (group C) using the Type 6 stem from March 2003 to September 2004 and 46 hips (group T) using the Exeter stem from October 2004 to February 2006. The mean postoperative observation period was 12.4 years in group C and 10.8 years in group T. Bone grafting was used for bone defects of the acetabulum, and all polyethylene cups (Aeonian, Kyocera) and each stem were cemented using the IBBC and third-generation cement techniques, respectively. The immediate postoperative cementing grades of the femoral side were A in 35 and 33 hips in group C and group T, respectively, and B in 8 hips in group C and T. The survival rate at 10 years when the revision surgery for all causes was set as the endpoint was 96% in group C and 98% in group T, with no significant difference between both groups (log-rank test, $p = 0.49$). No radiolucent lines, focal osteolysis, or cement fractures were observed in either group. Cortical hypertrophy (CH) was observed in eight hips (18.6%) in group C and seven hips (17.1%) in group T (NS). CH located on the medial side of the femur was significantly less common in group C and more common in group T. The peak location of CH was significantly different between groups C (100.4%) and T (84.3%). Cancellization was observed in 21 (44.8%) hips in group C and 16 (39.0%) hips in group T (NS). The long-term results of cemented THA using IBBC were excellent, regardless of differences in stem characteristics. CH and cancellization were observed in both groups; however, no significant difference in the frequency of occurrence was observed. CH occurred more distally and laterally in group C and more proximally and medially in group T. The difference in the location of the CH in groups C and T indicated that the composite beam Type 6 stem transmitted the load more distally and laterally than the taper slip Exeter stem.

5 Conclusion

Today's THA has evolved greatly owing to the relentless efforts of our predecessors over the past 60 years. Rehabilitation can be started immediately after surgery owing to improvements in implants, preoperative planning, approach, fixation methods, and postoperative therapy. Day surgery is also common in developed countries. Additionally, some implants have been reported to have good outcomes for >30 years. However, complications such as infection and fracture are increasing due to the diversity of coexisting lesions and population aging. Surgeons are required to accurately judge models without relying on the flood of information available in the market. When selecting a model, using one that has been reported to have long-term results and learning directly from experts in the field through literature, seminars, and surgical observation are important. The cemented stems of the composite beam type include all cemented stems other than the taper slip type; therefore, discussing them in general is difficult. Therefore, the concept and results of each stem should be carefully examined before its use.

Finally, "primary THA is the beginning of THA revision."

Acknowledgments I am deeply grateful to Dr. Makoto Kobori and Dr. Kaori Kobori, who performed the operation 30 years ago and provided the images.

References

1. Jingushi S, et al. Osteoarthritis hip joints in Japan: involvement of acetabular dysplasia. J Orthop Sci. 2011;16:156–64.
2. Petheram TG, et al. The Exeter universal cemented femoral stem at 20 to 25 years: a report of 382 hips. Bone Joint J. 2016;98-B:1441–9.
3. Fujita H, et al. Multi-center study of the use of the Exeter stem in Japan: evaluation of 1000 primary THA. J Orthop Sci. 2012;17:370–6.
4. Shen G, Femoral stem fixation. An engineering interpretation of the long-term outcome of Charnley and Exeter stems. J Bone Joint Surg Br. 1998;80:754–6.
5. Alfaro-Adrián J, et al. Cement migration after THR. A comparison of Charnley elite and Exeter femoral stems using RSA. J Bone Joint Surg Br. 1999;81:130–4.
6. Nivbrant B, et al. Increased migration of the SHP prosthesis: radiostereometric comparison with the Lubinus SP2 design in 40 cases. Acta Orthop Scand. 1999;70:569–77.
7. Cassar-Gheiti AJ, et al. Current concepts and outcomes in cemented femoral stem design and cementation techniques: the argument for a new classification system. EFORT Open Rev. 2020;5:241–52.
8. Hirose S, et al. Outcomes of Charnley total hip arthroplasty using improved cementing with so-called second- and third-generation techniques. Orthop Sci. 2012;17:118–23.
9. Goto K, et al. Long-term results of total hip arthroplasty using Charnley Elite-Plus stem and the effect of stem geometry on radiographic distal femoral cortical hypertrophy. J Orthop Sci. 2018;23:365–70.
10. Hinrichs F, et al. A comparative clinical outcome evaluation of smooth (10-13 year results) versus rough surface finish (5-8 year results) in an otherwise identically designed cemented titanium alloy stem. Arch Orthop Trauma Surg. 2003;123:268–72.
11. Kovac S, et al. Long-term survival of a cemented titanium-aluminium-vanadium alloy straight-stem femoral component. J Bone Joint Surg Br. 2006;88:1567–73.
12. Oonishi H, et al. THA with hydroxyapatite granules at the cement-bone interface: 15- to 20-year results. Clin Orthop Relat Res. 2008;466:373–9.
13. Oonishi H, et al. Total hip arthroplasty around the inception of the interface bioactive bone cement technique. Clin Orthop Surg. 2016;8:237–42.
14. Okutani Y, et al. Long-term outcome of cemented total hip arthroplasty with a Charnley-type femoral stem made of titanium alloy. J Orthop Sci. 2019;24:1047–52.
15. Takaoka Y, et al. The long-term results of total hip arthroplasty with a cemented β-titanium stem. J Arthroplast. 2020;35:2167–72.
16. Junnila M, et al. Implant survival of the most common cemented total hip devices from the Nordic Arthroplasty Register Association database. Acta Orthop. 2016;87:546–53.
17. Hashimoto Y, et al. The effects of rasp oversize on the clinical and radiographic outcomes of total hip arthroplasty with a collared satin-finished. J Arthroplast. 2021;36:2055–61.
18. Palan J, et al. The influence of cemented femoral stem choice on the incidence of revision for periprosthetic fracture after primary total hip arthroplasty: an analysis of national joint registry data. Bone Joint J. 2016;98-B:1347–54.
19. Fujita H, et al. Multi-centre study of cement-in-cement and in-cement femoral revision total hip arthroplasty using polished, stainless steel stems. J Orthop Sci. 2021;S0949-2658(21):00214-1.
20. Fujita H, et al. Prospective, comparative study of cemented, smooth-surfaced titanium stems and polish-surfaced, stainless steel stems at a minimum follow-up of 10 years. Eur J Orthop Surg Traumatol. 2020;30:501–12.

Part VII

Principles in Surgical Procedures

Hip Center

43

Tomohiro Goto

1 Cup Installation Height

In principle, the goal of the cup placement height is the true anatomical position. Previous biomechanical studies have shown that the true anatomical position is the most efficient position for the abductor muscles to exert an abduction force with the least stress on the cup [1]. This is also understandable from the viewpoint of the postoperative range of motion and impingement. Regardless of whether a cemented or cementless cup is used, the concept of aiming for the true anatomical position to obtain sufficient bony coverage and strong initial fixation is commonly accepted. However, in Japan, most patients with osteoarthritis of the hip joint who are candidates for total hip arthroplasty (THA) have secondary osteoarthritis of the hip joint due to acetabular dysplasia. In dysplastic hips (especially Crowe grade 2 or higher), acetabular deformity and hypoplasia may result in inadequate bony coverage of the acetabular cup in the true anatomical position. Insufficient bony coverage of the cup is an important problem that affects initial fixation and cup laxity [2, 3]. In highly dysplastic hips, placing a cup in the true anatomical position results in a large residual bone defect above the cup, necessitating augmentation or the use of a massive bone graft to augment the cup fixation. When a cementless cup is used, care must be taken because the risks of cup laxity, migration, and collapse of the grafted bone are high if a large, massive bone graft is placed with insufficient bony coverage of the cup by the matrix bone [4]. Elevation of the cup position is a technique that can be used to increase contact with the host bone and obtain sufficient bony coverage by raising the cup height in severe dysplastic hips (Fig. 43.1). In the case of cemented cups, it has been reported that high cup placement results in a high rate of implant loosening [5–7], In principle, the primary cup is placed in the primary position using bone grafts, such as massive bone grafts, impaction bone grafts, and KT plates [8–10]. However, in the case of cementless cups, the concept of high cup placement has been widely accepted in recent years because of its good long-term results and the evolution of cup surface treatment [11]. However, there are concerns that elevated cups may increase stress on the cups and decrease abductor muscle strength; moreover, there is still room for debate as to what level of elevation is acceptable from the perspective of joint biomechanics and what level of bone covering is necessary. Previous biomechanical tests using cadaveric bones reported that the stress on the cup did not increase significantly even when the cup was placed 3.7 cm above the teardrop line [12, 13]. In both studies, lateral placement of the cup was cited as a defective factor. The results of other basic and clinical studies have also reported many problems with lateral placement of the cup, and caution should be exercised because it may have a negative effect on treatment outcomes [1, 2, 14]. When considering the appropriate cup height, it is important to fully understand the advantages and disadvantages of elevation (Table 43.1). The installation should be made close to the original mortar site to avoid unnecessary cup elevation.

T. Goto (✉)
Tokushima University, Tokushima, Japan

Y. Nakashima (ed.), *Advances in Total Hip Arthroplasty*, https://doi.org/10.1007/978-981-97-2461-1_43

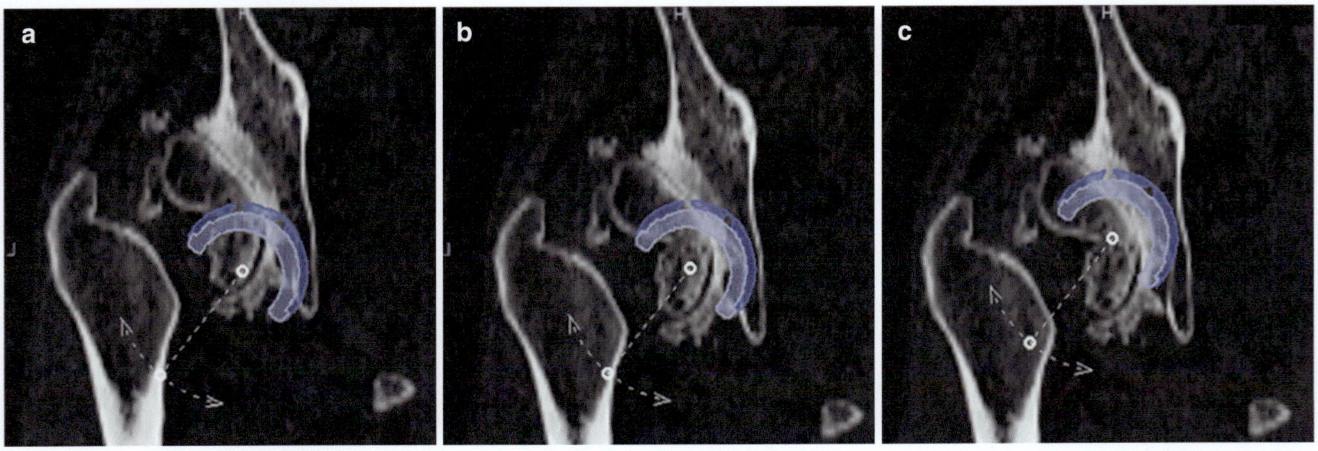

Fig. 43.1 Simulation of cup height placement using a 3D template in the Crowe 3 dysplastic hip. (**a**) 15 mm above the teardrop line. (**b**) 25 mm above the teardrop line. (**c**) 35 mm above the teardrop line

Table 43.1 Advantages and disadvantages of high cup placement

Advantages	Disadvantages
Increased contact area with the host bone	Increased risk of cup loosening in lateral placement of the cup
Bony covering (cup CE angle) may increase	Risk of weakness of postoperative abductor muscle strength
Relatively easy procedure	Reduced postoperative range of motion Risk of postoperative dislocation
No (less) bone grafting required	Less bone stock recovery

2 Preoperative Planning

Cup height is generally evaluated by the distance between the vertical and horizontal positions of the center of the femoral head in the frontal view of both hip joints, using the teardrop line and its perpendicular line as the reference (Fig. 43.2). In terms of cup placement height, we examined the anatomical location of the femoral head center (the original hip joint position) by assessing a total of 200 joints in Japanese women, 100 joints without dysplasia, and 100 hip joints with early stage secondary hip osteoarthritis caused by acetabular dysplasia. The results show that the position of the femoral head center averaged 14.6 mm in normal joints and 16.8 mm in early stage of osteoarthritis secondary to acetabular dysplasia in the vertical direction from the tear drop line. Based on the variance of these measurements, the height of the true anatomical hip center position in Japanese women with dysplasia was defined as 11–22 mm from the teardrop line [15]. In addition, a study of the height of the femoral head center in 98 Japanese patients without abnormal hip morphology showed that the average height of the femoral head center was 15.2 mm from teardrop line [16]. When making a preoperative plan, the cup height should be compared with that of the healthy side in unilateral cases; however, in bilateral cases, it is reasonable to consider 15–20 mm from the tear drop line as the primary acetabular position. Generally, the boney coverage of the cup was evaluated using the cup center edge (CE) angle (Fig. 43.3). Although the cup CE angle itself can be evaluated using simple radiographs, in actual clinical practice, it is important to understand the condition of the acetabular bone, location of the cyst, contact area with the cup, thus, preoperative planning using CT is recommended (Fig. 43.4). In particular, the condition of the anterior and posterior wall of the acetabulum must be checked because it is a very important factor in the initial fixation of the cup [17]. The authors determined

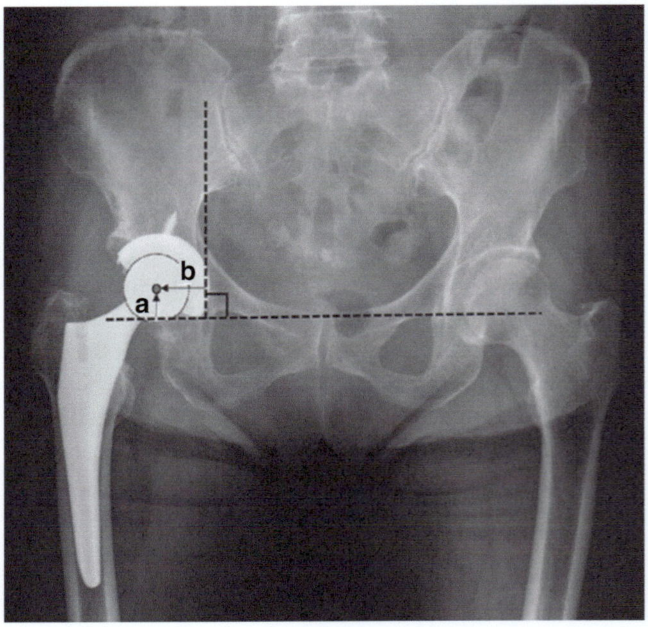

Fig. 43.2 Evaluation of cementless cup installation location. (**a**) Vertical distance from teardrop line. (**b**) Horizontal distance from perpendicular to the teardrop line

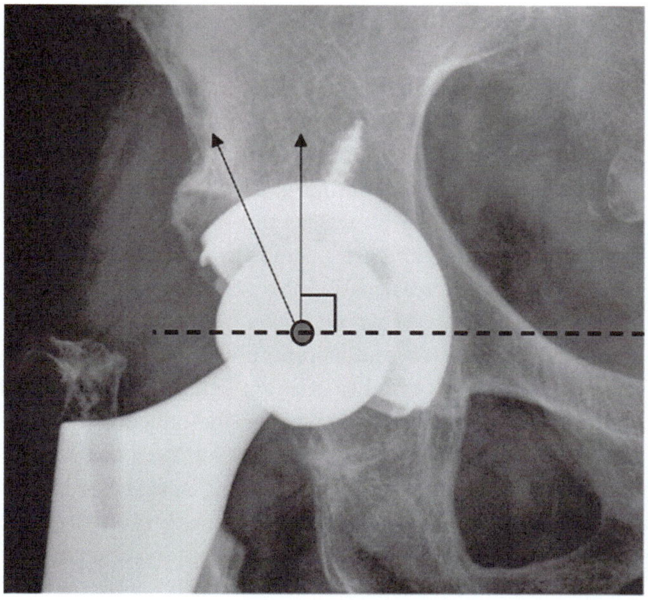

Fig. 43.3 Cup CE Angle. Angle between the perpendicular line of the teardrop line and the lateral edge of the acetabulum in contact with the cup from the center of the epiphysis (dotted line: horizontal line with teardrop line)

the height of cup placement to ensure a cup CE angle of at least 0° to 10° in the coronal plane through the center of the head. The results of the validation of cases Crowe 2 and 3 using this policy showed that cup placement simulation using a 3D template system showed that the cups of both Crowe 2 and 3 should be placed an average of 20 mm from the tear drop line to ensure a cup CE angle of at least 0°, and an average of 25 mm from the tear drop line to ensure a cup CE angle of at least 10°. In the postoperative image evaluation of the actual surgery, the average height of cup placement from the tear drop line was 26.6 mm for Crowe 2 and 27.6 mm for Crowe 3, and the maximum height was 31.4 mm for Crowe 2 and 36.7 mm for Crowe 3 [18]. Based on these results, the current policy of the authors is to secure a cup CE angle of at least 0° to 10°, and the maximum allowable cup height position is 30 mm from the tear drop line. When the cup was placed to fit the anteroposterior aspect of the acetabulum, the mean cup size of Japanese crows 2 and 3 was the largest when the cup height was 20 mm from the tear drop line (Crowe 2: 50.7 mm, Crowe 3: 50.1 mm) and decreased as the height of the cup increased. It would be very

Fig. 43.4 Preoperative 3D template using CT images

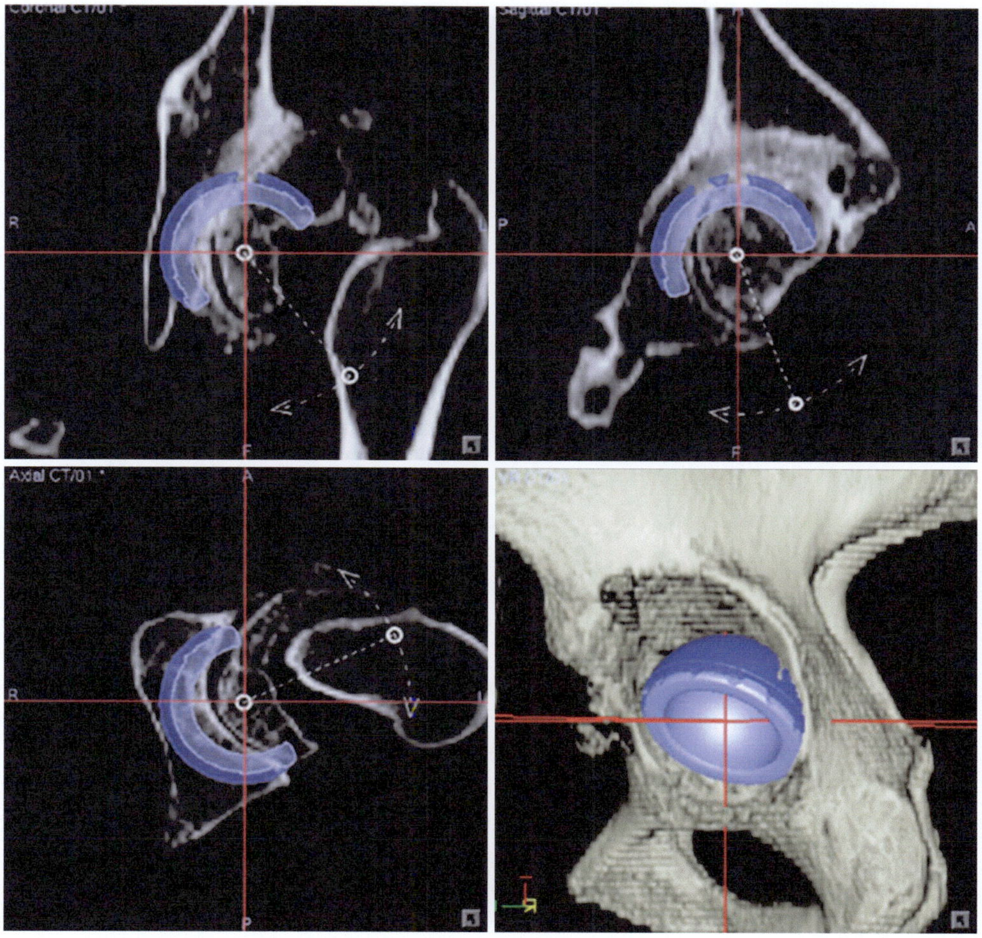

useful for surgeons to know the mean cup height and cup size in Japanese patients with dysplastic hips to make appropriate preoperative plans.

3 Intraoperative Evaluation

In dysplastic hips, it is difficult to accurately determine the orientation of the acetabulum and place reaming implants at the intended location. In patients with severe deformities, the double floor of the acetabulum and large anterior osteophytes make it difficult for even the most experienced hip joint surgeon to accurately determine the center of the acetabulum and to place the implant appropriately. For accurate intraoperative judgment, it is very important to have a detailed understanding of the relationship between the cup and acetabular bone, on preoperative CT, and intraoperative imaging and radiography are recommended to confirm the height of the cup placement. Surgical assistive technologies, such as navigation and robotic arms, are extremely useful, enabling intraoperative three-dimensional positioning and reproduction of the preoperative plan with high accuracy, not only in terms of implant placement angle but also in terms of placement position (Fig. 43.4). According to a report on the accuracy of cup placement using navigation in Japanese patients with hip osteoarthritis secondary to acetabular dysplasia, the average error from the preoperative plan in cup alignment was 1.5° for the cup inclination angle and 2.1° for the cup anteversion angle, indicating a high accuracy. The cup placement position was reproducible with an average error of approximately 2 mm in all directions [19]. In addition, it has been shown that cup placement can be performed with high accuracy in patients with severe deformities including severe dislocation of the hip, ankylosis, and bone destruction without significant differences from a control group with normal hip morphology [20]. Recently, the accuracy of implant placement using a robotic arm has been reported as the latest technology, and even more accurate cup placement has been achieved compared with navigation [21]. In actual surgical settings, it is difficult to determine the final implant position based on the cup alone, and it is necessary to make a comprehensive judgment of the degree of joint contracture and amount of leg extension. In particular, patients with severely dysplastic hips with Crowe 2 or higher often have abnormalities in rotation and morphology on the femoral side. Owing to the anatomical limitations of cup placement, it is important to consider stem selection with a wide range of variations in offset, neck length, and anteversion angle adjustment. Once adequate initial cup fixation is achieved, a bone graft is usually unnecessary. The authors have never used a massive bone graft and have reported good results with reamer bone grafted above the cup.

4 Impact on Postoperative Outcomes

The results of cemented cups combined with bone grafting for cases of acetabular dysplasia with large bone defects have been reported. After a mean follow-up period of 11 years, incorporation of the grafted bone into the host bone was observed in all cases, and there was no case of loosening [8]. Regarding the postoperative results of acetabular reconstruction using KT plates with autologous or allogeneic bone, all patients underwent initial surgery with 100% implant survival and incorporation of the grafted at an average follow-up period of 8 years [9]. In a report of 40 cases of THA with impaction bone grafts for acetabular dysplasia and large bone defects, reconstruction in the original position was achieved without loosening in all cases except in one case of acute infection at an average follow-up time of 7.5 years [10]. Although these surgical procedures require a skilled technique, acetabular reconstruction with a cement cup, performed with a correct understanding of the concept and correct surgical technique, is excellent for reconstruction of the hip in the original position and recovery of bone stock and may be theoretically one of the ideal treatment options.

However, the postoperative mid- to long-term results of high cup placement using cementless cups have recently been reported to be excellent, with a cup survival rate comparable to that of conventional cementless cups. When a cup was placed using only a press-fit cup with a CE angle of at least 10°, the bone coverage of the cup surface was 76% or higher and no case of cup loosening was observed at an average follow-up time of 7.4 years [17]. In addition, when a cup was placed ≥20 mm above the tear drop line to ensure at least 75% bone coverage of the cup surface for 30 dysplastic hip joints, no case of cup loosening was reported after an average follow-up time of 15 years and 2 months [3]. In addition, the postoperative results of high cup placement for relatively severe dysplastic hips (Crowe 2 and 3) showed that the average height of the cup was 26.6 mm from the tear drop line in Crowe 2 and 27.6 mm in Crowe 3, with no case of cup loosening after a follow-up time of more than 7 years (7–11 years) postoperatively [18]. In another study, the 13-year results of THA for cases of acetabular dysplasia with Crowe 2 or higher, with an average cup height of 30.9 mm from the tear drop line, showed that there were no cases of cup loosening and that the Harris hip score exceeded 80 in 83% of cases [11].

A notable problem to be aware of is polyethylene wear in THA with high cup placement. Previously, it was reported that aseptic loosening due to increased polyethylene wear occurs at a high rate in cases of cup elevation using conventional polyethylene [5, 7]. However, recent studies have shown that cup elevation has a negative effect on polyethylene wear [11], and some authors have suggested that conventional polyethylene does not increase polyethylene wear if the medialization of the cup is properly performed [2].

In addition, there have been varying reports of postoperative deterioration of the abductor muscle strength due to elevated cup placement. In the results of the evaluation of the Trendelenburg sign and abductor muscle strength in 50 Japanese patients with unilateral OA, the positive rate of the Trendelenburg sign was significantly higher in the high placement group, and abductor muscle strength (compared to the healthy side) was 7.7% lower in the high placement group [22]. In addition, a study on postoperative abduction muscle strength in 30 patients with unilateral OA showed a negative correlation between cup elevation and postoperative abduction muscle strength [23]. In a postoperative abductor muscle survey of 100 THA patients with dysplastic hip, recovery of abductor muscle strength was significantly delayed at 6 months in patients in whom the cup was placed >15 mm compared with the healthy side, but recovery was observed at 12 months [24]. However, it has also been suggested that the effect of cup placement height on postoperative abduction muscle strength is small. A comparison of 100 cases with and without the Trendelenburg sign showed that there was no significant difference in postoperative Trendelenburg sign and that cup placement height within 30 mm of the tear drop line had no effect on abduction muscle strength [14]. In addition, if the femoral offset and abductor lever arms were obtained appropriately, cup height placement >30 mm had no effect on postoperative patient-standing clinical assessment [25, 26]. The causal relationship between cup height and postoperative abductor muscle strength remains controversial and requires further investigation.

The effect of cup elevation on the postoperative range of motion is a matter of concern. A simulation study using CT images of dysplastic hips (Crowe 2 and 3) showed that flexion and internal rotation range of motion decreased with increasing cup height and that cup height should be limited to 35 mm from the tear drop line [27]. A clinical study of THA for 1079 dysplastic hips showed that high cup placement (>23.9 mm) could be a risk factor for postoperative dislocation [28]. Furthermore, a CT-based simulation study reported the effects of the shape of the inferior anterior iliac spine (AIIS) and cup height on the range of motion and impingement, especially in the Hetsuroni classification of AIIS, showing that cup height decreases the range of motion in types II and III [29]. We should be fully aware that elevated cups decrease the postoperative range of motion and increase the risk of impingement, and intraoperative range-of-motion checks and dislocation tests should be performed carefully.

In recent years, with the development of surface processing of cups, new implants, such as 3D porous cups made of titanium and tantalum, have been developed and clinically introduced. These developments are expected to improve the initial fixation and promote bone ingrowth by increasing the scratching effect; furthermore, the currently considered acceptable range of cup placement positions may change as cup fixation improves. In addition, the development of surgical support technologies, such as robots, is progressing, and their widespread clinical use is expected to lead to the generalization of implant placement with higher accuracy and improved outcomes.

References

1. Johnston RC, et al. Reconstruction of the hip. A mathematical approach to determine optimum geometric relationships. J Bone Joint Surg Am. 1979;61:639–52.
2. Nawabi DH, et al. Durable fixation achieved with medialized, high hip center cementless THAs for Crowe II and III dysplasia. Clin Orthop Relat Res. 2014;472:630–6.
3. Kaneuji A, et al. Minimum ten-year results of a porous acetabular component for Crowe I to III hip dysplasia using an elevated hip center. J Arthroplasty. 2009;24:187–94.
4. Hendrich C, et al. Cementless acetabular reconstruction and structural bone-grafting in dysplastic hips. J Bone Joint Surg Am. 2006;88:387–94.
5. Callaghan JJ, et al. The results of revision for mechanical failure after cemented total hip replacement, 1979 to 1982. A two- to five-year follow-up. Joint Surg Am. 1985, 676:1074–85.
6. Russotti GM, et al. Proximal placement of the acetabular component in total hip arthroplasty: a long-term follow-up study. J Bone Joint Surg Am. 1991;73:587–92.
7. Pagnano MW, et al. The effect of superior placement of the acetabular component on the rate of loosening after total hip arthroplasty. J Bone Joint Surg Am. 1996;78:1004–14.
8. Oe K, et al. Long-term results of acetabular reconstruction using three bulk bone graft techniques in cemented total hip arthroplasty for developmental dysplasia. Int Orthop. 2016;40:1949–54.
9. Oe K, et al. Bone remodeling in acetabular reconstruction using a Kerboull-Type reinforcement device and structural bone-grafting in Total hip arthroplasty. J Arthroplast. 2017;32:908–14.
10. Iwase T, et al. Favorable results of primary total hip arthroplasty with acetabular impaction bone grafting for large segmental bone defects in dysplastic hips. J Arthroplast. 2016;31:2221–6.
11. Galea VP, et al. Developmental dysplasia treated with cementless total hip arthroplasty utilizing high hip center reconstruction: a minimum 13-year follow-up study. J Arthroplast. 2018;33:2899–905.
12. Doehring TC, et al. Effect of superior and superolateral relocations of the hip center on hip joint forces: an experimental and analytical analysis. J Arthroplasty. 1996;11:693–703.
13. Delp SL, et al. How superior placement of the joint center in hip arthroplasty affects the abductor muscles. Clin Orthop Relat Res. 1996;328:137–46.
14. Fukui K, et al. How far above the true anatomic position can the acetabular cup be placed in total hip arthroplasty? Hip Int. 2013;23:129–34.
15. Fukui K, et al. A radiological study of the true anatomical position of the acetabulum in Japanese women. Hip Int. 2011;21:311–6.
16. Takamatsu T, et al. Radiographic determination of hip rotation center and femoral offset in Japanese adults: a preliminary investigation toward the preoperative implications in Total hip arthroplasty. BioMed Research Int. 2015;2015:610763.

17. Takao M, et al. The results of a press-fit-only technique for acetabular fixation in hip dysplasia. J Arthroplast. 2011;26:562–8.
18. Takasago T, et al. Postoperative radiographic evaluation and simulation study for optimal cup placement in high-hip center total hip arthroplasty. Hip Int. 2021;31:335–41.
19. Tsutsui T, et al. Efficacy of a computed tomography-based navigation system for placement of the acetabular component in total hip arthroplasty for developmental dysplasia of the hip. J Orthop Surg (Hong Kong). 2017;25:2309499017727954.
20. Kajino Y, et al. Does degree of pelvic deformity affect the accuracy of computed tomography-based hip navigation? J Arthroplast. 2012;27:1651–7.
21. Ando W, et al. Comparison of the accuracy of the cup position and orientation in total hip arthroplasty for osteoarthritis secondary to developmental dysplasia of the hip between the Mako robotic arm-assisted system and computed tomography-based navigation. Int Orthop. 2021;45:1719–25.
22. Kiyama T, et al. Effect of superior placement of the hip center on abductor muscle strength in total hip arthroplasty. J Arthroplast. 2009;24(24):240–5.
23. Asayama I, et al. Reconstructed hip joint position and abductor muscle strength after total hip arthroplasty. J Arthroplast. 2005;20:414–20.
24. Fukushi J, et al. Does hip center location affect the recovery of the abductor moment after total hip arthroplasty? Orthop Traumatol Surg Res. 2018;104:1149–53.
25. Traina F, et al. The influence of the center of rotation on implant survival using a modular stem hip prosthesis. Int Orthop. 2009;33:1513–8.
26. Traina F, et al. Modular neck prostheses in DDH patients: 11-year results. J Orthop Sci. 2011;16:14–20.
27. Komiyama K, et al. Does high hip center decrease range of motion in total hip arthroplasty? A computer simulation study. J Arthroplasty. 2016;31:2342–7.
28. Komiyama K, et al. Does high hip center affect dislocation after total hip arthroplasty for developmental dysplasia of the hip? Int Orthop. 2019;43:2057–63.
29. Tabata T, et al. Influence of hip center position, anterior inferior iliac spine morphology, and ball head diameter on range of motion in total hip arthroplasty. Orthop Traumatol Surg Res. 2019;105:23–8.

Offset

44

Tomohiro Shimizu

1 Offsets in Total Hip Arthroplasty

Determining the appropriate offset in total hip arthroplasty (THA) is important from the perspective of the effect on the joint forces and range of motion applied to the hip by the acquisition of dislocation resistance and definition of the lever arm of the abductor muscle. In addition to femoral and acetabular offsets, the global offset, which is the distance from the pubic symphysis to the femoral axis, is evaluated for the hip offset (Fig. 44.1). In the past, preoperative and postoperative evaluations were often performed using two-dimensional (2D) frontal X-ray images of the hip joint. However, in recent years, three-dimensional (3D) evaluation using computed tomography (CT) (Fig. 44.2) has improved our understanding of the anatomy of the hip joint and enabled detailed preoperative and postoperative planning and evaluation. Factors that determine the offset include the thickness of the acetabular base, presence or absence of a double floor, joint congruity, deformation of the femoral head, and anteversion and cervical angles of the femur. The position of the

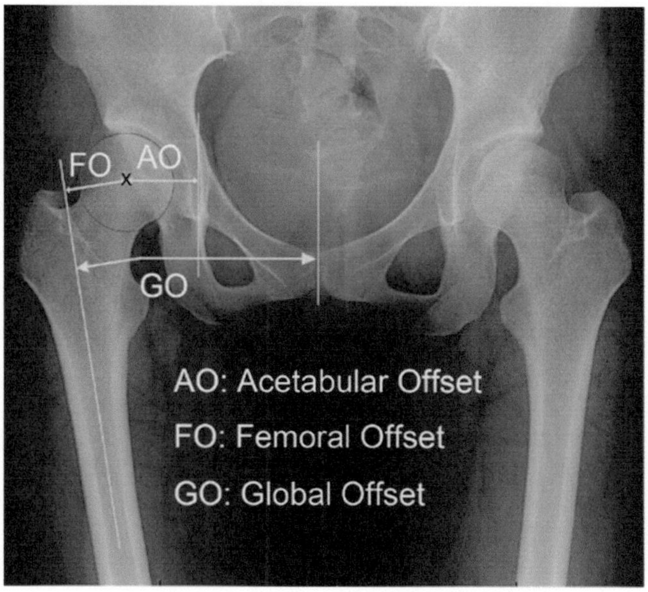

Fig. 44.1 Simple frontal X-ray image of the hip joint. *AO* acetabular offset, *FO* femoral offset, *GO* global offset

T. Shimizu (✉)
Department of Orthopaedic Surgery, Faculty of Medicine and Graduate School of Medicine, Hokkaido University, Sapporo, Hokkaido, Japan
e-mail: simitom@wg8.so-net.ne.jp

Y. Nakashima (ed.), *Advances in Total Hip Arthroplasty*, https://doi.org/10.1007/978-981-97-2461-1_44

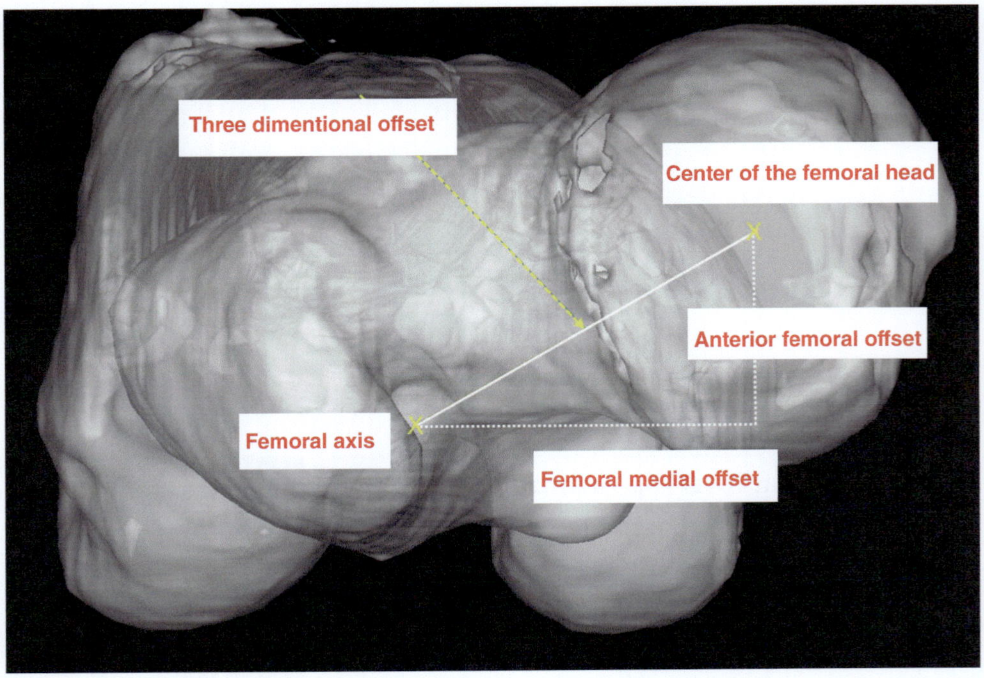

Fig. 44.2 Three-dimensional reconstructed image of the femur. The three-dimensional femoral offset is divided into the medial and anterior offsets

hip center on the affected and healthy sides should be accurately evaluated in two and three dimensions using simple radiographs and CT, and the offset should be determined from the viewpoint of biomechanics (kinematics and kinetics) during movement.

2 Preoperative Planning

Two-dimensional preoperative planning using simple radiographic images is still widely used because of its simplicity and low cost. Using the line connecting the bilateral tear drops as a reference and referring to the height of the contralateral hip center and the position of the inner plate, the cup is installed in a position where the target cup abduction angle of 40°–45° and cup center of edge (CE) angle (10° or more) can be achieved. In cases where a double floor is present, the cup should be internalized and placed on the outer wall of the lacrimal scar. This determines the acetabular offset, and the implant is selected by setting the offset size to obtain an appropriate femoral offset, referring to the global offset of the contralateral side (Fig. 44.3). The left–right difference in the postoperative leg offset defines soft-tissue tension, and minimizing the left–right difference in the postoperative leg offset is reportedly important for obtaining adequate soft-tissue tension over a wide range of flexion angles [1]. In cases with unilateral disease or without severe degeneration or deformity, it is advisable to plan cup placement by considering the hip center and femoral offset, using the healthy or contralateral side as the index. In cases with strong bilateral degeneration and deformity, a study of Japanese adults

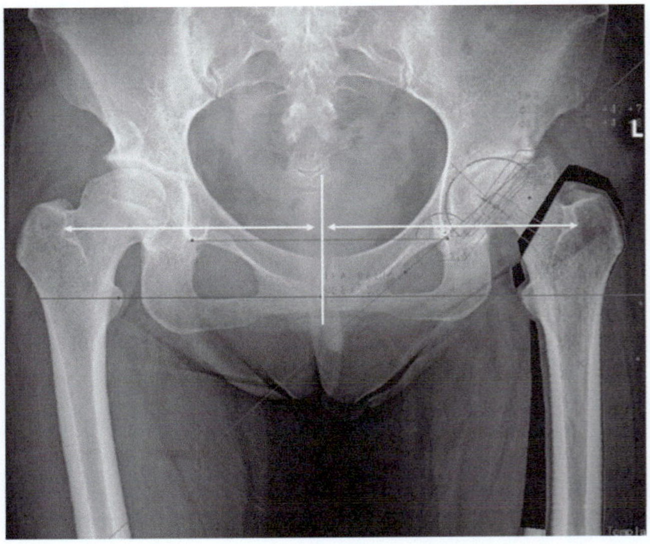

Fig. 44.3 Two-dimensional preoperative planning using software. The global offset and leg length are drawn to match those on the healthy side

reported that the width of the left and right superior anterior iliac spine is related to the acetabular offset and can serve as a reference for setting an appropriate hip center when placing a THA cup [2]. However, evaluating the anteversion angle of the femur is difficult using simple radiographic evaluation alone. Although the stem size offset is defined to some extent by the medullary cavity width of the femur, the degree of femoral anteversion may alter the medullary cavity width and result in an underestimation of the femoral offset (Fig. 44.4). In fact, the agreement rate with the actual stem size used in preoperative planning of THA using simple radi-

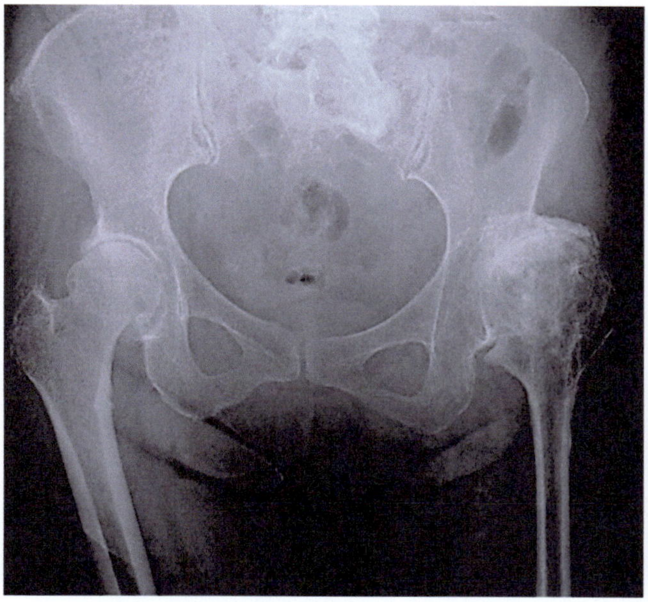

Fig. 44.4 Simple frontal X-ray of the hip joint in two-dimensional evaluation of a case. The left femur was externally rotated, and the medullary cavity diameter was narrow

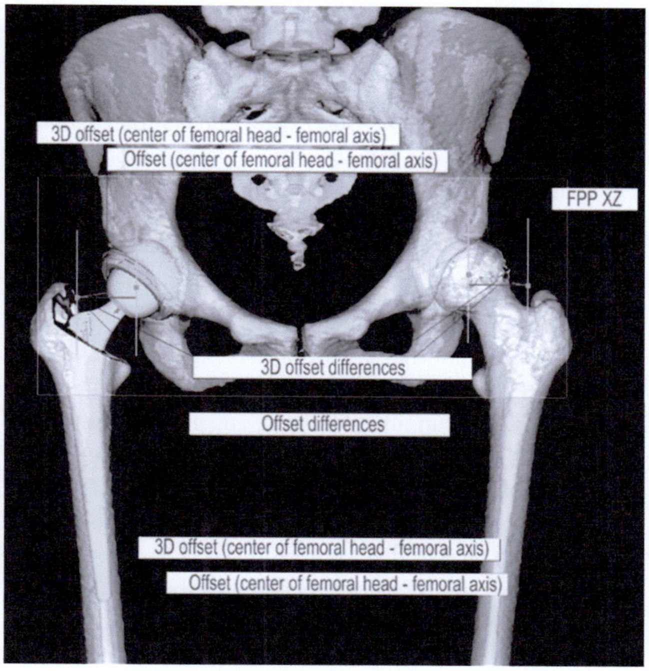

Fig. 44.5 Three-dimensional preoperative planning using software. Preoperative planning should be performed to achieve the same leg length and offset as those on the healthy side

ography is reportedly 32–79% [3], and, especially in cases of severe femoral deformity, there are limitations to evaluating only 2D images in preoperative planning.

Although 3D preoperative planning requires careful consideration of the increased radiation exposure with the additional CT scan and the cost of software, it has many advantages in terms of the anatomical information obtained and the simulation performed. Even if the software is unavailable, the information obtained from CT is useful for 2D drawing because it allows us to evaluate the width of the acetabular base, presence of a double floor, anteversion angle of the femur, and medullary cavity diameter. During 3D preoperative planning, it is possible to consider the anteroposterior offset in particular (Fig. 44.5). In a simulation study, when the hip center was placed posteriorly in patients with acetabular dysplasia, the bony coverage of the cup improved, but the range of motion of flexion and internal rotation decreased [4]. In addition, similar to the medical femoral offset, the anterior femoral offset has been shown to be related to dislocation resistance and range of motion [5]; therefore, the planning must consider the anterior–posterior offset. Simulations show that the stem offset length of THA that meets the conditions of 110°flexion, 90° flexion, 30° flexion, 30° internal rotation, 30° external rotation, and 30° extension has a medial femoral offset of at least 32.1 mm and an anteroposterior femoral offset of 15–25 mm on average.

Three-dimensional simulation studies have shown that the stem offset and stem cervical body angle are related to dislocation resistance after THA, and selecting an appropriate implant design during preoperative planning is important [6]. Generally, the offset length is defined by the stem size, which is limited by the medullary cavity size. There are designs that extend the offset by reducing the cervical body angle even with the same stem size; therefore, when selecting an implant, the medullary cavity size and global offset must be considered. In cases with femoral morphology abnormalities, the offset size should be increased when the anteroposterior torsion of the stem is adjusted from the original anatomical shape in consideration of combined anteversion. Although stems with changeable necks can be easily offset in the anteroposterior direction, their associated revision surgery rate is reported to be higher than that of conventional stems [7, 8]; therefore, age and indications must be considered when using them. A study determined that the incidence of lateral femoral cutaneous neuropathy was significantly increased in patients with a small femoral offset when THA was performed using a direct anterior approach [9]; therefore, anatomical characteristics should be considered when planning the approach.

3 Intraoperative Evaluation

Experimental models have demonstrated that the least error in the intraoperative evaluation of leg length and offsets occurs when maintaining the limb in the mid-position in the adduction and abduction directions [10]. Recently, in minimally invasive THA with myotendon preservation, the use of navigation and intraoperative imaging is recommended because of the lack of a sufficient field of view [11]. However, owing to the high cost of purchasing a navigation system and considering radiation exposure, using navigation in all centers and patients is not realistic. Considerations differ depending on the surgical technique (cup-first or stem-first) and the approach. This section describes the intraoperative evaluation of the cup-first posterior approach for THA.

Regarding cup placement, in cases of acetabular dysplasia (DDH), the cup center reportedly tends to shift medially, upward, and posteriorly from the normal hip center on the normal side [12], and the medialization of the cup affects the global offset. For cup internalization, i.e., global offset reduction associated with acetabular offset reduction, compensating with a femoral offset is an effective strategy. The reduction of the global offset with cup internalization, i.e., acetabular offset reduction, can be effectively compensated with a femoral offset. The anteversion of stem insertion must be evaluated prior to intraoperative determination of the femoral offset. Intraoperative evaluation of the anteversion angle of the stem is performed using a goniometer, and a goniometer integrated with a rasp and stem insertor can be used for a more accurate evaluation (Fig. 44.6) [13]. There are models with different cervical body angles for the neck trial (increasing the offset size by decreasing the cervical body angle) and models with offset size variations (Fig. 44.7), and adjustments are made by considering the range of motion related to stability, soft-tissue tension, leg length, and dislocation resistance after test preparation. Intraoperative evaluation of the posterior approach using a goniometer has shown that a large offset size increases the range of motion of internal rotation and dislocation resistance, and the offset size must be in accordance with intraoperative stability [14]. In contrast, intraoperative evaluation using an anterolateral approach with a navigation system showed no differences other than the range of motion of external rotation, depending on the stem offset length. Thus, it is necessary to recognize that the range of motion of the joint affected by the offset may differ depending on the approach [15].

After cup and stem insertion, test repositioning is performed, and if the dislocation resistance is weak, adjustments can be made by changing the head offset. While extending the head offset allows the global offset to be extended, care should be taken as this can cause leg overextension. In addition, the global offset can be increased using an offset liner (Figs. 44.8 and 44.9). However, while no incidence of osteolysis or revision 5 years postoperatively has been reported, the offset liner experiences more wear than the normal liner, and so the indication should be decided by considering long-term results [16].

Intraoperative precautions should be taken to ensure that the global offset changes according to the degree of cup medialization and the internal or external insertion of the stem. Therefore, if soft tissue tension deviates from the preoperative plan, intraoperative radiographic confirmation may be considered, accounting for the possibility of implant placement problems.

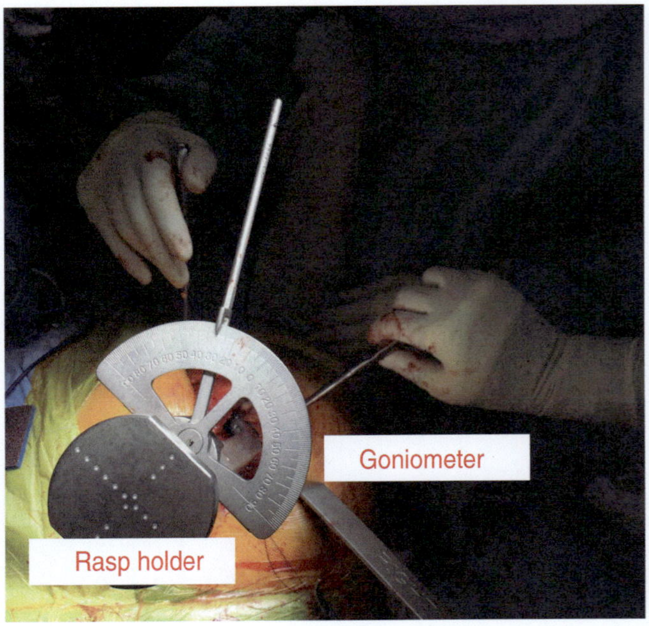

Fig. 44.6 Goniometers that can be attached to the rasp holder. The angle of intraoperative rasping and torsion angle before stem insertion can be accurately evaluated (VLIAN stem standard instrument; Teijin Nakashima Medical Ltd.)

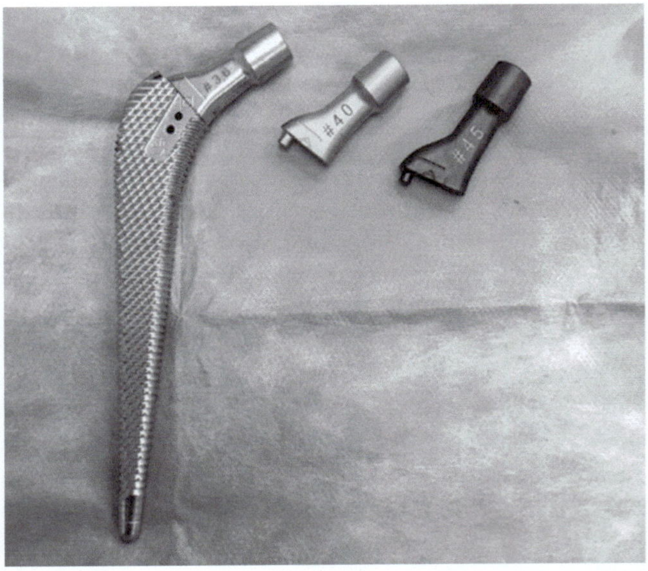

Fig. 44.7 Variation in the neck trial. An appropriate offset size can be evaluated after rasp insertion (VLIAN stem standard instrument; Teijin Nakashima Medical)

Fig. 44.8 Variation of the liner. An offset liner (right) (Stryker) can be used to increase the acetabular offset

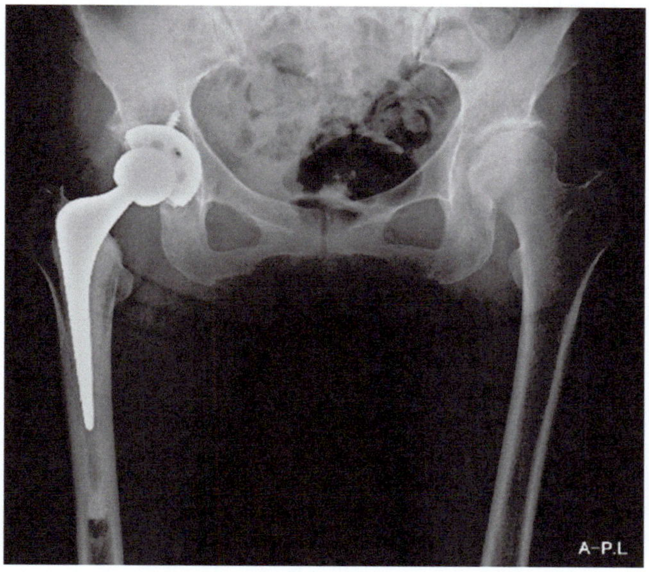

Fig. 44.9 Case of redisplacement with an offset liner. The patient had frequent dislocations after revision with dual mobility; therefore, he underwent revision with an offset liner and a long offset stem

4 Impact on Postoperative Outcomes

In Japan, good 10-year results of cementless THA using a high-offset stem have been reported in Crowe group I or II patients with DDH [17]. A longer offset decreases the hip joint union and increases resistance to dislocation due to increased afferentity with less bony impingement but may cause pain in the greater trochanter. In a randomized controlled trial, the Western Ontario and McMaster Universities Osteoarthritis Index pain score after THA showed that pain scores increased with a longer offset compared with a normal offset [18]. Care should be taken to avoid over-elongation of the offset and leg. In addition, an inappropriate offset reportedly limits the joint range of motion [19], and an appropriate offset should be reconstructed from the viewpoint of kinematics.

With regard to post-THA muscle strength, the moment arms of the middle and small gluteus maximus muscles increase with an increasing global offset [20, 21], and the middle and small gluteus maximus muscles also increase with increasing global offset [22]. In the posterior [23] and minimally invasive anterolateral and lateral approaches, the femoral offset and abductor muscle strength were shown to correlate [24], and the femoral offset and abductor muscle strength were also found to be related in different approaches. Three-dimensional gait analysis revealed that abductor muscle strength decreased when the sum of the femoral and acetabular offsets was shortened by 5 mm or more compared with the contralateral side after surgery, whereas the abductor muscle strength and SLR were not affected up to a leg length difference of 20 mm or less [25]. Therefore, the offset setting may be an important prognostic factor in THA from the viewpoint of biomechanics as well as lower limb length. In addition, cup-head translation was not related to the offset or abductor muscle strength in THA reconstructed in the appropriate position [26], which is also important from the viewpoint of long-term outcomes.

References

1. Takao M, et al. Postoperative limb-offset discrepancy notably affects soft-tissue tension in total hip arthroplasty. J Bone Joint Surg Am. 2016;98:1548–54.
2. Takamatsu T, et al. Radiographic determination of hip rotation center and femoral offset in Japanese adults: a preliminary investigation toward. Biomed Res Int. 2015;2015:610763.
3. Hananouchi T, et al. Preoperative templating of femoral components on plain X-rays. Rotational evaluation with synthetic X-rays on an ORTHODOC. Arch Orthop Trauma Surg. 2007;127:381–5.
4. Sakemi Y, et al. How does anteroposterior cup placement affect bone coverage and range of motion in primary total hip arthroplasty for hip dysplasia? J Orthop Sci. 2019;24:269–74.
5. Hirata M, et al. Optimal anterior femoral offset for functional range of motion in total hip arthroplasty – a computer simulation study. Int Orthop. 2015;39:645–51.
6. Shoji T, et al. The influence of stem offset and neck shaft angles on the range of motion in total hip arthroplasty. Int Orthop. 2016;40:245–53.
7. Graves SE, et al. The use of femoral stems with exchangeable necks in primary total hip arthroplasty increases the rate of revision. Bone Joint J. 2017;99-B:766–73.
8. Colas S, et al. Exchangeable femoral neck (dual-modular) THA prostheses have poorer survivorship than other designs: a nationwide cohort of 324,108 patients. Clin Orthop Relat Res. 2017;475:2046–59.
9. Ozaki Y, et al. Small femoral offset is a risk factor for lateral femoral cutaneous nerve injury during total hip arthroplasty using a direct anterior. Orthop Traumatol Surg Res. 2016;102:1043–7.

10. Kawamura H, et al. Effects of lower limb and pelvic pin positions on leg length and offset measurement errors in experimental total hip arthroplasty. J Orthop Surg Res. 2021;16:193.
11. Weber M, et al. Fluoroscopy and imageless navigation enable equivalent reconstruction of leg length and global and femoral offset in THA. Clin Orthop Relat Res. 2014;472:3150–8.
12. Bonnin MP, et al. Do we medialize the hip center of rotation in total hip arthroplasty? Influence of acetabular offset and surgical technique. Hip Int. 2012;22:371–8.
13. Mitsutake R, et al. A simple angle-measuring instrument for measuring cemented stem anteversion during total hip arthroplasty. BMC Musculoskelet Disord. 2020;21:113.
14. Jinno T, et al. Intraoperative evaluation of the effects of femoral component offset and head size on joint stability in total hip arthroplasty. J Orthop Surg (Hong Kong). 2017;25:2309499016684298.
15. Hayashi S, et al. Excessive femoral offset does not affect the range of motion after total hip arthroplasty. Int Orthop. 2013;37:1233–7.
16. Chapman RM, et al. Higher offset cross-linked polyethylene acetabular liners: is wear a significant clinical concern? Hip Int. 2019;29:652–9.
17. Nishino T, et al. Ten-year results of 55 dysplasia hips of hip offset and leg length reconstruction in total hip arthroplasty with cementless tapered. J Orthop Surg (Hong Kong). 2020;28:2309499020909499.
18. Liebs TR, et al. The influence of femoral offset on health-related quality of life after total hip replacement. Bone Joint J. 2014;96-B:36–42.
19. Weber M, et al. Inaccurate offset restoration in total hip arthroplasty results in reduced range of motion. Sci Rep. 2020;10:13208.
20. Mahmood SS, et al. Association between changes in global femoral offset after total hip arthroplasty and function, quality of life, and abductor muscle strength. A prospective cohort study of 222 patients. Acta Orthop. 2016;87:36–41.
21. Rüdiger HA, et al. Effect of changes of femoral offset on abductor and joint reaction forces in total hip arthroplasty. Arch Orthop Trauma Surg. 2017;137:1579–85.
22. Kiyama T, et al. Hip abductor strengths after total hip arthroplasty via lateral and posterolateral approaches. J Arthroplast. 2010;25:76–80.
23. Yamaguchi T, et al. Total hip arthroplasty: the relationship between posterolateral reconstruction, abductor muscle strength, and femoral. J Orthop Surg (Hong Kong). 2004;12:164–7.
24. Tezuka T, et al. Effects of hip joint center location and femoral offset on abductor muscle strength after total hip arthroplasty. Mod Rheumatol. 2015;25:630–6.
25. Sato H, et al. Effect of femoral offset and limb length discrepancy on hip joint muscle strength and gait trajectory after total hip. Gait Posture. 2020;77:276–82.
26. Kiyohara M, et al. Do component position and muscle strength affect the cup-head translation during gait after total hip arthroplasty? Eur J Orthop Surg Traumatol. 2019;29:1263–9.

Safe Zone

45

Hidetoshi Hamada

1 History of Safe Zone Recommendations

Suboptimal implant placement is one of the main causes of total hip replacement (THR) failure. It has been reported that a steep cup inclination angle of 50° or more causes edge loading, which leads to increased liner wear and liner fracture rate. Suboptimal cup alignment further causes prosthetic impingement and associated complications, such as dislocation, implant fracture, metallosis, and early loosening [1–4]. A suboptimal cup angle also influences postoperative outcomes after THR. In recent years, with improvements in materials of the bearing parts, the larger diameter of the artificial femoral head has increased the prosthetic range of motion (ROM). If the cup placement angle with regards to stem anteversion is appropriate, dislocation and complications associated with impingement can be avoided without postoperative restriction [5], and the postoperative quality of life of the patient is reported to be high [6].

The acceptable range of alignment for THR implants, especially regarding the cup alignment, which avoids complications associated with impingement is called the safe zone, and various recommendations have been made historically regarding this zone. In the 1970s and the 1980s, Charnley et al. recommended a cup inclination angle of 45° and cup anteversion angle of 0° [7], and Harris et al. recommended a cup inclination angle of 30° and cup anteversion of 20° [8] based on their practice, but these recommendations were not universal. The safe zone with a 40° ± 10° cup inclination and 15° ± 10° anteversion reported by Lewinnek et al. in 1978 is well known. It was based on the evidence that the dislocation rate was high outside the rectangular zone measured in the postoperative plain anteroposterior radiograph [9]. However, the recommended zone analyzed by their method is not truly considered a safe zone due to the prevalence of dislocation cases even inside the safe zone. Their radiographic evaluation is also not universal as they correct the patient's posture such that the anterior pelvic plane (APP) including the bilateral anterior superior iliac spine and the pubic tuberosity is horizontal. Further, stem anteversion angle, the factor essential for prosthetic impingement, is not evaluated.

For a safe zone considering the stem anteversion angle, Ranawat et al. proposed that the combined anteversion (CA), which is the sum of the cup anteversion angle and the stem anteversion angle, should be in the range of 25° to 35° in males and 25° to 45° in females [10]. Since then, there have been reports recommending that the cup be placed such that the CA is a fixed value, but the reported values widely range from 25 to 60 [11–13].

Since the 1990s, there have been reports of the use of simulations to determine safe zones. This method mathematically calculates the combination of implant angles to avoid prosthetic impingement so that the prosthetic ROM exceeds the predefined ROM required in daily activities. In these simulations, the predefined ROM, the prosthetic ROM determined by the implant geometry, the stem placement angle in the zero position, and the pelvis and femur reference coordinates affect the calculated cup angle. Seki et al. set 90° of flexion and 30° of extension as the required ROM and recommended 40° ± 10° of cup inclination and 20° ± 10° of anteversion as the safe zone [14]. Widmer et al. set the required ROM at 130° of flexion, 40° of extension, 50° of abduction, 40° of external rotation, and 80° of internal rotation and recommended the cup anteversion angle that summed 0.7 times the stem anteversion angle to 37.3° [15]. However, this report should be interpreted with care as the ROM did not include the flexion and internal rotation motions necessary for daily activities, and a cup inclination angle of up to 50° is allowed. Yoshimine et al. set the required ROM at 125° of flexion, 40° of extension, 40° of abduction, 40° of external rotation, 90° of internal rotation, and 45° of internal rotation in flexion and recommended a cup inclination and anteversion that summed 0.77 times the stem

H. Hamada (✉)
Osaka University, Suita, Osaka, Japan
e-mail: hide-hamada@umin.ac.jp

Y. Nakashima (ed.), *Advances in Total Hip Arthroplasty*, https://doi.org/10.1007/978-981-97-2461-1_45

anteversion angle to 84.3° [16]. Moreover, Ohmori et al. proposed that the optimal combination to avoid prosthetic impingement as well as bony impingement within the required ROM is 20° of stem anteversion and 23.3° of cup anteversion angle [17]. In their 2020 report, Widmer et al. recommended the cup anteversion angle that summed 0.68 times the stem anteversion to 31.3° [18]. The reason why the CA recommended in their current report was lower than that in their previous reports may be because the current required ROM was set with not only a single direction of extension and external rotation but also with a combined extension motion of 20° and external rotation of 25°.

The first step in defining a clinically relevant safe zone is to accurately measure the required ROM of the hip joint. Accurate measurement of hip ROM as a relative angle between the pelvic and femoral reference coordinates is needed, instead of a measurement using a goniometer on the appearance. The following studies were performed: (1) maximum ROM was intraoperatively measured using a computed tomography (CT)-based navigation system after implantation of THR 5], (2) ROM was measured by open magnetic resonance imaging in a deep flexion posture during daily activities of a young woman without hip pathology [19], (3) ROMs during the daily activities of the patients after THR were measured by motion analysis [20], (4) ROM determined by bony impingement was measured using a bone model of patients without hip pathology [21], (5) ROM was measured by motion analysis during deep flexion activities by fluoroscopic image matching with a CAD model of the THR implant [22]. These studies concluded that hip ROM for daily living could be covered by flexion at 120°, extension at 40°, external rotation at 40°, and internal rotation at 40° and 90° of the flexion, which were determined as clinically relevant and required ROMs of the hip joint.

Next, a combination of cup inclination and cup anteversion to exceed the required ROM was simulated for each 5° of stem anteversion, using the CAD model of the implant. The clinical models (Trident cup 50 mm, polyethylene flat liner inner diameter 32 mm, Centpillar stem #5, neck-shaft angle 127°, 32 mm head) was used and the original stem position was set to 10° flexion and 5° adduction in the reference coordinate of the stem to simulate implantation in the proximal femur (Fig. 45.1). The inside of the triangular zone is the cup angle that can avoid prosthetic impingement, but if the cup is placed at an inclination of 45° or less, the white area inclined at 45° or less is the safe zone that avoids edge loading. The distribution and extent of the safe zone differed by 5° of stem anteversion. The safe zone is relatively broad for 30° and 35° of stem anteversion, very narrow for 50° of stem anteversion, and does not exist for 10° of stem anteversion. These data indicate that an optimal CA is not a fixed value. There are concerns that cup inclination

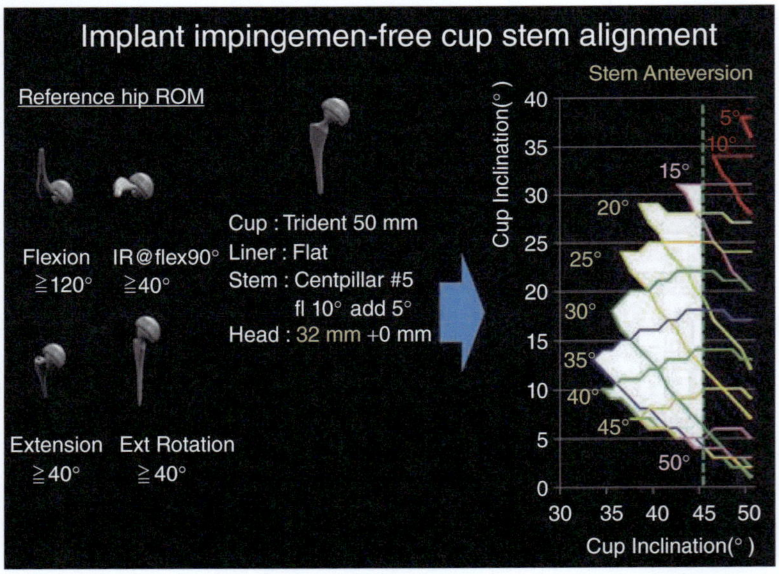

Fig. 45.1 Combination of the cup angle without impingement at 120° of flexion, 40° of extension, 40° of external rotation, and 40° of internal rotation at 90° of flexion in hip range of motion. The results were obtained using a flat liner with a 32 mm femoral head offset of 0 mm and a Centpillar stem placed at 10° of flexion and 5° of adduction in the reference coordinate of the stem for each 5° of stem anteversion (from 5° to 50°). The vertical and horizontal axes indicate the cup anteversion and inclination angles, respectively. The inside of the triangular zone is the cup angle that can avoid impingement. If the cup anteversion is placed to 45° or less to avoid edge loading, a safe zone does not exist for a stem anteversion angle of 10° and is very narrow for an angle of 50°. Adequate combined anteversion is not a fixed value, and safe zones do not exist for excessive low and high stem anteversion. The distribution and extent of the safe zone differ for stem anteversion. The safe zone is broad for 30° anteversion, but the cup anteversion could only be plus or minus 4°, around 17° at 40° inclination. (Cited from Sugano N, et al. J. Jpn. Orthop. Assoc. 2018; 92: 987–995)

of less than 35° significantly narrows the prosthetic ROM, and an inclination of 45° or more increases the risk of edge loading in patients with posterior pelvic tilt in the standing position.

When 40° of cup inclination was targeted, even if the safe zone was relatively wide for stem anteversion of 30°, the accepted cup anteversion angle ranged ±4° centered at 17°, which indicates that precision and accuracy for the targeted angle are needed intraoperatively. In addition, studies on the effect of bone and soft tissue on hip ROM have previously reported that impingement at the anterior acetabulum and anterior capsule limits ROM [23, 24]. Resection of the osteophytes around the acetabulum and thickened anterior capsule may therefore be considered.

To determine the cup placement angle, it is necessary to define the reference coordinates for the pelvis. APP is an anatomical coordinate that is not always parallel to the body axis in the supine or standing position. In a previous study in subjects without hip pathology, the mean value of the sagittal tilt of the APP in the supine position was determined to be 3.8° posteriorly, with a standard deviation of 6.1° [21]. In the supine position, it was reported that the APP was not always in a horizontal plane, and the APP was not recommended as a standard reference coordinate for the pelvis [25]. It is recommended that the functional pelvic plane (FPP), reflecting the sagittal pelvic tilt in the supine position, be used as a reference posture for hip joint ROM measurement [26, 27]. Since the change in sagittal tilt of the APP is less than 10° between the supine and standing positions in majority of the patients [28], it is reasonable to set the supine FPP as the reference coordinate of the pelvis to align the cup avoiding prosthetic impingement when moving within the required hip ROM. Three-dimensional optical motion analysis combining skin markers with the models of the bone and implant showed that patients after THR with implantation into the aforementioned safe zone with reference coordinates of the supine FPP were able to perform both Japanese and Western-style daily activities without prosthetic impingement [20].

2 Safe Zone for Patients with Posterior Pelvic Tilt in the Standing Position

Setting the pelvic reference coordinate in patients with a posterior pelvic tilt more than 10° from the supine to the standing position is a challenging problem. In such patients, even if the cup is accurately placed in the reference coordinate of the supine FPP, the inclination and anteversion angles increase while standing, which may increase the risk of anterior dislocation. This is due to posterior impingement during extension and external rotation motions and may cause accelerated wear of bearing parts due to edge loading. As previously reported, approximately 10% of patients with THR had a posterior pelvic tilt of 10° or more from the supine to standing position [28], and the incidence rate has increased to approximately 20% according to a recent report [29–31]. The factors associated with more than 10° of posterior pelvic tilt from supine to standing position were found to be older age and primary hip osteoarthritis or rapidly destructive hip arthrosis [29]. In a multivariate analysis including sagittal alignment of the spine, compression vertebral fracture, lumbar spondylolisthesis, smaller anterior tilt in the superior aspect of the first sacral vertebra, and older age were the four factors associated with the posterior pelvic tilt of 10° or more from the supine to standing position [32]. The results of this study suggest that back muscle atrophy associated with aging as well as deformation of the spinal column itself affects posterior pelvic tilt. This suggests that pelvic tilt changes with age after THR. In a 10-year follow-up study of pelvic tilt in THR patients, the pelvis was found to be tilted 3° posteriorly at 1 year postoperatively but did not change until 10 years thereafter. In contrast, the pelvis in the standing position was reported to tilt posteriorly, 5 years postoperatively by an average of 11° over 10 years [33]. Even if the pelvis tilts posteriorly by 30°, the cup initially placed at an inclination of 40° with the reference coordinate of the supine FPP can avoid 50° or more of the inclination, causing edge loading. Intraoperatively, the inclination should be set to 45° or less considering future changes, even in patients without spinal degeneration [29].

Using the same motion analysis method as described above [20], the deep hip flexion and extension movements of patients after THR and the maximum angles of hip flexion, extension, and external rotation with reference coordinates of the supine FPP were determined. There was no significant difference in the maximum angles of hip flexion, extension, and external rotation between patients with a posterior tilt angle of 10° or more from the supine to standing position and those with a posterior tilt angle of less than 10° [34]. Furthermore, a study measuring the maximum ROM intraoperatively by a CT-based navigation system after implantation of THR found no significant difference in the maximum angle of extension between patients with 10° or more of posterior tilt from the supine to standing position and those with a posterior tilt of less than 10° [35]. These reports indicate that, even in patients with a large posterior pelvic tilt in the standing position, the hip extension angle measured from the supine FPP as the neutral position is not large. Further, the cup angle should be adequately aligned in the reference coordinate of the supine FPP, similar to that of the other patients.

There was no difference in the required ROM between the two groups, which indicated that the reduced cup anteversion in the reference coordinate of the supine FPP would cause posterior dislocation with anterior impingement in the flexion motion. This is due to concerns about posterior pelvic tilt in the standing position. It has been reported that the cup needs to be aligned with the reference coordinate of the supine FPP to satisfy the required hip angles of 120° flexion, 40° extension, 40° external rotation, and 40° internal rotation at 90° flexion [34, 35]. In addition, when considering changes over time, 43% of patients had ≥10° of posterior pelvic tilt from the supine to the standing position 10 years postoperatively. Although a change in posterior tilt in the standing position with aging is unavoidable, there was no significant change in pelvic tilt in the supine position, and no case of anterior dislocation associated with posterior tilt in the standing position was reported [33]. Adequate preoperative planning using the reference coordinates of the supine FPP was validated (Fig. 45.2).

As another trend, sagittal spinopelvic alignment has been evaluated using lateral radiographs in the standing and sitting positions since 2004 [36], and the relationship between individual variations in sagittal spinopelvic mobility and postoperative dislocation after THR has been reported. The THR cup safe zone was recommended for each type according to sagittal spinopelvic parameters, where the patients with a large pelvic posterior tilt in the standing position, patients with a 10° or more difference between pelvic incidence and lumbar lordosis, and patients with 10° or less of posterior pelvic tilt from the standing to sitting position were at high risk of dislocation [37, 38]. It has been suggested that inadequate implant alignment is more likely to cause prosthetic impingement in cases with less posterior pelvic tilt from the standing to sitting position. In cases with more posterior pelvic tilt in the standing position, these abnormal spinopelvic characteristics aggravate the risk of dislocation due to inadequate implant alignment. However, this recommended safe zone needs to be interpreted in light of the fact that stem anteversion was not evaluated and the pelvic stiffness varied postoperatively [31].

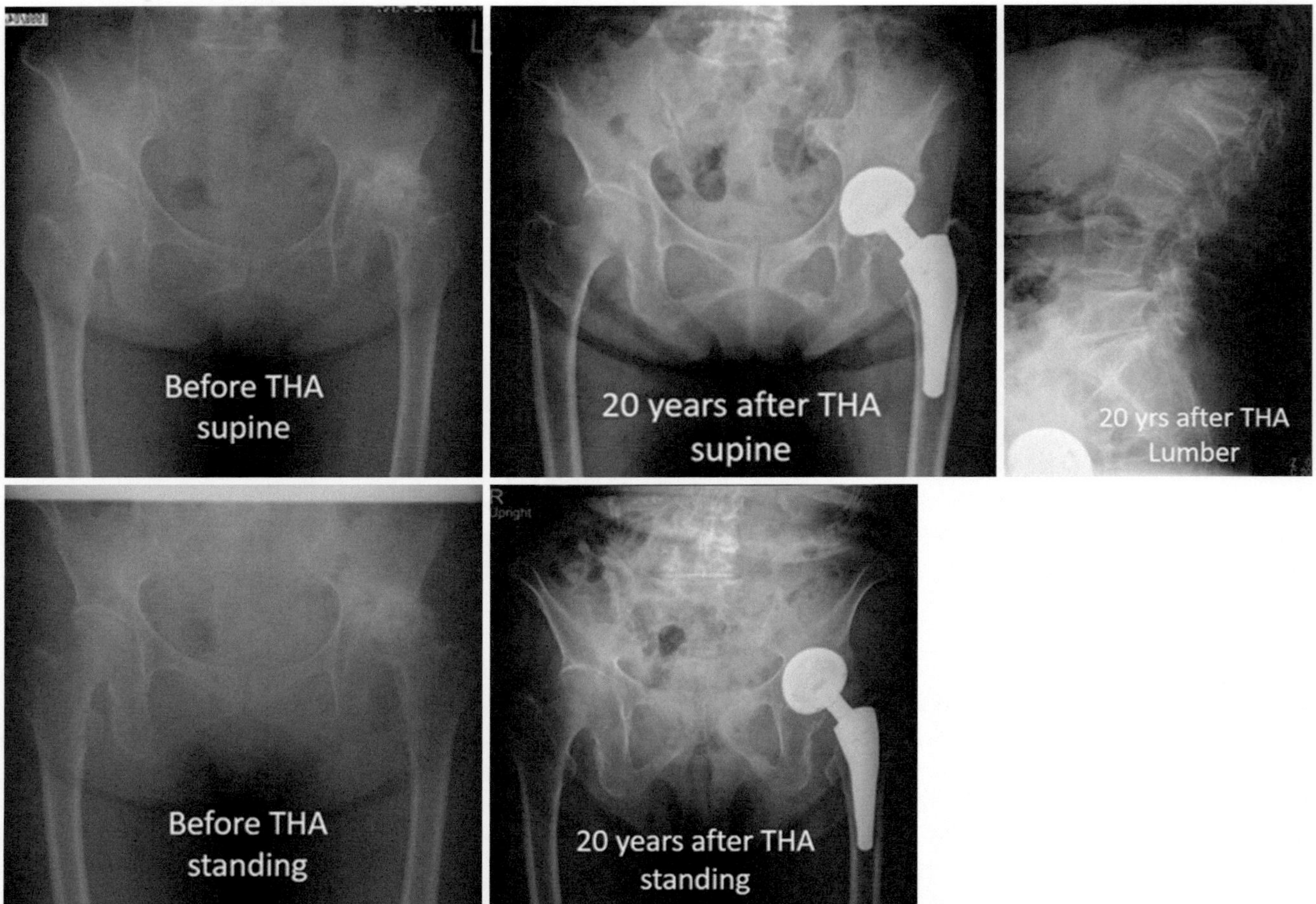

Fig. 45.2 Pelvic tilt in the standing position increases with age, but there is no significant change in pelvic tilt in the supine position. It is recommended that the supine functional pelvic plane is used as the standard reference coordinate for preoperative planning. The pelvis tilt 18° posteriorly from the supine to standing 20 years after THA, but there was no hip pain or dislocation. Lumbar vertebral compression fracture, lumbar spondylolisthesis, and smaller anterior tilt in the superior aspect of the first sacral vertebra is known to be associated with the posterior tilt of the pelvis in the standing position

3 Safe Zone for Patients with Spinal Fusion Surgery

It has been reported that spinal fusion surgery for adult spinal deformities also affects pelvic tilt, and abnormal cup anteversion in the standing position caused by spinal degeneration has been shown to improve after spinal fusion surgery [39]. In contrast, and reoperation and poor clinical outcomes [40–42]. In cases of spinal fusion surgery, pain due to degeneration of the intervertebral spaces adjacent to the fusion and neurological symptoms in the lower extremities caused by spinal canal stenosis may negatively impact the walking ability and the quality of life. After spinal fusion, spinal mobility is reduced, requiring an increase in hip flexion angle during forward bending from a sitting position. There is also an increase in the risk of anterior impingement, leading to an increase in the rate of dislocation and revision surgery for THR with suboptimal cup alignment or small prosthetic ROM [43].

In patients who underwent lumbosacral fusion surgery for kyphoscoliosis, the median pelvic tilt in the supine position was −8° both pre- and postoperatively. While in the standing position, it was −21° preoperatively and − 12° postoperatively, where lumbar lordosis was corrected to 14° on average of the difference between pelvic incidence and lumbar lordosis [44]. The pelvic tilt in the supine position did not change; however, the pelvic posterior tilt in the standing position improved through spinal fusion (Figs. 45.3 and 45.4). Even if spinal deformity occurs over time and spinal fusion surgery is subsequently performed, there is little change in supine FPP over time; therefore, there is no need to change the original reference coordinate of the supine FPP.

In modern THR, where the prosthetic ROM has increased owing to the improvement in the materials of the bearing part, it is important to align the cup in the reference coordinate of the supine FPP. While accurately aligning the cup angle, it is important to consider the stem anteversion to reduce the risk of complications associated with impingement.

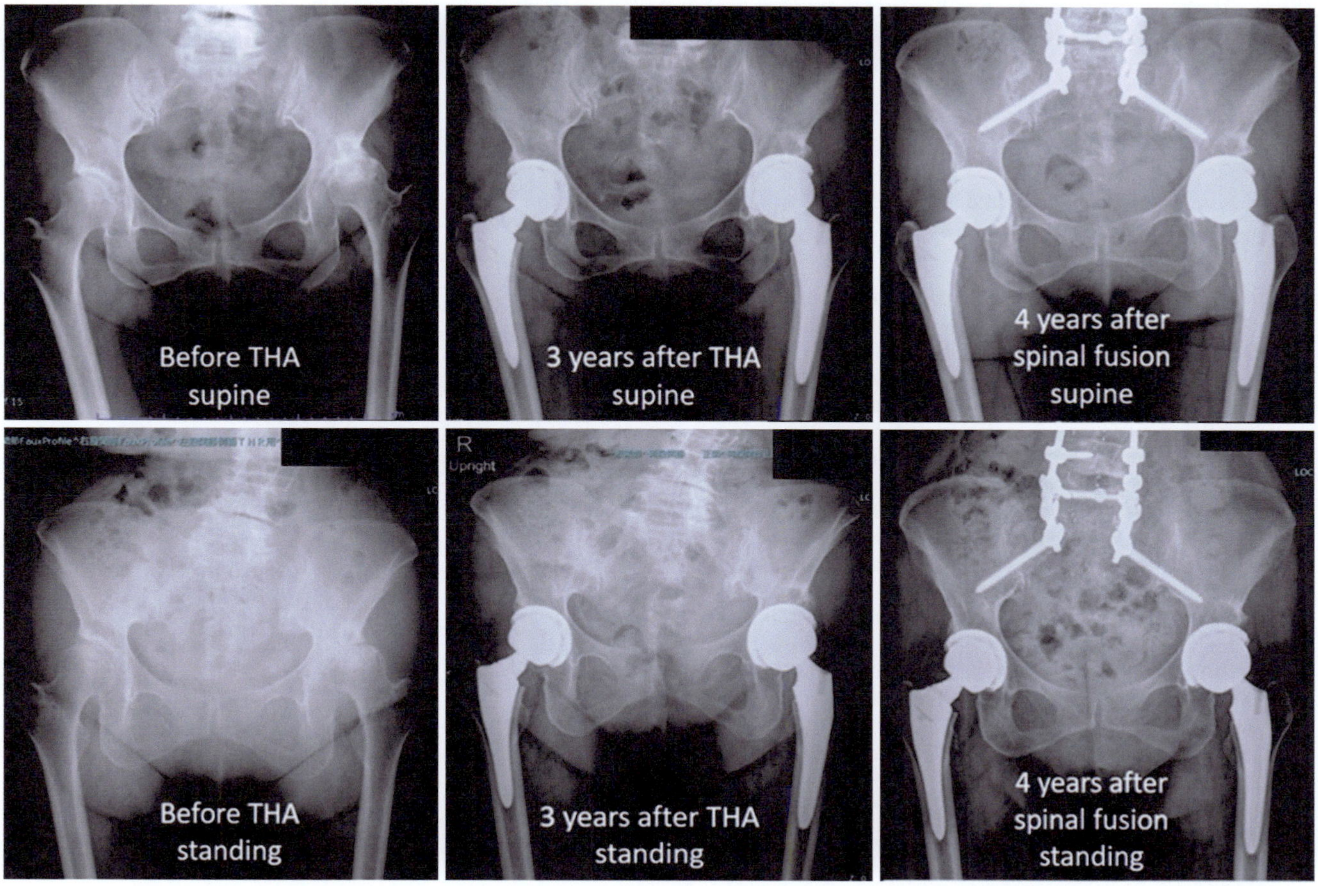

Fig. 45.3 The patient had a posterior pelvic tilt of 10° from the supine to standing position. Three years after THA, she had no hip pain or dislocation but had difficulty walking due to back pain. The abnormal posture caused by spinal deformity underwent corrective fusion from the thoracic spine to the pelvis. After spinal fusion surgery, the posterior pelvic tilt in the standing position improved, and THA did not cause any problems

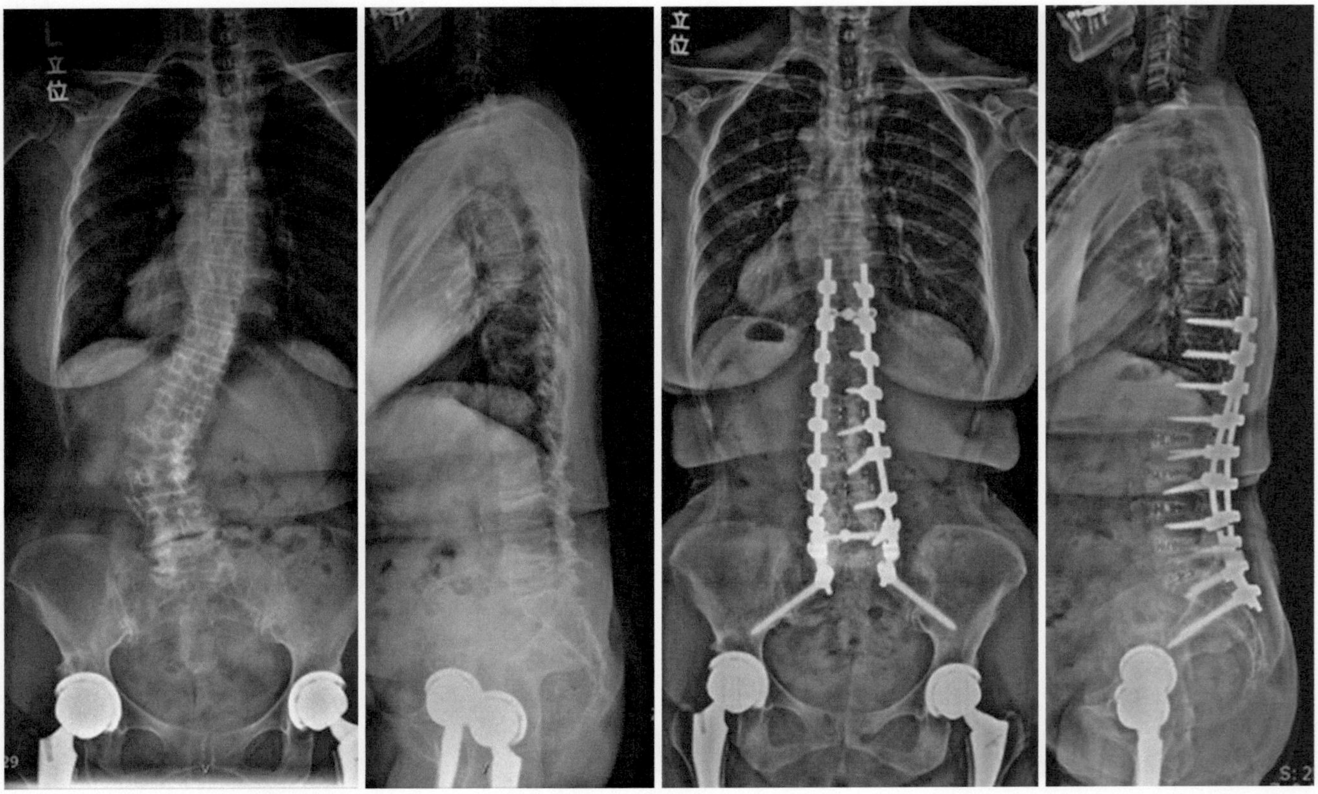

Fig. 45.4 Anteroposterior and lateral radiographs of the whole spine in the standing position. The left images were taken before spinal surgery for kyphosis, and the right ones were taken after spinal surgery, showing improved spinal alignment

References

1. Tower SS, et al. Rim cracking of the cross-linked longevity polyethylene acetabular liner after total hip arthroplasty. J Bone Joint Surg Am. 2007;89:2212–7.
2. Magnissalis EA, et al. Wear of retrieved ceramic THA components--Four matched pairs retrieved after 5–13 years in service. J Biomed Mater Res. 2001;58:593–8.
3. De Haan R, et al. Correlation between inclination of the acetabular component and metal ion levels in metal-on-metal hip resurfacing replacement. Erratum in: J Bone Joint Surg Br. 2009;91:700.
4. Shon WY, et al. Impingement in total hip arthroplasty a study of retrieved acetabular components. J Arthroplast. 2005;20:427–35.
5. Miki H, et al. Anatomic hip range of motion after implantation during total hip arthroplasty as measured by a navigation system. J Arthroplast. 2007;22:946–52.
6. Maeda Y, et al. Improvement of activities of daily living after total hip arthroplasty using a computed tomography-based navigation system. J Artif Organs. 2017;20:152–7.
7. Charnley J. Total hip replacement by low-friction arthroplasty. Clin Orthop Relat Res. 1970;72:7–21.
8. Harris WH. Advances in surgical technique for total hip replacement: without and with osteotomy of the greater trochanter. Clin Orthop Relat Res. 1980;146:188–204.
9. Lewinnek GE, et al. Dislocations after total hip-replacement arthroplasties. J Bone Joint Surg Am. 1978;60:217–20.
10. Ranawat CS, et al. Modern techniques of cemented total hip arthroplasty. Tech Orthop. 1991;6:17–25.
11. Nakashima Y, et al. Combined anteversion technique reduced the dislocation in cementless total hip arthroplasty. Int Orthop. 2014;38:27–32.
12. Fujishiro T, et al. The effect of acetabular and femoral component version on dislocation in primary total hip arthroplasty. Int Orthop. 2016;40:697–702.
13. Imai H, et al. Preoperative planning and postoperative evaluation of total hip arthroplasty that takes combined anteversion. Eur J Orthop Surg Traumatol. 2016;26:493–500.
14. Seki M, et al. Analysis of optimal range of socket orientations in total hip arthroplasty with use of computer-aided design simulation. J Orthop Res. 1998;16:513–7.
15. Widmer KH, et al. Compliant positioning of total hip components for optimal range of motion. J Orthop Res. 2004;22:815–21.
16. Yoshimine F. The safe-zones for combined cup and neck anteversions that fulfill the essential range of motion and their optimum combination in total. J Biomech. 2006;39:1315–23.
17. Ohmori T, et al. The optimal combined anteversion pattern to achieve a favorable impingement-free angle in total hip arthroplasty. J Orthop Sci. 2019;24:474–81.
18. Widmer KH. The implementation-free, prosthesis specific, and anatomy-adjusted combined target zone for component positioning in THA depends on the design and implantation parameters of both components. Clin Orthop Relat Res. 2020;478:1904–18.
19. Yamamura M, et al. Open-configuration MRI study of femoroacetabular impingement. J Orthop Res. 2007;25:1582–8.
20. Sugano N, et al. Dynamic measurements of hip movement in deep bending activities after total hip arthroplasty using a 4-dimensional motion analysis. J Arthroplast. 2012;27:1562–8.
21. Nakahara I, et al. Gender differences in 3D morphology and bony impingement of human hips. J Orthop Res. 2011;29:333–9.
22. Koyanagi J, et al. In vivo kinematic analysis of squatting after total hip arthroplasty. Clin Biomech (Bristol, Avon). 2011;26:477–83.

23. Han S, et al. The continuum of hip range of motion: from soft-tissue restriction to bony impingement. J Orthop Res. 2020;38:1779–86.
24. Nakamura N, et al. Effect of soft-tissue impingement on range of motion during posterior approach Total Hip Arthroplasty: an in vivo measurement study. Comput Assist Surg (Abingdon). 2016;21:132–6.
25. Lembeck B, et al. Pelvic tilt makes acetabular cup navigation inaccurate. Acta Orthop. 2005;76:517–23.
26. Sugano N. Computer-assisted orthopedic surgery. J Orthop Sci. 2003;8:442–8.
27. Sugano N, et al. Mid-term results of cementless total hip replacement using a ceramic-on-ceramic bearing with and without computer navigation. Bone Joint Surg Br. 2007;89:455–60.
28. Nishihara S, et al. Measurements of pelvic flexion angle using three-dimensional computed tomography. Clin Orthop Relat Res. 2003;411:140–51.
29. Kyo T, et al. Factors predicting change in pelvic posterior tilt after THA. Orthopedics. 2013;36:e753–9.
30. Uemura K, et al. Change in pelvic sagittal inclination from supine to standing position before hip arthroplasty. J Arthroplast. 2017;32:2568–73.
31. Hamada H, et al. What changes in pelvic sagittal tilt occur 20 years after THA? Clin Orthop Relat Res. 2022;481:690–9. https://doi.org/10.1097/CORR.0000000000002382.
32. Tamura S, et al. Spinal factors influencing change in pelvic sagittal inclination from supine to standing position in patients before total. J Arthroplast. 2014;29:2294–7.
33. Tamura S, et al. Does pelvic sagittal inclination in the supine and standing positions change over 10 years of follow-up after total hip arthroplasty? J Arthroplast. 2017;32:877–82.
34. Tamura S, et al. Hip range of motion during daily activities in patients with posterior pelvic tilt from supine to standing position. J Orthop Res. 2015;33(4):542–7.
35. Miki H, et al. Anatomical hip range of motion after implantation during total hip arthroplasty with a large change in pelvic inclination. J Arthroplast. 2012;27:1641–1650.e1.
36. Lazennec JY, et al. Hip-spine relationship: a radio-anatomical study for optimization in acetabular cup positioning. Surg Radiol Anat. 2004;26:136–44.
37. Vigdorchik JM, et al. Otto Aufranc Award: a simple hip-spine classification for total hip arthroplasty: validation and a large multicenter. Bone Joint J. 2021;103-B(7 Supple B):17–24.
38. Vigdorchik J, et al. Evaluation of the spine is critical in the workup of recurrent instability after total hip arthroplasty. Bone Joint J. 2019;101-B:817–23.
39. Buckland AJ, et al. Acetabular anteversion changes due to spinal deformity correction: bridging the gap between hip and spine surgeons. J Bone Joint Surg Am. 2015;97:1913–20.
40. Eneqvist T, et al. Lumbar surgery prior to total hip arthroplasty is associated with worse patient-reported outcomes. Bone Joint J. 2017;99-B:759–65.
41. An VVG, et al. Prior lumbar spinal fusion is associated with an increased risk of dislocation and revision in total hip arthroplasty: a meta-analysis. J Arthroplast. 2018;33:297–300.
42. Salib CG, et al. Lumbar fusion involving the sacrum increases dislocation risk in primary total hip arthroplasty. Bone Joint J. 2019;101-B:198–206.
43. Lum ZC, et al. The current knowledge on spinopelvic mobility. J Arthroplast. 2018;33:291–6.
44. Hamada H, et al. Change in pelvic tilt from supine to standing position before and after spinopelvic fixation, 19th annual meeting of CAOS international, New York, 2019.

Soft-Tissue Evaluation

46

Toshihiko Hara

1 Intraoperative Evaluation

(a) Significance of intraoperative soft-tissue tension assessment

The significance of assessing and adjusting soft-tissue tension during total hip arthroplasty (THA) is to achieve a good range of motion (ROM) without dislocation. An accurate leg length adjustment is required for THA; however, there are cases in which leg length and soft-tissue tone are at odds. An extremely high soft-tissue tension impairs ROM. If the surrounding soft-tissue tension is extremely low, the joint continues to move without braking and bony or implant impingement can occur, causing dislocation. There is also the possibility of significant separation of the head from the liner without impingement [1]. The evaluation and adjustment of soft-tissue tension in THA assumes an accurate implant placement angle and surrounding bone and soft tissue treatment to achieve the proper ROM, and adjustment of the implant placement position, neck length, offset, and soft-tissue tension must consider leg length. Soft-tissue adjustment is an important technique to achieve appropriate ROM and braking and avoid inappropriate head release.

Even in patients with significantly loose soft-tissue tension, if prolonged postoperative rest is maintained and scar tissue forms around the implant, it may be possible to achieve stability without dislocation. However, not all patients develop appropriate scar tissue during early weight-bearing gait training. Thus, stable hip reconstruction as the initial condition is very important.

(b) Intraoperative soft-tissue tension assessment and preoperative planning

Preoperative planning for THA can be two- or three-dimensional, but the latter is more accurate [2, 3]. The first step in three-dimensional planning is to determine the pelvic coordinate system using the anterior pelvic plane or functional pelvic plane (FPP), although the latter is currently widely used [4]. The coronal plane is adjusted using the reference lines between the superior anterior iliac spine, teardrop, and inferior end of the sciatic bone; however, in many cases, these lines do not coincide [5] (Fig. 46.1).

There are two methods for determining femoral length: one is to use the functional axis, while the other is to use the femoral axis length without considering the offset. There is no clear definition of the distal reference point for determining the femoral length, which can be the intercondylar apex of the knee or a point on the articular contact line. The femoral anteversion angle can be measured using the posterior condylar plane coordinate system [6] or the International Society of Biomechanics coordinate system [7]. As described above, the reference coordinates and points used in preoperative planning are not standardized. Even if the same reference coordinates are used, differences may occur among planning physicians. In addition, there are no clear indicators for setting leg length and offset in the preoperative planning of bilateral end-stage hips. Furthermore, even if leg length and offset are restored to those of the normal contralateral hip, hip stability may be insufficient. In the normal hip joint, vacuum sealing is performed by the labrum to ensure stability of the maximum ROM [8], but this is not expected in THA. This is suggested by the fact that there have been cases of dislocation despite accurate implant placement using robot-assisted surgery [9]. Surgeons should adjust the balance between leg length and soft-tissue tension while referring to their own preoperative plan while considering that a completely correct preoperative plan does not exist.

(c) Intraoperative soft-tissue evaluation

Traditional methods of assessing reconstructive hip stability during THA surgery include shuck and dropkick tests, maximum hip extension external rotation, hip flexion, and internal rotation [10]. The shuck test is a soft-tissue tension

T. Hara (✉)
Iizuka Hospital, Iizuka, Fukuoka, Japan

Y. Nakashima (ed.), *Advances in Total Hip Arthroplasty*, https://doi.org/10.1007/978-981-97-2461-1_46

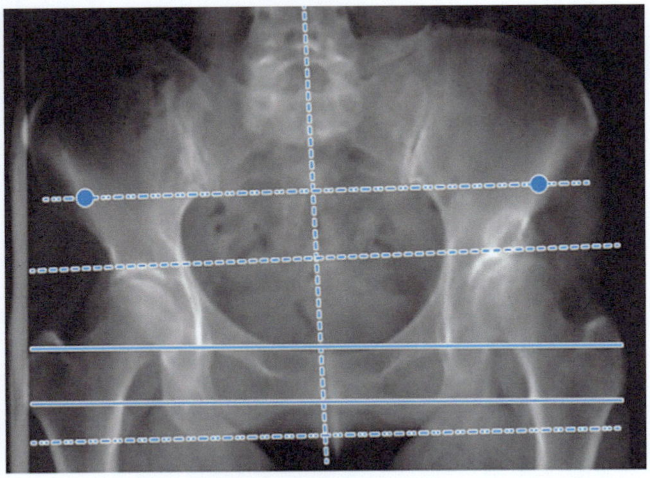

Fig. 46.1 Frontal view of the pelvis with various reference lines. Teardrop line is consistent with lines between the superior borders and inferior borders of obturator foramen, but not consistent with lines between inferior ends of the ischial tuberosities, tops of acetabulum or the superior anterior iliac spines (●). Because the amount of adjustment for leg length depends on the adopted reference, there could be differences in condition of soft tissue tension among each target implant position

evaluation method in which an assistant pulls the lower limb in the distal direction and observes the release of the head [11]. However, this test is not objective because rules are lacking for traction strength or hip and knee flexion angles during the evaluation. In the dropkick test, the patient is held in hip extension and knee flexion at 90° and the lower leg swings in the direction of extension when the hand is released. A positive result suggests strong tension of the extension mechanism. However, this test can also be positive for normal hip joints.

Evaluations of hip extension and external rotation include the presence or absence of anterior dislocation, assessment of maximum external rotation angle, presence or absence of posterior impingement, and anterior soft-tissue tension. The posterior approach usually yields a hard endpoint; however, the evaluation may differ depending on the external rotation force. For the hip flexion and internal rotation evaluation, the maximum ROM of internal rotation without dislocation, presence of anterior impingement, and posterior soft-tissue tension and braking (anterior and posterior approaches after soft-tissue reconstruction) should be observed. An internal rotation of 67° or more is associated with a low risk of postoperative dislocation [12]. These methods are used to intraoperatively evaluate reconstructed hip joint stability; however, they require an accurate leg length evaluation. An accurate evaluation of the anteroposterior pelvic tilt is also necessary because the hip flexion angle significantly impacts the evaluation. Therefore, proper measurement of soft-tissue tension requires a navigation system, a leg length measuring device, and adjustment of the pelvic reference plane by radiography after preoperative positional immobilization. Thus, the current evaluation of intraoperative stability has limited objectivity and accuracy.

Attempts have been made to evaluate reconstructed hip joint stability by quantitatively measuring the periprosthetic tension. In one study, an L-shaped plate with hooks was attached to the lower leg and plantar foot with a bandage and the plate was pulled at 40% of the body weight with the knee in extension. The maximum cup–head separation was observed at a traction angle of 15° of flexion, and the anterolateral approach showed greater separation at flexion angles up to 30° than the posterior approach [13]. In our own study, we attempted to measure the surrounding soft-tissue tension by continuously extending the neck length during surgery. In each case, the pelvic FPP was aligned with the laser line passing through the greater trochanter before surgery to reproduce an accurate hip flexion angle, and the hip flexion angle was measured based on the laser line (laser-guided system) [14]. In the posterior approach, soft-tissue tension was most relaxed at 60° of hip flexion, and the rising point of tension was shifted in the direction of the neck axis by a mean 4.4 mm versus extension. The lower leg axis was parallel to the pelvic reference plane, and measurements were made in the natural limb position [15].

Consensus is currently lacking on soft-tissue tension assessments during hip reconstruction, such as hip and knee flexion angle, traction/force load, amount of unstable cup–head separation, and appropriate soft-tissue tension; thus, these factors are left to the surgeon's discretion.

2 How to Ensure Proper Tension

(a) Approach and soft-tissue tension

The anterior approach involves dissecting part of the iliofemoral ligament but preserving the ischiofemoral ligament and short external rotator muscle, which provides excellent internal rotation braking. However, the iliofemoral ligament should be minimally transected with a good implant placement technique because it may be impaired in the extended external rotation position. In the posterior approach, the short external rotators and ischiofemoral ligament are transected while the iliofemoral ligament is preserved since it is superior in braking during external rotation but not during flexion or internal rotation. Care should be taken to avoid intraoperative failure to adjust the soft-tissue tension of the iliofemoral ligament during extension and external rotation, which can result in an extremely unstable hip reconstruction.

(b) Offset and soft-tissue tension

Increasing the femoral offset increases soft-tissue tension, prevents bony impingement [16], and improves abductor muscle strength by increasing the lever arm [17]. Methods to increase the femoral offset include neck length extension, a neck shaft angle reduction, and stem design selection, all of which affect leg length. Care should be taken when decreasing the femoral offset as it relaxes soft-tissue tension and increases the likelihood of bony impingement. Increasing the acetabular offset increases soft-tissue tension without affecting leg length, methods of which include lateral cup placement and offset liner use. Lateral cup placement should be determined during preoperative planning. The offset liner has slight influence on leg length depending on the cup inclination. Offset liner use should be carefully considered because of its reported high revision rate [18]. When evaluating and adjusting soft-tissue tension, the global offset should be considered along with the femoral and acetabular offsets rather than each individually [13, 19]. Maintaining the interosseous distance by offset is also associated with maintaining soft-tissue tension. Restoration of the global offset plays an important role in hip stability.

(c) Cup medialization and soft-tissue tension

Cup medialization relaxes the soft tissues without affecting leg length but requires attention to impingement because the acetabular limb and osteophyte are elevated relative to the hip center. Cup medialization has historically been recommended based on the theory that moving the center of rotation inward reduces joint stress [20]. However, excessive cup medialization beyond the femoral offset adjustment range can result in a significantly unstable hip reconstruction due to soft-tissue laxity, decreased abduction muscle strength, and bony impingement [16] (Fig. 46.2). Considering the good clinical results of the currently used highly cross-linked polyethylene, cup medialization should be performed with caution.

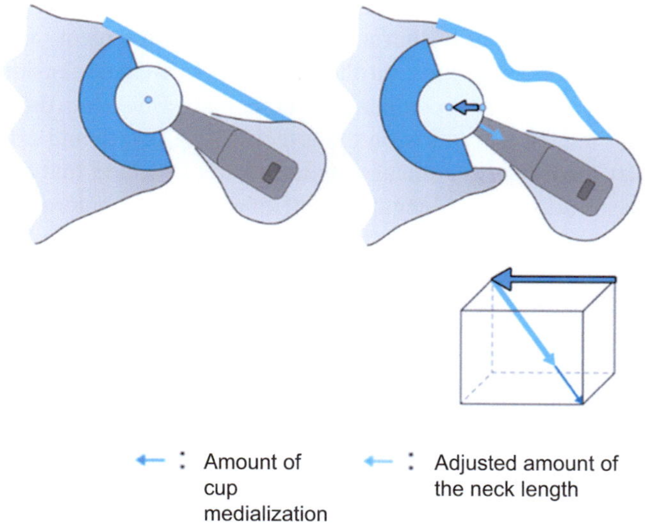

Fig. 46.2 Partial recovery of decreased acetabular offset by extending the neck length. Recovery of decreased acetabular offset requires longer amount of neck extension compared to the amount of cup medialization. Excessive cup medialization with decreased soft tissue tension and global offset could not be salvaged by the neck length adjustment

(d) Leg length and soft-tissue tension

Leg lengthening increases soft-tissue tension. The most important priority in THA is generally reconstructing a stable joint, and the leg length tends to be extended. THA is performed in various situations, and it is difficult to achieve both stability and perfect leg length equality in all cases [21]. However, if leg length and the surrounding soft-tissue tension can be accurately assessed, leg length discrepancies can be minimized by allocating excess or deficient tension to the offset and leg length. If there is a significant discrepancy between the preoperative plan and intraoperative soft-tissue tension, there may be a leg length discrepancy before degeneration or the preoperative plan itself may be problematic, in which case the final placement status should be reviewed. Leg length can be adjusted by changing of the stem level, cup level, or neck length. The stem anteversion angle and femoral offset also has a subtle influence on leg length.

(e) Soft-tissue tension adjustment in various situations

Five situations in which the soft tissue and leg length require adjustment during the trial are assumed, and the adjustment method is discussed for each situation as follows (Fig. 46.3): (i) strong soft-tissue tension and long leg length; (ii) weak soft-tissue tension and long leg length; (iii) strong soft-tissue tension and short leg length; (iv) weak soft-tissue tension and short leg length; and (v) leg length matching and tension problems.

(i) Strong soft-tissue tension and long leg length

The basic strategy in such cases is to reduce soft-tissue tension while shortening the leg length and neck length, lowering the stem level, and replacing the cup upward (in cases of cementless cups). In cases of neck shortening, bony impingement due to offset reduction should be considered.

(ii) Weak soft-tissue tension and long leg length

This is a situation encountered in patients who originally had a long leg on the affected side. In such cases, the degree of degeneration and degree of leg lengthening are irrational and can be predicted preoperatively. Forced leg length matching results in unstable hip reconstruction with significant soft-tissue laxity, thereby requiring a strategy to minimize leg lengthening while improving soft-tissue tension. If anticipated during planning, the surgeon can minimize cup medialization and ensure an acetabular offset. Intraoperative measures include the use of high-offset stems, neck lengthening, and offset liners. If these measures to increase soft-tissue tension are ineffective, a leg length discrepancy should be allowed to favor hip stability. However, this is often a predegenerative condition.

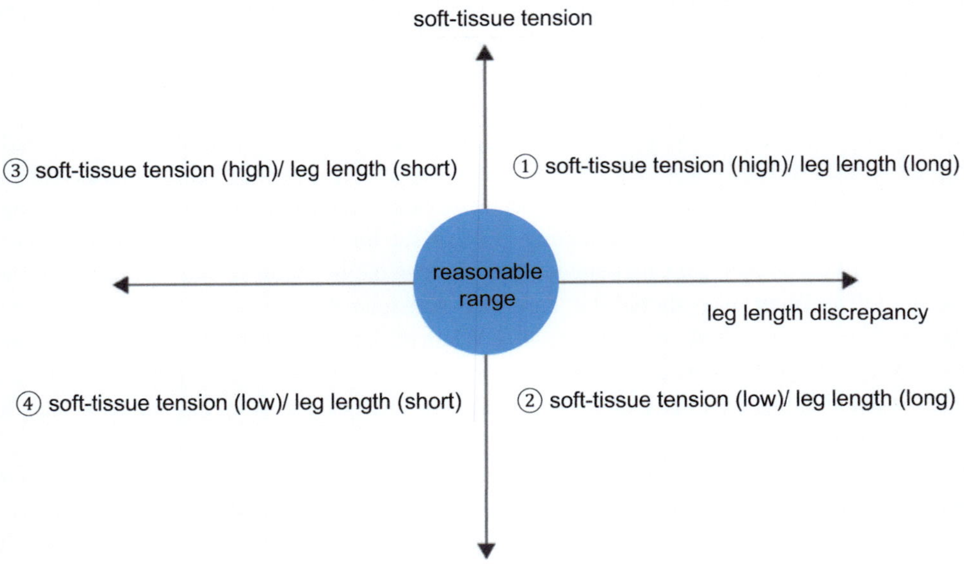

Fig. 46.3 Various situations of soft-tissue tension and leg length discrepancy

(iii) Strong soft-tissue tension and short leg length

This is assumed to be the case in cases of leg shortening from childhood, such as coxa plana or a secondary acetabulum. The goal is to reduce the global offset and soft-tissue tension to lengthen the leg by dissecting the periprosthetic soft tissues. For example, the maximum cup medialization can be added to place the stem higher and lengthen the neck. Dissection of the surrounding soft tissue should be minimized and performed with a trial because of the potential for significant soft-tissue imbalance.

(iv) Weak soft-tissue tension and short leg length

As limb lengthening increases soft-tissue tension, the strategy is to extend the neck (and increase the offset) and position the stem higher. Since the cup was in place during the trial, a lower cup placement was out of strategy.

(v) Leg length matching and tension problems

Because the leg lengths are matched, soft-tissue tension adjustment at the offset is the primary task. Adjustment of the neck shaft angle and neck length involves changes in leg length. If tension is high, the global offset may be lowered by adding cup medialization, shortening the neck length, lowering the stem offset, or lowering the stem placement level to allow leg shortening. Overstressed soft tissues may be released, but fine adjustment is difficult. If tension is loose, the global offset may be increased using neck lengthening, the use of a high-offset stem, and the placement of an offset liner but with leg lengthening. Cup lateralization is already impossible in the trial.

These are options for adjusting soft-tissue tension, but an accurate assessment of the implant placement status and soft-tissue evaluation at accurate hip flexion and extension angles are prerequisites for these adjustments. Cemented stems are more flexible than cementless stems in terms of changes in the stem height. In the cementless type, the tapered wedge and modular type stems are easier to adjust than the fit-and-fill type.

3 Effects on Postoperative Outcomes

Currently, there is no quantitative method or instrumentation for measuring intraoperative soft-tissue tension, nor are there any reports on the effects of soft-tissue tension adjustments on postoperative outcomes. However, there are numerous reports on the effects of leg length and offset recovery on clinical outcomes. It is assumed that the restoration of leg length and offset restores the interosseous distance along with the soft-tissue tension attached to the interosseous fragments. A decrease in postoperative dislocation [22, 23], improvement in abduction muscle strength [17, 24] and clinical scores [19], and a decrease in polyethylene wear [25, 26] reportedly occur. Dennis et al. reported a mean 3.3 mm of bone head separation in mid-to-abduction hip motion on three-dimensional matching using fluoroscopy [1]. In fact, the adjustment of soft-tissue tension and its effect on the hip joint is a subject for future study.

References

1. Dennis DA, et al. "In vivo" determination of hip joint separation and the forces generated due to impact loading conditions. J Biomech. 2001;34:623–9.
2. Sugano N, et al. Computed-tomography-based computer preoperative planning for total hip arthroplasty. Comput Aided Surg. 1998;3:320–4.
3. Sariali E, et al. Accuracy of reconstruction of the hip using computerised three-dimensional pre-operative planning and a cementless modular neck. J Bone Joint Surg Br. 2009;91:333–40.
4. Nishihara S, et al. Measurements of pelvic flexion angle using three-dimensional computed tomography. Clin Orthop Relat Res. 2003;411:140–51.
5. Yoshino S, et al. Anterior superior iliac spine is not always reliable as a pelvic reference axis. 3D study of pelvic axis. -3D study of pelvic axis. J Orthop Sci. 2022;S0949-2658(22):00024-0.
6. Sugano N, et al. The morphology of the femur in developmental dysplasia of the hip. J Bone Joint Surg Br. 1998;80:711–9.
7. Ge W, et al. ISB recommendation on definitions of joint coordinate system of various joints for the reporting of human joint motion – part I: ankle, hip, and spine. J Biomech. 2002;35:543–8.
8. Crawford MJ, et al. The 2007 Frank Stinchfield Award. The biomechanics of the hip labrum and the stability of the hip. Clin Orthop Relat Res. 2007;465:16–22.
9. Domb BG, et al. Minimum 5-year outcomes of robotic-assisted primary total hip arthroplasty with a nested comparison against manual primary total hip arthroplasty: a propensity score-matched study. J Am Acad Orthop Surg. 2020;28:847–56.
10. Boe RA, et al. Soft-tissue balancing in total hip arthroplasty. JBJS Rev. 2021;9:e20.00116.
11. Takao M, et al. [Reproducibility of the shuck test in total hip arthroplasty.] Cent Jpn J Orthop Traumat 2011;54:148–62. Japanese.
12. Tanino H, et al. Hip stability after total hip arthroplasty predicted by intraoperative stability test and range of motion: a cross-sectional study. BMC Musculoskelet Disord. 2018;19:1–7.
13. Takao M, et al. Postoperative limb-offset discrepancy notably affects soft-tissue tension in total hip arthroplasty. J Bone Joint Surg Am. 2016;98:1548–54.
14. Kohno Y, et al. A laser guide technique: a novel method for accurate acetabular cup alignment in total hip arthroplasty. Sci Rep. 2022;12:1–9.
15. Hara T. [Toward optimal total hip replacement placement – a consideration of soft tissue balance.] Orthoped Surg 2019;75:210–213. Japanese.
16. Kurtz WB, et al. Factors affecting bony impingement in hip arthroplasty. J Arthroplast. 2010;25:624–34.e1-2.
17. Mcgrory BJ, et al. Effect abductor of femoral muscle offset strength on total motion and after total hip arthroplasty. J Bone Joint Surg Br. 1995;77:865–9.
18. Archibeck MJ, et al. Acetabular loosening using an extended offset polyethylene liner. Clin Orthop Relat Res. 2009;467:188–93.
19. Clement ND, et al. Total hip replacement: increasing femoral offset improves functional outcome. Archiv Orthop Trauma Surg. 2016;136:1317–23.
20. Charles MN, et al. Soft-tissue balancing of the hip: the role of femoral offset restoration. J Bone Joint Surg Am. 2004;86:1078–88.
21. Austin MS, et al. Stability and leg length equality in total hip arthroplasty. J Arthroplast. 2003;18:88–90.
22. Ogawa T, et al. Soft tissue tension is four times lower in unstable primary total hip arthroplasty: does this comparison make sense? Int Orthop. 2018;42:2067–8.
23. Jinno T, et al. Intraoperative evaluation of the effects of femoral component offset and head size on joint stability in total hip arthroplasty. J Orthop Surg (Hong Kong). 2017;25:2309499016684298.
24. Asayama I, et al. Relationship between radiographic measurements of reconstructed hip joint position and the Trendelenburg sign. J Arthroplasty. 2002;17:747–51.
25. Devane PA, et al. Assessment of polyethylene wear in total hip replacement. Clin Orthop Relate Res. 1999;369:59–72.
26. Sakalkale DP, et al. Effect of femoral component offset on polyethylene wear in total hip arthroplasty. Clin Orthop Relate Res. 2001;388:125–34.

Leg Length Discrepancy

47

Yoshitomo Kajino

1 What Is Leg Length Discrepancy?

Leg length discrepancy refers to the difference between the lengths of the left and right lower limbs. The residual leg length difference after total hip arthroplasty (THA) significantly affects not only the clinical outcomes, such as hip instability, pain in the contralateral hip or adjacent joints, back pain, gait function, neurological deficits, and long-term outcomes of the implant, but also patient satisfaction with the surgery. Therefore, residual postoperative leg length difference is one of the major causes of medical litigation after THA in Europe and the United States [1].

There are two types of leg length differences that must be considered in THA: the spina malleolar distance (SMD), which is the length between the superior anterior iliac spine and the medial ankle joint, a characteristic point on the body surface, and the actual leg length difference measured from images, such as radiographs and computed tomography (CT) (objective leg length difference). The other is the subjective leg length difference perceived by the patient. The difference in length between the line connecting the right and left inter-teardrop line and the innermost edge of the left and right lesser trochanter in an anteroposterior (AP) hip radiograph is usually measured (Fig. 47.1). It is recommended to use a scale or metal ball as reference for the length to reduce the measurement error due to the magnification of the radiograph image. In patients with hip or knee joint contractures, measurement errors are likely to occur; therefore, measurement using CT and 3D templating software is effective. On the other hand, the subjective leg length difference is generally measured by the block test, in which a 5-mm-thick plate is placed under the shortened leg on the affected side one by one, and the thickness of the plate is measured based on the patient's feeling of equal length (Fig. 47.2). Notably, care should be taken because accurate measurement is not possible in patients with joint contractures. Due to the presence of joint contractures caused by long-term hip deformities and changes in the spinal and pelvic alignment, the subjective leg length difference may not always match the objective leg length difference [2]. For example, there are cases in which patients themselves complain that the leg on the operated side is longer than the leg on the isometric radiographs, and vice versa, in which patients themselves do not complain of the leg length difference at all even though the radiographs are not isometric.

As the postoperative leg length in THA depends on the position and alignment of the acetabular and femoral implants in relation to the bone, the surgeon needs to know the perceived leg length difference in each patient during the

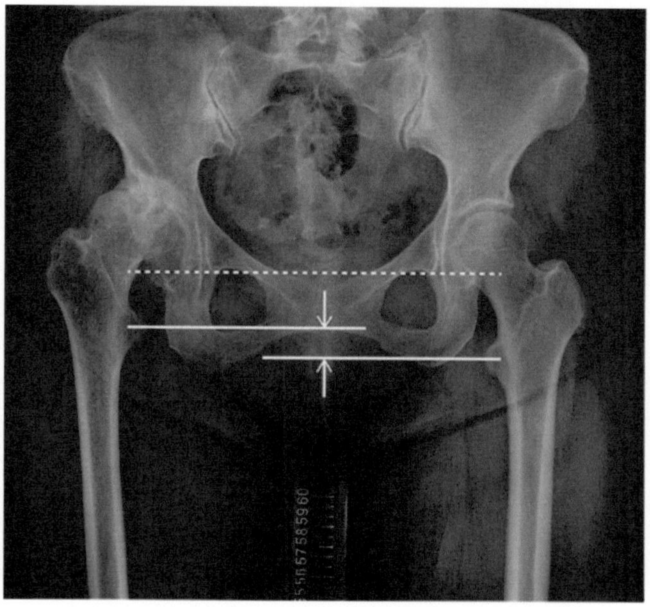

Fig. 47.1 Measurement of the leg length difference in the radiograph. The difference in length between the inter-teardrop line, which is the line connecting the lower edge of the right and left teardrops, and the innermost margin of the right and left lesser trochanter

Y. Kajino (✉)
Department of Orthopedic Surgery, Kanazawa University, Kanazawa, Ishikawa, Japan
e-mail: yoshitomokajino@staff.kanazawa-u.ac.jp

Y. Nakashima (ed.), *Advances in Total Hip Arthroplasty*, https://doi.org/10.1007/978-981-97-2461-1_47

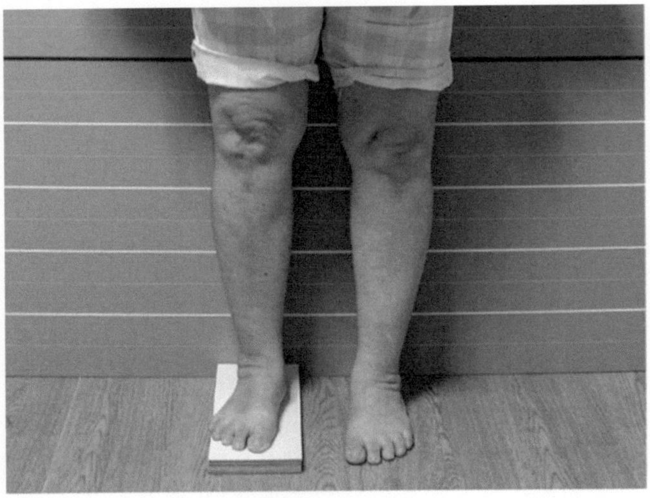

Fig. 47.2 Measurement of subjective leg length difference by the block test. A 5-mm-thick plate is placed under the foot of the affected (right) side one by one in turn and the measurement is taken

preoperative planning phase and determine the approximate amount of leg length correction during surgery. The surgeon must determine the approximate amount of leg length correction to be carried out during surgery. The surgeon must accurately follow the preoperative plan during surgery. It has been reported that surgeons with less experience in surgery have a higher risk of a residual leg length difference of 5 mm or more after surgery than surgeons with more experience in surgery, and surgeons with less experience in surgery especially need careful preoperative planning and surgery [3].

2 Factors Affecting the Leg Length Difference

In Japan, there have been several studies on the factors influencing the subjective leg length difference, methods to determine the amount of leg length correction during surgery, the amount of leg lengthening that can be corrected, and the goals of leg length correction.

The factors influencing both subjective and objective leg length differences have been investigated, including bone length asymmetry, lumbar spine mobility, lateral pelvic tilt, and lower limb alignment.

In the case of acetabular dysplasia, which is common in Japan, asymmetry in the length of the right and left femur has been reported, and an absolute difference in femur length of 5 mm or more was observed in approximately one quarter (24%) of cases with unilateral disease [4]. Asymmetry is related to the difference in the length between the apex of the greater trochanter and the most medial margin of the lesser trochanter (inter-trochanteric length) between the right and left hip joints. Preoperative measurement of the leg length difference by full-length radiographs or computed tomography is recommended for patients who complain of subjective leg length difference even though the frontal radiograph of the hip joint appears to show no leg length difference or in cases where a left-right difference in the inter-trochanteric length is observed.

In a study of preoperative lumbar lateral flexion mobility on the affected side and subjective leg length difference, a significant difference was found between the subjective leg length difference and the objective leg length difference in the group with lateral pelvic tilt to the hip on the affected side in the preoperative standing AP hip radiograph [5]. Patients with high lumbar lateral flexion mobility were less likely to experience a subjective leg length difference because of postoperative leveling of the pelvis with leg extension, whereas patients with low lumbar lateral flexion mobility had significantly greater values of postoperative subjective leg length difference. In patients who experienced a subjective leg-length difference postoperatively, the amount of leg lengthening by surgery was 5.9 mm longer than the subjective leg-length difference, suggesting that a subjective leg-length difference plus 5 mm may be appropriate as the target amount of leg lengthening during the surgery.

In another study, THA patients at an average of 29 months postoperatively who had undergone the procedure with the goal of eliminating the objective leg length difference on imaging were examined [6]. A subjective leg length difference was observed in 9% of all patients, of whom 83.3% felt that the affected leg was longer and 16.7% felt that the affected leg was shorter. In patients with a postoperative subjective leg length difference of less than 5 mm, 98.3% of the patients reported that the leg length difference did not inconvenience them, while 66.7% of the patients with a leg length difference between 7.5 and 10 mm reported that the subjective leg length difference inconvenienced them. The two most common factors that contributed to the subjective leg length difference being longer than the radiographic leg length difference were a preoperative lateral pelvic tilt to the affected side and preoperative lumbar scoliosis on the affected side.

In addition, it has been reported that the alignment of the entire lower limb, with a focus on the knee joint, also affects the leg length difference [7].

3 Methods for Ascertaining the Amount of Leg Length Correction During Surgery

Several methods have been reported to determine the amount of leg lengthening that is performed during THA surgery. As the simplest method that does not require special instruments, the distance between the outer edge of the cup and the apex of the greater trochanter is measured after the trial implant is placed and compared with the preoperative plan [8], and the distance between the posterior acetabular wall and a line drawn at the greater trochanter is measured before and after the placement of the implant [9]. Subsequently, a method to measure the distance between the pins inserted into the iliac and femoral bones using a simple device has been reported [10–12] (Fig. 47.3). These methods require attention because the measurement error becomes large depending on the slack of the pin and the position of the lower limb at the time of measurement.

Recently, the measurement accuracy of intraoperative leg length correction using a simple portable navigation system and a CT-based navigation system has been reported [11, 13–15] (Fig. 47.4). In both cases, the average error between the postoperative image evaluation and the intraoperative measurements was reported to be <2–3 mm. These navigation tools are useful because the implant alignment can also be measured intraoperatively. The utility of intraoperative radiographic imaging during the trial phase to confirm whether the implant placement and leg length correction are appropriate has also been reported [16].

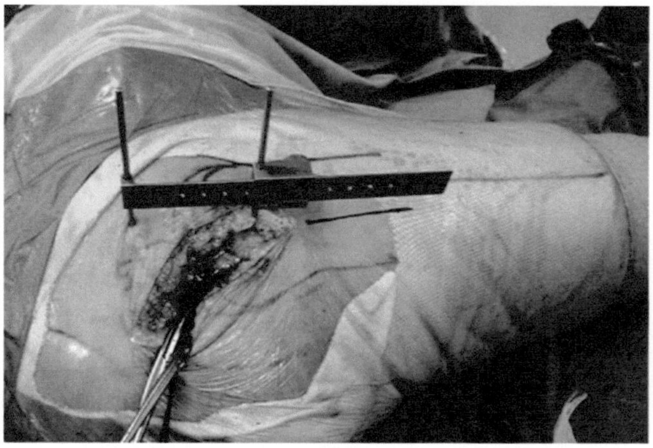

Fig. 47.3 Method for ascertaining the amount of leg length correction during surgery. The distance between the pins inserted into the iliac and femoral bones, respectively, is measured with a simple device (right hip joint, posterior approach)

4 Correctable Leg Extension

There are studies that have been performed to determine the amount of leg lengthening that can be corrected by surgery in relation to the preoperative measured objective leg length difference and the patient's own subjective leg length difference.

In a study of cases, including Crowe II-IV subtrochanteric shortening osteotomies, the hip flexion angle under preoperative anesthesia and the presence of ischiofemoral osteoarthritis (joint formation between the sciatic bone and femur) were correlated with the amount of leg length correction (= distance of possible abscissa drawdown) during surgery [17, 18].

In terms of neuropathy due to leg lengthening, it is generally reported that a primary leg lengthening of 3 cm or more is associated with a high risk of neuropathy; however, there are also reports that 5 cm or more is a risk factor and that 3.5% or more of the height to length ratio or 8.7% or more of the femur length ratio is a risk factor [19, 20]. For the details regarding nerve palsy associated with leg lengthening, please refer to the section on complications.

Fig. 47.4 Example of intraoperative leg length measurement using a simple portable navigation system. (Modified from Anjiki [15])

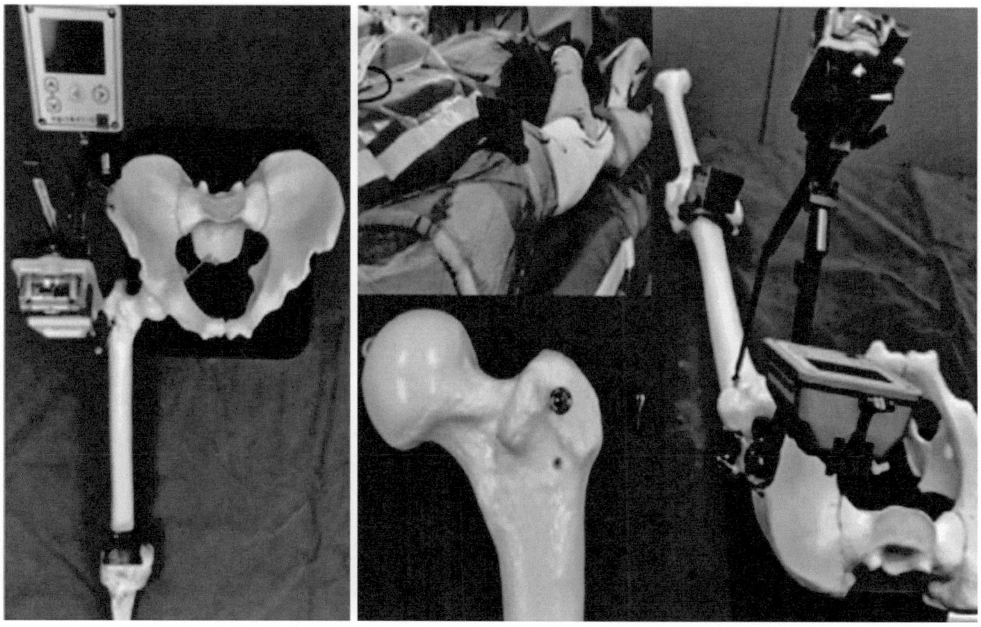

5 Leg Length Correction Targets

In patients without joint contractures, anatomic skeletal asymmetry, or spinal or lower extremity alignment abnormalities, there is no significant discrepancy between the subjective and objective leg length differences. However, in a previous study, it was reported that several postoperative patients after THA have a subjective feeling of having a longer leg on the affected side during the early postoperative period; subsequently, the subjective leg length difference decreases over time. Such changes in the subjective leg length difference should be informed to patients at the time of preoperative informed consent [5, 21]. However, subjective leg length difference was reported to have a more significant effect on postoperative pain, lower limb function, and patient satisfaction than objective leg length difference [7, 21, 22]. The subjective leg length difference is considered to be unchanged at 3 months and 1 year postoperatively, and patients who have experienced subjective leg length difference for a long period of time should be considered for the use of heel lifts or brace [21].

Although it was once accepted that a residual leg length difference of up to 10 mm was acceptable after THA, previous studies have shown that the patient's subjective leg length difference should be taken into account during preoperative planning and surgical manipulation.

References

1. Patterson DC, et al. Lawsuits after primary and revision total hip arthroplasties: a malpractice claims analysis. J Arthroplast. 2017;32:2958–62.
2. Maloney WJ, et al. Leg length discrepancy after total hip arthroplasty. J Arthroplast. 2004;19(4 Suppl 1):108–10.
3. Kishimoto Y, et al. A low-volume surgeon is an independent risk factor for leg length discrepancy after primary total hip arthroplasty: a case–control study. Int Orthop. 2020;44:445–51.
4. Tamura K, et al. Femoral morphology asymmetry in hip dysplasia makes radiological leg length measurement inaccurate. Bone Joint J. 2019;101-B:297–302.
5. Koga D, et al. The effect of preoperative lateral flexibility of the lumbar spine on perceived leg length discrepancy after total hip arthroplasty. J Med Dent Sci. 2009;56:69–77.
6. Fujita K, et al. Optimizing leg length correction in total hip arthroplasty. Int Orthop. 2020;44:437–43.
7. Fujimaki H, et al. Leg length discrepancy and lower limb alignment after total hip arthroplasty in unilateral hip osteoarthritis patients. Sciences. 2013;18:969–76.
8. Murayama T, et al. A novel non-invasive mechanical technique of cup and stem placement and leg length adjustment in total hip arthroplasty for dysplastic hips. Int Orthop. 2015;39:1057–64.
9. Tagomori H, et al. A new and simple intraoperative method for correction of leg-length discrepancy in total hip arthroplasty. J Orthop. 2019;16:405–8.
10. Shiramizu K, et al. L-shaped caliper for limb length measurement during total hip arthroplasty. J Bone Joint Surg Br. 2004;86:966–9.
11. Ogawa K, et al. Accurate leg length measurement in total hip arthroplasty: a comparison of computer navigation and a simple manual measurement device. Clin Orthop Surg. 2014;6:153–8.
12. Kawamura H, et al. Effects of lower limb and pelvic pin positions on leg length and offset measurement errors in experimental total hip arthroplasty. J Orthop Surg Res. 2021;16:193.
13. Kitada M, et al. Evaluation of the accuracy of computed tomography-based navigation for femoral stem orientation and leg length discrepancy. J Arthroplasty. 2011;26:674–9.
14. Tanino H, et al. Accuracy of a portable accelerometer-based navigation system for cup placement and intraoperative leg length measurement in total hip arthroplasty: a cross-sectional study. BMC Musculoskelet Disord. 2021;22:299.
15. Anjiki K, et al. Effectiveness of an accelerometer-based portable navigation for intraoperative adjustment of leg length discrepancy in total hip arthroplasty in the supine position. J Orthop Sci. 2022;27:169–75.
16. Kuroda K, et al. Do we need intraoperative radiographs for positioning the femoral component in total hip arthroplasty? Arch Orthop Trauma Surg. 2014;134:727–33.
17. Fujishiro T, et al. Leg length change in total hip arthroplasty with subtrochanteric femoral shortening osteotomy for Crowe type IV developmental hip dysplasia. J Arthroplast. 2012;27:1019–22.
18. Fujishiro T, et al. Predicting leg-length change after total hip arthroplasty by measuring preoperative hip flexion under general anaesthesia. J Orthop Surg (Hong Kong). 2012;20:327–30.
19. Higuchi Y, et al. Leg lengthening of more than 5 cm is a risk factor for sciatic nerve injury after total hip arthroplasty for adult hip dislocation. Nagoya J Med Sci. 2015;77:455–63.
20. Kabata T, et al. Safety range for acute limb lengthening in primary total hip arthroplasty. Int Orthop. 2019;43:2047–56.
21. Iwakiri K, et al. Changes in patient-perceived leg length discrepancy following total hip arthroplasty. Eur J Orthop Surg Traumatol. 2021;31:1355–61.
22. Nakanowatari T, et al. True or apparent leg length discrepancy: which is a better predictor of short-term functional outcomes after total hip arthroplasty? J Geriatr Phys Ther. 2013;36:169–74.

Part VIII

Preoperative Planning

48 Preoperative Planning

Tetsuya Tachibana

Preoperative planning is essential for a successful total hip arthroplasty (THA). It is difficult to perform a successful THA without adequate preoperative planning, and unforeseen complications can occur if preoperative planning is not appropriate. Preoperative planning includes appreciating the proximal femur geometry that varies in patients, selecting the appropriate acetabular and femoral components, determining the component size and positioning, estimating the postoperative leg length and global offset, identifying potential intraoperative problems and their remedies, and preparing extra implants for backup. The general goal of preoperative planning includes placing the cup in the native hip center as much as possible and restoring the femoral offset and leg length discrepancy.

Preoperative planning can be divided into two major types: two-dimensional templating using plain anteroposterior pelvic radiographs, and three-dimensional templating based on pelvic CT data. Conventionally, two-dimensional templating is performed by outputting films of pelvic radiographs, determining the acetabular and femoral component size by overlapping implant templates on the films, and then copying them on tracing paper. However, currently, to eliminate film storage and take advantage of electronic medical records, two-dimensional templating is generally performed using digital templating software on the electronic medical record [1] or applying the implant template onto the computer screen directly [2]. However, opinions differ regarding whether these digital templating methods are more accurate than conventional templating with tracing papers [3–5]. On the other hand, although there are issues of radiation exposure and cost, three-dimensional templating for THA using preoperative CT data is highly accurate and useful for preventing poor initial fixation, insufficient offset, and impingement [6]. Three-dimensional templating is rapidly becoming popular and has enhanced functions that enable the estimation of postoperative range of motion and impingement, and aids in performing robot-assisted surgery.

T. Tachibana (✉)
Dokkyo Medical University Saitama Medical Center, Koshigaya, Saitama, Japan
e-mail: t-tachi@dokkyomed.ac.jp

1 Two-Dimensional Preoperative Planning

(a) Appropriate radiograph in the hip

The most important prerequisite for two-dimensional templating is to obtain appropriate preoperative pelvic radiographs. Appropriate pelvic radiographs are characterized by symmetrical bilateral obturator foramina and bilateral acetabular teardrops, as well as the coincidence of the midsacral line and the pubic symphysis. The standard position of the femur in pelvic radiographs is slightly rotated internally (15°–20°) [7]. Since normal femoral anteversion is generally about 20°, slight internal rotation of the femur makes it possible to obtain proximal femoral radiographs parallel to the femoral neck, which enables accurate femoral stem templating (Fig. 48.1). Since a dysplastic hip is often associated with excessive anteversion of the femur greater than 40°, slight internal rotation of the femur cannot produce adequate AP femoral radiographs. Although there is no method to accurately evaluate the internal rotation of the femur on pelvic radiographs, the images are appropriate for preoperative planning if the width of the lesser trochanter is within 5 mm in Japanese patients [8] (Fig. 48.2).

Two-dimensional templating is difficult when appropriate pelvic radiographs cannot be obtained, as in cases with flexion contracture of the hip or knee joint, or with external rotation contracture of the hip joint. In these contracture cases, appropriate femur radiographs can be taken in a semi-sitting or prone position (Fig. 48.3). If the femur shape is similar bilaterally, two-dimensional stem templating should be performed on both the affected and unaffected sides.

Although preoperative templating is generally performed with pelvic radiographs in the supine position, pelvic radio-

Y. Nakashima (ed.), *Advances in Total Hip Arthroplasty*, https://doi.org/10.1007/978-981-97-2461-1_48

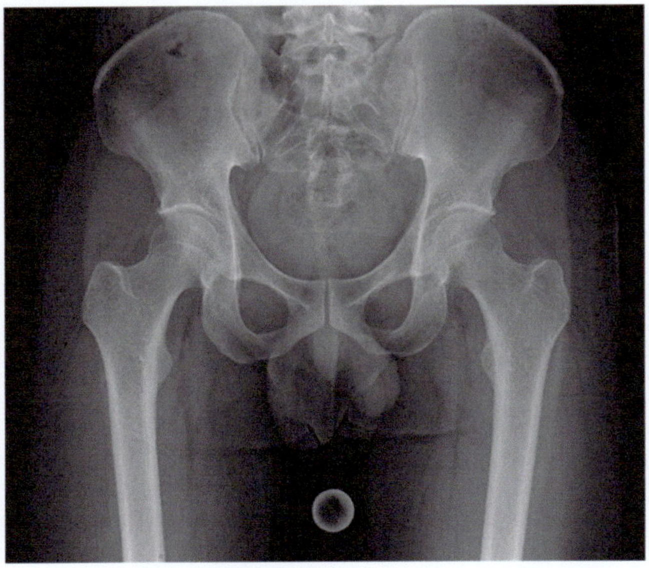

Fig. 48.1 Anteroposterior pelvic radiograph. The shapes of the bilateral obturator foramina are symmetrical, and the centers of the pubic symphysis and mid-sacral line are aligned

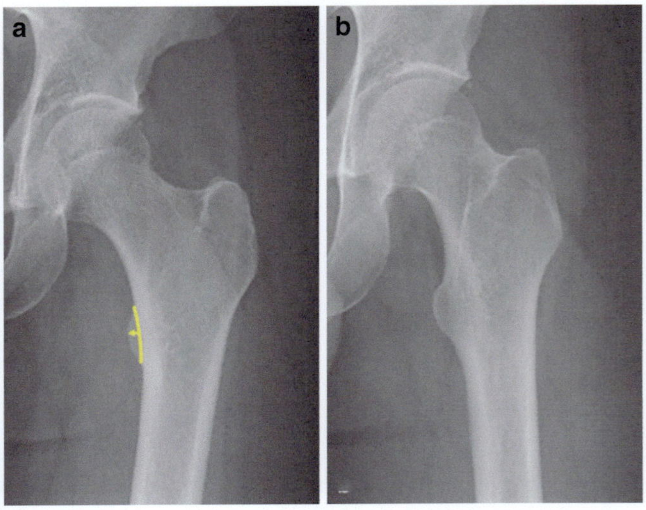

Fig. 48.2 (**a**) Appropriate pelvic radiograph. The width of the lesser trochanter is less than 5 mm (yellow arrow). (**b**) Radiograph showing external rotation of the left hip in the same patient. Preoperative planning for accurate stem placement is difficult due to the enlarged lesser trochanter, increased femoral neck-shaft angle, and decreased femoral offset

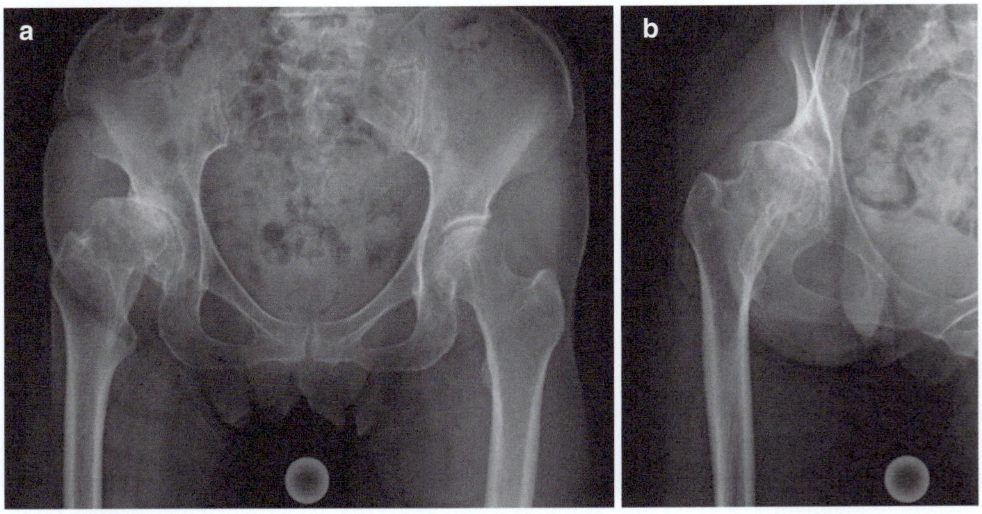

Fig. 48.3 Method of appropriate radiographs for external rotation contracture of the hip. (**a**) Plain radiographs of both hips under mild internal rotation. In the left hip, an appropriate anteroposterior radiograph of the proximal femur was obtained, but the right hip could not be templated accurately because it had an external rotation contracture. (**b**) Posteroanterior pelvic radiograph of the right hip in the prone position. The size of the lesser trochanter is appropriate, and a more accurate templating of the femoral stem is possible

graphs in the standing position are also required to confirm the change in pelvic tilt. The shape of the pelvic foramina and the vertical distance between the pubic tubercle and the sacroiliac joint can be easily measured as a substitute for pelvic tilt. Acetabular component planning should be performed more carefully in patients with a large change in posterior pelvic tilt from the supine to the standing position because of the high risk of dislocation [9].

Accurate calibration of magnification using radiographic markers is necessary to determine the component size. Errors are reduced by placing radiographic markers or measures in contact with the skin at the height of the greater trochanter during radiography [10] (Fig. 48.4). In cases in which contralateral THA has already been performed, calibrating the magnification according to the known diameter of the femoral head is useful.

(b) Acetabular component templating

In patients with primary hip osteoarthritis or mild acetabular deformity, the optimal acetabular component size can be

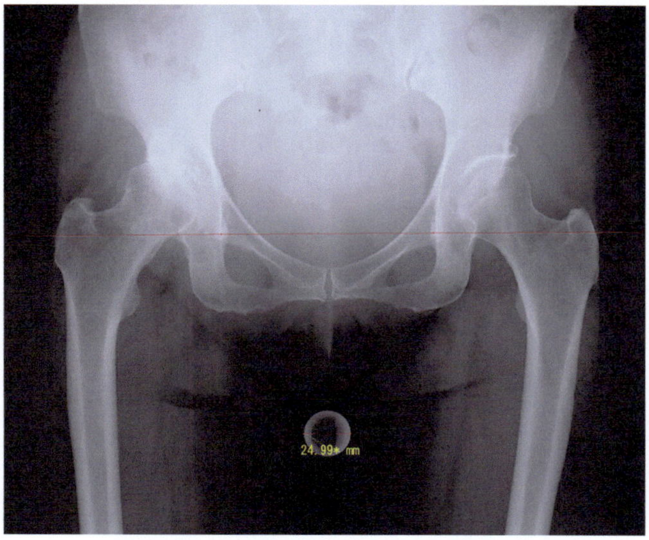

Fig. 48.4 Calibration of magnification and setting of the horizontal axis. A known marker (25 mm diameter in this Figure) was placed at the height of the greater trochanter. The horizontal axis is set by a straight line connecting the lower edges of the bilateral teardrops (red line)

estimated by placing the hemispherical acetabular template just lateral to the lateral wall of the tear drop and along the acetabular rim on the pelvic radiographs. However, patients with dysplastic hips often have subluxation of the femoral head in a superolateral direction and a lateralized acetabulum with excessive osteophytes in the acetabular fossa. In such cases, the aspherical shape of the femoral head and shallow acetabulum make it difficult to estimate the acetabular component size using the spherical cup template, and the only reference is the diameter of the unaffected acetabulum or femoral head. Therefore, the accuracy rate of cup size for two-dimensional templating is not high (approximately 40%), and the accuracy within one size of the preoperative plan is reported to be approximately 60–80% [3, 5]. If possible, CT should be performed to measure the anteroposterior acetabular diameter and acetabular anteversion to help determine the acetabular component size and cup position. It should be noted that the cup size determines the maximum diameter of the femoral head that can be used.

For the coordinate system in two-dimensional templating, the horizontal axis of the pelvis is a straight line connecting the lower edges of the bilateral tear drops. A straight line connecting the lower ends of the bilateral ischial bones may also be used as the horizontal axis, and both horizontal axes should be referenced if the acetabulum is severely deformed [11].

The ideal position for the acetabular component is generally achieved by placing the inferomedial edge of the cup adjacent to the lower edge of the lateral wall of the teardrop and along the acetabular rim. In most patients, this will anatomically restore the native hip center and enable medialization of the center of hip rotation. The advantages of medialization of the hip center include an increase in the lever arm of the abductor muscle, an improvement in hip stability, and a decrease in load transmission through prosthetic bearing surfaces [12]. Additionally, in most cases leg shortening is anatomically corrected by restoring the hip center. When the contralateral hip is normal or unaffected, the ideal position of the acetabular component was considered to be the contralateral hip center. In cases of bilateral hip deformities, the normal hip center is unknown; however, it is estimated to be 13–15 mm above the lower edge of the tear drop [13, 14]. The actual position of the acetabular component should be placed with reference to the normal center of rotation of the hip; furthermore, it should be placed superiorly or medially as necessary to achieve adequate initial fixation.

The superior bony coverage of the acetabular cup is generally measured as the angle between the vertical line and the line drawn from the cup center to the lateral edge of the acetabulum, which is called the cup center-edge angle (cup-CE angle) [15]. To obtain long-term cup stability, a cup-CE angle greater than 0° is acceptable with screw fixation [16], and with only press-fit fixation, the cup-CE angle is needed is greater than 10° [17].

If the cup CE angle is less than 0° when the cup is placed in the anatomically desirable position, stable cup fixation should be obtained by shifting the cup medially or superiorly so that the cup CE angle is greater than 0°.

Since cup medialization is biomechanically beneficial, priority is given to medialization along the lateral wall of the tear drop rather than shifting superiorly. If further cup medialization is required to obtain superior bony coverage, the lateral wall of the tear drop may be reamed, and the cup is placed adjacent to the medial cortex of the tear drop (Fig. 48.5). However, in this case, the global offset tends to decrease; therefore, it is necessary to compensate for the increase in the femoral offset appropriately.

It is not recommended to shift the center of rotation of the hip superiorly due to the risk of limping owing to decreased abductor muscle strength. However, since it has been reported that a 2 cm superior positioning of the hip center without lateral placement does not have major adverse effects on abduction moment arms [18]; hence, shifting the cup superiorly, by up to 2 cm until the cup CE angle is greater than 0°, is acceptable.

In the actual preoperative planning, the cup is first shifted medially with reference to the lateral wall of the tear drop. If the cup-CE angle is insufficient, the cup is shifted upward to avoid perforation of the acetabulum. Finally, leg lengthening and global offset are comprehensively determined in combination with stem planning.

However, in patients with high hip subluxation, such as Crowe's classification groups 3 or 4, the cup-CE angle may be insufficient. In such cases, there are two methods for achieving stable cup fixation: (1) to place the cup in an ana-

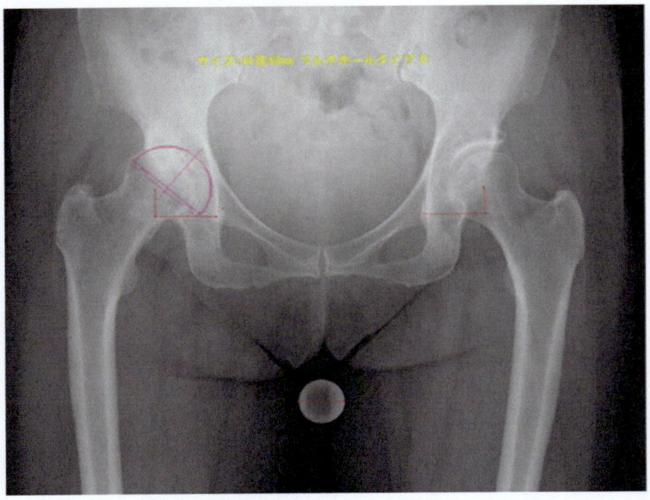

Fig. 48.5 Acetabular cup templating. As a goal of the native hip center, the distances from the lower edge of the tear drop to the center of the femoral head on the unaffected side (left hip) are measured vertically and horizontally (red line). The cup center should be determined so that it is symmetrical on the operative side and the native hip center is established. In this case, because medialization of the acetabular cup is beneficial, the cup is planned by placing it more medially than the native hip center. The cup CE angle is 24°, which is considered sufficient for stable initial fixation

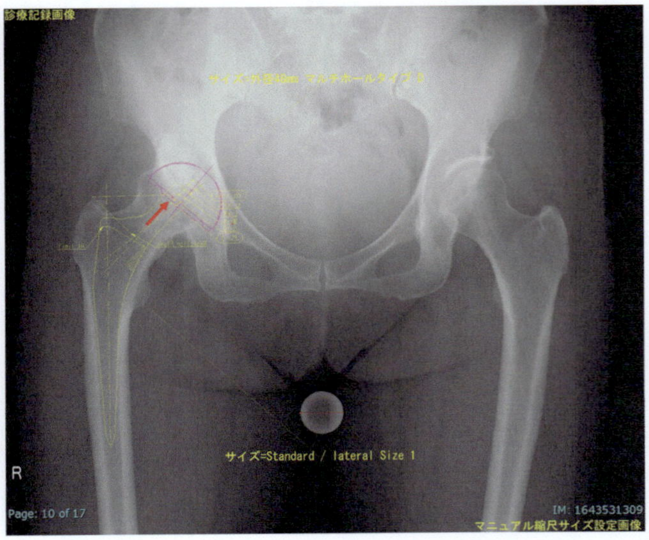

Fig. 48.6 Femoral stem templating. The stem should restore the center of the femoral head (point indicated by the red arrow). The femoral offset often tends to be underestimated in anteroposterior pelvic radiographs; therefore, care must be taken not to reduce it

tomically normal hip center, bulk bone allografts, metal augmentations, or a cemented cup with acetabular reinforcement rings can be used; (2) the small-diameter cup can be placed superiorly by up to 2 cm so that the cup CE angle is greater than 0°.

Studies have shown that Lewinnek's safe zone (inclination angle within 40° ± 10°; anteversion angle within 15° ± 10°) minimizes dislocation after THA [19]. However, more recently, the combined anteversion theory, in which the sum of cup and stem anteversion should be considered, is now becoming accepted as a more useful technique to reduce the risk of dislocation [20].

(c) Femoral component templating

In principle, the femoral component should be placed to restore the center of the femoral head. Because of the great variety of femoral stem designs, unlike acetabular cup designs, which have almost the same hemispherical shape, an appropriate femoral stem must be selected depending on each desirable primary fixation style. During preoperative planning, the patient's femur should be evaluated for bone quality (presence of osteoporosis), neck-shaft angle, femoral offset, proximal femur canal geometry, and femoral neck version. For example, if the patient has a wide canal diameter in the proximal femur (stovepipe type) with poor bone quality, distal fixation cementless stems, such as a rectangular stem or a cemented stem, should be considered. However, when the patient has a narrow diaphyseal canal with thick and distinct cortical bone (champagne flute type), proximal fixation cementless stems, such as a single wedge stem, should be considered. Furthermore, surgeons should be aware of an excessively narrow diaphyseal diameter as it causes the proximal fixation stem to be engaged in the diaphysis before maximizing its fit in the metaphysis. This causes rotational instability during initial fixation. If a patient has anatomical abnormalities and rotational malalignments in the proximal femur, adjustment for anteversion is performed using modular stems, changeable neck stems, and Wagner conical stems.

To reconstruct the femoral head center, the vertical height and femoral offset are estimated by reference to the contralateral unaffected femoral head center if the femoral head center of the affected side is unclear. Regarding the vertical height of the femoral head center, it has been reported that it is often 5 mm proximal to the height of the femoral neck saddle [21]. The femoral offset should be restored equally or increased by a few millimeters compared to the unaffected side, considering that the femoral offset tends to be underestimated on pelvic radiographs due to femoral anteversion. Decreasing femoral offset postoperatively is particularly to be avoided because decreasing offset has a number of negative effects that have been associated with decreased abductor muscle, increased limp, and instability [22, 23] (Fig. 48.6). As a guide to the stem depth, the distance from the top of the lesser trochanter to the medial aspect of the femoral stem and the distance from the top of the greater trochanter to the lateral aspect of the stem are measured to be reproduced intraoperatively (Fig. 48.7).

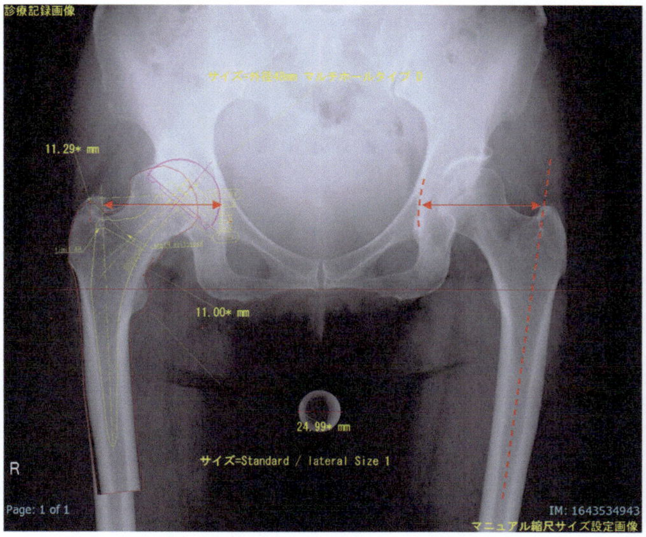

Fig. 48.7 Finished templating. Complete templating is performed after the stem is repositioned at the center of the cup. Leg length is the same as that of the unaffected side, based on the height of the apex of the lesser trochanter. The global offset (red arrow) is not smaller than that on the unaffected side. The depth of the stem from the apex of the greater trochanter and height from the lesser trochanter are measured to facilitate intraoperative confirmation

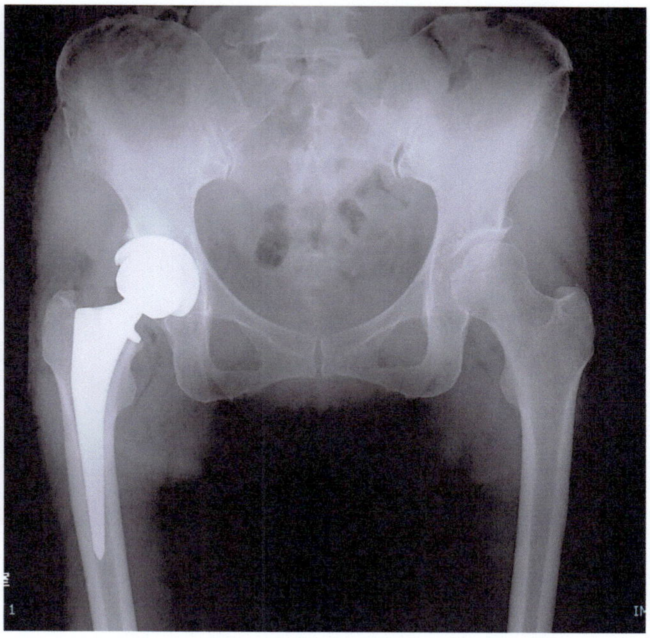

Fig. 48.8 Postoperative pelvic radiograph. The acetabular and femoral components were placed according to the preoperative plan

(d) Adjustment of leg length and global offset

Preoperative leg length discrepancy is measured in pelvic radiographs. First, the trans-teardrop line is drawn, and bilateral anatomical landmarks. Such as the apex of the lesser trochanter, are marked. Finally, the vertical distance between the trans-teardrop line and the apex of the lesser trochanter is measured. The difference between these distances in both hips shows the preoperative leg length discrepancy [11]. Patient-perceived leg length discrepancy is measured by placing an insert of a known thickness under the shorter leg. If the leg length discrepancy on the radiograph is almost equal to the patient-perceived leg length discrepancy, the leg lengthening that should be corrected surgically is the same. However, if the radiographic leg length discrepancy and patient-perceived leg discrepancy are significantly different, spinal degeneration or adduction or abduction contracture of the hip joint should be considered.

Global offset is defined as the horizontal distance between the central axis of the femur and the medial wall of the teardrop or the central pelvic axis connecting the pubic symphysis and the bilateral sacroiliac joint center, while femoral offset is defined as the horizontal distance between the center of the femoral head and the central axis of the femur [24, 25]. Global offset is measured using the unaffected side (or preoperative) as a reference. A large global offset increases abductor muscle strength and stability; however, an excessively large global offset restricts the range of adduction due to abductor tightness and greater trochanteric pain owing to bursitis. On the other hand, decreased offset has more negative effects, such as an increased risk of dislocation, decreased abductor muscle strength, increased limp, and impingement. In particular, a decrease in global offset greater than 5 mm is reported to significantly reduce abductor muscle strength and increase the risk of developing a limp (Fig. 48.8). [26].

The head should be selected to reproduce the leg lengths and offsets described above; however, it is desirable to avoid selecting the shortest or longest neck in preoperative planning to preserve intraoperative options. If necessary, changes in stem size should be considered.

2 Three-Dimensional Templating

Recently, three-dimensional templating based on CT has become widely used. Although three-dimensional templating is not the gold standard owing to radiation exposure caused by CT, some reports indicate that preoperative CT for THA to depict a limited area is associated with a slight and justifiable increase in radiation exposure in comparison to conventional radiographs, and its cost is not very high [27]. The advantages of three-dimensional templating include the following: (1) to determine the acetabular cup diameter and position considering the three-dimensional bony contact of the acetabulum; (2) to select the femoral stem design and size considering the femoral version and proximal femur geometry; (3) to obtain stable primary fixation using an appropriate stem design; (4) to obtain accurate leg lengthening, femoral offset, and global offset after surgery; and (5) to

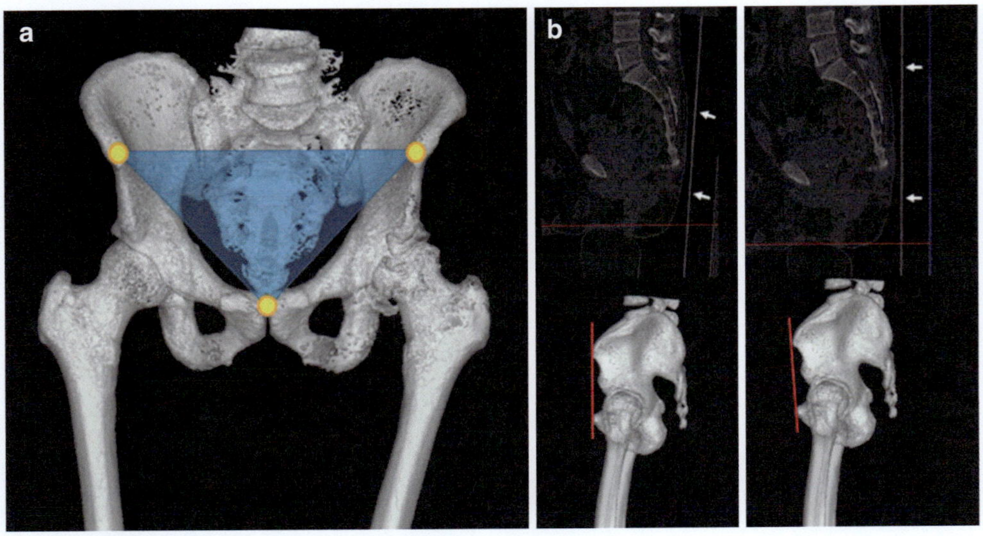

Fig. 48.9 Pelvic coordinate systems for three-dimensional templating. (**a**) Anterior pelvic plane (APP) comprising the plane connecting the bilateral anterior superior iliac spines and the midpoints of the bilateral pubic tuberosities (blue triangle). (**b**) Coordinate system based on APP. The CT table (white arrow) is tilted to make the APP (red line) vertical. (**c**) Coordinate system based on FPP. The CT table (white arrow) is vertical, and the pelvic tilt is measured by the tilt of the APP (red line). In this case, the pelvis was tilted by 4° anteriorly

prevent impingement and maximize postoperative range of motion [6, 24, 28]. In addition, visualization of the relationship between the acetabulum and cup and between the entry point of the femoral stem and osteotomized neck is very effective as a preoperative simulation.

(a) Setting of the pelvic and femur coordinate systems

Setting the pelvic and femur coordinate systems is an essential procedure for three-dimensional templating. There are two main pelvic coordinate systems in the hip joint: the anterior pelvic plane (APP), which is an anatomical coordinate system based on the plane created with the bilateral anterior superior iliac spines and the midpoint of the bilateral pubic tuberosities, and the functional pelvic plane (FPP), which is a functional coordinate system based on the pelvis in the supine position [29].

The coordinate system used for three-dimensional cup templating is the FPP that reproduces the pelvic tilt in the supine position unique to each patient (Fig. 48.9). This is because the pelvic tilt in the supine position does not change significantly over 10 years after THA, is stable compared to the pelvic tilt in the standing position, and is less affected by spinal fusion surgery [30]. In contrast, the femoral coordinate system is based on the posterior condylar plane, which is formed by the medial and lateral posterior condyles of the femur and posterior point of the greater trochanter (Fig. 48.10).

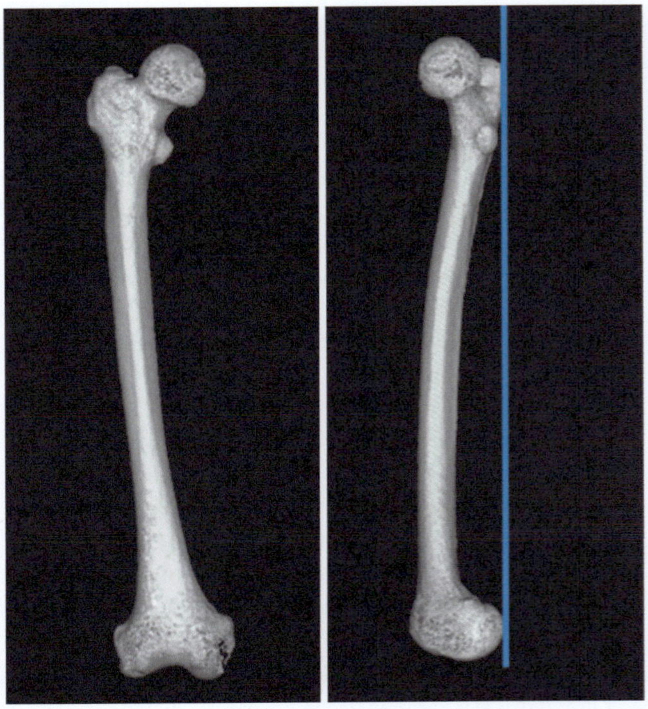

Fig. 48.10 Femur coordinate system based on the posterior condylar plane. The femur coordinate system is based on the plane formed when the femur is placed on the table (the plane composed of the medial and lateral posterior condyles and posterior aspect of the greater trochanter, indicated by the blue line)

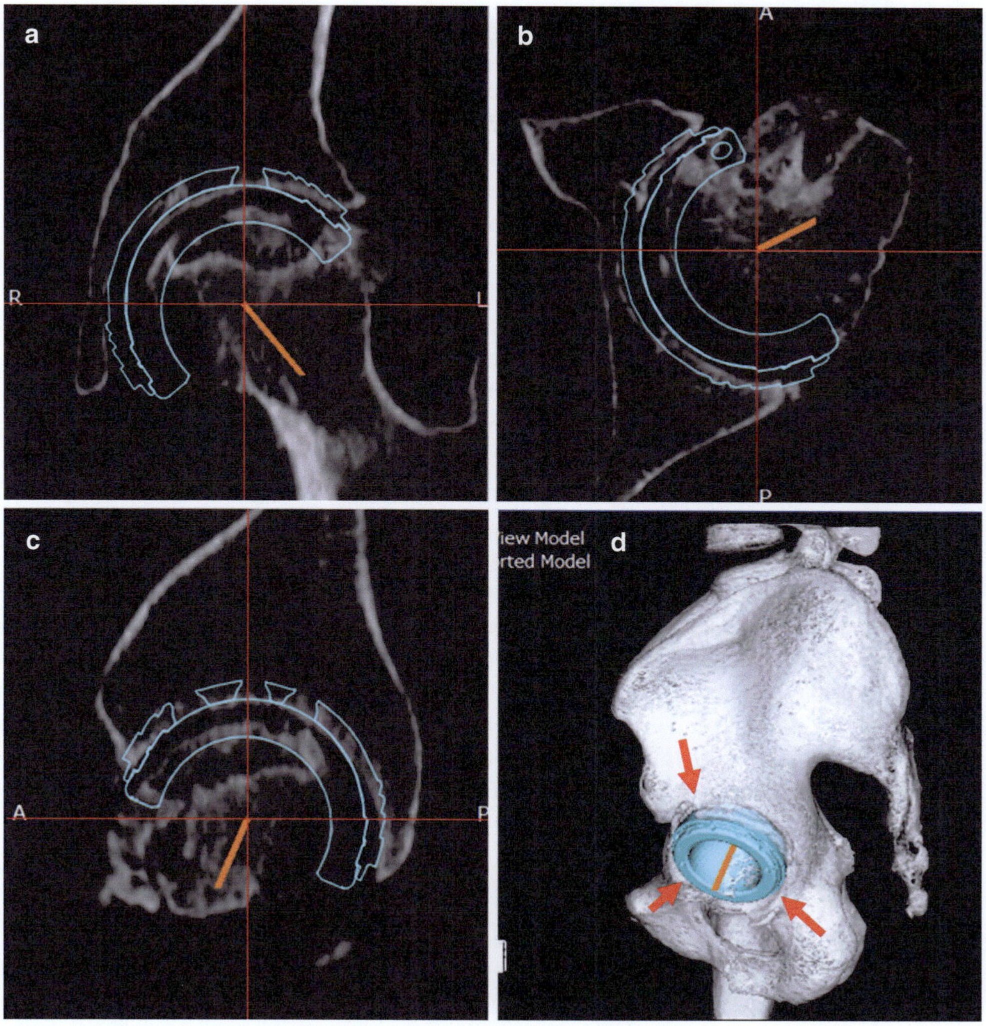

Fig. 48.11 Cup templating in three-dimensional templating. (**a**) Coronal plane. (**b**) Axial plane. In this case, the anteversion angle is 15°. (**c**) Sagittal plane. (**d**) Three-dimensional image. Press fit was obtained at three points around the anterior inferior iliac spine, pubic bone base, and ischial bone (red arrows)

(b) Acetabular cup templating

The cup size was determined to obtain an interference fit between the anterior and posterior walls of the acetabulum in the axial plane at the height of the native hip center. The accuracy of the cup size with three-dimensional templating is greater than 90% [31]. Regarding the position of the cup, as in two-dimensional templating, placing the cup superiorly (by up to 2 cm) according to the cup CE angle is acceptable to obtain a sufficient contact area with the acetabular bone. Three-dimensional acetabular cup templating allows simultaneous viewing in the coronal, axial, and sagittal planes. The cup position is adjusted in detail with medialization and elevation, taking care not to penetrate the medial wall of the acetabulum in each plane (Fig. 48.11).

Press fit fixation is a secure initial fixation method for cementless acetabular cups. Press fit fixation needs to be in contact with three points: the iliac bone, pubic bone, and ischial bone [32]. In cases where the cup CE angle is greater than 10°, the acetabular cup rim is fixed at these three points [17].

The Lewinneck's safe zone is a goal for acetabular cup inclination and anteversion; however, this safe zone alone cannot prevent dislocations, especially in cases with highly rotational malalignments of proximal femurs [33]. Therefore, the combined anteversion theory of Widmer et al. (which aims to ensure that the sum of the cup and stem anteversion is a constant value), mathematically examines the position of the cup and stem to avoid implant impingement and to obtain a sufficient range of motion, and has become widely accepted more recently [20]. The combined anteversion theory significantly reduces the dislocation rate [34] and a simplified and refined combined anteversion theory has been reported [35].

However, the combined anteversion theory is not applicable to cases of extremely high femoral anteversion. For example, when femoral anteversion is greater than 60°, anterior dislocation can occur due to posterior bone-implant impingement, even if cup anteversion is reduced. Therefore, in these extreme cases, a modular stem or changeable neck stem adjustable to femoral anteversion should be considered.

(c) Femoral stem templating

The goal of three-dimensional femoral stem templating, as in two-dimensional templating, is to restore the center of the femoral head. First, the stem design was selected with reference to proximal hip geometry and bone quality. Next, stem size was determined according to the proximal metaphyseal shape for the proximal fixation stem or the diaphyseal shape for the distal fixation stem in the coronal plane. Stem alignment should be appropriate for the stem fixation concept, and varus or valgus malalignment in the coronal plane and flexion or extension malalignment in the sagittal plane should be avoided. Usually, the desired alignment is in a neutral position.

In the axial plane, anteversion of the femoral stem should be adjusted to the anteversion of the proximal femur at the neck-cut level (Fig. 48.12). In a tapered wedge stem, femoral anteversion can be adjusted to some extent because of its thin shape in the proximal part. On the other hand, in the fit-and-fill type stem, femoral anteversion cannot be changed because its proximal body fills the femoral canal in the coronal and sagittal planes to achieve high canal occupancy [36]. However, in both cases, it is difficult to control stem anteversion at will because stem anteversion is affected by stem sagittal alignment (flexion or extension) [37]. If the stem anteversion is less than 10° or greater than 50° in preoperative planning, a modular stem, conical stem, or changeable neck stem should be considered. The accuracy of the three-dimensional femoral templating was 86%, which is slightly less than that of the cup. This is because the actual stem anteversion or sagittal alignment is often slightly different from the preoperative plan. Creating a cross-sectional view of the femur at the neck-cut level would be helpful for the insertion of the stem into the appropriate entry point (Fig. 48.12).

(d) Leg length and offset adjustment

One of the advantages of three-dimensional templating is the visibility of the global offset. Because the center of rotation of the hip is often more medial than that preoperatively, the femoral offset should be greater than the preoperative offset to maintain the global offset. The global offset based on three dimensional templating is accurate for estimating impingement because it considers femoral version [24]. In addition, three-dimensional templating is helpful in confirming leg lengthening after THA with aligned adduction and internal rotation of the bilateral hips (Fig. 48.13).

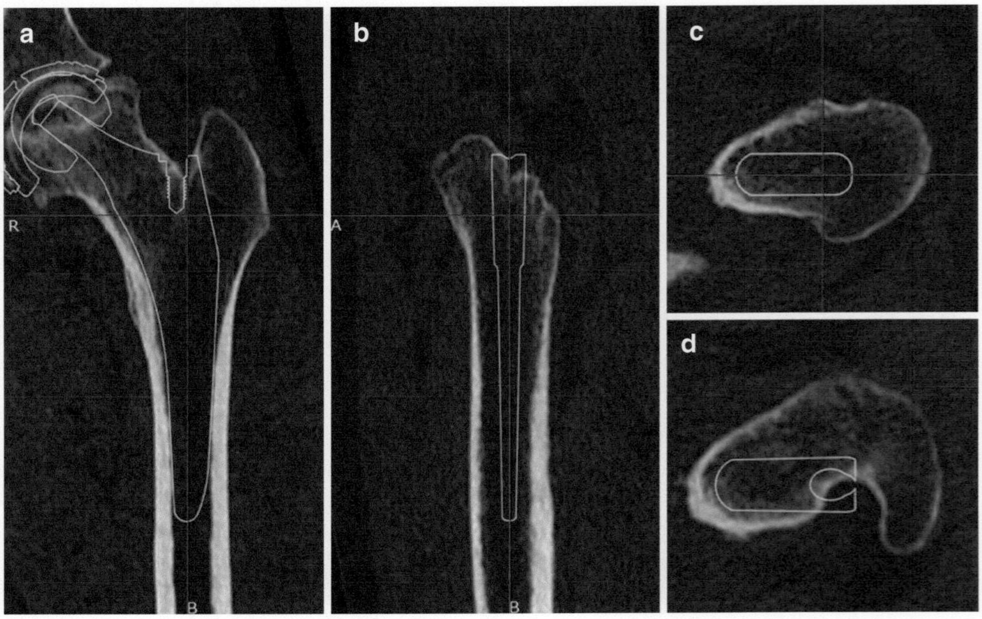

Fig. 48.12 Stem templating in three-dimensional templating. (**a**) Coronal plane. In this case, proximal fixation was planned and good contact with the metaphysis was obtained. (**b**) Sagittal plane. The neutral position is planned preoperatively. (**c**) Axial plane. The medial side of the stem (particularly the posterior medial side) was almost in contact with the bone cortex. (**d**) Cross-sectional plane of the neck cut. It is useful for determining the entry point of the femoral stem

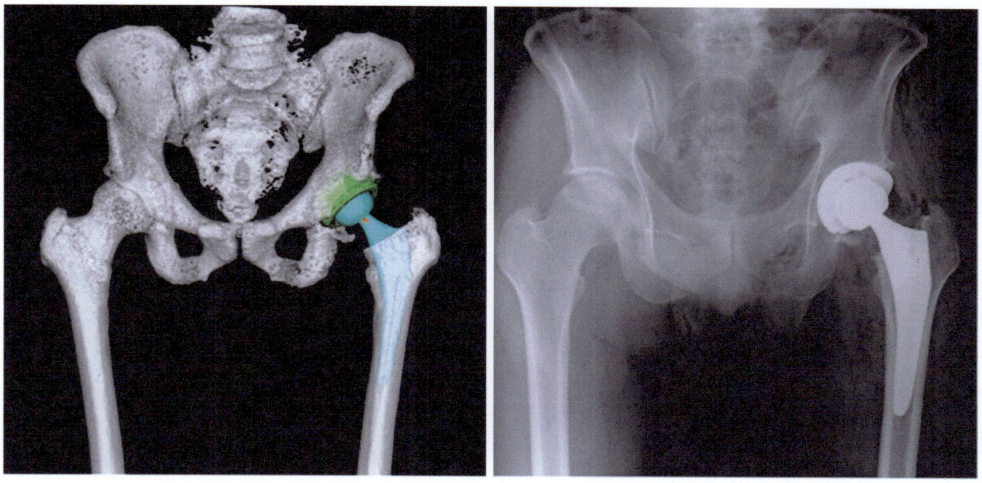

Fig. 48.13 Three-dimensional preoperative templating and postoperative pelvic radiograph. The inclination and anteversion of the cup and the stem's anteversion are restored during preoperative planning. Leg lengthening and global offset were also in good agreement with the preoperative planning

References

1. Bono JV. Digital templating in total hip arthroplasty. J Bone Joint Surg Am. 2004;86-A Suppl 2:118–22.
2. Oddy MJ, et al. Assessment of reproducibility and accuracy in templating hybrid total hip arthroplasty using digital radiographs. J Bone Joint Surg Br. 2006;88:581–5.
3. Gamble P, et al. The accuracy of digital templating in uncemented total hip arthroplasty. J Arthroplast. 2010;25:529–32.
4. Petretta R, et al. Acetate templating on digital images is more accurate than computer-based templating for total hip arthroplasty. Clin Orthop Relat Res. 2015;473:3752–9.
5. Iorio R, et al. A comparison of acetate versus digital templating for preoperative planning of total hip arthroplasty: is digital templating accurate and safe? J Arthroplast. 2009;24:175–9.
6. Sugano N, et al. Computed-tomography-based computer preoperative planning for total hip arthroplasty. Comput Aided Surg. 1998;3:320–4.
7. Clohisy JC, et al. A systematic approach to the plain radiographic evaluation of the young adult hip. J Bone Joint Surg Am. 2008;90(Suppl 4):47–66.
8. Hananouchi T, et al. Preoperative templating of femoral components on plain X-rays. rotational evaluation with synthetic X-rays on ORTHODOC. Arch Orthop Trauma Surg. 2007;127:381–5.
9. Inaba Y, et al. Preoperative planning for implant placement with consideration of pelvic tilt in total hip arthroplasty: postoperative efficacy. BMC Musculoskelet Disord. 2016;17:280. evaluation
10. Bayne CO, et al. Evaluation of the accuracy and use of x-ray markers in digital templating for total hip arthroplasty. J Arthroplast. 2009;24:407–13.
11. Meermans G, et al. Preoperative radiographic assessment of limb-length discrepancy in total hip arthroplasty. Clin Orthop Relat Res. 2011;469:1677–82.
12. Charnley J. Low friction principle. In: Charnley J, editor. Low friction arthroplasty of the hip: theory and practice. Heidelberg: Springer; 1979. p. 3–15.
13. Fujii M, et al. Can the hip joint center be estimated from pelvic dimensions in dysplastic hips? J Orthop Sci. 2017;22:1089–95.
14. Fukui K, et al. A radiological study of the true anatomical position of the acetabulum in Japanese women. Hip Int. 2011;21:311–6.
15. Sugano N, et al. Polyethylene sockets and alumina ceramic heads in cemented total hip arthroplasty. A ten-year study. J Bone Joint Surg Br. 1995;77:548–56.
16. Fujii M, et al. Minimum lateral bone coverage required for securing fixation of cementless acetabular components in hip dysplasia. Biomed Res Int. 2017;2017:4937151–7.
17. Takao M, et al. The results of a press-fit-only technique for acetabular fixation in hip dysplasia. J Arthroplast. 2011;26:562–8.
18. Delp SL, et al. How superior placement of the joint center in hip arthroplasty affects the abductor muscles. Clin Orthop Relat Res. 1996;328:137–46.
19. Lewinnek GE, et al. Dislocations after total hip-replacement arthroplasties. J Bone Joint Surg Am. 1978;60:217–20.
20. Widmer K-H, et al. Compliant positioning of total hip components for optimal range of motion. J Orthop Res. 2004;22:815–21.
21. Sugano N, et al. Predicting the position of the femoral head center. J Arthroplast. 1999;14:102–7.
22. Yamaguchi T, et al. Total hip arthroplasty: the relationship between posterolateral reconstruction, abductor muscle strength, and femoral offset. J Orthop Surg (Hong Kong). 2004;12:164–7.
23. Asayama I, et al. Reconstructed hip joint position and abductor muscle strength after total hip arthroplasty. J Arthroplast. 2005;20:414–20.
24. Lecerf G, et al. Femoral offset: an anatomical concept, definition, assessment, and implications for preoperative templating and hip arthroplasty. Orthop Traumatol Surg Res. 2009;95:210–9.
25. Kjellberg M, et al. A new radiographic method of measuring femoral offset. the Sundsvall method. Hip Int. 2009;19:377–81.
26. Mahmood SS, et al. Association between changes in global femoral offset after total hip arthroplasty and function, quality of life, and abductor muscle strength. A prospective cohort study of 222 patients. Acta Orthop. 2016;87:36–41.
27. Huppertz A, et al. Computed tomography for preoperative planning in minimal-invasive total hip arthroplasty: radiation exposure and cost. Eur J Radiol. 2011;78:406–13.
28. Sato Y, et al. Intraoperative simulation and planning using a combined acetabular and femoral (CAF) navigation system for total hip replacement. In: Delp SL, et al., editors. Medical image computing and computer-assisted intervention – MICCAI 2000. Heidelberg: Springer; 2000. p. 1114–25.
29. Takao M, et al. Pelvic and femoral coordinates and implant alignment representations in THA. In: Sugano N, editor. Computer assisted orthopaedic surgery for hip and knee: Current state of the art in clinical application and basic research. Singapore: Springer; 2018. p. 75–88.
30. Tamura S, et al. Does pelvic sagittal inclination in the supine and standing positions change over 10 years of follow-up after total hip arthroplasty? J Arthroplast. 2017;32:877–82.

31. Ogawa T, et al. Factors related to disagreement in implant size between preoperative CT-based planning and the actual implants used intraoperatively for total hip arthroplasty. Int J Comput Assist Radiol Surg. 2018;13:551–62.
32. Widmer K-H, et al. Load transfer and fixation mode of press-fit acetabular sockets. J Arthroplast. 2002;17:926–35.
33. Matsushita A, et al. Modular necks improve the range of hip motion in cases with excessively anteverted or retroverted femurs in THA. Clin Orthop Relat Res. 2010;468:3342–7.
34. Nakashima Y, et al. Combined anteversion technique reduced the dislocation in cementless total hip arthroplasty. Int Orthop. 2014;38:27–32.
35. Sugano N, et al. Dynamic measurements of hip movement in deep bending activities after total hip arthroplasty using a 4-dimensional motion analysis system. J Arthroplast. 2012;27:1562–8.
36. Taniguchi N, et al. Cementless hip stem anteversion in the dysplastic hip: a comparison of tapered wedge vs metaphyseal filling. J Arthroplast. 2017;32:1547–52.
37. Imai H, et al. Preoperative planning and operative techniques of the shorter tapered stem compared to the metaphyseal fit stem in cementless total hip arthroplasty. J Arthroplast. 2017;32:1192–9.

Risk Assessment (Comorbidities)

49

Gaku Koyano

1 Risk Assessment by Medical Interview

(a) Current medical history

The most important risk factor in the current medical history is a history of previous surgical treatment, such as osteotomy or myotomy. If possible, the previous surgical site and technique, including the surgical approach should be confirmed. If the surgery was performed at another hospital, the details of the surgery, including imaging information, should be ascertained.

(b) History, comorbidities

Examples of commonly asked questions related to medical history and comorbidities are shown in Table 49.1.

(c) Regularly used drugs

Regularly used medications may need to be discontinued or changed during the perioperative period, or additional prescriptions may be needed. Certain dietary supplements may also affect treatment (Table 49.2). Consult the relevant department regarding the duration of drug withdrawal, alternative medications, and measures to take in the event of symptom worsening.

(d) Previous anesthesia or surgery

This comprises previous surgical history and adverse events (malignant hyperthermia, drug allergy, etc.).

(e) Smoking and drinking history

Smokers have significantly longer hospital stays, higher total hospitalization costs, and significantly higher postoperative complications and mortality than non-smokers [1]. A meta-analysis examining the relationship between the duration of preoperative smoking cessation and postoperative complications showed that postoperative complications decreased with each week of smoking cessation, with a statistically significant difference at 4 weeks [2]. Therefore, it is recommended that patients stop smoking even for a short period of time. Although there is currently insufficient evidence to determine the safety and efficacy of e-cigarettes in perioperative smoking cessation, smoking cessation should be recommended in the same manner as regular cigarettes [3].

Heavy alcohol consumption (>60 g/day of pure alcohol) increases the risk of perioperative complications, including delirium [4]. Alcohol abstinence for 4–8 weeks before surgery decreases the risk of perioperative complications (wound infection, respiratory and circulatory complications, and ICU admis-

G. Koyano (✉)
Department of Orthopedic Surgery, Dokkyo Medical University Saimtama Medical Center, Koshigaya, Saitama, Japan
e-mail: g-koyano@dokkyomed.ac.jp

Y. Nakashima (ed.), *Advances in Total Hip Arthroplasty*, https://doi.org/10.1007/978-981-97-2461-1_49

Table 49.1 Examples of medical history and comorbidities that require consultation with other departments

Disease	Medical history and comorbidities
Circulatory system disease	Ischemic heart disease, valvular disease, heart failure, pacemaker, history of cardiovascular surgery, etc.
Respiratory disease	Bronchial asthma, chronic obstructive pulmonary disease, interstitial pneumonia, tuberculosis, home oxygen therapy, etc.
Renal disease	Chronic renal failure, with or without dialysis [40]
Liver disease	Abnormal liver function due to cirrhosis, etc.
Metabolic and endocrine disorders	Diabetes, thyroid disease, hormone replacement therapy for pituitary disease, etc.
Rheumatoid arthritis and collagen disease	Anti-rheumatic drugs, immunosuppressive drugs, steroid use, etc.
Malignant tumor	The patient's general condition and the presence or absence of metastasis to the surgical site must be confirmed
Allergy	Drug allergy, hypersensitivity to antiseptics, metal allergy, etc.
Hematological and hematopoietic diseases	Check for the presence of coagulation abnormalities, including the presence or absence of hemorrhagic tendency.
Neuromuscular disease	Epilepsy, spastic paralysis, etc.
Obesity	BMI>35

Table 49.2 Regular medications to be aware of in the perioperative period

Drugs	Check points
Anticoagulant, antiplatelet agent	Withdrawal of medication to prevent complications due to bleeding (note duration of withdrawal). Heparinization if withdrawal is not possible
Hormone drug	Female hormone preparations are associated with a high risk of thrombosis. Be aware that this includes osteoporosis drugs (SERMs) and oral contraceptives.
Antineoplastic drug	Molecularly targeted drugs and other drugs have the risk of delayed wound healing [41]
Anti-rheumatic drug	Withdrawal periods vary depending on the drug [42]
Immunosuppressant (drug)	Immunosuppressive effect may increase the risk of infection
Steroid	Beware of adrenal crisis due to perioperative drug withdrawal. Consider steroid coverage depending on duration and amount of oral medication
Diabetes drug	Patients with type 1 diabetes and other low autologous insulin secretion should be careful. Withdraw biguanides due to risk of lactic acidosis. Be careful as it may be contained in compounded products. Sliding scale with insulin for response
Antihypertensive	Discontinuation may worsen the condition
Antiparkinsonian drug	Use of intravenous levodopa if it cannot be taken orally in the perioperative period
Plaster	The patch using conductive metal as a support may cause burns on the applied area when an electrocautery is used.
Supplements	Fish oil (EPA/DHA), ginseng, ginger, garlic: Inhibition of platelet aggregation. Saw palmetto: Cyclooxygenase inhibitor. Ginkgo biloba: Suppression of platelet-activating factor. Kava, Valerian, St. John's Wort: Sedative

sion). Alcohol withdrawal symptoms begin at six to 48 hours and continue for approximately 1 week [5]. Patients are evaluated using the Clinical Institute Withdrawal Assessment for Alcohol Scale score, and benzodiazepines are administered prophylactically if the score is eight or higher [5].

2 Risk Assessment by Preoperative Examination

Preoperative examination generally includes blood and urine tests for the items listed in Table 49.3; however, laboratory tests should be performed according to the standards of each institution. Additional laboratory tests should be performed according to the patient's medical history to confirm that the comorbidities currently being treated are under control.

Infectious disease testing is important to ensure the safety of medical staff and other hospitalized patients. Traditionally, testing for HBV, HCV, HIV, and syphilis has been performed, and since the coronavirus disease 2019 (COVID-19) pandemic in 2020, more hospitals have been performing polymerase chain reaction testing for COVID-19 before admission.

Table 49.3 Examples of preoperative examinations

Examinations	Cases	Evaluation contents/applications
Blood and urine tests	All cases	Anemia, bacterial infection, inflammatory disease, renal dysfunction, hepatic dysfunction, easy bleeding, thrombosis, blood group, etc.
ECG	All cases	Circulatory disease
Echocardiography	Electrocardiogram abnormalities, elderly patients, patients with a history of cardiac disease, etc.	Cardiac function
Respiratory function test	Patients with respiratory disease, elderly, heavy smokers, etc.	Restrictive disease, obstructive disease, etc.
Simple X-ray image of the hip joint	All cases	Preoperative planning
Simple X-ray image of lumbar spine	All cases	Lumbar spine disease, see lumbar anesthesia
Simple X-ray image of the chest	All cases	Respiratory disease, circulatory disease, see intubation
Lumbar spine functional radiographs, full length lower extremity radiographs	Lumbar scoliosis, large leg length discrepancy, etc.	Evaluation of leg lengthening
Bone mineral density	Osteoporosis cases	Lumbar spine, femoral neck
CT	All cases	Bone geometry, lesion identification, 3D preoperative planning, and preparation for navigation system
MRI	Pyogenic hip arthritis, osteonecrosis of femoral head, subchondral fragile bone fracture, tumor, etc.	Identification of bone geometry, tumors, inflammation and other lesions
Infectious disease	All cases	HBV, HCV, HIV, syphilis, COVID-19, etc.

3 Preoperative Evaluation of Comorbidities and Disorders

(a) Evaluation of the whole body

Consultations with the relevant department and anesthesiology department should be made based on the medical history, both present and past, comorbidities, and test results. The duration of surgery, position (supine/lateral decubitus), amount of blood loss (e.g., preparation for autotransfusion), and anesthesia (general anesthesia/lumbar anesthesia) should be described in detail.

The American Society of Anesthesiologists (ASA) physical status classification (Table 49.4) and Rx-risk score are predictive indicators of serious perioperative complications and have been shown to be highly correlated with 30-day mortality after THA [6]. The complication and mortality rates after non-cardiac surgery in patients with ASA $\geq$ 3 were significantly higher than those in patients with ASA 1 or 2 [7].

Patients with limited range of motion or instability of the cervical spine are expected to have difficulty with intubation. Patients with limited mouth opening due to temporomandibular joint disorders also experience difficulty with intubation. Both conditions are common in patients with rheumatoid arthritis; therefore, and special attention should be given. Thus, early consultation with anesthesiologists is necessary.

If a patient has a malignant tumor that is currently being treated, we inquire about the patient's general condition, prognosis, whether surgery is indicated, and perioperative precautions.

(b) Cardiac function assessment

Cardiac function is a predictor of perioperative and long-term cardiovascular events, and ejection fraction is commonly assessed by echocardiography. An ejection fraction of <40% has been reported to increase mortality in noncardiac surgery, requiring caution [8]. The 2014 edition of the American College of Cardiology (ACC) and American Heart Association (AHA) guidelines use metabolic equivalents (METs) to assess cardiac reserve [8]. Metabolic equivalents are accurately determined by exercise testing but can also be estimated using a simple questionnaire (Duke Activity Status Index) [9]. The ACC and AHA guidelines allow surgery without further testing if the current metabolic equivalents are 4 METs or higher, and allow evaluation of myocardial ischemia if the current metabolic equivalents are 4 METs or lower. For example, playing a musical instrument and walking at a speed of approximately 3–5 km/h should be considered to be <4 METs, whereas climbing stairs, walking on a slope, and walking at a speed of 6 km/h on level ground should be considered to be >4 METs.

(c) Respiratory function assessment

Although hip arthroplasty is not considered a risk factor for postoperative respiratory complications [10], it is impor-

Table 49.4 American Society of Anesthesiologists (ASA) physical status classification

Class	Definition	Adult examples
I	A normal healthy patient	Healthy, nonsmoking, no alcohol use
II	A patient with mild systemic disease	Current smoker, social alcohol drinker, pregnancy, obesity (30<BMI<40), well-controlled diabetes mellitus/hyper tension mild lung disease
III	A patient with severe systemic disease	Poorly controlled diabetes mellitus or hyper tension, COPD, morbid obesity (BMI ≥ 40), active hepatitis, alcohol dependence or abuse, implanted pacemaker, moderate reduction of ejection fraction, end-stage renal disease undergoing regularly, MI, cerebral vascular accident, transient ischemic attack, coronary artery disease/stents (> 3 months).
IV	A patient with severe systemic disease that is a constant threat to life	Recent (< 3 months) MI, cerebral vascular accident, transient ischemic attack, coronary artery disease/stents, ongoing cardiac ischemia or severe valve dysfunction, severe reduction of ejection fraction, shock, sepsis, DIC, acute respiratory disease, not undergoing regularly scheduled dialysis.
V	A moribund patient who is not expected to survive without the operation	Ruptured abdominal/thoracic aneurysm, massive trauma, intracranial bleed with mass effect, ischemic bowel in the face of significant cardiac pathology or multiple organ/system dysfunction.
VI	A declared brain-dead patient whose organs are being removed for donor purposes.	–

Adapted in part from Horvath et al. [7]

tant to evaluate respiratory function associated with general anesthesia. Evaluation using spirometry should be performed in the presence of comorbidities and smoking habits. Generally, %FEV1 (forced expiratory volume in 1 s) <70% is considered obstructive ventilatory impairment and %VC (vital capacity) <80% is considered restrictive impairment.

(d) Liver function evaluation

The Child-Pugh classification is used to evaluate liver function in cirrhosis, but its use in clinical practice may be difficult for orthopedic surgeons because of the need to subjectively evaluate ascites and encephalopathy, which orthopedic surgeons are not familiar with, and it is practical to ask an internist to perform the evaluation. The model for end-stage liver disease (MELD) score [11] is based on laboratory values and can be evaluated by orthopedic surgeons.

MELD score = 11.2 × loge (PT-INR) + 9.57 × loge (serum creatinine level (mg/dL)) + 3.78 × loge (serum total bilirubin level (mg/dL)) + 6.43 × (constant)

Each test value is calculated as 1 if it is less than 1 mg/dL, and the serum creatinine value is calculated as 4 mg/dL in patients undergoing hemodialysis more than twice a week.

Constant: 0 for alcoholic or biliary congestion, 1 otherwise

Cirrhotic patients are reported to have a longer hospital stay for THA and TKA, discharge to a nursing home, readmission within 90 days, urinary tract infection, renal failure, blood transfusion, gastrointestinal bleeding, dislocation, infection, and higher 1-year and long-term mortality [12]. If the MELD score is 10 or higher, the risk of post-THA and TKA complications is three times higher, and the risk of death is 4.1 times higher [12].

(e) Renal function assessment

Increased mortality and complications after revision surgery have been reported in patients with poor renal function compared with those in non-infected cases [13]. Renal function is often assessed using creatinine clearance (CCr) or estimated glomerular filtration rate (eGFR), which can be calculated using the Cockcroft-Gault formula for CCr and the modification of diet in the renal disease (MDRD) formula for eGFR. CCr can be calculated using the Cockcroft-Gault formula.

CCr (male) = (140 − age) × weight/(72 × serum creatinine level),

CCr (female) = (140 − age) × weight × 0.85/(72 × serum creatinine level)

eGFR cre (male) = 194 × (serum creatinine level (mg/dL)) − 1.094 × (age) − 0.287.

eGFR cre (female) = eGFR cre (male) × 0.739

Perioperative fluid control and dose of drug should be considered after consultation with a nephrologist, in patients with severely impaired renal function such as a CCr < 30 mL/min. Dialysis should be considered if necessary.

(f) Allergy

Patients with a history of drug allergies, sensitivity to antiseptics, or metal allergies should consult a dermatologist for further evaluation. Patients allergic to mango, avocado, banana, kiwi, papaya, fig, melon, etc. are more likely to develop latex allergies and require caution during anesthesia and surgery.

Although there are no reports of randomized trials for metal allergies, good postoperative outcomes have been reported in patients with implants that do not contain the offending metal [14, 15]. Therefore, preoperative evaluation is important. Although patch tests and lymphocyte stimulation tests are commonly used to test for metal allergies, patch tests are most commonly used because of their simplicity [16]. Although it is not recommended to routinely screen all patients for metal allergies because of false-positive and false-negative results, the patch test is performed in patients with a history of metal allergies to jewelry [16, 17]. Confirmation with the manufacturer of implants is necessary regarding the trace metal content. Cementless implants are generally made of titanium alloys (Ti-6Al-4 V) and patients should be tested for titanium, aluminum, and vanadium. Metal heads are alloys of cobalt, chromium, and nickel, which are frequently allergic. Ceramic heads are mainly made of mixed oxides of aluminum and zirconium and may contain trace elements; however, in general, oxides are chemically stable, and metal ions do not leach out easily, making them less likely to cause allergies.

(g) Thromboembolism

There is no evidence regarding whether preoperative deep venin thrombosis (DVT) screening with D-dimer measurement or lower-extremity venous echocardiography can reduce symptomatic or fatal PTE, and the Guidelines for Prevention of Symptomatic Venous Thromboembolism 2017 neither explicitly recommend nor reject preoperative testing [18]. However, preoperative evaluation is recommended for patients with risk factors for thrombosis such as a history of thrombosis, malignancy, or antiphospholipid antibody syndrome. Although there are many false-positive D-dimer values in preoperative blood tests and no clear cutoff values have been shown, D-dimer values have been reported to be higher in patients with positive preoperative lower-extremity ultrasound for thrombosis than in those with negative results [19]. If D-dimer levels are high, lower-extremity venous echo or contrast-enhanced CT should be considered. The physical findings of DVT include subjective symptoms such as pain, edema, and swelling in the affected limb, and other findings such as the Homans sign (peroneal pain induced by dorsiflexion of the ankle joint) and the Lowenberg sign (grasping pain in the perineal region). These findings are often subclinical and should be treated as supplemental method for diagnosis.

(h) Abnormal coagulation

Patients with coagulation disorders, such as thrombocytopenia, hemophilia, and congenital plasminogen activator inhibitor 1 deficiency, are at a high risk of intraoperative bleeding. If platelets or coagulation factor products can be replenished in advance, bleeding can be managed [20–22]. Additionally, it is necessary to prepare more blood for transfusion than usual. Epsilon-aminocaproic acid and tranexamic acid are also effective in preventing bleeding [23].

4 Comorbidities Requiring Attention for Hip Joint Prosthesis Placement

(a) Bone fragility

Bone fragility is associated with an increased risk of intraoperative and postoperative fractures and loosening of prostheses owing to poor bone quality. Comorbidities that may indicate bone fragility include osteoporosis, rheumatoid arthritis, steroid use, and previous radiotherapy. Bone fragility increases the risk of intraoperative fractures; therefore, the selection of a stem (cementless or cemented, proximal or distal fixation, etc.), wiring plates, and support rings should be considered. In addition, it has been reported that osteoporosis drugs such as alendronate, suppress the decrease in bone mineral density due to postoperative stress shielding [24, 25] and are expected to be effective in the prevention of long-term fractures.

(b) Obesity

Although obese patients are prone to inaccurate cup placement [26], it is possible to place the cup with the same accuracy as non-obese patients using CT navigation [27]. Severely obese patients are at a high risk of soft tissue impingement and dislocation [28]; therefore, measures to prevent dislocation, as described below, should be taken. In addition, it has been reported that the risk of avulsion fracture of the greater trochanter is three times higher in supine surgery [29], even in the absence of bony fragility.

(c) Short stature

Short stature is caused by a variety of factors, including endocrine disorders, chromosomal abnormalities, bone system diseases, inborn metabolic errors, chronic diseases, and nutritional disorders. Many patients with short stature have accompanying bone and joint deformities, and THA is indicated in a small number of patients at a young age [30]. THA in patients with short stature makes it difficult to match the leg length of the operative side with that of the healthy side [31]. It has also been reported that the risk of intraoperative fracture is high owing to advanced deformity, poor implant fit, and poor bone quality [30, 32]. On the other hand, good results have been reported for THA using conventional implants [33]. In actual surgery, it is advisable to perform a detailed study using CT and three-dimensional preoperative planning to control leg length and evaluate the suitability of the implants.

(d) Metastatic bone tumors

In case of metastatic tumors, a conventional hip prosthesis may be placed after surgical curettage of the metastatic lesion to improve symptoms and activity without radical surgery. If the metastatic lesion extends beyond the metaphyseal end, it is resected in a single lump, and a tumor prosthesis is used. In the cases of metastasis on the acetabular side, the tumor is resected and a cup is placed on the sacrum [34] or curettage is performed and fixed with a cement cup or support ring [35]. In addition, metastatic bone tumors, especially renal cell carcinoma, may bleed frequently, and adequate blood transfusions and hemostatic devices should be prepared. In the case of osteogenic tumors such as breast and prostate cancers, rasping may be difficult.

(e) Infection

In the presence of active pyogenic arthritis of the hip joint, primary THA is contraindicated because the infection is difficult to control and may not heal or may worsen after the placement of a hip prosthesis. Infection of the hip prosthesis can also cause osteolysis, leading to decreased fixation and loosening of the prosthesis. The presence of infection is evaluated using a combination of local findings, history, blood tests (inflammatory response, erythrocyte sedimentation rate), culture studies, and imaging studies (MRI, CT, bone scintigraphy, inflammation scintigraphy, etc.).

(f) Cases with a high risk of dislocation

Patients at high risk of postoperative dislocation include those whose preoperative plan does not allow for leg lengthening, those with a large preoperative range of motion, and those with severe obesity. Patients with lumbar spine disease or postoperative lumbar spine deformity, reduced lumbar range of motion, or a strong posterior pelvic tilt are also at a high risk for postoperative THA dislocation. In such patients, we should consider countermeasures against dislocation, such as the surgical approach, preservation of muscle tendons by using a short stem, and preparation of a dual mobility system.

Patients with spastic paraplegia have very large forces applied to the joints and bones owing to muscle spasticity, which may cause hip dislocation or cup failure [36]. To prevent this, consider the reduction of tension by myotomy around hip joint, such as the adductor muscle, iliopsoas muscle, gluteus maximus tendon, piriformis muscle, and short external rotator muscle groups.

(g) Leg length discrepancy

The case with large preoperative leg length discrepancy requires large leg lengthening, and the risk of nerve palsy is high [37]. If the leg lengthening exceeds 3 cm, intraoperative nerve monitoring is useful [38, 39]. The leg length discrepancy may be evaluated by the position of the lesser trochanter on an anteroposterior view of hip radiograph; however, since there may be a difference in the femur and lower leg length between the right and left sides, a simple radiograph of the entire lower extremity should be taken to evaluate the difference.

(h) Patients with previous surgery and abnormal bone morphology

Bone deformities, defects, and soft tissue conditions vary from case to case, especially in patients with a history of hip arthroplasty, and should be evaluated in detail by CT, MRI, and other examinations. Perioperative complications, such as intraoperative fracture, dislocation, nerve palsy, and vascular injury, should be anticipated, the type of prosthesis and instruments used should be selected carefully, and back-up wires, plates, and support rings should be prepared.

References

1. Debbi EM, et al. Smoking and total hip arthroplasty: increased inpatient complications, costs, and length of stay. J Arthroplast. 2019;34:1736–9.
2. Mills E, et al. Smoking cessation reduces postoperative complications: a systematic review and meta-analysis. Am J Med. 2011;124:144–54.
3. Rusy DA, et al. Vaping and e-cigarette use in children and adolescents: implications on perioperative care from the American Society of Anesthesiologists Committee on Pediatric Anesthesia, Society for Pediatric Anesthesia, and American Academy of Pediatrics Section on Anesthesiology and pain medicine. Anesth Analg. 2021;133:562–8.
4. Egholm JWM, et al. Perioperative alcohol cessation intervention for postoperative complications. Cochrane Database Syst Rev. 2018;11:CD008343.
5. Ungur AL, et al. Perioperative management of alcohol withdrawal syndrome. Visc Med. 2020;36:160–6.
6. Kerr MM, et al. Does a prescription-based comorbidity index correlate with the American Society of Anesthesiologists physical status score and a registry study. Clin Orthop Relat Res. 2020;479:2181–90.
7. Horvath B, et al. The evolution, current value, and future of the American Society of Anesthesiologists physical status classification system. Anesthesiology. 2021;135:904–19.
8. Fleisher LA, et al. 2014 ACC/AHA guideline on perioperative cardiovascular evaluation and management of patients undergoing

noncardiac surgery: a report of the American College of Cardiology/American Heart Association Task Force on practice guidelines. J Am Coll Cardiol. 2014;64:e77–e137.
9. Hlatky MA, et al. A brief self-administered questionnaire to determine functional capacity (the Duke Activity Status Index). Am J Cardiol. 1989;64:651–4.
10. Smetana GW, et al. Preoperative pulmonary risk stratification for noncardiothoracic surgery: systematic review for the American College of Physicians. Ann Intern Med. 2006;144:581–95.
11. Kamath PS, et al. A model to predict survival in patients with end-stage liver disease. Hepatology. 2001;33:464–70.
12. Tiberi JV 3rd, et al. Increased complication rates after hip and knee arthroplasty in patients with cirrhosis of the liver. Clin Orthop Relat Res. 2014;472:2774–8.
13. Chen AZ, et al. Preoperative estimated glomerular filtration rate is a marker for postoperative complications following aseptic revision total hip. Eur J Orthop Surg Traumatol. 2021;31:175–82.
14. Kręcisz B, et al. Allergy to orthopedic metal implants – a prospective study. Int J Occup Med Environ Health. 2012;25:463–9.
15. Atanaskova Mesinkovska N, et al. The effect of patch testing on surgical practices and outcomes in orthopedic patients with metal implants. Dermatol. 2012;148:687–93.
16. Schalock PC, et al. Patch testing for evaluation of hypersensitivity to implanted metal devices: a perspective from the American Contact Dermatitis Society. Dermatitis. 2016;27:241–7.
17. Pacheco KA. Allergy to surgical implants. Clin Rev Allergy Immunol. 2019;2019(56):72–85.
18. The Japanese Orthopaedic Association Clinical Practice Guidelines Committee, et al. Routine screening for asymptomatic DVT. In: Guidelines for the prevention of symptomatic venous thromboembolism 2017. Tokyo: Nanko-do; 2017. pp. 32–33. Japanese.
19. Sato K, et al. Preoperative prevalence of deep vein thrombosis in patients scheduled to have surgery for degenerative musculoskeletal disorders. BMC Musculoskelet Disord. 2021;22:1–8.
20. Tateiwa T, et al. Perioperative management of hemophilia patients receiving total hip and knee arthroplasty: a complication report of two cases. Ther Clin Risk Manag. 2015;11:1383–9.
21. Takedani H, et al. Major orthopaedic surgeries for haemophilia with inhibitors using rFVIIa. Haemophilia. 2010;16:290–5.
22. Uprichard J, et al. Factor IX replacement to cover total knee replacement surgery in haemophilia B: a single-centre experience, 2000-2010. Haemophilia. 2012;18:46–9.
23. Hirose J, et al. Total hip arthroplasty and total knee arthroplasty in a patient with congenital deficiency of plasminogen activator inhibitor-1. Haemophilia. 2016;22:e237–9.
24. Yamaguchi K, et al. Cyclic therapy with etidronate has a therapeutic effect against local osteoporosis after cementless total hip arthroplasty. Bone. 2003;33:144–9.
25. Iwamoto N, et al. The effectiveness of mono or combined osteoporosis drug therapy against bone mineral density loss around femoral implants after. J Bone Miner Metab. 2014;32:539–44.
26. Chee YH, et al. Total hip replacement in morbidly obese patients with osteoarthritis: results of a prospectively matched study. J Bone Joint Surg. 2010;92:1066–71.
27. Imai N, et al. Accuracy of acetabular cup placement using CT-based navigation in total hip arthroplasty: comparison between obese and non-obese patients. J Orthop Sci. 2019;24:482–7.
28. Hayashi S, et al. Obese patients may have more soft tissue impingement following primary total hip arthroplasty. Int Orthop. 2012;36:2419–23.
29. Iwata H, et al. Total hip arthroplasty via an anterolateral supine approach for obese patients increases the risk of greater trochanteric fracture. J Orthop. 2018;15:379–83.
30. Chiavetta JB, et al. Total hip arthroplasty in patients with dwarfism. J Bone Joint Surg Am. 2004;86:298–304.
31. Fujishiro T, et al. Predicting leg-length change after total hip arthroplasty by measuring preoperative hip flexion under general anaesthesia. Orthop Surg (Hong Kong). 2012;20:327–30.
32. Heleniud I, et al. Total hip arthroplasty in diastrophic dysplasia. J Bone Joint Surg Am. 2003;85:441–7.
33. Oba M, et al. Conventional cementless total hip arthroplasty in patients with dwarfism with height less than 140 cm and minimum 10-year follow-up: a. J Orthop Sci. 2021;26:128–34.
34. Yoshitani K, et al. Pelvic metastasis of breast synovial sarcoma. J Orthop Sci. 2009;14:219–23.
35. Ochi H, et al. Total hip arthroplasty with reconstruction of the acetabulum through direct anterior approach for metastatic bone disease of the acetabulum: a case report. Trauma Case Rep. 2021;32:100447.
36. Fukaya H, et al. A case of bone ingrowth of an artificial acetabulum dislocated by spastic paraplegia due to cerebral hemorrhage. Hokkaido J Orthop Traumatol. 2007;49:44–7. Japanese
37. Kabata T, et al. Safety range for acute limb lengthening in primary total hip arthroplasty. Int Orthop. 2019;43:2047–56.
38. Hoshino C, et al. Femoral nerve palsy following primary total hip arthroplasty with a direct anterior approach. PLoS One. 2019;14:e0217068.
39. Kinoshita S, et al. Femoral nerve status during the anterolateral approach for total hip arthroplasty: motor-evoked potential analysis and an influencing factor. J Orthop Sci. 2021;S0949-2658(21):00334-1. https://doi.org/10.1016/j.jos.2021.09.007.
40. Inoue D, et al. Comparison of postoperative complications and survivorship of total hip and knee arthroplasty in dialysis and renal transplantation. J Arthroplast. 2020;35:971–5.
41. Suyama K, et al. Lenvatinib: a promising molecular targeted agent for multiple cancers. Cancer Control. 2018;25(1):1073274818789361.
42. Goodman SM, et al. American College of Rheumatology/American Association of Hip and Knee Surgeons Guideline for the perioperative management of antirheumatic medication in patients with rheumatic diseases undergoing elective total hip or total knee arthroplasty. J Arthroplast. 2017;2071(32):2628–38.

Part IX

Surgical Techniques: Approaches

Direct Anterior Approach (DAA)

50

Junichi Nakamura

1 Surgical Techniques

The direct anterior approach (DAA) is a true anatomical approach allowing superficial entry through the intermuscular and internervous planes between the sartorius muscle, innervated by the femoral nerve, and the tensor fascia lata muscle, innervated by the superior gluteal nerve, as well as deep entry through the intermuscular and internervous planes between the rectus femoris muscle, innervated by the femoral nerve, and the gluteus medius and minimus muscles, innervated by the superior gluteal nerve. Total hip arthroplasty (THA) via DAA is performed in the supine position, with or without a traction operating table.

2 DAA Without a Traction Operating Table

In the DAA without a traction operating table, the femur is raised anteriorly using the extension function of a standard operating table, and an assistant holds the leg of the patient after handwashing.

The DAA was first reported by Carl Hueter in 1883 and established by Marius Nygaard Smith–Petersen in 1949. In Japan, it was introduced as a minimally invasive procedure using the distal part of the Smith–Petersen approach in 2004 by Oinuma and Shiratsuchi. Additionally, detailed surgical techniques and short-term results were reported in 2007 [1]. Since then, DAAs have been the mainstay of DAA in Japan without the use of a traction operating table.

In this method, the patient is placed in the supine position on an operating table with an extension function. To extend the hip joint during femoral manipulation, the axis of extension of the operating table should be aligned with that of the greater trochanter. The non-operative side of the hip joint should be kept at 20° abduction to allow adduction of the operative side of the hip joint during femoral manipulation. The position of the bilateral superior anterior iliac spines is confirmed; additionally, the pelvic tilt is corrected by placing a towel over the buttocks. The skin incision is made parallel 2 cm outside the line connecting the superior anterior iliac spine and the head of the fibula, with the proximal end at the intersection with the extension of the inguinal groove, 6–9 cm distally depending on the patient's body size (Fig. 50.1). The length of the skin incision should not be too short to avoid overstressing the skin and delaying wound healing. In the original Smith–Petersen approach, an incision is made between the sartorius and tensor fascia lata muscles; however, in the DAA, the fascia of the tensor fascia lata muscle is incised 1 cm lateral to the intermuscular space to avoid damage to the lateral femoral cutaneous nerve. With the fin-

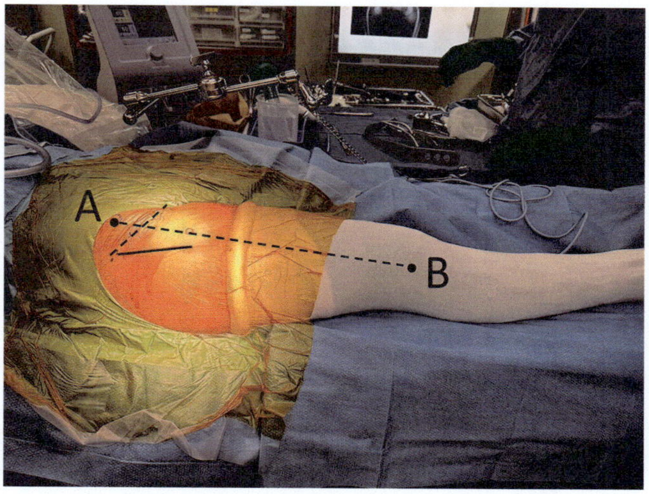

Fig. 50.1 Posture and skin incision line (for right hip joint). The proximal end is 2 cm lateral to the line (dashed line) connecting the anterior superior iliac spine (**a**) and the head of the fibula (**b**) and the intersection with the extension of the inguinal groove (dashed line). The skin incision line (solid line) is 6–9 cm distal to this point and lies on the belly of the tensor fascia latae. (By courtesy of Dr. Miura Yoko, Dr. Oinuma Kazuhiro, and Dr. Shiratsuchi Hideaki, Funabashi Orthopedic Hospital)

J. Nakamura (✉)
Chiba University, Chiba, Japan
e-mail: njonedr@chiba-u.jp

Y. Nakashima (ed.), *Advances in Total Hip Arthroplasty*, https://doi.org/10.1007/978-981-97-2461-1_50

gertips, the muscle belly of the tensor fascia lata muscle is bluntly exfoliated to the lateral side, and the entire anterior hip joint capsule encircling the femoral head is palpated. The first retractor is inserted into the superior lateral aspect of the capsule, deflecting the gluteus medius and minimus muscles. The second retractor is inserted lateral to the innominate tubercle of the greater trochanter, deflecting the tensor fascia lata muscle to the lateral side. Ligation or electrocoagulation and dissection of several lateral femoral circumflex arteries running under the fascia between the tensor fascia lata and sartorius muscles are performed, approximately in the middle of the first and second retractors. The third retractor is inserted inferiorly medial to the capsule to retract the rectus femoris muscle medially. While palpating the anterior acetabular wall, the fourth retractor is inserted medially in the direction of the femoral neck axis; the rectus femoris muscle is subsequently retracted medially (Fig. 50.2). This insertion position is in the distal third of the femoral head, where the Cobb accessory can be inserted without resistance by dissecting the space between the rectus femoris muscle and capsule. The fourth retractor is inserted because the femoral nerve, artery, and vein run medial to the rectus femoris muscle. The anterior capsule is incised, and a femoral neck osteotomy is performed using the superior margin of the head–neck junction as a landmark. The femoral neck is osteotomized at a slightly higher position, and after removal of the femoral head, it is osteotomized twice at the preoperatively planned height. If the osteotomy is too low, it may cause a chip fracture of the greater trochanter during femoral manipulation, and if it is too high, it may interfere with the insertion and removal of the acetabular reamers. The full circumference of the bony acetabulum should be developed to confirm the position and depth of the original acetabulum, considering the tendency of acetabular reaming in the DAA to overcut the posterior wall. During cup placement, the bilateral superior anterior iliac spines are palpated to ensure that the pelvis is not tilted by the retractor. Femoral manipulation is the most affected by the learning curve. The operating table is extended from 15° to 20°, the hip is hyperextended and slightly adducted, and the knee is externally rotated 90° until it points outward. The posterior capsule is separated, a sharp fork retractor is inserted behind the greater trochanter, and the proximal femur is raised anteriorly, while the osteotomy site is raised anteriorly with a blunt hook (Fig. 50.3). The restriction of external rotation is often caused by insufficient dissection of the posterior capsule, which detaches when external rotation is manually forced. If anterior elevation is inadequate despite the removal of external rotation restriction, the hip joint is further hyperextended. If the anterior elevation is still observed to be inadequate, a myotomy is performed. First, dissection of the attachments of the short external rotator muscles (superior gemellus, internal obturator, and inferior gemellus muscles) attached to the intertrochanteric fossa has to be carried out; second, dissection of the piriformis muscle attached to the posterior superior aspect of the greater trochanter is carried out; third, the skin incision has to be extended proximally, and the attachment of the tensor fascia lata muscle has to be dissected from the anterior superior iliac spine by approximately 1 cm. For femoral rasping, position the affected leg in a Figure-4 position by crossing it over the front of the healthy knee. In the DAA, the anteversion of the stem tends to increase when the assis-

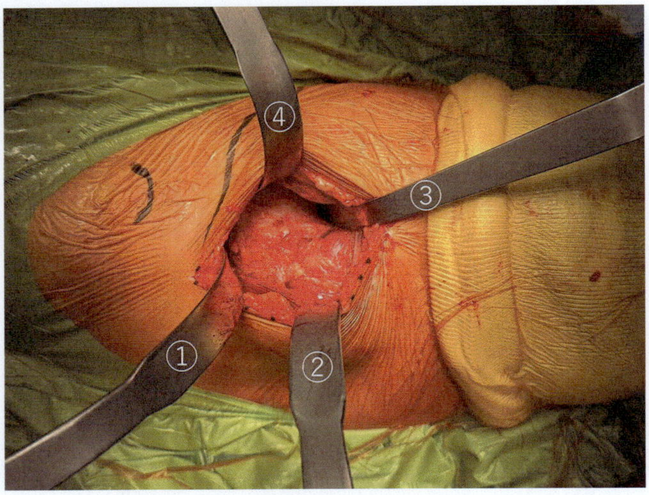

Fig. 50.2 Exposure of the anterior hip joint capsule. The anterior joint capsule is exposed with four retractors. The role of each retractor is to avoid the (1) gluteus medius and minimus muscles, (2) tensor fascia latae muscle, (3) rectus femoris muscle, and (4) rectus femoris muscle. (By courtesy of Dr. Miura Yoko, Dr. Oinuma Kazuhiro, and Dr. Shiratsuchi Hideaki, Funabashi Orthopedic Hospital)

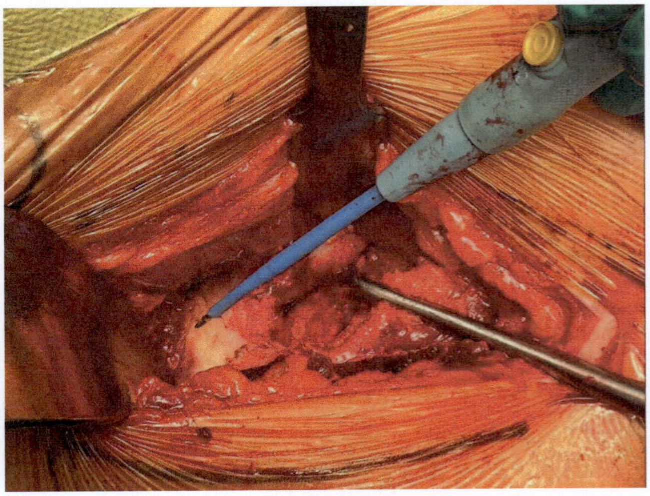

Fig. 50.3 Femoral manipulation. With the hip joint in hyperextension, mild adduction, and 90° external rotation, the posterior joint capsule is separated and lifted anteriorly while pulling the osteotomy forward with a single blunt hook. (By courtesy of Dr. Miura Yoko, Dr. Oinuma Kazuhiro, and Dr. Shiratsuchi Hideaki, Funabashi Orthopedic Hospital)

tant relaxes the external rotation of the femur. The wound is closed by suturing the fascia of the tensor fascia lata, taking care not to involve the lateral femoral cutaneous nerves. A hematoma should be prevented from spreading by applying mild compression with an elastic bandage.

3 DAA with a Traction Operating Table

In the DAA with a traction operating table, the special table holds the leg of the patient, and the femur is elevated using the extension function of the traction device.

The method of using a traction operating table was invented by the Judet School in France in the 1950s. It was not known to the English-speaking world until later and was recognized again as surgical support for the DAA in the 2000s. In Japan, the detailed surgical technique and 2-year postoperative results were reported in 2017 [2]. There are two types of traction operating tables: fixed and mobile. The mobile table is highly versatile because it can be installed on a standard operating table. The installation method differs depending on the type of product. When the footrest on the operative side is removed and the traction operating table is installed, the cephalocaudal direction is 180° opposite to that in the method without the traction operating table described above owing to the structure of the operating table. As with the traction operating table for fractures, a boot is placed on the patient's leg and connected to the traction operating table. In the case of the mobile traction operating table, only the leg on the operative side needs to be connected; however, in the case of bilateral simultaneous surgery or the fixed traction operating table, both legs should be connected by the boots.

The basic flow of the surgical procedure with the traction operating table is the same as that without; therefore, only the points that differ will be described. After positioning the patient, the fluoroscopy system is turned on to confirm the fluoroscopic range, height of the apex of the greater trochanter, and correction of the pelvic tilt. Owing to the narrow fluoroscopic range, an accurate anteroposterior image should be reproduced before performing the operation. The main points to be considered are the visualization of the tear drop, absence of obstructions around the hip joint, and confirmation of the apex of the greater trochanter. The pelvic tilt and tilt of the fluoroscopy system should be adjusted such that the contours of the obturator foramen coincide with the preoperative radiographs, and the Hilgenreiner line is leveled. The disinfection area should be rectangular in shape, with the umbilicus as the reference point, proximal and medial to the distal thigh, and disinfection of the entire lower extremity is not required. The anterior capsular ligament is incised in a V-shape with the iliocapsularis muscle as the landmark of the iliofemoral ligament [3], extending to the attachment site of the vastus lateralis muscle. Once the femoral neck is exposed, traction is applied with a traction device by externally rotating the affected lower limb by 45°, which opens the joint space and leads to subluxation. Because of traction force, the osteotomy site opens spontaneously when the femoral neck is osteotomized, and the femoral head can be easily extracted. The femoral side is manipulated by externally rotating the affected leg by 90° with a traction device, releasing the medial capsule, and a blunt fork retractor is inserted to move the osteotomy site outward. The joint capsule is dissected along the posterior cortical bone in a V-shape, with the arc to the intertrochanteric fossa and apex of the greater trochanter. A sharp fork retractor is inserted posteriorly at the greater trochanter, and a blunt hook is applied to the osteotomy site to raise the proximal femur anteriorly while lifting it. If the posterior aspect of the greater trochanter impinges on the posterior acetabular wall during external rotation, the osteotomy site will not be raised anteriorly, even if the hip is extended. In such cases, the posterior aspect of the greater trochanter should be exfoliated until it becomes easier to move outward. Subsequently, the hip is internally rotated once to release the impingement, and the osteotomy site is guided outward and forward, so that the posterior aspect of the greater trochanter overcomes the posterior acetabular wall; subsequently, it is externally rotated again. In addition, the extension function of the traction device is used to hyperextend the hip joint on the operative side to a maximum of 35° to elevate the proximal femur anteriorly.

4 Advantages and Disadvantages of Various Approaches

(a) Advantages of DAA
  (i) Being a true anatomical approach:

The greatest advantage of the DAA is that it is a true anatomic approach [1]. Theoretically, DAA is minimally invasive because it is an intermuscular and internervous plane approach. The muscles can be preserved independently of the length of the skin incision; in particular, there is no invasion of the gluteus medius and minimus muscles, which are the abductor muscles of the hip joint. Therefore, the following benefits can be expected: reduction of postoperative pain, early mobilization, acceleration of rehabilitation, early recovery of walking ability, shortening of hospital stay, and reduction of the postoperative dislocation rate.

The anterolateral approach (OCM or ALS) is an intermuscular approach; however, it is not an internervous approach because both the gluteus medius and tensor fascia latae muscles are innervated by the superior gluteal nerve. The lateral approach (direct lateral or mini-one anterolateral) cannot prevent injury to the gluteus medius muscle. Posterior approaches (posterolateral or conjoint tendon-preserving posterior) cannot prevent injury to the gluteus maximus muscle. In the past, the posterior approach was favored because long stems and large implants were preferred; however, in recent years, minimally invasive surgery has been favored because of smaller implant sizes and improved instrumentation, and the DAA is gaining prominence. In a comparison of 99 patients undergoing the DAA and 96 patients undergoing the posterior approach [4], the recovery of walking ability was faster with the DAA than with the posterior approach. Walking 200 m with one cane took 12.0 days after the DAA and 15.5 days after the posterior approach. Moreover, the required time for a single-leg stance of more than 5 s was 16.6 days after the DAA and 22.9 days after the posterior approach. In addition, the required time to recognize a negative Trendelenburg sign was 16.7 days after the DAA and 24.8 days after the posterior approach, and the residual Trendelenburg sign at 3 weeks postoperatively was observed in 29% of the cases after the DAA and 67% after the posterior approach. Notably, the 50-m walking time at 3 weeks postoperative was 52.3 s in the DAA and 74.5 s in posterior approach. Overall, 34% and 19% of patients were able to walk without cane at 3 weeks postoperatively after the DAA and the posterior approach, respectively. The length of hospitalization was 22.2 days after the DAA and 30.4 days after the posterior approach [4]. The walking ability was higher in patients after the DAA than after the posterior approach at 2 months postoperatively; however, the significant difference disappeared at 6 months postoperatively [4]. At 3 years postoperatively, the forgotten joint score-12 was better with the DAA than with the posterior approach [5]. The extent to which the joint capsule and short external rotator tendons can be preserved depends not only on the approach but also on comprehensive factors, including the patient background (e.g., physique, bone structure, and previous surgical history), implant model (e.g., cemented or cementless fixation style and design concept), and medical system (e.g., surgeon, staff, and facilities). The thickness of the internal obturator muscle increased 6 months postoperatively compared with preoperatively, when the posterior and anterior joint capsules were reconstructed [6].

(ii) Being in a supine position

The supine position is safe for anesthesia because of the ease of systemic control. Owing to the lack of positional changes, the airway can be secured even with a laryngeal mask, and the circulation is more stable. The surgical advantage of the supine position is that it is easy to visualize the cup placement. The supine position is consistent with the pelvic tilt on preoperative radiographs, and the pelvic position is stable throughout the surgery. The DAA has less angular variability (dispersion) than the posterior approach [7] and a higher safe-zone achievement rate (80–99% in the DAA vs. 40–91% in the posterior approach) [4, 7]. The use of CT-based navigation allows for more accurate cup placement than the use of freehand placement [8]. There was no difference in the accuracy of stem alignment between the DAA and the posterior approach [9]. In the supine position, the leg length discrepancy can be easily confirmed by the position of the knees and heels.

(iii) Ease of performing intraoperative fluoroscopy

In the DAA, implant placement can be confirmed in real time using intraoperative fluoroscopy. This allows even beginners to safely perform the DAA [10]. The International Commission on Radiological Protection has set the standard for radiation exposure at an average of 20 mSv/year for 5 years and no more than 50 mSv/year for any year. The fluoroscopic time and radiation dose for the DAA are much lesser than those for trauma surgery [11]. The fluoroscopic time and radiation dose for the DAA and osteosynthesis are 0.83 min and 8.91 min (6.14 min for the femoral neck fracture, 7.81 min for the intertrochanteric fracture, and 18.98 min for subtrochanteric fracture), respectively. The DAA exposure is 11.09 μSv (without a protector) for the thyroid, 0.21 μSv (with a protector) for the sternum, and 5.55 μSv (without a protector) for the eye lens. The fluoroscopy time for DAA is an average of approximately 2 min at the time of introduction, but is reduced to an average of less than 1 min after reaching 40 patients [10]. To reduce the radiation exposure, it is important to use pulsed fluoroscopy instead of continuous fluoroscopy and shield the lens and thyroid gland.

(iv) Shortest distance to the hip joint is directly reachable

The anterior portion of the groin is easily accessible even in patients with obesity because the subcutaneous fat is thinner in this portion than in the lateral and posterior portions [1]. However, patients with obesity tend to become unclean owing to the accumulation of plaque in the groove of the inguinal skin; therefore, it is better to avoid the crease and make a slight lateral skin incision to reduce the risk of infection.

(v) Ease of performing simultaneous bilateral surgery

The DAA does not require intraoperative positional changes, making it easy to perform simultaneous bilateral THA. According to a retrospective study of 325 patients with bilateral simultaneous THA [12], 0.3% of patients underwent blood transfusions, none experienced systemic postoperative complications, 0.2% experienced dislocation, 0.3% had cup migration, and 0.3% had periprosthetic femoral fracture at 2 years after surgery. Bilateral simultaneous THA is reportedly more satisfactory than unilateral THA [13].

(vi) Availability of surgical support with a traction operating table

A traction operating table is not essential for the DAA; however, it supports safer surgery through the following [1].

1. Holding the hip joint of the operative side in the desired position reduces the load of handwashing in assistants.
2. Hyperextension of the hip joint to a maximum of 35° facilitates anterior elevation of the proximal femur.
3. Because a perineal post holds the perineum, the pelvic position is stabilized even in hyperextension and traction movements.
4. Some traction operating tables release the traction force during extension to reduce the tension caused by hyperextension of the femoral nerve and to efficiently transmit the force of rasping while promoting anterior elevation of the osteotomy by fixing it in a position where pressure force is applied.
5. Manual reduction does not require the physical strength of the surgeon.
6. Radiation exposure can be reduced by shortening the intraoperative fluoroscopic time.

Upon comparing the learning curve of the DAA between the methods with the traction operating table and without [14], the surgical time (114.6 min vs. 120.5 min), fluoroscopy time (0.68 min vs. 1.12 min, $P < 0.05$) and blood loss (397 mL vs. 601 mL, $P < 0.05$) were lesser in the DAA with the traction operating table. When novice surgeons begin the DAA, the traction operating table allows them to safely perform the procedure, including femoral rasping, acetabular reaming, cup placement, and trial reduction with real-time fluoroscopic monitoring.

(vii) Restrictions are solved with proficiency.

All THA cases are eligible for the DAA, and the types of implants are not limited [1]. Hemiarthroplasty [15], single-cup THA, and dual-mobility cup THA [16]) for hip fractures with the DAA have been reported. In special cases, the DAA has been used for high dislocation of Crowe classification type IV [17], after hip fusion [18], acetabular reconstruction with a KT plate [19], pathological fracture due to bone metastasis of renal cancer [20], hemophilia [21], and Charcot joint [22].

(b) Disadvantages of DAA
(i) Learning curve

The learning curve is the greatest disadvantage of the DAA [1]. Compared with conventional approaches, this surgical technique is more difficult. Therefore, surgical indication and patient selection are important for beginners. For example, in men with muscularity and hard soft tissues, or in previously operated patients with adhesions or contractures, the DAA becomes extremely difficult. In patients with obesity, the route of entry to the tensor fascia lata muscle is often disoriented because of the thick subcutaneous fat, and especially if it strays medially, the sartorius muscle may be mistaken for the tensor fascia lata, and the femoral nerve or artery may be injured. In addition, if the bone morphology is highly deformed or if there is a mixture of osteosclerosis and fragility after osteotomy or trauma, perforation or intraoperative fracture is likely to occur during reaming, rasping, and implant placement. While the DAA is less likely to dislocate, pitfalls have been reported in which the patient is unable to dislocate after intraoperative trial reduction [23].

(ii) Difficulty in the femoral manipulation

Insufficient anterior elevation of the femur tends to cause the stem to enter the varus position, and insufficient lateral translation tends to cause the stem to enter the flexion position [24]. Intraoperative femoral fractures have been observed to be more common in the DAA approach than in the posterior approach [25]. The frequency of chip fractures of the greater trochanter was 29.4% (32 cases/109 cases) [26]. The longer the distance from the posterior margin of the femoral neck base to the posterior margin of the greater trochanter (depth of the greater trochanter), the more likely it is for the fracture to occur [26]. In cases of Perthes-like deformity, coxa breva, and overgrowth of the greater trochanter, the closer the distance between the acetabular margin and the femoral neck osteotomy site, the more difficult the manipulation becomes [27].

(iii) Anterior instability

The DAA carries the risk of anterior dislocation due to anterior entry. Soft tissue factors and a tendency to increase the anteversion of the cup have been pointed out as causes of anterior dislocation [24, 28]. In the DAA with simplified navigation, the pelvis tilts 7.6° anteriorly and 3.2° to the operative side during cup placement, and anteversion tends to be larger with a lower body mass index and a larger intraoperative hip internal rotation angle [29]. Even with CT-based navigation, the anteversion angle tends to be larger with the DAA than with the posterior approach [30].

(iv) Risk of lateral femoral cutaneous neuropathy

According to a questionnaire survey [31] of 122 patients at an average of 12.8 months postoperatively, lateral femoral cutaneous neuropathy occurred in 31.9% of the patients, with symptoms of hypoesthesia (dull sensation) in 46.2%, tingling or jolt-like sensation in 28.2%, and pain in 25% of the patients, respectively. Lateral femoral cutaneous neuropathy did not affect objective measures (Harris hip score [HHS] and JOA score) but decreased patient-oriented measures (forgotten joint score and JHEQ). A spontaneous recovery of lateral femoral cutaneous neuropathy occurred in 32% of the patients, with a mean recovery time of 6.4 months. According to a continuing survey, [32] at an average of 26.2 months postoperatively, lateral femoral cutaneous neuropathy remained in 11.2% of the patients; however, spontaneous recovery occurred in 96% of the patients, with complete resolution in 60% of them, and an improved quality of life in the remaining patients. A prospective randomized controlled study of 30 patients with bilateral THA who underwent DAA and ALS [33] showed that lateral femoral cutaneous neuropathy was present in 7 patients (23.3%) in the DAA group at 3 months postoperatively, but it resolved 1 year postoperatively. As a risk factor for lateral femoral cutaneous neuropathy, the shorter the offset of the femoral neck (the length of the perpendicular line drawn from the center of the femoral head to the femoral axis in a plain anteroposterior radiograph), the more likely it was to occur (23.9 mm for symptomatic vs. 27.7 mm for asymptomatic patients) [34]. The lateral femoral cutaneous nerve runs in a wide variety of locations, and when visualized by echo, [35] patients with the fan-type pattern were more likely to develop lateral femoral cutaneous neuropathy (positive predictive value, 90%), while asymptomatic patients were less likely to have the fan-type (specificity, 96.2%). In a cadaver study on the variation of the lateral femoral cutaneous nerve in 64 thighs, [36] the anterior branch type with anterior and thin branches running along the medial border of the tensor fascia lata muscle was found in 37% of cases, and the posterior branch type with a posterior branch as thick as the anterior branch was found in 63% of cases. The possibility of the lateral femoral cutaneous nerve crossing the DAA skin incision line is estimated to be 42%.

(v) Risk of femoral nerve palsy

The frequency of femoral nerve palsy is reportedly 1.1% (3/273 cases) [37]. The cause of femoral nerve palsy is attributed to the surgeon's learning curve, as it occurred in the first 20 cases and did not occur thereafter in one study [37], or to the insertion of a retractor into the anterior acetabular wall. Intraoperative monitoring of the femoral nerve (motor-evoked potential monitoring by transcranial electrical stimulation) [38] showed that in 17 of 22 cases (77%), the insertion of a retractor into the anterior acetabular wall caused a transient decrease in the potential. According to a cadaver study [39] of 84 limbs on femoral nerve travel, the femoral nerve is closest to the acetabular rim at 90° anteriorly (33.2 mm at 0°, 24.4 mm at 30°, 18.4 mm at 60°, 16.6 mm at 90°, 17.9 mm at 120°, and 23.2 mm at 150°), based on a line drawn from the superior anterior iliac spine to the acetabular center. Notably, the femoral nerve running anterior to the hip joint is stretched when the hip hyperextends during femoral manipulation.

(vi) The DAA does not cause zero damage to muscles

In real surgery, muscle damage can occur because of surgical manipulation, even in the DAA. An MRI comparison of the periprosthetic hip muscles before and 1 year after surgery [40] showed a decrease in the muscle cross-sectional area and fatty atrophy in the gluteus minimus, tensor fascia lata, and internal obturator muscles. A prospective randomized controlled trial [33] of 30 patients with bilateral simultaneous THA who underwent the DAA or ALS showed that the CT cross-sectional area of the gluteus medius muscles was unchanged in both groups, but that of the tensor fascia lata muscle decreased more from preoperatively to 3 months postoperatively for the DAA (78.8% for DAA versus 90.7% for ALS); moreover, MRI fatty degeneration at 1 year postoperatively was more prominent for DAA than for ALS. In addition, the CK values at 1, 4, 7, 10, and 14 days postoperatively were higher for the DAA than for the posterior approach [41].

(vii) An uncommon approach

As of 2011, a survey of 292 orthopedic surgeons in 57 countries worldwide found that the DAA was used only in 10% of cases, and in the Japanese Arthroplasty Registry as of 2012, it was used in 14.9% [10]. According to the Registry, as of March 31, 2018, in the 2 years from April 1, 2016, the DAA was estimated to have increased to 40%.

(viii) Few studies have reported long-term results

Because the DAA is still a new technique, the long-term outcome over 5 years is unknown. A retrospective comparative study of revision cases of primary THA performed since 2004 [25] showed that the main cause of revision was aseptic loosening of the femoral component, which was more common with the DAA than with other approaches (30% in 9 cases with DAA vs. 7% in 7 cases with non-DAA, $P = 0.002$). However, this result should be interpreted with caution because the analysis included all revision cases. In other words, DAA may not more prone to revision because the number of patients undergoing DAA was smaller. The time from the initial THA to revision was 3.0 years in the DAA and 3.6 years in the non-DAA group, with no significant difference. In a comparison of 5-year postoperative outcomes between 44 patients undergoing the DAA and 42 patients undergoing the posterior approach [42], the JOA score and HHS were good in both groups, with no significant difference. No revision occurred; however, dislocation occurred in no patients using the DAA and in 9.5% of patients using the posterior approach, and subsidence occurred in 2.2% of patients using the DAA.

References

1. Oinuma K, et al. Total hip arthroplasty by a minimally invasive, direct anterior approach. Oper Orthop Traumatol. 2007;19:310–26.
2. Nakamura J, et al. Direct anterior approach for total hip arthroplasty with a novel mobile traction table -a prospective cohort study. BMC Musculoskelet Disord. 2017 Jan 31;18(1):49.
3. Konno K, et al. An anatomic study of the relationship between the iliocapsularis muscle and the iliofemoral ligament in total hip arthroplasty. Arthroplast Today. 2021;12:57–61.
4. Nakata K, et al. A clinical comparative study of the direct anterior with mini-posterior approach: two consecutive series. J Arthroplast. 2009;24:698–704.
5. Ozaki Y, et al. Posterior versus direct anterior approach in total hip arthroplasty: difference in patient-reported outcomes measured with the forgotten joint score-12. SICOT J. 2018;4:54.
6. Kanda A, et al. Preservation of the articular capsule and short lateral rotator in direct anterior approach to total hip arthroplasty. Eur J Orthop Surg Traumatol. 2018;28:1111–6.
7. Kobayashi H, et al. Surgeons changing the approach for total hip arthroplasty from posterior to direct anterior with fluoroscopy should consider potential excessive cup anteversion and flexion implantation of the stem in their early experience. Int Orthop. 2016;40:1813–9.
8. Nishihara S, et al. Comparison between freehand technique and computed tomography-based navigation in acetabular cup placement through direct anterior approach for total hip arthroplasty. Arch Orthop Trauma Surg. 2021;142:323–9.
9. Abe H, et al. Difference in stem alignment between the direct anterior approach and the posterolateral approach in total hip arthroplasty. J Arthroplast. 2015;30:1761–6.
10. Homma Y, et al. Safety in early experience with a direct anterior approach using fluoroscopic guidance with manual leg control for primary total hip arthroplasty: a consecutive one hundred and twenty case series. Int Orthop. 2016;40:2487–94.
11. Jinnai Y, et al. Does a fluoro-assisted direct anterior approach for total hip arthroplasty pose an excessive risk of radiation exposure to the surgeon? SICOT J. 2020;6:6.
12. Tamaki T, et al. Perioperative complication rate of one-stage bilateral total hip arthroplasty using the direct anterior approach. J Orthop Sci. 2016;21:658–61.
13. Yoshii H, et al. Comparison of patient satisfaction after unilateral or simultaneous bilateral total hip arthroplasty through a direct anterior approach: evaluation using the Japanese Orthopaedic Association hip disease evaluation questionnaire. J Orthop Sci. 2016;21:332–5.
14. Banno S, et al. Use of traction table did not increase complications in total hip arthroplasty through direct anterior approach performed by novice surgeon. J Orthop Surg (Hong Kong). 2020;28:2309499020923093.
15. Ochi H, et al. Total hip arthroplasty via the direct anterior approach with a dual mobility cup for displaced femoral neck fracture in patients with a high risk of dislocation. SICOT J. 2017;3:56.
16. Homma Y, et al. In total hip arthroplasty via the direct anterior approach, a dual-mobility cup prevents dislocation as effectively in hip fracture as in osteoarthritis. Int Orthop. 2017;41:491–7.
17. Oinuma K, et al. Total hip arthroplasty with subtrochanteric shortening osteotomy for Crowe grade 4 dysplasia using the direct anterior approach. J Arthroplast. 2014;29:626–9.
18. Tamaki T, et al. Total hip arthroplasty through a direct anterior approach for fused hips. Hip Int. 2015;25:549–52.
19. Matsumoto M, et al. Kerboull-type plate in a direct anterior approach for severe bone defects at primary total hip arthroplasty: technical note. SICOT J. 2017;3:21.
20. Ochi H, et al. Total hip arthroplasty with reconstruction of acetabulum through direct anterior approach for metastatic bone disease of acetabulum combined with pathological proximal femoral neck fracture: a case report. Trauma Case Rep. 2021;32:100447.
21. Sano K, et al. Total hip arthroplasty via the direct anterior approach with Kerboull-type acetabular reinforcement device for an elderly female with factor XI deficiency. SICOT J. 2017;3:11.
22. Terakawa H, et al. Favorable outcome of a tractional direct anterior approach to total hip arthroplasty in a patient with Charcot arthropathy case report. Chiba Med J. 2018;94E:55–9.
23. Homma Y, et al. Benefit and risk in short term after total hip arthroplasty by direct anterior approach combined with dual mobility cup. Eur J Orthop Surg Traumatol. 2016;26:619–24.
24. Kawarai Y, et al. Does the surgical approach influence the implant alignment in total hip arthroplasty? Comparative study between the direct anterior and the anterolateral approaches in the supine position. Int Orthop. 2017;41:2487–93.
25. Eto S, et al. The direct anterior approach is associated with early revision total hip arthroplasty. J Arthroplast. 2017;32:1001–5.
26. Homma Y, et al. Greater trochanter chip fractures in the direct anterior approach for total hip arthroplasty. Eur J Orthop Surg Traumatol. 2016;26:605–11.
27. Tamaki T, et al. Radiographic factors to predict operation time of direct anterior total hip arthroplasty for dysplastic hips. Hip Int. 2021;31:90–6.
28. Yuasa T, et al. Safely transitioning to the direct anterior from posterior approach for total hip arthroplasty. J Orthop. 2018;15:420–3.
29. Okamoto M, et al. Effects of body mass index and range of motion on intraoperative change in pelvic tilt during total hip arthroplasty using the direct anterior approach. BMC Musculoskelet Disord. 2021;22:240.
30. Maeda Y, et al. The accuracy of a mechanical cup alignment guide in total hip arthroplasty (THA) through direct anterior and posterior approaches measured with CT-based navigation. J Arthroplast. 2015;30:1561–4.
31. Homma Y, et al. Lateral femoral cutaneous nerve injury with the direct anterior approach for total hip arthroplasty. Int Orthop. 2016;40:1587–93.
32. Ozaki Y, et al. Spontaneous healing of lateral femoral cutaneous nerve injury and improved quality of life after total hip arthroplasty via a direct anterior approach. J Orthop Surg (Hong Kong). 2017;25:2309499016684750.
33. Takada R, et al. Direct anterior versus anterolateral approach in one-stage supine total hip arthroplasty. Focused on nerve injury: a prospective, randomized, controlled trial. J Orthop Sci. 2018;23:783–7.
34. Ozaki Y, et al. Small femoral offset is a risk factor for lateral femoral cutaneous nerve injury during total hip arthroplasty using a direct anterior approach. Orthop Traumatol Surg Res. 2016;102:1043–7.
35. Ozaki Y, et al. Preoperative ultrasound to identify distribution of the lateral femoral cutaneous nerve in total hip arthroplasty using the direct anterior approach. SICOT J. 2018;4:42.
36. Sugano M, et al. Anatomical course of the lateral femoral cutaneous nerve with special reference to the direct anterior approach to total hip arthroplasty. Mod Rheumatol. 2020;30:752–7.
37. Hoshino C, et al. Femoral nerve palsy following primary total hip arthroplasty with the direct anterior approach. PLoS One. 2019;14:e0217068.
38. Ishimatsu T, et al. Motor-evoked potential analysis of femoral nerve status during the direct anterior approach for total hip arthroplasty. J Bone Joint Surg Am. 2018;100:572–7.

39. Yoshino K, et al. Anatomical implications regarding femoral nerve palsy during a direct anterior approach to total hip arthroplasty: a cadaveric study. J Bone Joint Surg Am. 2020;102:137–42.
40. Kawasaki M, et al. Muscle damage after total hip arthroplasty through the direct anterior approach for developmental dysplasia of the hip. J Arthroplast. 2017;32:2466–73.
41. Maezawa K, et al. Changes in serum creatine kinase and C-reactive protein after posterior and direct anterior approaches in total hip arthroplasty. Hip Int. 2020:1120700020978643. Online ahead of print
42. Yuasa T, et al. Comparison of direct anterior approach and posterior approach total hip arthroplasty: more than 5-year follow-up. J Orthop. 2021;24:271–3.

Anterolateral OCM Approach

51

Naoyuki Hirasawa

1 OCM Approach

The OCM approach was initiated by Röttinger et al. at the Orthopädische Chirurgie of München University [1], and has been used under various names, including the minimally invasive anterolateral, anterolateral, and modified Watson–Jones approaches.

The OCM approach enters the tensor fascia lata and vastus medialis muscles from the intermuscular space in the lateral decubitus position; it has the advantage of reducing the risk of dislocation due to preservation of the superior and posterior capsules and short external rotators and early postoperative recovery due to intermuscular entry [2]. The supine anterolateral (ALS) approach is a similar approach, but the OCM approach is performed in the lateral decubitus position, which has the advantages of confirming intraoperative stability [3] and easy femoral manipulation; however, the OCM approach has disadvantages, such as difficulty in confirming the accuracy of cup placement angle and leg length discrepancy [4].

2 Surgical Indication and Preoperative Planning

In principle, the OCM approach is indicated in almost all THA cases, including hip osteoarthritis, idiopathic osteonecrosis of the femoral head, rheumatoid arthritis of the hip, rapidly destructive coxopathy of the hip, and proximal femoral fractures. Caution should be exercised in cases of significant flexion contracture, anterior instability due to posterior pelvic tilt, and a highly dislocated hip that requires subtrochanteric femoral osteotomy or has a history of osteotomy or arthrodesis. Furthermore, in cases of hip osteoarthritis after pelvic fracture and screw removal from the posterior wall, the posterolateral approach must be selected.

In recent years, the long-term results of implants have improved, and there have been many cases of THA in adolescent patients wishing to engage in sports activities [5]. In these cases, patients often have relatively mild deformities and good preoperative range of motion, and a wide range of motion is required postoperatively. The OCM approach is well suited for this case because this approach makes it possible to complete preservation of the musculotendons and sufficient intraoperative confirmation of dislocation stability.

The lateral decubitus position is more prone to inaccurate cup placement than the supine position. Therefore, it is advisable to use a three-dimensional template for preoperative planning to obtain a three-dimensional view of the pelvic shape, or to use a navigation or robotic system.

3 Merits and Demerits of the OCM Approach Compared to Other Approaches

(a) Merits

Notably, the advantages of the OCM approach and other anterior muscle tendon-sparing approaches are prevention of dislocation and early recovery [2]. Unlike other anterior approaches, such as the direct anterior approach (DAA) and ALS approach, the OCM approach is performed in the lateral decubitus position, which is relatively easy for surgeons who have used posterior and lateral approaches to date [6]. Furthermore, compared with other anterior approaches, the femoral appearance is considered relatively better, making it easier for physicians with limited surgical experience to perform the procedure. Additionally, an easy view of the femur allows for the insertion of any stem shape, including a cement stem.

N. Hirasawa (✉)
Hokusuikan Memorial Hospital, Mito, Ibaraki, Japan
e-mail: hirari3@orion.ocn.ne.jp

Y. Nakashima (ed.), *Advances in Total Hip Arthroplasty*, https://doi.org/10.1007/978-981-97-2461-1_51

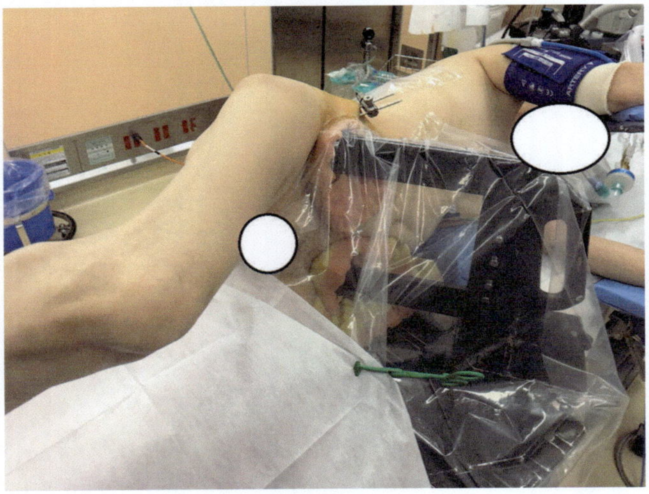

Fig. 51.1 Using a special lateral positioner that can hold both in place from an oblique angle and make possible to evaluate hip deep flexion intraoperatively

Furthermore, it was possible to perform a strict intraoperative stability check in the lateral decubitus position. The use of a special lateral positioner, as shown in Fig. 51.1, allows for adequate assessment of flexion stability in the lateral recumbent position and facilitates assessment in the hyperextended position. In contrast, in the supine position, even when the bed is extended by 20°, the stability evaluation of hyperextension is insufficient [3]. In recent years, THA has been required not only for pain relief but also for forgotten joints [7], and there is a growing trend to eliminate postoperative movement restrictions [8]. As described below, the number of dislocation cases is very small, and this is the greatest advantage of the OCM approach, which allows evaluation of sufficient intraoperative stability.

(b) Demerits

The major disadvantages of the OCM approach are that the cup placement angle tends to be inaccurate compared to the supine position, and that the difference in the length of the left leg cannot be adequately confirmed intraoperatively [9]. Therefore, to perform highly accurate operations using the OCM approach, it is necessary to devise a three-dimensional preoperative plan [10], and if possible, use navigation [11, 12]. Additionally, the high risk of intraoperative fracture when insertion is performed with a poorly appearing femur is a common disadvantage of all anterior approaches [13]. The OCM approach usually requires two assistants (although it is possible to manage with one assistant if the surgeon is skilled), which is a major economic disadvantage compared to the supine approach.

4 Surgical Techniques

(a) Anesthesia and position

Anesthesia can be either general or spinal. However, some patients may have been taking anticoagulants or antiplatelet agents before surgery, and if they are to undergo surgery without withdrawal of these drugs, general anesthesia should be used.

The patient was placed in the lateral recumbent position. The most important advantage of the lateral decubitus position is that the intraoperative stability can be adequately verified. Recently, a special lateral positioner that can be held in place from an oblique angle has become available and is highly useful (Fig. 51.1). This device allows intraoperative assessment of deep flexion equivalent to that in the supine position.

One leg should be removed at a time from the operating table because the affected leg must be lowered posteromedially during the operation. Once the patient is covered with a cloth and the field was cleaned, a bag was attached to keep the lower limbs clean.

(b) Skin incision and intermuscular development

The affected limb was held in mild flexion and mid-position during internal and external rotation. A skin groove at the border between the tensor fascia femoris and anterior border of the gluteus medius was palpated 6–7 cm distal to the top of the greater trochanter, and a straight skin incision, 8–10 cm long, was made from there to a point 5 cm posterior to the anterior superior iliac spine along the iliac crest. In obese patients, the skin incision should be placed slightly anteriorly, whereas in slender patients, it should be placed slightly posteriorly to facilitate smooth access to the intermuscular space.

The fascia was revealed when the subcutis was unfolded. The fascia covering the tensor fascia femoris is thin and pinkish, as shown in Fig. 51.2. The intermuscular plane is slightly posterior to the tensor fascia femoris; therefore, it is invaded by the white fascia posterior to the pink fascia. The fascia is easily identified when developed distally because of the overlap of the gluteus medius and tensor fascia femoris muscles in the proximal part. The intermuscular space is easily identified because of the fatty layer that extends between the muscles. In patients with a highly dislocated hip, the vastus medialis runs laterally and often overlaps with the tensor fascia lata muscle in many cases. In patients with muscular disorders, the muscles are often difficult to identify because of the lack of fat between them; therefore, these patients should be treated with skills.

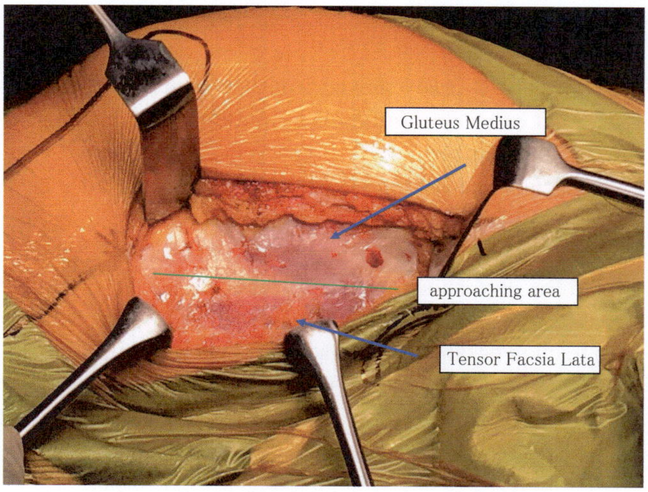

Fig. 51.2 The fascia covering the tensor fascia femoris is thin and pinkish

Abundant blood vessels run between the muscles. Thus, hemostasis should be secured adequately; however, ligation is unnecessary. The superior gluteus nerve can sometimes be identified between the tensor fascia femoris and vastus medialis. Invasion of this nerve may cause postoperative atrophy of the tensor fascia femoris muscle; therefore, it is desirable to preserve it, as much as possible [14, 15]. However, even if the nerve is damaged, it does not present with strong clinical symptoms and rarely leads to subjective symptoms in patients [16]. The anterior iliofemoral ligament is easily accessible when the intermuscular space develops.

(c) Articular ligament deployment

Once the iliofemoral ligament was identified, a retractor was inserted into the proximal femoral neck, and the gluteus medius and minimus muscles were retracted posteriorly. Similarly, a retractor was applied to the medial part of the femoral neck (medial calcar) to retract the rectus femoris anteriorly and expose the iliofemoral ligament. A longitudinal incision of the iliofemoral ligament allows preservation of the longitudinal fibers of the iliofemoral ligament; however, if the patient is unfamiliar with the procedure, the anterior iliofemoral ligament may be resected to allow sufficient expansion.

The proximal portion of the superior ligament was incised until it extended to the acetabular edge. The inferior margin of the ligament incision is the medial calcar of the femur for 7–8 cm (if the head is large, the ligamentum flavum incision should be extended medially and inferiorly because it cannot be removed if the incision is small).

(d) Femoral neck osteotomy and femoral head removal

After a thorough incision of the iliofemoral ligament, the femoral neck was osteotomized and the femoral head was extracted. Unlike the posterior approach, the lesser trochanter cannot be identified; therefore, the osteotomy line should be measured preoperatively using the apex of the femoral head and the saddle between the femoral head and greater trochanter. The position of the apex of the femoral head was confirmed by inserting a folded elevatorium into the joint. Subsequently, the distance to the femoral osteotomy line measured in the preoperative plan was used to determine the osteotomy line. A straight elevator was inserted into the saddle, and the osteotomy line was rechecked. The distance to the planned femoral osteotomy line, which was also measured preoperatively, was measured to confirm if the osteotomy line was correct. The femoral neck was osteotomized, focusing on the direction of the femoral axis. If a surgeon is not familiar with the technique, the osteotomy line is made proximally and a double cut can be made after the small trochanter is palpated, or a calcar reamer can be used to reach the planned osteotomy position.

Once the osteotomy was completed, the femoral head was extracted anteriorly and superiorly with the insertion of a bone head extractor.

(e) Soft tissue release around the greater trochanter and proximal femoral elevation

After removing the femoral head, the lower limb was placed in the pouch with the hip in a hyperextended external rotation. A two-pronged elevator was applied to the medial aspect of the femoral neck. However, the broach insertion site is deep in the operative field and not deployable; therefore, the superior capsule should be dissected. A retractor was inserted between the superior capsule and gluteus minimus, and the tip of the central retractor was applied over the apex of the greater trochanter (a shallow application may result in fracture or cutout of the greater trochanter; therefore, deep application is important) to expose the superior capsule. The superior capsule was incised towards the tip of the retractor and central apex of the greater trochanter. The osteotomy surface was gradually raised to allow broaching manipulation (Fig. 51.3). If the appearance of the femoral bone is insufficient, raising it will improve by separating the posterior capsule, piriformis muscle attachment area, external obturator muscle attachment area, and conjoint tendon. It is desirable for surgeons who are unfamiliar with the OCM approach to perform an osteotomy after sufficient deployment to prevent fractures.

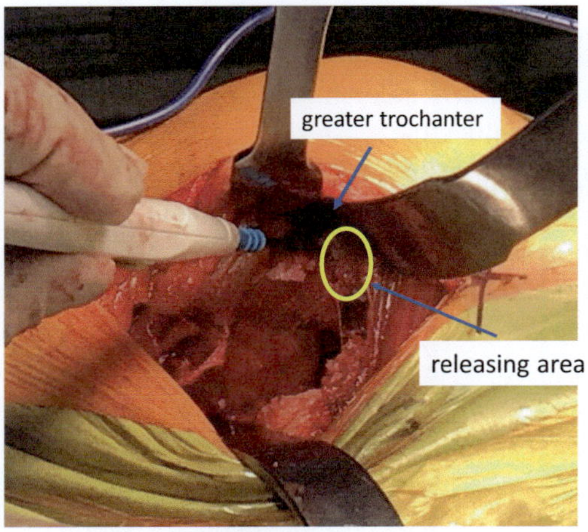

Fig. 51.3 The superior capsule is incised toward the tip of the retractor the central apex of the greater trochanter

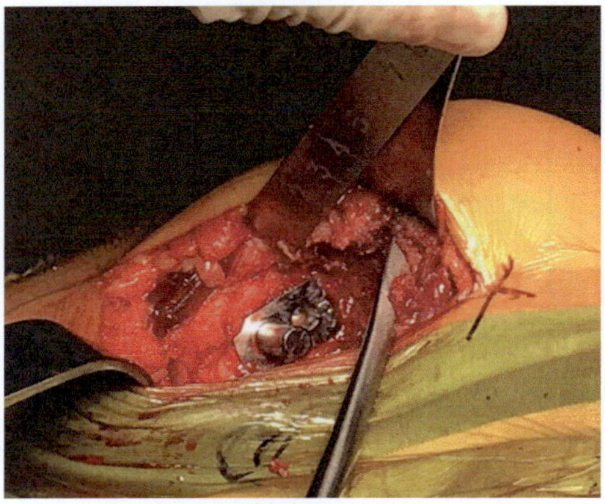

Fig. 51.4 The depth of broaching was defined as the distance from the apex of the greater trochanter to the shoulder of the broach

(f) Broaching operation

Broaching was performed with the lower limbs in the pouch in the same position (hip extension, 20°; internal rotation, 30°; and external rotation, 90°). When a straight stem is used, the insertion site is positioned posterolateral to the osteotomy site. Unlike the supine position, the view during femoral broaching was good if the soft tissue release was sufficient, making it an easy approach for inexperienced surgeons. Broaching depth was defined as the distance from the apex of the greater trochanter to the shoulder of the broach (Fig. 51.4).

(g) Acetabular deployment and reaming

The lower limb was removed from the pouch, and hip and knee joint extensions, mild external rotation, and mild traction were performed.

Retractors can be applied to the acetabulum at four locations: posterior, anterior, inferior (foramen magnum), and superior positions. The greater trochanter is retracted posteriorly to the acetabulum. Once the retractors are in place, the acetabulum can be easily manipulated without requiring lower limb traction or other manipulations (Fig. 51.5). If the anterior acetabular retractor is deep, there is a risk of neurovascular injury; however, this risk is lower than with the DAA approach because of the intervening tensor fascia lata muscles. Reaming was initiated after a full circumferential resection of the acetabular labrum. If the surgeon is not familiar with the technique, a small reamer should be used, and the reaming should be gradually made slightly larger. The distances from the superior acetabular rim to the superior margin of the excavation site and from the anterior acetabular rim to the anterior margin of the excavation site were checked visually. The distance from the medial margin of the excavation site to the medial wall cortex was checked using a sharp spoon. The distance from the posterior acetabular margin to the posterior margin of the excavation site was determined by palpation. If the posterior labrum is not resected well, it is difficult to confirm the posterior distance; therefore, it must be removed sufficiently in advance.

As mentioned above, the femur is displaced posteriorly, and when a straight reamer is used, the excavation tends to be pushed anteriorly by the femur because of the seesaw phenomenon [17]. If the anterior margin of the acetabulum is excavated, the cup may come in contact with the iliopsoas

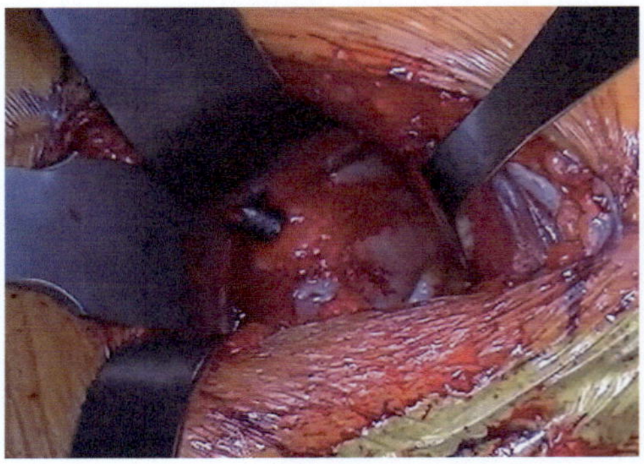

Fig. 51.5 Once the retractors are in place, the acetabulum can be easily manipulated without the need for lower limb traction or other manipulation

muscle, thereby causing impingement. However, if the posterior margin of the acetabulum is excavated with too much awareness, the initial fixation of the cup will be problematic. A careful check of whether the cup is reamed in the acetabulum should be performed before proceeding with the reaming.

(h) Cup installation and dome screw insertion

The femur often interferes with the cup inserter during cup placement; therefore, it is preferable to use an offset handle. Many surgeons use the cup handle as a guide to determine the angle of the cup placement; however, this is not recommended for THA in the lateral decubitus position because pelvic tilt may occur during the procedure. The distance from the anterior margin of the acetabulum to the anterior margin of the cup was palpated and visualized, and the distance from the superior margin of the cup to the point of contact between the cup and acetabulum was visualized. If the cup is drilled to the preoperatively planned position and if these two distances are as planned, then a terrible malposition in the placement angle is unlikely to occur.

When a dome screw is inserted, it must be placed posterior to the cup in order to prevent intraoperative vascular injury. However, the safety zone should be limited to the posteroinferior to posterosuperior region because the apparent and anatomical positions differ due to the anterior-posterior tilt of the pelvis. With the anterior approach, it is easier to place the screw posteriorly.

(i) Intraoperative trial

A trial liner and trial neck were placed, and the gauze was grasped by hooking it to the trial neck to aid in the repositioning and dislocation (Fig. 51.6).

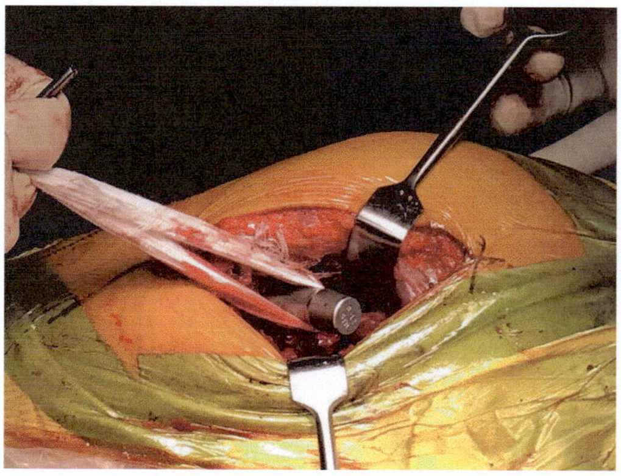

Fig. 51.6 Gauze was grasped by hooking the gauze to the trial neck to aid in the reposition and dislocation

A good view of the neck can be obtained by pulling the gauze over the neck upward and positioning the patient in hip-knee extension, hip adduction at 20°, and hyper-external rotation. If this is insufficient to raise the neck, the neck may be raised by slightly flexing the hip joint while maintaining knee extension, hip adduction, and hyper-external rotation. Both these positions are recommended.

The hip joint is repositioned by pushing the outer head into the acetabulum with a head-driving device, while the patient is in mild hip flexion and mid-internal/external rotation, with the lower limb in traction. Care was taken to avoid entrapment of the anterior capsule.

Once the trial head is repositioned, flexion internal rotation should be performed as much as possible, followed by hyperextension adduction and external rotation to confirm that the dislocation stability is satisfactory. If the patient is unstable only anteriorly or posteriorly despite the expected leg length and head diameter, we should consider changing the cup placement angle or using an elevated liner. If possible, a larger head diameter may be considered, as long as the OCM approach uses a head diameter of 32 mm, which when placed correctly, provides sufficient intraoperative anterior and posterior stability to eliminate postoperative ADL limitations. In other words, it is important not to compromise the stability of dislocations.

A Shuck test should be performed to ensure that the tension is not too high, and if there are no problems, a real implant placement should be performed. The patient was dislocated by pulling the aforementioned gauze over the neck in the direction of the surgeon, with mild traction of the lower limb and the hip in adduction, mild flexion, and mild external rotation.

(j) Implant placement and wound suture

The head and neck were removed, a retractor was applied to the acetabulum, and a liner was placed. Next, the lower limb was placed in a pouch, and the same position, as in the broach operation, was used again, after which the broach was removed. If necessary, the medullary cavity was cleaned, and the implant was placed. Usually, cementless implants are inserted at the same depth as the broach; however, the depth of insertion may vary, depending on bone quality, stem coating, and broach shape. With the OCM approach, the femoral bone appears well; therefore, a cemented stem can be placed relatively easily. If there were no problems, the head was placed, and a final repositioning was performed. After irrigation, the fascia and subcutis were sutured and the operation was completed.

5 Clinical Results

Several authors reported satisfactory short-term results using the OCM approach [16]. However, to the best of our knowledge, no report has summarized the mid- to long-term outcomes of the OCM approach. At the authors' institution, among the 4520 initial THA cases (excluding THA with subtrochanteric osteotomy), 65 (1.4%) had deep infections, 128 (2.8%) had perioperative fractures (including intraoperative wiring), and 7 (0.15%) had postoperative dislocations. The low dislocation rate was due to the ability to check the dislocation stability in the lateral decubitus position, as described above. However, it should also be emphasized that this low dislocation rate was due to the high precision of the operation using a three-dimensional template and navigation, which was not obtained solely by using the OCM approach.

We would like to thank Editage (www.editage.com) for English language editing.

References

1. Röttinger H. The MIS anterolateral approach for THA. Orthopade. 2006;35(708):710–5.
2. Mjaaland KE, et al. Implant survival after minimally invasive anterior or anterolateral vs. conventional posterior or direct lateral approach: an analysis of 21,860 total hip arthroplasties from the Norwegian arthroplasty register (2008 to 2013). J Bone Joint Surg Am. 2017;99:840–7.
3. Ogawa H, et al. Intraoperative stability evaluation of total hip arthroplasty: a comparative study using significance navigation in the lateral and supine positions, respectively. Hip Joint. 2017;43:328–30. Japanese
4. Takada R, et al. Supine versus lateral position for accurate positioning of the acetabular cup in total hip arthroplasty using the modified Watson-Jones approach: a randomized single-blind controlled trial. Orthop Traumatol Surg Res. 2019;105:915–22.
5. Sowers CB, et al. Return to sports after Total hip arthroplasty: an umbrella review for consensus guidelines. Am J Sports Med. 2023;51:271–8.
6. Coomber R, et al. Total hip replacement for hip fracture: surgical techniques and concepts. Injury. 2016;47:2060–4.
7. Behrend H, et al. The "forgotten joint" as the ultimate goal in joint arthroplasty: validation of a new patient-reported outcome measure. J Arthroplast. 2012;27:430–6.
8. Crompton J, et al. Do hip precautions after posterior-approach total hip arthroplasty affect dislocation rates? A systematic review of 7 studies of 6,900 patients was performed. Acta Orthop. 2020;91:687–92.
9. Grammatopoulos G, et al. Pelvic positioning in the supine position leads to a more consistent orientation of the acetabular component after total hip arthroplasty. Bone Joint J. 2018;100-B:1280–8.
10. Ito T, et al. Development concept of three-dimensional preoperative planning software. J Joint Surg. 2020;39:604–8. Japanese
11. Iwana D, et al. Accuracy of angle and position of the cup using computed tomography-based navigation systems in total hip arthroplasty. Comput Aided Surg. 2013;18:187–94.
12. Ogawa H, et al. Does an augmented reality-based portable navigation system improve the accuracy of acetabular component orientation during THA? Randomized controlled trial. Clin Orthop Relat Res. 2020;478:935–43.
13. Greidanus NV, et al. Outcomes of minimally invasive anterolateral THA are not superior to those of minimally invasive direct lateral and posterolateral THA. Clin Orthop Relat Res. 2013;471:463–71.
14. Takada R, et al. Incidence of tensor fascia lata muscle atrophy after using the modified Watson-Jones anterolateral approach in total hip arthroplasty. Eur J Orthop Surg Traumatol. 2021;31:533–40.
15. Müller M, et al. Evidence of reduced muscle trauma through a minimally invasive anterolateral approach by means of MRI. Clin Orthop Relat Res. 2010;468:3192–200.
16. Inaba Y, et al. Little clinical advantage of modified Watson-Jones approach over modified mini-incision direct lateral approach in primary total hip arthroplasty. J Arthroplast. 2011;26:1117–22.
17. Nakasone S, et al. Study of iliopsoas impingement by cup protrusion after anterior approach THA. Hip Joint. 2017;43:363–6. Japanese

Anterolateral-Supine (AL-S) Approach

52

Arihiko Kanaji

1 Basic Concept of the Anterolateral-Supine Approach

There are four major total hip arthroplasty (THA) entry methods: anterior, anterolateral, lateral, and posterior. Since Crockett et al. proposed minimally invasive THA (MIS-THA) in 1998, many attempts have been made to reduce the physical and social burden on patients with osteoarthritis of the hip [1]. THA using minimally invasive procedures has been widely used in Japan, where dysplastic hips are common. MIS-THA using the anterolateral-supine (AL-S) approach (the modified Watson-Jones approach), which was reported by Pflüger et al. in 2007 [2], has become more common in Japanese hip arthroplasty registries in recent years (Figs. 52.1 and 52.2). The AL approach is an intermuscular approach that has been widely used to preserve not only the gluteus medius muscle but also the external rotator muscles, tensor fascia femoris, and posterior capsule.

The surgeon's surgical proficiency and ease of dealing with anatomical abnormalities, such as dislocated hip and neck hyper anteversion, low complication rate (e.g., no dislocation or femoral fracture), and minimally invasive soft tissues, including muscles and tendons, should be considered when deciding the approach for THA. During THA, surgeons must plan preoperatively for a variety of patient background factors, including age, sex, activity, bone morphology, and bone strength. The AL-S approach is an intermuscular entry technique performed between the tensor fascia femoris and gluteus medius muscles. Its advantages are as follows: 1. early recovery due to intermuscular entry, deafferentation, and complete preservation of muscle–tendon attachments; 2. stability of the pelvic plane in the supine position; 3. easy confirmation of leg length difference in the supine position; 4. confirmation of extra-articular indicators such as both superior anterior iliac spines and pubic symphysis during cup placement; 5. easy image use in the supine position; 6. direct visualization and selective preservation of the pisiform

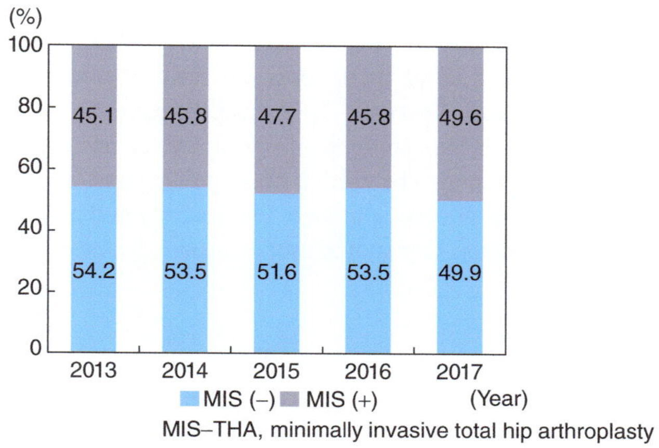

Fig. 52.1 Trends in MIS-THA ratio. (Modified from the 2013–2017 Japanese Orthopedic Association National Registry [JOANR])

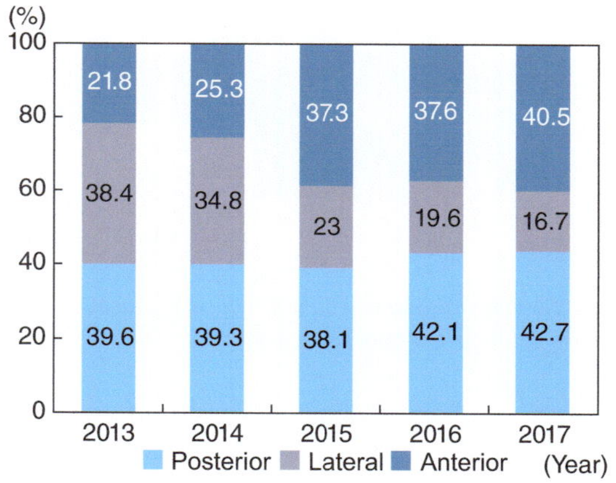

Fig. 52.2 Trends in approach selection ratios. (Modified from the 2013–2017 Japanese Orthopedic Association National Registry [JOANR])

A. Kanaji (✉)
Orthopedic Surgery Restorative Medicine of Neuro-Musculoskeletal System, Fujita Health University, Nagoya, Aichi, Japan

Fujita Health University, Bantane Hospital, Nagoya, Aichi, Japan
e-mail: arihiko.kanaji@fujita-hu.ac.jp

Y. Nakashima (ed.), *Advances in Total Hip Arthroplasty*, https://doi.org/10.1007/978-981-97-2461-1_52

muscle and common tendon with the crank cut; 7. high cup placement accuracy and good navigation system placement accuracy in the supine position; 8. the cup placement accuracy is excellent and frequency of dislocation is low because the external rotators and ischiofemoral ligament joint capsule are preserved; and 9. the capsular ligament, especially the iliofemoral ligament, can be preserved owing to direct visualization.

2 AL-S Surgical Technique and Its Features

AL-S is an intermuscular approach that is performed between the tensior fashiae latae muscle and vastus medialis muscles and is performed in the supine position, similar to the direct anterior approach (DAA) [2–7]. The AL-S approach can be performed on a conventional operating table; however, a traction table can also be used to preserve the capsular ligaments, as in the anterior approach. Fluoroscopy and portable navigation systems (for supine use) can easily be used intraoperatively. A skin incision is made 5 cm posterior to the superior anterior iliac spine and the distal end of the femoral tuberosity, with a 4 cm central and 4–6 cm peripheral incision line centered at the apex of the greater trochanter. In the AL-S approach, after the skin incision, the subcutaneous fat tissue and iliotibial tract are also incised at approximately the same level, and the anterior edges of the gluteus medius and gluteus minimus are retracted posteriorly to expand the periarticular area. As a rule, the joint capsule is cut and resected appropriately according to each case, while preserving the posterior tissues, including the longitudinal fibrous bundle of the iliofemoral and ischiofemoral ligaments, to avoid dislocation and maintain good soft tissue tension (Fig. 52.3). The principle of femoral manipulation for accurate broaching and stem insertion is to externalize the femoral insertion site by adduction of the affected limb; however, compared with anterolateral entry in the supine position, excessive adduction and external rotation cannot be achieved, and thus requires skill in femoral manipulation. Insufficient femoral manipulation for accurate broaching and stem insertion may result in femoral fracture [6].

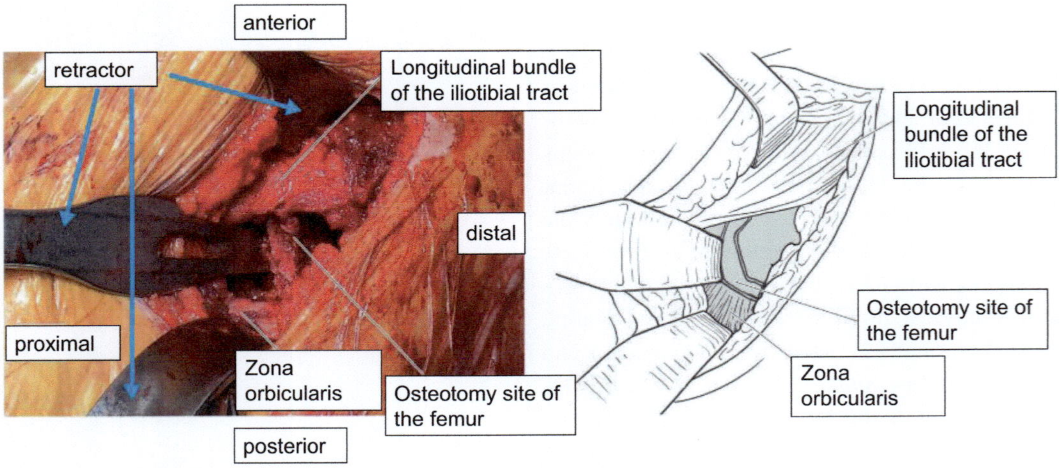

Fig. 52.3 Supine arthrodesis with arthrodesis ligament sparing total hip arthroplasty (right femoral manipulation). Preservation of the longitudinal fibers of the iliopsoas femoral ligament

3 Advantages of the AL-S Approach

(a) Selective and strategic arthrodesis ligament preservation

The hip joint has capsular ligaments that maintain static stability, and the preservation of these ligaments has been advocated in THA. The three periarticular ligaments of the hip joint are the iliofemoral (superior and inferior bands), ischiofemoral, and pubofemoral ligaments [8, 9]. Capsular ligaments are often difficult to identify in the operative field; therefore, understanding the anatomical location of the ligaments is important.

The superior band of the iliofemoral ligament originates from the superior border of the acetabulum and inferior anterior iliac spine and stops at the greater trochanter from the lateral intertrochanteric area, mainly braking adduction and external rotation. The inferior band of the iliofemoral ligament originates from the superior border of the acetabulum and inferior anterior iliac spine, and stops at the intertrochanteric line, mainly controlling the extension and external rotation. The ischiofemoral ligament originates from the posterior inferior surface of the acetabulum and is widely attached to the medial aspect of the greater trochanter, together with the articular capsule. It is the primary braking factor for hip abduction and internal rotation. The pubofemoral ligament originates from the pubic portion of the acetabulum and the suprapubic branch and stops slightly posterior to the longitudinal fibrous bundle of the iliopsoas femoral ligament. It also restricts external rotation, abduction, and extension, and it may be a factor that inhibits dislocation in the anterior approach. The capsular ligament plays an important role in supporting and stabilizing the hip joint, which is a spherical joint that moves in all three dimensions [8, 9]. However, whether the capsular ligament can be preserved in THA depends on the entry method used.

In THA using the posterior approach and in cases with significant contracture or deformity, the ischiofemoral ligament is often transected simultaneously with the joint tendon,. The anterior capsule, which is likely to involve the iliofemoral ligament, may also be incised and resected. The DAA is an intermuscular approach that involves entry between the sartorius and tensior fashiae latae muscles. The DAA requires resection of the anterior and posterior capsules because the medullary cavity is manipulated by anterior elevation of the femur. In many cases, resection of the inferior band of the iliofemoral ligament, which antagonizes anterior elevation of the femur, is needed. In addition, the ischiofemoral ligament and external rotator muscle tendon may need to be transected to improve manipulation of the anterior elevation of the femur.

The AL-S approach has the notable advantage of selectively and strategically preserving the capsular ligaments and external rotation muscles, and THA, which preserves the capsular ligaments, has the advantages of avoiding postoperative dislocation, ensuring good stability and range of motion (ROM) of the hip, a three-dimensional (3D) mobile ball joint, and maintaining adequate soft tissue tone.

(b) Good posterior dislocation resistance

The posterior approach is still widely used in THA because it is easy to deploy and treat in all cases. In contrast, THA using the posterior approach may decrease resistance to posterior dislocation due to posterior soft tissue invasion [10]. In general, posterior dislocation is caused by various impingements due to hyperflexion or hyper-internal rotation of the hip joint, and the risk of posterior dislocation is particularly high in patients with osteoarthritis of the hip, with minimal deformity and good preoperative ROM. In THA, accurate implant placement should be achieved, to avoid implant impingement, because the safe zone for implant placement is very narrow. Recently, the importance of using computer-assisted orthopedic surgery (CAOS) techniques, such as navigation systems, has been reported [8, 9], and we should use the CAOS technique for such high-risk cases; however, in cases where CAOS techniques are not available or in THA for patients with a good preoperative ROM, we strongly recommend performing THA with the AL-S approach, using the complete preservation technique of the ischiofemoral ligament. The AL-S THA technique allows a high rate of simultaneous preservation of all short external rotators and the ischiofemoral ligament, which is advantageous for maintaining resistance to posterior dislocation.

(c) Simultaneous maintenance of anterior and posterior dislocation resistance

In patients with no preoperative leg length discrepancy or posterior pelvic tilt, preservation of the anterior capsular ligament, especially the inferior band of the iliofemoral ligament, is required. The inferior band of the iliofemoral ligament has extension and external rotation control of the hip, and ligament resection increases the risk of anterior dislocation [8, 9]. To avoid anterior dislocation, preserving the inferior band of the iliofemoral ligament is important. The AL-S THA technique allows a high rate of simultaneous preservation of all short external rotators, the ischiofemoral ligament, and the inferior band of the iliofemoral ligament, thus reducing the risk of anterior and posterior dislocation resistance.

(d) Preservation of selective short external rotator muscle groups

Although AL-S is an anterior approach, it is characterized by good visibility of the tendon stop of short external rotators. In particular, patients with a small ROM have a lower risk of postoperative dislocation than those with a large ROM because various impingements are less likely to occur. However, patients with a reduced ROM after THA may not be fully satisfied with their ambulation and activities of daily living. Accurately predicting the intraoperative and postoperative ROM based on the preoperative ROM in individual patients is difficult. Therefore, if full preservation of the short external rotator muscles in patients with hip contractures does not provide sufficient intraoperative flexion and internal rotation ROM, selective resection of the short external rotator muscles is recommended to provide a strategically adequate ROM while ensuring resistance to dislocation. The AL-S approach, which is superior in selective preservation of the capsular ligament and short external rotator muscles, can reduce the risk of dislocation and limitation of motion and may lead to improved patient satisfaction.

(e) Accuracy of cup installation

In the supine THA technique, pelvic anatomical indexes such as the bilateral superior anterior iliac spines and pubic symphysis can be easily grasped, and fluoroscopy and portable navigation systems (for the supine position) can be used intraoperatively. Good cup placement within 3° has been reported for supine THA using portable navigation [11–15].

4 Disadvantages of the AL-S Approach

The disadvantages of the AL-S approach are 1. the need for skill, 2. risk of femoral fractures, and 3. stem selection [6]. These problems are not unique to the AL-S approach but are considered common disadvantages of supine anterior advancement. Notably, the principle of femoral manipulation for stem insertion in the AL-S approach is externalization of the femoral insertion site by adduction of the affected limb, and that insufficient femoral manipulation for accurate broaching and stem insertion may result in femoral fracture [6]. Moreover, femoral fracture may occur if femoral manipulation for accurate broaching and stem insertion is inadequate. If the angle of internal rotation is 45° or greater, the rotator muscles should be preserved as much as possible; however, if the angle of internal rotation is 10° or less, contracture is often strong and externalization is difficult, and the amount of joint capsule resection and partial removal of the common muscle tendon should be considered before surgery. Preoperative computed tomography should be performed to determine the degree of internal meniscal atrophy. Atrophy of the internal meniscus muscle suggests the presence of contracture, and release of the conjoint tendon should be considered.

Short stems tend to be easier to use, whereas rectangular stems with relatively thicker anteroposterior diameters and longer stem lengths tend to be more difficult [7]. To address this problem, a 3D template system should be used to select a stem that matches preoperative bone morphology [7].

References

1. Crockett HC, et al. Mini-incision for total hip arthroplasty. New Orleans: AAOS; 1998.
2. Pflüger G, et al. Minimally invasive total hip replacement via the anterolateral approach in the supine position. Int Orthop. 2007;31(Suppl 1):S7–S11.
3. Martin R, et al. Anterolateral minimally invasive total hip arthroplasty: a prospective randomized controlled study with a follow-up of 1 year. J Arthroplast. 2011;26:1362–72.
4. Takada R, et al. Direct anterior versus anterolateral approach in one-stage supine total hip arthroplasty: Focused on nerve injury: a prospective, randomized, controlled trial. J Orthop Sci. 2018;23:783–7.
5. Nishimura M, et al. Comparison of gait motion including post-operative trunk deflection between direct lateral and anterolateral approaches in supine total hip arthroplasty. J Arthroplast. 2016;31:1603–8.
6. Nakai T, et al. Early complications of primary total hip arthroplasty in the supine position with a modified Watson-Jones anterolateral approach. J Orthop. 2014;11:166–9.
7. Kutzner KP, et al. One-stage bilateral implantation of a calcar-guided short-stem in total hip arthroplasty: minimally invasive modified anterolateral approach in supine position. Oper Orthop Traumatol. 2017;29:180–92.
8. Martin HD, et al. The function of the hip capsular ligament.
9. Hidaka E, et al. Ligament strain on the iliofemoral, pubofemoral, and ischiofemoral ligaments in cadaver specimens: biomechanical measurement and anatomical observation. Clin Anat. 2014;27:1068–75.
10. Hodges PW, et al. Insight into the function of the obturator internus muscle in humans: observations with development and validation of an electromyography recording technique. J Electromyogr Kinesiol. 2014;24:489–96.
11. Anjiki K, et al. Effectiveness of an accelerometer-based portable navigation for intraoperative adjustment of leg length discrepancy in total hip. J Orthop Sci. 2022;27:169–75.
12. Hasegawa M, et al. Accuracy of acetabular cup insertion in an anterolateral supine approach using an accelerometer-based portable navigation system. J Artif Organs. 2021;24:82–9.
13. Okamoto M, et al. Comparison of accuracy of cup position using portable navigation versus alignment guide in total hip arthroplasty in the supine position. Hip Int. 2021;31:492–9.
14. Takada R, et al. Portable imageless navigation system and surgeon's estimate for accurate evaluation of acetabular cup orientation during total hip. Eur J Orthop Surg Traumatol. 2020;30:707–12.
15. Hayashi S, et al. Evaluation of the accuracy of acetabular cup orientation using the accelerometer-based portable navigation system. J Orthop Sci. 2020;25:612–7.

Mini-One Anterolateral Approach

53

Masahiko Mihara and Kazuo Hirakawa

1 Mini-One Anterolateral Approach

Mini-one anterolateral approach [1–6] is a minimally invasive surgery (MIS) that is easy to learn for beginners because of its good visual field on both the acetabular and femoral sides.

2 Indications for Surgery

The indications for surgery are wide-ranging, and almost all cases of primary THA can be treated.

(a) Primary disease, age, bone quality

The acetabular and femoral views are excellent and easy to manipulate, so the indications of surgery are wide-ranging and can be used in almost all cases. The most common indications are osteoarthritis of the hip (primary OA, developmental dysplasia of the hip, Crowe's classification type I and II) and osteonecrosis of the femoral head. Restrictions based on age or bone quality are not present.

(b) Difficult cases

Highly obese, cases of varus with short neck (e.g., Perthes like deformity) are not contraindications; however, requires a learning curve.

3 Preoperative Planning

The images should be prepared by X-ray and CT so that they can be checked at any time during the operation. The use of 3D template is also considered for patients with severe deformities.

(a) X-ray image (standing and lying A-P view, lateral view)

Both hip joints should be in the middle position of internal/external rotation and adduction/adduction, and the pelvis should be in a position in which the midline of the sacrum and the line of the pubic symphysis are aligned as much as possible. In the standing position, the presence or absence of posterior pelvic tilt should be checked. In the axial position, the cup size and the anteversion can be determined to some extent. The cup size and anteversion should be reconfirmed on CT.

(b) CT image

The amount of osteophytes, thickness of the acetabular base, and anteversion angle of the femoral neck should be measured by photographing from the pelvis to the knee joint. The amount of osteophytes protruding from the acetabulum anteriorly and posteriorly when the cup is placed should be determined. The anteversion angle of the insertion stem should be determined to some extent, and the anteversion angle of the cup should be determined accordingly to ensure that the cup fits within the safe zone.

4 Problems of the Mini-One Anterolateral Approach

Although this approach has a better view and a wider range of surgical indications than other MIS approaches, it is necessary to consider the effect on the patient for releasing the anterior 1/3rd of the gluteus medius muscle from the greater trochanter. Yukizawa et al. [7] compared the muscle-sparing OCM approach with the present approach and reported that the present approach had a disadvantage in hip abductor muscle strength up to 6 weeks postoperatively; however, any difference thereafter was not observed.

M. Mihara (✉) · K. Hirakawa
Shonan Kamakura Joint Reconstruction Center, Kamakura, Kanagawa, Japan

Y. Nakashima (ed.), *Advances in Total Hip Arthroplasty*, https://doi.org/10.1007/978-981-97-2461-1_53

5 Surgical Techniques

(a) Preparation of body position, marking, etc. (Fig. 53.1)

The patient is positioned in the lateral position and secured at the bilateral superior anterior iliac spines and sacrum using a pelvic immobilizer. To prevent peroneal nerve palsy, sponges are placed on the thigh and lower leg, and the head of the peroneal bone is elevated. The surgeon marks the midline of the body axis at the shoulder position, and the anesthesiologist indicates this position intra-operatively. A surgeon uses an alignment guide to determine the anterior cup opening angle.

The surgeon standing in front of the patient begins the operation. The first assistant stands behind the patient, and the second assistant stands next to the surgeon (near the patient's head). When the surgeon starts the femoral side operation (stem), the surgeon moves backward and changes places with the first assistant.

During the operation, the leg must be lowered from the operating table during manipulation of the femur (stem), so it is convenient to use a drape with leg pouches on both sides. If the preparation is difficult, the drape can be folded in half and used as a leg bag.

(b) Skin incision (Fig. 53.2)

The length of the skin incision is determined according to the body shape and the degree of deformity, which is usually 8 cm. If the patient is tall, obese, or muscular, or if the deformity is severe and a leg lengthening of >2 cm is planned, the length should be extended up to 10 cm. The femur is traced over the iodine drape in the lateral position, and the median line of the bony axis is drawn. A skin incision line is drawn approximately 2.5 cm distal to the apex of the greater trochanter at 20° to 30° angle to the midline of the shaft.

(c) Procedure of fascia and gluteus minimus and medius (Figs. 53.3 and 53.4)

The skin (dermis) and subcutaneous fat are hemostatically removed. The fascia is incised parallel to the femur axis, and the proximal gluteus maximus muscle is divided manually while the hip joint is kept in an intermediate position of internal and external rotation. The fascia is then opened anteriorly and posteriorly with a Charnley wound opener. The gluteus

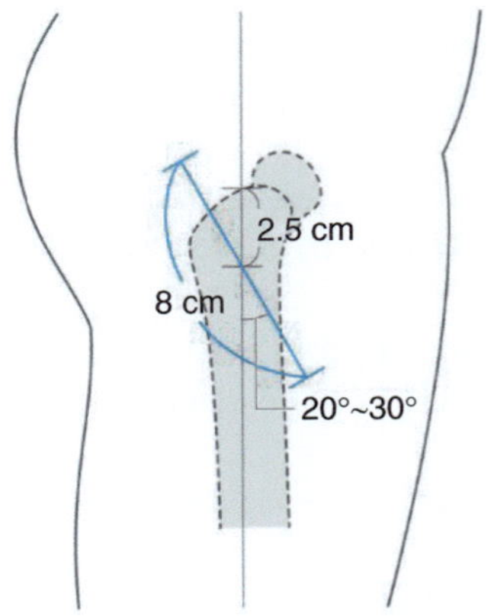

Fig. 53.2 Skin incision. The skin incision was 8 cm straight line centered 2.5 cm distal from the tip of the greater trochanter and inclined approximately 20–30 degrees from the femoral axis

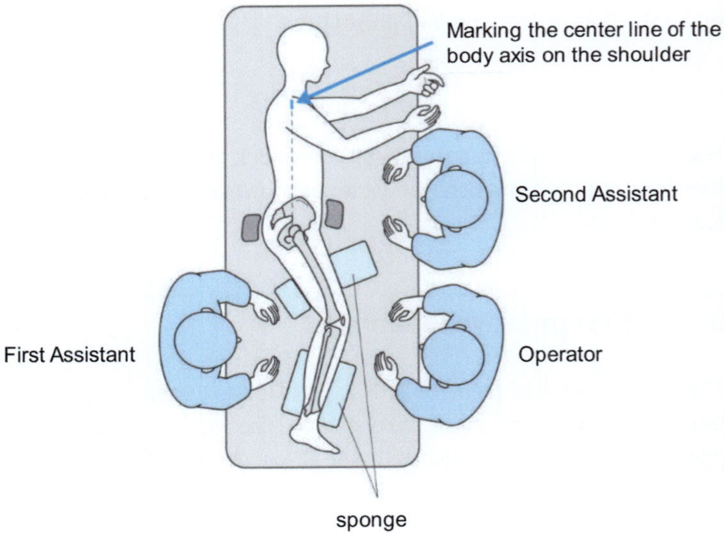

Fig. 53.1 Preparation of posture, marking, etc. The patient is placed in the lateral position and secured at the bilateral superior anterior iliac spines and sacrum using a pelvic fixator

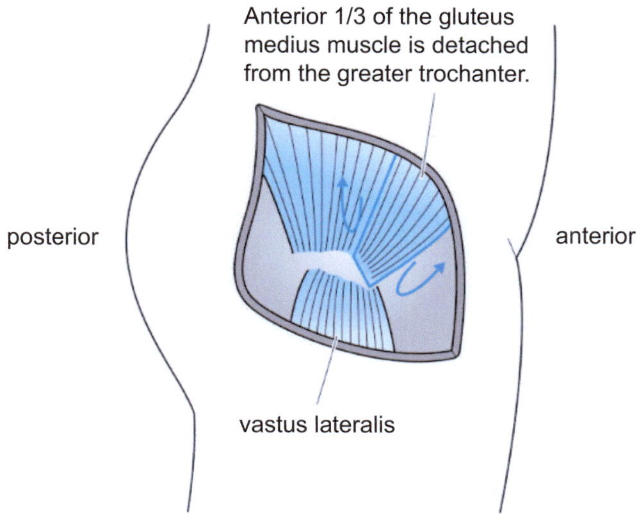

Fig. 53.3 Operation of the gluteus medius muscle. The anterior margin of the gluteus medius and the vastus lateralis are identified, and the anterior one-third of the gluteus medius is flap-exposed from the greater trochanter

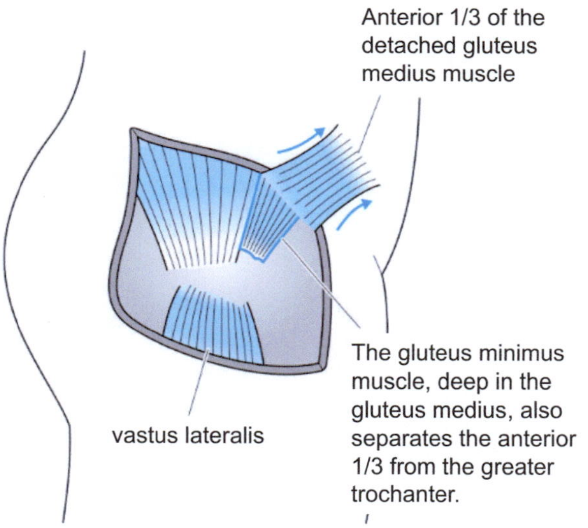

Fig. 53.4 Operation of the gluteus minimus muscle. The gluteus minimus muscle also separates anteriorly from the greater trochanter in a flap-like fashion

medius and vastus lateralis muscles can be identified by resection of the bursa directly beneath them. The anterior margin of the gluteus medius and vastus lateralis are identified, and the anterior and inferior one-third of the gluteus medius is dissected as a flap from the greater trochanter. The inferior border of the gluteus medialis muscle overlaps the vastus lateralis muscle in part, so care should be taken not to damage the vastus lateralis muscle when exfoliating the anterolateral inferior border of the vastus medialis muscle. In patients with short-necked, varus neck, or severe deformity of the head of the bone, the gluteus medius muscle runs anteriorly; therefore, care should be taken. The gluteus minimus can be identified by excising the fatty layer immediately deep to the gluteus medius and bluntly dissecting it, so the small gluteus medius is also dissected anteriorly in the lower one-third in a flap shape from its attachment to the greater trochanter. Since it is often attached to the joint capsule, the muscular belly of the gluteus medius and minimus are carefully dissected using an electric scalpel or Cobb raspatorium.

(d) Periarticular operation (Fig. 53.5).

The hip joint is placed in mild external rotation, and a Hohmann retractor is applied to the three points: above the lateral side of the femoral neck, anteriorly below the femoral head, and axially along the neck to further release the gluteus minimus muscle attached to the joint capsule while applying force to secure vision.

Although the capsule may be preserved in some cases, it may cause postoperative impingement, impair leg length compensation, and impair range of motion. The anterior capsule is resected by making a T-shaped incision with an electrocautery scalpel that reaches the inter-trochanteric ridge and acetabular capsule.

(e) Dislocation and femoral neck osteotomy (Fig. 53.6)

The hip joint is flexed, externally rotated, and adducted, the knee is flexed at 90°, and the lower leg is allowed to droop. If it is difficult to dislocate, resection of the residual bursa inside the neck is recommended. At this point, the lesser trochanter should be identified. The neck is osteotomized

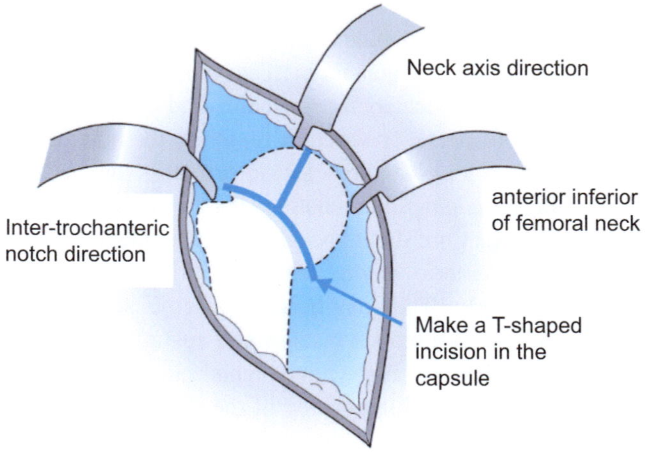

Fig. 53.5 Operation around capsules. The Hohmann retractor is used to make three T-shaped incisions with an electric scalpel over the lateral aspect of the femoral neck, antero-inferior aspect of the femoral head, and axial aspect of the neck, reaching the inter-trochanteric notch and acetabular cap

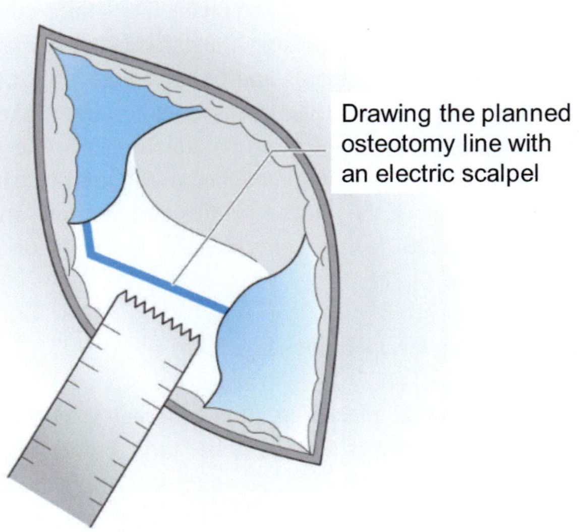

Fig. 53.6 Femoral neck osteotomy. The osteotomy line is determined by referring to the distance from the lesser trochanter and the distance from the inter-trochanteric notch and is marked with an electrocautery scalpel. Osteotomy of the neck was performed with a bone saw

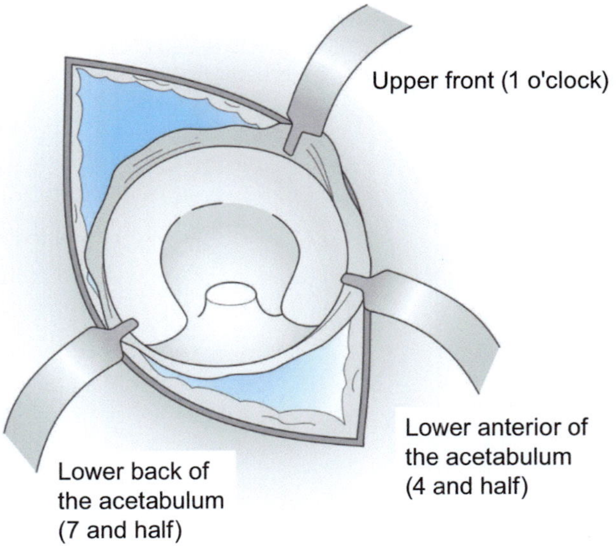

Fig. 53.7 Operation of the acetabulum. One Hohmann retractor is applied posteriorly downward to the acetabulum, one anteriorly downward to the acetabulum, and one anteriorly upward (at 1 o'clock, 4:30, and 7:30) to allow the entire acetabulum to be deployed

according to the preoperative plan, using the distance from the lesser trochanter and inter- trochanteric notch as a guide to mark the osteotomy line. Osteotomy of the neck is performed with a bone saw. Osteotomy is performed avoiding damage to the vastus lateralis and gluteus medius muscle by using a retractor.

(f) Acetabular exposure (Fig. 53.7)

The lower leg is returned, the hip is placed in mild flexion and external rotation, and the femur is moved posteriorly and inferiorly with one Hohmann retractor applied posteriorly and inferiorly to the acetabulum, and one anteriorly and superiorly to the acetabulum (at 1:00, 4:30, and 7:30) to expand the entire acetabulum. The retractor should be inserted along the acetabular rim bone, and care should be taken to avoid nerve or vascular injury to the soft tissues deep within. A long-handled electric scalpel is then used to remove the remaining inflammatory synovium and labrum of the acetabulum. Care should be taken not to damage the femoral nerve and arterio-venous veins anteriorly and the sciatic nerve posteriorly. The long-handled electric scalpel may energize nerves and blood vessels when treating the residual joint capsule, labrum, and synovial membrane.

The planned insertion position of the cup is marked with an electric scalpel on the X-ray or CT image to be used as an index for reaming.

(g) Acetabular reaming

The hip joint is mildly flexed and externally rotated to move the femur firmly posteriorly, and reaming is initiated using a

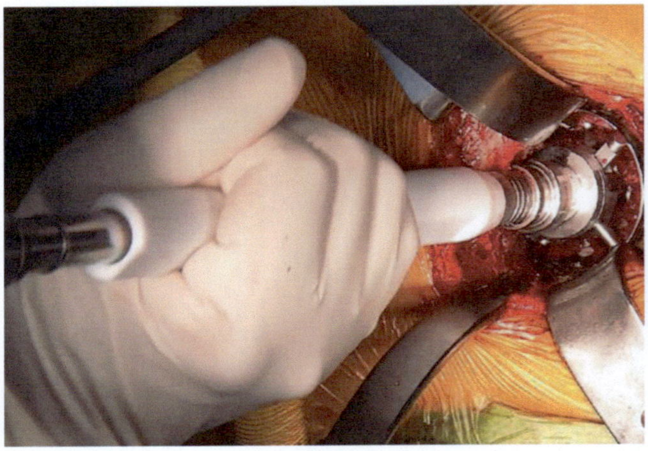

Fig. 53.8 Acetabular reaming (final reaming). The shaft of the reamer gradually collapses after inwardization

reamer. The shaft of the reamer cannot be tipped over if it is obstructed by soft tissues; therefore, at the start of reaming, the reamer shaft should be tipped over gradually after medialization (Fig. 53.8), and reaming can be performed without being obstructed by skin or soft tissues.

(h) Cup insertion (Fig. 53.9)

After cleaning, the cup is inserted after confirming that the surrounding soft tissues are not entrapped. The cup is inserted while firmly grasping the cup holder and checking the alignment guide, body axis, pelvic tilt, amount of overhang of the osteophyte, and amount of overhang of the cup so that the alignment is not disturbed by the skin, femur, or soft tissues.

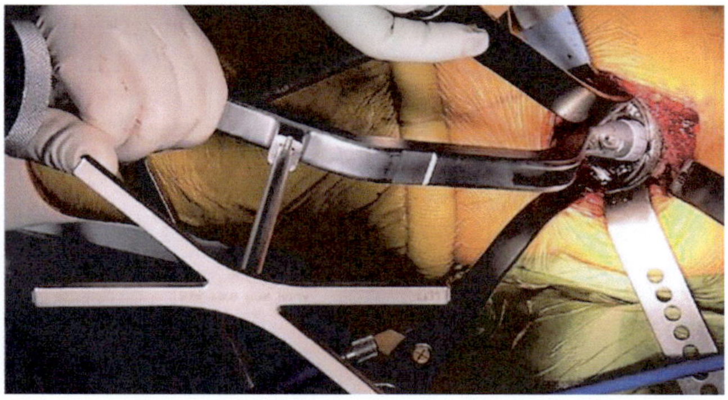

Fig. 53.9 Installation of cups. The alignment guide, body axis, pelvic tilt, amount of osteophyte overhang, and amount of cup overhang should be checked while hammering in

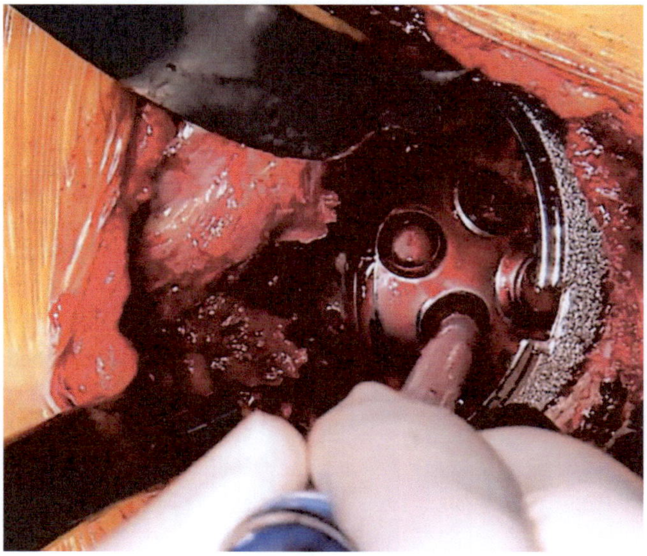

Fig. 53.10 Confirmation of cup installation using screw holes. By using a multi-hole cup and observing the acetabulum from all screw holes, it is possible to confirm that the acetabulum is contacted circumferentially without gaps

The anesthesiologist should be informed about the preoperative shoulder marking position (midline of the body axis), which will be used as an indicator. The transverse acetabular ligament also serves as an indicator of anterior acetabular anteversion and inferior acetabular rim.

When a multi-hole cup is used, it is possible to confirm that the acetabulum is contacted circumferentially without gaps by observing the acetabulum from all screw holes (Fig. 53.10). In case of a gap, it should be hammered in again. A flexible drill is used with a guide to fix one to three screws from posterior to supra-posterior to the acetabulum. At least one screw should be inserted to obtain rotational stability.

The anterior superior and posterior inferior osteophytes of the acetabulum are resected with a folding chisel. This is an important operation to prevent impingement and dislocation. In the case of internalization, large osteophytes often protrude, so it is important to resect the surrounding osteophytes firmly. Careful resection of the inferior osteophyte should be performed because the chisel can easily enter more distally, with a risk of hemorrhage. After ensuring that any soft tissue is not left behind, a liner is inserted.

(i) Femur exposure

The surgeon moves behind the patient and changes places with the first assistant. The hip is flexed, externally rotated, and adducted, the knee is flexed 90° and the lower leg is lowered. A wide, highly deformed obturator retractor, two-pronged retractor to raise the femoral neck, and Hohmann retractor medial to the neck are applied to expose the stem entry site (Fig. 53.11). The remaining bursa is resected from anterior to posterior of the osteotomy surface to make the stem insertion site easier to observe. Care is taken not to damage the short external rotator muscle group. The intertrochanteric notch is resected using a box cutter or wells, and the insert site is prepared. The postero-lateral side of the stem is cut using a curved rasp to prevent the stem from setting inwardly.

For an anterolateral approach, the stem tends to enter posteriorly from the anterior side (flexed insertion); therefore, rasping should be performed with the posterior lateral side in mind.

After confirming the direction of the medullary cavity with a starter reamer, rasping is performed in order of decreasing number. At this time as well, care must be taken not to turn the rasp from anterior to posterior. The lateral cortex should also be considered to prevent the rasp from being placed inwardly. The preoperative anteversion angle of the stem is determined to some extent by preoperative CT or

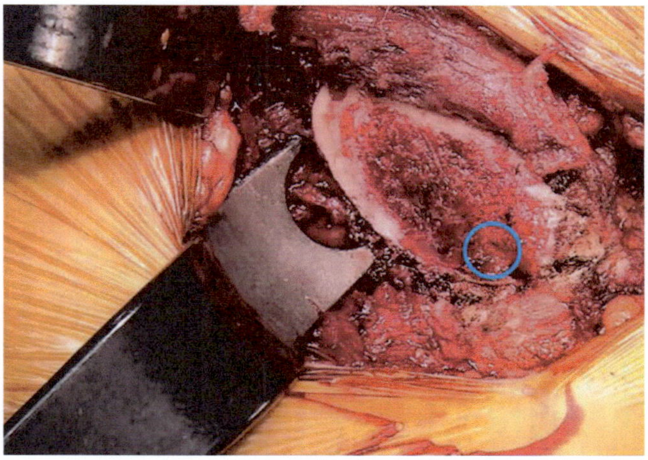

Fig. 53.11 Exposure of the stem penetration site. Residual capsules from anterior to posterior of the osteotomy surface should be resected to facilitate visualization of the stem insertion site. Care should be taken not to damage the short rotator muscles

3D template, and rasping is performed while maintaining this angle. Although the strength of the hammer should be adjusted to avoid fracture, it should be tapped firmly and repeatedly until the rasp does not sink down. When a trial rasp is inserted, perform a test repair according to the case.

(j) Stem insertion

After cleaning, it is struck patronizingly, aiming to fit in the same position as the final rasp.

(k) Reposition

A trial head (or neck in the case of a modular head) is placed, and a trial reposition is performed. The surgeon grasps the neck with gauze and has an assistant perform traction and internal rotation of the lower extremity in the direction of mild flexion. Leg length and dislocation stability are evaluated (stability is confirmed in flexion- adduction- internal rotation and extension- external rotation - and internal rotation). In a trouble free situation, a real head (or neck in the case of a modular neck) is applied.

(l) Repair of the gluteus medius and minimus

After placement of all implants and reconfirmation of leg length and dislocation stability, the gluteus medius and minimus muscles are repaired that has been separated during approach. The hip joint is externally rotated to locate the anterior component of the transected gluteus minimus. No. 5 Tycron is sewn onto the tendon component of the gluteus minimus according to the Kirchmayer technique. The hip joint is abducted, the tendon is relaxed, and then the posterior component of the untranslated gluteus medius is sewn onto the posterior component (Fig. 53.12).

The anterior component of the gluteus medius is sutured with a mattress suture after threading three, No. 5 Tycron, through the anterior separation of the greater trochanter. As shown in Fig. 53.13, the hip joint is placed in abduction and mild internal rotation, and the muscle tendons are loosened to facilitate manipulation.

(m) Subcuticular closure

The fascia, subcutaneous fat, and subcutaneous skin are carefully sutured and skin is covered with a dressing.

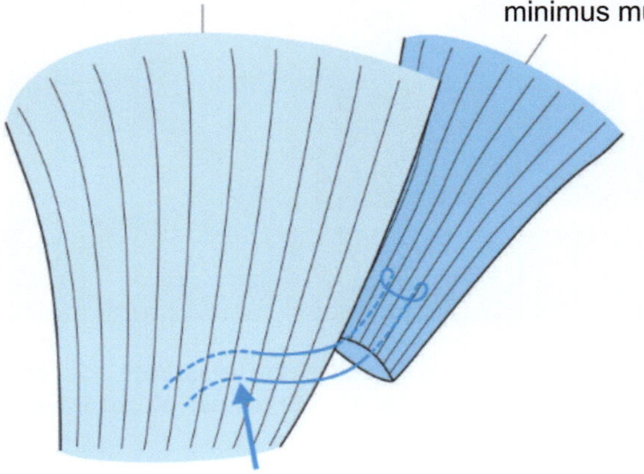

Fig. 53.12 Repair of the gluteus minimus muscle. A No. 5 Tycron is sewn onto the tendon component of the gluteus minimus muscle according to the Kirchmayer technique. The hip is placed in abduction and mild internal rotation, and the tendon is relaxed and then sewn onto the posterior component of the gluteus medius muscle, which has not been transected

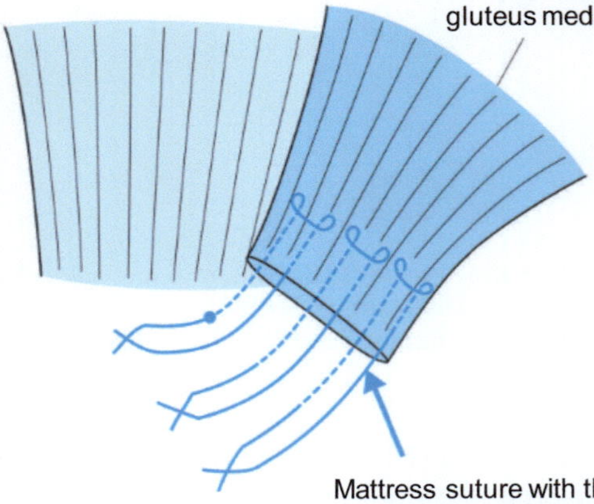

Fig. 53.13 Repair of the gluteus medius muscle. The anterior portion of the gluteus medius is sutured with a mattress suture after three No. 5 Tycron are threaded through the anterior portion of the greater trochanter

6 Rehabilitation

On the evening of the day of surgery, if the patient's vital signs are stable, with no discomfort (nausea, vomiting, etc.), and with good pain control, the patient should perform standing and stepping training with a physical therapist. In case the patient has no intraoperative fractures, he is allowed to walk with full weight bearing, and gait training using a pick-up walker, which is usually started the day after surgery. On the second and third postoperative days, gait training with a T-cane is started, followed by stair climbing training on the fourth and fifth postoperative days, and patients are discharged home on the fifth postoperative day under the normal clinical path.

References

1. Berger RA. Mini-incision total hip replacement using an anterolateral approach; technique and results. Orthop Clin North Am. 2004;35:143–51.
2. Hirakawa K. Clinical outcome of mini-one incision THA: to establish early discharge. J Musculoskelet Syst. 2004;17:1309–14. [Japanese]
3. Tsuji K, et al. Mini-one anterolateral THA. In: Iwamoto Y, et al., editors. OS NOW instruction no. 9 hip arthroplasty: techniques for MIS to revision. Tokyo: Medical View; 2009. p. 74–86. Japanese.
4. Hirakawa K. 2nd Mini-one anterolateral approach hip arthroplasty for hip OA due to acetabular dysplasia. J Joint Surg. 2011;30:135–44. Japanese
5. Hirakawa K. The anterolateral intrusion method. In: Matsuno T, Ito H, editors. All about total hip arthroplasty [THA]. Revised 2nd ed. Tokyo: Medical View; 2015. pp. 184–200. Japanese.
6. Ikeda T, et al. Relationship between motor function before mini-one antero-lateral incision and postoperative gait expansion. Sci Orient Med Physiother. 2009;24:127–30. Japanese
7. Yukizawa Y, et al. Effect of partial gluteal muscle dissection on postoperative outcomes in MIS-THA. Hip Joint. 2010;36:185–9. Japanese

Posterolateral Approach

54

Daisuke Takahashi

1 Advantages and Disadvantages of the Posterolateral Approach

The posterolateral approach for total hip arthroplasty (THA) is the most commonly used worldwide. The acetabular view is good, and the femoral stem insertion is simple; therefore, the procedure can be performed smoothly and is applicable to almost all THA cases.

The advantages of the posterolateral approach are as follows: (1) smooth deployment without dissection of the greater trochanter or damage to the abductor muscle group; (2) very low risk of intraoperative femoral fracture due to a good visual field; (3) ability to use any type of stem; (4) very low risk of sciatic nerve palsy due to easy nerve protection; and (5) minimal patient burden, owing to the short operative time.

The risk of postoperative dislocation following use of the posterolateral approach has been reported to be slightly higher than that associated with other approaches [1–4]. The risk of dislocation can be minimized by accurate implant placement, evaluation of stability with intraoperative trial repositioning, and improvement in range of motion with impingement release. In addition, it has been reported that the risk of postoperative dislocation is similar to that of other approaches after repairing the short external rotator muscle group and joint capsule [5–7].

2 Indications and Limitations of the Posterolateral Approach

The posterolateral approach is applicable to almost all THA cases, including Crowe type IV THA with subtrochanteric femoral shortening osteotomy and revision THA, as well as initial THA due to various diseases. The posterolateral approach is also suitable for contracted hips with a severely limited range of motion; very rarely can cases not be treated with the posterolateral approach.

3 THA by Posterolateral Approach

(a) Body position

Surgery is performed in the lateral recumbent position. For accurate cup placement in the lateral recumbent position, the pelvis must be firmly fixed, such that the straight line connecting the two superior anterior iliac spines is perpendicular to the floor to minimize pelvic movement during the operation (Fig. 54.1). After fixation, the bed tilt is adjusted with fluoroscopic confirmation to position the pelvic tilt completely perpendicular to the floor (Fig. 54.2) [8].

(b) Skin incision, arthrodesis, and osteotomy

The skin incision should be a straight line, 10–12 cm from the innominate tubercle through the posterosuperior border of the greater trochanter (Fig. 54.3a). The fascia is incised in the same direction as the skin incision, and proximally, the greater trochanter is identified by the blunt division of the gluteus maximus muscle in the direction of the fibers (Fig. 54.3b). The hip joint is held in extension and mild internal rotation. The sciatic nerve runs in the dorsal fatty tissue and is retracted using a muscle hook or other means as gently as possible. Short external rotators are identified by dissecting the bursa and fatty tissue from the posterior border of the greater trochanter. First, the piriformis muscle, which is the most proximal muscle with a long tendon (often white), must be identified. The elevatorium is then inserted distally from the boundary between the piriformis minor- and piriformis muscles, and the short external rotators are sutured with stay sutures before dissection (Fig. 54.4a). The quadriceps femoris muscle should be avoided as much as possible because a branch of the medial femoral rotator artery runs within the mus-

D. Takahashi (✉)
Faculty of Medicine and Graduate School of Medicine, Department of Orthopaedic Surgery, Hokkaido University, Sapporo, Hokkaido, Japan
e-mail: rainbow-quest@pop02.odn.ne.jp

Y. Nakashima (ed.), *Advances in Total Hip Arthroplasty*, https://doi.org/10.1007/978-981-97-2461-1_54

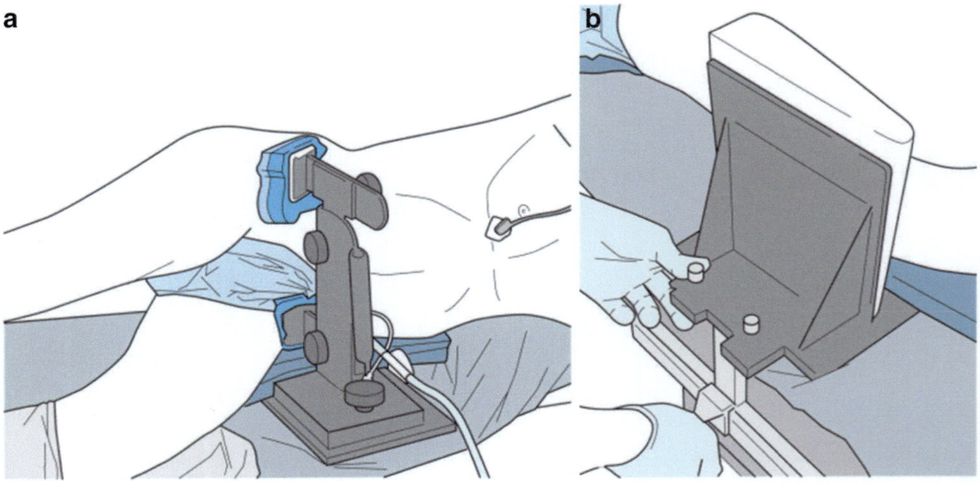

Fig. 54.1 Positions. The patient is placed in a full lateral supine position with the pelvis fixed at the three points of the bilateral superior anterior iliac spines and the sacrum

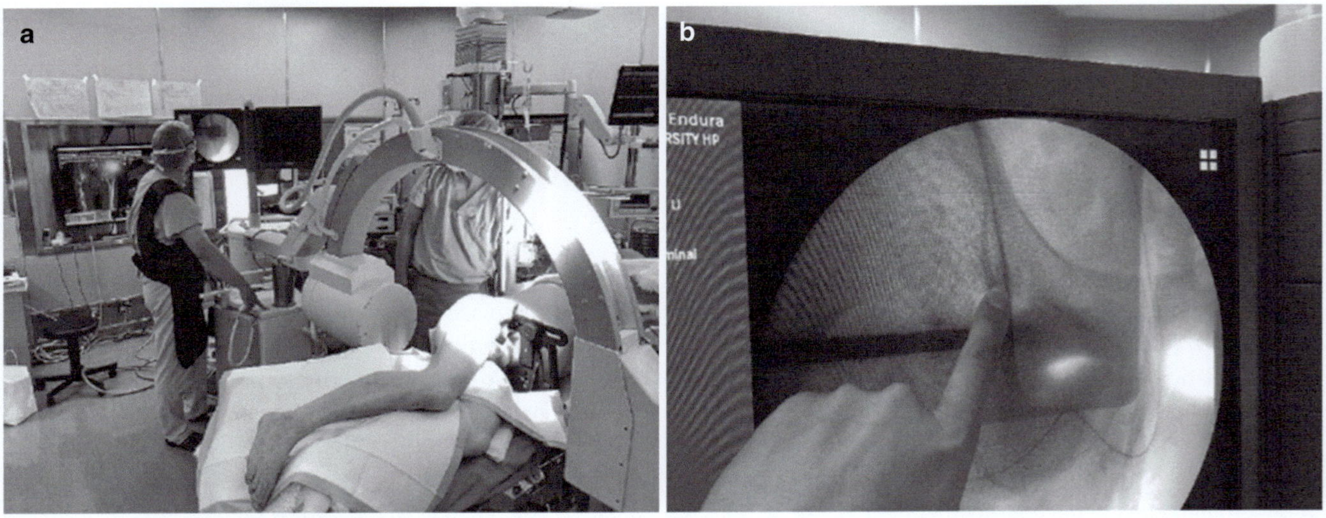

Fig. 54.2 Confirmation and adjustment of pelvic tilt under fluoroscopic observation. The bed tilt is adjusted to make the pelvic tilt perpendicular to the floor, while checking the lower edge of the lacrimal scar under fluoroscopic guidance

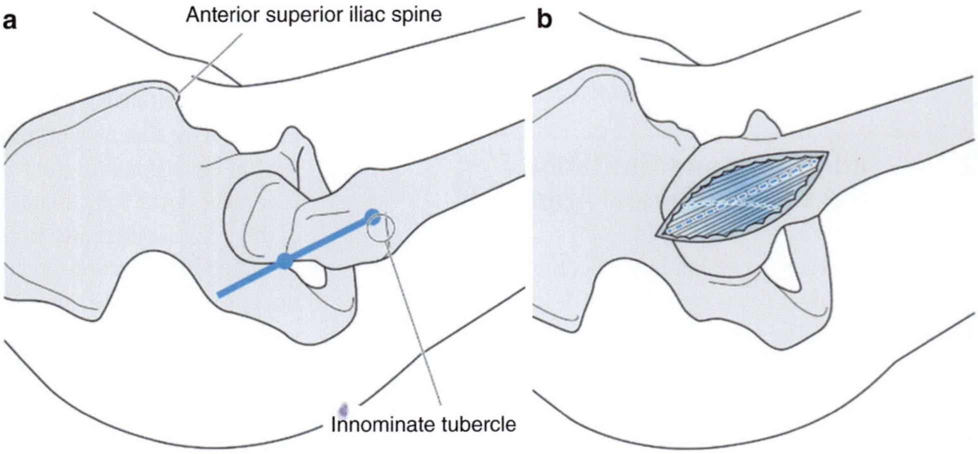

Fig. 54.3 Skin incision and fasciotomy. (**a**) A skin incision is made at the proximal end of the greater trochanter. (**b**) The fascia is incised in the same direction as the skin incision, and proximally, the gluteus maximus is bluntly divided in the direction of the fibers

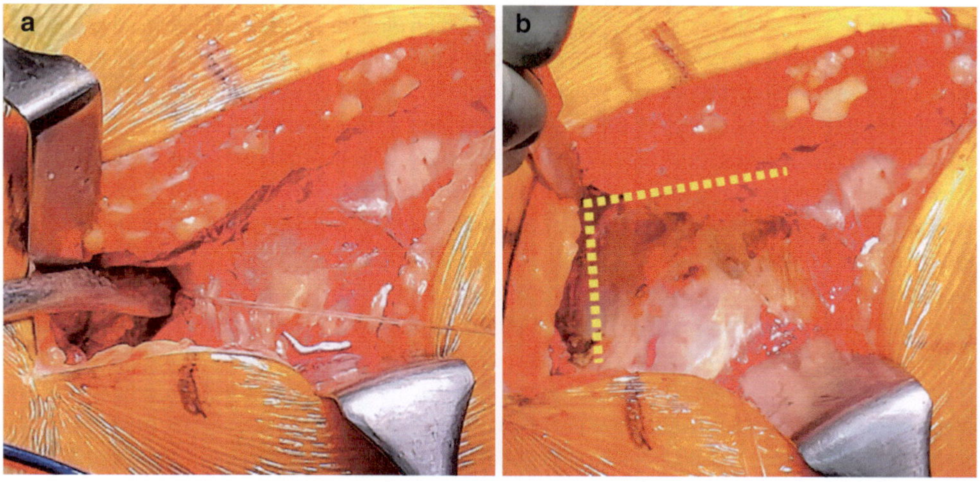

Fig. 54.4 L-shaped incision of the short external rotator muscle group and arthrodesis. (**a**) Insertion of the elevatorium proximal to the piriformis muscle and dissection of the short external rotator muscle group; (**b**) L-shaped incision of the glenoid arthrodesis

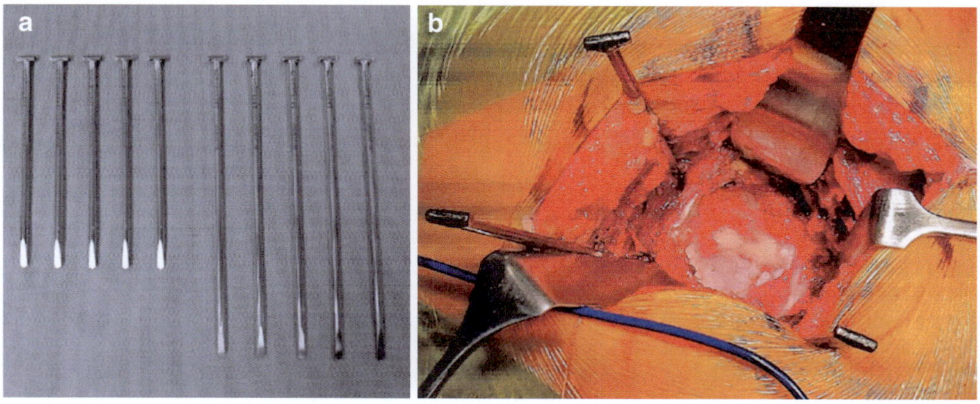

Fig. 54.5 Development of the acetabulum. (**a**) Two types of Steinmann pins (125 mm and 175 mm in length) are used, depending on the patients' physiques. (Teijin Nakashima Medical Ltd.). (**b**) The femur is retracted anteriorly using a Cobra retractor (Zimmer Biomet) to obtain a good field of view

cle. The gluteus minimus and gluteus medius are dissected with the raspatorium, and the gluteus minimus is retracted proximally, while the gluteus medius is incised in an L-shape for preservation (Fig. 54.4b). After inserting the elevatorium distal to the femoral neck, an assistant flexes, adducts, and internally rotates the hip joint to dislocate the femoral head posteriorly.

The femoral head is elevated with a retractor, the elevator is inserted into the proximal aspect of the femoral neck, and osteotomy is performed with a bone saw. In principle, the osteotomy should be positioned at the level of the lesser trochanter; however, if the lesser trochanter is difficult to reach, the apex of the greater trochanter or the base of the femoral head may be used as an index. Large bone forceps are used to remove the head of the bone.

(c) Acetabular deployment and cup placement

Three Steinmann pins (Teijin Nakashima Medical) (Fig. 54.5a) and a Cobra retractor (Zimmer Biomet) are used for the acetabular deployment. An assistant places a short blunt graphene hook over the femoral neck osteotomy and gently pulls up the femur to facilitate the deployment operation. The posterior capsule is grasped with large Kocher forceps and pulled posteriorly, where it is dissected distally posterior to the acetabulum with an electrocautery scalpel. Once the sciatic position is confirmed, the first Steinmann pin is placed distally posteriorly and obliquely, the second Steinmann pin is placed supraacetabularly, and the third Steinmann pin is placed anteriorly supraacetabularly (often with a metal hammer). Next, the labrum is resected for exposure. If the thickened anterior capsule extends into the acetabulum, it should be resected to thin the capsule as much as possible. Next, a portion of the anterior capsule is incised at the anterior margin of the acetabulum using an electrocautery scalpel, and a Cobra retractor is applied to obtain a good field of view (Fig. 54.5b). A Cobra retractor can be easily applied if the elevatorium is inserted first.

After reaming, the cup must be placed at an appropriate angle and fixed with two screws, following which the trial liner is finally placed.

(d) Femoral deployment and test repair

An assistant must hold the hip joint in 60° flexion, 30° adduction, and 90° internal rotation. The proximal femur is elevated using a special retractor, and the elevatorium is deployed over the lesser and greater trochanters. The soft tissues are resected and the pisiform fossa is identified. Then, the trabecular bone is resected using a femoral box chisel, and a starter reamer is inserted into the medullary cavity (Fig. 54.6). The medial aspect of the greater trochanter is reamed with a greater trochanter reamer to allow intermediate insertion of the stem. After rasping, the anteroposterior angle of the stem is checked with an angiometer (Teijin Nakashima Medical Ltd.) and fine-tuned based on the total anteversion (Fig. 54.7). Tests are performed to check for stability, soft tissue tension, and differences in leg length. Indicators of stability are as follows: (1) no posterior dislocation in deep flexion, (2) no posterior dislocation in flexion 70° + internal rotation 60°, and (3) no anterior dislocation in extension external rotation (Fig. 54.8).

If posterior dislocation occurs due to impingement at a shallow angle of internal rotation during the trial, one must release impingement and adjust the offset and head length. If no improvement is observed after performing the operations described previously, the angle of cup placement should be checked fluoroscopically because if the cup is not placed properly, reinstallation should be considered. One may also consider increasing the range of motion using a dual-mobility head. Finally, a cement plug is placed after the medullary cavity is thoroughly flushed.

(e) Installation of liner stem

A real liner is placed at this stage. After the medullary cavity is dried, cementing is performed, and the stem is inserted and retained to the expected level. After confirming that the cement has hardened sufficiently, test restoration is performed again with a trial head (Fig. 54.9). After cleaning, the head is inserted, and restoration is performed.

(f) Repair and closure of short external rotator muscle groups and joint capsule

Two non-absorbable threads are placed over the short external rotator muscles and bursa, which were preserved during

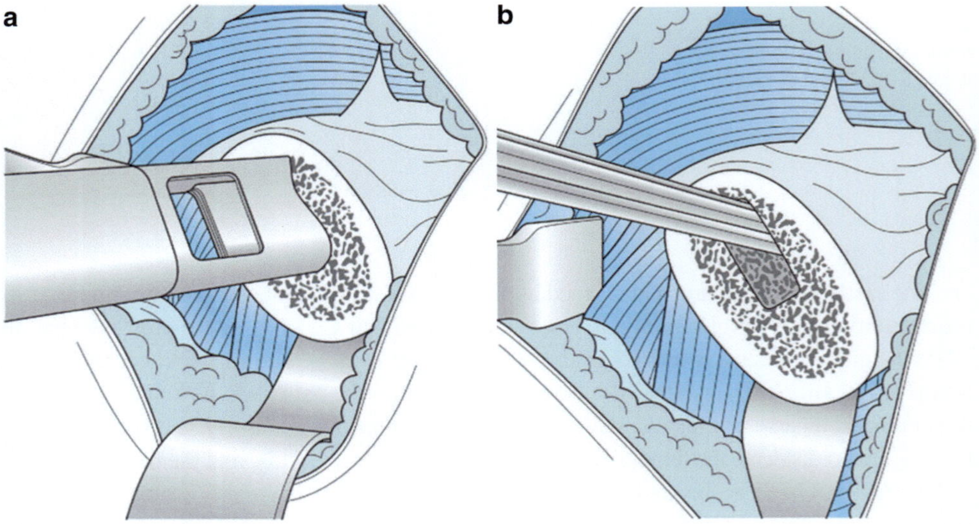

Fig. 54.6 Femur development and treatment. (**a**) Bone resection with a box chisel. (**b**) Insertion of a starter reamer into the medullary cavity while preserving midline muscle attachment

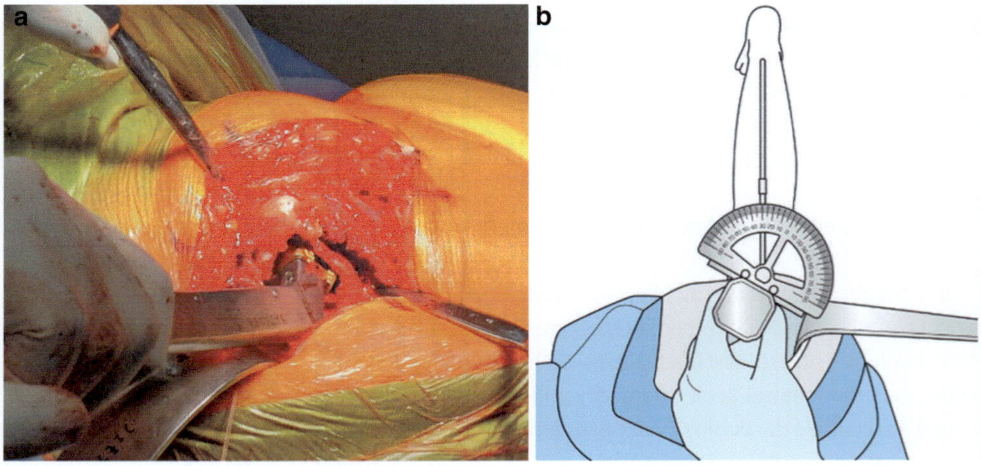

Fig. 54.7 Confirmation of rasping and front twist angle. (**a**) Rasping is performed until gyration stability is obtained. (**b**) An angle meter is placed on the rasp to check the anterior torsional angle, using the lower leg axis as an indicator

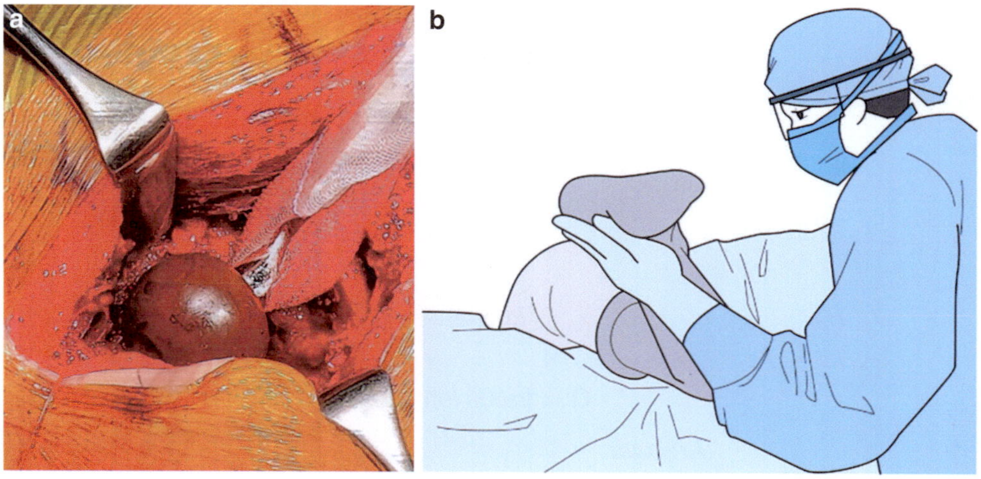

Fig. 54.8 Confirmation of stability by test shaping. (**a**) Using an alignment shaft, the depth of stem insertion can be confirmed, even at the height of the center of the head and greater trochanter. (**b**) Test repositioning using a trial neck and head to check factors such as range of motion and leg length difference

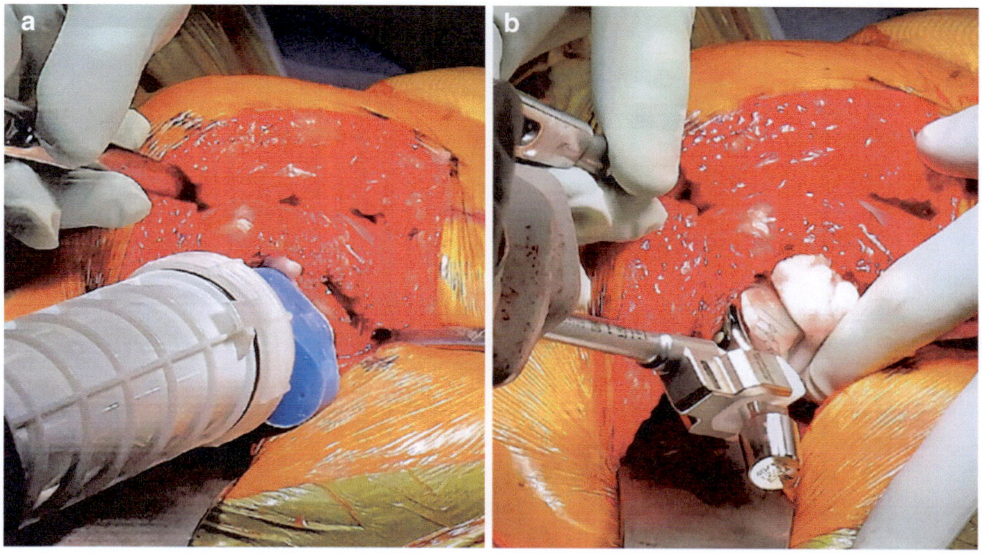

Fig. 54.9 Cementing and stem insertion and installation. (**a**) Cement is pressed inside with a cement gun. (**b**) Excess cement is removed while holding the stem with a stem insertor until cement hardens

deployment. Two holes are made posterior to the greater trochanter using a 2.4-mm diameter Kirschner steel wire, and two absorbable threads are threaded using a special hook and sutured using the double-loop sliding knot technique (Fig. 54.10). Each layer is sutured, and the wound is ultimately closed.

(g) Posttreatment

The day after surgery, the patient begins full-load standing training and gait training using a walker or cane. Patients can be discharged from the hospital when they are stable while walking and climbing stairs 2–3 weeks after surgery.

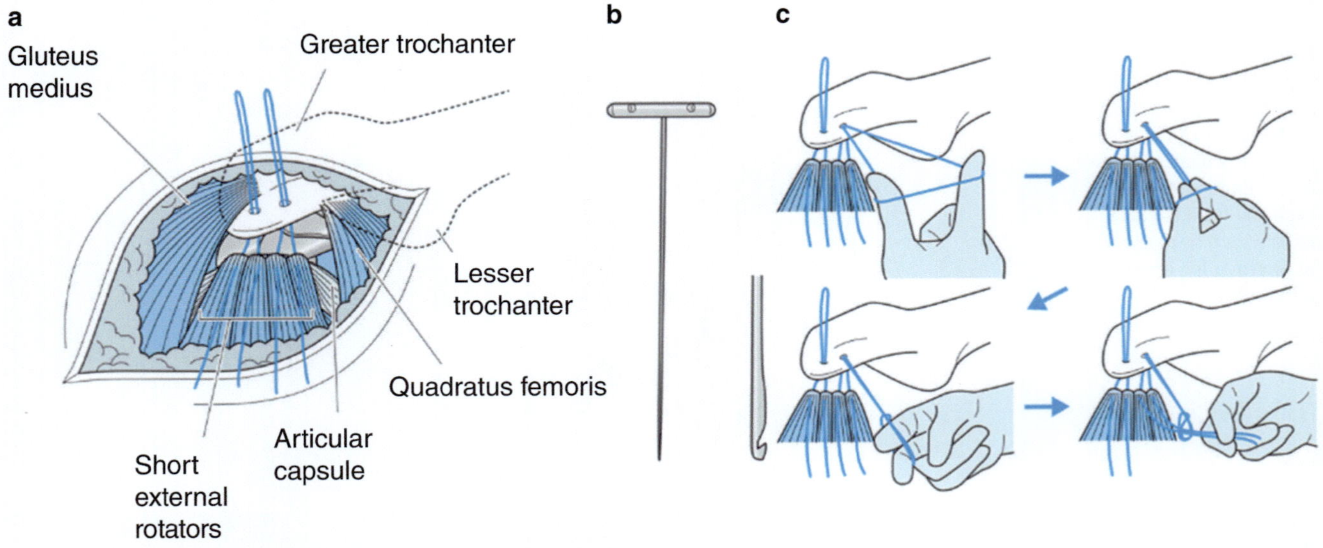

Fig. 54.10 Soft tissue repair. (**a**) Two nonabsorbable threads are looped over the short external rotator muscle group and posterior glenoid and then threaded through the bony foramen posterior to the greater trochanter using a special hook. (**b**) Special hook. The suture is placed on the hook at its tip. (**c**) Double-loop sliding knot method in practice

4 Cemented Stems

In a registry of various countries, cemented stems have been reported to have better long-term results than cementless stems, especially in elderly patients with osteoporosis [9]. Cemented stems are classified into composite beam stems and polished tapered stems. The composite beam theory is based on the idea that the stem and cement provide strong fixation. The taper slip theory, based on a polished tapered stem, suggests that the shear stress at the bone-cement interface is reduced by slight settling in the cement mantle and that a large compressive stress provides long-term stability [10]. In recent years, the mainstay of the taper slip technique has been the polished tapered stem, represented by the Exeter stem (Stryker), SC stem (Kyocera), and VLIAN stem (Teijin Nakashima Medical), because of its excellent long-term results (Fig. 54.11).

The factors involved in postoperative satisfaction with THA include appropriate leg length and offset adjustment. However, in many of the polished tapered stems available in Japan, the offset is based on body standards for stem size. The offset is determined by the diameter of the medullary cavity, which increases with age; it is often difficult to reproduce the original appropriate offset in patients. However, the Exeter stem and VLIAN stem can reproduce the optimal leg length and offset for almost all patients because of the size variation of each offset, which is ideal for the construction of correct hip biomechanics.

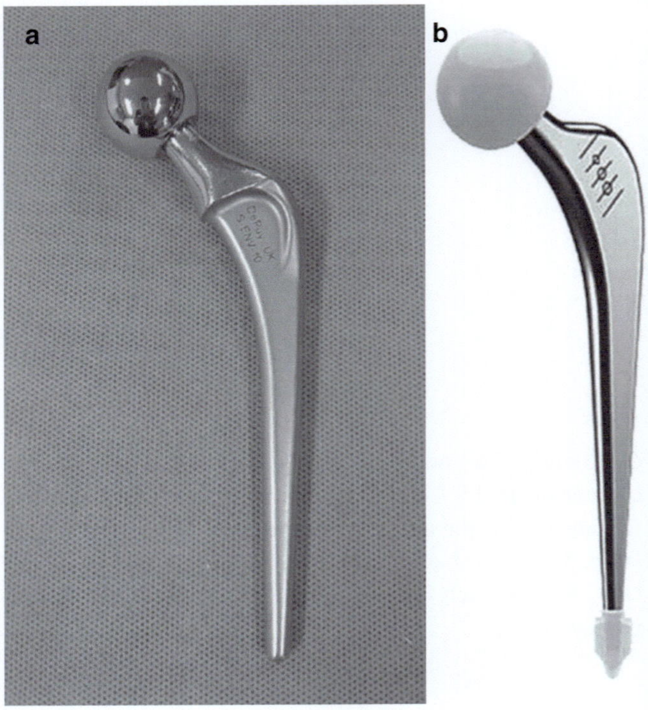

Fig. 54.11 Composite beam stem and polished tapered stem. (**a**) Elite plus stem (Depuy Synthes), a typical composite beam stem. (**b**) VLIAN stem, a typical polished tapered stem. (Teijin Nakashima Medical). (Image courtesy of Depuy Synthes, Teijin Nakashima Medical Ltd)

The advantages of polished tapered stems are as follows: (1) good ultralong-term results, (2) reproduction of correct biomechanics (leg length, offset, and anteversion), (3) bone preservation of the proximal femur owing to less stress shielding, (4) very low risk of intraoperative fractures, (5) easy removal during revision surgery (cement removal is not necessary, even in infected cases, if there is no loosening), and (6) a single cement stem can be used in almost all cases. The disadvantages of cementation are as follows: (1) It is necessary to learn the cementation technique, but it does not require much time to master this method under the guidance of a well-trained physician. (2) It was previously used for the cementation of cemented prostheses. However, this risk can be significantly reduced by taking the following precautions: (a) secure closure of the medullary cavity with a plug; (b) jet washing of the bone marrow and blood; and (c) cementing at the appropriate time, as cementing should begin after the cement is no longer attached to the fingers.

We believe that all hip surgeons should learn cementation techniques because cemented stems can be used in cases that are difficult or impossible to treat with cementless stems.

References

1. Morrey BF. Instability after total hip arthroplasty. Orthop Clin North Am. 1992;23:237–48.
2. Roberts JM, et al. A comparison of posterolateral and anterolateral approaches to total hip arthroplasty. Clin Orthop Relat Res. 1984;187:205–10.
3. Turner RS. Postoperative total hip prosthetic femoral head dislocations. Incidence, etiologic factors, and management. Clin Orthop Relat Res. 1994;301:196–204.
4. Vicar AJ, et al. A comparison of anterolateral, transtrochanteric, and posterior surgical approaches in primary total hip arthroplasty. Orthop Relat Res. 1984;188:152–9.
5. Maratt JD, et al. No difference in dislocation seen in anterior vs posterior approach total hip arthroplasty. J Arthroplast. 2016;31(9 Suppl):127–30.
6. Ritter MA, et al. A clinical comparison of anterolateral and posterolateral approaches to the hip. Clin Orthop Relat Res. 2001;385:95–9.
7. Pellicci PM, et al. Posterior approach to total hip replacement using enhanced posterior soft tissue repair. Clin Orthop Relat Res. 1998;355:224–8.
8. Hayashi T, et al. The accuracy of socket orientation using a pelvic fixation device for total hip arthroplasty with the patient in the lateral decubitus position in comparison to the vacuum bag fixation method. Fukuoka Igaku Zasshi. 2007;98:19–25.
9. Troelsen A, et al. A review of current fixation use and registry outcomes in total hip arthroplasty: the uncemented paradox. Clin Orthop Relat Res. 2013;471:2052–9.
10. Shen G. Femoral stem fixation. An engineering interpretation of the long-term outcome of Charnley and Exeter stem. J Bone Joint Br. 1998;80:754–6.

55 Conjoined Tendon-Preserving Posterior (CPP) Approach

Taro Mawatari

1 What Is the CPP Approach?

In total hip arthroplasty (THA), the posterior approach is the most commonly used, accounting for 42% of initial THAs and 58% of revisions, according to THA registry statistics from 2018 [1]. The posterior approach is particularly user-friendly for young surgeons learning both BHA and THA, due to the ease with which the femoral side can be managed. Additionally, it facilitates handling complex cases, such as severe deformities, bone defects, and revisions, which further contributes to its widespread use.

However, the posterior approach leaves the posterior soft tissues of the hip joint in the entry path vulnerable to dislocation, which is a major complication of THA. The dislocation rate can be reduced by repairing the incised soft tissue after implant placement. However, even if the incised short external rotator muscles and joint capsule are firmly repaired, failure may still occur immediately after surgery, with a reported 70–90% failure rate at 3 months postoperatively [2].

The hip joint has a joint capsule in the deep layers of the periarticular muscles, and it has been reported that the capsular ligament plays a very important role as a static stabilizer of the joint. In particular, the anterior iliofemoral ligament and posterior ischiofemoral ligament are considered crucial for maintaining the tension of the hip joint and preventing dislocation [3].

The conjoined tendon preserving posterior (CPP) approach preserves the short external rotators, including the piriformis and the conjoined tendon composed of the superior gemellus, obturator internus, and inferior gemellus, as well as the underlying joint capsule, all of which were completely dissected in the conventional posterior approach, while the obturator externus and proximal part of the quadratus femoris muscle are resected. An L-shaped incision is made down to the inferior part of the joint capsule.

The CPP approach is considered a "capsular ligament-preserving" posterior approach because it preserves all three primary capsular ligaments: the iliofemoral ligament, ischiofemoral ligament, and pubofemoral ligament, which are largely transected in the conventional posterior approach.

The CPP approach has been applied to BHA [4, 5], and we have applied this approach to THA since 2017. To the best of our knowledge, this is the first detailed explanation of CPP-THA in the English literature.

2 Difference Between the CPP Approach and Other Posterior Approaches

In addition to the conventional posterior approach (conventional PL), new posterior approaches, such as piriformis muscle preservation [6], external rotator preservation procedure (ERP), supercapsular percutaneously assisted total hip (SuperPATH), and direct superior approach (DSA), have been proposed recently. The invasiveness of the CPP and other posterior approaches to the short rotator muscles and capsular ligaments is shown in Fig. 55.1.

In the CPP approach, the piriformis and conjoined tendon are preserved, while the obturator externus and the proximal part of the quadratus femoris are transected and then repaired. The iliofemoral and pubofemoral ligaments, located anterior to the femoral head, can be preserved in the posterior approach, while the posterior ischiofemoral ligament, located behind the femoral head, can be preserved only in the CPP approach. Invasion of the ischiofemoral ligament is inevitable in piriformis-sparing and ERP procedures. Preservation of the inferior gemellus is crucial because if this muscle is transected, the underlying ischiofemoral ligament is also transected simultaneously.

T. Mawatari (✉)
Department of Orthopaedic Surgery, KKR Hamanomachi Hospital, Fukuoka, Japan

Y. Nakashima (ed.), *Advances in Total Hip Arthroplasty*, https://doi.org/10.1007/978-981-97-2461-1_55

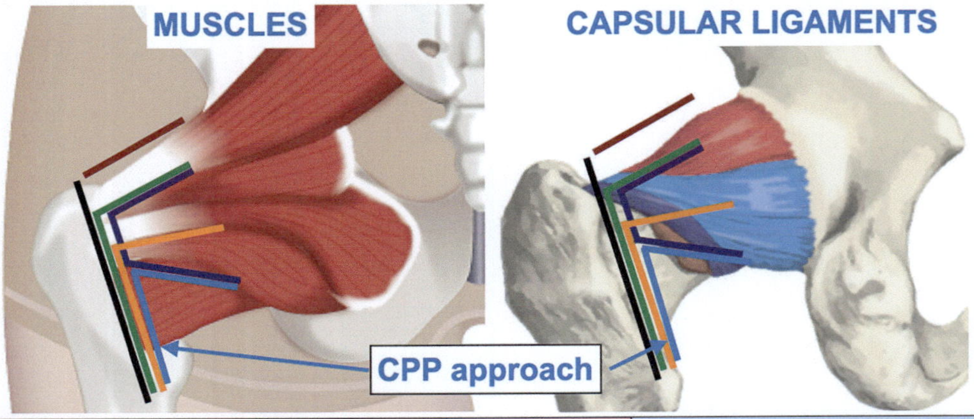

APPROACH	MUSCLES			Capsular ligaments
	Piriformis	Conjoined Tendon	Obturator Externus	
Conventional	Transect & Repair	Transect & Repair	Transect & Repair	Ischio-femoral lig. damage
Piroformis-sparing	Preserve	Transect & Repair	Transect & Repair	Ischio-femoral lig. damage
ERP	Preserve	Transect Inferior Gemellus & Repair	Transect & Repair	Ischio-femoral lig. damage
DSA	Transect & Repair / Preserve	Transect & Repair	Preserve (Difficult Stem insertion)	Ischio-femoral lig. damage
CPP	Preserve	Preserve	Transect & Repair	Minimal damage
SuperPATH	Preserve	Preserve	Preserve (Difficult Stem insertion)	Ilio-femoral / Ischio-femoral lig. damage

Fig. 55.1 Surgical invasiveness of muscles and capsular ligaments in various posterior approaches

3 Pros and Cons of the CPP Approach

(a) Pros

The advantages of the CPP approach include the following points:

- The three capsular ligaments, the ischiofemoral ligament, iliofemoral ligament, and pubofemoral ligament, each serving distinct functional roles in stabilizing the joint, can be preserved.
- The short external rotators, which act as dynamic stabilizers, can be preserved, except for the obturator externus and the proximal part of the quadratus femoris. The soft muscle belly of the inferior gemellus may be contused by acetabular reaming or cup placement; however, even in such cases, the underlying joint capsule is resilient and typically preserved, and the remaining short rotator muscles are almost always kept intact.
- Our experience suggests that resistance to dislocation can be enhanced by preserving the static and dynamic stabilizers mentioned above.
- Early recovery is expected due to the minimal invasiveness of the procedure.
- Hip joint tension is maintained, reducing the likelihood of requiring leg lengthening for joint stability compared with the conventional PL approach.
- In cases of intraoperative difficulty, it is easy to transect the preserved muscles and joint capsule to convert to the conventional PL approach.
- Stem placement is relatively straightforward and comparable to the conventional PL approach. The preserved muscles and joint capsule can be easily retracted and usually do not interfere with the procedure; however, increasing the internal rotation of the femur can further reduce interference with the preserved conjoined tendon. A variety of implant designs, including the Zweymüller stem, can be used, and CPP-THA can be performed regardless of whether cement is used.
- Once the surgeon becomes accustomed to the procedure, the operative time is not significantly different from that of the conventional PL approach, and blood loss is expected to be reduced.

- Good compatibility with the portable navigation system is also expected, as described below.
- Additionally, in many revision cases, a strong secondary joint capsule-like solid soft tissue forms around the joint. It is often possible to perform revision surgery through an L-shaped incision of the posterior secondary joint capsule, similar to the CPP approach, which is very helpful in maintaining hip joint stability.

(b) Cons
The possible challenges associated with the CPP approach are as follows.

- Proper technique is required for dislocating the hip joint; in addition to flexion and internal rotation, applying abduction is important, and the use of a specially designed dislocation spatula can be helpful.
- Direct visualization of the acetabulum is difficult, so finger palpation is necessary to confirm the presence of the labrum or other residual soft tissues, as well as the depth of acetabular reaming.
- The use of a specially designed conjoined tendon retractor to retract muscles and joint capsules along with a helmet equipped with a light, is also beneficial.
- Pelvic tilt may occur during intraoperative manipulation, which is a particular concern for THA performed in the lateral position, regardless of the approach. Intraoperative X-ray radiography is recommended, and a portable navigation system can help address this issue.
- When the conjoined tendon is preserved, the sciatic nerve is very close to the acetabulum, especially with the hip in flexion. Careful attention must be paid to the sciatic nerve throughout the surgery.
- The use of a specially designed sciatic nerve retractor to physically protect the nerve during acetabular cup preparation is recommended.

4 Limitations and Indications of the CPP Approach in THA

In our experience, approximately 95% of patients underwent THA using the CPP approach, and we believe that this approach can be applied to most THA cases. If leg lengthening is necessary to match the length of the contralateral healthy side, it can be accomplished by dissecting the anterior iliofemoral ligament, which has a greater influence on joint tension than the ischiofemoral ligament.

In contrast, for highly subluxed hips of Crowe classification type III or higher, which require bone graft or metallic augmentation on the acetabular side, the bony acetabular rim should be exposed to confirm the stability of the reconstruction. These cases may not be suitable for the CPP approach and may require wider exposure. Cases with ankylosis or severely contracted hips with severe deformities are also not considered suitable for the CPP approach.

5 Surgical Techniques for CPP-THA [7]

The following steps should be considered by surgeons who have performed THA using the conventional posterior approach or by surgeons who have performed CPP-BHA:

- Identification of the short external rotator muscles and application of an L-shaped incision down to the joint capsule
- Hip joint dislocation maneuver
- Total resection of the labrum
- Reaming maneuver
- Acetabular cup placement

In both CPP-BHA and CPP-THA, many surgical procedures are performed with the hip in flexion; therefore, care must be taken to avoid sciatic nerve injuries.

(a) Surgical position

The patient is placed in the lateral decubitus position and stabilized using a padded pelvic fixator. Preoperative radiography is helpful in confirming the lateral tilt, rotation, and anteversion of the pelvis, which can be adjusted in advance by tuning the patient or fixator position and tilting the operating table to obtain the true lateral position. The position of both knees on the operating table and the leg length discrepancy on X-ray radiography should be confirmed before surgery begins.

(b) Exposure

Identification of the short external rotator muscle group and an L-shaped incision of the muscle and joint capsule (Fig. 55.2).

The skin incision follows the same path as that used in the conventional posterior approach. The incision begins approximately 4 cm distal to the tip of the greater trochanter (GT) at the posterior border of the GT and extends proximally for another 4 cm. The skin, subcutaneous fat, fascia lata, and iliotibial band are incised longitudinally, splitting along the fibers of the gluteus maximus. The hip is then internally rotated, and the piriformis and other short external rotators are identified. The landmarks include the piriformis and the obturator internus, which have a white tendinous component, and palpating the inferior border of the femoral head is also

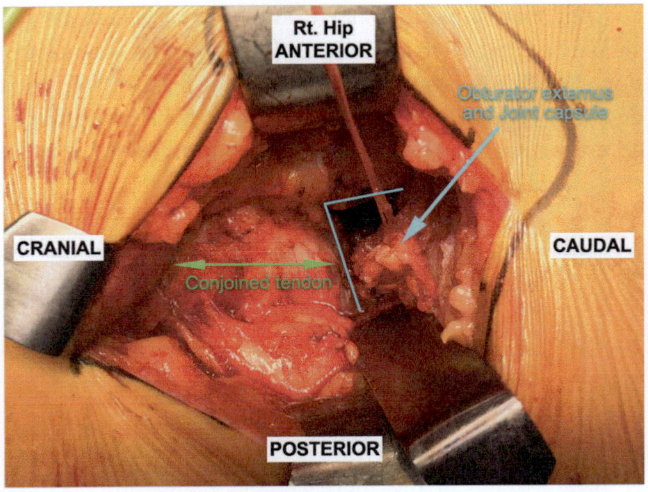

Fig. 55.2 L-shaped incision. The hip joint is flexed and internally rotated to expose the short external rotators. First, the proximal part of the quadratus femoris is incised, the obturator externus was stay-sutured and then incised, and the joint capsule is exposed. An L-shaped incision is made down to the posterior part of the joint capsule, as shown in the figure. The level of the incision should be at the inferior border of the inferior gemellus, which coincides with the inferior border of the ischiofemoral ligament

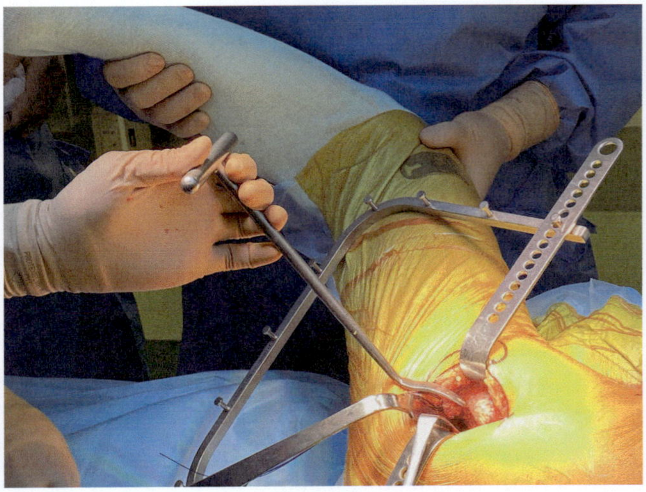

Fig. 55.3 Dislocation maneuver. Dislocation is performed with the hip in flexion, abduction, and internal rotation. Applying abduction is the key. A specially developed dislocating spatula is useful for this step

helpful, as it coincides with the inferior border of the inferior gemellus.

The proximal part of the quadratus femoris is dissected to expose the obturator externus. Attention should be paid to the medial femoral circumflex artery located at the proximal part of the quadratus femoris, which may require hemostasis using an electrocoagulator. At the surface of the inferior joint capsule, the obturator externus, which also has a white tendinous component, is identified, stay sutures are applied, and it is then dissected. Subsequently, the joint capsule is incised using a scalpel at the inferior border of the inferior gemellus. After applying stay sutures to the inferior part of the joint capsule, a longitudinal incision of the joint capsule along the femur is made to complete the L-shaped incision. During these procedures, the hip joint should be kept in extension because the sciatic nerve moves closer to the acetabulum when the hip joint is in flexion.

(c) Dislocation (Fig. 55.3)

The dislocation maneuver differs between BHA and THA, and it is also a critical aspect of CPP-THA, distinct from the conventional PL approach. A dislocation spatula is inserted into the joint space, and the hip is dislocated by flexion, abduction, and internal rotation (Fig. 55.4). Abduction can loosen the tension of the joint capsule. Care must be taken when using a dislocation spatula, especially in cases of osteoporosis, as improper insertion of the spatula into the bone of the femoral head can complicate manipulation. If the

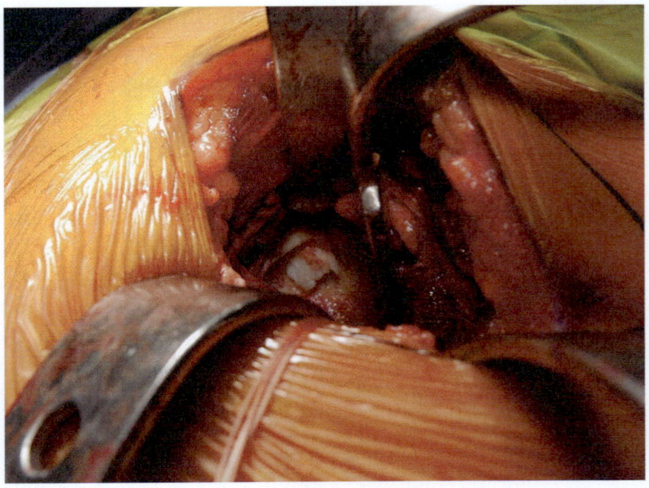

Fig. 55.4 Total resection of the labrum. The labrum should be completely resected. If any portion of the labrum is left behind, it may impair reaming and potentially cause postoperative pain. The use of a rongeur is helpful for the complete resection of the labrum

osteophyte at the inferior acetablum obstructs the dislocation maneuver, it should be removed with a chisel beforehand. Once the suregeon becomes familiar with the technique, this dislocation maneuver is relatively straightforward.

(d) Femoral neck osteotomy

Femoral neck osteotomy is performed using a reciprocal saw or oscillating saw. When using a reciprocating saw, care must be taken to avoid damaging the skin on the dorsal side of the blade. The osteotomy level can be determined relatively easily by palpating the lesser trochanter, as with the conventional posterior approach. When considering combined anteversion, femoral stem preparation can be performed prior to acetabular cup preparation, particularly in cases with severe deformities.

(e) Acetabular cup preparation

This is one of the most important steps in CPP-THA. The steps include deployment with retractors, complete labral resection, reaming, and placement of a cup and liner. The sciatic nerve is pulled by the preserved muscles and is much closer to the acetabulum than if the short external rotator muscles are completely transected. The position of the sciatic nerve must be checked throughout the acetabular cup preparation, especially with the hip is in the flexed position.

(i) Deployment with Retractors

Various types of retractors, including newly developed sciatic nerve and conjoined tendon retractors, are used. As in the conventional posterior approach, the first retractor is placed above the anterior wall of the acetabulum to move the femur anteriorly. A custom-designed sciatic nerve retractor positioned beneath the transverse acetabular ligament is useful for protecting the sciatic nerve during reaming and cup placement. A custom-designed conjoined tendon retractor is applied above the superior wall of the acetabulum to further expose the acetabulum covered by the conjoined tendon and joint capsule. This retractor should be placed between the joint capsule and the labrum. Usually, the width of the preserved joint capsule from the bony edge of the acetabulum is approximately 30–35 mm.

(ii) Labrum excision (Fig. 55.4)

In THA, the labrum should be completely resected, unlike in BHA, where the labrum is preserved. In CPP-THA, a surgical instrument such as a rongeur is helpful for the complete excision of the labrum, pulvinar, and loose soft tissues.

(iii) Reaming (Fig. 55.5)

The reamer head is first pointed distally and inserted by sliding it over the sciatic nerve retractor toward the acetabulum, followed by reaming after flipping the reamer head toward the acetabulum. The reaming depth needs to be confirmed by palpation with the fingers, by examining the amount of resected bone in the reamer, or, if necessary, by direct observation using the conjoined tendon retractor. If it is difficult to insert the reamer even after adjusting the limb position, the osteotomy level of the femoral neck should be checked, as difficulty in inserting the reamer is often due to the residual neck being too long.

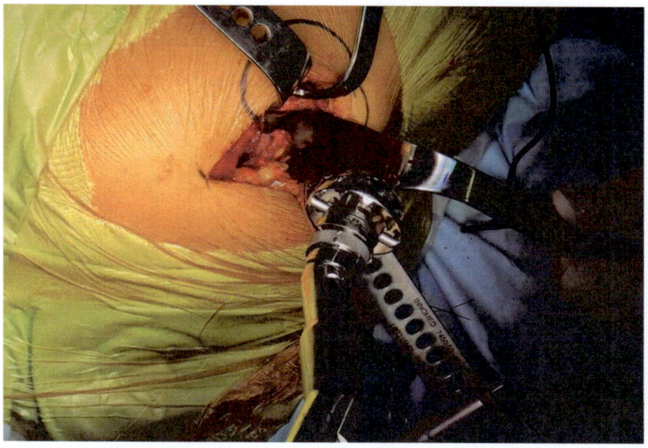

Fig. 55.5 Reaming maneuver. The reamer is first pointed distally and slid downward into the acetabulum over the custom-made sciatic nerve retractor, then reaming is performed after flipping the reamer head toward the acetabulum

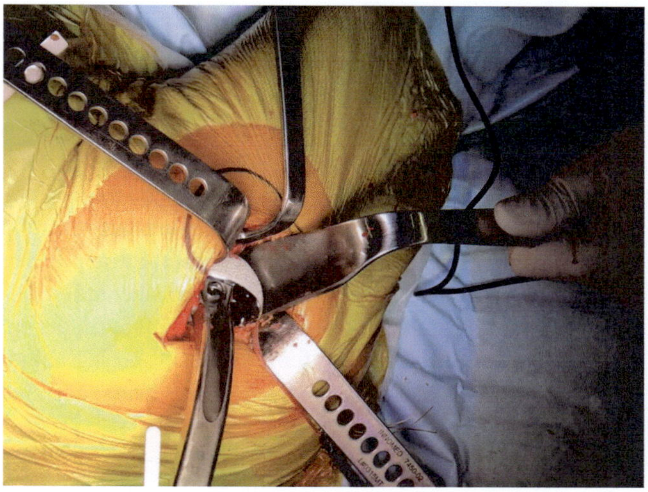

Fig. 55.6 Acetabular cup installation. Similar to reamer insertion, the cup is placed by sliding it downward into the acetabulum over the sciatic nerve retractor

(iv) Cup installation (Fig. 55.6)

During installation, the cup is first pointed distally and inserted by sliding it over the sciatic nerve retractor, as in the reamer insertion. After flipping the cup toward the acetabulum, an angle guide is used for precise cup placement, as in the conventional posterior approach. The acetabular cup is fixed with one or two screws if necessary, especially in cases of displastic hips. Once the first screw is fixed at the easiest location and the cup is stabilized, the following screw can be fixed by applying the conjoined tendon retractor above the superior edge of the cup. If necessary, osteophytes around the cup should be resected to avoid potential impingement.

(v) Liner trial

The liner trial should be installed after the cup is fully exposed by retracting surrounding soft tissues.

(f) Femoral side procedure (Fig. 55.7)

The femoral-side procedure of CPP-THA is the same as that in the conventional posterior approach. A major advantage of this technique is that it can be applied regardless of stem design, with or without cement. With the hip joint in flexion and internal rotation, a blunt femoral elevator, a curved Hohmann retractor at the lesser trochanter, and an elevatrium at the greater trochanter are used to retract soft tissues. Routine femoral preparation and broaching can then be performed. If the preserved soft tissues interfere with the procedure, applying slightly more internal rotation should help expose the broaching area.

(g) Trial reposition, intraoperative radiograph, stem size, and neck length determination

After placing the trial neck and head, the repositioned hip joint is tested for joint tension, stability, and leg length discrepancy. The size, offset of the stem, and neck length of the head are then determined. Care should be taken not to entrap the sciatic nerve when repositioning the joint. Intraoperative radiography is recommended to confirm the size and installation angle of the cup, possible changes in pelvic tilt during surgery, size and offset of the femoral stem, and leg length discrepancy. Safe dislocation is facilitated using a single blunt hook on the neck.

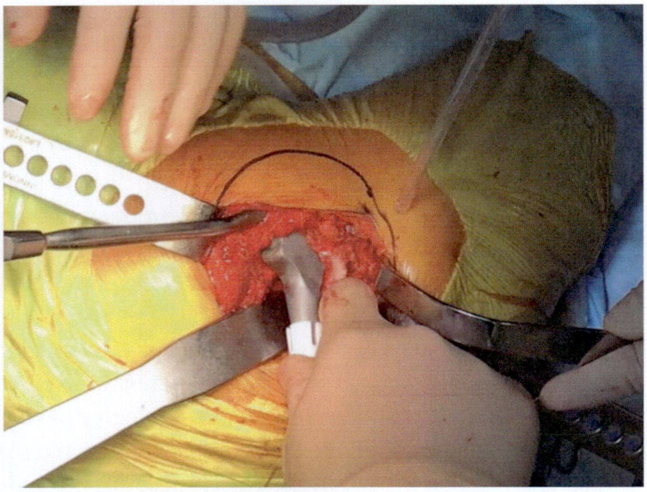

Fig. 55.7 Stem installation. The procedure on the femoral side is the same as that of the conventional PL approach, and the stem insertion can be done relatively easily

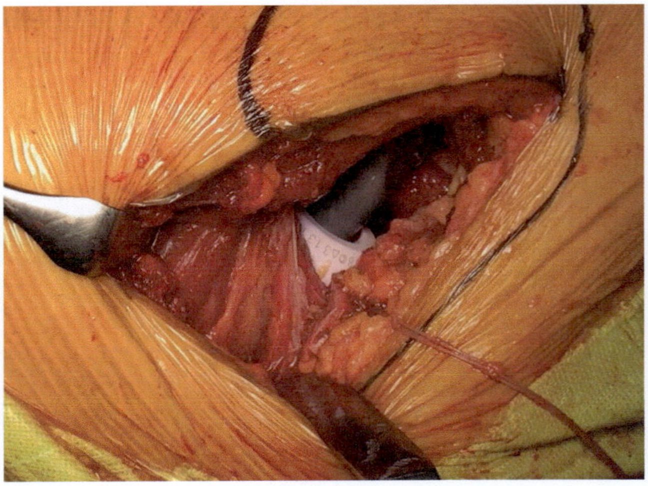

Fig. 55.8 Representative condition of the conjoined tendon just after the final implant placement. The figure shows the condition of the hip joint after total hip implant placement, before the repair of the incised inferior joint capsule and muscles. The acetabular cup completely covered by the conjoined tendon and joint capsule, stabilizing the hip joint in flexion and internal rotation

(h) Final component placement, repair of the joint capsule and obturator externus, and wound closure

After removing the trial implants and thorough cleaning, the final liner, stem, and head are placed. The stability of the hip in combined flexion and internal rotation, and combined extension and external rotation, presence of impingement, and soft tissue balance are checked again. Figure 55.8 shows the intraoperative condition immediately after final implant placement. The cup is completely covered by the unexcised conjoined tendon and capsule, and the hip joint is stabilized. The inferior joint capsule is sutured side by side with the proximal portion. The obturator externus is sutured to the tendinous part of the gluteus medius or anchored to the greater trochanter. Locking sutures are useful for closing the fascia lata, ITB, and gluteus maximus. Drainage is usually unnecessary.

(i) Rehabilitation

Rehabilitation is not significantly different from conventional THA and can be started on the same day. The schedule is determined based on a comprehensive evaluation of the patient's age, muscle strength, bone quality, implant fixation, and stability.

6 Clinical Results of CPP-THA [7]

The clinical outcomes of 148 self-experienced patients with 162 hips (14 males and 148 females, mean age 66.9 years [range 35–89 years]) were investigated. Ambulatory patients with osteoarthritis, rheumatoid arthritis, or osteonecrosis of the femoral head were included, while those with ankylosis or severely dislocated hips (Crowe classification type III or higher) were excluded. The Zimmer Biomet G7® Socket and Taperloc® stem were primarily used, and the Zimmer Biomet G7® Socket and Alloclassic® stem were used in some patients with Dall type C femurs. The mean operative time from skin incision to closure was 63.5 ± 14.7 min, and the mean blood loss was 135.3 ± 206.1 g. X-ray radiographs were consistently taken preoperatively and intraoperatively. One patient experienced incomplete sciatic nerve palsy (peroneal nerve area), which resolved within 6 months postoperatively. No other nerve palsies, including lateral femoral cutaneous nerve or femoral nerve involvement, were observed. No intraoperative fractures occurred. There were no cases of deep infection, and no revision surgery was required during the 12-month follow-up period. Radiographs showed that the cup was placed at an average angle of 39.3° ± 4.0° for radiographic lateral inclination and 21.2° ± 2.5° for radiographic anteversion, with 95.1% of cups positioned within the Lewinnek safe zone. All patients began full weight-bearing gait training on the same day or the day after surgery. The improvement rates of walking speed and timed up and go (TUG) test at 2 weeks postoperatively were significantly better than those of the conventional posterior approach, with functional recovery occurring approximately 1 week faster.

7 Preservation of the Obturator Internus by CPP-THA [8] (Fig. 55.9)

The obturator internus preserved by CPP-THA is one of the pelvic floor muscles, and its relationship with abdominal pressure, urinary incontinence, and sexual function has recently been reported. We retrospectively evaluated this muscle in 102 hips of 85 patients who underwent THA at least 1 year postoperatively using computed tomography (CT). Seventy-six hips in 67 patients underwent the CPP approach, while the remaining 26 hips in 23 patients underwent the conventional PL approach. In both the CPP and PL groups, the transected tissues were repaired at the end of the surgery. The results showed that the pre- and postoperative ratios of the obturator internus width on axial CT images (in some cases, the ratio was calculated using the contralateral healthy side) were 98.6% in the CPP-THA group and 65.5% in the PL group. The percentage of patients with obturator internus atrophy (width less than 90% compared with the preoperative state) was 5.3% in the CPP group and 100% in the PL group. These results clearly indicate that the CPP approach is useful in preserving the obturator internus, at least morphologically, while it seems difficult to preserve this muscle with the conventional PL approach.

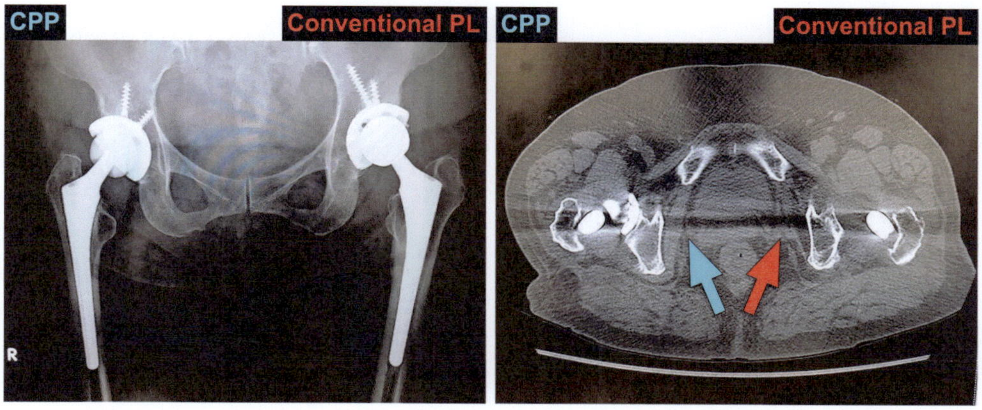

Fig. 55.9 CT observation of the obturator internus in patients who underwent THA through the CPP approach and conventional PL approach more than 1 year ago. The observed obturator internus (arrows) on axial CT images was very atrophic in THA performed through the conventional PL approach (left hip) and preserved in CPP-THA (right hip)

8 CPP-THA and Portable Navigation System [9] (Fig. 55.10)

The CPP approach has been introduced in an increasing number of institutions for BHA [4, 5]. In CPP-THA, the acetabulum is covered with the conjoined tendon and joint capsule, making it difficult to visualize directly during surgery. Additionally, in THA performed with the patient in the lateral decubitus position, it has been reported that the pelvis is not always in the true lateral position even before surgery, and the pelvic position often changes during surgery. In recent years, navigation systems and robotic arm-assisted surgery have been introduced for THA to improve the accuracy of implant placement. However, depending on the approach and soft tissue preservation, such instruments using markers or robotic arms may interfere with surgical exposure, making the procedure challenging. We adopted a portable navigation system (HipAlign New Lateral®, Zimmer Biomet) that can trace intraoperative pelvic movements and confirmed that the CPP-THA procedure can be performed without any interference, which is useful for improving the accuracy of implant placement [9].

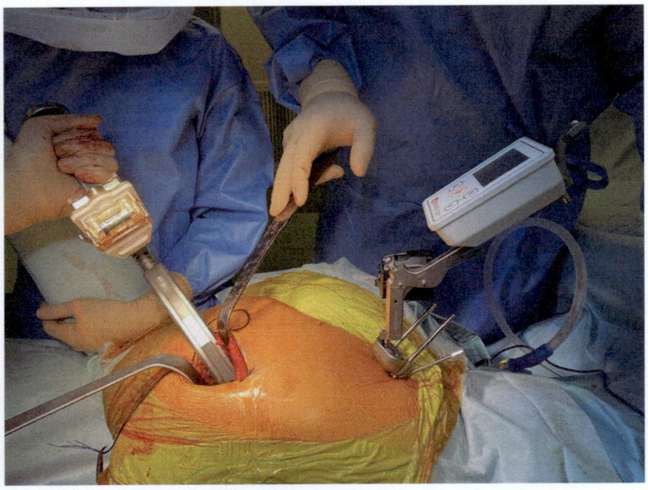

Fig. 55.10 CPP-THA with HipAlign New Lateral®. CPP-THA can be performed using a portable navigation system (HipAlign New Lateral®, Zimmer Biomet) without any difficulty

References

1. Committee for Japanese Arthroplasty Register. The Japanese Society for Replacement Arthroplasty. THA Registry Statistics (February 2006–March 2018). 2018 [in Japanese].
2. Loiba V, et al. The analysis of posterior soft tissue repair durability after total hip arthroplasty in primary osteoarthritis patients. Hip Int. 2015;25:420–3.
3. Takao M, et al. The posterior capsular ligamentous complex contributes to hip joint stability in distraction. J Arthroplast. 2018;33:919–24.
4. Nakamura T, et al. Conjoined tendon preserving posterior approach in hemiarthroplasty for femoral neck fractures: a prospective multicenter clinical study of 322 patients. J Orthop Surg. 2021;29(3):1–8.
5. Tetsunaga T, et al. Bipolar hip arthroplasty using a conjoined tendon-preserving posterior approach in geriatric patients. Acta Med Okayama. 2021;75:25–30.
6. Minokawa S, et al. Preservation technique of the piriformis tendon is superior to reattachment technique in terms of contiguity and muscle atrophy. Hip Int. 2018;28:559–905.
7. Mawatari T, et al. Conjoined tendon preserving posterior (CPP) approach. Orthop Surg Traumatol. 2020;63:1037–46. [in Japanese]
8. Ayabe Y, et al. CT evaluation of the obturator interunus in CPP-THA. Hip Joint. 2021;47:261–4. [in Japanese]
9. Ayabe Y, et al. Changes in pelvic tilt angle in THA performed in lateral position and experience with HipAlign new lateral®. Hip Joint. 2012;47:276–81. [in Japanese]

Part X

Computer Assisted Surgery

CT-Based Navigation

56

Kunihiko Tokunaga

1 Principle and History of Computed Tomography (CT)-Based Navigation

In total hip arthroplasty (THA) using CT-based navigation, three-dimensional (3D) bone models are created based on preoperative CT images imported into the navigation system. A computer-aided design (CAD) of the implant is placed at the optimal position on the bone coordinates selected by the surgeon. Range of motion (ROM) is simulated to confirm the presence or absence of implant-to-implant collisions in the ROM required for activities of daily living.

To achieve preoperative planning, registration of the bone model to the actual bone is required intraoperatively. Infrared trackers are placed on the iliac wing, distal frontal aspect of the femoral diaphysis, and surgical instruments. The position and angle of the bone, instruments, and implants in 3D coordinates are determined in real-time using an infrared camera. The Stryker navigation system (Stryker, US) is the only CT-based navigation system currently available for clinical use in Japan. Registration is performed using the surface-matching method. After 6–7 reference points are taken on the surface of the pelvis and femur, 30–40 arbitrary points are taken on the bone surface. Registration is completed by superimposing the intraoperatively created surface profile on the surface profile of the preoperative bone model (Figs. 56.1 and 56.2).

In 1997, Osaka University independently developed a CT-based navigation system for THA. Stryker's CT-based navigation system for THA, based on Osaka University's system, was approved by pharmaceutical affairs bodies in 2004 [1]. Since then, the accuracy of infrared cameras has improved, and some parts of the software have been updated.

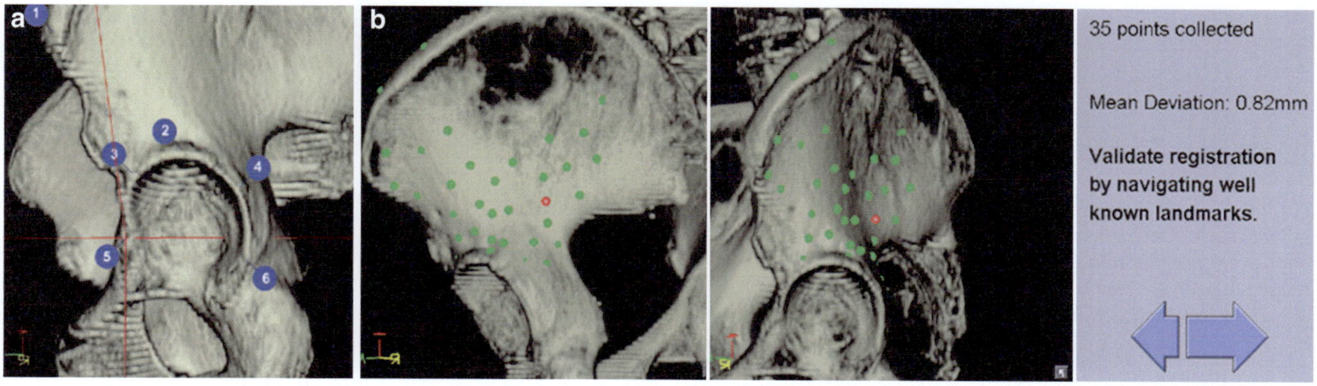

Fig. 56.1 Surface-matching registration of the pelvis. (**a**) Prior to surface matching, six reference points should be set at the characteristic sites around the superior anterior iliac spine and acetabulum. (**b**) For surface matching, 30–40 points should be taken at the sciatic plane, including the sciatic notch, anterior site, including the anterior pelvic column, and posterior iliac site, by creating three planes of different directions in three-dimensions; 90% of the points should be adopted, and the mean deviation should be less than 1 mm. The accuracy of registration should be checked by palpating the bony surface with a pointer after registration, and we should not put too much confidence in the mean deviation values

K. Tokunaga (✉)
Niigata Hip Center, Kameda Daiichi Hospital, Niigata, Japan

Y. Nakashima (ed.), *Advances in Total Hip Arthroplasty*, https://doi.org/10.1007/978-981-97-2461-1_56

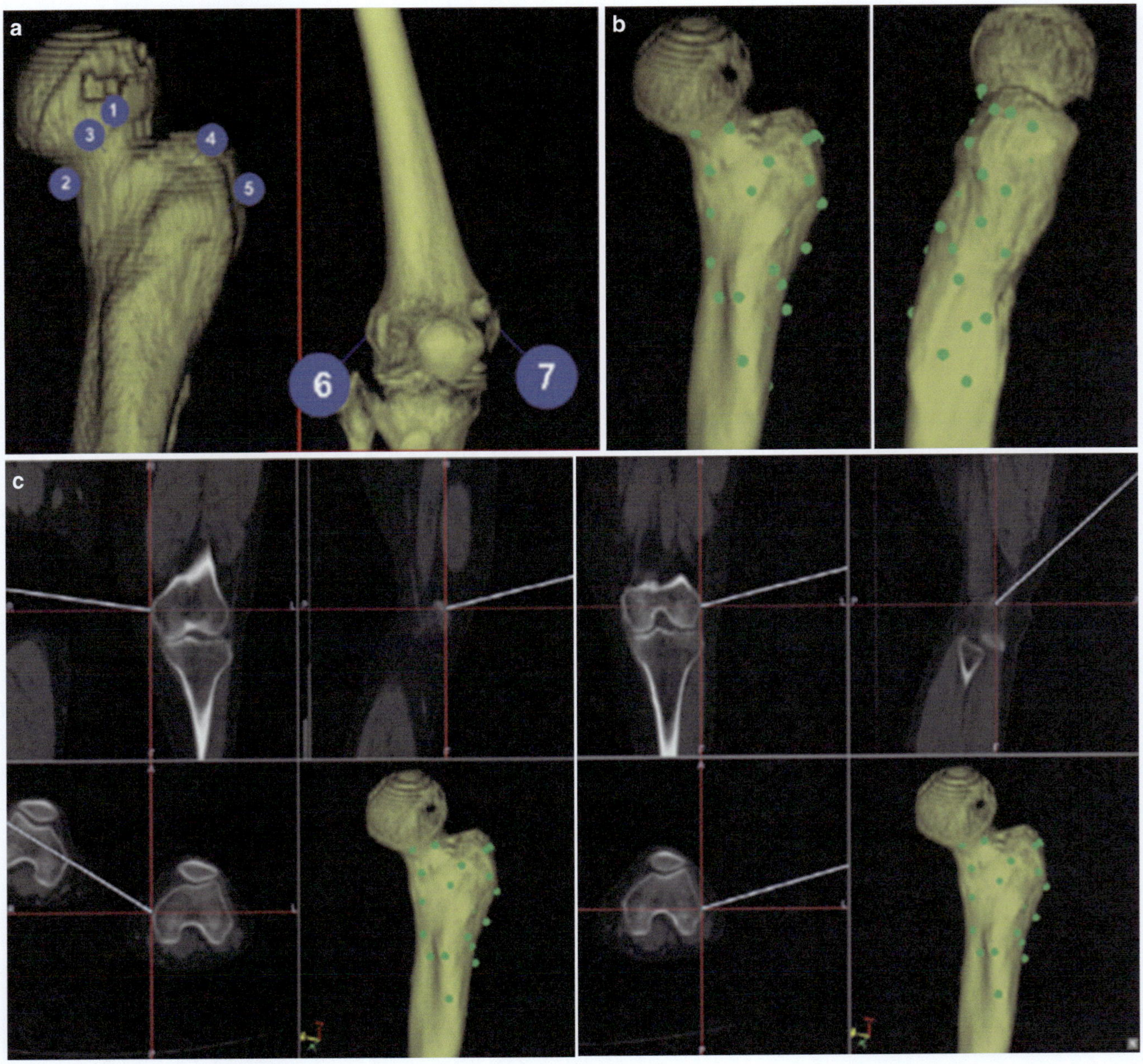

Fig. 56.2 Surface-matching registration of the femur. (**a**) A reference point is taken at a characteristic site on the femur and the skin is pierced with an 18G needle at the distal medial and lateral femoral condyles. A reference point is obtained while palpating the bony surface. (**b**) Thirty points for surface matching on the proximal femur surface. A modified Watson–Jones with L-shaped fasciotomy is then performed. (**c**) Do not put too much confidence in the mean deviation values displayed on the screen. The bone surface is touched with the pointer after registration and the accuracy of the distal part is confirmed by directly touching the bony surface of the medial and lateral femoral condyles

2 Preoperative Planning for CT-Based Navigation

In principle, a cementless cup is selected as the acetabular cup, which is placed in the original position of the acetabulum. Fine adjustments are made according to the offset and leg length using navigation software. If there is a bone defect in the upper part of the acetabulum, bone grafting is performed using the resected femoral head without hesitation. Stems should be selected to accommodate the planned offset, optimal angles (anteversion, valgus/vasus, and flexion/extension angles), and intramedullary geometry; ROM simulation is used to confirm the optimal combination of position, angle, and fine-tuning. A new navigation software for the Exeter stem, a cemented stem with high global reliability, has been developed and is available, enabling accurate placement of cement stems (Fig. 56.3).

If we use navigation software alone for preoperative planning, we can only choose Stryker implants that do not always match the patient's bone geometry. ZedHip (LEXI, Japan), a 3D image analysis software with a large amount of clinically applicable implant information, can be used to identify the

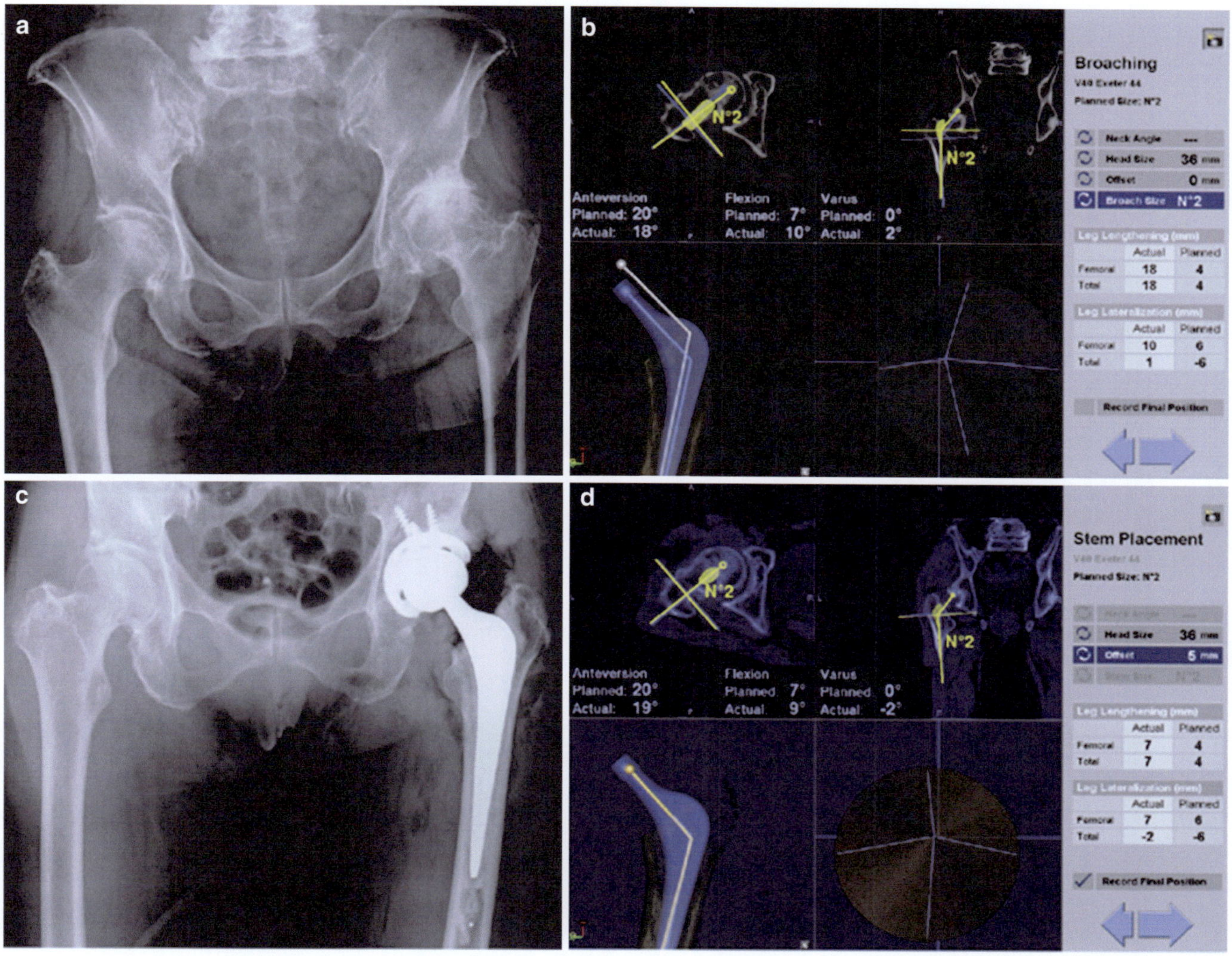

Fig. 56.3 Insertion of the Exeter stem with navigation assistance. (**a**) Preoperative radiographic hip joint frontal view; (**b**) Navigation operation screen during rasp insertion; (**c**) Immediate postoperative X-ray hip joint frontal view; (**d**) Navigation operation screen during stem insertion

most appropriate implant for a 3D bone model. Furthermore, ZedHip's ROM simulation can verify not only implant-to-implant but also bone-to-implant and bone-to-bone collisions.

3 Intraoperative Registration

The Stryker navigation for THA was originally developed for the posterolateral approach. It is important to take reference points for surface-matching registration around the posterior pelvic aspect, including the sciatic notch. Even in the anterolateral approach [2], where the sciatic notch is difficult to expose, an L-shaped incision of the femoral fascia allows the sciatic notch area and posterior parts of the femur to be exposed, which facilitates surface-matching registration for both the pelvis and femur (Fig. 56.4).

Prior to the surface-matching method, six to seven reference points on the bone, which correspond to the reference points on the 3D bone model, were registered (Figs. 56.1 and 56.2). The accuracy of the positions of these reference points is important for the next surface-matching method [3]. Surface matching becomes easier when the average distance error of the reference points is less than 2 mm.

Even if we perform femoral registration at the proximal portion only and obtain a distance error of less than 2 mm, registration of the distal femur may still be inaccurate (Fig. 56.5). To increase the accuracy of femoral registration, we recommend using a reference point on the femoral condylar surface. It is important to check the accuracy of registration by touching several points on the bone surface with a pointer rather than relying on the distance errors displayed on the navigation screen.

In the pelvic surface-matching method, we collected as many points as possible over a wide area, and 30–40 points were collected on the pelvis with awareness of three different bony surfaces, namely the sciatic notch area and the anterior and posterior portions of the anterior pelvic column (Fig. 56.1) but not within the acetabulum. In end-stage osteoarthritis, the acetabular surface cannot be accurately reproduced after segmentation between the pelvis and

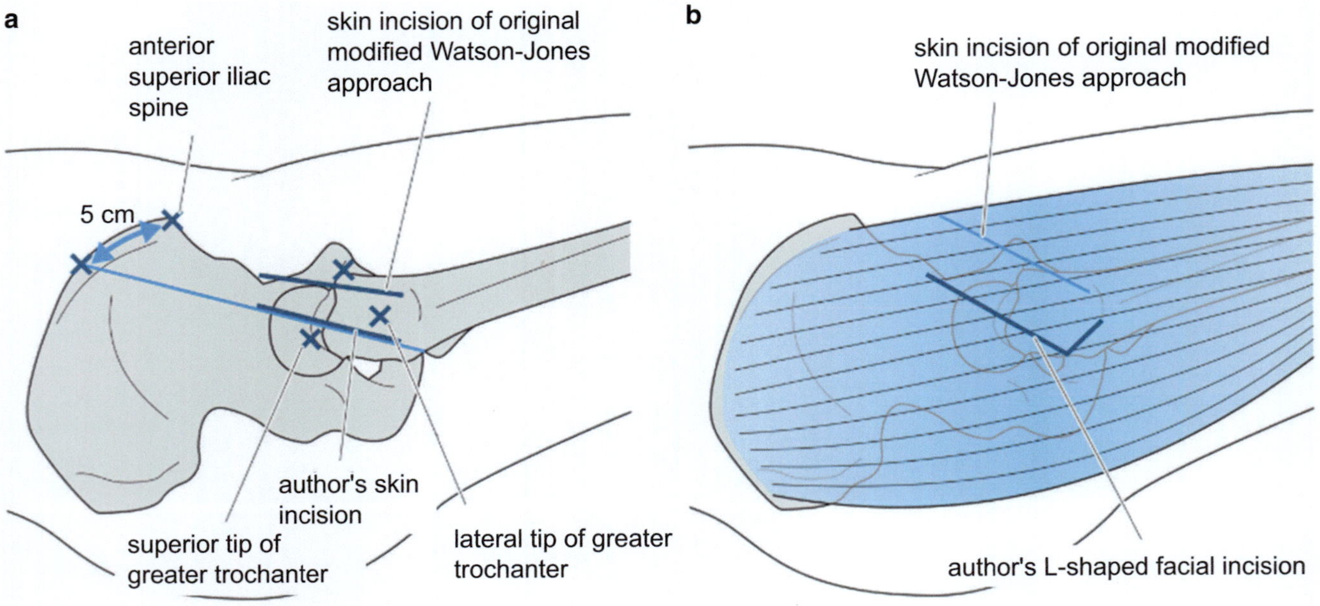

Fig. 56.4 Modified Watson-Jones antero-lateral approach. (**a**) The skin incision is placed posterior to the original modified Watson-Jones approach to reach the posterior border of the greater trochanter. (**b**) The fascia is incised in an L-shape around the apex and posterior aspect of the greater trochanter

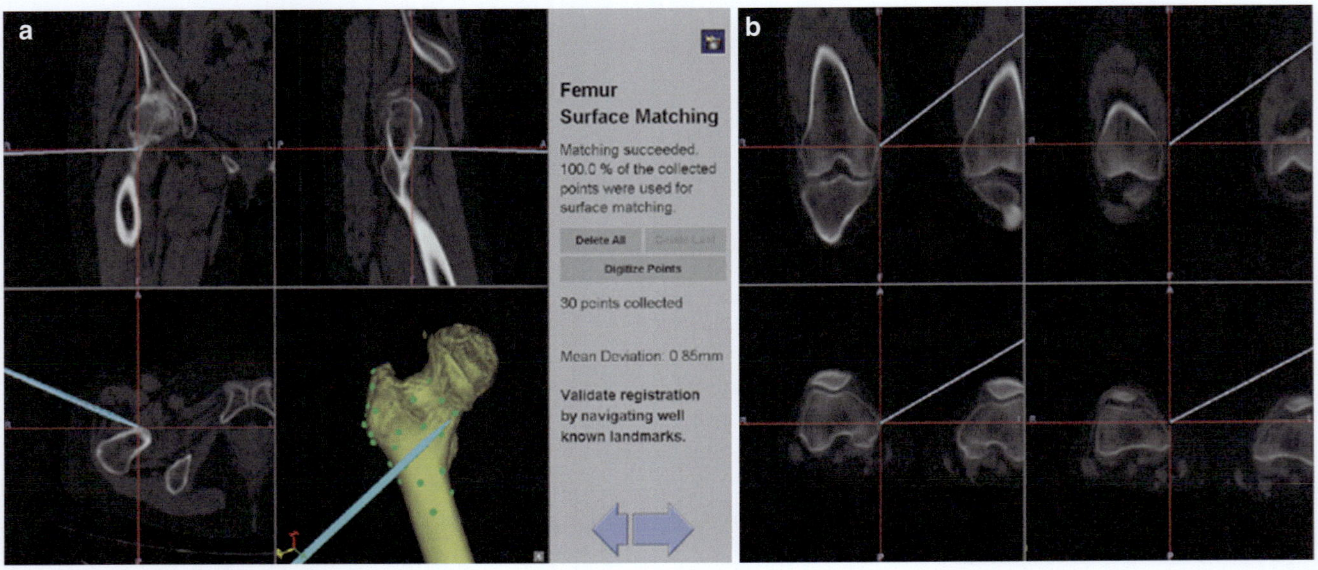

Fig. 56.5 Accuracy at the distal femur. (**a**) Thirty points were obtained from the femoral surface. 100% of the palpated points were used at that time. The mean distance deviation was 0.85 mm, and the accuracy was investigated well in the proximal femur using the pointer. (**b**) In the distal femur, the pointer appeared accurate at the medial condyle; however, when the pointer was shifted proximally, the pointer tip was separated from the bony surface and the registration was inaccurate

femur. Stryker's navigation system semiautomatically separates the femur and pelvis, which often results in a 3D bone model that differs from the actual acetabular surface. In addition, if the acetabular surface is always used as a registration point, one must register the case with hip ankylosis and arthrodesis using an unfamiliar registration method, which results in an inaccurate navigation procedure.

In the early days of CT-based navigation, the optimal CT slice width was 3 mm [5], but today's CTs are capable of rapid 1-mm width imaging with low radiation exposure. Metal halation elimination software is also available, and accurate registration can be achieved by taking a reference point on the implant surface in revision cases in which the implant is stable (Fig. 56.6). In cases of periprosthetic femoral stem fractures, accurate registration can be achieved through preoperative planning prior to fracture CT, if available (Fig. 56.7).

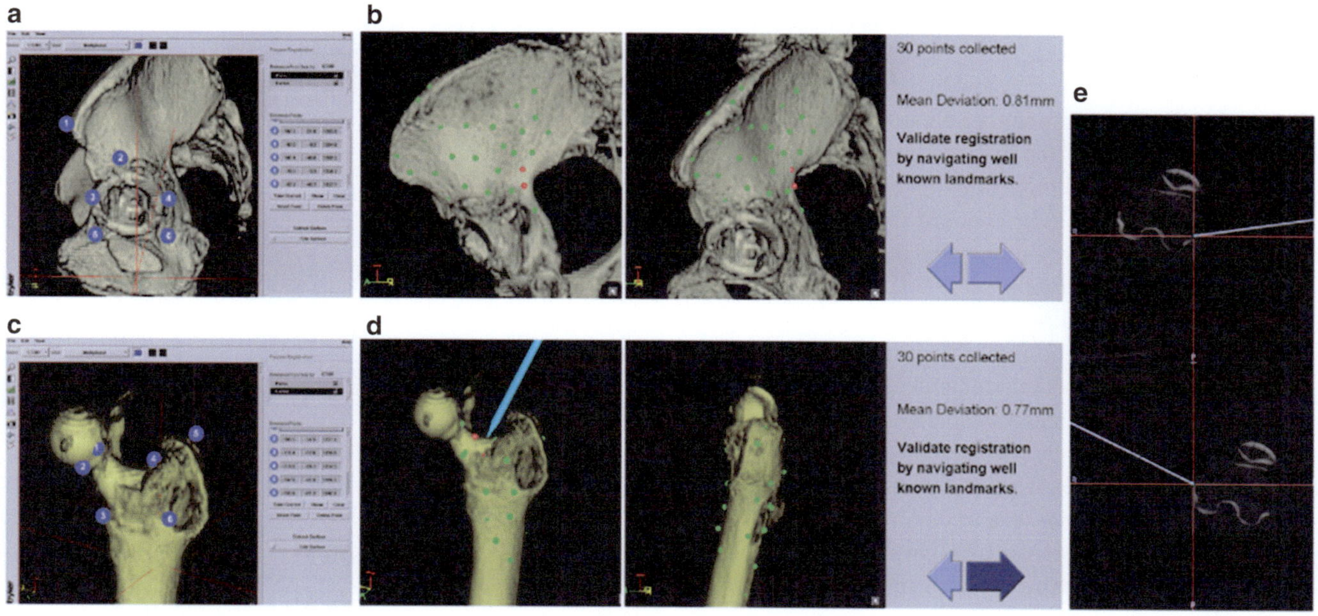

Fig. 56.6 Surface matching in revision surgery. (**a**) Pelvic reference points are placed on the margin of the existing acetabular cup. (**b**) Surface matching of the pelvis. As usual, we did not use points in the acetabular cup. (**c**) Femoral reference points are placed on the margin of the existing femoral neck and proximal bony surface. (**d**) Surface matching of the femur. (**e**) After registration was completed, a pointer was inserted through the distal femur through the skin wound using 18G needle to palpate directly on the bone to check the accuracy

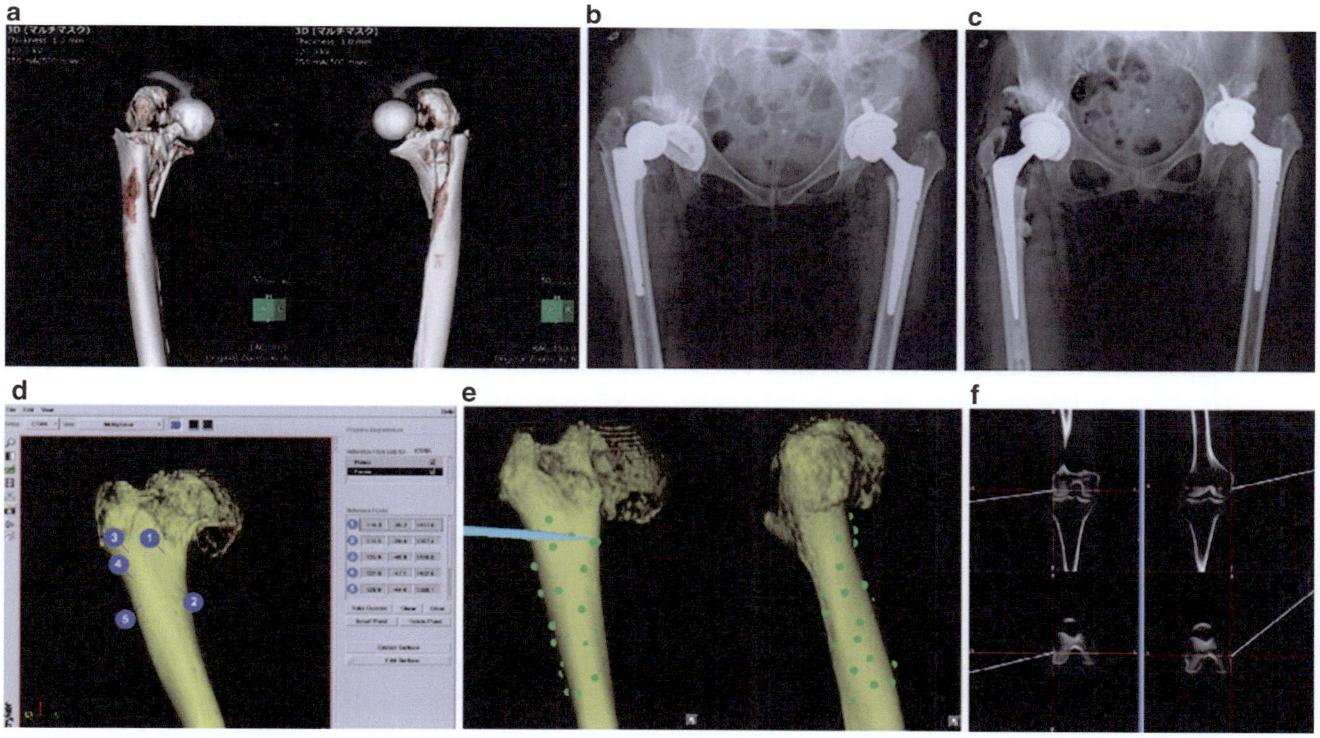

Fig. 56.7 Surface-matching method for periprosthetic femoral fracture. (**a**) Three-dimensional (3D) image constructed from the computed tomography (CT) image of the femur at the time of periprosthetic femoral fracture. (**b**) Posterior dislocation due to periprosthetic fracture of the right femur; C: Repair and fixation of the fracture using a Nesplon cable (Alfresa Pharma Corporation, Japan) followed by revision surgery using a cemented stem; (**d**) Preoperative bone model was created using the preoperative CT before the fracture, and a reference point was planned for the area without fracture by referring to the 3D image in (**a**). (**e**) Thirty points were taken for surface matching widely outside the fracture area; (**f**) After registration was completed, a pointer was inserted through the skin at the distal femur with an 18G to directly palpate the bony surface to confirm accuracy

4 Intraoperative Cup and Stem Placement

Both cup and stem operations are controlled by adjusting the positions and angles of the tools displayed on the screen in real time to match the preoperative plan. If any discrepancy exists, the tracker should be checked for slack or tool damage. If no abnormality is found, the registration should be repeated without hesitation.

For cup insertion, a straight reamer-cup holder can be used for accurate insertion via the posterolateral approach. However, in anterior and anterolateral approaches, an instrument with an offset is preferable to avoid the femur. Under navigation assistance, the holder must be tapped several times for accurate placement, and it is necessary to have a fixation that does not sag even after frequent tapping. The offset cup impactor (Precimed, US) is easy to use because it has a support mechanism that reinforces screwed-in fixation.

Although THA using CT-based navigation is becoming popular, very few surgeons have used navigation for stem insertion. It is difficult to reproduce the preoperative planning angle of the stem with manual insertion when advocating the combined anteversion theory. While cup navigation requires adjustment of three factors: position, inclination, and anteversion, and the stem requires adjustment of four factors: position, anteversion, varus/valgus, and extension/flexion. However, the current navigation system requires extraordinary time for preoperative planning and intraoperative manipulation to navigate the stem and cup. We hope that the software will be improved to shorten preoperative preparation and intraoperative procedures.

5 Accuracy and Precision of CT-Based Navigation and its Clinical Effects

The accuracy and precision of CT-based navigation have been reported to be excellent [4, 5] and highly accurate not only in primary THA, but also in revision THA and in patients with significant pelvic deformities [6, 7]. In addition, CT-based navigation demonstrated high accuracy in THA in obese patients, in whom image-less navigation could not achieve an accurate procedure [8]. The clinical results of patients who underwent CT-based navigation THA at Osaka University for >10 years showed significantly better results than those who did not use navigation [9], and the postoperative quality of life evaluation was also superior to that of patients using conventional methods [10].

The dislocation rate in 600 self-experimental patients using Striker navigation was 0.2% [11], which was comparable to the results reported in more than 1000 patients using Striker navigation [12]. Scatter plots of the cup placement angles for the 703 recent self-experimental cases using striker navigation are shown in Fig. 56.8. The absolute errors of the cup and stem angles were 2.5° ± 2.0° for the cup with inclination, 3.0° ± 2.6°for anteversion, and 3.8° ± 3.4°for stem antetorsion, which were larger than those reported previously. When the outlier was defined as an angle of 5°or more, it accounted for 12.2% of the cup inclination, 17.4% of the anteversion, 26.9% of the combined inclination and anteversion, and 25.8% of the stem antetorsion. The outlier rate for the combined cup and stem was 46.0% and approximately half of the implantation angles were incorrect. The clinical results of THA using BrainLab's (Brainlab, Germany) CT-based navigation over 10 years showed no difference in dislocation rate, imaging findings, or hip function compared with the mid-term results of conventional THA using the same implants [13]. Although THA using CT-based navigation allows for accurate implant placement according to the preoperative plan, the current system is still not perfect, and improvements are needed to achieve more reproducible accuracy and precision. Furthermore, the complexity of preoperative planning and intraoperative procedures, as well as the high cost of purchasing the system, have hindered its widespread use, and improvements in these issues are desired.

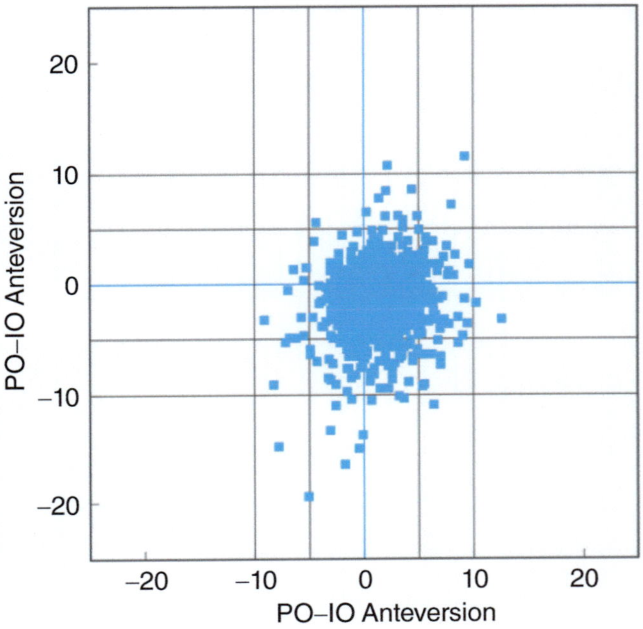

Fig. 56.8 Scatter diagram of angle errors in cup implantation. Most of the points were within the range of less than 5°; however, there were a few points with an error of more than 5° or more than 10°. The error was larger for anteversion angles than for inclination angles. Solid line: 0° error; large dashed line: 5° error; fine dashed line: 10° error

References

1. Sugano N. History of surgical navigation in orthopaedics. J Joint Surg. 2015;34:113–6. Japanese
2. Bertin KC, et al. Anterolateral mini-incision hip replacement surgery: a modified Watson-Jones approach. Clin Orthop Relat Res. 2004;429:248–55.
3. Fujii Y, et al. An analysis of the characteristics and improved use of a newly developed CT-based navigation system in total hip arthroplasty. Acta Med Okayama. 2017;71:279–89.
4. Iwana D, et al. Accuracy of angle and position of the cup using a computed tomography-based navigation system in total hip arthroplasty. Comput Aided Surg. 2013;18:187–94.
5. Nakahara I, et al. Effect of improved navigation performance on the accuracy of implant placement in total hip arthroplasty with a CT-based navigation. J Artif Organs. 2018;21:340–7.
6. Nakamura N, et al. Application of computed tomography-based navigation for revision total hip arthroplasty. J Arthroplast. 2013;28:1806–10.
7. Kajino Y, et al. Does degree of pelvic deformity affect the accuracy of computed tomography-based hip navigation? J Arthroplast. 2012;27:1651–7.
8. Imai N, et al. Accuracy of acetabular cup placement using CT-based navigation in total hip arthroplasty: comparison between obese and nonobese patients. J Orthop Sci. 2019;24:482–7.
9. Sugano N, et al. Does CT-based navigation improve the long-term survival in ceramic-on-ceramic THA? Clin Orthop Relat Res. 2012;470:3054–9.
10. Maeda Y, et al. Improvement of activities of daily living after total hip arthroplasty using a computed tomography-based navigation system. J Artif Organs. 2017;20:152–7.
11. Tokunaga K, et al. Navigation total hip arthroplasty. J Joint Surg. 2020;39:631–41. Japanese
12. Nakahara I, et al. Results of THA cases performed using navigation system on both cup and stem sides. Hip Joint. 2020;46:231–5. Japanese
13. Tokunaga K, et al. Clinical results of CT-based navigation assisted THA at a minimum 10 year-follow up. Hip Joint. 2020;46:236–41. Japanese

Image-Free Navigation

57

Masahiro Hasegawa

1 Navigation in Total Hip Arthroplasty

Navigation in total hip arthroplasty (THA) was first reported in 1998, to enable implant placement according to preoperative plans [1, 2]. In Japan, computed tomography (CT)-based navigation was approved in 2001 and has since been used clinically. Image-free navigation, which does not require preoperative and intraoperative imaging, was approved in 2009 and is available for use in THA. Portable navigation using gyroscopic and accelerometer sensors was approved and used in 2016.

For navigation, tracker pins were inserted into the pelvis, and a reference marker was fixed (Fig. 57.1a). The marker was tracked using an infrared camera. In CT-based navigation, CT is used to make the system aware of the position of the reference marker and the pelvic reference plane. Image-free navigation can be used without preoperative and intraoperative imaging information and allows intraoperative registration of the anterior superior iliac spine (ASIS). The pubic symphysis was registered, and virtual anterior pelvic plane (APP) coordinates were set according to reference markers. After registration of the internal surface of the acetabulum, the inclination and anteversion angles of the cup with respect to the APP reference were displayed in conjunction with a hammering handle and reference markers (Fig. 57.1b). In image-free navigation, inaccuracies may occur because of the thickness of the subcutaneous tissue on the pubic surface [3]. Using echocardiography, when the soft tissue thicknesses on the ASIS surface and the pubic symphysis surface were measured, the ASIS surface was on average 7.9 mm and the pubic symphysis surface was on average 13.6 mm, with the latter being 5.7 mm thicker. When the inclination angle was 45°, the anteversion angle was 2.8°greater [4]. In Europe and the United States, image-free navigation is the mainstay because of the cost and radiation exposure associated with CT imaging.

1.1 Cup Angle in Navigation

There are three definitions of cup angles: radiographic, anatomical, and operative [5]. The radiographic definition corresponds to cup angles measured on a simple radiograph. The radiographic inclination (RI) is the angle between the long axis of the patient and the projected axis of the polar axis of the cup onto the coronal plane. The radiographic anteversion (RA) is the angle between the polar axis of the cup and the coronal plane. The anatomical anteversion (AA) is the angle between the transverse axis and the projected axis of the polar axis of the cup onto the transverse plane, as seen in the CT images. The anatomical inclination (AI) is the angle between the polar axis of the cup and the long axis of the patient. The operative definition is the angle from the surgeon's perspective. The operative inclination (OI) is the angle between the polar axis of the cup and the sagittal plane, and the operative anteversion (OA) is the angle between the long axis of the patient and the projected axis of the polar axis of the cup onto the sagittal plane [5]. The CAOS Japan recommends a radiographic definition. The Brainlab navigation system uses an operative definition. The 40° OI and 20° OA are equivalent to 42° RI and 15° RA, respectively.

1.2 Method of Registration

The Brainlab CT-based navigation system is no longer supported and cannot be used after failure, and the Brainlab navigation system uses image-free Hip 6.0. The Brainlab image-free navigation system can be used for registration of the fifth lumbar spinous process and unilateral ASIS in the lateral decubitus position to avoid registration errors at the pubic symphysis. However, owing to poor accuracy [6], bilateral ASIS registration was performed in the supine

M. Hasegawa (✉)
Department of Orthopaedic Surgery, Mie University Graduate School of Medicine, Edobashi Tsu City, Mie, Japan
e-mail: masahase@clin.medic.mie-u.ac.jp

Y. Nakashima (ed.), *Advances in Total Hip Arthroplasty*, https://doi.org/10.1007/978-981-97-2461-1_57

a b

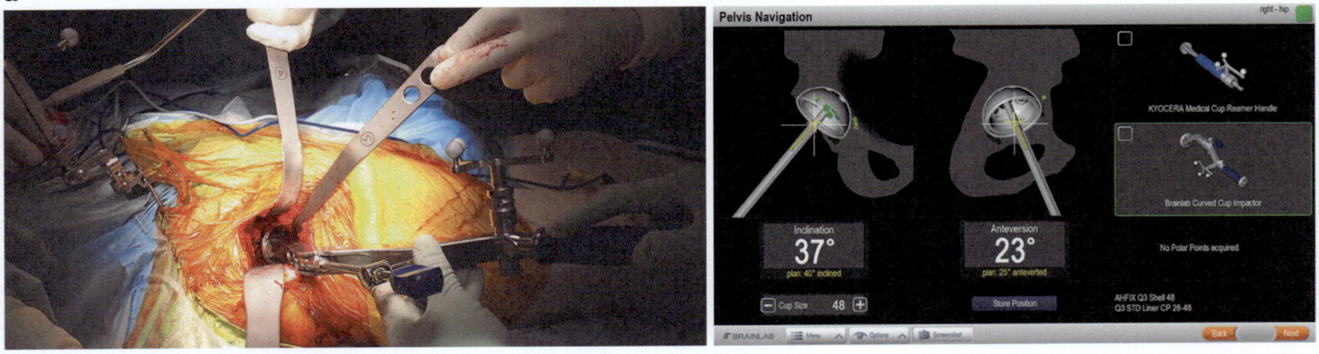

Fig. 57.1 Image-free navigation. (**a**) Intraoperative photograph. (**b**) Screenshot after intraoperative cup placement. The preoperative plan was OI 40° and OA 25°, and OI 37° and OA 23° after implantation. *OI* operative inclination, *OA* operative anteversion

position, and the lumbar spine was not used. The registration of the pubic symphysis is not necessary; instead, registration of the landmark at the anterior rim of the acetabulum (psoas valley) is used. The pelvic tilt must be determined on preoperative images to place the cup according to the supine functional pelvic plane (FPP). If the APP and FPP are consistent, the preoperative pelvic tilt can be ignored, and a constant 15° or 20° anteversion on the intraoperative view is acceptable. However, such cases are extremely rare. For example, in simple preoperative planning, if the pelvis is tilted 5° anteriorly, the anteversion angle is increased by 5° (Fig. 57.1b), and if it is tilted 5° posteriorly, it is decreased by 5°. The portable navigation system had a gyro sensor and an acceleration sensor, and the FPP standard angle was displayed. Using portable navigation, the intraoperative display of the anteversion angle can be kept constant at 15° or 20°, ignoring the preoperative pelvic tilt.

2 Accuracy of Image-Free Navigation

2.1 Historical Accuracy Reporting

In the meta-analysis, there was no difference in the inclination or anteversion angles in cases with image-free navigation compared to controls, but there were significantly more cases within Lewinnek's safe zone [7] in cases with navigation [8]. In a meta-analysis of cases with image-free navigation, there was no difference in the inclination angle between cases with navigation and controls, but the anteversion angle had a smaller error in cases with navigation [9]. In another meta-analysis, it was reported that there was less error in the inclination and anteversion angles of the cup in the case of image-free navigation than in the control [10]. The accuracy of cup placement for image-free navigation is 1.8–4.4° for the inclination angle and 2.0–11.6° for the anteversion angle, and there are reports that the accuracy of the anteversion angle is poor (Table 57.1) [3, 6, 11–21]. THA with OrthoPilot (B. Braun-Aesculap) has been reported to have an error of 5.8–5.9° in stem anteversion (Table 57.1) [18–20]. Considering combined anteversion and navigation with the stem-first technique, 92% of the cases were within the target range of 37 ± 5°, but only 42% of the cases were within the range of the cup-first technique [15]. However, when the anteversion angle of the cup was determined using the stem-first technique in cases of developmental dysplasia of the hip, the anteversion of the stem was greater and the anteversion angle of the cup was smaller. The cup protruded more than 3 mm anterior to the pelvis in 58% of cases, raising concerns regarding the occurrence of iliopsoas impingement [20].

2.2 Examination of Factors Affecting Accuracy

The effect of obesity on cup placement was not correlated with the body mass index [16], and there was no difference in cup placement accuracy between obese and non-obese cases [19, 21]. However, it has been reported that the cup anteversion angle error is significantly larger in obese cases [3, 17]. The presence of developmental dysplasia of the hip did not affect cup placement error [17], but it was reported that the cup anteversion angle error was significantly larger in Crowe groups 2–4 than in group 1 [21]. Soft tissue thickness measured by CT between the pubic symphysis surface and the skin surface had no correlation with cup placement error. However, in 77% cases, the postoperatively measured anteversion angle was greater than the intraoperative navigation value [16]. When the intraoperative and interoperative intraclass correlation coefficients were determined, the reliability of the cup placement angle was higher in the supine position than in the lateral decubitus position [22]. However, it has been reported that there is no difference in cup placement accuracy between the supine and lateral decubitus positions [19, 21].

Table 57.1 Absolute error of image-free navigation

	Reporter	Company	Cup		Stem
			Inclination (°)	Anteversion (°)	Anteversion (°)
Overseas	Kalteis T et al. [11]	Brainlab[a]	2.9 ± 2.2	4.2 ± 3.3	
	Ybinger T et al. [12]	Plus Orthopedics[a]	3.5 ± 4.4	6.5 ± 7.3	
	Lass R et al. [13]	ORTHOsoft[a]	3.2 ± 2.4	6.5 ± 3.7	
	Parratte S et al. [3]	Praxim Medivision[a]	4.0 ± 2.8	3.4 ± 3.6	
			3.3 ± 3.1[b]	11.6 ± 6.1[b]	
	Ryan JA et al. [14]	Orthopaedics Inc., Brainlab[a]	1.8 ± 1.2	2.0 ± 2.0	
Japan	Fukunishi S et al. [15]	B. Braun Aesculap[a]	1.9 ± 1.9	2.6 ± 1.6	
	Fukui T et al. [16]	B. Braun Aesculap[a]	1.2	2.3	
	Tsukada S et al. [17]	B. Braun Aesculap[a]	2.4 ± 2.0	3.7 ± 2.3	
	Fukunishi S et al. [18]	B. Braun Aesculap[a]	3.0 ± 2.6	5.0 ± 3.5	5.9 ± 4.2
	Takeda Y et al. [19]	B. Braun Aesculap[a]	3.6 ± 2.7	6.9 ± 3.6	5.8 ± 4.6
	Okada T et al. [20]	B. Braun Aesculap[a]	4.0 ± 3.2	5.5 ± 4.6	5.8 ± 4.4
	Hasegawa M et al. [6]	Brainlab[c]	2.1 ± 1.8	4.1 ± 3.6	
		Brainlab[d]	4.4 ± 4.2	7.1 ± 6.3	
	Naito Y et al. [21]	Brainlab[c]	3.3 ± 2.8	5.8 ± 4.9	

[a]Registration of the bilateral anterior superior iliac spine and pubic
[b]Body mass index of ≥27 kg/m^2
[c]Registration of the bilateral anterior superior iliac spine
[d]Registration of the anterior superior iliac spine and fifth lumbar spinous process on the affected side

3 Impact on Clinical Outcomes

3.1 Dislocation and Revision Rates in Cases of Navigation Use

In a 10-year study, there were no cases of dislocation or revision due to loosening in 30 cases with or without navigation, and there was no difference in clinical scores [23]. Of 803,732 THA cases, navigation was used in 14,540 (1.8%), and the mean follow-up period was 2.6 and 3.0 years in THA cases with and without navigation, respectively. The revision rate for cup loosening was also significantly lower in THA cases with navigation [24].

3.2 Overseas Registry Data

In the National Joint Registry of England, Wales, Northern Ireland, and the Isle of Man, navigation was used in 871 cases (2%), and the 10-year revision rate was 1.1% and 3.9% in cases with and without navigation, respectively. However, there was no difference in the Oxford hip score or EuroQol-5 dimension (EQ-5D) 6 months postoperatively. The satisfaction rate was higher in cases who used navigation [25]. In the Australian registry data, navigation use increased from 1.9% in 2009 to 4.4% in 2019; however, revision rates did not differ with or without navigation use. However, the 10-year revision rate due to dislocation was significantly lower in the navigation-use cases. In the five highly used combination implants, the 10-year revision rates for all reasons were also significantly lower in navigation-use cases [26].

3.3 THA Navigation Usage Rate in Japan

According to the Japanese Arthroplasty Society, THA navigation has increased from 6% in 2013 to 14% in 2017. We hope that navigation will increase the number of cases of accurate cup placement and improve long-term outcomes.

References

1. DiGioia AM, et al. The Otto Aufranc Award. Image guided navigation system to measure intraoperatively acetabular implant alignment. Clin Orthop Relat Res. 1998;355:8–22.
2. Jaramaz B, et al. Computer assisted measurement of cup placement in total hip replacement. Clin Orthop Relat Res. 1998;354:70–81.
3. Parratte S, et al. Validation and usefulness of a computer-assisted cup-positioning system in total hip arthroplasty. A prospective, randomized, controlled study. J Bone Joint Surg Am. 2007;89:494–9.
4. Richolt JA, et al. How does soft tissue distribution affect the anteversion accuracy of the palpation procedure in image-free acetabular cup navigation? An ultrasonographic assessment. Comput Aided Surg. 2005;10:87–92.
5. Murray DW. The definition and measurement of acetabular orientation. J Bone Joint Surg Br. 1993;75:228–32.
6. Hasegawa M, et al. Comparison of the accuracies of computed tomography-based navigation and image-free navigation for acetabular cup insertion in total hip arthroplasty in the lateral decubitus position. Comput Assist Surg (Abingdon). 2021;26:69–76.
7. Lewinnek GE, et al. Dislocations after total hip-replacement arthroplasties. J Bone Joint Surg Am. 1978;60:217–20.
8. Beckmann J, et al. Navigated cup implantation in hip arthroplasty. Acta Orthop. 2009;80:538–44.
9. Liu Z, et al. Imageless navigation versus traditional method in total hip arthroplasty: a meta-analysis. Int J Surg. 2015;21:122–7.
10. Snijders T, et al. Precision and accuracy of imageless navigation versus freehand implantation of total hip arthroplasty: a systematic

review and meta-analysis. Int J Med Robot. 2017;13:10.1002/rcs.1843.
11. Kalteis T, et al. Imageless navigation for insertion of the acetabular component in total hip arthroplasty: is it as accurate as CT-based navigation? J Bone Joint Surg Br. 2006;88:163–7.
12. Ybinger T, et al. Accuracy of navigation-assisted acetabular component positioning studied by computed tomography measurements: methods and results. J Arthroplast. 2007;22:812–7.
13. Lass R, et al. Total hip arthroplasty using imageless computer-assisted hip navigation: a prospective randomized study. J Arthroplast. 2014;29:786–91.
14. Ryan JA, et al. Accuracy of computer navigation for acetabular component placement in THA. Clin Orthop Relat Res. 2010;468:169–77.
15. Fukunishi S, et al. Assessment of accuracy of acetabular cup orientation in CT-free navigated total hip arthroplasty. Orthopedics. 2008;31. orthosupersite.com/view.asp?rID=31515
16. Fukui T, et al. Use of image-free navigation in the determination of acetabular cup orientation: analysis of factors affecting precision. Orthopedics. 2010;33(10 Suppl):38–42.
17. Tsukada S, et al. Decreased accuracy of acetabular cup placement for imageless navigation in obese patients. J Orthop Sci. 2010;15:758–63.
18. Fukunishi S, et al. Accuracy of combined anteversion in image-free navigated total hip arthroplasty: stem-first or cup-first technique? Int Orthop. 2016;40:9–13.
19. Takeda Y, et al. Accuracy of component orientation and leg length adjustment in total hip arthroplasty using image-free navigation. Open Orthop J. 2017;11:1432–9.
20. Okada T, et al. Total hip arthroplasty using stem-first technique with navigation: the potential of achievement of optimal combined anteversion being a risk factor for anterior cup protrusion. Eur J Orthop Surg Traumatol. 2019;29:807–12.
21. Naito Y, et al. The accuracy of acetabular cup placement in primary total hip arthroplasty using an image-free navigation system. BMC Musculoskelet Musculoskeletal Disord. 2021;22:1016.
22. Ohashi H, et al. Intra- and intersurgeon variability in image-free navigation system for THA. Clin Orthop Relat Res. 2009;467:2305–9.
23. Parratte S, et al. No benefit after THA performed with computer-assisted cup placement: 10-year results of a randomized controlled study. Clin Orthop Relat Res. 2016;474:2085–93.
24. Bohl DD, et al. Computer-assisted navigation is associated with reductions in the rates of dislocation and acetabular component revision following primary total hip arthroplasty. J Bone Joint Surg Am. 2019;101:250–6.
25. Davis ET, et al. Reduced risk of revision with computer-guided versus non-computer-guided THA: an analysis of manufacturer-specific data from the National Joint Registry of England, Wales, Northern Ireland and the Isle of Man. JBJS Open Access. 2021;6:e21.00006.
26. Agarwal S, et al. The use of computer navigation in total hip arthroplasty is associated with a reduced rate of revision for dislocation: a study of 6,912 navigated THA procedures from the Australian Orthopaedic Association National Joint Replacement Registry. J Bone Joint Surg Am. 2021;103:1900–5.

58 Portable Navigation

Kentaro Iwakiri

1 Usefulness and Characteristics of Hip Replacement with Portable Navigation

In total hip arthroplasty (THA), following the preoperative plan for placement of the implant in the proper place and at the proper angle is an important because it reduces post-THA dislocation and polyethylene wear and improves range of motion, thus, resulting in a good long-term postoperative outcome [1–3]. So-called robotic (such as Mako) and navigation systems (CT-based navigation and image-free navigation) can enable very precise implant placement (placement angle and position) [4], but they are very expensive. Additionally, they are not available to all surgeons and institutions because they require additional time for preoperative planning and intraoperative work and personnel skilled in navigation work. Therefore, portable navigation has recently become popular as a simple, relatively inexpensive, and highly accurate tool for acetabular component placement.

The following points should be considered when using portable navigation: (1) surgical position, (2) reference plane, (3) invasiveness, (4) cost, (5) preoperative planning, (6) availability of implants, and (7) accuracy.

1. Surgical position: there are three types of models, i.e., those that can be used only in the supine or lateral recumbent position, and those that can be used in both positions.
2. The reference plane in surgery: the functional pelvic plane (FPP, parallel to the bed in the supine position) and anterior pelvic plane (APP, consisting of the three points of the bilateral anterior superior iliac spines and pubic symphysis) are used to guide the implant. The APP is used to guide implant placement.
3. Invasiveness: some models require pin insertion into the patient's body (pelvis), while others do not require pin insertion at all.
4. Cost of use: there are multiple- and single-use models (the latter costs money each time).
5. Necessity of preoperative planning: some models require preoperative planning using CT imaging, while others require only simple radiographic imaging.
6. Available implants: There are models that have restrictions on the implants that can be used depending on the distributor, and there are those that can be used in any of the models that the surgeon wishes to use.
7. Accuracy: evaluation by acetabular component placement angle, radiographic inclination (RI), and radiographic anteversion (RA).

This section outlines the features and accuracy of various portable navigation systems that are currently available in Japan.

2 Urushitani Molar Socket Guide (Fig. 58.1; Table 58.1)

This model was first reported in the late 1990s and can be considered as the conceptual beginning of portable navigation, which prevents poor acetabular component placement due to intraoperative pelvic movement [5].

K. Iwakiri (✉)
Shiraniwa Hospital, Nara, Japan

Y. Nakashima (ed.), *Advances in Total Hip Arthroplasty*, https://doi.org/10.1007/978-981-97-2461-1_58

Table 58.1 Points to consider when using portable navigation systems available in Japan

	Urushitani-shiki socket guide	HipCOMPASS®	APP Lateral Positioner®	iFlex G®	HipAlign®Lateral	HipAlign®Supine	NaviSWISS	AR-HIP	intellijoint HIP®	iFlexS (the only portable navigation for stem)
<1> Surgical positions	Many reports of use in the lateral decubitus position. Possible be also used in supine position	Used in supine position only	Used in lateral position only	Used in lateral position only. Must be used with APP Lateral Positioner	Used in lateral position only	Used in supine position only	Used in supine and lateral position, both	Used in supine and lateral position, both	Used in supine and lateral position, both	Reported in lateral position only
<2> Reference plane at surgery	FPP	FPP and APP, both	APP (if preoperative planning for FPP criteria is done, conversion to the APP plane is required. Details are described in <5>Preoperative plan)	APP (if preoperative planning for FPP criteria is done, conversion to the APP plane is required. Details are described in <5>Preoperative plan)	Planes are defined from the direction of gravity and the long axis of the spinal column. Since the long axis of the spine is the axis that the surgeon points to during the procedure, the reference plane is neither the FPP nor the APP, so accuracy in RA remains an issue. <Fig. 58.5a>	FPP <Fig. 58.5b>	FPP. Tips for the lateral position: While supine, define the FPP plane by placing the tag approximately 2 cm posterior to the affected anterior superior iliac spine and then pointing to both anterior superior iliac spines. The FPP plane can be defined in the lateral position by going from supine to lateral position with the tag in place on the pelvis. When going from supine to lateral recumbency with the tag placed in the pelvis, the pin entry site and tag may become unclean. To avoid this, there is a mode in which the reference plane is set after lateral position, but in this mode, the longitudinal axis of the spinal column is the axis pointed to by the surgeon during the surgery, so the reference plane is neither FPP nor APP, and accuracy in RA remains an issue. <Fig. 58.6b>	FPP define the FPP and APP planes by pointing to both anterior superior iliac spines and the pubic symphysis after placing a tag several centimeters posterior to either anterior superior iliac spine while supine. The FPP plane can be defined in the lateral position by going from supine to lateral with the tag in place on the pelvis	[Lateral] Lateral position: Planes are defined from the direction of gravity and the long axis of the spinal column. Same as in the HipAlign lateral, the long axis of the spine is the axis pointed to by the surgeon intraoperatively, so the reference plane is strictly neither FPP nor APP. Accuracy in RA remains an issue [Supine] Supine position: FPP	The posterior femoral condylar axis, required to measure the stem antetorsion angle, cannot be palpated. Then, the tibial ridge is palpated intraoperatively and used. The stem antetorsion angle is measured by having the assistant hold the tibial ridge parallel to the direction of gravity and estimating the posterior femoral condylar axis intraoperatively, reflecting the angle between the tibial ridge and the posterior femoral condylar axis (Kanekasu view) that was measured preoperatively. <Fig. 58.9b>

(continued)

Table 58.1 (continued)

	Urushitani-shiki socket guide	HipCOMPASS®	APP Lateral Positioner®	iFlex G®	HipAlign®Lateral	HipAlign®Supine	NaviSWISS	AR-HIP	intellijoint HIP®	iFlexS (the only portable navigation for stem)
<3> invasiveness of portable navigation	One thread pin (3.2–4.0 mm diameter) is inserted into each bilateral anterior superior iliac spine to be perpendicular to the bed in the supine position. The accuracy of cup placement angle is affected by whether or not these pins are perpendicular to the bed	One pin (1.5 mm diameter) is inserted into each bilateral anterior superior iliac spine	No pin insertion into the body. No invasiveness to the patient	No pin insertion into the body. No invasiveness to the patient. To be installed and fixed on the APP Lateral Positioner from the top of the surgical drape	Insert 2 to 3 pins (2 with a diameter of 4.5 mm and an additional pin with a diameter of 3.2 mm) several centimeters posteriorly from the anterior superior iliac spine on the affected side		Two pins (3 mm diameter) are inserted several centimeters posteriorly from the anterior superior iliac spine on the affected side	Two pins (3.2 mm diameter) are inserted several centimeters posteriorly from the anterior superior iliac spine on the affected side	Two pins (6.5 mm diameter) are inserted several centimeters posteriorly from the anterior superior iliac spine on the affected side	Possible to be easily attached to any model rasp handle. No pin insertion into the body, so there is no invasiveness to the patient. <Fig. 58.9c>
<4> Cost for use	Reusable (pins only; single use)	Reusable (available at LEXI Co. Ltd. Tokyo)	Reusable (available at LEXI Co. Ltd., ZimmerBiomet, OrthoDevelopment, DepuySynthes, Smith & Nephew and FUJIFLEX)	Reusable (available at LEXI Co. Ltd., OrthoDevelopment, DepuySynthes, and FUJIFLEX)	Single-use only. Costly each time. (Available at ZimmerBiomet)		Single-use only. Costly each time. (Available at KYOCERA)	Single-use only. Costly each time.(Available at ZimmerBiomet)	Single-use only. Costly each time.(Available at OrthoDevelopment, Japan MDM)	Reusable (available at FUJIFLEX Co. Ltd., OrthoDevelopment, and Smith& Nephew)
<5> Preoperative plan	Possible with only X-ray	Preoperative planning is performed using 3D template software (CT imaging required) to measure Operative Inclination (OI) and Operative Anteversion (OA) to the reference plane before surgery	Preoperative planning is performed using 3D template software (CT imaging required) to measure Operative Inclination (OI) and Operative Anteversion (OA) to the reference plane before surgery. If the angle between FPP and APP can be measured by simple x-ray of the lateral pelvis, FPP-based RI and RA can be converted to APP-based OI and OA. This is a pelvic fixator, and good acetabular component placement accuracy can be obtained when used with portable navigation, including iFlexG	Preoperative planning is performed using 3D template software (CT imaging required) to measure Operative inclination (OI) and Operative Anteversion (OA) to the reference plane before surgery. If the angle between FPP and APP can be measured by simple x-ray of the lateral pelvis, FPP-based RI and RA can be converted to APP-based OI and OA	Possible with only X-ray		Possible with only X-ray	Possible with only X-ray	Possible with only X-ray	Preoperative planning by using 3D template software (CT imaging required) before surgery. However, among the actual stem placement angles, the stem anteversion angle has been reported to have a greater variability than the stem internal/ external or flexion/ extension angles, compared to the preoperative plan. This tool is more useful as an intraoperative antetorsion angle measuring device than as a navigation device

(continued)

Table 58.1 (continued)

	Urushitani-shiki socket guide	HipCOMPASS®	APP Lateral Positioner®	iFlex G®	HipAlign®Lateral	HipAlign®Supine	NaviSWISS	AR-HIP	intellijoint HIP®		iFlexS (the only portable navigation for stem)
<6> Available Implants	Currently handled by KYOCERA	Possible for any model that the surgeon wishes to use	Possible for any model that the surgeon wishes to use		ZimmerBiomet products only in Japan		Kyocera products only in Japan.	ZimmerBiomet products only in Japan	JapanMDM products only in Japan		Possible for any model that the surgeon wishes to use
<7> Accuracy	No data available	RI 2.9 ± 2.3° RA 2.9 ± 2.1°	RI 2.3 ± 2.3° RA 3.2 ± 2.7°		RI 3.7 ± 3.0° RA 6.0 ± 4.5° RA accuracy is less accurate than that in supine version. The reason for the error was that the spine long axis was the axis that the surgeon pointed to intraoperatively when setting the reference plane, and the reference plane was neither FPP nor APP	RI 3.3 ± 2.7° RA 3.8 ± 3.4°	Lateral position RI 2.7° RA 2.6° Supine position RI 2.2° RA 2.2°	Supine position RI 1.9 ± 1.3° RA 2.8 ± 2.2°	Lateral position RI 2.5 ± 2.9° RA 6.8 ± 4.3° The RA accuracy is less accurate than the supine position because the error is caused by the fact that the reference plane is neither FPP nor APP, since the long axis of the spine is the axis that the surgeon pointed to during the surgery when setting the reference plane	Supine position RI 4.0 ± 3.6° RA 3.4 ± 4.1°	Absolute value error between intraoperative measured antetorsion stem angle and actual antetorsion stem angle (calculated by using postoperative CT). 4.9 ± 3.9°(with Kanekasu view Xp imaging) 8.9 ± 5.4°(without Kanekasu view Xp imaging)

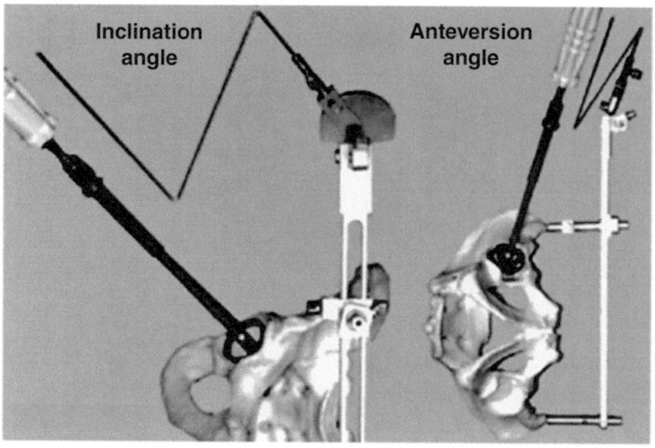

Fig. 58.1 Urushitani acetabular socket guide

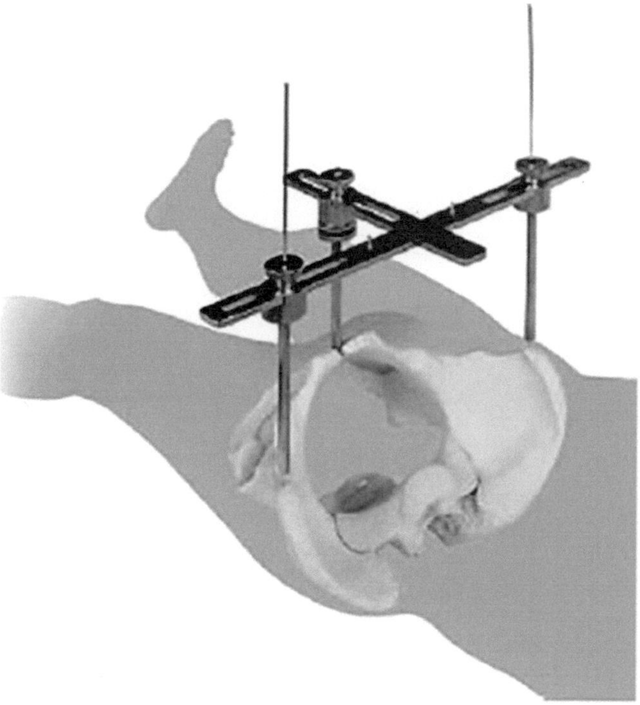

Fig. 58.2 HipCOMPASS®

3 HipCOMPASS® (Fig. 58.2; Table 58.1)

Along with the development of a three-dimensional (3D) preoperative planning template software (ZedHip®: LEXI), a model reported in 2016 was developed to reflect 3D preoperative planning intraoperatively [6].

4 APP Lateral Positioner® (Fig. 58.3; Table 58.1)

This pelvic fixator suppresses the pelvic tilt in any direction (especially in the sagittal and transverse planes), which is a problem in lateral supine THA, and creates an APP plane by holding both anterior superior iliac spines and the pubic symphysis in the same plane during pelvic fixation and surgery. The part that holds the pubic symphysis is removable; therefore, it can also be used as a conventional pelvic fixator (fixation at the bilateral superior anterior iliac spines and sacrum). To reflect 3D preoperative planning intraoperatively, the author developed a tool for supine THA reported in 2017 as the first model of its kind in Japan [7].

5 iFlex G® (Fig. 58.4; Table 58.1)

This is a tool to guide the target acetabular component placement angle from the APP plane using the APP Lateral Positioner® as a reference and to easily attach it to the APP Lateral Positioner® from the top of the surgical drape during surgery. iFlex G® and APP Lateral Positioner® were developed by the author in Japan and reported in 2017 [8].

Fig. 58.3 APP Lateral Positioner®

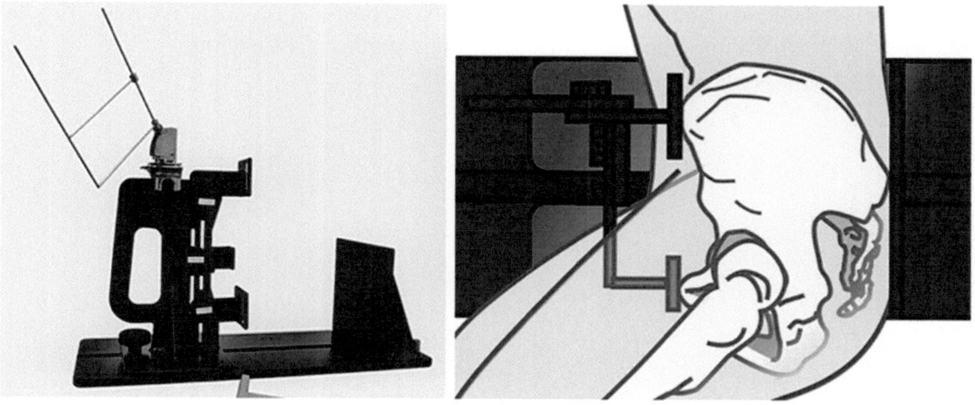

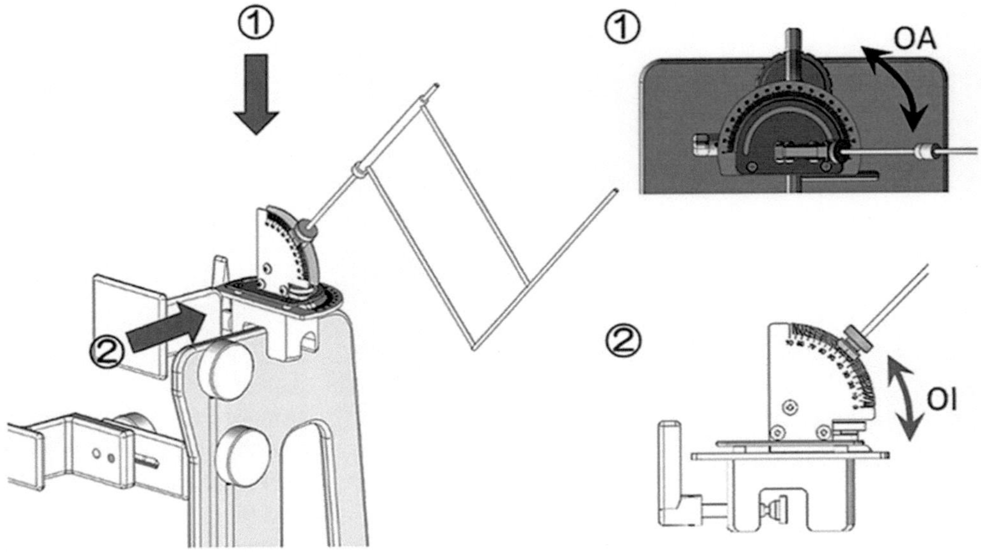

Fig. 58.4 iFlex G®

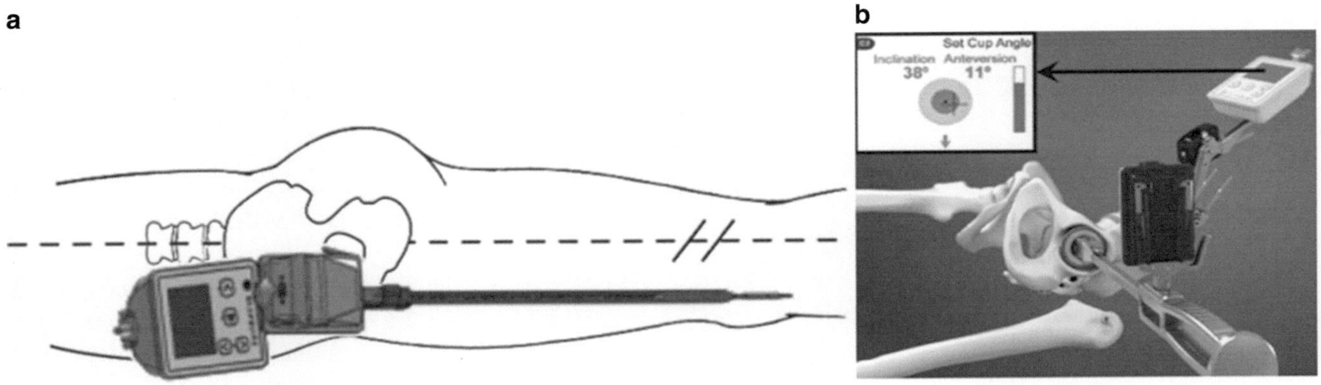

Fig. 58.5 HipAlign®. (**a**) In the lateral recumbent position, the long axis of the spinal column is defined intraoperatively. (**b**) For supine position

6 HipAlign® (Fig. 58.5; Table 58.1)

The navigation unit contains an accelerometer and an angular rate meter, and it can be used for preoperative planning based only on simple X-ray images. The navigation unit was developed by OrthoAlign, Inc. in the United States, and it was made available in Japan in 2018 for the side-lying position and in 2019 for the supine position [9–11]. Another function is the measurement of leg extension.

7 NaviSWISS (Fig. 58.6; Table 58.1)

A model that uses an infrared sensor to recognize a small tag device and can be used in both supine and lateral recumbent positions. It was developed by NaviSWISS, Switzerland, and has been available in Japan since 2021 [12, 13]. Another function is the measurement of leg extension.

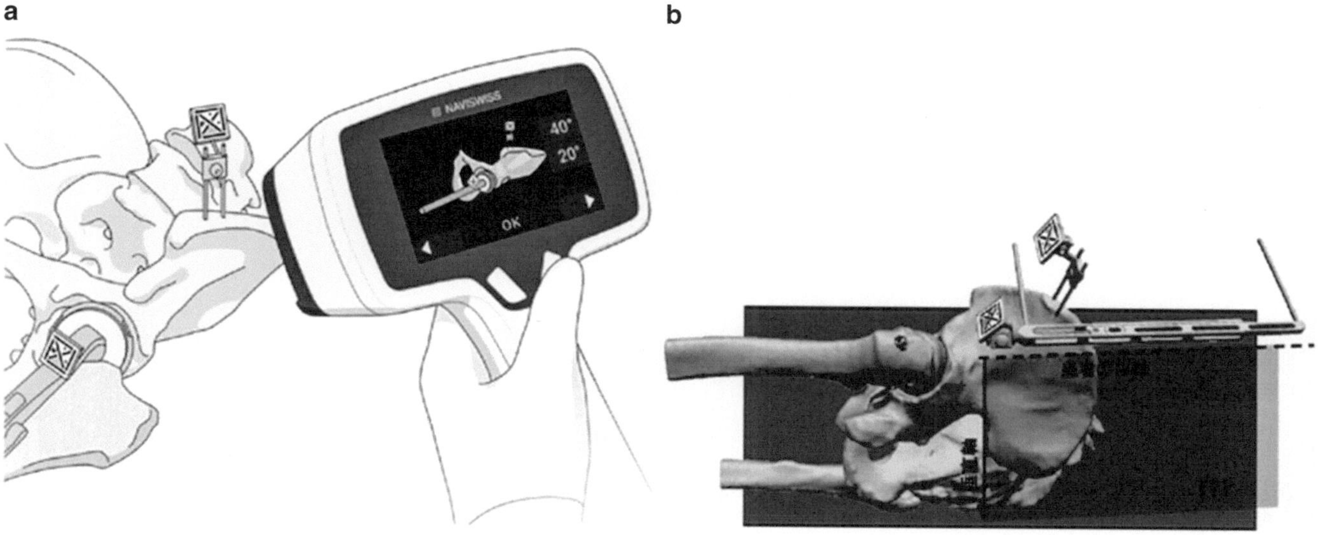

Fig. 58.6 NaviSWISS. (**a**) NaviSWISS usage diagram. (**b**) Setting the reference plane after fixation in the lateral recumbent position. The body axis must be defined intraoperatively

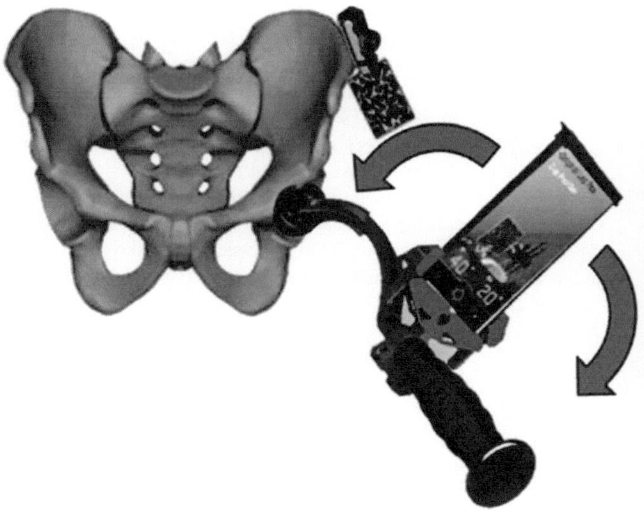

Fig. 58.7 AR-HIP

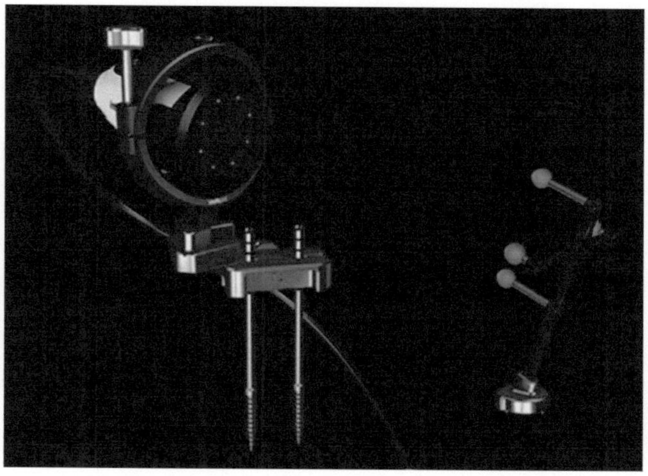

Fig. 58.8 Intellijoint HIP

8 AR-HIP (Fig. 58.7; Table 58.1)

Navigation using augmented reality (AR) with a dedicated smartphone application can be used in both supine and lateral recumbent positions by recognizing the FPP and APP planes with dedicated markers pinned to the pelvis, bilateral anterior superior iliac spines, and pubic symphysis in the dedicated application. The device can be used in both the supine and lateral recumbent positions. It was developed by Dr. Hiroyuki Ogawa of Hokusuikai Memorial Hospital, and is a Japanese model that has been available in Japan since 2021 [14].

9 Intellijoint HIP® (Fig. 58.8; Table 58.1)

This model uses an infrared camera placed on the iliac bone with two pin fixations to recognize four small metal ball markers (passive balls), guide the appropriate placement angle, and can be used in both supine and lateral recumbent positions. It was developed by Intellijoint Surgical, Canada, in 2013, and has been available in Japan since 2021 [15, 16]. This device can measure the femoral offset and leg length difference as additional functions.

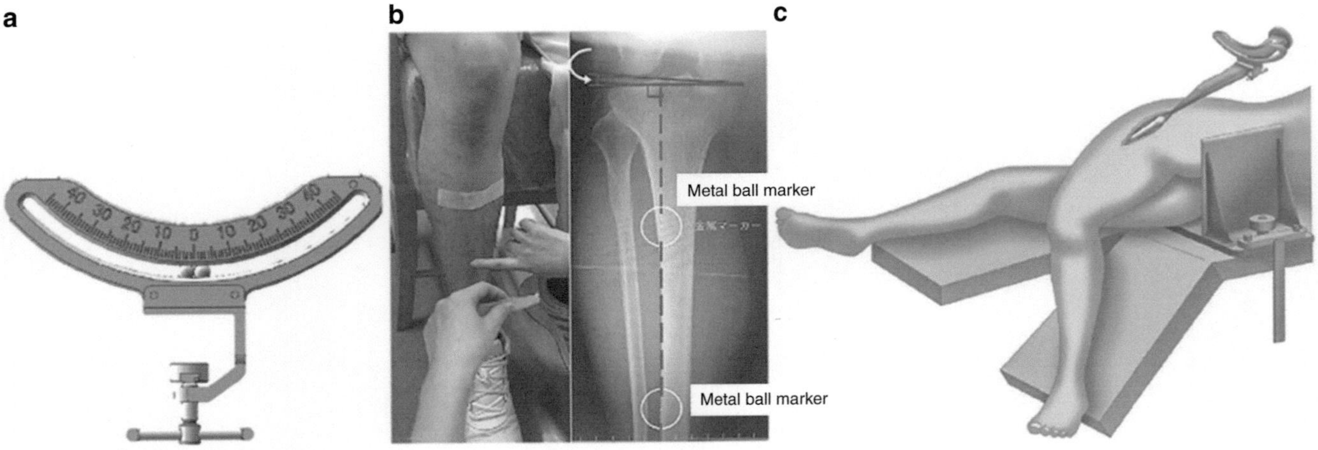

Fig. 58.9 iFlexS femoral anteversion angle measuring device. (**a**) iFlexS. (**b**) Measurements of the angle between the tibial ridge line and posterior femoral condylar axis. (**c**) Easily fixed to the rasp handle [18]

10 iFlexS (Fig. 58.9; Table 58.1)

This is the only portable navigation method for the femoral component described in this chapter. The anterior torsion angle of the stem is the angle between the neck axis and the femoral posterior condylar axis (measured by projection onto the XY plane). The angle between the tibial ridge and the femoral posterior condylar axis was measured preoperatively (Kanekasu view) [17], and the femoral posterior condylar axis was estimated intraoperatively to enable the measurement of the anterior stem torsion angle [18].

11 Future Tasks

Portable navigation is the only tool for improving the accuracy of acetabular component placement angles, and it is difficult to improve the position of acetabular component placement, such as robotic (e.g., Mako) or navigation (CT-based navigation) systems. Therefore, in Japan, where there are many cases of acetabular dysplasia and strong deformities, it is necessary to use CT in preoperative planning to improve the accuracy of the placement position by performing 3D preoperative planning, including the size of the surrounding osteophytes and distance from the acetabulum. Currently, most portable navigation tools can be used only for acetabular component placement, and iFlexS is the only tool that can be used for femoral component placement (anteversion angle of the stem). To take into account the combined anteversion theory [19] intraoperatively, it would be desirable to use a tool such as the iFlexS, which is useful for femoral component placement.

References

1. Lewinnek GE, et al. Dislocations after total hip-replacement arthroplasties. J Bone Joint Surg Am. 1978;60:217–20.
2. Kennedy JG, et al. Effect of acetabular component orientation on recurrent dislocation, pelvic osteolysis, polyethylene wear, and component. J Arthroplast. 1998;13:530–4.
3. Kummer FJ, et al. The effect of acetabular cup orientations on limiting hip rotation. J Arthroplast. 1999;14:509–13.
4. Ando W, et al. Comparison of the accuracy of the cup position and orientation in total hip arthroplasty for osteoarthritis secondary to developmental dysplasia of the hip between the Mako robotic arm-assisted system and computed tomography-based navigation. Int Orthop. 2021 July;45(7):1719–25.
5. Kawaguchi H, et al. Prevention of postoperative dislocation by external rotator muscle groups and arthrodesis repair during posterior THA entry. Cent Jpn J Orthop Traumat. 2000;43:369–70. Japanese
6. Suda K, et al. Cup implantation accuracy using the HipCOMPASS mechanical intraoperative support device. Springerplus. 2016;5:784.
7. Iwakiri K, et al. Efficacy of the anatomical-pelvic-plane positioner in total hip arthroplasty in the lateral decubitus position. J Arthroplast. 2017;32:1520–4.
8. Iwakiri K, et al. Efficacy of a pelvic lateral positioner with a mechanical cup navigator based on the anatomical pelvic plane in total hip arthroplasty. J Arthroplast. 2017;32:3659–64.
9. Tanino H, et al. Portable accelerometer-based navigation system for cup placement of total hip arthroplasty: a prospective, randomized, controlled study. J Arthroplast. 2020;35:172–7.
10. Takada R, et al. Portable imageless navigation system and surgeon's estimate for accurate evaluation of acetabular cup orientation during total hip arthroplasty in supine position. Eur J Orthop Surg Traumatol. 2020;30:707–12.
11. Hasegawa M, et al. Accuracy of acetabular cup insertion in an anterolateral supine approach using an accelerometer-based portable navigation system. J Artif Organs. 2021;24:82–9.
12. Ohyama Y, et al. A new accelerometer-based portable navigation system provides high accuracy of acetabular cup placement in total hip arthroplasty in both the lateral decubitus and supine positions. Arch Orthop Trauma Surg. 2023;143(7):4473–4480.

13. Hasegawa M, et al. Accuracy of a novel accelerometer-based navigation (Naviswiss) for total hip arthroplasty in the supine position. BMC Musculoskelet Disord. 2022;23(1):537.
14. Ogawa H, et al. Does an augmented reality-based portable navigation system improve the accuracy of acetabular component orientation during THA? Randomized controlled trial. Clin Orthop Relat Res. 2020;478:935–43.
15. Jacob I, et al. Acetabular positioning is more consistent with the use of a novel miniature computer-assisted device. Int Orthop. 2020;44:429–35.
16. Bradley MP, et al. Accuracy of acetabular component positioning using computer-assisted navigation in direct anterior total hip arthroplasty. Cureus. 2019;11:e4478.
17. Kanekasu K, et al. Axial radiography of the distal femur to assess rotational alignment in total knee arthroplasty. Clin Orthop Relat Res. 2005;434:193–7.
18. Iwakiri K, Ohta Y, Ueno S, Minoda Y, Kobayashi A, Hiroaki N. A novel device for intraoperative measurement of stem anteversion angle in total hip arthroplasty. Arthroplasty Today. 2024;20:28:101458.
19. Widmer KH, et al. Compliant positioning of total hip components for optimal range of motion. J Orthop Res. 2004;22:815–21.

Robotic Surgery

59

Yutaka Inaba and Hiroyuki Ike

Modern surgical treatment often involves minimally invasive techniques and early postoperative recovery. The clinical introduction of robot-assisted surgery, represented by the da Vinci surgical system (Intuitive Surgical, Sunnyvale, CA, USA), continues to progress in a variety of clinical fields. In the realm of orthopedic surgery, new robotic surgical systems have been developed for spine and joint surgery that are expected to improve surgical accuracy, making surgeries safer than ever before.

In total hip arthroplasty (THA), implant placement position and angle are associated with complications such as wear of the articulating surface and dislocation and important for achieving a good postoperative range of motion. Appropriate implant placement also leads to good clinical outcomes and lower revision rates; thus, further dissemination and promotion of computer-assisted surgery, including robotic surgery, is desirable [1].

This section outlines the platforms, clinical outcomes, future challenges, and prospects of robotic surgery for THA.

1 Robotic Surgery

Robotic surgery is performed using computer-controlled equipment that performs part or all of the surgery based on the input anatomical information. The navigation system is used as an auxiliary tool to inform the surgeon of a patient's anatomy and the relative position and angle of the implant. The use of a navigation system improves the accuracy of implant positioning and angulation and reduces outliers [2]. Although navigation systems are useful for surgeons to recognize bone geometry and implant location, they are nevertheless passive assistive devices; as such, they cannot prevent failures due to human error, a surgeon's hand tremor, or insufficient skill. In robotic surgery, a robot controls the surgical instruments and implants, which allows osteotomy, reaming, and implant placement to be performed more accurately according to a preoperative plan than in navigation-assisted surgery.

Robotic surgery is classified into two types: active and semiactive [3]. In the active type, surgery is performed without surgeon involvement. The surgeon confirms safety by monitoring and operating the emergency stop button in case of an emergency but does not directly operate the robot. In the semiactive type, also called the "haptic system," a surgeon operates a robot to perform the surgery while receiving information regarding preoperative planning and intraoperative data. The haptic system conveys information to the surgeon via warning sounds (auditory), vibration and resistance (tactile), and color changes on display (visual) to prevent excessive tissue removal and implant malposition [4].

Robotic surgery often requires computed tomography (CT) images to create a patient-specific three-dimensional (3D) model. The preoperative planning workstation determines the appropriate implant size and position by referencing 3D images and CT slices in the transverse, coronal, and sagittal planes (Fig. 59.1). During surgery, registration was used to align the patient's anatomy with preoperative CT images. A registration problem is generally viewed as an optimization problem in which a transformation parameter is adjusted to obtain the result most similar to the rest of the data when a certain transformation is applied to the target data (image or point position) and is formulated by transformation, similarity, and regularization functions [5]. After confirming registration accuracy, osteotomy, reaming, and implant placement are performed using a robot based on the preoperative plan.

Y. Inaba (✉) · H. Ike
Department of Orthopaedic Surgery, Yokohama City University School of Medicine, Yokohama, Kanagawa, Japan
e-mail: yute0131@yokohama-cu.ac.jp; yute@yhc.att.ne.jp

Y. Nakashima (ed.), *Advances in Total Hip Arthroplasty*, https://doi.org/10.1007/978-981-97-2461-1_59

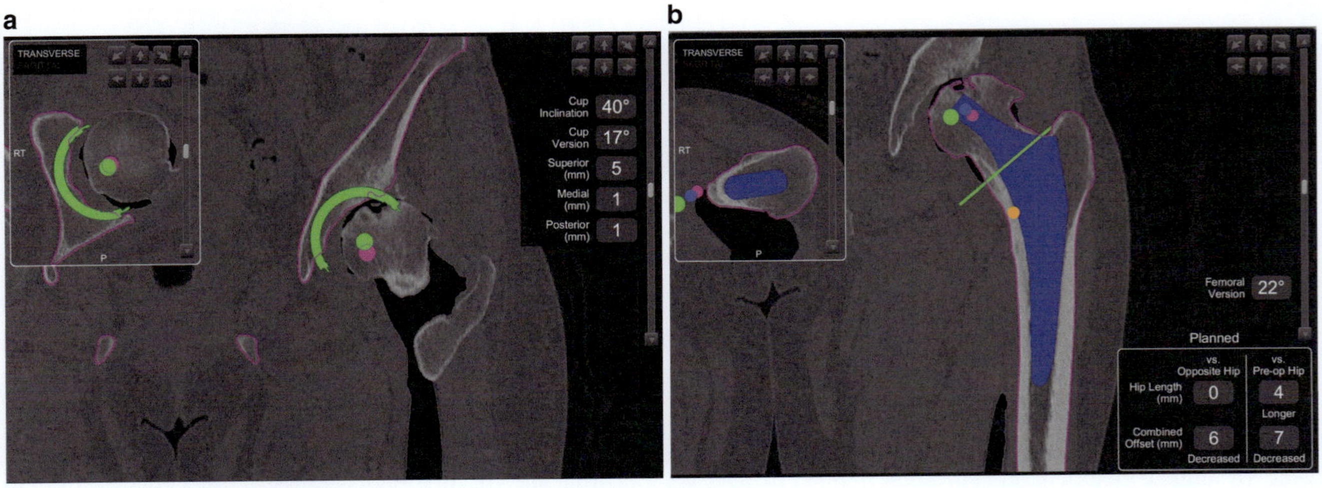

Fig. 59.1 Preoperative planning using the Mako®. (**a**) acetabular side, (**b**) femoral side

2 Platform in THA

The current platforms used for THA include the TSolution One® (Think Surgical Inc., Fremont, CA, USA) and Mako Robotic-Arm Assisted Surgery® (Stryker, Fort Lauderdale, FL, USA).

2.1 ROBODOC®, TSolution One®

The ROBODOC® (Integrated Surgical Systems, Davis, CA, USA) was the first surgical assistance robot designed for THA. The ORTHODOC® is a preoperative planning workstation that plans stem location in 3D using CT image data that selects the appropriate model and size while considering the compatibility of the implant with the medullary cavity and the occupancy of the medullary cavity in each case (Fig. 59.2). The ROBODOC is an active robot with a high-speed cutter controlled by a robotic arm that excavates the femoral bone marrow cavity. When the bone motion monitor detects femoral movement, the machine stops. There are three compatible registration methods: insertion of two marker pins into the femur, surface profile matching (Fig. 59.3), and two-dimensional–3D or 3D-3D registration using intraoperative fluoroscopic images [6]. This system is available only for the femoral side and cannot perform reaming or cup placement on the acetabulum.

The TSolution One®, the successor of the ROBODOC®, is also an active robot approved by the U.S. Food and Drug Administration in 2014 (Figs. 59.4 and 59.5) [8]. This open-platform system is compatible with implants of various companies and designs. In addition, the TCAT® arm enables acetabulum reaming and cup placement while maintaining the angle.

2.2 Mako®

New robotic surgery systems have recently been developed and used in the field of orthopedic surgery in the United States, Europe, and other countries. Japan is no exception, where the clinical application of the Mako®, a semi-active robot, has already begun.

Similar to CT-based navigation, the Mako® creates an individual preoperative plan for each patient using preoperative CT images to refer to information such as implant size and placement location, acetabular osteotomy extent, femoral neck osteotomy position, and leg length and offset [9]. A pelvic tracker is placed on the iliac crest or pelvis outside the acetabulum using two or three pins, and a femoral tracker is placed on the femur using a screw. Checkpoint screws are inserted into the femur and pelvis. During femoral and acetabular registration, landmark matching is performed first, followed by surface matching and verification (Fig. 59.6). To prevent manual manipulation errors, a surgeon performs reaming and cup placement using the robotic arm (robotic arm interactive orthopedic system [RIO®]). Reaming can be performed only once using the final size or multiple times by increasing the size in steps from smaller to larger. If the reamer advances more than 2 mm deeper than the preoperative plan, the fail-safe function ensures that it automatically stops (Fig. 59.7) [10]. Cup position and impactor angle are also constrained by the robotic arm during cup placement, and the cup is placed according to the preoperative plan while verifying the anteversion, inclination, and rotation center position displayed on the monitor. After a femoral broach is inserted, the anteversion angle, leg length, and offset are measured and compared with the preoperative values, and the neck length, head diameter, and stem size are selected accordingly. These values are all measured after stem placement.

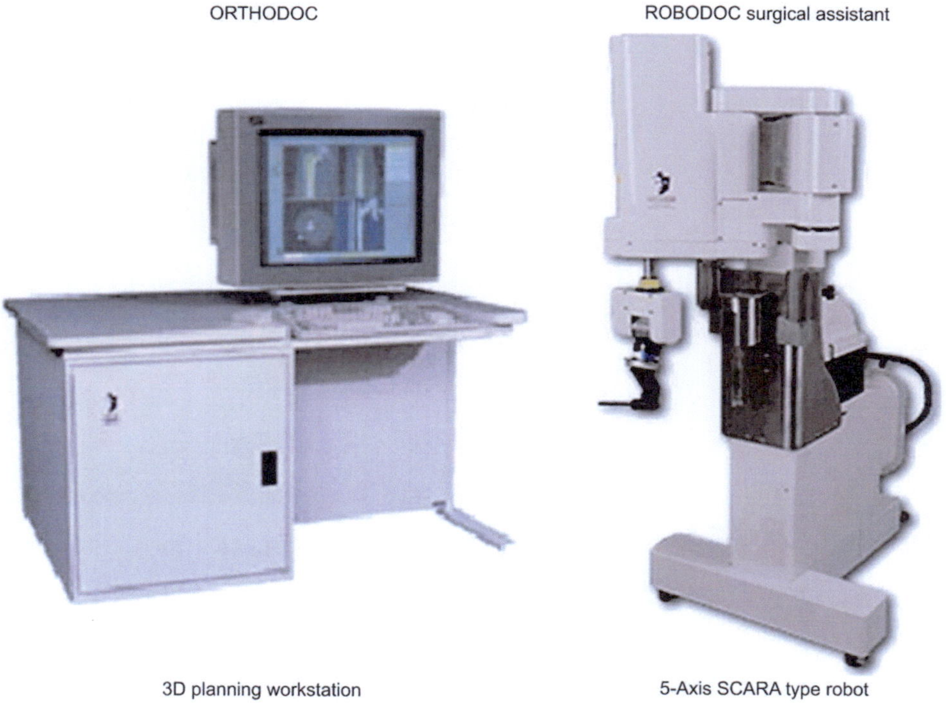

Fig. 59.2 The ROBODOC® and ORTHODOC® systems. (Adapted from Sugano[6]. Fig. 1)

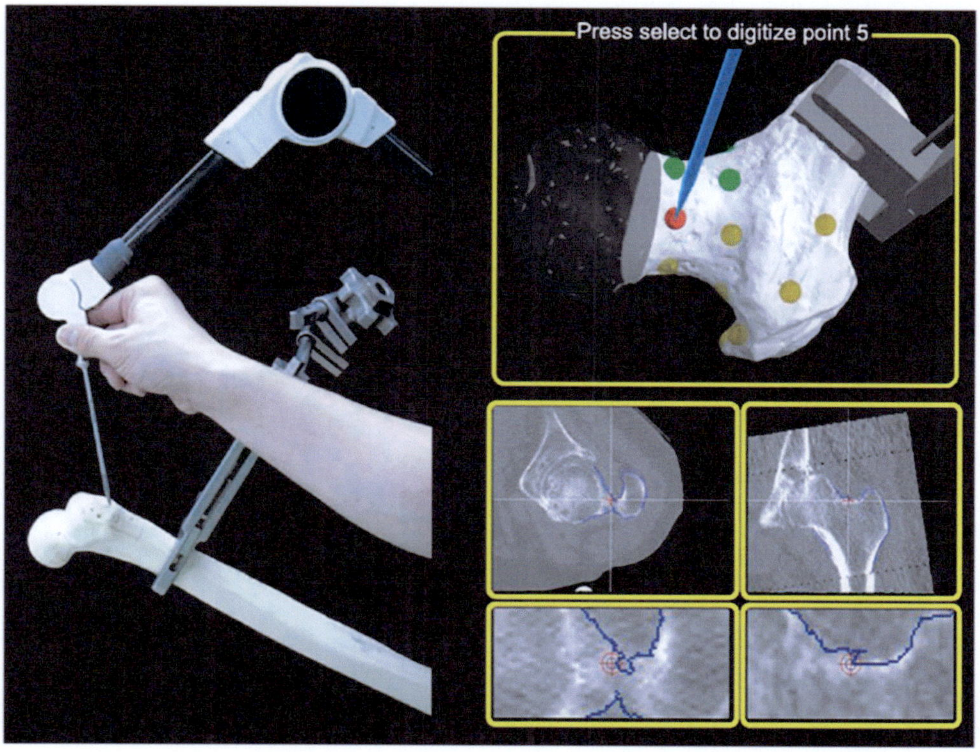

Fig. 59.3 Registration by surface profile matching with the ORTHODOC®. (Adapted from Sugano [6]. Fig. 3)

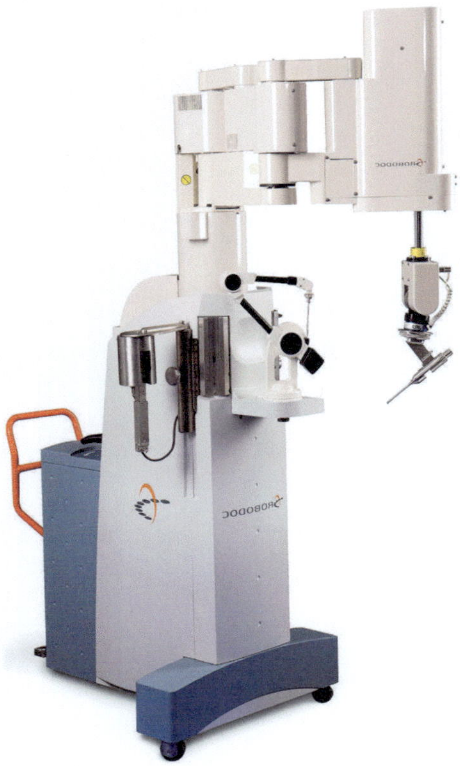

Fig. 59.4 The TSolution-One® system. (Adapted from Liow [7]. d)

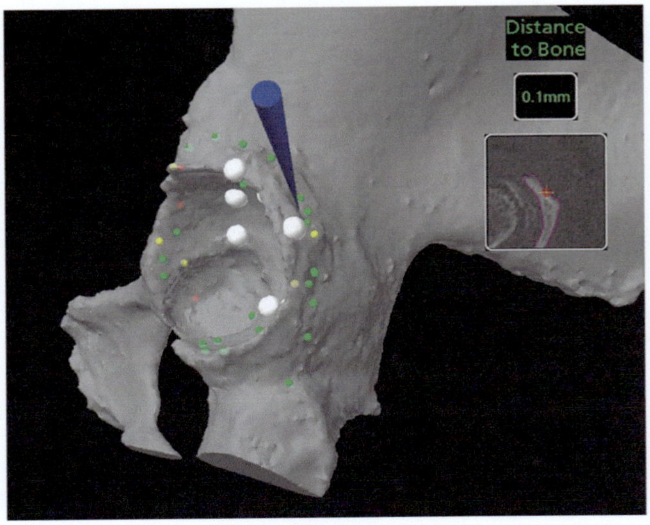

Fig. 59.6 Registration in the Mako® system

Fig. 59.5 Femoral drilling with the TSolution One® system. TSolution One® is an active robot that actively excavates the femoral bone marrow cavity using a rotating bar. (Adapted from Dungy [8]. Fig. 2)

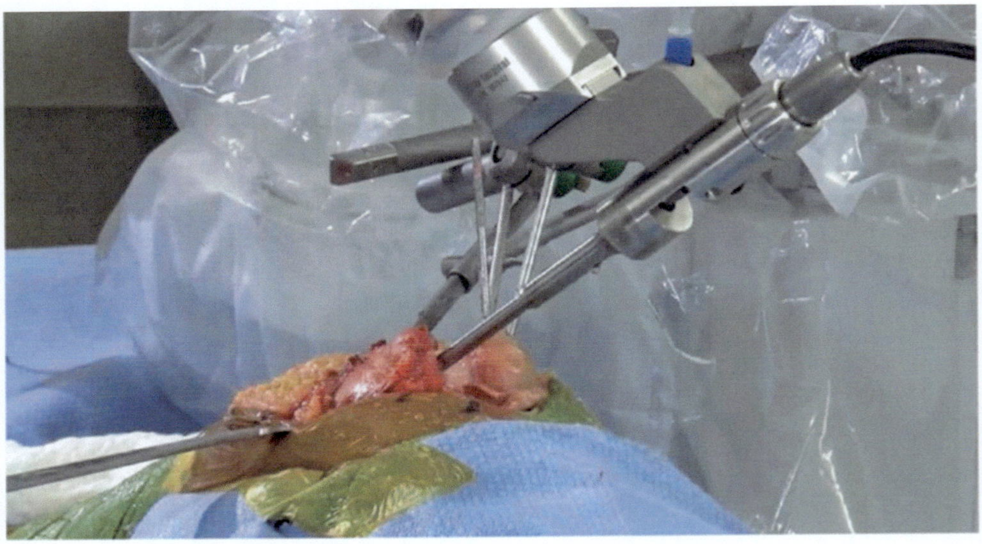

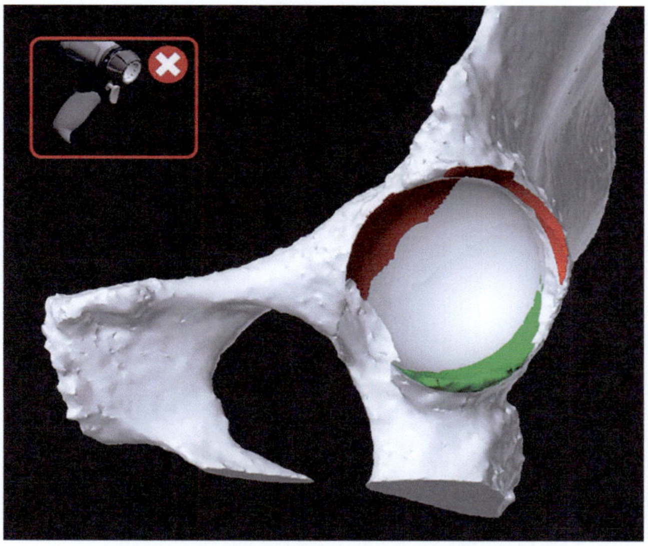

Fig. 59.7 Reaming with the Mako® system. The area to be excavated in the acetabulum is shown in green, the area excavated according to the preoperative plan is shown in white, and the area excavated more than 1.0 mm deeper than the preoperative plan is shown in red

3 Clinical Results

A study in Japan reported that the ROBODOC®, which was clinically introduced in 1992, is superior to conventional surgery in terms of stem alignment, stem and femoral marrow cavity compatibility, and Merle D'Aubigne hip score at 2 years postoperative with fewer intraoperative fractures. However, the time required for surgery and femoral bone marrow cavity processing is reportedly longer [11]. Although a reduction in stress shielding was observed at 5 years postoperative compared to conventional surgery, there was no significant difference in the Japanese Orthopedic Association hip disease function criteria (JOA hip score) [12]. In addition, no significant difference in JOA hip score was observed in cases for 10 years postoperative (97 ± 5 points in the ROBODOC® group vs. 96 ± 7 points in the conventional surgery group, p = 0.159). There were no cases of aseptic loosening or revision surgery in either group, and the degree of heterotopic ossification was not significantly different between them (p = 0.186) [13]. However, other studies reported high dislocation (18%) and revision (15%) rates presumably caused by a gluteus medius muscle injury [14], although the ROBODOC® improved stem alignment and compatibility. Therefore, preoperative planning and intraoperative soft tissue protection are important.

In a study comparing the accuracy of cup placement between conventional and robotic surgery with Mako®, the accuracy of cup placement within the Lewinnek safe zone (30–50°of inclination, 5–25°of anteversion) was significantly higher in the Mako® group (50/50 cases [100%]) than in the conventional surgery group (40/50 cases [80%]). The accuracy of cup placement in the Mako® group was significantly higher than that in the conventional surgery group (p = 0.001) (Fig. 59.8) [15]. A comparative study of navigation and robotic surgery using the Mako® reported a lower mean numerical rating scale (NRS) score and significantly shorter time to independent walking (p < 0.001) at 1 to 2 weeks postoperative in the robotic surgery group. In addition, robotic surgery demonstrated superior early postoperative recovery with significantly higher Harris hip score at discharge (85.3 ± 3.2 points in the robotic group vs. 81.0 ± 8.5 points in the navigation group, p = 0.01) [16]. In another report, the absolute error between the intraoperative display angle and postoperative placement angle was less than 5° in 88% of inclination and 84% of anteversion, and the distance of the center of rotation before and after surgery was within 3 mm vertically and within 5 mm horizontally in 81.5% of cases. The superiority of the Mako® over navigation has thus been reported to improve the accuracy of the position of the center of rotation in cup placement [10]. In addition, another study in Japan has reported the accuracy of cup placement position and angle in acetabular dysplasia cases as superior to that of robotic surgery with the Mako® system [17].

Although these reports demonstrated the high accuracy of cup placement using the Mako®, an anterior approach (odds ratio, 5.1) and posterior pelvic tilt (odds ratio, 1.1) were identified as predictors of inaccurate cup placement [18]; thus, this should be performed with caution, especially when using an anterior approach with a small skin incision that may decrease the accuracy of pelvic registration.

4 Future Issues and Prospects

The advantages of robotic surgery are as follows: (1) improved implant placement accuracy; (2) good joint range of motion and stability; and (3) improved safety due to the fail-safe functionality. However, it also features several disadvantages, including: (1) high initial installation cost and continuing need for maintenance and disposable devices; (2) prolonged operation time; (3) radiation exposure and cost of preoperative CT; (4) time required for preoperative planning; and (5) existence of a learning curve. In addition, owing to the lack of tactile sensors in robotic devices, surgeons must be very careful to secure the surgical field and avoid soft tissue injuries.

As robotic surgery becomes more popular and smaller devices become available, initial installation and maintenance costs are expected to decrease. Proficiency is required to perform registration work quickly with high accuracy and affect the accuracy of implant placement. A systematic review by Sweet et al. reported that robot-assisted THA increased the operative time by 12–25 min compared with

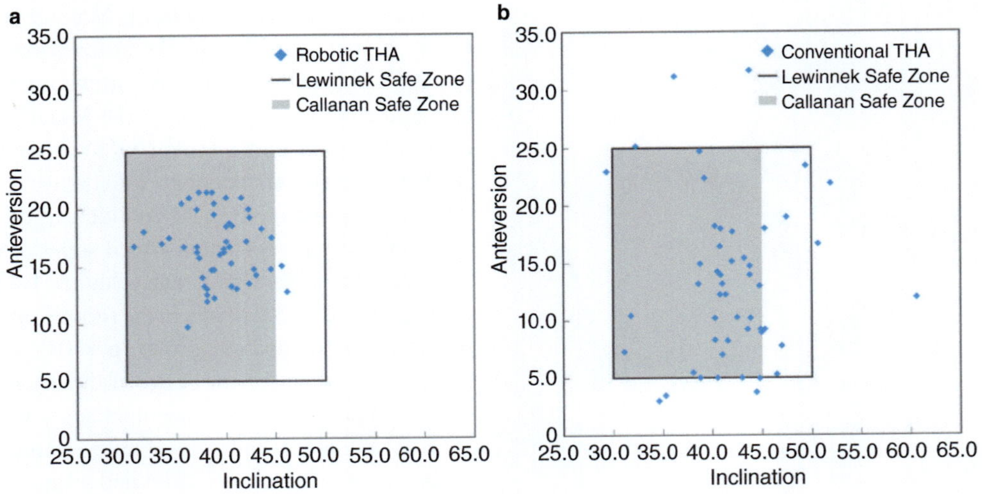

Fig. 59.8 Cup placement angle in robotic surgery using the Mako® versus conventional surgery. (**a**) Angle of cup installation in the Mako group. (**b**) Angle of cup installation in the conventional surgery group. The target cup angles for both groups were 40° for the inclination and 20° for the anteversion. One hundred percent (50/50) of cups in the robotic THA group were in the safe zone of Lewinnek et al. [26], compared with 80% (40/50) of cups in the conventional THA group (p = 0.001). Ninety-two percent (46/50) of cups in the robotic THA group were in the safe zone of Callanan et al. [5], compared with 62% (31/50) of cups in the conventional THA group (p = 0.001). (Adapted from Domb [15]. Figs. 4-A, 4-B)

conventional THA, further increasing perioperative complications including surgical site infection [19].

Robotic surgery for THA can be performed using an open or closed platform. Open platforms allow for the use of implants with designs other than those of the robot's manufacturer. Although a wide variety of implants can be used, the functionality of the robot and/or the preoperative planning workstation is often limited. Closed platforms do not allow the use of implants other than specified products, which may limit the surgeon's choices. Therefore, the advantages and disadvantages of each platform should be recognized and evaluated.

Industry 4.0 refers to the fourth industrial revolution, which aims to transform the industrial structure by utilizing innovative technologies such as the Internet of Things (IoT), big data, and machine learning. In the medical field, efforts are being made to digitize surgery by sharing, integrating, and utilizing information obtained during the perioperative period using new technologies and methodologies such as IoT, process modeling, and deep learning to improve surgical quality, efficiency, and performance (Fig. 59.9) [20]. This is the essential concept of Surgery 4.0, in which surgical robots are expected to play an important role.

To improve surgical quality, efficiency, and outcomes, all surgical operations, including approach and closure, may be automated in the future, possibly eliminating the need for surgeons. To automate the entire surgical process, it is necessary to develop sensing technology that recognizes the surgical field using a machine. In addition, it is important to clarify where responsibility lies in the case of medical accidents or complications in robotic surgery and ensure system safety and reliability, as mechanical system errors in surgery can be life-threatening. To realize remote surgery and automation of the entire surgical process by robotic surgery, it is necessary to solve technical problems, establish laws and systems, and build a social consensus on the risk of liability in the event of medical accidents caused by robotic surgery. In other words, surgeons should not only aim to develop technological innovations and resolve legal issues through medical–engineering collaboration, they should also consider current issues and promote discussion about the ideal state of robotic surgery.

Although the use of surgical robots improves implant placement accuracy, its use rate remains low owing to problems such as installation costs. Moreover, reports on the mid- and long-term outcomes of robot-assisted THA are limited and further evidence is expected. There is also a lack of evidence on whether improved implant placement accuracy leads to improved clinical outcomes and reduced complications, and an evaluation from the cost-effectiveness perspective is also necessary.

Just as the rapid development of information and communication technology has caused structural changes in the industry, a paradigm shift is also occurring in the field of orthopedics owing to the introduction of robotics. Further development and dissemination of computer-assisted surgery, including robotic surgery, is expected.

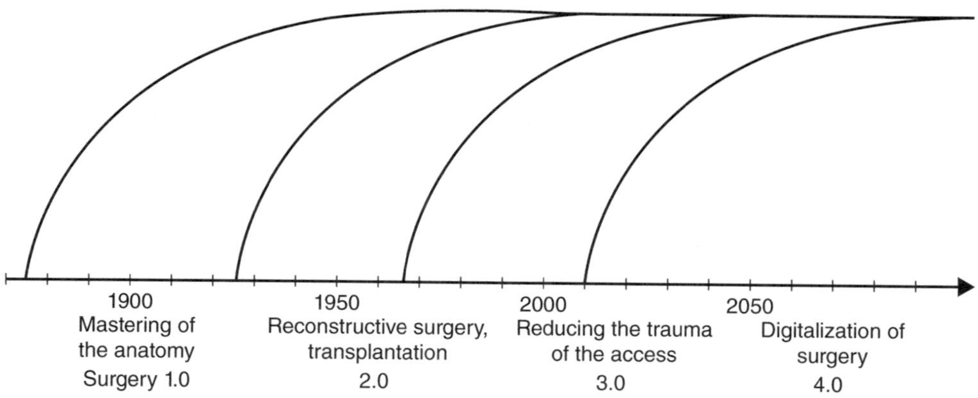

Fig. 59.9 Development of surgery. Surgery 1.0: Development in anatomy. 2.0: Development of reconstructive surgery and transplantation. 3.0: Shift to minimally invasive surgery (MIS), reduced functional impairment, and postoperative pain. 4.0: Transformation in surgical digitization. (Adapted from Feußner [20]. Fig. 2)

References

1. Inaba Y, et al. Modern navigation systems and robotic surgery. Joint Surg. 2017;36:1190–7. Japanese
2. Sugano N, et al. Mid-term results of cementless total hip replacement using a ceramic-on-ceramic bearing with and without computer navigation. Joint Surg Br. 2007;89:455–60.
3. Wasterlain AS, et al. Navigation and robotics in total hip arthroplasty. JBJS Rev. 2017;5:e2.
4. Jacofsky DJ, et al. Robotics in arthroplasty: a comprehensive review. J Arthroplast. 2016;31:2353–63.
5. Ohtake Y. Registration in medical image analysis and surgical support systems. Med Imag Tech. 2017;35:11–7. Japanese
6. Sugano N. Computer-assisted orthopedic surgery and robotic surgery in total hip arthroplasty. Clin Orthop Surg. 2013;5:1–9.
7. Liow MHL, et al. THINK surgical TSolution-One® (Robodoc) total knee arthroplasty. SICOT J. 2017;3:63.
8. Dungy DS, et al. Active robotics for total hip arthroplasty. Am J Orthop (Belle Mead NJ). 2016;45:256–9.
9. Tarwala R, et al. Robotic assisted total hip arthroplasty using the MAKO platform. Curr Rev Musculoskelet Med. 2011;4:151–6.
10. Kanawade V, et al. Precision of robotic guided instrumentation for acetabular component positioning. J Arthroplast. 2015;30:392–7.
11. Nishihara S, et al. Comparison between hand rasping and robotic milling for stem implantation in cementless total hip arthroplasty. J Arthroplast. 2006;21:957–66.
12. Nakamura N, et al. A comparison between robotic-assisted and manual implantation of cementless total hip arthroplasty. Clin Orthop Relat Res. 2010;468:1072–81.
13. Nakamura N, et al. Does robotic milling for stem implantation in cementless THA result in improved outcomes scores or survivorship compared with hand results of a randomized trial at 10 years. Clin Orthop Relat Res. 2018;476:2169–73.
14. Honl M, et al. Comparison of robotic-assisted and manual implantation of a primary total hip replacement: a prospective study. J Bone Joint Surg Am. 2003;85:1470–8.
15. Domb BG, et al. Comparison of robotic-assisted and conventional acetabular cup placement in THA: a matched-pair controlled study. Clin Orthop Relat Res. 2014;472:329–36.
16. Shibanuma N, et al. Early postoperative clinical recovery of robotic arm-assisted vs. image-based navigated total hip arthroplasty. BMC Musculoskelet Disord. 2021;22:314.
17. Ando W, et al. Comparison of the accuracy of the cup position and orientation in total hip arthroplasty for osteoarthritis secondary to developmental dysplasia of the hip between the Mako robotic arm-assisted system and computed tomography-based navigation. Int Orthop. 2021;45:1719–25.
18. Hayashi S, et al. Accuracy of cup position following robot-assisted total hip arthroplasty may be associated with surgical approach and pelvic tilt. Sci Rep. 2021;11:7578.
19. Sweet MC, et al. Comparison of outcomes after robotic-assisted or conventional total hip arthroplasty at a minimum 2-year follow-up: a systematic review. JBJS Rev. 2021;9 https://doi.org/10.2106/JBJS.RVW.20.00144.
20. Feußner H, et al. Surgery 4.0: the natural culmination of the industrial revolution? Innov Surg Sci. 2017;2:105–8.

60 Patient-Specific Instruments

Junya Shimizu and Satoshi Nagoya

1 Accuracy of Patient-Specific Instrument (PSI)

1.1 Basic Concept of PSI

PSI is a technology that uses preoperative computed tomography (CT) and magnetic resonance imaging (MRI) data to create a three-dimensional (3D) structure using a 3D printer based on a simulation of an arbitrary shape that fits the bone surface of each patient. PSI has the potential to indicate the direction and depth of an arbitrary 3D structure of a bone or joint in the surgical field. In total hip arthroplasty (THA), intraoperative guides for acetabular cup placement, femoral osteotomy position, stem placement depth, internal/external rotation, and anteversion angle have been created, which are effective for more accurate implant placement [1, 2]. Although more accurate placement is possible using robots and navigation systems, the introduction of such systems in hospitals requires time and excessive costs [3–5]. A preoperative bone model can be fabricated using a 3D printer at a relatively low cost and in a short time, and THA using a PSI fabricated from a bone model has been reported in Japan and overseas [6, 7].

1.2 Application to Cups

Accurate cup placement is important to reduce not only the dislocation rate but also the wear of the polyethylene liner and acetabular rim loading [8, 9]. In manual cup placement without surgical assistance, such as navigation or PSI (conventional method), the probability of acceptable acetabular cup placement is reportedly 20–50% [3, 4, 10–13]. Meanwhile, when PSI is applied to cup installation, the installation accuracy is higher than with conventional methods (Fig. 60.1). Small et al. [14] have reported that the average error of radiological inclination (RI) of the cup was −2.0° ± 7.3° and the average error of radiological anteversion (RA) was −2.0° ± 6.9°. The errors in the conventional method were RI 1.3° ± 9.1° and RA −6.9 ± 8.9°, the average error was reduced by using PSI. Spencer-Gardner et al. [15] have also reported errors of placement angles in 100 cases using PSIs created from preoperative 3DCT. The average error of RI was 3.9° ± 2.9°, and the average of RA was 3.6° ± 3.2°. In 91% of cases, both RI and RA were placed within 10° from the planned angle. Hananouchi et al. [1] reported an error of 3.2° ± 2.3° for RI and 3.7° ± 2.7° for RA.

In a report on PSI produced based on MRI, Inoue et al. [16] have reported that the mean absolute error when PSI was used was 3.7° ± 2.2° for RI and 4.5° ± 3.9° for RA. Of the 14 patients, the absolute errors in the 7 patients who underwent the first half and the 7 patients who underwent the second half were 4.7° ± 2.1°, 2.8° ± 1.8° for RI, and 6.1° ± 4.0°, 3.2° ± 2.9 for RA, respectively. The errors in the seven cases performed in the second half were smaller. This indicates that a learning curve exists in the use of PSI. Wang et al. [17] used a PSI-based guide for subluxated hip arthroplasty with Crowe classification type II/III. And they have reported that the preoperative planning was RA 15.1° and RI 44.7° and postoperative CT showed an RA of 15.3° and an RI of 45.6° with placement errors of 4.2° ± 2.5° for RI and 2.7° ± 2.0° for RA and indicated more accurate placement of the cup was possible.

1.3 Application to Stem

Osteotomy of the femoral neck is important in THA. PSIs that indicate osteotomy sites have also been developed. Sakai et al. [18] have reported in a cadaveric study that the absolute

J. Shimizu
Department of Orthopedic Surgery, Sapporo Medical University, Sapporo, Hokkaido, Japan

S. Nagoya (✉)
Department of Orthopaedic Surgery, Sapporo Kojinkai Memorial Hospital, Sapporo, Hokkaido, Japan
e-mail: nagoya@sapmed.ac.jp

Y. Nakashima (ed.), *Advances in Total Hip Arthroplasty*, https://doi.org/10.1007/978-981-97-2461-1_60

Fig. 60.1 Previous reports of installation angle errors of cups using PSI

Author (year)	Imaging Modality	Cases	Absolute Error	
			RI (°)	RA (°)
Small et al. (2014)[14)]	CT	18	−2.0±7.3	−2.0±6.9
Spencer–Gardner et al. (2016)[15)]	CT	100	3.9±2.9	3.6±3.2
Hananouchi et al. (2010)[1)]	CT	31	3.2±2.3	3.7±2.7
Inoue et al. (2019)[16)]	MRI	14	3.7±2.2	4.5±3.9
Wang et al. (2019)[17)]	CT	20	4.2±2.5	2.7±2.0

error between the preoperative planning and postoperative implant insertion position was 0.6° ± 0.6° in the coronal section, 1.0° ± 0.9° in the sagittal section, and 1.0 ± 1.1 mm for the medial placement height, all of which were reported to have small errors. Schneider et al. [19] have reported that 29 of 30 patients achieved osteotomy within 3 mm of the preoperative plan. Ito et al. [2] reported that the accuracy of PSI for femoral stem placement was 2.1° ± 4.1° for anterior bending, 1.0° ± 0.7° for varus/valgus, and 4.7° ± 1.2° for anteversion.

2 Impact of PSI on Clinical Outcomes

PSI has high accuracy, and accurate implant placement is possible. However, clinical outcomes regarding the reduction of dislocation rate, polyethylene wear, and acetabular rim loading, which were the original objectives of PSI, have not yet been reported. Naturally, dislocation may occur when soft tissue is sacrificed even if the implant is placed in an appropriate bony position and angle, and postoperative pain is also a problem. Many of the PSIs currently in use are based on the assumption that they are fitted to the bone and that soft tissues must be sacrificed to some extent for accurate placement. For example, PSI placed in a cup may require a larger expansion of the joint capsule and muscles. In the future, the issue is how to preserve soft tissues in addition to more accurate implant placement.

References

1. Hananouchi T, et al. Tailor-made surgical guide reduces incidence of outlier of cup placement. Clin Orthop Relat Res. 2010;468:1088–95.
2. Ito H, et al. A patient-specific instrument for femoral stem placement during total hip arthroplasty. Orthopedics. 2017;40:e374–7.
3. DiGioia AM 3rd, et al. Comparison of a mechanical acetabular alignment guide with computer placement of the socket. J Arthroplast. 2002;17:359–64.
4. Saxler G, et al. The accuracy of free-hand cup positioning—a CT based measurement of cup placement in 105 total hip arthroplasties. Int Orthop. 2004;28:198–201.
5. Kalteis T, et al. Imageless navigation for insertion of the acetabular component in total hip arthroplasty: is it as accurate as CT-based navigation? Bone Joint Surg Br. 2006;88:163–7.
6. Parratte S, et al. Validation and usefulness of a computer-assisted cup-positioning system in total hip arthroplasty. J Bone Joint Surg Am. 2007;89:494–9.
7. Sugano N, et al. Mid-term results of cementless total hip replacement using a ceramic-on-ceramic bearing with and without computer navigation. Joint Surg Br. 2007;89:455–60.
8. Goyal P, et al. Effect of acetabular component positioning on functional outcomes in primary total hip arthroplasty. J Arthroplast. 2017;32:843–8.
9. Callanan MC, et al. The John Charnley Award: risk factors for cup malpositioning: quality improvement through a joint registry at a tertiary hospital. Clin Orthop Relat Res. 2011;469:319–29.
10. Conn KS, et al. A simple guide to determine the magnification of radiographs and to improve the accuracy of preoperative templating. J Bone Joint Surg Br. 2002;84:269–72.
11. Mainard D, et al. Accuracy and reproducibility of preoperative three-dimensional planning for total hip arthroplasty using biplanar low-dose radiographs: a pilot study. Orthop Traumatol Surg Res. 2017;103:531–6.
12. Minoda Y, et al. Are acetabular component alignment guides for total hip arthroplasty accurate? J Arthroplast. 2010;25:986–9.
13. Lei P, et al. Greater trochanter osteotomy with cementless THA for Crowe type IV DDH. Orthopedics. 2013;36(5):e601–5.
14. Small T, et al. Comparison of acetabular shell position using patient specific instruments versus standard surgical instruments: a randomized clinical trial. J Arthroplast. 2014;29:1030–307.
15. Spencer-Gardner L, et al. Patient-specific instrumentation improves the accuracy of acetabular component placement in total hip arthroplasty. Bone Joint J. 2016;98-B:1342–6.
16. Inoue D, et al. A prospective clinical trial to assess the accuracy of an MRI-based patient-specific acetabular instrument guide in total hip arthroplasty. Eur J Orthop Surg Traumatol. 2019;29:65–71.
17. Wang C, et al. Accuracy and practicability of a patient-specific guide using acetabular superolateral rim during THA in Crowe II/III DDH patients: a retrospective study. J Orthop Surg Res. 2019;14:19.
18. Sakai T, Hamada H, Takao M, et al. Validation of patient-specific surgical guides for femoral neck cutting in total hip arthroplasty through anterolateral approach. Int J Med Robot. 2017;13:e1830.
19. Schneider AK, et al. Clinical accuracy of a patient-specific femoral osteotomy guide in minimally invasive posterior hip arthroplasty. Hip Int. 2018;28:636–41.

Part XI

Postoperative Rehabilitation

Range of Motion, Walking Function, and Muscle Strength

61

Naonobu Takahira

1 Basics of Postoperative Rehabilitation

Postoperative rehabilitation should be performed with the aim of early release from bed and early return to society. Scientific evidence suggests that pre- and postoperative rehabilitation in total hip arthroplasty (THA) is useful in improving walking ability, muscle strength, range of motion (ROM), and psychological status, such as improved self-efficacy (grade B) [1]. In addition, differences in the effects of improvement in muscle strength, functional score, and muscle volume were observed depending on the program content, such as water training, bicycle ergometer use, and resistance exercise strengthening. However, the medium- to long-term effects over 1 year after surgery are not clear [1].

Preoperative, intraoperative, and postoperative hip joint assessments are helpful in rehabilitation. In addition, the operated and non-operated hip joints, adjacent knee joint, lumbar spine, ankle joint, upper limb, and trunk functions are important. In addition, evaluation of cognitive function should be considered when implementing rehabilitation.

1.1 Types of THA

In recent years, minimally invasive surgery (MIS) with muscle–tendon preservation has become the mainstay primary THA, which is a significant paradigm shift from previous postoperative rehabilitation. Additionally, continuous postoperative ROM training was performed with continuous passive motion. Gait function training was also performed immediately after surgery, starting with partial loading in stages. However, these techniques are no longer required in most cases. However, rehabilitation is not always the same for cemented or cementless THA and primary or revision THA. In addition, the rehabilitation program for patients who have undergone previous osteotomies or other surgical procedures differs individually because of the difficulty in THA conversion.

N. Takahira (✉)
Kitasato University Graduate School of Medical Sciences, Sagamihara, Kanagawa, Japan
e-mail: takahira@med.kitasato-u.ac.jp

1.2 Patient's Condition

The rehabilitation program differs depending on the patient's disease status before surgery. For example, the period between disease onset and THA is often long in hip osteoarthritis (OA), and the ROM is often extremely reduced. Therefore, postoperative ROM training is essential to maintain the degree of ROM obtained intraoperatively. The patient should be carefully evaluated during surgery, considering that the patient is under anesthesia during surgery. Conversely, for patients with idiopathic osteonecrosis of the femoral head and femoral neck fractures, the duration between onset and THA is relatively short, and such patients originally have a low degree of limitation in ROM. Excessive ROM training may be a risk factor for dislocation and should be carefully considered based on the patient's condition.

1.3 Postoperative Conditions and Complications

The degree of improvement in postoperative pain compared with preoperative pain should be evaluated. Specifically, the extent of the pain site, type of pain, and circumstances under which the pain occurs should be confirmed, and a plan for postoperative rehabilitation made and used as a reference for future evaluation of activities of daily living (ADLs).

Postoperative complications include leg length difference (LLD), dislocation, fractures, especially peri-prosthetic fractures, bleeding, infection, venous thromboembolism (VTE) due to deep vein thrombosis (DVT) and pulmonary thromboembolism (PTE), nerve palsy, and loosening. Avoidance of these complications is essential; however, evaluating them is important when they occur, and rehabilitation should be

Y. Nakashima (ed.), *Advances in Total Hip Arthroplasty*, https://doi.org/10.1007/978-981-97-2461-1_61

planned accordingly. For example, if postoperative LLD is present, the presence or absence of apparent LLD should be evaluated by measuring SMD and reconfirmed using simple radiographic imaging. In the case of apparent LLD, rehabilitation aimed at improving the pelvic tilt may be useful, and in the case of true LLD, a prosthetic orthosis, such as an insole, may be an option.

1.4 Team Medicine

Physical therapists (PTs) play a central role in treatment, and team-based medicine is important. Communicating closely with not only nurses but also PTs is effective in sharing sufficient information, respecting their opinions, and clearly dividing their roles in the rehabilitation process.

2 Patient Education

Patient education is an important component of rehabilitation. To prevent dislocation, information should be shared between medical staff and patients regarding the surgical approach, implants used, cup placement angles, femoral head diameter, soft tissue repair, and intraoperative assessment of the risk of dislocation. Information is also provided on the need for early release from bed, active ankle joint exercise, and physical prophylaxis with an intermittent pneumatic compression device in addition to pharmacologic prophylaxis as preventive measures for VTE.

2.1 Commencement Period

Postoperative rehabilitation depends on the surgical technique and severity of the patient's pain. Recently, muscle–tendon sparing or an intermuscular approach, such as MIS, is being used, and the duration of postoperative rehabilitation is shorter compared with the past. For example, in the postoperative plan for MIS-THA, active ankle joint exercise, muscle strength training around the hip joint, and ROM training were performed immediately after surgery, and by 7 days, patients were trained to walk with a parallel bar, use a single cane on one side, climb stairs, and use the toilet. It is common for patients to be discharged within 1–2 weeks [2]. In the United States, patients are discharged from the hospital the day after THA surgery, and rehabilitation should be started as early as possible after THA surgery if the implant placement is stable and there are no problems with the patient's general condition. However, the situation in Japan is not necessarily the same as that in other countries because of the introduction of a comprehensive medical payment system (DPC) and other differences among facilities.

2.2 Intervention Method

Providing patients with the necessary information before surgery is advisable; hospitals with DPC have the option of continuous rehabilitation from acute-care to a convalescent hospital when a certain time has been reached. Home exercise is useful after discharge from the hospital [3]. Implementation of a clinical path as a rehabilitation plan from the preoperative period can facilitate rehabilitation by allowing patients to know what to do next. The clinical path is discussed in detail in the following section.

2.3 Medical and Welfare Services

In many cases, medical–social workers (MSWs) provide medical and welfare services, such as applications for physical disability certificates and long-term care insurance before hospitalization. In addition, MSWs deal with a wide range of issues, such as medical expenses, public systems, social reintegration, nursing care, medical care life, and medical care location.

3 Range of Motion Training

3.1 Commencement Period

Immediately after THA surgery, a hematoma may form around the hip joint, and performing THA as soon as possible is important to avoid soft tissue adhesion and joint contracture, depending on the amount of blood loss and presence of a drain. Conversely, anticoagulants may be used to prevent DVT, and careful measures should be taken to prevent bleeding.

3.2 Intervention Method

Postoperative ROM training should be performed according to the surgical approach, which varies depending on whether the entry is through an incision in the muscle, an intermuscular approach, or a partial osteotomy and reattachment of the greater trochanter. For example, with the intermuscular approach, little damage to the muscles occurs, and active ROM training can be started on the same day or the day after surgery. In some cases, if postoperative pain is very mild, the patient may actively move the muscle. Conversely, if muscles or tendons have been sutured, it is advisable to refrain from active ROM training at least until 3 weeks after suturing to avoid retearing. During this time, it is advisable to perform ROM training within the limits of pain control and take protective care of the sutured muscles and tendons.

Compensatory movements may occur undetected. It is important to consider these situations when implementing countermeasures.

3.3 Dislocation

The postoperative ROM after THA is rarely lower than the preoperative ROM and usually satisfactory. However, postoperative dislocation must be avoided, and more attention must be paid to avoid it. This is influenced by the surgical technique or approach, and the ROM at risk for dislocation should be determined by checking the intraoperative ROM and bony or soft tissue impingement. Generally, after THA using a conventional posterior approach, hip flexion of 90° or more, internal rotation, and adduction are dangerous postures for posterior dislocation. Conversely, hyperextension, adduction, and external rotation of the hip joint are often dangerous postures for anterior dislocation after THA using an anterior approach. In addition, reconfirming the findings of plain radiography, such as the shape of the implant, angle of cup placement, anteversion angle of the stem, and offset distance, is important. Abduction is a safe position with low risk of dislocation after THA. Therefore, hip ROM training is safer when performed during mild abduction. Thus, rehabilitation that considers posture is important.

When THA placement is unstable or complications such as periprosthetic fracture occur, careful guidance is necessary for repositioning, bed movement, and transfer. If instability is present, using an abduction pillow while sleeping may be an option. If a risk of dislocation due to hyperflexion exists, we should also consider providing support for ADLs through the use of self-help devices like reachers and sock aids. Further, it is useful to teach the patient certain ADLs such as how to use the toilet, bathe, put on and take off socks, and pick up things from the floor.

4 Gait Function Training

4.1 Commencement Period

Proceeding in a stepwise fashion is important according to the individual loading status of each patient from the immediate postoperative period until the patient can perform standing movements.

4.2 Intervention Method

After the patient is able to move from a bed to a wheelchair, he or she can begin gait training using parallel bars or a four-point walker. The patient subsequently progresses from walking with a double cane to walking with a single cane. Patients with a good postoperative course and early stabilization of pain, muscle strength, and balance may be discharged from the hospital without the need for a cane, such as patients with high mobility who have always played sports, those with rapidly progressive disease, and those who underwent muscle–tendon-sparing surgery using an intermuscular approach. Conversely, if claudication persists despite sufficient recovery of muscle strength and ROM, considering a decrease in the sense of position is necessary, which is an intrinsic sensation of the hip joint, due to surgical invasion. In such cases, rehabilitation that considers the recovery of sensory function is necessary because it may lead to the prevention of dislocation.

Alignment may change when leg length is corrected by THA, and this may affect the pelvis, spine, knee joint, ankle joint, and foot.

Effective interventions for functional gait training include step exercises [4], gait exercises with reduced ankle push-off [5], and loading exercises [6]. Assessments useful for predicting prognosis include gait changes [7], pelvic and trunk tilt [8, 9], maximum walking speed [10], preoperative hip abduction muscle strength, and history of falls [11].

5 Muscle Strengthening Training

5.1 Commencement Period

Quickly restoring muscle strength through muscle strengthening training is important to avoid postoperative muscle weakness due to the surgical invasiveness of THA, muscle atrophy, pain caused by preoperative disuse, and anxiety and fear.

A later starting date may be considered if a risk of overloading muscle strengthening training exists with loading, such as after bone grafting for acetabular or femoral defects.

5.2 Intervention Method

Although it is important to perform an assessment prior to intervention, it is often difficult to perform muscle strength assessment immediately after surgery when there is pain or muscle damage around the hip joint or when there is a hematoma in the joint. Muscle strength should be assessed when the hip joint is stabilized. Thigh and lower leg circumference should also be measured to assess the degree of muscle atrophy, edema, and periarticular swelling.

If the preoperative hip disease of patients is osteoarthritis, the muscle volume may have decreased over time, and muscle weakness may have occurred. If the disease is acute, such as idiopathic osteonecrosis of the femoral head, muscle atrophy and weakness may not be severe.

Table 61.1 Evidence for effective gait function training interventions and prognostically useful assessments

Evidence for effective interventions for functional gait training
Step exercise
Walking exercise with reduced push-off of ankle joint
Load-bearing exercise
Assessment useful for prognostication
Gait change
Pelvic and trunk tilt
Maximum walking speed
Preoperative hip abduction strength and history of falls

Table 61.2 Evidence of muscle strengthening training effectiveness interventions

Reconstruction of the posterolateral part of the hip joint with total hip arthroplasty
Placement of the hip center
Interval walking training at home
Training of external rotator of the hip joint
Training of the gluteus medius muscles
Training for hip extensor strength and knee extensor muscle strength
Training for hip abduction and knee extension
Training for hip abductor muscle strength during hip joint flexion
Training for muscle strength on the nonsurgical side
Preoperative malnutrition is a risk factor for delayed functional recovery
Combining branched-chain amino acid supplements with physical exercise
Improvement of anemia

Immediately after surgery, the intervention methods differ depending on the surgical approach, such as muscle incision, intermuscular approach, or partial osteotomy and reattachment of the greater trochanter and other bones. In addition to the operated hip joint, training of the contralateral non-operated lower limb, trunk, and upper limb should be performed.

Therefore, it is important to consider the effects of muscle strengthening training on the adjacent knee joint and lumbar spine and rehabilitate the entire body with an awareness of the kinetic chain. For example, open kinetic chain training is performed immediately after the start of rehabilitation, concentrating on certain muscles, such as the gluteus medius and quadriceps, and closed kinetic chain training is performed when a certain level of applied movement is possible.

Evidence for interventions with strength training benefits include posterolateral reconstruction of the hip joint at THA [12, 13], placement of the hip center [14], interval walking training at home [15], training of the hip external rotators [16], training of the gluteus medius muscles [17], strength training of the hip extensors and knee extensors [18], strength training of the hip abductor muscles in flexion and knee extension [19], hip abduction strength training during hip joint flexion [20], strength training on the nonoperative side [21], preoperative malnutrition as a risk factor for delayed functional recovery [22], branched-chain amino acid supplements combined with physical exercise [23], and improving anemia [24]. (Tables 61.1 and 61.2).

6 Evaluation

Methods for assessing hip function and patient satisfaction are important in postoperative rehabilitation. In addition, rehabilitation with feedback from evidence-based assessment methods mentioned below is important. These include the Japanese Orthopaedic Association Hip Function Criteria and the Harris hip score, which are specific to osteoarthritis, the Medical Outcome Study 36-item short-form health survey, Western Ontario and McMaster Universities Osteoarthritis Index, and Japanese Orthopedic Association Hip-Disease Evaluation Questionnaire. Recently, new evaluation methods, such as the forgotten joint score [25], have also been considered because of the ceiling effect in the postoperative evaluation of THA.

References

1. Japanese Orthopedic Association Corporation Committee, et al., editors. Clinical Question 10. Is rehabilitation useful for THA? In: Osteoarthritis clinical practice guideline 2016. Revised 2nd ed. Tokyo: Nankodo; 2016. p. 188–90. Japanese.
2. Takahira N, et al. DAA, modified Watson-Jones, ALS, and other anterior system approaches. J Musculoskeletal Med. 2017;28:350–5. Japanese
3. Sashika H, et al. Home program of physical therapy: effect on disabilities of patients with total hip arthroplasty. Arch Phys Med Rehabil. 1996;77:273–7.
4. Tsukagoshi R, et al. Stepping exercises improve muscle strength in the early postoperative phase after total hip arthroplasty: a retrospective study. Am J Phys Med Rehabil. 2012;91:43–52.
5. Tateuchi H, et al. Immediate effects of different ankle pushoff instructions during walking exercise on hip kinematics and kinetics in individuals. Gait Posture. 2011;33:609–14.
6. Tsukagoshi R, et al. Functional performance of female patients more than 6 months after total hip arthroplasty shows greater improvement with randomized controlled trial. Eur J Phys Rehabil Med. 2014;50:665–75.
7. Gomi M, et al. Early clinical evaluation of total hip arthroplasty by three-dimensional gait analysis and muscle strength testing. Gait Posture. 2018;66:214–20.
8. Kamimura A, et al. Pelvic inclination angle and hip abductor muscle strength after total hip arthroplasty. J Phys Ther Sci. 2013;25:215–9.
9. Nishimura M, et al. Comparison of gait motion including postoperative trunk deflection between direct lateral and anterolateral approaches in supine total hip arthroplasty. J Arthroplast. 2016;31:1603–8.
10. Shibuya M, et al. The maximal gait speed is a simple and useful prognostic indicator for functional recovery after total hip arthroplasty. BMC Musculoskelet Disord. 2020;21:84.

11. Ninomiya K, et al. Predictors of falls in patients during the first year after total hip arthroplasty: a prospective cohort study. Health Sci Rep. 2020;3:e184.
12. Yamaguchi T, et al. The effect of posterolateral reconstruction on range of motion and muscle strength in total hip arthroplasty. J Arthroplast. 2003;18:347–51.
13. Yamaguchi T, et al. Total hip arthroplasty: the relationship between posterolateral reconstruction, abductor muscle strength, and femoral offset. J Orthop Surg (Hong Kong). 2004;12:164–7.
14. Kiyama T, et al. Effect of superior placement of the hip center on abductor muscle strength in total hip arthroplasty. J Arthroplast. 2009;24(24):240–5.
15. Morishima Y, et al. Effects of home-based interval walking training on thigh muscle strength and aerobic capacity in female total hip arthroplasty patients: a randomized, controlled pilot study. PLoS One. 2014;9:e108690.
16. Nankaku M, et al. Hip external rotator exercise contributes to improving physical functions in the early stage after total hip arthroplasty using an anterolateral approach: a randomized controlled trial. Disabil Rehabil. 2016;38:2178–83.
17. Matsumoto H, et al. Gluteus medius muscle activity and recovery of gait function in total hip arthroplasty patients in the early postoperative period. Sci Orient Med Physiother. 2009;24:665–8. Japanese
18. Arima Y, et al. Relationship between vertical ground reaction force and muscle strength while climbing stairs after total hip arthroplasty. Clin Biomech (Bristol, Avon). 2020;78:105088.
19. Sekita J, et al. A predictive model for hip abductor strength and knee extensor strength 12 months after total hip arthroplasty with an interaction term. BMC Musculoskelet Disord. 2021;22:827.
20. Miura T. A study on the transition of lower limb muscle strength at 2 weeks after total hip arthroplasty by the anterior approach method. Sci Orient Med Physiother. 2019;34:587–91. Japanese
21. Fukumoto Y, et al. Changes in hip and knee muscle strength in patients following total hip arthroplasty. J Jpn Phys Ther Assoc. 2013;16(1):22–7.
22. Nanri Y, et al. Preoperative malnutrition is a risk factor for delayed recovery of mobilization after total hip arthroplasty. PM R. 2021;13:1131–9.
23. Ikeda T, et al. Effect of exercise therapy combined with branched-chain amino acid supplementation on muscle strength in elderly women after total hip arthroplasty. Asia Pac J Clin Nutr. 2019;28:720–6.
24. Maezawa K, et al. Postoperative hemoglobin and recovery of hip muscle strength after total hip arthroplasty. J Orthop. 2018;15:886–8.
25. Behrend H, et al. The "forgotten joint" as the ultimate goal in joint arthroplasty: validation of a new patient-reported outcome measure. Arthroplasty. 2012;27:430–6.

Clinical Path

62

Satoshi Nakasone

1 Clinical Path

Clinical pathways were first introduced to medical care in the 1980s as a care map based on the production process management system used in factories in the United States [1]. It is a medical treatment method that aims to improve the quality of patient care and the efficiency of hospital stay adjustment by standardizing and increasing the efficiency of medical treatment mainly through process management and was introduced in the United States in 1983 with the DRG/PPS (diagnosis related groups/prospective payment system). It has been utilized as a method for management improvement and has spread worldwide [1, 2].

Japanese Society for Clinical Pathway defines a clinical path as "a standard medical care plan that includes patient status, goals of medical treatment, and evaluation and recording, and a method to improve the quality of medical care by analyzing the deviation from the standard (variance)" [3]. It is mainly made in the form that the time axis from admission to discharge is taken as the horizontal axis and diagnosis, examination, surgery, medication, diet, rehabilitation, guidance, nursing care, etc., are taken as the vertical axis, and the schedule of medical treatment activities for patients in a hospital is summarized in a table. Medical treatment includes not only examinations and medications but also nursing and clerical work. The medical staff and patients can share the information by clarifying the tasks and instructions among the various professions, and the medicine team can be performed smoothly. This allows nonspecialized staff to participate in care without anxiety, and as a result, risk management and informed consent from multiple professions become possible, and patient education can be easily introduced. Patients can also share the schedule, which allows them to know the examination schedule, the number of hospitalization days, and the estimated medical costs. Moreover, by analyzing patients who deviate from the schedule, it is possible to review the standard of care, such as the contents of tests and drugs. Thus, the clinical pathway is a system that can provide safe medical care to patients and improve the quality of medical care.

2 Effectiveness of the Clinical Path

Total hip arthroplasty (THA) is a highly satisfactory surgical procedure that reduces pain and improves function, and perioperative management using a clinical pathway is considered useful for early discharge. In the United States, the average length of hospital stay was reported to have decreased from 9.3 days to 3.1 days between 1996 and 2008 [4]. In Japan, a clinical pathway was introduced for THA in approximately 2000, resulting in a reduction in hospital stay from an average of 50 days to approximately 3 weeks [5]. In 2013, a new system of diagnosis procedure combination (DPC) was introduced to set different reimbursement points according to the length of hospitalization, clinical paths were used in many hospitals for THA, and early rehabilitation and measures against complications were reported to be effective in reducing the length of hospital stay and medical costs [6, 7].

In recent years, electronic medical records have been introduced in many hospitals, and prescriptions, injections, and image orders can be easily set, making the clinical path for THA even more efficient [8]. In our hospital, patients are briefed with a simple sheet, and the clinical path is operated on the electronic medical record. Doctors can reliably administer pre- and postoperative medications, injections, rehabilitation instructions, and imaging tests according to the clinical path, thereby eliminating order errors. Nurses can eliminate errors in the calculation of procedures and various additional fees, as well as errors in preoperative nursing care, postoperative vital measurements, and detailed postoperative observations. The clinical pathway for THA in our hospital was designed with detailed nursing checkpoints. Since nurses actually spend most of their time with patients, the inclusion of many observation items in the clinical pathway allows

S. Nakasone (✉)
Graduate school of medicine, University of the Ryukyus, Nishihara, Okinawa, Japan

Y. Nakashima (ed.), *Advances in Total Hip Arthroplasty*, https://doi.org/10.1007/978-981-97-2461-1_62

Table 62.1 Nursing Care Items for the Early Detection of Complications in the THA Clinical Pathway

Measurement: Weight, height, pulse rate, respiration rate, blood pressure, SpO_2
Intake: Main and side meals
Meals: Setting, serving, and discharging
Excretion: Urinary frequency, stool frequency, nature of excretion
IN: Peripheral line, central line
OUT: Foley urine, drain effluent volume monitoring
Cleanliness: Showering, morning care, bed sheet change, perineal care
Infection survey: Presence of indwelling ureter, nature of urine, wound (redness, pain, tenderness, heat)
Observation of infection: Peripheral catheter insertion site, urinary tract, wound assessment (redness, pain, tenderness, heat, induration)
Education and guidance: Movement guidance, fall and tumble risk assessment, nursing guidance, and nutritional guidance
Observations (day of admission, up to 2 days postoperatively, and 3 days or more postoperatively): ankle assessment (movement and circulatory disturbance), foot assessment (hypesthesia, use of elastic stockings, use of air compression pump, plantar dorsiflexion movement, skin condition, color tone), taste disorder evaluation, pain control, postoperative anesthesia arousal, left-right difference in both lower extremities, numbness assessment, epidural tube management, headache monitoring, bedsore assessment and pressure ulcer prevention, transfer assistance, oral care, feeding assistance, dressing/undressing, and abnormalities in limb position, etc.
Resting level and activity: Transferring movement (talking to, looking after, supporting body, all assistance)

detailed observation without compromising the quality of nursing care and leads to the early detection of complications (Table 62.1).

3 Advantages of Clinical Pathways in THA

There are two major advantages of using a clinical pathway in THA. The first is that it promotes early rehabilitation and early discharge from the hospital. The development of minimally invasive surgery in THA has clarified the goal of early recovery of patients, and early release from the hospital, measures against complications, and pain control are now performed by multiple disciplines. It was reported that prevention of postoperative anemia by bleeding control and alleviation of postoperative pain by drug therapy both enable early rehabilitation and that rehabilitation has a synergistic effect in preventing serious complications such as deep vein thrombosis (DVT)/pulmonary embolism (PE) [9]. The use of a clinical path enables on-site organization of team medicine involving multiple professions, such as physicians, nurses, pharmacists, and dietitians, and safety management with standardization of medical care and clarification of responsibilities. It is also thought to be possible to prevent errors and accidents caused by omissions of checks by not only patients and doctors but also doctors' instructions and the medical team and to manage management by adjusting the length of hospital stay and medical resource investment and reimbursement.

The second advantage is the ability to provide feedback on perioperative management through analysis of variances. The clinical pathway can provide feedback on perioperative management by analyzing the factors of cases that deviate from clinical plan [2]. By examining the reasons for deviations, the roles and duties of multiple professions during the hospitalization period can be reviewed. Factors that have been reported to prolong the length of hospital stay for THA include older age (60 years or older), female sex, BMI over 40, and high complication rates [10–12], and factors that shorten the length of hospital stay include strong preoperative abductor muscle strength [11], rehabilitation from the day of surgery [13], and strict pain control [14]. Variance analysis of the clinical pathway will allow us to focus on such deviating factors and improve preoperative patient education and complication management by multiple disciplines. The same clinical path should not be applied to all patients, but it is necessary to select patients for the clinical path. In our department, the achievement level of rehabilitation after 7 days postoperatively varies depending on age, muscle strength, and the presence of complications.

4 Future Prospects

In other countries, the team system comprising this clinical path has been effectively functioning not only during hospitalization but also before admission and after discharge [15]. In addition, not only orthopedic surgeons but also anesthesiologists and internists participate in the path [15, 16]. Before hospitalization, patients are educated about dislocation prevention, postoperative rehabilitation, and pain control, and medical control is provided for glucose metabolism and renal function protection. After discharge, wound management, rehabilitation, and DVT prevention are provided at home and in outpatient settings. In Japan, these services are provided during the hospitalization period, but it is expected that the same services can be provided in the outpatient setting if the medical staff form a team based on the pathway.

References

1. Zander KS. Collaborative care: two effective strategies for positive outcomes. In: Zander KS, editor. Managing outcomes through collaborative care: the application of caremaping and case management. Chicago: American Hospital Publishing; 1995.
2. Konishi T. Clinical pathways. J Jpn Soc Intern Med. 2007;96:2325–9. Japanese

3. Website of the Japanese Society for Clinical Pathway: definition of a clinical path: http://www.jscp.gr.jp/about.html#sub02. Japanese. Accessed in August 2024.
4. Cram P, et al. Clinical characteristics and outcomes of Medicare patients undergoing total hip arthroplasty, 1991–2008. JAMA. 2011;305:1560–7.
5. Mitsuyasu S, et al. Relationship between total arthroplasty case volume and patient outcome in an acute care payment system in Japan. J Arthroplast. 2006;2006(21):656–63.
6. Ninomiya K, et al. The comparison study of postoperative short term outcome between 3-days length of clinical pathway and conventional after total hip. Clin Orthop Surg. 2016;51:977–81. Japanese
7. Sasamori M. A study of short-term clinical path after total hip arthroplasty. Hip Joint. 2017;43:S1–4. Japanese
8. Funada C. Clinical pathways and nursing record efficiency. Clin Nurs Rec. 2019;29:2–6. Japanese
9. Matsubara M, et al. Postoperative deep vein thrombosis in THA with DVT pharmacotherapy vs. nonpharmacotherapy physical therapy and early weaning and loading program. Nichi Jinko Kansetsu Kaishi. 2013;43:15–6. Japanese
10. Otero JE, et al. Length of hospitalization after joint arthroplasty: does early discharge affect complications and readmission rates? J Arthroplast. 2016;31:2714–25.
11. Oosting E, et al. The influence of muscle weakness on the association between obesity and inpatient recovery from total hip arthroplasty. J Arthroplast. 2017;32:1918–22.
12. Sibia US, et al. Predictors of hospital length of stay in an enhanced recovery after surgery program for primary total hip arthroplasty. J Arthroplast. 2016;31:2119–23.
13. Okamoto T, et al. Day-of-surgery mobilization reduces the length of stay after elective hip arthroplasty. J Arthroplast. 2016;31:2227–30.
14. Sculco PK, et al. Perioperative solutions for rapid recovery joint arthroplasty: get ahead and stay ahead. J Arthroplast. 2015;30:518–20.
15. Stambough JB, et al. Rapid recovery protocols for primary total hip arthroplasty can safely reduce length of stay without increasing readmissions. J Arthroplast. 2015;30:521–6.
16. Tait MA, et al. Preoperative patient education for hip and knee arthroplasty: financial benefit? Bone Joint J. 2013;95B:246–51.

Part XII

Postoperative Function and Patient Satisfaction

63 Functional Evaluation

Shinya Hayashi

1 Functional Evaluation

In Japan, the Japanese Orthopedic Association (JOA) hip function assessment criteria (JOA score) are most commonly used to evaluate hip function before and after total hip arthroplasty (THA). Although these criteria are widely used by medical professionals or researchers, in recent years, patient-oriented assessment methods in which the patients themselves fill questionnaires have become more important. The Oxford hip score (OHS), hip disability and osteoarthritis outcome score (HOOS), and generalized quality of life (QOL) scores, such as the short-form 36-item health survey (SF-36) and EuroQol 5 dimensions (EQ. 5-D), have been used in both clinical practice and research. The details of the JOA score, OHS, HOOS, EQ. 5-D, and JHEQ have been described in separate sections (Chap. 12, Sects. 2 and 3), Therefore, this section describes other methods of functional evaluation and patient satisfaction evaluation before and after THA. Functional evaluation methods can be classified into those that use questionnaires and those that measure motor function and activity levels. The following is a brief description of each method.

1.1 Questionnaire Method

- The University of California, Los Angeles (UCLA) activity rating score (Table 63.1), Tegner activity score, activity rating scale (ARS), and International Physical Activity Questionnaire (IPAQ) have been used to evaluate the amount of activity in daily life and sports activities.
- The Falls Efficacy Scale (Table 63.2) and Activities-Specific Balance Confidence scale were used to assess the risk of falls [2].

S. Hayashi (✉)
Kobe University Orthopedic Surgery, Kobe, Hyogo, Japan

Table 63.1 University of California, Los Angeles activity rating score [11]

Level	Activity
10	Regular participation in impact sports such as tennis, skiing, acrobatics, ballet, heavy labour or backpacking
9	Sometimes participates in impact sports
8	Regular participation in very active events, such as bowling or golf
7	Regular participation in active events, such as bicycling
6	Regular participation in moderate activities, such as swimming and unlimited housework and shopping
5	Sometimes participates in moderate activities
4	Regular participation in mild activities, such as walking, limited house work, and limited shopping
3	Sometimes participates in mild activities
2	Mostly inactive: restricted to minimal activities of daily living
1	Wholly inactive: dependent on others; cannot leave residence

Table 63.2 Overview of the fall efficacy scale [12]

Activity:	Score: 1 = very confident 10 = not confident at all
Take a bath or shower	
Reach into cabinets or closets	
Walk around the house	
Prepare meals not requiring carrying	
Heavy or hot objects	
Get in and out of bed	
Answer the door or telephone	
Get in and out of a chair	
Getting dressed and undressed	
Personal grooming (i.e. washing your face)	
Getting on and off of the toilet	
Total score	

1.2 Measuring Athletic Ability and Activity

1.2.1 Measuring Motor Function

Stair climbing test (time required to climb a set number of stairs), 30-s chair stand test (30 sCST) (number of times the participant could repeatedly stand up and sit down from a

Y. Nakashima (ed.), *Advances in Total Hip Arthroplasty*, https://doi.org/10.1007/978-981-97-2461-1_63

chair in 30 s), 40 m fast-paced walk test (40 mFPWT) (time required to walk 40 m as fast as possible), timed-up-and-go test (TUG) (time required to stand up from a chair, walk 3 m, and sit down), and 6-min walk test (6 MWT) (walking distance in 6 min). According to the Osteoarthritis Research Society International, the 30 sCST, 40 mFPWT, 6 MWT, and 10 mFPWT are recommended as evaluation methods for motor function in patients with hip osteoarthritis [3–5].

In addition, there are also methods for measuring walking speed, such as instructing patients to walk 16 m, measuring the normal walking speed and maximum walking speed at the passage of 10 m [6], and measuring walking speed using an optical sensor [7]. Other methods for measuring motor function include direct measurement of hip abductor muscle strength with a muscle strength-measuring device and measuring the amount of muscle activity during maximum voluntary contraction and walking using surface electromyography [8].

1.2.2 Measuring Physical Activity

Measurement of physical activity can be performed with a variety of different methods, such as having patients wear accelerometers for more than 10 h a day and measuring the amount of activity in daily life according to the measured acceleration and duration [9].

2 Results of THA Using Each Functional Assessment Method

Previous reports have demonstrated both the preoperative and postoperative results of THA using various functional assessment methods. In a study of 65 patients with periprosthetic bone mineral density after THA with tapered wedge stems, bone mineral density at 6 and 12 months postoperatively decreased in Gruen zones 1 and 7, while at 24 months postoperatively, bone mineral decreased in Gruen zone 7. The analysis of factors affecting periprosthetic bone mineral density after THA showed a positive correlation between the UCLA activity rating score at 24 months and the change in periprosthetic bone mineral density in Gruen zones 1, 6, and 7. These results suggest that higher activity is associated with better periprosthetic proximal bone density [1].

In a study on the risk of falls after THA in 214 patients, the Falls Efficacy Scale was modified to include 12 items that took into account the Japanese lifestyle (e.g., the need to commonly get up from the floor), and the association between the Falls Efficacy Scale and other items was investigated using multivariate analysis [2].

A study of functional assessment of exercise capacity and activity measures in patients undergoing THA with a 10-m walking speed showed that the time to stand on one leg on the contralateral side before surgery was significantly correlated with comfortable walking speed at 10 m after surgery in 91 patients in a multivariate analysis of preoperative factors affecting walking speed after THA surgery [3]. In a study evaluating TUG, 28 patients were randomized to receive postoperative external rotator strength training after THA surgery, and abductor muscle strength and TUG were compared between the two groups. The results showed that the intervention group (i.e., the group that received external rotation strength training) showed significant improvement in both the items [4]. In addition, the 10-m walking speed was measured in this study. Furthermore, in a study that measured 10-m walking speed, receiver operating characteristic curve analysis of preoperative factors affecting early functional recovery after THA surgery in 317 patients showed that the preoperative 10 mFPWT was significantly higher than the preoperative 10-m comfortable walking speed when performing 50-m independent walking on the fifth day after THA surgery was used as the outcome. The preoperative 10 mFPWT correlated with 50-m independent walking 5 days after THA surgery more favorably than the preoperative 10-m comfortable walking speed [6]. In a study on motion analysis using an optical sensor, 17 patients who underwent THA were fitted with a three-dimensional optical sensor (Vicon system) at 1, 3, 6, and 12 months after surgery, and their walking speed, range of motion of each joint, and period of motion were measured and analyzed. Walking speed improved to the average Japanese walking speed by 12 months postoperatively. Conversely, the maximum hip flexion angle during walking was significantly decreased on the operated side before surgery compared to the healthy side, and the flexion angle on the operated side was still significantly decreased compared to the healthy side at 12 months postoperatively [7]. In a study that directly measured muscle strength and activity, abductor muscle strength was measured postoperatively using a muscle testing device, and muscle activity was measured during maximal voluntary contraction and walking using surface electromyography of the midline muscles in seven patients who underwent THA using an anterolateral approach, and the results were evaluated over time. The hip abductor muscle strength was lower than that before surgery at 3–4 weeks postoperatively, but recovered to the same level as before surgery at 5 weeks postoperatively, and the activity of the gluteus medius muscle did not change before or after surgery. This suggests that organic dysfunction of the vastus medialis muscle does not improve in the early postoperative period [8]. In another study, we investigated the effects of supine posture during hip abduction on postoperative function during postoperative THA rehabilitation. Twenty patients who underwent THA were randomized into two groups: supine abduction and supine training. The abductor muscle strength of the hip was measured using manual electromyography, while muscle activity was measured using surface electromyography on the 10th day after surgery [9]. In a study that evaluated activity using accelerometers, 153 patients who underwent continuous observation for 1 year after THA were included, and all data were analyzed in 131 patients who could be traced. The number of steps per day and the duration of moderate-to-vigorous physical activity per week were measured preoperatively and at 1 and 3 years postoperatively, showing a significant increase at 1 and 3 years postoperatively compared with preoperatively. Furthermore, multivariate analysis of age, body mass index, Oxford hip score, SF-8 mental component, employment status, daily number of steps, and weekly duration of moderate-to-vigorous-intensity physical activity showed that the number of daily steps tended to increase with decreasing age, and that the duration of moderate-to-vigorous-intensity physical activity per week increased with decreasing age [10].

In the future, the use of patient-oriented functional assessment and measurement of motor function and activity as a functional assessment method after THA surgery is expected to improve the postoperative outcomes of THA.

References

1. Hayashi S, et al. Daily activity and initial bone mineral density are associated with periprosthetic bone mineral density after total hip. Hip Int. 2016;26:169–74.
2. Nagai K, et al. Fear of falling during activities of daily living after total hip arthroplasty in Japanese women: a cross-sectional study. Physiotherapy. 2014;100:325–30.
3. Ohmori T, et al. Contralateral lower-limb functional status before total hip arthroplasty: an important indicator for postoperative gait speed. Bone Joint Surg Am. 2021;103:1093–103.
4. Nankaku M, et al. Hip external rotator exercise contributes to improving physical functions in the early stage after total hip arthroplasty using an anterolateral approach: a randomized controlled trial. Disabil Rehabil. 2016;38:2178–83.
5. Dobson F, et al. Reliability and measurement error of the osteoarthritis research society international (OARSI) recommended performance-based tests of physical function in people with hip and knee osteoarthritis. Osteoarthr Cartil. 2017;25:1792–6.
6. Shibuya M, et al. The maximal gait speed is a simple and useful prognostic indicator for functional recovery after total hip arthroplasty. BMC Musculoskelet Disord. 2020;21:84.
7. Miki H, et al. Recovery of walking speed and symmetrical movement of the pelvis and lower extremity joints after unilateral THA. J Biomech. 2004;37:443–55.
8. Matsumoto H, et al. Early postoperative activity of the medialis muscle and recovery of gait function in patients undergoing total hip arthroplasty. J Phys Ther Sci. 2009;24:665–8. Japanese
9. Ikematsu K, et al. Effects of different supine postures during hip abduction exercises on postoperative function after total hip replacement. J Phys Ther Sci. 2018;33:959–64. Japanese
10. Matsunaga Y, et al. Three-year follow-up study of physical activity, physical function, and health-related quality of life after total hip. J Arthroplast. 2020;35:198–203.
11. Fisher NE, et al. Sporting and physical activity following hip resurfacing. Int Orthop. 2011;35:977–80.
12. Tinetti ME, et al. Falls efficacy as a measure of fear of falling. J Gerontol. 1990;45:P239–43.

JOA Hip Score, JOA Hip Disease Evaluation Questionnaire, and Locomotive Syndrome

64

Takeyuki Tanaka

1 Japanese Orthopaedic Association Hip Function Criteria

1.1 Score Characteristics

The Japanese Orthopaedic Association hip score (JOA score) is the leading physician-completed, disease-specific measure of hip function in Japan. The JOA score consists of the following four items: pain (40 points), range of motion (20 points), ability to walk (20 points), and activities of daily living (20 points). It was developed in 1971, revised in 1995, and is still being widely used. In 2010, the reliability and validity of the JOA score for hip function were reported by Kuribayashi et al. [1]. The Harris hip score (HHS) is the most commonly used physician-completed, disease-specific scale worldwide; moreover, Fujisawa et al. revealed the significant correlation between the JOA score and HHS [2].

To improve convenience, it would be helpful if the JOA score form can be filled in an spreadsheet so that the score can be automatically calculated (Fig. 64.1).

It should also be noted that there are challenges with physician-completed scales, as they are evaluated only by healthcare professionals, and there is a risk of discrepancy due to patients reporting subjective symptoms.

1.2 Results in Total Hip Arthroplasty

As described above, the reliability and validity of the JOA scores are particularly useful in evaluating patients' outcome after total hip arthroplasty (THA), especially in Japanese patients. Moreover, several papers that used the JOA score to evaluate patients' outcomes have been accepted and published in international journals [3].

T. Tanaka (✉)
Orthopaedic Surgery, Sensory and Motor System Medicine, Surgical Sciences, Graduate School of Medicine, The University of Tokyo, Bunkyo-ku, Tokyo, Japan
e-mail: tanakata-ort@h.u-tokyo.ac.jp

Y. Nakashima (ed.), *Advances in Total Hip Arthroplasty*, https://doi.org/10.1007/978-981-97-2461-1_64

Appendix 1. Japan Orthopaedic Association hip score

	Full mark = 40 points	
A. Pain	Right	Left
A. None[a]	40	40
B. Ignores[b]	35	35
C. Slight[c]	30	30
D. Moderate[d]	20	20
E. Severe[e]	10	10
F. Unbearable[f]	0	0

[a]No pain and/or no complaints relating to hip joint
[b]No pain. Inconstant symptoms including weary feeling or dullness
[c]No spontaneous pain. Some pain when walking (including slight pain when starting to walk or after walking for long distance)
[d]No spontaneous pain. Some pain when walking but which disappears quickly after a short rest
[e]Spontaneous pain. Pain is severe when attempting to walk; it decreases after a rest
[f]Continuous pain during rest and/or at night

	Full mark = 20 points	
B. Range of motion	Right	Left
Flexion arc	0–12	0–12
Abduction arc	0–8	0–8

Scores are detemined by multiplying 10° of motion in each arc by 1 point in flexion and 2 points in abduction. Range of contracture should be subtracted
A flexion are more than 120° is detemined as 12 points and the abduction are more than 30° as 8 points
Either flexion and abduction is measured in neutral position on rotation and described by its arc by passive motion

C. Ability to walk	Full mark = 20 points
A. Normal[a]	20
B. Slight limp[b]	18
C. Mild limp[c]	15
D. Severe limp[d]	10
E. Difficult to walk[e]	5
F. Impossible[f]	0

[a]Able to walk long distance without limp. Able to walk fast
[b]Able to walk long distance including walking with a slight limp. Able to walk fast
[c]Able to walk 30 min or 2 km without support. Mild limp
[d]Able to walk 10–15 min or 50 m without support
[e]Able to do household activities. Difficult to do outdoor activities. Difficult to walk outdoors without bilateral supports
[f]Impossible or almost impossible to walk

	Full mark = 20 points		
D. Activities of daily life	Normal	Difficult	Impossible
A. Sitting on chair	4	2	0
B. Standing work (including housework)[a]	4	2	0
C. Squatting, standing up from sitting on the floor[b]	4	2	0
D. Going up and down stairs[c]	4	2	0
E. Getting into car or entering public transport	4	2	0

[a]Able to continue longer than 30 min. Needs to take a rest — score as "difficult." Unable to continue longer than 5 min — score as "impossible"
[b]Support needed — score as "difficult"
[c]Handrail needed — score as "difficult"

Fig. 64.1 Hip examination chart

2 Hip Disease Evaluation Questionnaire (Fig. 64.2, JHEQ Form from the Website of the Japanese Hip Society)

2.1 Score Characteristics

The importance of patient-reported outcomes (PROs) is often discussed, and the Westan Ontario and McMaster Universities Osteoarthritis Index (WOMAC) and the MOS 36-item short-form health survey (SF-36) have been widely used as representative international PROs. On the other hand, the Japanese Orthopaedic Association Hip Disease Evaluation Questionnaire (JHEQ) is the first PRO of hip function devised in Japanese and was presented at the Japanese Hip Society meeting in 2012 [4]. The JHEQ consists of three subscales: pain, movement, and mental health, with a visual analog scale (VAS). It has six questions for pain and seven questions for movement and mental health. The maximum score is 84, and the higher the score, the higher the quality of life (QOL). As another indicator, the higher the score in the VAS evaluation, the stronger the patient's dissatisfaction. The questionnaire can be downloaded by anyone from the website of the Japanese Hip Society, and an English version is also available. Modifying the questionnaire is prohibited. PROs have the advantage of detecting subtle subjective symptoms, but they also have disadvantages, such as the need to check for incomplete answers each time and the possibility of detecting symptoms other than hip joint diseases. Furthermore, it is often combined with a physician-completed disease-specific measure, such as the JOA score.

2.2 Results in Total Hip Arthroplasty

The JHEQ has been used in many reports on the outcomes of THA in Japan. Although we cannot describe all of these studies, some are introduced below.

Takegami et al. used the JHEQ to evaluate whether patients who underwent unilateral THA had improved pain and function in the contralateral hip [5]. On the other hand, Yoshii et al. evaluated the results of unilateral THA and bilateral same-day THA with an anterior approach using the

Japanese Orthopedic Association Hip Disease Evaluation Questionnaire

The following questionnaire asks you to describe your hip-joint condition, including difficulties you may encounter in daily life. Your honest appraisals will help us improve treatment for patients in the future. Thank you for taking the time to fill out this questionnaire.

Guidelines

① Please make sure that your answers to the questions pertain to your hip-joint condition within the past 3 months.

② Please refer to the examples below in answering the questions.

③ Please try answer all questions. However, if you find some questions difficult to answer, you may skip them.

④ In this questionnaire, there are two types of questions:

I. The first type asks you to place an X on a line to indicate your answer.

II. The second type asks you place a check mark in the box that applies to you.

Fig. 64.2 Japanese Orthopaedic Association Hip Disease Evaluation Questionnaire. (Cited from the website of the Japanese Hip Society)

Type I Question Example.

<Important Point>

Place an X on the line that corresponds to your answer.

Good Examples

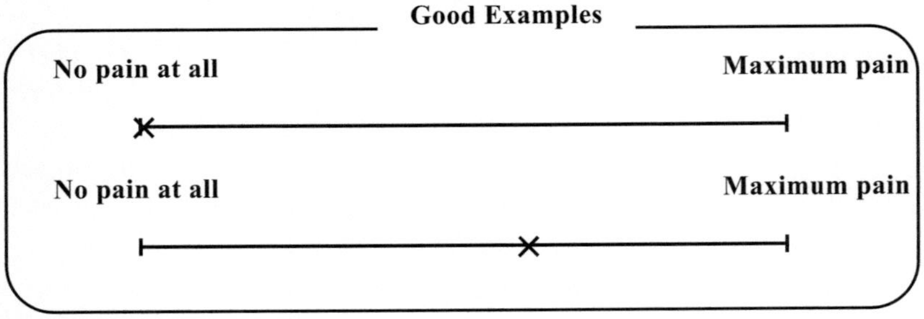

In this case the X is not on the line, but beside or above it.

Bad Examples

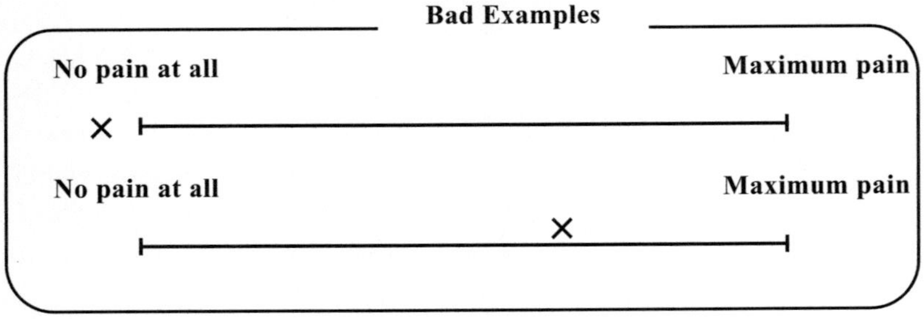

Type II Question Example

<Important Point>

Please check ONE box that most applies to you.

Good example:
1 box is checked.

Bad example:
2 boxes are checked.

		Strongly Agree	Agree	Uncertain	Disagree	Strongly Disagree
Good example	Even when I am at rest, my hip is painful.	☐	☑	☐	☐	☐
Bad example	Even when I am at rest, my hip is painful.	☑	☑	☐	☐	☐

Fig. 64.2 (continued)

The Questionnaire Begins Here

*Please respond regarding your current (within the last 3 months) hip condition.

1. In general, rate your current level of satisfaction as it pertains to your hip condition.

Place an X on the continuum, where the left side represents complete satisfaction and the right represents complete dissatisfaction.

Complete Satisfaction **Complete Dissatisfaction**

2. How severe is your hip pain?

Place an X on the continuum of pain, where the left side represents no pain at all and the right side represents maximum pain.

<Right Side Hip>

No Pain at all **Maximum Pain**

<Left Side Hip>

No Pain at all **Maximum Pain**

Fig. 64.2 (continued)

Please place a check mark in the box for each question that applies to you.

		Strongly Agree	Agree	Uncertain	Disagree	Strongly Disagree
1. Even when I am at rest, my hip is painful.	**Right**	☐	☐	☐	☐	☐
	Left	☐	☐	☐	☐	☐
2. My hip is painful when I sit in a chair.	**Right**	☐	☐	☐	☐	☐
	Left	☐	☐	☐	☐	☐
3. I feel pain in my hip when I start to move.	**Right**	☐	☐	☐	☐	☐
	Left	☐	☐	☐	☐	☐
4. I cannot move my hip joint freely because of the pain.	**Right**	☐	☐	☐	☐	☐
	Left	☐	☐	☐	☐	☐
5. The pain in my hip joint prevents me from moving with strength.	**Right**	☐	☐	☐	☐	☐
	Left	☐	☐	☐	☐	☐
6. Because of hip-joint pain, I occasionally cannot sleep well.	**Right**	☐	☐	☐	☐	☐
	Left	☐	☐	☐	☐	☐
7. It is difficult for me to climb up and down stairs.		☐	☐	☐	☐	☐
8. It is difficult for me to get up from the floor and tatami.		☐	☐	☐	☐	☐
9. It is difficult for me to squat.		☐	☐	☐	☐	☐
10. It is difficult for me to use a Japanese-style toilet.		☐	☐	☐	☐	☐

Fig. 64.2 (continued)

		Strongly Agree	Agree	Uncertain	Disagree	Strongly Disagree
11. It is difficult to get in and out of a bathtub.		□	□	□	□	□
12. It is difficult to cut my toenails.	**Right**	□	□	□	□	□
	Left	□	□	□	□	□
13. It is difficult to put on my socks.	**Right**	□	□	□	□	□
	Left	□	□	□	□	□
14. Because of hip-joint disease, I sometimes get irritated or feel nervous.		□	□	□	□	□
15. Because of hip-joint disease, I feel dispirited and avoid going out.		□	□	□	□	□
16. Because of hip-joint disease, I feel anxiety about my livelihood/daily life.		□	□	□	□	□
17. Because of hip-joint disease, I feel dissatisfied with my health.		□	□	□	□	□
18. My hip-joint condition deeply affects my well-being.		□	□	□	□	□
19. Because of hip joint disease, it is difficult for me to actively undertake various things.		□	□	□	□	□
20. Because of hip-joint pain, sometimes participation in local events and neighborhood relationships does not go smoothly for me.		□	□	□	□	□

Thank you for completing this questionnaire.

Fig. 64.2 (continued)

JHEQ and found that the outcomes of bilateral same-day THA were superior [6].

Murofushi et al. found a correlation between health status and JHEQ at 6 months after THA, which is valuable data that suggests the involvement of various factors in improving QOL after THA from the viewpoint of physical therapy [7].

Fukui et al. used the JHEQ and found that mental factors influenced the outcomes of THA after 6 months [8].

These reports have shown that the JHEQ is useful after THA in Japan, and it is expected that the JHEQ will be used not only for QOL assessment but also to assess several other outcomes.

3 Locomotive Syndrome

3.1 Score Characteristics

Locomotive syndrome (locomo) is a condition in which mobility is impaired due to disorders of the locomotor system, a concept proposed by Nakamura in 2008 [9]. With the advent of a super-aged society, various academic societies, such as the Japanese Orthopaedic Association, are conducting educational activities on locomotive syndrome, which is a condition that should be recognized accurately in the future.

The diagnosis of locomotive syndrome is based on the results of three tests: the first is a "stand-up test" to see whether a person can stand upright on one or both legs; the second is a "two-step test" to measure the distance between two steps taken with the thighs as wide as possible; and the third is the "locomo 25 (25-question geriatric locomotive function scale: GLFS-25)" which has 25 questions regarding musculoskeletal disorders. The details of the GLFS-25 can be accessed by anyone from the website of the Japanese Orthopaedic Association.

Although the age range of patients with hip joint disease is assumed to be diverse, it should be understood that THA is expected to be performed in an increasing number of patients with locomotive syndrome in the future super-aged society.

3.2 Results in THA

Although locomotive syndrome is a relatively new concept, reports on it and THA have been published in recent years.

Ninomiya et al. evaluated the prevalence of locomotive syndrome in 593 patients more than 10 years after THA. The results showed that 27.7% of patients who had undergone THA more than 10 years prior had locomotive syndrome, and this figure was equivalent to that of the elderly generation [10].

Taniguchi et al. followed 88 patients with THA and locomotive syndrome for 2 years. They found that the degree of locomotion improved in 56.8% of these patients [11].

Miyazaki et al. analyzed 105 patients with preoperative locomotion level 3 who underwent THA. They found that 46.7% of the patients had improved locomotion at 3 months postoperatively, and the standing test and GLFS-25 were particularly associated with improvement in locomotion [12].

Ohmori et al. analyzed 72 patients with preoperative locomotive syndrome for 1 year. The results showed that 65% of the patients had improvement in locomotion, especially within the first 6 months postoperatively [13].

Maezawa et al. evaluated GLFS-25 scores and straight leg raising (SLR) muscle strength in 12 male and 75 female patients undergoing THA. Ninety percent of patients had hip disability, which decreased to 30% at 3 months postoperatively, indicating the importance of improving SLR muscle strength.

Although we have not been able to comprehensively discuss all of the above, we hope that the relationship between locomotive syndrome and THA will become clearer in the future [14].

References

1. Kuribayashi Y, et al. Reliability and validity of the Japanese Orthopaedic Association hip score. J Orthop Sci. 2010;15:452–8.
2. Fujisawa M, et al. A comparative study of hip joint functional scoring systems between Japanese Orthopaedic Association hip score and Harris hip score [in Japanese]. Seikeigeka. 2001;52:628–33.
3. Tanaka T, et al. Comparison between cup implantations during total hip arthroplasty with or without a history of rotational acetabular osteotomy. Arch Orthop Trauma Surg. 2022;142:3539–47.
4. Matsumoto T, et al. Japanese Orthopaedic Association Hip Disease Evaluation Questionnaire (JHEQ): a patient-based evaluation tool for hip-joint disease. The Subcommittee on Hip Disease Evaluation of the Clinical Outcome Committee of the Japanese Orthopaedic Association. J Orthop Sci. 2012;17:25–38.
5. Takegami Y, et al. Does unilateral total hip arthroplasty improve pain and function in the nonoperative hip joint? Eur J Orthop Surg Traumatol. 2020;30:1411–6.
6. Yoshii H, et al. Comparison of patient satisfaction after unilateral or simultaneous bilateral total hip arthroplasty through a direct anterior approach: evaluation using the Japanese Orthopaedic Association Hip Disease Evaluation Questionnaire. J Orthop Sci. 2016;21:332–5.
7. Murofushi Y, et al. The relationship between physical condition and JHEQ at 6 months after total hip arthroplasty. Rigakuryoho Kagaku. 2016;31:469–72.
8. Fukui K, et al. Clinical assessment after total hip arthroplasty using the Japanese Orthopedic Association Hip-Disease Evaluation Questionnaire. J Orthop. 2015;12(Suppl 1):S31–6.
9. Nakamura K. A "super-aged" society and the "locomotive syndrome". J Orthop Sci. 2008;13:1–2.
10. Ninomiya K, et al. Prevalence of locomotive syndrome in Japanese patients more than 10 years after total hip arthroplasty: a cross-sectional cohort study. J Orthop Sci. 2022;27:176–80.
11. Taniguchi N, et al. Improvement of locomotive syndrome after total hip arthroplasty: a two-year longitudinal cohort study. Mod Rheumatol. 2021;31:1050–8.
12. Miyazaki S, et al. Effect of total hip arthroplasty on improving locomotive syndrome in hip disease patients: a prospective cohort study focused on total clinical decision limits stage 3. J Orthop Sci. 2022;27:408–13.
13. Ohmori T, et al. The efficacy of total hip arthroplasty on locomotive syndrome and its related physical function in patients with hip osteoarthritis. J Orthop Sci. 2021;26:389–95.
14. Maezawa K, et al. Early clinical results of total hip arthroplasty assessed with the 25-question Geriatric Locomotive Function Scale and muscle strength testing. J Orthop Sci. 2018;23:538–41.

Short Form Survey, EQ5D, Oxford Hip Score, Hip Disability and Osteoarthritis Outcome Score, Forgotten Joint Score, and Patient Joint Perception

65

Yasuhiko Takegami

Total hip arthroplasty (THA) is one of the most successful orthopedic procedures. With improvements in implants and surgical techniques, stable outcomes have been achieved from the early postoperative period to more than 10 years postoperatively. Recently, the importance of patients' quality of life (QOL) and satisfaction with the procedure has become more important in the postoperative evaluation of THA.

Traditionally, physicians have evaluated improvements in postoperative pain and physical function (physician-reported outcomes). In recent years, several patient-reported outcomes (PROs) have been developed to allow patients to evaluate the effectiveness of treatment, such as activities of daily living and satisfaction. In addition to physician-reported outcomes, PROs are increasingly being used worldwide for the postoperative evaluation of THA [1].

PROs can be classified using comprehensive and disease-specific scales. Comprehensive measures are used to assess health-related QOL in all individuals. Comprehensive scales allow comparisons with the general population and between populations with certain diseases. The MOS short-form 36-item Health Survey (SF-36), SF-12, SF-8, and EQ-5D are examples of comprehensive scales. Disease-specific scales measure and assess the impact of disease-specific problems on daily life and patient satisfaction with surgery. The Japanese Orthopedic Association Hip-Disease Evaluation Questionnaire (JHEQ), Oxford hip score (OHS), and Hip Disability and Osteoarthritis Outcome Score (HOOS) are disease-specific scales related to hip disease. The forgotten joint score (FJS) and patient joint perception are available for THA-specific PROs.

When using PRO, improvements in statistical values do not mean clinical improvements in patients [2]. Therefore, it is necessary to examine whether improvements in PRO values are clinically meaningful. The concept of minimal clinically important difference (MCID) is the smallest change in PRO that can be interpreted as beneficial to the patient [3].

1 MOS 36-Item Health Survey (SF-36), SF-12, and SF-8

The SF-36 is one of the most widely used international comprehensive scales. It can be used to compare the QOL measured using the SF-36 between patients with different diseases and to compare the health status of patients with that of the general population. The Japanese version of the scale has also been validated [4, 5].

The SF-36 has eight subscales of health concepts: physical functioning, role-physical, role-emotional, general health, social functioning (SF), bodily pain, vitality (VT), and mental health. It can also calculate the following summary scores: physical component summary (PCS), mental component summary (MCS), and role/social component summary [6]. The SF-36 consists of 36 questions and is highly burdensome to answer. Therefore, the SF-12 and SF-8 were developed as simplified versions [7, 8]. The SF-12 and SF-8 are also capable of calculating eight subscale- and summary scores. However, SF-12 and SF-8 do not evaluate changes in health status over a 1-year period as effectively as SF-36. The SF-8 cannot be compared with the national standard; therefore, it should be used with knowledge of the limitations of the scale.

Table 65.1 shows the MCIDs before and after THA for the eight subcomponents of the SF-36. (Table 65.1), with MCIDs of 4.15–14.11 for PCS and 2.88–5.91 for MCS for SF-12 in THA [9].

Y. Takegami (✉)
Department of Orthopedics/Rheumatology/Hand Surgery, Nagoya University Graduate School of Medicine, Nagoya, Aichi, Japan
e-mail: takegami@med.nagoya-u.ac.jp

Y. Nakashima (ed.), *Advances in Total Hip Arthroplasty*, https://doi.org/10.1007/978-981-97-2461-1_65

Table 65.1 MCID of SF-36 subscales 6 months after THA surgery

	MCID (standard deviation)
Physical functioning	20.4 (20.1)
Role-physical	10.8 (31.0)
Role-emotional	−6.5 (60.5)
General health	0.4 (18.8)
Social functioning	8.6 (26.0)
Bodily pain	14.7 (26.5)
Vitality	10.1 (23.5)
Mental health	9.0 (22.4)

MCID minimal clinically important difference, *SF-36* MOS short-form 36-item Health Survey, *THA* total hip arthroplasty

2 EQ-5D

The EQ-5D is a standardized instrument developed by the EurolQol group for use as a measure of health outcomes. It is applicable to a variety of illnesses and treatments, and provides a simple descriptive profile and a single index value for health status. The five dimensions of the EQ-5D are mobility, self-care, usual activities, pain/discomfort, and anxiety/depression. Each dimension has three levels (no problems, some problems, major problems) and together defines 243 health states (3 to the power of 5 gives the 243 possible combination), to which have been added "unconscious" and "dead" leading to a total of 245 health states. QOL values are obtained using a tariff [10].

The most important feature of the EQ-5D is that it can provide quality of life values to calculate quality-adjusted life years, which are used in the economic evaluation of health technology [11].

EQ-5D calculates QOL values by answering only five questions, but does not calculate sub-items, like the SF-36.

The use of the EQ-5D is cost-free for research purposes, but registration and permission with the EuroEQoL group are needed [12].

The EQ-5D has also been used to evaluate THA, and its MCID has been reported to range from 0.22 to 0.41 [13].

3 Oxford Hip Score

The OHS is a PRO that evaluates the efficacy of surgical treatment for hip arthroplasty, including THA, and consists of 12 questions on a 5-point Likert scale that assesses pain (6 questions) and function (6 questions). The OHS was developed by Dawson et al. in 1996 with a total score ranging from 12 to 60, and a lower score indicates a better outcome [14]; however, in 2007, it was revised by Murray with a total score ranging from 0 to 48, with a higher score indicating a better outcome [15]. The validity of the Japanese version of the OHS was also demonstrated by Uesugi et al. [16, 17]. As of 2021, registration and application [18] to the OHS organization are required for new users of OHS, but the use of OHS for academic research purposes is cost-free.

The OHS has a good response rate compared to common scales, such as the SF-36, EQ-5D, and Western Ontario and McMaster Universities Osteoarthritis Index (WOMAC) [19]. Because of its ease of response, this PRO is also used in the National Joint Registry in the UK and the Prosthetic Joint Registry in the Netherlands and has been used internationally in many cases [20]. However, it should be noted that OHS does not support Japanese-style daily activities (e.g., use of Japanese-style toilets), and using it to evaluate mental aspects is difficult [18].

For the MCID of OHS, the 2015 study by Beard et al. gave a score of 5 [21].

4 Hip Disability and Osteoarthritis Outcome Score

HOOS is a PRO developed in Denmark by Roos et al. based on the Western Ontario and McMaster Universities Osteoarthritis Index (WOMAC) and the opinions of patients with hip disease [22, 23].

The HOOS consists of 40 questions in 5 domains: pain (10 questions), other symptoms (5 questions), activities of daily living (17 questions), sports and recreational activities (4 questions), and hip-related quality of life (4 questions). Each question is answered on a 5-point Likert scale from 0 to 4, with scores ranging from 0 to 100, where a higher score indicates a better condition. Three variations of the HOOS include the HOOS-JR, which measures pain and daily living [24], the HOOS-PS, which measures physical function assessment [25], and the HOOS-12, which is a shortened version of the HOOS [26]. The HOOS has been translated into 14 languages, including Japanese, and is widely used internationally, including in the Norwegian registry, after reliability and validity evaluations [27]. As of 2021, the paper version of the Japanese version of HOOS could be downloaded as a PDF file from the Web [28].

HOOS is unique in that it can be used not only for THA but also to assess the outcomes of osteotomies for dysplastic hip arthroplasty in young patients [26–28] and hip arthroscopy [29] and evaluate postoperative pain management [20]. However, it is time-consuming to answer and tabulate the results of the 40 questions.

The MCID in THA of the HOOS was 15.1 for pain, 10.5 for symptoms, 9.6 for functional ADLs, 15.5 for participation in sports and recreation, and 16.2 for quality of life [30].

5 Forgotten Joint Score

FJS is a questionnaire developed in 2012 for individuals with artificial joints with the ultimate goal being "to be able to carry out daily life forgetting that you are after arthroplasty" [31]. The FJS consists of 12 questions, each of which asks whether the patient can forget about the joints in 12 different situations of daily life, using a 5-point scale. The FJS consists of 12 questions; the total score is divided by the number of items answered, and the average score is multiplied by 25 to obtain a total score of 0–100. One hundred points are subtracted from this score, and the higher the score, the greater the patient's ability to forget their artificial joints.

The FJS is an index that has demonstrated reliability and validity postoperatively in patients who have undergone THA [32]. Its conversion to Japanese and validity have been assured [33, 34].

The FJS assesses patient awareness and perception of prostheses. In other words, the FJS is a PRO with different aspects from the JHEQ, WOMAC, OHS, and HOOS, which ask about activities of daily living and function. Additionally, the FJS has only been internally and externally validated in patients who undergo THA. Therefore, it should be used with caution in other hip surgeries (osteotomies, arthroscopy, etc.). The MCID of the FJS in THA has been reported to be 17.5 [34].

6 Patient Joint Perception (PJP)

PJP is an assessment method developed in 2012 that asks patients who underwent THA whether their operated joint feels like a 'native or natural joint' [35]. In the PJP, patients are asked to respond to the question "How do you feel about your operated hip joint?" by choosing the following five responses: "natural joint," "no limitations but feels artificial," "little limitations and feels artificial," "great limitations and feels artificial," or "not a functioning joint".

It can be easily used and assessed in a busy outpatient setting because it is possible to obtain a patient's perception using a single simple question. In addition, Puliero et al. showed that the five questions in the PJP were correlated with WOMAC scores and FJS values, indicating the possibility of external validity [36].

Interestingly, in this study, approximately 30% of patients with perfect scores for the FJS, which asks about forgetting the artificial joint, and WOMAC, which asks about pain and function of the hip joint in general, answered "not a natural joint," while approximately 20% of patients who answered "the operated joint is my own or a natural joint" did not have perfect scores on the FJS or WOMAC [36]. In other words, it can be said that there is a difference between forgetting that a joint that has undergone THA is an artificial joint in daily life, having no pain and having no functional impairment, and feeling that a joint that has undergone THA is one's own joint.

Another limitation of this questionnaire is that it does not allow us to know what the patient is dissatisfied with, why he/she feels dysfunctional, or why he/she is not satisfied because the questions are simple.

PROs allow for a more multifaceted and patient-centered evaluation of THA. It is hoped that postoperative THA evaluations will include both physician evaluations and PROs.

References

1. Siljander MP, et al. Current trends in patient-reported outcome measures in total joint arthroplasty: a study of 4 major orthopedic journals. J Arthroplast. 2018;33:3416–21.
2. Paulsen A, et al. Minimal clinically important improvement (MCII) and patient-acceptable symptom state (PASS) in total hip arthroplasty (THA) patients 1 year postoperatively. Acta Orthop. 2014;85:39–48.
3. Kvien TK, et al. Minimal clinically important improvement/difference (MCII/MCID) and patient acceptable symptom state (PASS): what do these concepts mean? Ann Rheum Dis. 2007;66(Suppl 3):iii40–1.
4. Fukuhara S, et al. Translation, adaptation, and validation of the SF-36 Health Survey for use in Japan. J Clin Epidemiol. 1998;51(11):1037–44.
5. Fukuhara S, et al. Psychometric and clinical tests of validity of the Japanese SF-36 Health Survey. J Clin Epidemiol. 1998;51:1045–53.
6. Suzukamo Y, et al. Validation testing of a three-component model of Short Form-36 scores. J Clin Epidemiol. 2011;64:301–8.
7. Ware J Jr, et al. A 12-Item Short-Form Health Survey: construction of scales and preliminary tests of reliability and validity. Med Care. 1996;34:220–33.
8. Fukuhara S, et al. Health-related quality of life scales – SF-8 and SF-36. J Clin Exp Med. 2005;213:133–6. Japanese
9. Conner-Spady BL, et al. Comparing the validity and responsiveness of the EQ-5D-5 L to the Oxford hip and knee scores and SF-12 in osteoarthritis patients 1 year following total joint replacement. Qual Life Res. 2018;27:1311–22.
10. Tsuchiya A, et al. Estimating an EQ-5D population value set: the case of Japan. Health Econ. 2002;11:341–53.
11. Brooks R. EuroQol: the current state of play. Health Policy. 1996;37:53–72.
12. EQ-5D Registration Form. https://euroqol.org/eq-5d-registration-form/
13. Quintana JM, et al. Predictors of health-related quality-of-life change after total hip arthroplasty. Clin Orthop Relat Res. 2009;467:2886–94.
14. Dawson J, et al. Questionnaire on the perceptions of patients about total hip replacement. J Bone Joint Surg Br. 1996;78:185–90.
15. Murray DW, et al. The use of the Oxford hip and knee scores. J Bone Joint Surg Br. 2007;89(8):1010–4.
16. Uesugi Y, et al. Validity and responsiveness of the Oxford hip score in a prospective study with Japanese total hip arthroplasty patients. J Orthop Sci. 2009;14:35–9.
17. Uesugi Y, et al. Quality of life of total hip arthroplasty patients. J Jpn Soc Nurs Res. 2006;29:81–7. Japanese
18. The Oxford Hip Score (OHS). https://innovation.ox.ac.uk/outcome-measures/oxford-hip-score-ohs/

19. Weick JW, et al. Measures of hip function and symptoms. Arthritis Care Res. 2020;72(Suppl 10):200–18.
20. Rolfson O, et al. Patient-reported outcome measures in arthroplasty registries. Acta Orthop. 2016;87(Suppl 1):3–8.
21. Beard DJ, et al. Meaningful changes for Oxford hip and knee scores after joint replacement surgery. J Clin Epidemiol. 2015;68:73–9.
22. Klässbo M, et al. Hip disability and osteoarthritis outcome score. An extension of the Western Ontario and McMaster universities osteoarthritis an extension of the Western Ontario and McMaster universities osteoarthritis index. Scand J Rheumatol. 2003;32:46–51.
23. Nilsdotter AK, et al. Hip disability and osteoarthritis outcome score (HOOS)—validity and responsiveness in total hip replacement. Responsiveness in total hip replacement. BMC Musculoskelet Disord. 2003;4:10.
24. Davis AM, et al. Comparative, validity and responsiveness of the HOOS-PS and KOOS-PS to the WOMAC physical function subscale in total joint replacement for osteoarthritis. Osteoarthr Cartil. 2009;17:843–7.
25. Lyman S, et al. Validation of the HOOS, JR: a Short-form Hip Replacement Survey. Clin Orthop Relat Res. 2016;474:1472–82.
26. Gandek B, et al. A 12-item short form of the Hip Disability and Osteoarthritis Outcome Score (HOOS-12): tests of reliability, validity and responsiveness. Osteoarthr Cartil. 2019;27:754–61.
27. Satoh M, et al. Cross-cultural adaptation and validation reliability, validity of the Japanese version of the hip disability and osteoarthritis outcome score (HOOS) in patients with hip osteoarthritis. Osteoarthr Cartil. 2013;21:570–3; Larsen JB, et al. 14-year hip survivorship after periacetabular osteotomy: a follow-up study on 1,385 hips. Acta Orthop 2020; 91: 299–305
28. HOOS (Hip Disability and Osteoarthritis Outcome Score) Japanese version. http://plaza.umin.ac.jp/~msatoh-ken/hoos.html
29. Grammatopoulos G, et al. Does severity of acetabular dysplasia influence clinical outcomes after periacetabular osteotomy?–a Case-Control Study. J Arthroplast. 2018;33(7S):S66–70.
30. McClincy MP, et al. Periacetabular osteotomy improves pain and function in patients with lateral center-edge angle between 18° and 25°, but are these hips truly borderline dysplastic? Clin Orthop Relat Res. 2019;477:1145–53.
31. Flores SE, et al. When do patients improve after hip arthroscopy for femoroacetabular impingement? A prospective cohort analysis. Am J Sports Med. 2018;46:3111–8.
32. Padilla JA, et al. A novel opioid-sparing pain management protocol following total hip arthroplasty: effects on opioid consumption, pain severity, and patient-reported outcomes. J Arthroplast. 2019;34:2669–75.
33. Ornetti P, et al. Cross-cultural adaptation and validation of the French version of the Hip Disability and Osteoarthritis Outcome Score (HOOS) in hip osteoarthritis patients. Osteoarthr Cartil. 2010;18:522–9.
34. Behrend H, et al. The "forgotten joint" as the ultimate goal in joint arthroplasty: validation of a new patient-reported outcome measure. J Arthroplast. 2012;27:430–436.e1.
35. Furuya H, et al. Reproducibility and validity of the Japanese version of the forgotten joint score in postarthroplasty patients. Jpn J Phys Ther. 2019;53:742–50. Japanese
36. Matsumoto M, et al. Validation study of the Forgotten Joint Score-12 as a universal patient-reported outcome measure. Eur J Orthop Surg Traumatol. 2015;25:1141–5.

Postoperative Sports Activities

66

Satoshi Hamai

1 Sports After Total Hip Arthroplasty

Total hip arthroplasty (THA) is an effective treatment that improves the quality of life (QOL) of patients. In recent years, interest in sports participation after THA has increased owing to the younger age at surgery and changing lifestyles. Postoperative sports participation is discussed on the premise that implants are placed in the correct alignment and that there is good soft tissue tension and muscle strength recovery. Sports are expected to improve physical function, such as muscle strength, agility, and balance; control weight and blood pressure; prevent lifestyle-related diseases and dementia; and have a positive psychological impact. In addition, moderate mechanical stimulation through sports may improve bone quality. Various regular sports activities are important for preventing the decline of physical and mental functions, enjoying hobbies, and leading vigorous lives.

Problems associated with sports after THA include polyethylene wear, breakage, loosening, dislocation, and periprosthetic fractures. From a survey of Hip & Knee Society members in the United States, squash, jogging, football, basketball, soccer, and baseball were listed as sports that were not recommended after THA [1]. Sports that are permitted for experienced players include skiing, weightlifting, and skating. Sports, such as golf, swimming, walking, aerobics, and ellipticals, which have low load intensity and can be done at one's own pace without competition, are considered desirable. There are still few reports on the impact of postoperative sports participation on mid- to long-term implant outcomes. However, recent reports of favorable patient-oriented evaluation of THA and good mid- to long-term implant survival rates with revision as the endpoint, tend to ease restrictions on postoperative sports participation.

S. Hamai (✉)
Department of Medical-Engineering Collaboration for Healthy Longevity, Kyushu University, Fukuoka, Japan
e-mail: hamai.satoshi.075@m.kyushu-u.ac.jp

2 Sports Participation Rate

The sports participation rate before and after THA in Japanese patients with hip osteoarthritis doubled from 16% preoperatively to 31% postoperatively, in most low-impact sports, such as walking, swimming, aerobics, bowling, and golf, while patients also participated in high-impact sports, such as jogging, baseball, and tennis [2]. The factors that prevented patients from returning to sports were often not the hip symptoms themselves, such as pain, limited range of motion (RoM), and muscle weakness, but included patients' own sense of anxiety, lack of physician recommendation, other joint diseases, and medical comorbidities. Patients who are more active preoperatively tend to be more active postoperatively, but postoperative activity itself correlated with patient-oriented hip function scores and satisfaction. It has been reported that it is advisable to actively encourage moderate physical activity from the perspective of increasing patient self-evaluation and satisfaction.

3 Highly Cross-Linked Polyethylene

The development and use of highly crosslinked polyethylene (XLPE) in the 1990s led to many reports on the excellent wear resistance of XLPE liners. In discussing the extent to which sports should be permitted after THA, it is necessary to report the effects of the type and frequency of sports on XLPE liner wear and loosening.

Jogging is classified as a high-impact sport because the peak contact force on the hip is 5.5 times the body weight and 1.6 times the force experienced during walking. The participation rate of jogging in Japan ranges from 5.4% before THA to 3.8% after THA, and the rate was reported to be 7.5 km/h for 29 minutes at 3.6 km, four times a week after surgery [3]. The reasons for not jogging despite an interest in jogging were anxiety in 61% of the patients; pain, limited RoM, and muscle weakness in 24% of patients; and low back and knee joint pain in 15% of patients. The results showed no

Y. Nakashima (ed.), *Advances in Total Hip Arthroplasty*, https://doi.org/10.1007/978-981-97-2461-1_66

obvious implant laxity, migration, XLPE liner wear, or increased blood metal ion concentrations in resurfaced metal-on-metal THA after 5 years of postoperative follow-up. In a multivariate analysis, the authors stated that the factors associated with postoperative jogging were male sex and preoperative jogging cases and that patient factors, such as willingness to resume jogging were more influential than implant type.

A study investigated whether postoperative sports participation affects XLPE wear in primary cementless THA performed with a first-generation XLPE liner, using a questionnaire survey and wear measurements [4]. The sports participation rate was 39% postoperatively, and low impact sports, such as walking, swimming, aerobics, bowling, and golf were played by 75% of patients, while high-impact sports, such as jogging and baseball were played by 13% of patients. The steady-state wear rate was 0.005 mm/year during the 13-year postoperative follow-up period, indicating extremely good wear resistance (Fig. 66.1). There was no difference of postoperative sports participation in steady-state wear, and the multivariate analysis showed no effect of sports participation or other factors on steady-wear rate. No difference in steady-state wear rate was found between the low- and intermediate-impact groups and the high-impact group in the sub-analysis. There were no cases of revision (polyethylene wear, breakage, loosening, dislocation, or peri-prosthetic fracture) due to sports participation among THA patients during the study period. This study was mainly performed in middle-aged patients and is useful in advising patients on sports participation after THA. Further evidence of the effects of THA on long-term endurance and appropriate mechanical stimulation is warranted.

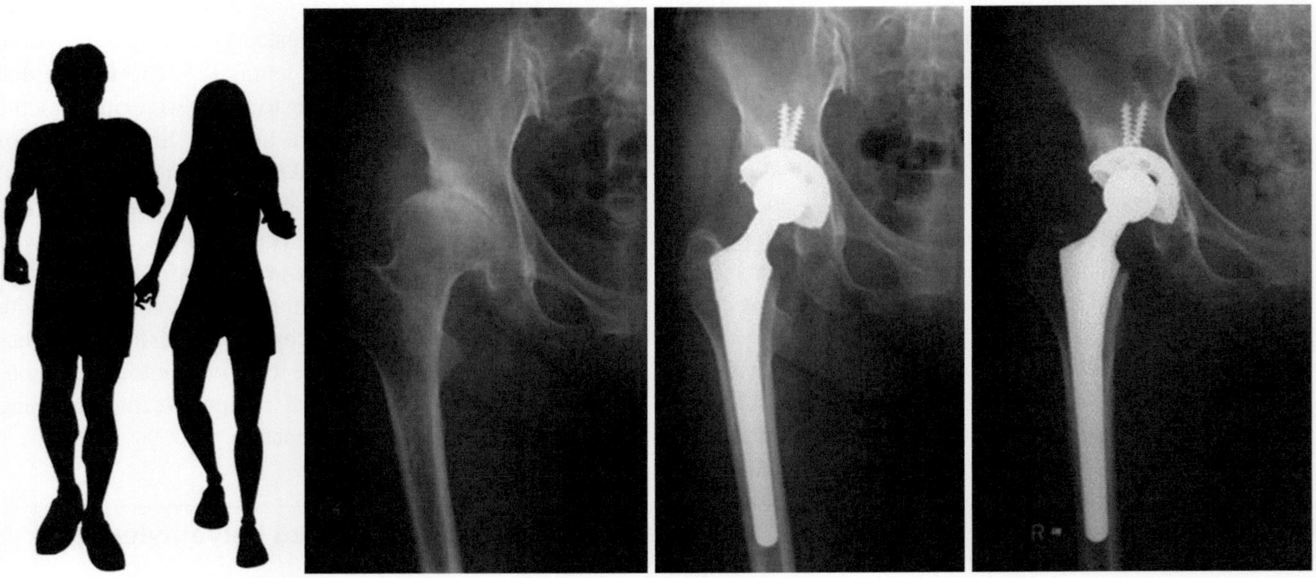

Fig. 66.1 Jogging after total hip arthroplasty for osteoarthritis of the hip (age at surgery: 54 years, female) (left: before surgery, center: immediately after surgery) (5 times a week, 30 min per session, for 12 years). No obvious crosslinked polyethylene wear or implant loosening was observed at the final follow-up (right panel)

4 Sports Motion Analysis

There have been few reports on the analysis of implant behavior during sports. Clarification of the dynamics of different types of sports is expected to lead to feedback on sports coaching and surgical procedures. By superimposing digitally reconstructed radiographs (DRRs) from CT voxel values or three-dimensional (3D) computer-aided design (CAD) models on X-ray images, the 3D dynamics of physiological and artificial joints can be analyzed. It has been reported that it is possible to precisely measure the coordinated motion of the pelvis and femur (Fig. 66.2), and the relative relationship of the stem to the cup and liner during various movements, including sports [5–7].

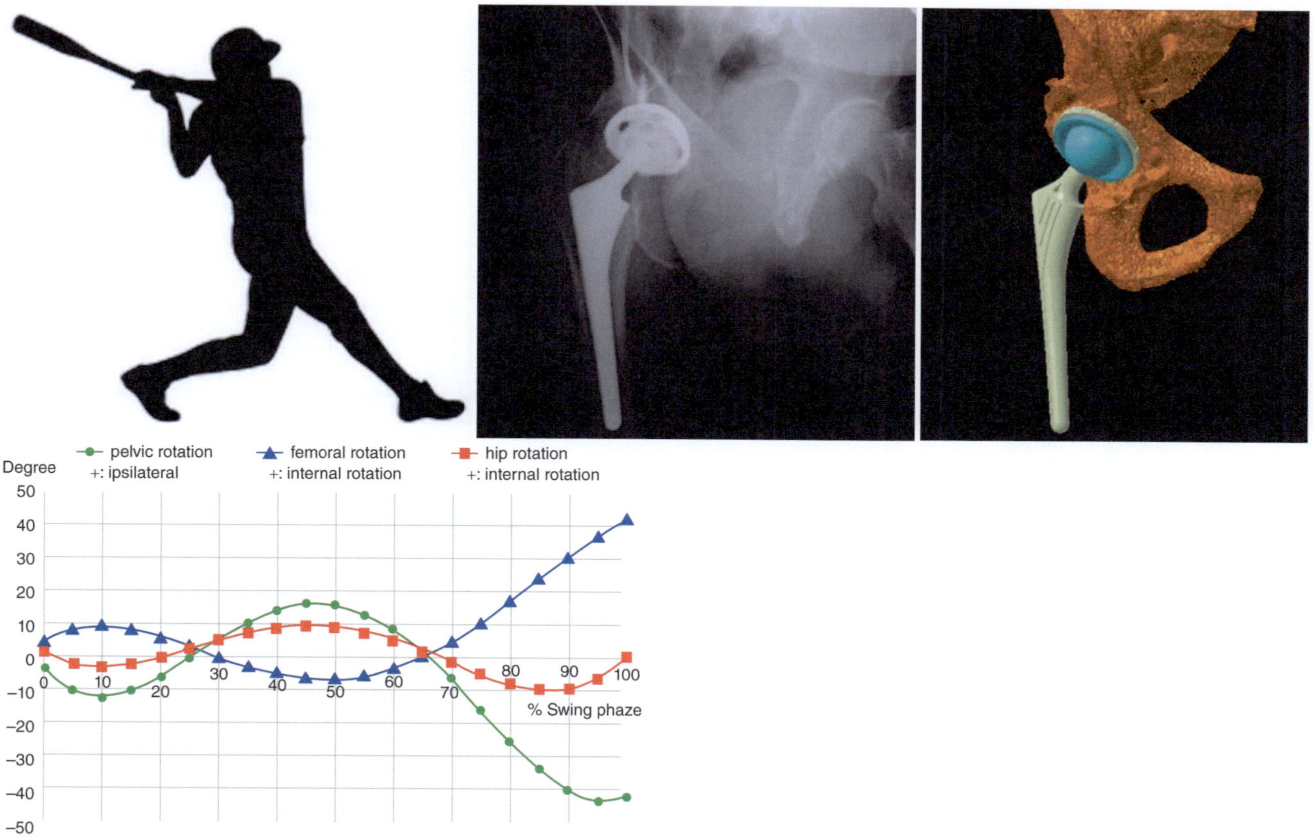

Fig. 66.2 Analysis of batting dynamics after total hip arthroplasty. In the present case, the pelvis and femur were rotated in a coordinated manner; the rotation of the hip joint was approximately 10°, and no contact between the neck and liner was observed

5 Golf Swing Dynamics

3D motion analysis during golf swings after THA have been reported [5]. The implant placement angles were 40° cup abduction, 17° anteversion, 33° stem anteversion, and 49° combined anteversion (CA); the liner was flat in one case and elevated in 10 cases. The rotation from the address to the top of the backswing (right: 22° internal rotation, left: 27° external rotation) (Fig. 66.3), and from the top to the end (right: 23° external rotation; left: 23° internal rotation), the total rotation was approximately 50°. Impingement was observed between the neck and elevated liner during maximal external rotation in four hips (two right and two left). The impingement group had a greater maximum external rotation (impingement group: 38°, no impingement group: 21°) than did the no impingement group; however, no significant differences was found in the change in rotation (impingement group: 43°, no impingement group: 53°), flexion angle (impingement group: −12°, no impingement group: −1°), and abduction angle at maximum external rotation (impingement group: 6°, no impingement group: 5°). There was no difference in the angle of abduction (impingement group: 6°; no impingement group: 5°). The cup anteversion (impingement group: 27°, no impingement group: 12°) was larger, but there were no differences in CA (impingement group: 45°, no impingement group: 51°) or cup-head translation (impingement group: 1 mm, no impingement group: 2 mm). Impingement and dislocation risks associated with rotation were assumed in the golf swing after THA. In the present report, the authors stated that no excessive rotation or instability was observed during swinging compared to the normal hip, and it was permitted as a postoperative sports activity. However, contact between the elevated liner and neck at maximum external rotation was observed in four out of 11 cases, and it was reported that care should be taken with cup anteversion and the use of the elevated liner.

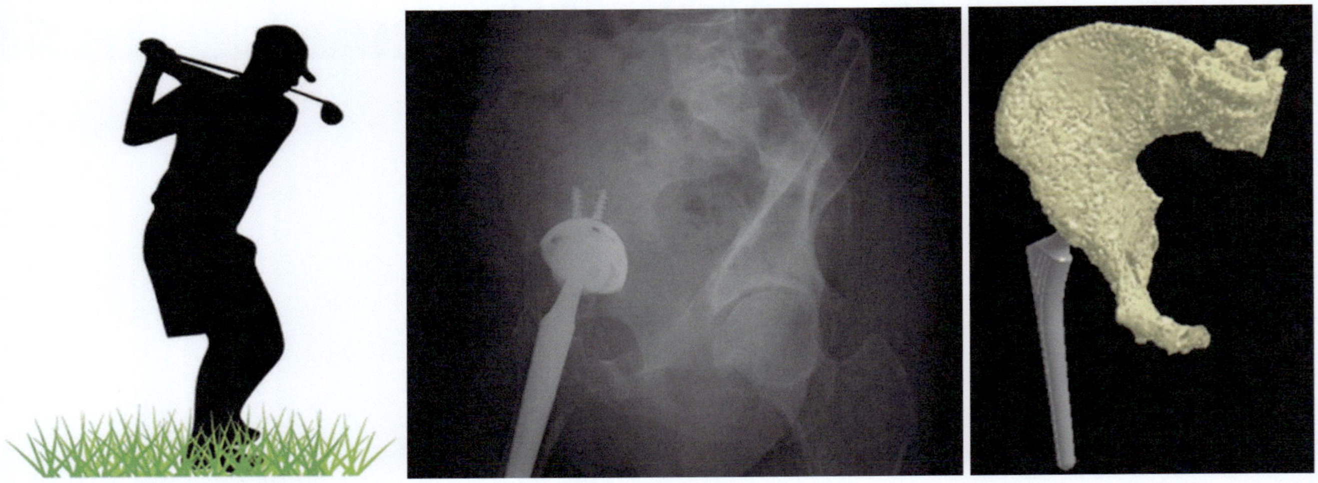

Fig. 66.3 Analysis of golf swing dynamics after total hip arthroplasty

6 Aerobike Dynamics

Dynamic analysis during aerobike exercise, which is a frequently performed postoperative sport in Japan following walking, swimming, and gymnastics [2, 4], has been reported (Fig. 66.4) [6]. Cup abduction, cup anteversion, stem anteversion, and CA were 40°, 22°, 36°, and 58°, respectively. The pelvis was almost fixed at a 27° posterior tilt, and hip flexion varied between 19° and 59°, showing no excessive RoM. Hip abduction and rotation were almost unchanged at 9° abduction and 4° external rotation, and no dislocated limb was observed. Aerobiking is considered a sport that can be safely performed after THA because of the low load acting on the joints. The closest distance between the liner and neck is 14 mm at maximum flexion in the anterior direction and 9 mm at maximum extension in the posterior direction, indicating significant clearance in the anterior direction. Furthermore, in a simulation analysis of the safe RoM for cup placement, it was reported that even in aerobike motion without excessive RoM, there was a risk of contact between the posterior implants in the most extended position when using the elevated liner, and that the cup placement condition to avoid contact was an anterior opening angle of 30° or less.

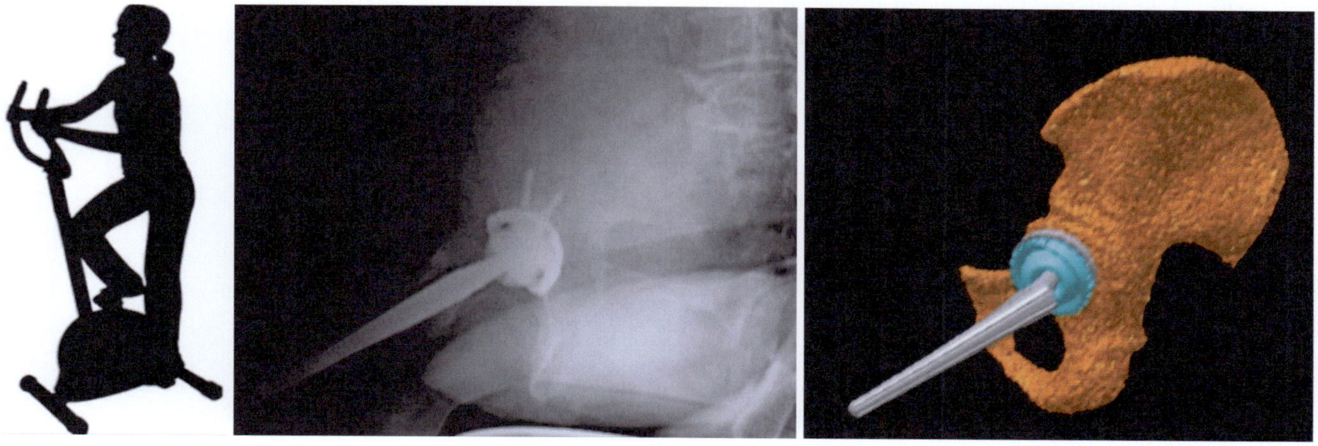

Fig. 66.4 Aerobike kinematics analysis after total hip arthroplasty

7 Dance Dynamics

Ninety-four percent of hip surgeons recommend dancing (ballroom, jazz, and square dancing) as a sport after THA [1]. Dancing is effective in improving balance by shifting the center of gravity to the front, back, left, and right, in emphasizing plantar sensation, as well as in maintaining and strengthening the muscles of the trunk and hip joint through the half-squat posture. The hula (Kaholo and Kao) and classical ballet (plié and développé from the basic first position), can be performed as hobbies after THA; moreover, 3D motion analysis during ballet (plié and développé from the basic position) have been reported in a case report (Figs. 66.5 and 66.6) [7]. In both cases, no sudden change in hip angle, excessive RoM, or contact between the liner and the neck was observed. Hip flexion, abduction, and external rotation angles of the plié were 34°, 29°, and 43°, respectively. In the développé, the femur was flexed 75° to the ground, with a simultaneous 28° pelvic retroversion and 48° hip flexion. The report noted that further case analysis would provide deeper insights into the dynamics of the hulas and ballet, as there are many different step poses in hula and ballet, and the dynamics vary widely in individual's flexibility and skill. In the motion analysis of a jazz dance instructor using a motion capture system, the maximum flexion angles were 100° for squatting and 119° for long-sitting flexion of the lower limbs. Reports have also shown that real-time data confirmation is useful in objectively evaluating functional recovery, confirming whether the limb position is safe, and reliving anxiety [8].

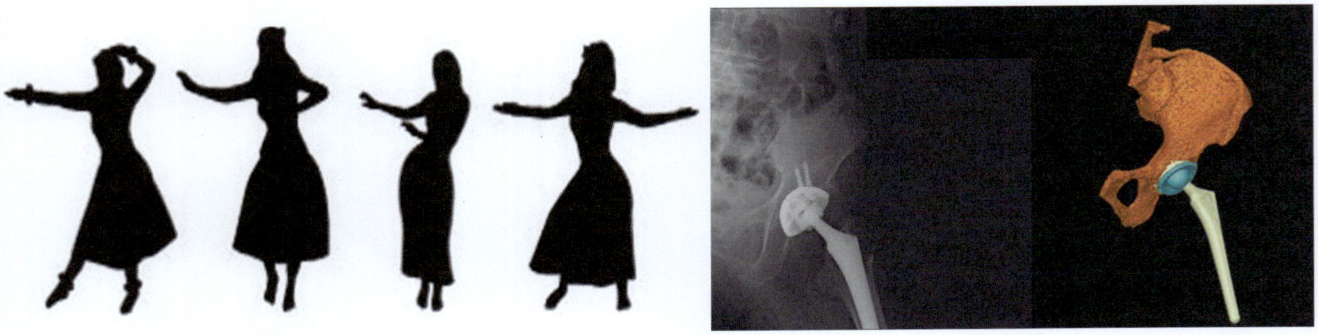

Fig. 66.5 Hula dance (chao) dynamics analysis after total hip arthroplasty. (Adapted from Komiyama [7])

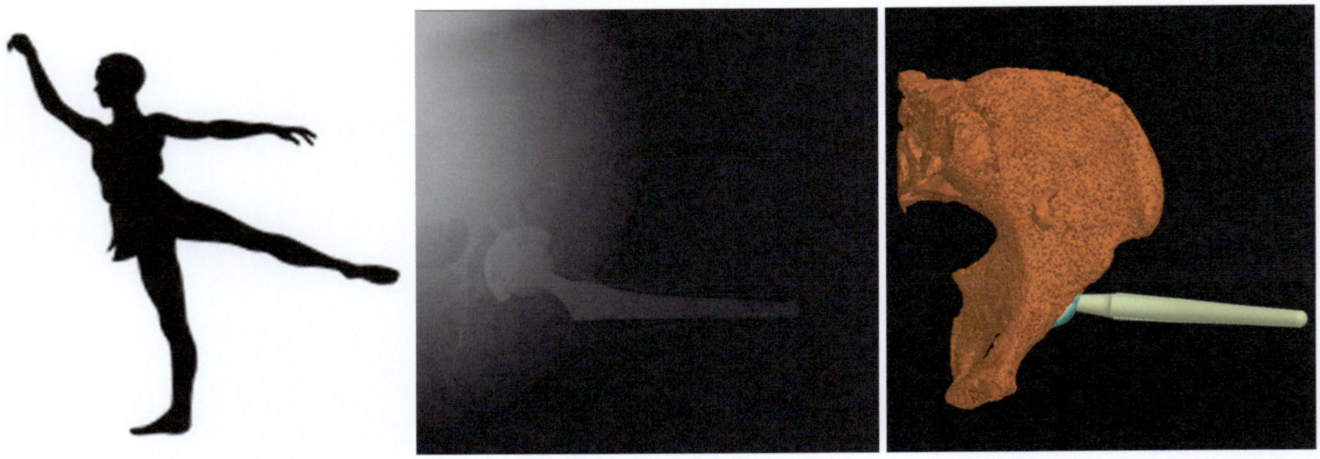

Fig. 66.6 Analysis of ballet (développé) dynamics after total hip arthroplasty. (Adapted from Komiyama [7])

References

1. Klein GR, et al. Return to athletic activity after total hip arthroplasty. Consensus guidelines based on a survey of the Hip Society and American Association of Hip and Knee Surgeons. J Arthroplast. 2007;22:171–5.
2. Hara D, et al. Sports participation in patients after total hip arthroplasty vs periacetabular osteotomy: a propensity score-matched Asian cohort study. J Arthroplast. 2018;33:423–30.
3. Abe H, et al. Jogging after total hip arthroplasty. Am J Sports Med. 2014;42:131–7.
4. Harada S, et al. Wear analysis of the first-generation cross-linked polyethylene at minimum 10 years follow-up after THA: no significant effect of sports participation. J Artif Organs. 2022;25:140–7.
5. Hara D, et al. Dynamic hip kinematics during the golf swing after total hip arthroplasty. Am J Sports Med. 2016;44:1801–9.
6. Komiyama K, et al. In vivo kinematic analysis of replaced hip during stationary cycling and computer simulation of optimal cup positioning against prosthetic impingement. Clin Biomech. 2019;68:175–81.
7. Komiyama K, et al. Dynamic hip kinematics during recreational classical ballet and hula dance after total hip arthroplasty: two case reports. J Med Case Rep. 2019;13:11.
8. Mizoguchi Y, et al. Safer return to jazz dance instruction after simultaneous bilateral total hip arthroplasty. BMJ Case Rep. 2021;14:e243214.

Keys to Improving Patient Satisfaction

67

Keiichiro Ueshima

THA is an effective treatment for advanced and end-stage hip arthroplasty that can improve hip function and quality of life (QOL) and is also considered an economically effective medical treatment [1, 2]. In recent years, improvements in the materials used for sliding surfaces of implants and advances in surgical techniques have led to excellent long-term results, and THA has been increasingly indicated for use in younger patients. In addition, the elderly are becoming increasingly aware of the need to extend their healthy life expectancy and their desire to participate in social activities, such as work and hobbies. Even after THA surgery, there has been a demand for increased physical activity, including recreational and sports activities among patients, with the expectation that this will improve QOL and mental health. To improve patient satisfaction after THA surgery, it is important for physicians to understand the pathophysiology of hip arthroplasty before surgery, as well as the expectations and wishes of patients and how to meet these by providing appropriate treatment.

1 Patient Satisfaction

Donabedian reported that patient satisfaction is an important evaluation of the quality of medical care from the patient's point of view [3]. In other words, improving patient satisfaction is linked to improving quality of medical care. It has been proposed that the quality of medical care should be evaluated from three aspects: structure, process, and outcome [3]. "Structure" is a factor that constitutes the conditions under which medical care is provided, such as facilities, equipment, expertise, and qualifications of medical staff. "Process" includes the medical care itself (e.g., diagnosis, treatment, nursing, surgery, rehabilitation, hospitality, and participation of patients and their families in the medical care) and shows how the medical care was provided. "Outcome" is the effect of the medical care provided (health outcomes, presence or absence of complications, changes in QOL, and patient satisfaction) (Fig. 67.1). A good structure leads to a good process, and a good process leads to a good outcome. Improving patient satisfaction is an issue that physicians and all other healthcare providers involved with patients should always work on.

K. Ueshima (✉)
Kyoto Interdisciplinary Institute Hospital of Community Medicine, Kyoto, Japan
e-mail: ueshima@koto.kpu-m.ac.jp

Y. Nakashima (ed.), *Advances in Total Hip Arthroplasty*, https://doi.org/10.1007/978-981-97-2461-1_67

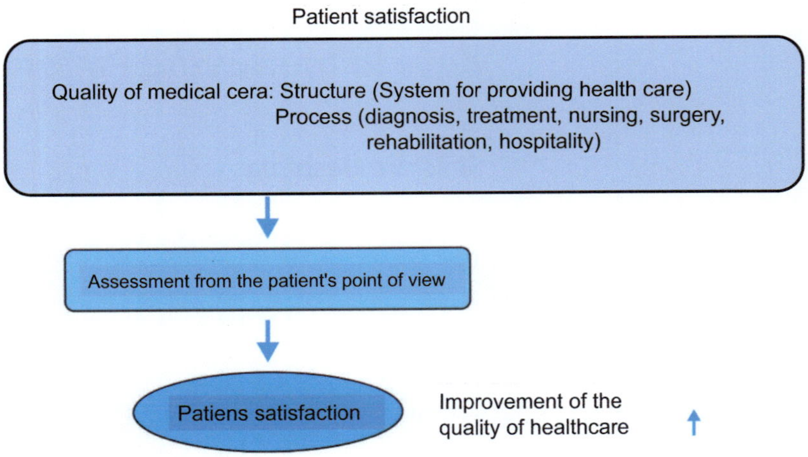

Fig. 67.1 Quality of medical care and patient satisfaction

2 THA and Patient Satisfaction

In recent years, THA has been established as an effective treatment modality for advanced and end-stage hip arthroplasty, and good long-term clinical outcomes have been reported [1, 2]. However, it has been reported that 7–15% of patients who underwent THA are dissatisfied with their postoperative course [4, 5]. It has been pointed out that residual pain and limitation of physical function lead to dissatisfaction after surgery even in the absence of surgical complications.

Many patients with hip disease expect THA to relieve pain and improve their daily activities. Specific expectations include improved gait, range of motion, recovery time, happiness, social interaction, leisure, sports and recreational participation, and improved QOL (Fig. 67.2). Although there have been few reports on how patients' preoperative status relates to their postoperative expectations, recent studies have investigated the impact of their preoperative functional status on these expectations. A US study reported that the mean percentages of preoperative expectations fulfilled at 1 year postoperatively were 59% for total hip replacement and 72–73% for THA [5, 6]. Tilbury et al. found that the percentage of preoperative expectations for sports and exercise fulfilled was higher for THA than for TKA, THA was associated with preoperative expectations, and the achievement of those expectations was higher with THA than with TKA [7]. The fulfillment of patients' preoperative expectations is also associated with improved postoperative health-related QOL.

Although there have been recent studies on the association between patients' assessments of preoperative expectations and self-reported outcomes, it is important to note that there are significant differences in expectations for THA across countries and practices.

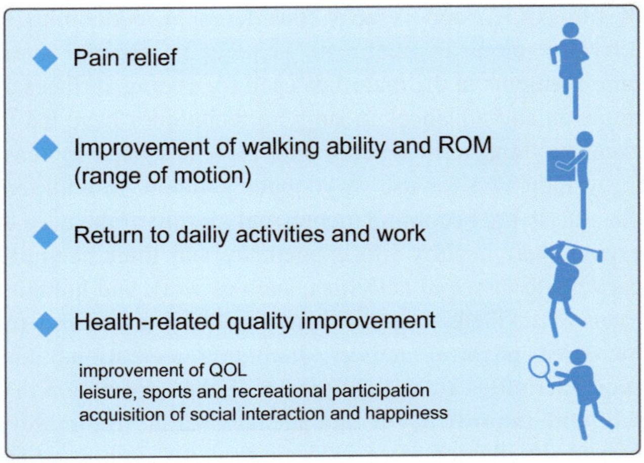

Fig. 67.2 Patient expectations for THA

3 Patient-Reported Outcome Measures and Patient Satisfaction

Patient-reported outcome measures (PROMs) are increasingly being used to assess the effectiveness of surgical procedures, in addition to the traditionally used physician-completed disease-specific measures. Representative PROMs include the Western Ontario and McMaster Universities Osteoarthritis Index (WOMAC) as a disease-specific measure, hip disability and osteoarthritis outcome score (HOOS), and Oxford hip score. The Japanese Orthopaedic Association Hip Disease Evaluation Questionnaire is also used in Japan. Comprehensive measures used to assess health status across different diseases include the SF-36 and SF-12. In addition to their clinical importance, PROMs and patient satisfaction have entered the field of healthcare payment restructuring in the United States.

A relationship between preoperative THA expectations and PROMs has also been reported, with higher preoperative THA expectations predicting better PROMs. Jain et al. reported that a younger age at THA surgery and poorer preoperative hip function predicted higher preoperative expectations. They also reported that higher preoperative expectations predicted better PROMs and higher levels of patient satisfaction and fulfillment of expectations [8]. The importance of positive patient encouragement during preoperative counseling was reported by Gandhi et al. who showed that patients expecting no postoperative pain had better postoperative pain and PROMs and expecting more pain relief predicted greater improvement in pain scores [9]. One of the reasons why high expectations before THA surgery have a positive impact on the outcome is that high expectations improve motivation for rehabilitation and lead to good outcomes, whereas low expectations make it more likely that rehabilitation will not proceed smoothly and the patient will not fully benefit from the surgery [8].

4 Toward Improving Patient Satisfaction

It is understood that physicians must make efforts to improve their diagnostic and therapeutic abilities and expertise in their daily practice. Efforts to improve communication skills with patients are also important. A better understanding of the patient's expectations before surgery facilitates a shared decision-making process between the patient and physician and provides the patient with specific information about which preoperative expectations are most likely to be met. Preoperative counseling is expected to motivate patients to set their expectations as high as possible, and a higher sense of accomplishment after surgery is expected to increase patient satisfaction (Fig. 67.3). However, if patients have unrealistic postoperative expectations, it is possible to modify them in advance through preoperative counseling.

At present, the influence of preoperative counseling on expectations and postoperative PROMs and satisfaction is not clear, and PROMs, such as WOMAC, HOOS, and SF-36, are considered to be an issue for future study. Validation of their effectiveness and relationship with patient satisfaction is expected. Recently, wearable devices, such as smartphones and smartwatches, have enabled more objective measurements of changes in physical function and activity. It has been reported that good results of functional recovery after THA surgery were obtained by self-checking using a wearable device [10]. Future clinical studies should clarify the relationship between the results obtained using wearable devices and PROMs and patient satisfaction. In addition, it is expected that the wearable device will enable the patients themselves to feel the improvement in function after THA surgery, which will lead to further improvement in their motivation for activity and patient satisfaction.

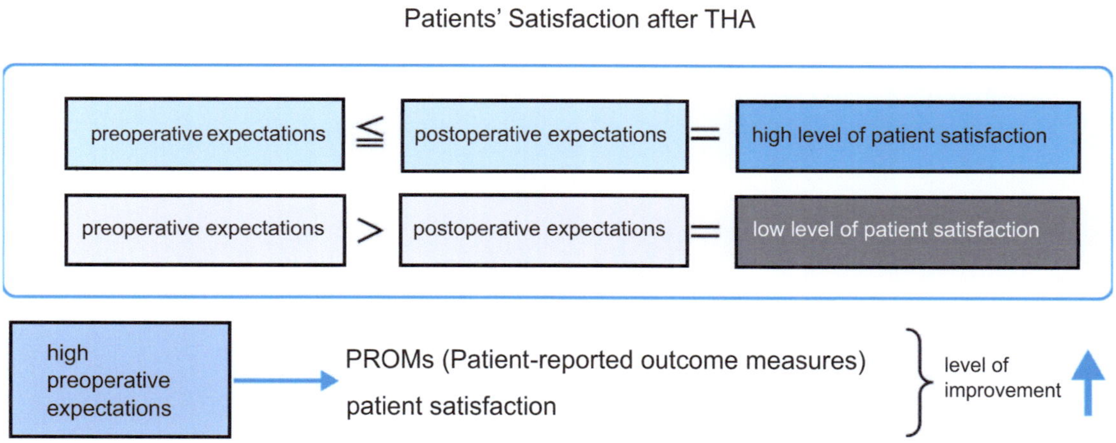

Fig. 67.3 Preoperative expectations and patient satisfaction

References

1. Learmonth ID, et al. The operation of the century: total hip replacement. Lancet. 2007;370:1508–19.
2. Kamaruzaman H, et al. Cost-effectiveness of surgical interventions for the management of osteoarthritis: a systematic review of the literature. BMC Musculoskelet Disord. 2017;18:183.
3. Donabedian A. Explorations in quality assessment and monitoring, vol. 1. The definition of quality and approaches to its assessment. Ann Arbor: Health Administration Press; 1980. p. 77–124.
4. Palazzo C, et al. Determinants of satisfaction 1 year after total hip arthroplasty: the role of expectations fulfilment. BMC Musculoskelet Disord. 2014;15:53.
5. Anakwe RE, et al. Predicting dissatisfaction after total hip arthroplasty: a study of 850 patients. J Arthroplast. 2011;26:209–13.
6. Scott CE, et al. Patient expectations of arthroplasty of the hip and knee. J Bone Joint Surg Br. 2012;94:974–81.
7. Tilbury C, et al. Unfulfilled expectations after total hip and knee arthroplasty surgery: there is a need for better preoperative patient information and education. J Arthroplast. 2016;31:2139–45.
8. Jain D, et al. Do patient expectations influence patient-reported outcomes and satisfaction in total hip arthroplasty? A prospective, multicenter study. J Arthroplast. 2017;32:3322–7.
9. Gandhi R, et al. Patient expectations predict greater pain relief with joint arthroplasty. J Arthroplast. 2009;24:716–21.
10. Crawford DA, et al. Early outcomes of primary total hip arthroplasty with use of a smartphone-based care platform: a prospective randomized controlled trial. Bone Joint J. 2021;103-B(7 Supple B):91–7.

Part XIII

Clinical Results

Young People

68

Yusuke Endo and Shigeru Mitani

Total hip arthroplasty (THA) provides excellent pain relief and functional recovery in advanced hip arthritis. On the other hand, if revision is required for any reason, the incidence of complications is observed to increase, unlike initial replacement. For this reason, some consider that patients should be kept on a conservative waiting list even if they have pain until after the age of 60 years. Hip arthroplasty has been perform for more than 50 years, and the results of hip arthroplasty are stable. The definition of young patients for THA is not clear; however, several overseas studies indicate that the age of THA patients is less than 30 years old. In this section, we discuss the clinical outcomes of THA in relatively young patients in Japan and in patients under 30 years of age in Europe and the United States.

1 Indication for Young People

Of the 2881 primary THAs performed at our hospital between April 2010 and April 2021, 9 cases and 12 hips (3 bilateral cases) were under 30 years old. In all cases, the patients were asked for their preference between joint preservation by osteotomy or THA, and THA was selected. The most common reason for choosing THA was the desire for early and reliable pain relief and the ability to return to society.

The mean age at surgery was 27 (19–30) years. The causative disease was osteoarthritis in 3 patients (1 with DDH treatment, 1 after pyogenic arthritis of the hip, and 1 postoperative pseudarthrosis of the femoral neck fracture on dialysis). There were 6 cases of osteonecrosis of the femoral head in 9 hips, steroid-related osteonecrosis in 4 cases and 7 hips (2 cases of SLE, 1 case of anti-SRP antibody-related necrotizing myopathy, and 1 case of traumatic sudden hearing loss), alcohol-related osteonecrosis in 1 case and idiopathic osteonecrosis in 1 case. Cemented stem was used in only one case after infectious arthritis, and a 32-mm diameter head was used in all the cases. The mean postoperative hospital stay was 15 days (8–20 days), except for one patient with hemodialysis (30 days), and there were no perioperative complications in all patients. The mean postoperative observation period was 5.2 (1–9) years, and good clinical results were obtained in all patients until the last survey.

2 Key Points of Surgery

In patients with no previous surgery and moderate deformity, an intermuscular approach is preferable. The preoperative spinal and pelvic mobility and dynamic posture (standing and sitting position) should be evaluated, and hypermobility should be considered, especially in young women. The angle of cup placement is particularly important to avoid implant impingement and liner wear; additionally, we use preoperative evaluation by computed software and intraoperative navigation in our institute. In young patients, the range of motion in daily life is high; therefore, we carefully perform a dislocation test during the trial, often use a high-offset stem as appropriate to obtain sufficient tension, and adjust the leg length as required. Basically, we used a combination of a cementless cup, a high cross-linked polyethylene flat liner,

Y. Endo (✉) · S. Mitani
Department of Orthopedic Surgery, Kawasaki Medical School, Kurashiki, Okayama, Japan

Y. Nakashima (ed.), *Advances in Total Hip Arthroplasty*, https://doi.org/10.1007/978-981-97-2461-1_68

and a 32 mm ceramic head with a cementless short taper wedged stem. Dual mobility is not currently used due to concerns regarding long-term results.

In patients with a history of previous surgery, we reviewed the previous operative record before surgery, confirmed the previous scar, and discussed the approaches and skin incision to avoid adding new wounds as much as possible. However, if the patient has undergone multiple surgeries and severe adhesion is recognized in subcutaneous tissue, there is a risk of delayed wound healing. Therefore, a new skin incision should be made along with the approach, and the wound should be enlarged, if necessary. In young patients with THA, it has been reported that the antetorsion of the femoral neck is often large; notably, excessive antetorsion may increase the risk of anterior dislocation, requiring caution especially in the anterior approach [1]. In cases where the antetorsion of the femoral neck is more than 40°or posteriorly distorted on the table top plane, an adjustment of the anteversion of the stem should be considered, taking the combined anteversion into account.

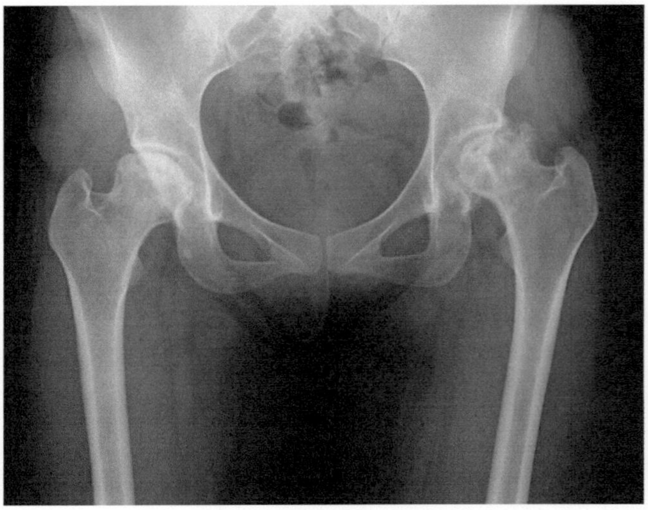

Fig. 68.1 A preoperative radiograph shows both avascular necrosis of femoral heads in a 25-year-old woman. She received steroid therapy for systematic lupus erythematosus

2.1 Cases

A 25-year-old woman was diagnosed with systematic lupus erythematosus and received steroid medication. She experienced hip pain 6 months after taking the medication and was followed up conservatively due to her family circumstances at the first visit. One and a half years later, the femoral head was observed to progressively collapse (Fig. 68.1). At 27 years of age, THA was performed on the left hip using a cementless cup and taper wedged stem with modified Watson-Jones approach, and THA of the right hip was performed in the same manner 1 year later (Fig. 68.2). Five years after surgery, there were no clinical hip symptoms; however, the stem of the right hip was distally fixed, and a clear zone was observed proximally (Fig. 68.3).

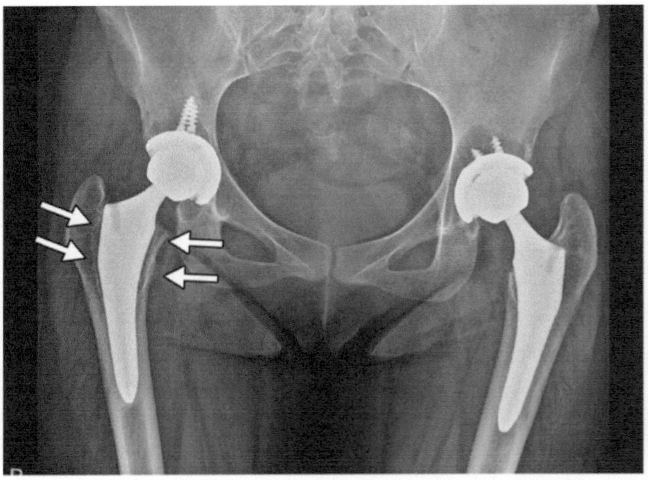

Fig. 68.3 A radiograph taken at the 7-year after THA. Left hip: 4 years postoperatively, Right hip: 5 years postoperatively

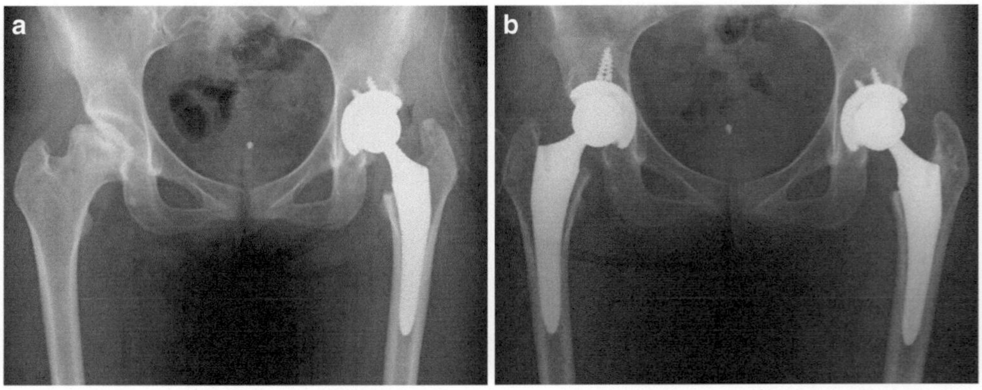

Fig. 68.2 A postoperative radiograph after cementless THA

3 Investigation of THA Results in Young Patients

Most reports in Japan indicate that THA in young patients was performed at the age of 40 years or younger. In cemented THA, the results at an average of 14 years (10–20 years) were reported in 55 hips (follow-up rate, 83.3%) in patients younger than 50 years at the time of surgery and in 273 hips (follow-up rate, 42.9%) in patients older than 50 years [2]. The postoperative 16-year cup survival rates were 68% for patients younger than 50 years and 88.7% for patients older than 50 years, and the stem survival rates were 96.4% for patients younger than 50 years and 94.1% for patients older than 50 years, showing comparable results for the stem side but inferior results for the cup side in patients younger than 50 years. In a recent report of cementless THA, XLPE polyethylene liners and zirconia heads were used in 19 hips under 40 years of age, and the annual wear rate was 0.015 mm with no cases of revision after more than 10 years of follow-up [3]. Regarding the postoperative activity after THA, the results were similar in patients under the age of 60 years, especially in men [4]. It is important to select a material with low wear resistance and to place the implant in younger patients who are likely to engage in sports activities.

In Europe and the United States, THA is frequently performed in younger patients because of fewer cases of joint preserving surgery when compared to Japan. Thus, there are several reports of THA in patients younger than 30 years of age. THA patients under 30 years of age have a relatively high incidence of idiopathic osteonecrosis, developmental dysplasia of the hip, juvenile rheumatoid arthritis (JRA), pyogenic hip arthritis, and trauma. A systematic review of THA in patients under 30 years of age reported no significant difference in the clinical outcomes before and after 1988, i.e., no improvement in THA outcomes in younger patients [5].

Overall, 975 cases of first THA in patients under 21–30 years of age from the Australian registry between 1999 and 2012 was investigated, and revision surgery are performed for an average of only 4.6 years after the initial replacement.

In another study, the risk of revision was investigated in 55 of 680 patients under 30 years of age who underwent THA revision was 77 hips [6]. The results have shown that the reasons for revision were aseptic loosening in 51%, liner wear in 24%, infection in 8%, frequent dislocation in 6%, implant breakage in 4%, and osteolysis in 7% of the patients. The statistical preoperative risk factors for patients who required revision included a history of pediatric hip disease, infection, inflammatory diseases, such as JRA, and postoperative risk factors, including dislocation within 3 months, intraoperative fracture, and postoperative nerve palsy [7].

Hip resurfacing arthroplasty (HRA) has been used to achieve both bone preservation in young patients and dislocation resistance by using a metal-on-metal bearing. A multicenter study of HRA in patients aged 50 years or younger reported a 22-year survival rate of 88.9% after an average follow-up of 7.6 years. Alternatively, poor results in certain models and an inferior survival rate of 78.3% at 20 years in DDH cases have been reported [8].

While a report has shown that the pregnancy and delivery rate is lower in women after THA surgery [9], another report found no difference in the WOMAC pain and function scores between patients who had experienced pregnancy and delivery before and after THA surgery and those who had not [10]. There is also a report that a history of pregnancy and childbirth does not cause a significant difference in the subsequent implant survival rates [11]. Although these reports indicate that there are no issues with pregnancy and delivery in young women even after THA, there is a case report in which high concentrations of chromium and cobalt were detected in the blood of a newborn baby born to a woman who developed pseudotumors with bilateral metal on metal [12]. We require careful selection regarding the bearing material in young women of childbearing potential.

Notably, it is still debated whether cemented or cementless THA is the better choice. A cemented stem has the advantage that the cement mantle can be preserved and a cement-in-cement technique can be performed if it is not broken at the time of revision. Cementless THA has the problem that even a well-fixed implant has to be removed when the femoral head cannot be procured due to the manufacturer's circumstances when revision occurs after a long period of time.

The problem with THA in young patients is that early wear and breakage, instability, and dislocation are common due to high activity. Recently, a trend has been observed in Japan to not restrict sports activities after THA; thus, it is difficult to restrict activities in young patients. It is necessary to perform surgery aimed at muscle tendinous tissue and bone preservation as much as possible in the initial THA and to achieve both long-term stable implant fixation and dislocation resistance.

Regardless of which approach is used or which implant is selected, THA in young patients is still observed to have an unstable outcome, and it is advisable to have continuous long-term follow-up at the institute in which the surgeons who have detailed knowledge about revision surgery are employed.

References

1. Watanabe K, et al. An increase in the risk of excessive femoral anteversion for relatively younger age and types of femoral morphology in total. J Orthop Surg (Hong Kong). 2019;27:2309499019836816.
2. Kobayashi S, et al. Comparative study of total hip arthroplasty between younger and older patients. Clin Orthop Relat Res. 1997;339:140–51.
3. Sonohata M, et al. Wear of XLPE liner against zirconium heads in cementless total hip arthroplasty for patients under 40 years of age. Hip Int. 2017;27:532–6.
4. Takeuchi K, et al. Recovery of activity level following total hip arthroplasty in patients less than 60 years of age. Hip Int. 2020;31:632–6.
5. Adelani A, et al. Has total hip arthroplasty in patients 30 years or younger improved? A systematic review. Clin Orthop Relat Res. 2013;471:2595–601.
6. Sedrakyan A, et al. Survivorship of hip and knee implants in pediatric and young adult populations: analysis of registry and published data. J Bone Joint Surg Am. 2014;96(Suppl 1):73–8.
7. Girard J, et al. Risk factors for revision of hip arthroplasties in patients younger than 30 years. Clin Orthop Relat Res. 2011;469:1141–7.
8. Van Der Straeten C, et al. Hip resurfacing arthroplasty in young patients: international high-volume centers' report on the outcome of 11,382 metal-on-metal hip resurfacing arthroplasties in patients ≥50 years at surgery. Hip Int. 2022;32:353–62.
9. Artama M, et al. Lower birth rate in patients with total hip replacement A nationwide population-based study in Finland. Acta Orthop. 2006;87:492–6.
10. Lally L, et al. Pregnancy does not adversely affect postoperative pain and function in women with total hip arthroplasty. J Clin Rheumatol. 2005;21:323–5.
11. Kuitunen I, et al. No effect of delivery on total hip replacement survival: a nationwide register study in Finland. Acta Orthop. 2019;90:433–8.
12. Fritzsche J, et al. High chromium and cobalt levels in a pregnant patient with bilateral metal-on-metal hip arthroplasties. Clin Orthop Relat Res. 2012;470:2325–31.

69 Elderly People

Akihiro Sudo

1 Clinical Outcomes of THA in the Elderly

A retrospective review of 149 primary total hip arthroplasties (THAs), performed from 1997 to 2017 over 90 years of age (mean age, 92 years), was reported in 2021 [1]. Mortality rates at 90 days and 1 year after surgery were 6% and 8%, respectively, with 5-year cumulative incidences of any revision and reoperation were 1% and 4%, respectively, and the Harris hip score (HHS) improved significantly. The most frequent postoperative complications were periprosthetic femoral fractures and early postoperative delirium.

A retrospective review of 10,860 THAs performed at a single institution in New York between January 2011 and May 2019 was reported in 2021 [2]. There were 609 patients aged ≥80 years (mean age, 84.3 years), and 10,251 patients aged <80 years (mean age, 61.6 years). Patients with fractures were excluded from this study. The risk of surgery, as determined by both American Society of Anesthesiologists (ASA) score and Charlson comorbidity index (CCI), was significantly higher along with longer postoperative hospital stay and lower rate of home discharge in patients aged ≥80 years. However, no significant differences were observed in 90-days emergency visits, myocardial infarction, periprosthetic joint infection, dislocation, or aseptic failure; moreover, no significant differences in the patient-oriented measures like FJS (forgotten joint score) and HOOS (hip disability and osteoarthritis outcome score) were observed between the two groups. Based on these results, we concluded that THA is safe and effective in patients aged >80 years.

A retrospective study of 2327 consecutive THAs performed at a single institution in New Hampshire from April 2011 to July 2016 was reported in 2020 [3]. There were 187 patients in the group aged ≥80 years, and 2140 patients in the group aged <80 years. Compared to the group aged <80 years, the preoperative CCI was significantly higher, and the physical component summary (PCS) was significantly lower in patients aged ≥80 years, but the improvement in PCS after surgery was similar between the two groups. On the other hand, length of hospital stay was significantly longer, and facility discharge rate was higher in the group aged >80 years. Based on these results, it was concluded that patients over 80 years of age had similar adjusted postoperative functional outcomes following THA but utilized substantially more resources in the form of increased time in the hospital and higher rates of facility discharges compared to younger patients.

2 Cementless or Cemented

A retrospective analysis Of the 301 hips or Of them hemiarthroplasty cases performed with plasma-sprayed titanium-hydroxyapatite femoral stems in patients >70 years (70–93 years old; mean age, 78 years) from 2008 to 2016 at three hospitals in Tuscany, Italy, was reported in 2021 [4]. Of the 301 hips, 35 were hip hemiarthroplasties for displaced femoral neck fractures and 266 were THAs, including 227 hips with osteoarthritis. The stem was implanted in femur Dorr type A, B, and C in 24, 124, and 153 cases, respectively. Intraoperative femoral fractures were observed in three patients (0.9%), and one of them was treated at 10 days with stem revision and cerclage of the femur. Postoperative periprosthetic fractures were observed in 11 patients (3.6%), all of which were associated with trauma. Periprosthetic infection and dislocation occurred in 12 (3.9%) and 9 (2.9%) patients, respectively. The cumulative survival of the stem was 98.5% with stem revision for any reason as the endpoint at the 10-year follow-up. Stem survival was 100%, with stem revision due to aseptic loosening as the endpoint. The mean FJS score was 98.7, and five patients reported a negative FJS score. The implants showed complete osseointegration, with slight stress-shielding signs in less than 10% of the hips.

A. Sudo (✉)
Mie University, Tsu, Mie, Japan
e-mail: a-sudou@med.mie-u.ac.jp

Y. Nakashima (ed.), *Advances in Total Hip Arthroplasty*, https://doi.org/10.1007/978-981-97-2461-1_69

Mean stem subsidence in the 11 patients with stem distal migration was 1.6 mm. These results suggested that the cementless plasma-sprayed porous Ti/HA femoral stem is useful in elderly patients.

A clinical and radiological comparative study of cementless THAs performed in patients aged ≥80 years (older group; mean, 83.1 years) and 60–69 years (control group; mean, 64.1 years) between 2001 and 2012 was reported in 2016 [5]. There were 30 patients in the older group, including 24 cases of osteoarthritis, 5 of hips with osteonecrosis of the femoral head, and 1 case of rapidly destructive coxopathy. All 27 female patients in the control group had osteoarthritis. The mean follow-up period was 5.5 years. A Converge cup and Natural Hip stem was used in 6 hips, and a Reflection cup and Synergy Select stem was used in 24 hips in each group. Comorbidities, such as hypertension and respiratory disease, were observed in 16 (53.3%) and 6 (20%) patients in the older and control group respectively, indicating a significant difference between the two groups. The number of patients who required allogeneic blood transfusions was significantly higher in the older group. Although the HHS improved in both groups, the score was significantly lower in the older group 1 year after the surgery and at final follow-up. In the radiological evaluation, there were significantly more patients with preoperative stovepipes in the older group (7, 23.3%) than in the control group (1, 3.3%), but no cases of postoperative sinking were reported in the older group. These results suggested that cementless THA in elderly patients is safe and durable at follow-up.

The analysis of 6540 THA cases aged 80 years or older (mean, 82.7 years) performed between 1980 and 2004, using data from the Finnish Arthroplasty Registry data, was reported in 2008 [6]. In 82% of cases, both components were fixed with cement, 12% underwent hybrid fixation (91% had a femoral component fixed with cement, 9% with an acetabular component fixed with cement), and 6% had cementless fixation of both. The difference was significant ($p < 0.05$) only when patients who underwent hybrid fixation were compared with those who underwent cementless fixation; survival at 5 years was 98% and 95%, respectively. A secondary analysis indicated that, for patients with cementless acetabular fixation, survival was better for those with a cemented femoral component than those without it. However, in patients with a cemented acetabular component, no significant differences in prosthesis survival were observed between those with a cemented femoral component and those with a cementless femoral component. Based on these results, we concluded that cementation of the femoral stem resulted in better long-term outcomes than cementless fixation.

A retrospective comparison study of 110 cemented THAs (mean age, 68.88 years) and 114 cementless THA (mean age, 68.52 years) performed between March 2010 and March 2015, including patients aged ≥65 years with femoral neck fractures associated with osteoporosis, was reported in 2020 [7]. The mean follow-up duration was 60 months. Cemented THA was performed using the Exeter system, and cementless THA was performed using a combination of a CORAIL stem and a REFLECTION cup. The HHS was significantly higher in the cemented THA group from 3 years after surgery to the final follow-up. No difference was observed in the frequency of dislocation. On the other hand, significant differences in revision, loosening, and periprosthetic fracture were observed between the cementless and cemented THA groups (revision: 14.0% vs. 5.5%, $p = 0.031$; loosening: 20.2% vs. 10.0%, $p = 0.034$; periprosthetic fracture: 12.3% vs. 4.5%, $p = 0.038$). These outcomes suggested that cemented THA results in higher functional scores and lower rates of hip-related complications than cementless THA in elderly patients with displaced osteoporotic femoral neck fractures.

A retrospective comparison of 184 cemented THAs and 182 cementless THAs operated between June 2006 and February 2013 in patients aged 60–80 years, performed using data from the South China Hip Arthroplasty Registry was reported in 2019 [8]. The primary diagnosis included osteoarthritis of the hip, femoral neck fracture, osteonecrosis of the femoral head, and rheumatoid arthritis, with associated osteoporosis in all the cases. The mean follow-up duration was 75 months. HHS was significantly higher in cemented THA from 3 months after surgery to the final follow-up. There was no difference in the frequency of periprosthetic fracture. The prosthetic loosening rates in the cementless and cemented groups were 26.4% and 16.8%, respectively, at a minimum of 5 years ($p = 0.027$). A significant difference was observed in the rate of early revision at the final follow-up (7.6% vs. 14.8%, $p = 0.028$). Approximately 39.2% of cementless THAs were revised for femoral loosening compared to 21% of cemented THAs ($p = 0.033$). These results suggested that cemented THA may be associated with superior outcomes in elderly patients with primary osteoporosis.

3 Countermeasures Against Osteoporosis

A retrospective study of 268 elderly patients (≥ 70 years) with end-stage osteoarthritis who underwent dual-energy X-ray absorptiometry (DXA) within 3 months prior to primary THA from 2018 to 2019 was reported in 2021 [9]. Out of 268, 90 patients (34%) were males and 178 (66%), females. Fifteen (5.6%) patients had previously experienced fragility fractures (24 fractures in total). Further, 14 out of 24 presented as vertebral fractures. Importantly, 49 out of the total 268 patients analyzed (18%) were diagnosed with osteoporosis, while 110 patients (41%) had T-scores in the range of osteopenia. Only 13 (27%) out of the 49 patients with osteoporosis had been diagnosed previously and 18 (37%) patients received vitamin D supplementation. Moreover, antiresorptive treatment was prescribed to only 11 patients (22%). These results suggested that a general assessment of BMD prior to THA should be considered in patients aged >70 years to optimize treatment and outcomes, including the prevention of complications such as intraoperative periprosthetic fractures and aseptic loosening.

In the review article titled "Total hip arthroplasty in patients with osteoporosis" that was published in 2020, it was stated that bone health should be evaluated and improved before patients undergo THA to ensure the most successful outcome, due to the high prevalence of osteoporosis in patients undergoing THA [10]. In osteoporotic hips, the use of bisphosphonates may improve implant osseointegration and reduce periprosthetic fractures. Furthermore, in patients with osteoporosis, a cemented stem can be chosen, but to choose a cementless stem suitable bone geometry is required. The surgical technique of stem fixation is of great importance since poor bone quality is an adverse risk factor for implant loosening, regardless of the fixation method.

References

1. Dagneaux L, et al. Contemporary mortality rate and outcomes in nonagenarians undergoing primary total hip arthroplasty. J Arthroplast. 2021;36:1373–9.
2. Zak SG, et al. Primary total hip arthroplasty outcomes in octogenarians. Bone Jt Open. 2021;2:535–9.
3. Austin DC, et al. Patient outcomes after total hip arthroplasty in extremely elderly patients older than 80 years. Hip Int. 2020;30:407–16.
4. Piolanti N, et al. Use of a plasma-sprayed titanium-hydroxyapatite femoral stem in hip arthroplasty in patients older than 70 years. Fixation a reliable option in the elderly? J Clin Med. 2021;10:4735.
5. Yuasa T, et al. Cementless total hip arthroplasty in patients aged ≥80 years. J Orthop. 2016;13:29–32.
6. Ogino D, et al. Total hip replacement in patients eighty years of age and older. J Bone Joint Surg Am. 2008;90:1884–90.
7. Zhou X, et al. Uncemented versus cemented total hip arthroplasty for displaced femoral neck fractures in elderly patients with osteoporosis: a retrospective analysis. J Int Med Res. 2020;48:300060520944663.
8. Yang C, et al. Cemented versus uncemented femoral component total hip arthroplasty in elderly patients with primary osteoporosis: retrospective analysis with 5-year follow-up. J Int Med Res. 2019;47:1610–9.
9. Delsmann MM, et al. High prevalence and undertreatment of osteoporosis in elderly patients undergoing total hip arthroplasty. Osteoporos Int. 2021;32:1661–8.
10. Karachalios TS, et al. Total hip arthroplasty in patients with osteoporosis. Hip Int. 2020;30:370–9.

Developmental Dysplasia of the Hip (DDH)

70

Gaku Koyano and Tetsuya Jinno

1 Degree of Dislocation in DDH

Developmental dysplasia of the hip (DDH) presents as dislocation, formerly known as congenital dislocation of the hip, and subluxation of the hip (see Anatomy 2: Pelvis and femur morphology in DDH). DDH also includes acetabular dysplasia. Osteoarthritis of the hip joint due to DDH is often associated with high dislocation or subluxation, and the degree of dislocation is often classified using Crowe's classification system [1] or the Hartofilakidis classification system [2] (Fig. 70.1). For a quick comparison of the two classification systems, note that Crowe I corresponds to Hartfilakidis dysplasia; Crowe II and III to Hartfilakidis low dislocation; and Crowe IV to Hartfilakidis high dislocation.

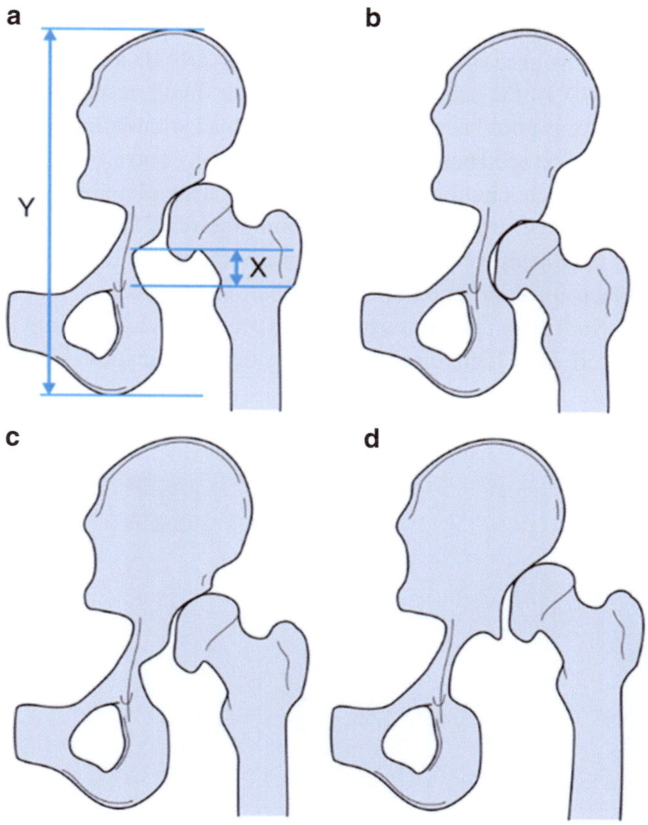

Fig. 70.1 Classification of hip joint dislocation. (**a**) Crowe classification; X: distance from the inter-teardrop line to the head-neck junction; Y: height of the pelvis. The normal head diameter is approximately equal to 1/5 of Y. Group I: $X/Y < 0.1$ (subluxation of the head less than 50%). Group II: $0.1 \leq X/Y < 0.15$ (subluxation of the head is between 50% and 75%). Group III: $0.15 \leq X/Y < 0.2$ (subluxation of the head is between 75% and 100%). Group IV: $0.2 \leq X/Y$ (dislocation of the bone head is more than 100%). (**b-d**) Hartofilakidis classification. (**b**) Dysplastic type: the head of the bone is in the primary acetabulum. (**c**) Low dislocation type; the secondary acetabulum overlaps the primary acetabulum. (**d**) High dislocation type; the secondary acetabulum is above the primary acetabulum

G. Koyano · T. Jinno (✉)
Department of Orthopedic Surgery, Dokkyo Medical University Saitama Medical Center, Koshigaya, Saitama, Japan
e-mail: jinnot@dokkyomed.ac.jp

Y. Nakashima (ed.), *Advances in Total Hip Arthroplasty*, https://doi.org/10.1007/978-981-97-2461-1_70

2 Cup Placement in DDH Cases

Patients with Crowe I and II rarely have difficulty with cup placement. However, patients with Crowe III (Fig. 70.2) and IV (Fig. 70.3) have a significantly different hip joint structure from the normal hip joint. Crowe III and IV have little bone stock because of the primary acetabulum and femur hypoplasia, making implant selection and placement difficult in some cases. In addition, the initial fixation may not be adequate because of poor bone quality caused by bone atrophy.

(a) Cup placement position

The cup should be placed in the primary acetabulum to align the center of rotation on the contralateral side, which may compensate for the leg length discrepancy. This cup placement in the primary acetabulum may result in the tensioning of the gluteus medius and gluteus minimus muscles to improve strength, and in less impingement. Even if the cup is placed in a high position in Crowe III cases without acetabular floor thickening, host bone coverage may be insufficient and initial fixation may not be achieved. In these cases, acetabular reconstruction using a bulk bone graft or metal augmentation is necessary.

In the case of Crowe IV (intramuscularly dislocated hip, Fig. 70.3), the acetabular bone mass is small, and the bone quality is poor because of the lack of load stimulation to the bone. The acetabulum has a narrow opening and a triangular rather than circular shape. In preoperative planning, it is advisable to use 3D-computed tomography (CT) image and 3D preoperative planning software to assess the three-dimensional structure of the acetabulum accurately.

No fixation problems have been reported when the cup is placed at a high hip center without lateralization [3, 4]. There is also evidence that cup placement at the primary acetabulum reduces the reaction force of the hip joint and prevents mechanical failure [5]. Regarding the range of motion of the hip joint, CT-based simulations have shown that raising the center of the hip joint improves the bone coverage of the cup but reduces the range of hip flexion and internal rotation [6]. In addition, the risk of dislocation increases when the center of rotation of the hip is placed more than 23.9 mm proximal to the inter-teardrop line [7]. If the cup must be placed higher than the level of the above-mentioned dislocation risk, it is better to prepare for dislocation, such as by using a large-diameter head, a dual-mobility cup, or a modular-type stem or cemented stem, which can adjust stem anteversion.

(b) Cementless cup

The cup center edge (CE) angle and cup coverage were used to evaluate cup placement stability (Fig. 70.4). The cup CE angle is generally used because the coverage estimated in two dimensions is larger than that evaluated in three dimensions [8]. In a study of cementless cups placed in dysplastic hips without screws (87 cases, 98 joints), bone-ingrowth was observed in all cases, and the minimum cup CE angle was reported to be 8.4° [9]. When screws were used for cementless cup fixation, bone ingrowth was observed, and bony fixation was obtained even in cases with a cup CE angle of 0° or less [10]. Considering the direction of the resultant force acting on the hip joint, it is recommended that the cup CE angle be 8° or more when screws are not used, and the cup CE angle should be 0° or more even when a screw is used. When the bone defect is covered with host bone, morselized bone obtained by acetabular reaming is

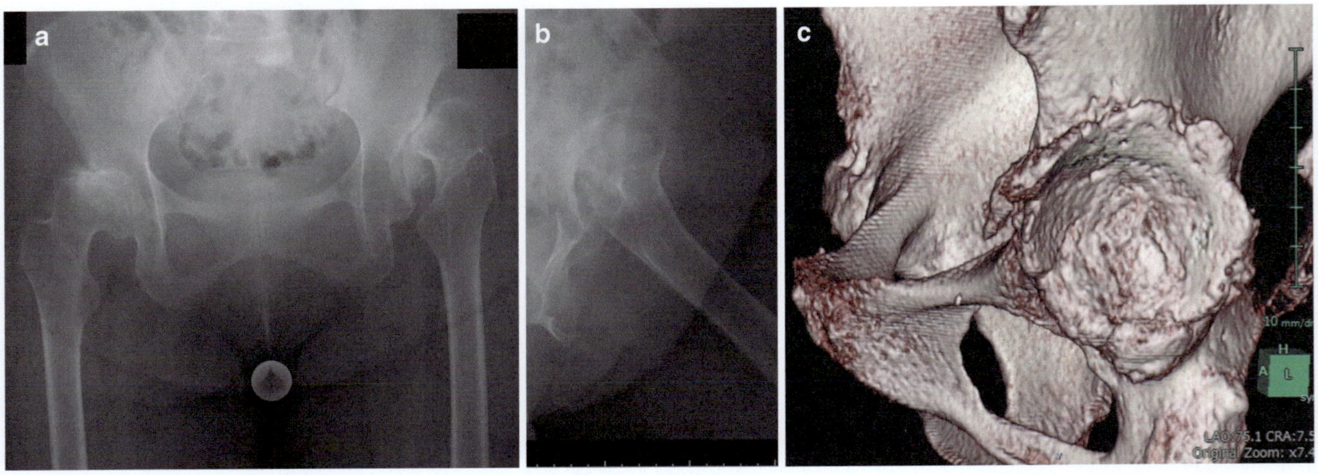

Fig. 70.2 X-ray images of a case of low dislocation. (**a**) Anteroposterior view, (**b**) Lauenstein view, (**c**) Left acetabular CT three-dimensional reconstruction image. Right hip; Crowe I, Hartofilakidis dysplasia type. Left hip; Crowe III, Hartofilakidis low dislocation type. The boundary between the secondary and primary acetabula was difficult to distinguish

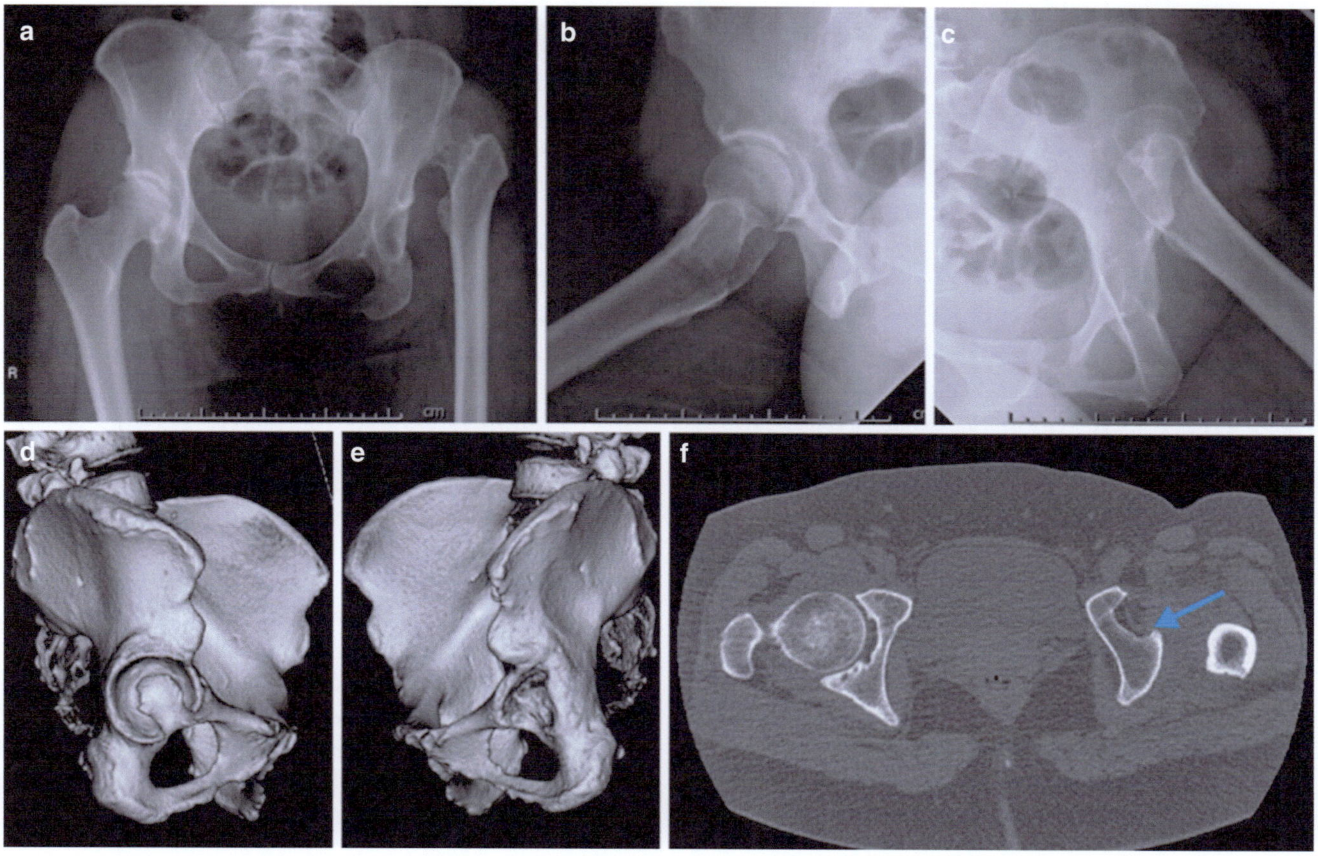

Fig. 70.3 X-ray images of a case of high dislocation. (**a**) Anteroposterior view; (**b**) Right Lauenstein view; (**c**) Left Lauenstein view; (**d**) Right acetabular CT three-dimensional reconstruction image; (**e**) Left acetabular CT three-dimensional reconstruction image; (**f**) Acetabular horizontal-section CT. Right hip; Crowe I with mild dysplasia (CE angle 22°) and Hartofilakidis dysplasia. The acetabulum was normal. Left hip; Crowe IV, Hartofilakidis high dislocation type. The acetabulum was atrophic and the posterior wall was thicker than the anterior wall. The center of the cup is located near the anterior margin of the posterior wall (arrow)

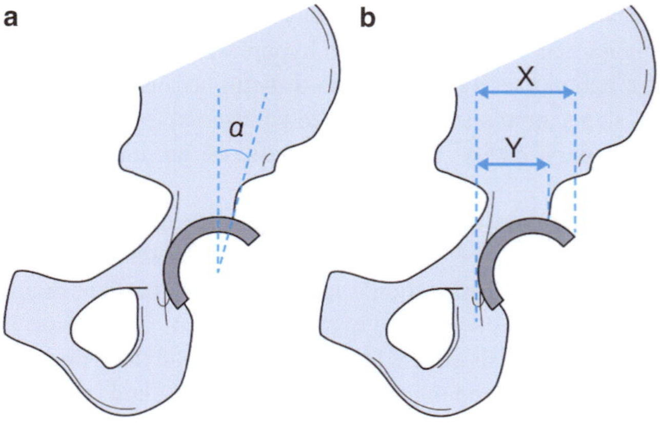

Fig. 70.4 Evaluation of the cup coverage by the host bone. (**a**) Cup CE angle. The angle between the vertical direction from the center of rotation to the lateral end covered by the host bone. (**b**) Cup coverage (= Y/X). X; Horizontal width of the cup. Y; Horizontal width of the covering host bone

suitable as a bone-grafting material for residual bone defects on the outer side of the cup, because it can be quickly incorporated [8–10].

Bulk bone grafts have been used in cases without sufficient host bone cup coverage. Cementless cups in conjunction with bulk femoral head autologous grafts have been reported to have good short- and mid-term results [11, 12]. The cup coverage of the bulk bone graft is usually evaluated using the surface coverage of the bone graft and lateral cup coverage (Fig. 70.5). There have been cases of early postoperative cup migration when a cementless cup was fixed in conjunction with a bulk bone graft [13, 14], and in these cases, the cup surface coverage of the bone graft was >40% [14]. This indicates that the coverage of the cup by the host bone should be greater than 60%. There is one report of a fractured metal cup when lateral coverage of the bone graft was 97% [11].

High hip center reconstruction is another method of obtaining sufficient coverage without bulk bone grafting. The survival rate 12 years after high hip center THA using revision surgery as an endpoint, was reported to be 96% for Crowe I, which is close to the result of primary acetabular THA. The 12-year survival rate for Crowe II and III using the high hip center technique was 97% [3].

In DDH cases, a large bone cyst may be observed in the acetabulum due to stress concentration near the acetabular

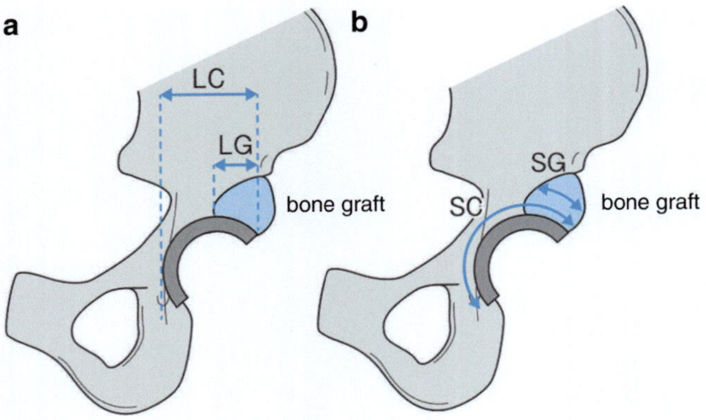

Fig. 70.5 Cup coverage with a bulk bone graft. (**a**) Lateral coverage of the cup by the bone graft. LG/LC: evaluated by the length projected horizontally. (**b**) Surface coverage of the cup by the bone graft. SG/SC; evaluated by the length of the arc along the cup

rim, and loosening of the cup due to the cyst may be worried. However, it has been reported that acetabular bone cysts do not influence long-term implant fixation if the cup is correctly placed [15]. In cases of acetabular bone cysts, curettage and transplantation of the morselized bone obtained during acetabular reaming are recommended.

(c) Cemented cup

It has been reported that a cemented cup should have at least 50% lateral coverage by the iliac crest to achieve stable fixation of the cup [16, 17]. When the cup is placed in the primary acetabular space and bone coverage of >50% is not achieved, proximal placement should be considered to improve coverage. Cemented cups with bulk bone grafting showed good results, [18–21] however, the 10-year survival rate of cemented cups with cup revision surgery as the endpoint decreased in Crowe III (76.1%) and IV (60.9%) cases, compared to Crowe I (94.1%) and II (91.7%) cases.

The long-term results of bulk bone grafting using autologous femoral heads are better than those of allograft, and the revision rate of cemented cups is reported to increase as the coverage of bone graft increases [17]. Seventy eight percent of hips that achieved long term fixation were supported by graft with the coverage less than 50%, while only 36% were so supported in those who underwent revision surgery [17]. Moreover, none of the patients underwent revision surgery in which the lateral coverage of the grafted bone was less than 30%, suggesting that the risk of revision increased as the coverage of the bone graft increased.

(d) For accurate cup placement

In Crowe III and IV DDH cases, the "strike zone" for cup placement is narrow, so preoperative planning for accurate cup placement is important [23]. Three-dimensional planning using CT is more accurate than two-dimensional planning [24]. Computer-assisted surgery (see Chap. 10, CAS) is useful in DDH cases because the deformity is advanced, orientation during surgery is difficult, and it is often hard to distinguish the boundary between the secondary and primary acetabula with the naked eye. A CT navigation system allows for more accurate cup placement than freehand surgery [24, 25]. In the group using the CT navigation system, 97.7% of cups were placed within the target zone, which was higher than that obtained using the mechanical guide (61.3%) [25]. Robot-assisted THA has less variation from the target position than that using the CT navigation system, resulting in more accurate cup placement [26]. Although these techniques are useful in Crowe III and IV cases with strong deformities, they are not yet widely used. It is important to perform detailed preoperative planning using CT, especially 3D reconstruction, to recognize bone shape [27]. As shown in Fig. 70.3, the center of cup placement is often located near the posterior wall in Crowe IV cases [27]. To accurately determine the height of cup placement intraoperatively, the distances from the superior rim of the secondary acetabulum to the rim of the primary acetabulum, the acetabular fossa, the transverse acetabular ligament, and the obturator foramen should be comprehensively evaluated [27–31].

3 Stem Placement in DDH Cases

It is reported that the anteversion angle of the femoral neck is large in cases of DDH [32]. In addition, Crowe III and IV cases have a narrow medullary cavity and poor bone quality because of decreased loading caused by leg length discrepancies and subluxation due to hypoplasia. Choosing a stem for patients with DDH is often difficult because some types of stems are not small enough to be inserted into the narrow medullary cavity and run the risk of an intraoperative fracture.

In the case of increased femoral neck anteversion, the combined anteversion does not settle in the safe zone [33], and the risk of dislocation is high; therefore, the stem should be inserted with reduced anteversion. Stem anteversion in dysplastic hips is affected by the approach and the type of the stem [34, 35], and thus preoperative planning and control of the anteversion are important. When mild control of stem anteversion is required, a neck-adjustable stem is an option [36–38]. On the other hand, a modular stem [39] or cone-type stem [40] is used when controlling a large angle. In the case of cemented stems, if the thickness is greater than a certain level, anteversion can be controlled intraoperatively [41].

In Crowe IV cases, leg length discrepancies are large, and THA with femoral shortening osteotomy is often performed (see 14-1. Crowe IV, with shortened osteotomy). THA with femoral shortening osteotomy allows both the appropriate positioning of the center of rotation of the hip joint and the prevention of nerve damage. It is also possible to compensate for excessive anteversion of the proximal femur, and both cementless and cemented stems are available. When a cementless stem is used, it is necessary to obtain sufficient rotational stability at both the proximal and distal ends of the osteotomy, so a modular stem with a fin structure at the distal end of the stem is used.

4 Leg Length Control

In cases of DDH, especially Crowe III and IV, leg lengthening is necessary. The ideal leg length correction is equal on both sides. In bilateral cases, it is easier to correct the leg length discrepancy than in unilateral cases because the center of rotation can be adjusted between the left and right sides. The position of the lesser trochanter on the anteroposterior radiograph of the hip joint is usually used as a landmark for leg length. The lengths of the femur and lower leg sometimes differ between the right and left sides in patients with DDH [42]. Therefore, it is necessary to confirm leg length not only by evaluating the anteroposterior X-ray image of the hip joint but also by evaluating the full-leg length X-ray or CT images. In addition, there are cases of patients feeling a subjective leg length discrepancy after THA, even if that on X-ray images disappeared. This usually occurs when there is deformity or disability in the knee or hip due to a preoperative leg length discrepancy, or when there is compensatory scoliosis, decreased flexibility of the spine, or hip contracture [43–45]. Therefore, appropriate leg lengthening is necessary. In preoperative planning, the subjective and objective leg length discrepancies were measured and used as references to determine the amount of leg lengthening required. The appropriate leg length is often determined between subjective and objective leg length discrepancies. Other factors to be considered include the presence of spinal deformity and flexibility of the spine, hip contracture, lower extremity alignment, height, and surgical history (e.g., soft tissue conditions).

If the preoperative leg length discrepancy is large, the planned leg lengthening will also be large; thus, the risk of nerve palsy will be high [46, 47]. Intraoperative nerve monitoring is another option when leg lengthening exceeds 3 cm, [46, 48, 49]. However, there are cases in which paralysis does not occur, even after a 3-cm or greater one-stage leg lengthening. Sometimes, the cause of paralysis is something other than leg lengthening, such as nerve damage caused by intraoperative retractors [46, 49]. A recent study using intraoperative neuromonitoring reported that an average lengthening of 22 mm (5% of the femoral length) was the upper limit of one-stage leg lengthening when a 50% decrease in the amplitude of motor or somatosensory evoked potentials or a 10% lngthening of latency was set as the limit [50].

The joint range of motion is reduced when muscle, tendon, and ligament tensions are increased by leg lengthening. In some cases, the patient may not be able to obtain a good limb position because of joint contracture. There is a case report of cup dislocation due to increased muscle tone after cerebral hemorrhage [51], and it is important to reduce tension by cutting or releasing the periarticular muscles and ligaments. If abduction is difficult, the pubic attachment of the adductor tendon should be released. When leg lengthening is required, several procedures should be considered, including dissection of the gluteus maximus attached posteriorly to the femur, dissection of the iliopsoas muscle from the lesser trochanter, incision of the iliotibial ligament, and dissection of the descending iliofemoral ligament. In the anterior approach, the piriformis and short external rotator muscles may be partially dissected from the greater trochanter. Too much dissection of these ligaments leads to posterior dislocation; thus, the dissection should be minimal.

5 Clinical Results

(a) Long-term durability

The survival rates of prostheses with revision surgery as an endpoint for DDH patients were 94–100% at 15 years with cemented stems [19, 20] and 99% with cementless stems [52] on the femoral side. Survival rates with revision surgery as endpoints after THA with femoral shortening osteotomy were 95.8–100% at 5 years with cemented stems [53, 54] and 87.8% at 10–15 years [53]. Good results for the survival rate after cementless THA with femoral shortening osteotomy were reported at 92% at 5 years [54].

The survival rate of cemented cups with revision surgery as an endpoint was reported to be 97–99% at 10 years [19, 20, 55] and 96–96.6% at 15 years [19, 20, 55]. In cementless cups, the survival rate was 91.6% after an average follow-up of 11 years, with revision surgery as the endpoint [56]. On the other hand, there are reports that the long-term outcome is lower in Crowe IV cases (60.9% at 10 years and 15.6% at 15 years [22]. A report of Crowe IV patients from Japan showed a good survival rate of 87.8% at 15 years, with implant revision for any reason as endpoints [53]. Good results can be expected with accurate preoperative planning and appropriate surgical techniques.

(b) Clinical symptoms

The Harris hip score (HHS, 100-point scale) and the Japanese Orthopedic Association hip function scale (JOA,

100-point scale) are physician-assessed hip function scales. In a report on cementless hip arthroplasty, the HHS score improved from 45–56 points preoperatively to 85–99 points at the latest postoperative observation [37, 57–59], and the JOA score improved from 43–45 points preoperatively to 87–91 points at the last postoperative observation [60, 61]. Similarly, in hybrid THA, the HHS score improved from 42–44.5 points preoperatively to 87.5–88 points at the latest postoperative observation [12, 58], showing excellent results. In the cases with cemented THA, the JOA score improved from 45.3 points preoperatively to 83.2 points at the latest postoperative observation [55]. The HHS score of patients who underwent THA with shortening osteotomy was reported to have improved from 53.1–56.1 points preoperatively to 80.8–84.5 points postoperatively [53, 57]. The mean preoperative range of motion of the hip joint was 71.5° of flexion and 14°of abduction, and the mean postoperative range of motion improved to 93.5° of flexion and 30° of abduction [37].

References

1. Crowe JF, et al. Total hip replacement in congenital dislocation and dysplasia of the hip. J Bone Joint Surg Am. 1979;61:15–23.
2. Hartofilakidis G, et al. Congenital hip disease in adults. Classification of acetabular deficiencies and operative treatment with acetabuloplasty. J Bone Joint Surg Am. 1996;78:683–92.
3. Nawabi DH, et al. Durable fixation achieved with medialized, high hip center cementless THAs for crowe II and III dysplasia. Clin Orthop Relat Res. 2014;472:630–6.
4. Galea VP, et al. Developmental dysplasia treated with cementless total hip arthroplasty utilizing high hip center reconstruction: a minimum 13-year follow-up study. J Arthroplast. 2018;33:2899–905.
5. Johnston RC, et al. Reconstruction of the hip. A mathematical approach to determine optimum geometric relationships. J Bone Joint Surg Am. 1979;61:639–52.
6. Komiyama K, et al. Does high hip center decrease range of motion in total hip arthroplasty? A computer simulation study. J Arthroplast. 2016;31:2342–7.
7. Komiyama K, et al. Does high hip centre affect dislocation after total hip arthroplasty for developmental dysplasia of the hip? Int Orthop. 2019;43:2057–63.
8. Ueno T, et al. Three-dimensional host bone coverage required in total hip arthroplasty for developmental dysplasia of the hip and its relationship with 2-dimensional coverage. J Arthroplast. 2019;34:93–101.
9. Takao M, et al. The results of a press-fit-only technique for acetabular fixation in hip dysplasia. J Arthroplast. 2011;26:562–8.
10. Fujii M, et al. Minimum lateral bone coverage required for securing fixation of cementless acetabular components in hip dysplasia. Biomed Res Int. 2017;2017:4937151.
11. Spangehl MJ, et al. Uncemented acetabular components with bulk femoral head autograft for acetabular reconstruction in developmental dysplasia of the. J Bone Joint Surg Am. 2001;83:1484–9.
12. Ito H, et al. Hybrid total hip arthroplasty using specifically-designed stems for patients with developmental dysplasia of the hip. Int Orthop. 2011;35:1289–94.
13. Ito H, et al. Intermediate-term results after hybrid total hip arthroplasty for the treatment of dysplastic hips. J Bone Joint Surg Am. 2003;85:1725–32.
14. Hintermann B, et al. Total hip replacement with solid autologous femoral head graft for hip dysplasia. Arch Orthop Trauma Surg. 1995;114:137–44.
15. Takada R, et al. Longitudinal morphological change of acetabular subchondral bone cyst after total hip arthroplasty in developmental dysplasia of the hip. Eur J Orthop Surg Traumatol. 2018;28:621–5.
16. Kobayashi S, et al. Total hip arthroplasty with bulk femoral head autograft for acetabular reconstruction in developmental dysplasia of the hip. Bone Joint Surg Am. 2003;85:615–21.
17. Shinar AA, et al. Bulk structural autogenous grafts and allografts for reconstruction of the acetabulum in total hip arthroplasty. J Bone Joint Surg Am. 1997;79:159–68.
18. Oe K, et al. Long-term results of acetabular reconstruction using three bulk bone graft techniques in cemented total hip arthroplasty for developmental dysplasia. Int Orthop. 2016;40:1949–54.
19. Akiyama H, et al. Long-term results of cemented total hip arthroplasty in developmental dysplasia with acetabular bulk bone grafts after improving. J Arthroplast. 2010;25:716–20.
20. Iida Y, et al. Cemented total hip arthroplasty with acetabular bone graft for developmental dysplasia. J Bone Joint Surg Br. 2000;82:176–84.
21. Inao S, et al. Cemented total hip arthroplasty with autogenous acetabular bone grafting for hips with developmental dysplasia in adults: the results. J Bone Joint Surg Br. 2000;82:375–7.
22. Chougle A, et al. Severity of hip dysplasia and loosening of the socket in cemented total hip replacement: a long-term follow-up. J Bone Joint Surg Br. 2005;87:16–20.
23. Gill TJ, et al. Total hip arthroplasty with use of an acetabular reinforcement ring in patients who have congenital dysplasia of the hip. J Bone Joint Surg Am. 1998;80:969–79.
24. Inoue D, et al. Value of computed tomography-based three-dimensional surgical preoperative planning software in total hip arthroplasty with developmental dysplasia of the hip. J Orthop Sci. 2015;20:340–6.
25. Tsutsui T, et al. Efficacy of a computed tomography-based navigation system for placement of the acetabular component in total hip arthroplasty for developmental dysplasia of the hip. J Orthop Surg (Hong Kong). 2017;25:2309499017727954.
26. Ando W, et al. Comparison of the accuracy of the cup position and orientation in total hip arthroplasty for osteoarthritis secondary to developmental dysplasia of the hip between the Mako robotic arm-assisted system and computed tomography-based navigation. Int Orthop. 2021;45:1719–25.
27. Yoshitani J, et al. Morphometric geometrical analysis to determine the centre of the acetabular component placement in Crowe type IV hips undergoing. Bone Joint J. 2019;101-B:189–97.
28. Shimodaira H, et al. The acetabular fossa may not be located at the true center of the acetabulum: a detailed analysis using preoperative CT images. J Bone Joint Surg Am. 2018;100:e27.
29. Beverland DE, et al. Placement of the acetabular component. Bone Joint J. 2016;98-B(1 Suppl A):37–43.
30. Fujita K, et al. The use of the transverse acetabular ligament in total hip replacement: an analysis of the orientation of the trial acetabular. Bone Joint J. 2014;96-B:306–11.
31. Sotereanos NG, et al. Using intraoperative pelvic landmarks for acetabular component placement in total hip arthroplasty. J Arthroplast. 2006;21:832–40.
32. Noble PC, et al. Three-dimensional shape of the dysplastic femur: implications for THR. Clin Orthop Relat Res. 2003;417:27–40.
33. Widmer K-H, et al. Compliant positioning of total hip components for optimal range of motion. J Orthop Res. 2004;22:815–21.

34. Taniguchi N, et al. Cementless hip stem anteversion in the dysplastic hip: a comparison of tapered wedge vs metaphyseal filling. J Arthroplast. 2017;32:1547–52.
35. Taniguchi N, et al. Comparative study of stem anteversion using a cementless tapered wedge stem in dysplastic hips between posterolateral and anterolateral approaches. Orthop Traumatol Surg Res. 2019;105:1271–6.
36. Miki H, et al. Modular neck for prevention of prosthetic impingement in cases with excessively anteverted femur. Clin Biomech (Bristol, Avon). 2011;26:944–9.
37. Sakai T, et al. A modular femoral neck and head system works well in cementless total hip replacement for patients with developmental dysplasia of the hip. J Bone Joint Surg Br. 2010;92:770–6.
38. Nakahara E, et al. Effect of a modular neck hip prosthesis on anteversion and hip rotation in total hip arthroplasty for developmental dysplasia of the hip. J Artif Organs. 2020;23:255–61.
39. Tamegai H, et al. A modified S-ROM stem in primary total hip arthroplasty for developmental dysplasia of the hip. J Arthroplast. 2013;28:1741–5.
40. Takahashi E, et al. Satisfactory outcomes in patients operated with primary total hip arthroplasty for perthes-like deformities: results from a surgical technique utilizing a conical stem, an elevated hip center, and no shortening femoral osteotomy. Arthroplast Today. 2020;7:29–36.
41. Scheerlinck T, et al. Hip implants can restore anatomical and medialized rotation centres in most cases: a 3D templating study comparing four implantation strategies. Bone Jt Open. 2021;2:476–85.
42. Tamura K, et al. Femoral morphology asymmetry in hip dysplasia makes radiological leg length measurement inaccurate. Bone Joint J. 2019;101-B:297–302.
43. Wylde V, et al. Prevalence and functional impact of patient-perceived leg length discrepancy after hip replacement. Int Orthop. 2009;33:905–9.
44. Koga D, et al. The effect of preoperative lateral flexibility of the lumbar spine on perceived leg length discrepancy after total hip arthroplasty. Med Dent Sci. 2009;56:69–77.
45. Fujita K, et al. Optimizing leg length correction in total hip arthroplasty. Int Orthop. 2020;44:437–43.
46. Hoshino C, et al. Femoral nerve palsy following primary total hip arthroplasty with a direct anterior approach. PLoS One. 2019;14:e0217068.
47. Kabata T, et al. Safety range for acute limb lengthening in primary total hip arthroplasty. Int Orthop. 2019;43:2047–56.
48. Ishimatsu T, et al. Motor-evoked potential analysis of femoral nerve status during the direct anterior approach for total hip arthroplasty. J Bone Joint Surg Am. 2018;100:572–7.
49. Kinoshita S, et al. Femoral nerve status during the anterolateral approach for total hip arthroplasty: motor-evoked potential analysis and an influencing factor. J Orthop Sci. 2021;S0949–2658(21):00334–1. On line ahead of print
50. Bayram S, et al. Critical limit of lower-extremity lengthening in total hip arthroplasty: an intraoperative neuromonitorization study. J Bone Joint Surg Am. 2020;102:664–73.
51. Fukaya H, et al. A case of dislocation of bone ingrowth prosthesis due to spastic paraplegia caused by cerebral hemorrhage. Hokkaido J Orthop Traumatol. 2007;49:44–7. Japanese
52. Fukunishi S, et al. Long term results of the Bicontact D stem in dysplastic osteoarthritis: a 10-15 year follow-up study. Arch Orthop Trauma Surg. 2021; https://doi.org/10.1007/s00402-021-04262-8. On line ahead of print
53. Kawai T, et al. Total hip arthroplasty combined with subtrochanteric transverse shortening osteotomy: factors associated with delayed union at the osteotomy site. J Am Acad Orthop Surg Glob Res Rev. 2020;4(e20):00056.
54. Inoue D, et al. Comparison of mid-term clinical results between cementless and cemented femoral stems in total hip arthroplasty with femoral. Arch Orthop Trauma Surg. 2021;141:1057–64.
55. Goto K, et al. Long-term results of cemented total hip arthroplasty for dysplasia, with structural autograft fixed with poly-L-lactic acid screws. Arthroplasty. 2009;24:1146–51.
56. Hendrich C, et al. Cementless acetabular reconstruction and structural bone-grafting in dysplastic hips. J Bone Joint Surg Am. 2006;88:387–94.
57. Hasegawa Y, et al. Total hip arthroplasty for crowe type IV developmental dysplasia. J Arthroplast. 2012;27:1629–35.
58. Harada Y, et al. Anatomically designed prosthesis without cement for the treatment of osteoarthritis due to developmental dysplasia of the hip: a 6- to to 13-year follow-up study. J Orthop Sci. 2007;12:127–33.
59. Kawasaki M, et al. Muscle damage after total hip arthroplasty through a direct anterior approach for developmental dysplasia of the hip. J Arthroplast. 2017;32:2466–73.
60. Hayama T, et al. Clinical results of a short stem with flat tapered wedge design in primary total hip arthroplasty for hip dysplasia in Asians. J Orthop Surg (Hong Kong). 2020;28:2309499020956742.
61. Ueoka K, et al. Patient-reported outcomes following primary total hip arthroplasty in Crowe type III or IV developmental dysplasia are comparable to those in Crowe type I: a case-control study of 96 hips with intermediate-term follow-up. BMC Musculoskelet Disord. 2020;21(1):344.

Idiopathic Osteonecrosis of the Femoral Head

71

Koichi Kinoshita and Takuaki Yamamoto

1 Features of Osteonecrosis of the Femoral Head

Idiopathic osteonecrosis of the femoral head (ONFH) is an ischemic disease characterized by a hip pain and limited range of motion, caused by a collapse of the femoral head. Preoperative planning should be made based on the width of the necrotic lesion as well as the patients' age and social environment.

2 Indications for Surgery

Asymptomatic patients need to be carefully followed up. Symptomatic patients of stage 3 or 4 are indicated for the surgical treatments, including varus or rotational osteotomy, bipolar head arthroplasty (BHA), or total hip arthroplasty (THA).

(a) Femoral osteotomy

When intact lesion remains in the femoral head, joint preserving procedures, including intertrochanteric curved varus osteotomy (CVO) [1] or transtrochanteric rotational osteotomy (TRO) [2] are considered. The principle of these procedures are to obtain more than 34% of the postoperative intact area [3]. Radiographs and magnetic resonance imaging (MRI) are useful for the preoperative planning. Contrast-enhanced MRI can be performed to prevent underestimation of necrotic areas if there are multiple low-intensity band lesions in the femoral head on standard MRI.

Consideration should be given to the patient's medical history and general health. Patients undergoing osteotomy need to be informed as to a longer periods to return work as compared to THA or BHA. When an osteotomy is performed based on the appropriate indication, postoperative results are equivalent to those of THA [4, 5].

(b) Bipolar head arthroplasty

BHA is indicated for patients with stage 3 who do not show joint space narrowing on radiographs. Acetabular cartilage damage is generally determined by the joint space narrowing based on radiographs or computed tomography. In acetabular dysplasia, the T2 value on preoperative MRI has been reported as a prognostic factor for progression of osteoarthritis after acetabular rotational osteotomy [6]. This may also help to determine the indications for BHA for ONFH.

In a report of BHA performed for ONFH around 1990 with an average follow-up period of 11.4 years, the 15-year postoperative survival rate was 70%, which was inferior to the results of THA [7]. However, with recent improvements in the neck diameter, polyethylene wear resistance, and antioxidant properties, the clinical results of BHA are expected to improve.

BHA is not indicated for stage 4 patients with joint space narrowing.

(c) Total hip arthroplasty

THA is indicated for stage 4 patients who show joint space narrowing on radiographs and CT images. The recent advent of bone-preserving implant geometries, improvement of initial and biological fixation with various surface treatments, and improvement of polyethylene wear and oxidation resistance, are all factors expected to improve implant survival rates.

K. Kinoshita (✉) · T. Yamamoto
Faculty of Medicine, Department of Orthopaedic surgery, Fukuoka University, Fukuoka, Japan
e-mail: kkouichi@adm.fukuoka-u.ac.jp; yamamotot@fukuoka-u.ac.jp

Y. Nakashima (ed.), *Advances in Total Hip Arthroplasty*, https://doi.org/10.1007/978-981-97-2461-1_71

3 Preoperative Planning

(a) Implant selection

In Japan, cementless THA is performed more frequently than cemented THA [8]. In both techniques, it is important to use the correct surgical technique. If the difference in leg length is small, it is advisable to prepare a stem with two or more types of lateral offset, one with a normal lateral offset and the other with a large lateral offset. This enables the appropriate tension to be obtained while minimizing the amount of leg extension. As a countermeasure for postoperative dislocation, models that can be transferred to a dual mobility system or restrained liner may be considered without cup revision.

When BHA is performed, it is recommended to prepare the acetabular instruments in cases of unexpected acetabulum cartilage damage identified intraoperatively.

(b) Approach selection

Obtaining adequate tension in THA for ONFH can be difficult and dislocation is a concern in such cases. The dislocation rate within 1 year following THA surgery was 2.1% for patients <65 years and 3.5% for patients >65 years in the ONFH group, and 1.4% for patients <65 years, and 2.4% for patients >65 years in the OA group, showing that the dislocation rate was significantly higher in patients with ONFH [9]. The use of a muscle-sparing approach helps to obtain adequate joint tension and good dislocation resistance (Fig. 71.1).

(c) Cup position

Cup placement in the primary acetabulum is relatively easy. Plan medialization to the extent that the global offset is not reduced.

(d) THA with previous femoral osteotomy

(i) Intertrochanteric curved varus osteotomy

If the medial cortex at the base of the neck is laterally shifted, a fit-and-fill type or short stem type may not fit to that lesion. A tapered wedge or cemented stem can then be inserted easily while preserving the overhang site of the greater trochanter. Moreover, intermediate to distal fixation-type stems can be used without any fixation problems even if the same site is osteotomized.

(ii) Transtrochanteric rotational osteotomy

When a leg-length discrepancy exists, leg-lengthening is necessary. As the patient may have excessive anteversion or retroversion as a result of osteotomy, the anteversion angle should be evaluated using CT. If adjustment is necessary, a cemented stem or a cementless stem with a tapered rectangle, cone, or variable neck should be used. The position of the greater trochanter may also be changed. The presence or absence of an anterior femoral bump or osteophyte should be determined, and the extent of bone removal should be estimated.

As THA after TRO significantly increases blood loss compared with standard THA, [10, 11] it is advisable to prepare preoperative deposits of autologous blood or an intraoperative/postoperative blood collection device. If the plate or screws used during osteotomy remain, preparation should be made in anticipation of removal difficulties.

4 Knack and Pitfalls During THA

(a) Pelvic side

If the subchondral bone does not undergo sclerotic changes, reaming with a force similar to that used for general

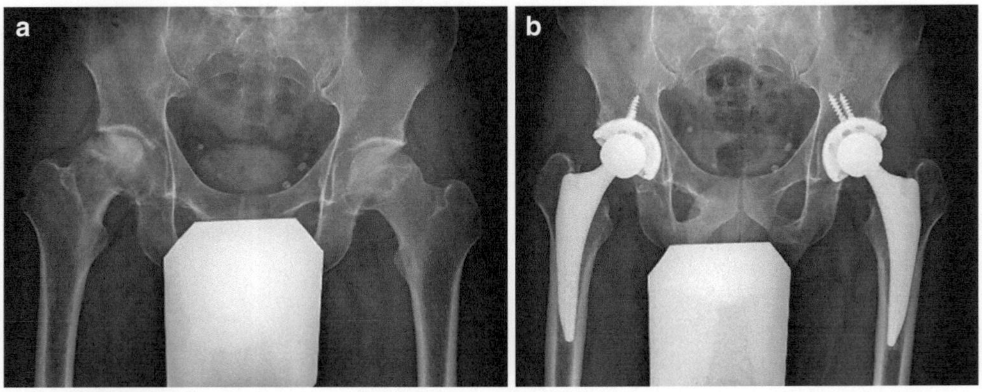

Fig. 71.1 (**a**) Preoperative anterior-posteriol (AP) radiograph. Bilateral alcohol-related osteonecrosis of the femoral head. Both patients had type C-2, stage 4 (right) and stage 3 (left). (**b**) Postoperative AP radiograph. First, a right total hip arthroplasty (THA) was performed, followed by a left THA 9 months later. Implant placement position and angle were appropriate. No postoperative leg length difference was observed because of almost no leg lengthening on either side, and no claudication was observed

OA will result in rapid bone loss, increasing medialization and proximalization of the cup position more than anticipated. The distance from the medial cortex should be measured using forceps to achieve the preoperative plan of medialization. If excessive medialization occurs, leg lengthening is required to achieve adequate tension.

(b) Femur side

No specific difficulties were encountered in THA without a previous femur osteotomy. If a previous femoral osteotomy has been performed, the osteophytes in impingement should be osteotomized according to the preoperative plan. If osteosclerosis is severe, a bone saw should be considered due to the risk of fracture when using a bone chisel or Riehl.

5 Postoperative Therapy

If implant fixation is satisfactory and there are no intraoperative fractures or other complications, postoperative therapy is performed as usual at each institution.

6 Clinical Results

In this section, the results obtained in Japan are discussed.

(a) Cementless THA
   (i) Polyethylene Liner
      The 10-year postoperative survival rate with revision as the endpoint in 14 joints of 11 patients who underwent cementless THA with polyethylene liners between 1994 and 2001 was reported to be 93% [12]. Crosslinked polyethylene became available after the year 2000 in Japan. Since then, the medium to long-term outcomes with revision as the endpoint are generally favorable: 91% at 10 years in 58 patients with 68 joints [13]; 97.5% at 10 years in 78 patients with 86 joints [14]; and 100% at 8 years in 71 patients with 73 joints [15]. The 8-year survival rate was 100% in 71 patients with 73 joints.
   (ii) Ceramic-on-ceramic bearing
      In a mean 7.1-year follow-up of 71 joints in 60 patients who underwent cementless THA with ceramic-on-ceramic between 1998 and 2001, 14 joints (20%) had squawking, no revision, implant laxity, dislocation, or ceramic failure [16].

(b) Cemented THA

When revision surgery was used as the endpoint, a report found the 10-year postoperative survival rate was 85.7% for 28 joints in 20 patients who underwent cemented THA between 1981 and 1985 [17]. Moreover, a report showed that the 15-year postoperative survival rate was 88.5% for 122 joints who underwent cemented THA between 1980 and 1990 [18], and an 18-year survival rate of 98% for the femoral side and 83% for the acetabular side following hybrid THA using a cementless cup [19].

There have been reports of lower revision rates in THA using second-generation cementation techniques compared to first-generation cementation techniques [17, 20]. Currently, cementation techniques are being improved to the fourth-generation; though, familiarity with the technique is essential for improving long-term outcomes.

(c) Resurfacing hip arthroplasty

The results of 16 joints in 12 patients who underwent resurfacing hip arthroplasty for ONFH between 2002 and 2006 (hemi-resurfacing arthroplasty (HRA) group; mean follow-up 6.5 years) and 16 joints in 10 patients who underwent THA (total hip resurfacing arthroplasty (TRA) group; mean follow-up 5.5 years) were compared. The mean Japanese Orthopedic Association hip score improved from a preoperative mean of 57 points to 84 points in the HRA group and a preoperative mean of 54 points to a postoperative mean of 96 points in the TRA group. In addition, THA was performed in five joints in the HRA group due to neck fracture, labral protrusion, progressive arthropathic changes, and groin pain [21].

(d) Dislocation

The Pearl Diver Database was used to search for patients who underwent THA for ONFH or OA between 2007 and 2011. A total of 45,002 patients with ONFH and 8429 patients with OA were studied. The dislocation rate within 1-year after surgery was 2.1% for ONFH, and 1.4% for OA in patients <65 years of age, 3.5% for ONFH, and 2.4% for OA in patients >65 years of age [9].

Comparing 111 joints in which a CT-based navigation system was used (navigation group) and 98 joints in which a navigation system was not used (non-navigation group), it was reported that the dislocation rate was significantly lower in the navigation group (2.7% versus 11.2% in the non-navigation group). The variance in cup inclination and anteversion angle was significantly smaller in the navigation group, and the use of a navigation system was described as a predictive factor for preventing dislocation [22].

Intraoperative radiography, image intensifiers, mechanical guides, and navigation systems should be used to obtain the target implant placement for THA for any disease.

(e) Comparison with OA

The 29 ONFH joints (ONFH group) and 63 OA joints (OA group) in which THA was performed between 1970 and

1984 were followed up for an average of 7 years. The revision rate was reported to be 6.4% in the OA group and 27.6% in the ONFH group [23]. The OA group consisted of three Charnley-Müllar joints and one Charnley joint on the femoral side, and the ONFH group consisted of two Trapezoidal-28 joints on the pelvic side, five Charnley-Müllar joints on the femoral side, and one Kyocera Bioceram joint on both sides. The implant survival rate was low for THA in older patients with ONFH.

A mean 10-year follow-up of 78 patients with 86 joints in the ONFH group and 78 patients with 86 joints in the age and sex-matched OA group who underwent cementless THA between 2000 and 2013 was conducted to determine the incidence of complications. The 10-year survival rate of implants, Harris hip score, Japanese Orthopedic Association Hip Disease Evaluation Questionnaire (JHEQ), physical component summary, and role/social component summary scores of the short form-36 (SF-36). No significant differences were found in the complication rate, 10-year implant survival rate, Harris hip score, physical component summary, and role/social component summary scores of the JHEQ (SF-36). Although no significant differences were found in the mental score of the JHEQ, the mental component summary score of the SF-36, and the visual analog scale score of satisfaction were reported to be significantly lower in the ONFH group than in the OA group [14]. Although recent years suggest that THA for OA has achieved comparable postoperative outcomes to THA for ONFH, a patient-based evaluation indicates that postoperative satisfaction with THA for ONFH is low.

(f) THA with previous femoral osteotomy

At a mean follow-up of 10.5 years for THA after TRO in 29 patients with 34 joints (post-TRO group), and 58 patients with 68 joints (primary group) matched for age and sex, the Harris hip score and pre- and postoperative range of motion at the last follow-up were significantly lower in the post-osteotomy group. Moreover, the short form-36, Oxford hip score, JHEQ, complication rate, and 10-year survival rate (81% in the post-TRO group and 91% in the primary group) with revision surgery as endpoints were not significantly different between the two groups [13]. A meta-analysis conducted in PubMed, Web of Science, and the Cochrane Library reported no significant difference in Harris hip score at the final follow-up compared with THA without previous osteotomy [12]. Although there was no difference in clinical outcomes, both studies reported a higher rate of stem malalignment in THA than in ONFH, which is technically demanding.

7 Future Tasks

There are two main issues: durability implants and dislocation. The implant survival rate has improved with the improvement of implant performance, including the sliding surface. Moreover, the dislocation rate has reduced with the introduction of the muscle-sparling approach and navigation systems. As patients with ONFH undergoing THA are often younger than patients undergoing THA with OA, improving the wear and tear resistance of sliding surfaces to withstand more intense exercise and labor is important. Moreover, selecting an approach and implant placement angle according to the individual patient to is essential to prevent dislocation.

References

1. Sakano S, et al. Curved intertrochanteric varus osteotomy for osteonecrosis of the femoral head. J Bone Joint Surg Br. 2004;86:359–65.
2. Sugioka Y. Transtrochanteric anterior rotational osteotomy of the femoral head in the treatment of osteonecrosis affecting the hip: a new osteotomy operation. Clin Orthop Relat Res. 1978;130:191–201.
3. Miyanishi K, et al. Prediction of the outcome of transtrochanteric rotational osteotomy for osteonecrosis of the femoral head. J Bone Joint Surg Br. 2000;82:512–6.
4. Seki T, et al. Quality of life following femoral osteotomy and total hip arthroplasty for nontraumatic osteonecrosis of the femoral head. J Orthop Sci. 2008;13:116–21.
5. Osawa Y, et al. Curved intertrochanteric varus osteotomy vs total hip arthroplasty for osteonecrosis of the femoral head in patients under 50 years old. J Arthroplast. 2020;35:1600–5.
6. Shoji T, et al. Preoperative T2 mapping MRI of articular cartilage values predicts postoperative osteoarthritis progression following rotational acetabular osteotomy. Bone Joint J. 2021;103:1472–8.
7. Ito H, et al. Bipolar hemiarthroplasty for osteonecrosis of the femoral head. A 7- to 18-year follow up. Clin Orthop Relat Res. 2000;374:201–11.
8. The Japanese Society for Replacement Arthroplasty. Artificial Joint Registry Survey, 2017 Report. https://jsra.info/data/pdf/report-2017.pdf. Japanese.
9. Scott Y, et al. Does osteonecrosis of the femoral head increase surgical and medical complication rates after total hip arthroplasty? A comprehensive analysis in the United States. Hip Int. 2015;25:237–44.
10. Shigemura T, et al. Total hip arthroplasty after failed transtrochanteric rotational osteotomy for osteonecrosis of the femoral head: a systematic review and meta-analysis. Orthop Traumatol Surg Res. 2018;104:1163–70.
11. Utsunomiya T, et al. The results of total hip arthroplasty after Sugioka transtrochanteric anterior rotational osteotomy for osteonecrosis. J Arthroplast. 2017;32:2768–73.
12. Shigemura T, et al. Cementless total hip arthroplasty for osteonecrosis of the femoral head in systemic lupus erythematosus: a study with 10-16 years of follow-up. Eur Orthop Traumatol. 2013;4:15–20.

13. Osawa Y, et al. Total hip arthroplasty after transtrochanteric rotational osteotomy for osteonecrosis of the femoral head: a mean 10-year follow-up. J Arthroplast. 2017;32:3088–92.
14. Osawa Y, et al. Cementless total hip arthroplasty for osteonecrosis and osteoarthritis produces similar results at ten years follow-up when matched for age and gender. Int Orthop. 2018;42:1683–8.
15. Kim YH, et al. Cementless total hip arthroplasty with alumina-on-highly cross-linked polyethylene bearing in young patients with femoral head osteonecrosis. J Arthroplast. 2011;26:218–23.
16. Baek SH, et al. Cementless total hip arthroplasty with alumina bearings in patients younger than fifty with femoral head osteonecrosis. J Bone Joint Surg Am. 2008;90:1314–20.
17. Kantor SG, et al. Cemented total hip arthroplasty in patients with osteonecrosis. A 6-year minimum follow-up study of second-generation cement techniques. J Arthroplast. 1996;11:267–71.
18. Christophe N, et al. Charnley-Kerboull total hip arthroplasty for osteonecrosis of the femoral head a minimal 10-year follow-up study. J Arthroplast. 2006;21:533–40.
19. Kim YH, et al. Contemporary total hip arthroplasty with and without cement in patients with osteonecrosis of the femoral head: a concise follow-up, at an average of seventeen years, of a previous report. J Bone Joint Surg Am. 2011;93:1806–10.
20. Garino JP, et al. Total hip arthroplasty in patients with avascular necrosis of the femoral head: a 2- to 10-year follow-up. Clin Orthop Relat Res. 1997;334:108–15.
21. Kabata T, et al. Hemi-resurfacing versus total resurfacing for osteonecrosis of the femoral head. J Orthop Surg. 2011;19:177–80.
22. Takashima K, et al. Does a computed tomography-based navigation system reduce the risk of dislocation after total hip arthroplasty in patients with osteonecrosis of the femoral head? A propensity score analysis. J Artif Organs. 2020;23:247–54.
23. Saito S, et al. Long-term results of total hip arthroplasty for osteonecrosis of the femoral head. A comparison with osteoarthritis. Relat Res. 1989;244:198–207.

Rheumatoid Arthritis

72

Isao Matsushita

1 Hip Joint Disorders in Rheumatoid Arthritis (RA)

RA is an autoimmune disease that causes synovitis in the joints throughout the body. Persistent synovitis causes bone and cartilage destruction via inflammatory cytokines and matrix metalloproteases, eventually leading to joint dysfunction. The hip joint is one of the synovial joints that can be destroyed, and the incidence of RA is reported to be between 13.0% and 19.8% [1, 2]. Once hip destruction appears in RA, it progresses relatively rapidly, and the patient's daily activities are significantly impaired. Recently, with advances in drug therapies for RA, the number of patients with RA complaining of hip disability has decreased, as has the frequency of surgery [3]. However, a fair number of patients with RA still have progressive destruction of the hip joints and undergo surgery.

2 Clinical Outcomes of Total Hip Arthroplasty in RA

Total hip arthroplasty (THA) is a surgical treatment for RA-related hip disorders. Clinical outcomes have been reported using the Harris hip score, with postoperative scores exceeding 80 points and a significant improvement in pain scores [4–11] (Table 72.1). Clinical studies using the Merle d'Aubigné and Postel score [12, 13] (Table 72.2) and the Japanese Orthopaedic Association score [14, 15] (Table 72.3) have shown very good improvement in postoperative clinical outcomes, especially in pain. THA for RA has shown promising clinical results.

In a questionnaire survey of 1156 members of the Japan Rheumatology Friendship Association, [16] 106 responses were obtained for THA. A total of 93.4% of the respondents answered that THA had the expected effect, and 89.6%

I. Matsushita (✉)
Department of Rehabilitation Medicine, Kanazawa Medical University, Kahoku, Ishikawa, Japan
e-mail: matsu-i@kanazawa-med.ac.jp

Y. Nakashima (ed.), *Advances in Total Hip Arthroplasty*, https://doi.org/10.1007/978-981-97-2461-1_72

Table 72.1 Harris hip score

Study	Report year	Number of joints	Observation period	Score
Tang [4]	2001	28	Mean 10.8 years	Improved from 25.3 points preoperatively to 88.9 postoperatively. Overall, 67.9% were completely pain-free postoperatively.
Thomason [5]	2001	98	Mean 7.4 years	Improved from 42 points preoperatively to 85 postoperatively. Postoperatively, 30.6% were excellent and 44.9% were good.
Katsimihas [6]	2003	65	Mean 9.1 years (6.8–14 years)	Improved from 21.5 points preoperatively to 81.5 postoperatively. Pain score improved from 5.8 points preoperatively to 40 postoperatively.
Yuasa [7]	2016	28	Mean 11.5 years (10–16 years)	Improved from 48.3 points preoperatively to 76.8 postoperatively. Pain score improved from 16.4 points preoperatively to 34.2 postoperatively. Gait score improved from 15.8 points preoperatively to 24.1 postoperatively.
Zhen [8]	2018	20	2.5–6 years	Improved from 55.3 points preoperatively to 92.2 postoperatively.

Table 72.2 Merle d'Aubigné and Postel score

Study	Report year	Number of joints	Observation period	Score
Garcia Araujo [9]	1998	33	Minimum 5 years	Improved from 7.2 points preoperatively to 16.6 postoperatively. Pain score improved from 1.6 points preoperatively to 5.6 postoperatively. Activity score improved from 2.9 points preoperatively to 5.7 postoperatively. Gait score improved from 2.1 points preoperatively to 5.3 postoperatively.
Jana [10]	2001	82	Mean 11 years	Pain score improved from 2.3 points preoperatively to 4.3 postoperatively (improvement of 91%). Gait score improved from 2.8 points preoperatively to 4.3 postoperatively (improvement of 61%).

Table 72.3 JOA score

Study	Report year	Number of joints	Observation period	Score
Matsushita [11]	2014	21	Minimum 10.5 years Mean 12.2 years	Improved from 36.9 points preoperatively to 73.4 postoperatively. Pain improved from 11.0 points preoperatively to 39.8 postoperatively. Range of motion improved from 13.9 points preoperatively to 16.0 postoperatively. Gait improved from 5.5 points preoperatively to 7.8 postoperatively. ADL improved from 6.5 points preoperatively to 9.9 postoperatively.
Haraguchi [12]	2017	191	Mean 12.6 years	Pain score improved from 13.5 points preoperatively to 39 postoperatively.

JOA Japanese Orthopaedic Association

answered that they were satisfied with THA, indicating that THA is a highly satisfactory treatment method that can meet patients' expectations.

In contrast, postoperative clinical outcomes of THA for patients with RA reportedly do not improve sufficiently when disease activity is high or when there are disorders in other joints of the lower limb [17]. It is important to control disease activity and provide comprehensive treatment that includes other joints.

3 Long-Term Survival Rate of Prostheses (Table 72.4) [10, 13–15, 18–20]

There have been concerns about the long-term durability of THA in patients with RA who have fragile bone quality. However, although there are some differences in long-term survival rates depending on the age of the patients and the type of prosthesis, many reports have shown that the 10-year

Table 72.4 Survival rate of prosthesis

Study	Report year	Number of joints	Survival rate
Jana [10]	2001	82	10-year survival rate with revision for all reasons as the endpoint Cementless stem: 98.1% (95% CI: 94.5–100), Porous-coated cementless cup: 93.8% (95% CI: 85.5–100), cemented cup: 84.3% (95% CI: 68.0–100)
Eskelinen [15]	2006	2557	10-year survival rate with revision by aseptic loosening as the endpoint Porous-coated cementless stem: 97% (95% CI: 96–99), cemented stem: 90% (95% CI: 88–92), press-fit porous-coated cementless cup: 92% (95% CI: 89–94), cemented cup: 91%(95% CI: 89–94)
Zwartele [16]	2008	77	10-year survival rate with failure due to aseptic loosening as the endpoint Cementless cup: 95% (95% CI: 0.90–1.0), Cementless stem: 100%
Makela [17]	2011	4019	10-year survival with revision by aseptic loosening as the endpoint Cementless THA: 96% (95% CI: 94–98), cemented THA (loaded taper stem): 94% (95% CI: 93–96), cemented THA (composite-beam stem): 90% (95% CI: 88–92)
Matsushita [11]	2014	21	14-year survival rate with revision by aseptic loosening as the endpoint Cementless stem: 100%, Cementless cup: 87.8%
Yuasa [7]	2016	28	16-year survival rate with aseptic loosening as the endpoint: 92.9%
Haraguchi [12]	2017	191	17-year survival rate with revision as the endpoint Cementless cup: 93.9%, Cementless stem: 99.5

or longer prosthesis survival rate is generally 90% or higher, with revision or aseptic loosening as the endpoint. Eskelinen et al. [18] analyzed 2557 THAs for RA and found that the 10-year survival rates of stems with revision due to aseptic loosening as the endpoint were 97% (95% CI: 96–99) for porous-coated cementless stems, 90% (95% CI: 88–92) for cemented stems, 92% (95% CI: 89–94) for porous-coated cementless cups, and 91% (95% CI: 89–94) for cemented cups. Zwartelé et al. [19] and Matsushita et al. [14] reported that no aseptic loosening of the stem was observed after >10 years of follow-up. Thus, THA for RA is a surgical technique with promising long-term durability.

4 Comparison with Osteoarthritis

Rud-Sørensen et al. [21] used the Danish registry to compare RA and osteoarthritis (OA) (cemented and cementless components were mixed). The maximum follow-up period was 14 years, and the relative risk rate of cup revision due to aseptic loosening of RA versus OA was 1.14 (95% CI: 0.77–1.68), with no difference between RA and OA. In contrast, the relative risk rate of stem revision due to aseptic loosening of RA versus OA was 0.54 (95% CI: 0.32–0.92), which was slightly better for RA than for OA.

Zwartelé et al. [22] compared RA and OA in cemented THA. The relative risk rate of cup revision for RA versus OA was 1.2 (95% CI: 0.4–3.3) in the Norwegian registry and 1.14 (95% CI: 0.77–1.68) in the Danish registry. The relative risk rate of stem revision for RA versus OA was 1.1 (95% CI: 0.6–2.0) in the Norwegian registry and 0.54 (95% CI: 0.32–0.92) in the Danish registry, indicating that RA and OA had similar outcomes.

Ravi et al. [23] compared revision rates in a meta-analysis of 40 clinical studies and reported that within 5 years, the odds ratio of revision with RA versus OA was 1.33 (95% CI: 1.03–1.71), but there was no difference between RA and OA at 6–10 years, and at >10 years, the risk of revision with RA was lower than that with OA.

5 Which Is Better for RA: Cemented or Cementless THA?

There is limited evidence comparing cementless and cemented THA for RA hip disorders. The 2020 update of the Japan College of Rheumatology clinical practice guidelines for the management of RA [16] conducted a meta-analysis of studies comparing cemented and cementless THA [7, 8, 13, 24] and found that the risk ratio of revision due to aseptic loosening of cementless to cemented cups was 0.17 (95% CI: 0.04–0.64). Cementless cups were superior to cemented cups. The risk ratio of revision due to aseptic loosening of cementless to cemented stems was 0.77 (95% CI: 0.06–9.12), confirming that the results were comparable (Fig. 72.1).

A systematic review by Zwartelé et al. [25] reported that the ratio of aseptic loosening of cementless to cemented prostheses was 0.6 (95% CI: 0.14–2.60) for cups and 0.71 (95% CI: 0.06–8.55) for stems, with no difference between them.

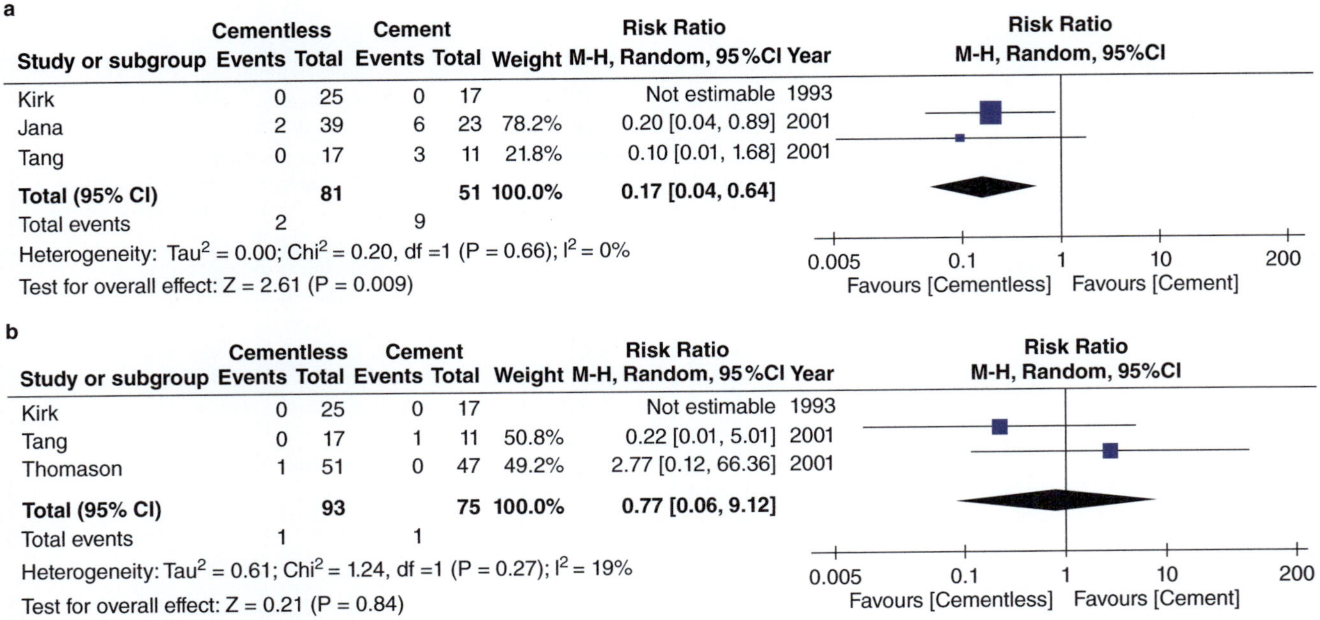

Fig. 72.1 Risk ratio of revision of cementless total hip arthroplasty (THA) to cemented THA in rheumatoid arthritis (RA) due to aseptic loosening. (**a**) Cup, (**b**) Stem. (Cited from [16])

6 Durability of Cementless Tapered Wedged Stem for RA Hips

Cementless tapered wedged stems have been increasingly used in patients with OA to preserve bone, but their fixation and durability in patients with RA have yet to be fully clarified. Keisu et al. [26] performed THA using a cementless tapered wedged stem in 50 hips of patients with RA and reported that no aseptic loosening was observed at a minimum follow-up of 5 years. Carl et al. [27] also performed THA using a cementless tapered wedged stem in 39 hips of patients with RA and reported no revision due to aseptic loosening at a minimum follow-up of 10 years. The authors compared the cementless short-tapered wedged stem for RA and OA and found no stem subsidence in patients with RA, confirming the same fixation as for OA [28].

7 Complications of THA in RA

A systematic review reported that the fracture incidence rates were 0–4% on the acetabular side and 0–21% on the femoral side [25]. The incidence of femoral fractures with cementless stems was slightly higher and required caution. In a systematic review summarizing cemented THA, the infection and dislocation rates were reported to be 0.7–3.7% and 9–11%, respectively [21].

According to Ravi et al. [23] patients with RA are at higher risk of dislocation after THA, with an adjusted odds ratio to OA of 2.16 (95% CI: 1.52–3.07). Compared with OA, patients with RA had a hazard ratio of 1.91 for dislocation within 2 years after THA [29]. Postoperative dislocation is a complication of particular concern in patients with RA who experience weakness of soft tissues and muscles. To reduce the dislocation rate, we should strive for more accurate intraoperative component placement, preservation of soft tissue as much as possible, and careful postoperative rehabilitation therapy.

8 Perioperative Drug Management

Patients with RA take various antirheumatic drugs to suppress disease activity, which must be managed during the perioperative period. Regarding methotrexate (MTX), Grennan et al. [30] reported that postoperative infections and complications did not increase when surgery was performed without MTX withdrawal. The 2020 update of the Japan College of Rheumatology clinical practice guidelines for the management of RA [16] also indicates that "it is recommended that MTX not be withdrawn during the perioperative period of orthopedic surgery."

A meta-analysis [31] of clinical studies comparing surgery with and without TNF inhibitor withdrawal in the perioperative period showed that the perioperative withdrawal of TNF inhibitors reduced the risk of postoperative surgical site infections (relative risk: 0.62) and complications (relative risk: 0.6). The 2020 update of the Japan College of Rheumatology clinical practice guidelines for the management of RA [16] recommends withdrawal of biological agents during the perioperative period of orthopedic surgery.

Goodman et al. [32] proposed that the preoperative drug withdrawal period should be the drug administration interval plus 1 week for biological agents and 4 days for JAK inhibitors. However, there is a lack of evidence for this, and future data accumulation and analysis are needed. Currently, the Japanese College of Rheumatology recommends that surgery be performed after a certain period of drug withdrawal based on the drug's half-life. The drug should be restarted after confirming that the postoperative wound has healed and that there are no signs of infection [33].

9 Conclusion

THA for RA offers excellent pain relief and long-term durability in cemented and cementless prostheses. Therefore, THA is a well-timed treatment for patients with RA who complain of difficulty walking due to hip dysfunction. However, dislocation occurs approximately twice as frequently as OA, and postoperative dislocation should be carefully monitored. Therefore, perioperative drug management should also be considered. We hope patients with RA who undergo THA will have good clinical outcomes with appropriate timing, accurate surgery, and careful perioperative management.

References

1. Drossaers-Bakker KW, et al. Radiographic damage of large joints in long-term rheumatoid arthritis and its relation to function. Rheumatology (Oxford). 2000;39:998–1003.
2. Dirven L, et al. Large-joint damage in patients with early rheumatoid arthritis and its association with treatment strategy and damage of the small joints. Rheumatology. 2012;51:2262–8.
3. Momohara S, et al. Recent trends in orthopedic surgery are aiming to improve quality of life for those with rheumatoid arthritis: data from a large observational cohort. J Rheumatol. 2014;41:862–6.
4. Lachiewicz P. Porous-coated total hip arthroplasty in rheumatoid arthritis. J Arthroplast. 1994;9:9–15.
5. Learmonth I, et al. Cementless replacement to the acetabulum at total hip arthroplasty in rheumatoid arthritis. J Orthop Rheumatol. 1996;9:33–6.
6. Loehr J, et al. Uncemented total hip arthroplasty in patients with rheumatoid arthritis. Clin Orthop. 1999;366:31–8.
7. Tang WM, et al. Primary total hip arthroplasty in patients with rheumatoid arthritis. Int Orthop. 2001;25:13–6.
8. Thomason H, et al. The influence of technique on the fixation of primary total hip arthroplasty in patients with rheumatoid arthritis. J Arthroplast. 2001;16:628–34.
9. Katsimihas M, et al. Cementless acetabular replacement in patients with rheumatoid arthritis. J Arthroplast. 2003;18:16–22.
10. Yuasa T, et al. Cementless total hip arthroplasty for patients with rheumatoid arthritis: a more than 10-year follow-up. Eur J Orthop Surg Traumatol. 2016;26:599–603.
11. Zhen P, et al. Total hip arthroplasty to treat acetabular protrusions secondary to rheumatoid arthritis. J Orthop Surg Res. 2018;13:92.
12. Araujo J, et al. Rheumatoid arthritis and hydroxyapatite-coated hip prostheses: five-year results. International ABG Study Group. J Arthroplasty. 1998;13:660–7.
13. Jana AK, et al. Total hip arthroplasty using porous-coated femoral components in patients with rheumatoid arthritis. J Bone Joint Surg Br. 2001;83:686–90.
14. Matsushita I, et al. Long-term clinical and radiographic results of cementless total hip arthroplasty for patients with rheumatoid arthritis: minimal 10-year follow-up. Mod Rheumatol. 2014;24:281–4.
15. Haraguchi A, et al. Minimum 10-year results of cementless total hip arthroplasty in patients with rheumatoid arthritis. Mod Rheumatol. 2017;27:598–604.
16. Japan College of Rheumatology ed. 2020 Update of the Japan College of Rheumatology clinical practice guidelines for the management of rheumatoid arthritis. Tokyo, Shindan to Chiryo Sha. Japanese.
17. Imagama T, et al. Risk factors associated with short-term clinical results after total hip arthroplasty for patients with rheumatoid arthritis. Orthopedics. 2018;41:e772–6.
18. Eskelinen A, et al. Total hip arthroplasty for rheumatoid arthritis in younger patients. Acta Orthop. 2006;77:853–65.
19. Zwartele R, et al. Long-term results of cementless primary total hip arthroplasty with a threaded cup and a tapered, rectangular titanium stem in rheumatoid arthritis and osteoarthritis. Int Orthop. 2008;32:581–7.
20. Mäkelä KT, et al. Cemented versus cementless total hip replacement in patients fifty-five years of age or older with rheumatoid arthritis. J Bone Joint Surg Am. 2011;93:178–86.
21. Rud-Sørensen C, et al. Survival of primary total hip arthroplasty in rheumatoid arthritis patients. Acta Orthop. 2010;81:60–5.
22. Zwartelé R, et al. Cemented total hip arthroplasty in rheumatoid arthritis: a systematic review of the literature. Hip Int. 2013;23:111–22.
23. Ravi B, et al. A systematic review and meta-analysis comparing complications following total joint arthroplasty for rheumatoid arthritis versus for osteoarthritis. Arthritis Rheum. 2012;64:3839–49.
24. Kirk P, et al. Total hip arthroplasty in rheumatoid arthritis: comparison of cemented and uncemented implants. Can J Surg. 1993;36:229–32.
25. Zwartelé RE, et al. Cementless total hip arthroplasty in rheumatoid arthritis: a systematic review of the literature. Arch Orthop Trauma Surg. 2012;132:535–46.
26. Keisu K, et al. Cementless femoral fixation in a rheumatoid patient undergoing total hip arthroplasty. J Arthroplast. 2001;16:415–21.
27. Carl HD, et al. Cementless total hip arthroplasty in patients with rheumatoid arthritis using a tapered designed titanium hip stem minimum: 10-year results. Rheumatol Int. 2011;31:353–9.
28. Matsushita I, et al. Clinical outcomes and radiological evaluation of cementless hip arthroplasty using short tapered wedged stem: a comparative study of RA and OA. Nichi Jinko Kansetsu Kaishi. 2015;45:833–4. Japanese
29. Ravi B, et al. Increased risk of complications following total joint arthroplasty in patients with rheumatoid arthritis. Arthritis Rheumatol. 2014;66:254–63.
30. Grennan DM, et al. Methotrexate and early postoperative complications in patients with rheumatoid arthritis undergoing elective orthopaedic surgery. Ann Rheum Dis. 2001;60:214–7.
31. Clay M, et al. Risk of postoperative infections and the discontinuation of TNF inhibitors in patients with rheumatoid arthritis: a meta-analysis. Joint Bone Spine. 2016;83:701–5.
32. Goodman SM, et al. 2022 American College of Rheumatology/American Association of hip and Knee Surgeons guideline for the perioperative management of antirheumatic medication in patients with rheumatic diseases undergoing elective total hip or total knee arthroplasty. Arthritis Care Res. 2022;74:1399–408.
33. Japan College of Rheumatology. Guidelines. https://www.ryumachi-jp.com/publish/guide/. Japanese.

Traumatic Arthropathy

73

Takeshi Sawaguchi

1 Causes of Traumatic Arthropathy

It is well known that acetabular fractures can cause traumatic arthropathy in the early stages if the displacement of the weight bearing dome remains, and currently, the results have improved by accurate reduction and stable fixation by surgery [1, 2]. However, traumatic arthropathy still occurs in approximately 17.6–20% of patients [3, 4]. The main causes of traumatic arthropathy are the severity of the fracture, poor reduction, femoral head injury, and femoral head avascular necrosis. In addition, traumatic arthropathy often occurs in femoral head fractures due to inadequate reduction by comminution and complications of femoral head avascular necrosis [5]. In patients with advanced traumatic arthropathy, joint replacement is an effective treatment for pain relief and functional recovery.

2 Precautions for Arthroplasty for Traumatic Arthropathy

Arthroplasty for traumatic arthropathy after femoral head fracture without acetabular fracture is not much different from arthroplasty for nontraumatic arthropathy. On the other hand, arthroplasty after acetabular fracture is technically more difficult than conventional arthroplasty because of pseudoarthrosis, pelvic malunion, bone loss, presence of internal fixation material, severe scar tissue, heterotopic ossification, infection (including occult infection) and sciatic nerve palsy [3, 6–9].

3 Preoperative Evaluation

In addition to the usual imaging evaluation, deformity, leg length discrepancy, joint range of motion, neurologic evaluation, heterotopic ossification, internal fixation material, and occult infection should be assessed.

4 Key Points of Surgery

If there is little deformity or heterotopic ossification and the internal fixation material does not interfere with the placement of the acetabular component, a conventional hip replacement can be performed. The posterior approach should be avoided in cases of sciatic nerve palsy. If posterior approach is necessary to remove the implant, nerve monitoring should be considered. If screws are present in the joint, they should be removed or reamed out as necessary. Inappropriately positioned plates should be removed or covered with a bone graft. For acetabular bone defects and pseudarthroses, bone grafting is performed using the resected femoral head (Fig. 73.1), or impaction grafting using allograft is performed to reinforce the acetabulum using an acetabular revision implant or mesh. The author mainly use K-T plates® for acetabular reconstruction. A large-diameter cup is also an option. In case of malunion with large medial displacement, a sufficient amount of bone graft is placed in the acetabular floor to lateralize the center of rotation, and a stem with a large offset is used to prevent dislocation. If the deformity is severe and the orientation of the acetabular

T. Sawaguchi (✉)
Fukushima Medical University/Shinyurigaoka General Hospital,
Kawasaki, Kanagawa, Japan
e-mail: sawaguch@mxq.mesh.ne.jp

Y. Nakashima (ed.), *Advances in Total Hip Arthroplasty*, https://doi.org/10.1007/978-981-97-2461-1_73

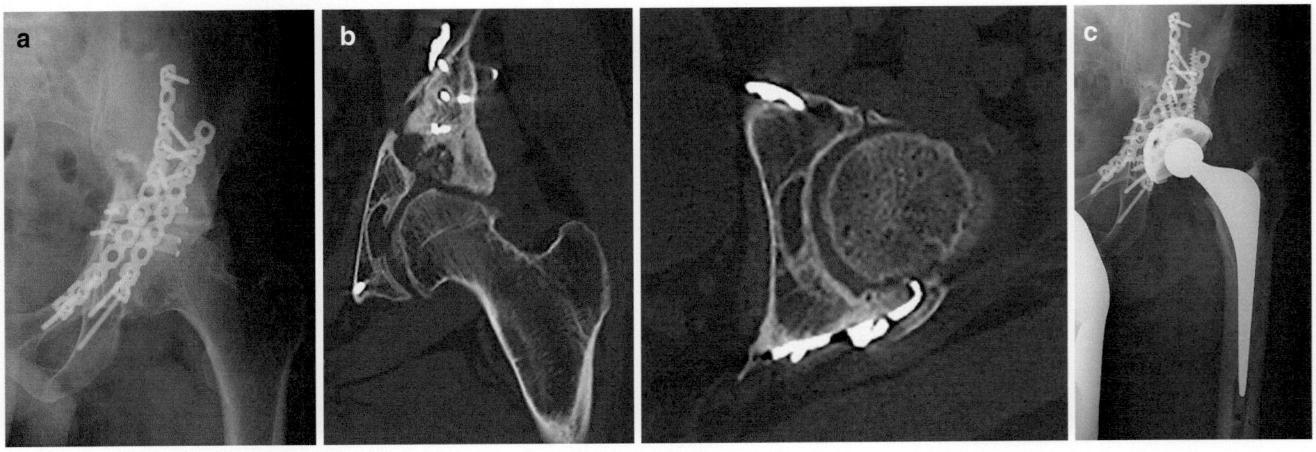

Fig. 73.1 A 45-year-old male. Left acetabular T-shaped fracture after osteosynthesis. (**a**) Arthropathy due to poor reduction 1 year after open reduction and internal fixation at another hospital. (**b**) CT image. The acetabular articular surface was poorly reduced, and nonunion of the acetabular side. There was a defect in the weight-bearing area of the femoral head, and the hook plate fixing the posterior wall damaged the head. (**c**) 2.5 years postoperatively. The hook plates were removed, and the resected head was grafted to the nonunion site. The acetabular side was a cementless cup, and the femoral stem was cemented with a high offset stem. There is no loosening on both acetabular and femoral side

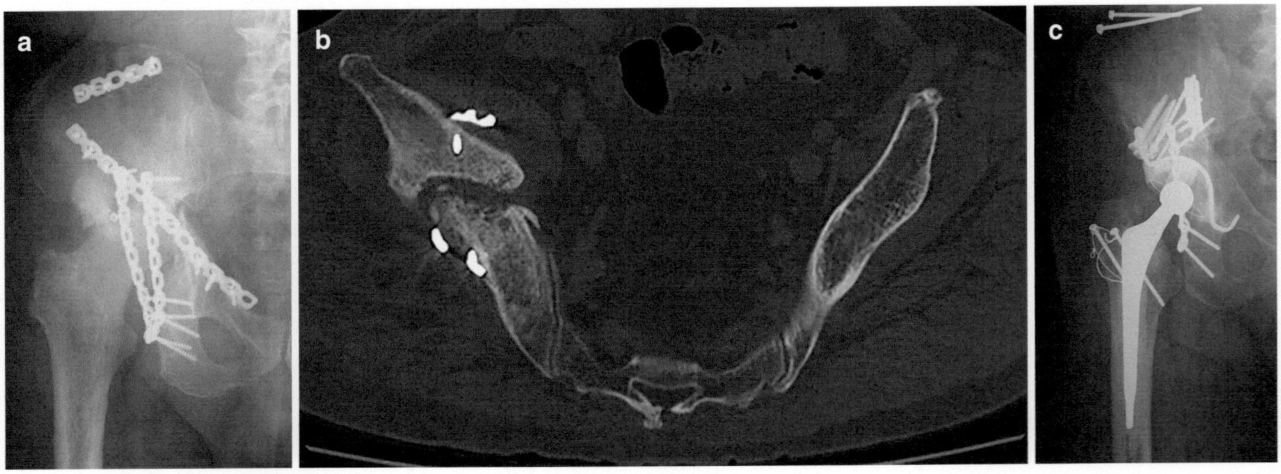

Fig. 73.2 68-year-old male. After osteosynthesis of the right acetabular both column fracture. (**a**) Arthropathy due to poor reduction 1 year after open reduction and internal fixation at another hospital. (**b**) CT image. The anterior and posterior columns are not united and the anterior column is rotated posteriorly. (**c**) The rotational deformity of the anterior column was corrected by osteotomy, the resected head was used for bone grafting to the nonunion and acetabular floor, and the acetabulum was reconstructed using a K-T plate followed by cement cup fixation. The stem was cemented. There is no loosening on both acetabular and femoral side 6 years after surgery

component placement is not adjustable, an osteotomy to correct the deformity may be necessary (Fig. 73.2). Heterotopic ossification that results in impaired range of motion should be resected. Radiation is effective in preventing recurrence, but the benefits and harms must be fully explained [9]. If infection is present or suspected, it is advisable to perform a two-stage joint replacement with debridement, femoral head resection and placement of an antibiotics loaded cement spacer at the initial surgery. And perform joint replacement at the second stage (Fig. 73.3).

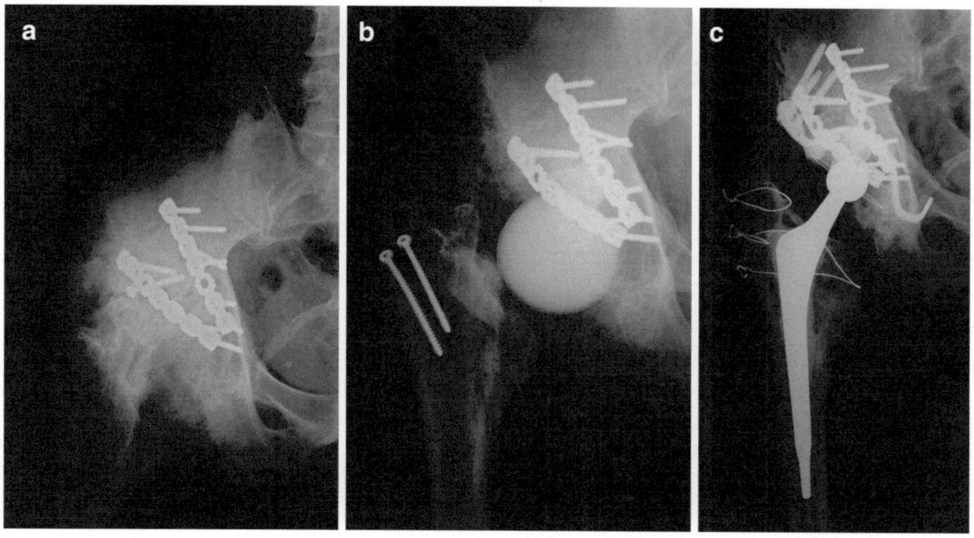

Fig. 73.3 57-year-old male. Right transverse fracture + posterior wall fracture after surgery. (**a**) Twenty years after open reduction and internal fixation, the patient is with end-stage arthropathy and he was scheduled for arthroplasty. However, 2 weeks before the admission, the patient had fever and pain. It turned out of infection. (**b**) The first-stage surgery included femoral head resection, intra-articular debridement, and placement of an antibiotics loaded cement spacer. (**c**) Eight weeks after the first stage of surgery, the acetabulum was reconstructed using an allogeneic bone graft and K-T plate, and a cement cup was fixed. The stem was cemented. There is no loosening on both acetabular and femoral side 4 years after surgery

5 Postoperative Management

Postoperative management includes antimicrobial agents, deep vein thromboprophylaxis, and rehabilitation under pain control, as in the case of conventional arthroplasty. When osteosynthesis and joint replacement are performed simultaneously for pseudoarthrosis or osteotomy for malunion, weight bearing should be applied gradually.

6 Clinical Outcome, Complications

Arthroplasty for traumatic arthrosis after acetabular fracture has been reported to be associated with a relatively early onset of aseptic loosening of both acetabular components and femoral stems, inferior long-term outcomes, and a higher frequency of revision compared with initial arthroplasty for nontraumatic cases [1, 6–9, 11–13].

Romness et al. reported a mean follow-up of 7.5 years in 55 hips with 53 cases of joint replacement after acetabular fractures mainly treated conservatively and found that 13.7% required acetabular revision with 52.9% radiographic loosening and 27.5% symptomatic loosening on the acetabular side and 7.8% required femoral revision with 29.4% radiographic loosening and 15.7% symptomatic loosening on the femoral side [11]. Weber and Roth et al. reported that in 66 cases of joint replacement after acetabular fracture with osteosynthesis, when the endpoint was revision due to loosening, the survival rate at 10 years after surgery was 78% without revision on both the acetabular and femoral sides (87% on the acetabular side only, 84% on the femoral side only). The survival rate at 20 years after surgery in 25 patients excluding subsequent deaths was 71% for the acetabular component with revision by loosening as the endpoint and 57% with all revisions as the endpoint [8, 12]. In a systematic review of 645 cases, Makridis et al. found that the 10-year survival rates for acetabular components and femoral stems with the endpoints of loosening, osteolysis or revision were 81% and 95%, respectively [6]. Morrison et al. reported the survival rates for acetabular components after acetabular fracture and for osteoarthritis or femoral head necrosis. In a case–control study of 74 cases each (cementless on all acetabular sides and cementless in 70 of 74 femoral side cases), the 10-year survival rate was 70% after acetabular fracture and 95% in the control group with revision as the endpoint. The results of arthroplasty after acetabular fracture were definitely inferior to the arthroplasty for nontraumatic osteoarthritis [1]. Compared with nontraumatic arthroplasty, there are more younger adults with traumatic arthropathy and there seems to be more burden on the arthroplasty after surgery [8]. And in elderly patients, abductor muscle weakness due to trauma and repeated falls are considered to be the cause of poor prognosis. Complications such as bleeding, prolonged surgery, sciatic nerve palsy, dislocation, infection, and heterotopic ossification, in addition to component loosening, are more frequent than in conventional joint replacement [1, 7–12]. Arthroplasty for traumatic arthropathy, especially after acetabular fracture, has more problems than conventional arthroplasty, and it is recommended to be performed at a facility with abundant experience in acetabular fracture osteosynthesis along with arthroplasty [1, 6].

References

1. Morison Z, et al. Total hip arthroplasty after acetabular fracture is associated with lower survivorship and more complications. Clin Orthop Relat Res. 2016;474:392–8.
2. Ziran N. Outcomes after surgical treatment of acetabular fractures: a review. Patient Saf Surg. 2019;16(13):16.
3. Kelly J, et al. Surgical management of acetabular fractures -a contemporary literature review. Injury. 2020;51:2267–77.
4. Giannoudis PV, et al. Operative treatment of displaced fractures of the acetabulum: a meta-analysis. J Bone Joint Surg Br. 2005;87:2–9.
5. Giannoudis PV, et al. Management, complications and clinical results of femoral head fractures. Injury. 2009;40:1245–51.
6. Makridis KG, et al. Total hip arthroplasty after acetabular fracture: incidence of complications, reoperation rates and functional outcomes: evidence today. J Arthroplast. 2014;29:1983–90.
7. Morita D, et al. Acetabular fracture nonunion with pelvic discontinuity treated with two-stage total hip arthroplasty after intra- and extra-articular plate fixation. Nagoya J Med Sci. 2019;81:113–9.
8. Weber M, et al. Total hip arthroplasty after operative treatment of an acetabular fracture. J Bone Joint Surg Am. 1998;80:1295–305.
9. Ranawat A. Total hip arthroplasty for posttraumatic arthritis after acetabular fracture. J Arthroplast. 2009;24:759–67.
10. Davis JA. Does postoperative radiation decrease heterotopic ossification after the Kocher-Langenbeck approach for acetabular fracture?
11. Romness DW, et al. Total hip arthroplasty after fracture of the acetabulum. Long-term results. J Bone Joint Surg Br. 1990;72:761–4.
12. Roth P, et al. Total hip arthroplasty after operatively treated acetabular fracture. J Bone Joint Surg Am. 2015;97:288–91.
13. Scott CEH. Cemented total hip arthroplasty following acetabular fracture. J Bone Joint Surg Br. 2017;99:1399–408.

Part XIV

Difficult Cases

74 THA with Subtrochanteric Shortening Osteotomy for Crowe Group IV Dislocated Hips

Masaki Takao

1 Total Hip Arthroplasty with Subtrochanteric Shortening Osteotomy

Total hip arthroplasty (THA) for Crowe group IV dislocated hips is a difficult procedure because of the high incidence of complications, such as cup loosening due to a hypoplastic acetabulum, stem loosening due to a narrow and straight femoral medullary canal, limping due to residual leg length difference or non-union of the osteotomized greater trochanter, nerve palsy due to intraoperative manipulation or excess leg lengthening, and postoperative dislocation [1]. In the 1990s, THA with subtrochanteric femoral shortening osteotomy was reported to achieve stable cup placement in the original acetabular position, prevent over-lengthening of the leg, produce good rotational stability of the stem by preserving the proximal femur, and offer appropriate restoration of the abductor muscle function by avoiding pseudoarthrosis of the greater trochanter. In Japan, this procedure has been widely performed with good mid- to long-term results (Table 74.1) [2–15]. Before the implementation of a nationwide screening for hip dislocation during infant health checkups in Japan in the 1970s, many cases of hip dislocation were left untreated. Thus, many clinical outcomes and surgical techniques for THA of highly dislocated hips have been reported. In the future, the number of cases is expected to decrease; however, the health examination system has become weaker with the decrease in the number of infant hip dislocation cases, whereas the number of diagnosed cases of hip dislocation after the start of walking is increasing. A 2-year nationwide survey conducted by the Japanese Pediatric Orthopedics Association Multi-Center Study Committee in 2011 revealed that nearly 100 cases were diagnosed annually after the start of walking [16]. Therefore, THA combined with subtrochanteric femoral shortening osteotomy is a necessary surgical technique for some patients with highly dislocated hips. However, imparting surgical techniques to the next generation could be difficult because of the decline in cases of hip dislocations. This section summarizes the indications, tips, key points, clinical outcomes, and implant selection for THA with subtrochanteric femoral shortening osteotomy.

M. Takao (✉)
Department of Orthopaedic Surgery, Graduate School of Medicine, Ehime University, Toon, Ehime, Japan

Department of Orthopaedic Surgery, Ehime University Graduate School of Medicine, Toon, Japan
e-mail: takao.masaki.ti@ehime-u.ac.jp

Y. Nakashima (ed.), *Advances in Total Hip Arthroplasty*, https://doi.org/10.1007/978-981-97-2461-1_74

Table 74.1 Reported clinical outcomes of total hip arthroplasty with subtrochanteric shortening osteotomy in Japan

Author	Reported year	Number of joints	Observation period (years)	Osteotomy design	Approach	Stem fixation (cementless/cemented)	Stem revision[a] (joint)	Stem loosening (joint)	Intraoperative fracture (joint)	Pseudojoint (joint)	Dislocation (joint)	Neuropathy (joint)
Onodera S	2006	14	5.1	Transverse	PL	14/0	1	1	6	1	1	0
Makita H	2007	11	5.4	Step-cut	Modified Watson–Jones	11/0	1	1	NA	0	1	2[b]
Nagoya S	2009	20	5.1	Transverse	PL	20/0	2	1	2	1	0	0
Takao M	2011	33	8	Step-cut	PL	33/0	1	1	8	0	2	0
Kawai T	2011	19	3.5	Transverse	Transtrochanteric	0/19	0	0	NA	0	0	0
Akiyama H	2011	15	6.3	Transverse	Dall	0/15	3	0	NA	3	2	0
Ogawa H	2011	6	8.1	Transverse	PL	6/0	0	0	NA	0	2	0
Sonohata M	2012	22	7.6	Double chevron	PL	22/0	1	1	4	1	5	0
Oe K	2013	34	5.2	Transverse	Dall	0/34	0	0	NA	1	3	0
Oinuma K	2014	12	3.7	Oblique	DAA	12/0	0	0	NA	0	1	0
Sonohata M	2016	34	7.3	Double chevron or V shaped	PL	34/0	2	1	2	2	3	0
Kawai T	2020	27	10	Transverse	Hardinge or Dall	17/10	2	1	NA	1	3	0
Inoue D	2021	26	5.7/3.5[c]	V-shaped	PL	13/13	0	0	NA	0	3/0[c]	0

PL posterolateral approach, *DAA* direct anterior approach, *NA* not available
[a]Stem replacement for all reasons
[b]Only numbness in the thigh was relieved in 3 weeks
[c]Cementless stem/cement stem, in order

2 Indications for Surgery

The indications for THA with subtrochanteric shortening osteotomy include Crowe group IV dislocated hips, cases in which the greater trochanter is elevated because of a Perthes deformity which require a leg extension of ≥3 cm, and hips with angular deformities of the femoral medullary cavity due to previous femoral osteotomies, such as Schanz osteotomy [13]. Primary diseases include developmental hip dysplasia, congenital bone diseases such as spondyloepiphyseal dysplasia and multiple epiphyseal dysplasia, and postpyogenic arthritis of the hip joint.

3 Tips for Surgical Planning

Preoperative planning is performed using simple anteroposterior and lateral radiographic images of the hip joint and computed tomography (CT). In principle, the cup should be placed in its original acetabular position. As the acetabulum is often hypoplastic, predicting the anteroposterior diameter of the acetabulum from an anteroposterior simple radiographic hip joint image can be difficult. The acetabulum is often triangular rather than hemispherical, and selecting the cup size is easier by checking the fit of the cup in the hypoplastic acetabulum using a three-dimensional template (Fig. 74.1).

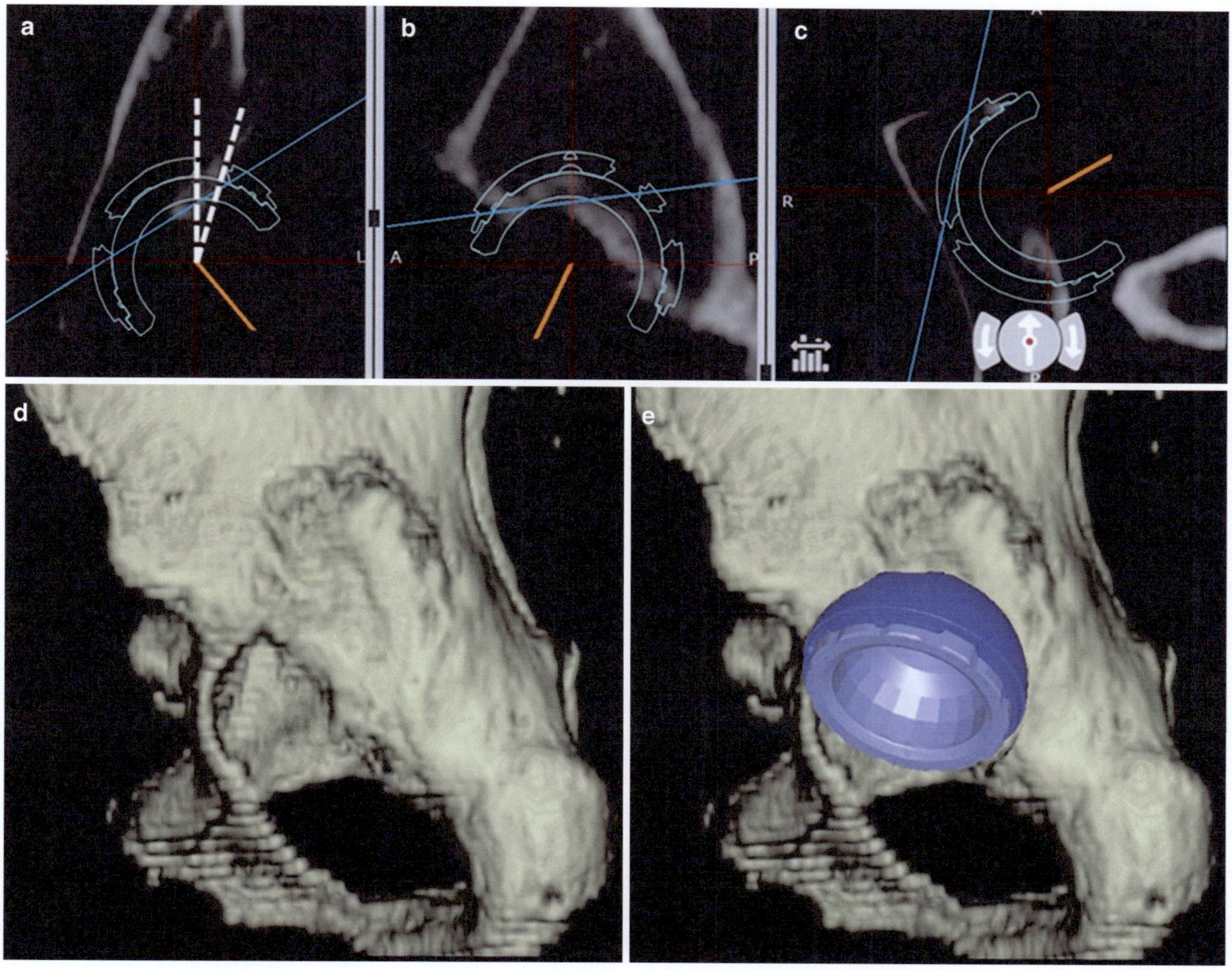

Fig. 74.1 Three-dimensional plan of cup placement. Coronal (**a**), sagittal (**b**), axial (**c**), and volume-rendering images (**d**, **e**). The position of the cup should be determined such that the cup center-edge angle (dashed line) is at least 10°. The acetabulum is often buried in the bone and should be identified by examining the closed foramen at the inferior margin of the acetabular fossa, base of the sciatic bone, and secondary acetabulum

The cup of the selected size is placed in the original acetabular position on an anteroposterior simple X-ray hip joint image or on a coronal image of a three-dimensional CT-based plan; the cup center-edge (CE) angle [17, 18] is measured as the angle between the vertical line through the cup center and the oblique line through the cup center and the lateral edge of the acetabulum (Fig. 74.1a) to evaluate bony coverage. The cup should be placed as far medially as possible to achieve a cup CE angle of ≥10°. If the cup CE angle is <10° in the preoperative planning, we consider bulk bone grafting using the resected femoral head (Fig. 74.1).

The stem brand is determined by checking the fit in both the proximal and distal medullary cavities after subtrochanteric shortening osteotomy. In the case of a proximal fixation cementless or cemented stem, femoral neck preservation should be checked to ensure rotational stability of the stem [15]. First, a stem template is placed in a position where the center of the femoral head is aligned with the tip of the greater trochanter. However, because the correlation between the tip of the greater trochanter and height of the femoral head center is never high, the placement position should be adjusted by prioritizing stem fitting in the proximal medullary cavity (Fig. 74.2) [19].

Preoperative methods for predicting the extent to which the greater trochanter can be moved distally at surgery include radiography in the flexed-adducted position to confirm how far distally the tip of the greater trochanter can be moved [9] and preoperative range of motion (ROM) [20]. In particular, patients with secondary acetabulum formation often have stronger joint contractures than those with complete dislocation [8, 21], thus making moving the greater trochanter distally difficult.

The short diameter of the oval-shaped medullary cavity was measured using CT. If the stem size is slightly larger than this diameter, it serves as an intramedullary nail to provide firm fixation of the osteotomy site (Fig. 74.2).

Subsequently, leg elongation is measured, and if it exceeds 3 cm, shortening osteotomy is performed at a position of 1–2 cm below the lesser trochanter. The designs of shortening osteotomy include step, oblique, horizontal, V-shaped, and inverted V-shaped (chevron) osteotomies [22]. Step, oblique, V-shaped, and inverted V-shaped osteotomies ensure rotational stability and increase the contact area of the osteotomy; however, they are more difficult to perform when the osteotomy site requires anteversion adjustment. Hotokebuchi et al. developed an instrument for V-shaped osteotomy and demonstrated its usefulness [14, 23].

After determining the design of the shortening osteotomy, the stem size is finally confirmed by checking the fit between the distal stem and distal medullary cavity. In cases in which the stem size is small because of a narrow medullary cavity, the femoral offset may be insufficient to ensure hip joint stability. Hence, we consider the use of a modular-type stem with modularity in the stem body or a stem with a modular neck.

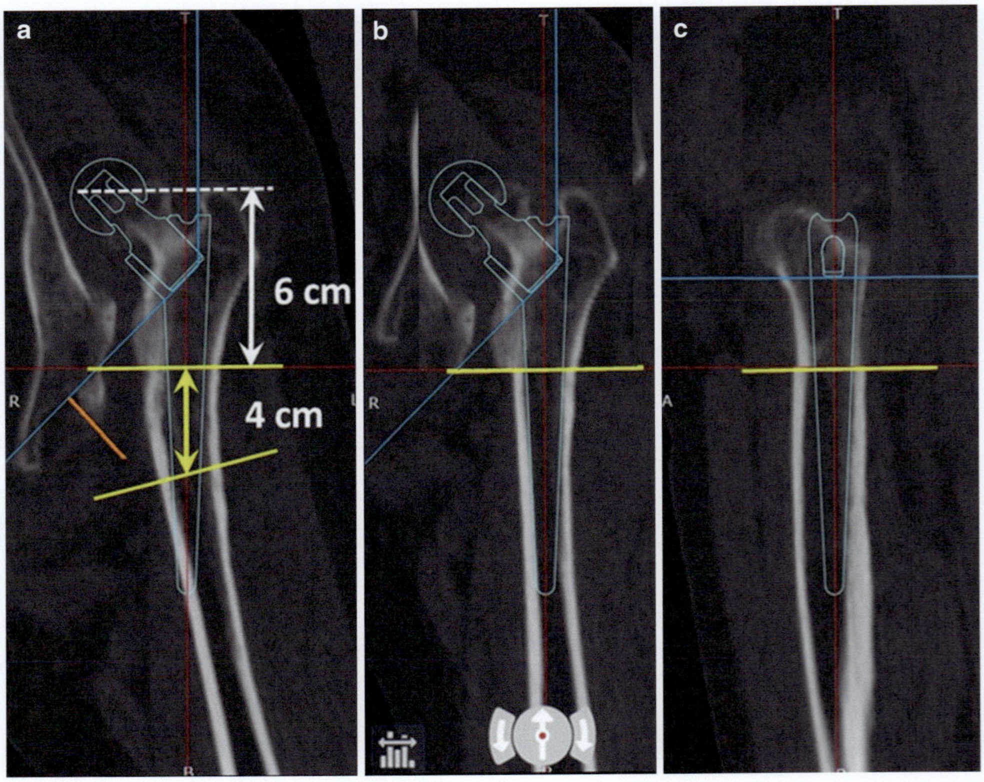

Fig. 74.2 Three-dimensional plan of stem placement. Coronal (**a**, **b**) and sagittal (**c**) views of the stem placement and shortening osteotomy. Plan of stem placement and shortening osteotomy for the proximal fragment (**a**), and the final plan after shortening osteotomy (**b**, **c**). This case is a post-Shanz osteotomy, so we planned to correct the femoral angular deformity with shortening. First, a stem template is placed such that the greater trochanter is aligned with the center of the artificial femoral head. The leg elongation is 7 cm; therefore, we decided to shorten the leg by 4 cm. The distance from the apex of the greater trochanter to the shortened osteotomy (white arrow) and the distance of the shortened area (yellow arrow) are measured to guide the intraoperative procedure

4 Tips for Surgical Techniques

(a) Approach

Although the posterolateral approach has been demonstrated in many cases, the use of the anterior, anterolateral, and lateral approaches has also been reported [11]. THA with subtrochanteric shortening osteotomy is a technically demanding procedure; therefore, surgeons may prefer using a familiar surgical approach. The elongated capsule is released from its bony attachment and detached from the surrounding musculature, and the original acetabulum is identified. If the elongated capsule prevents leg lengthening required for hip joint reduction, it should be removed completely. The acetabulum is often covered with soft tissue that is also resected. Subsequently, a neck osteotomy is performed. If a proximal fixation stem is used, the neck should be preserved to the maximum extent possible. Osteotomy at the base of the neck decreases the rotational stability of the stem [15].

(b) Reaming of the acetabulum and cup placement

The acetabulum is often observed as a slope, and identifying the reaming position is often difficult (Fig. 74.1). The original acetabular position is identified using the following landmarks as a guide: inferior margin of the acetabular fossa, base of the sciatic bone, and secondary acetabulum. The bone covering the original acetabulum is removed using a curved chisel. The reaming position is determined such that the superior iliac bone would be preserved. If a bulk femoral head bone graft is to be used, it should be fixed above the acetabulum with two AO screws and reamed.

(c) Preparation of the femoral medullary canal and subtrochanteric shortening osteotomy

In the absence of femoral deformity, such as after osteotomy, rasping and reaming of the femoral medullary canal are performed before shortening osteotomy. A longitudinal incision is made through the fascia of the vastus lateralis muscle, and the femur is osteotomized around the femoral attachment of the gluteus maximus with the planned shortening amount (Fig. 74.3). Hemostasis is performed as appropriate because of the proximity of the profunda femoris artery branches. The distal fragment is reamed at the depth determined in the preoperative plan.

(d) Stem placement

In the case of cementless stems, the femoral osteotomy is temporarily fixed with bone-holding forceps, and the stem is placed (Fig. 74.4). In the case of cemented stems, inserting the cemented stem is easier after the proximal and distal bone fragments are fixed with a bone or metal plate. The anteversion angle of the stem is adjusted to 30° on the basis of the lower leg axis, with the knee flexed at 90° during the operation. Moreover, assistants should pay careful attention not only to the changeability of the anteversion of the cemented stem until fixation but also to cementless stems during seating. If bone quality is poor, prophylactic wire clamping

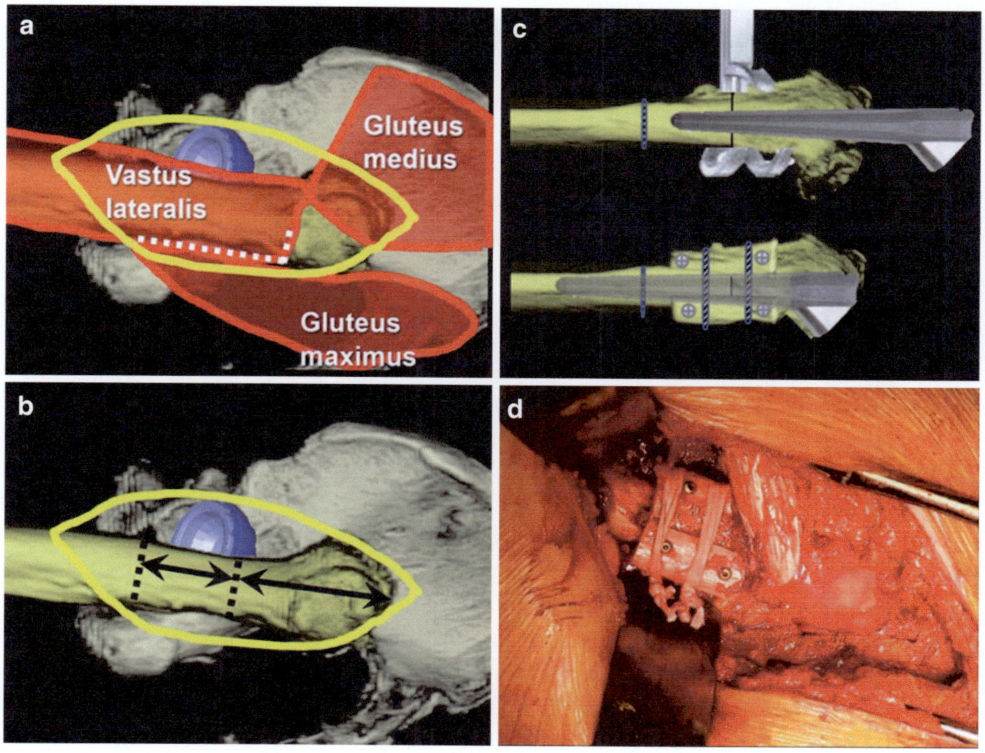

Fig. 74.3 Subtrochanteric shortening osteotomy and stem placement. (**a**) Exposure of the subtrochanteric femoral diaphysis for shortening osteotomy. The fascia of the greater trochanter attachment of the vastus lateralis muscle is incised posteriorly (white dashed line), and the vastus lateralis muscle is reflected anteriorly to expose the femoral diaphysis. (**b**) Marking the shortened osteotomy. Marking is on the basis of the distance from the apex of the greater trochanter, as measured in the surgical plan. (**c**) Stem placement and bone plate grafting. In the present case, a tapered cone-type stem is used because of the hypoplastic femoral neck. (**d**) Image of the surgical field of a bone plate graft

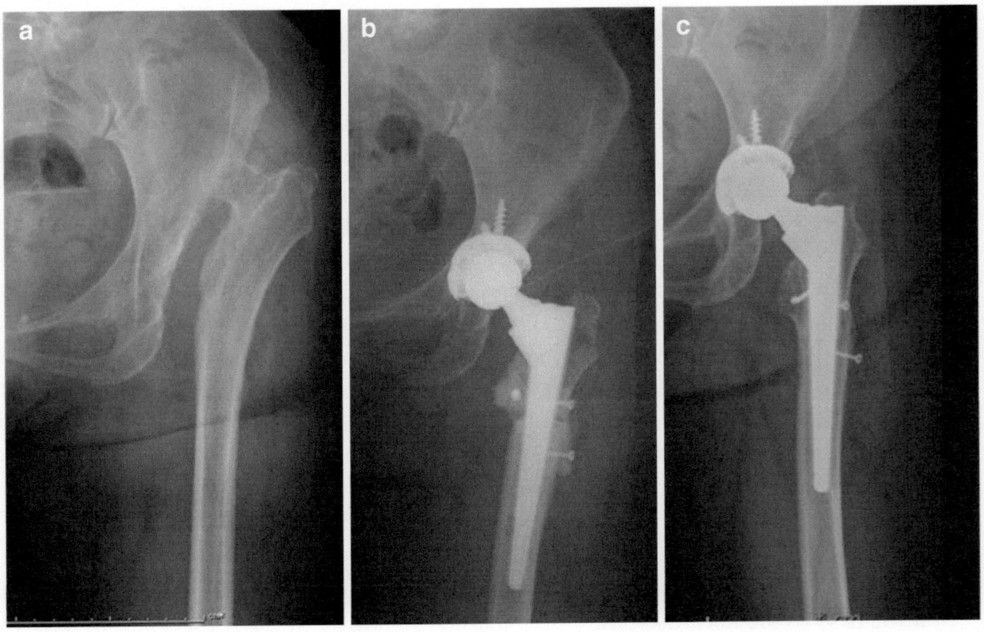

Fig. 74.4 Simple X-ray anteroposterior view of the hip joint preoperatively (**a**), immediately postoperatively (**b**), and 3 years postoperatively (**c**)

should be performed on the proximal and distal fragments. To promote bone formation, a morselized bone is grafted into the osteotomy gap or around the bone plate during the cup reaming. In the case of cemented stems, cement that protrudes from the osteotomy gap should be removed.

5 Clinical Results

(a) Clinical results and stem models

Good mid-term results using cementless or cemented stems have been reported in Japan (Table 74.1). Cementless stems have various types, such as modular, tapered cone, cylindrical fully coated, tapered round, and double-wedge metaphyseal filling types. A Crowe group IV femur is generally characterized by a narrow medullary canal and hypoplasia of the proximal portion, although many individual differences exist. Hence, optimizing the femoral offset and femoral neck anteversion, which are not correlated with the shape of the femoral medullary canal, is also necessary. Takao et al. performed cementless THA with subtrochanteric femoral shortening osteotomy in a step-cut design with a modular type S-ROM stem for a Crowe group IV dislocated hip. A good 10-year survival rate of 97% with an endpoint of revision of 33 joints in 25 patients has been reported [15]. Meanwhile, in one case, the femoral neck was hypoplastic, and the neck osteotomy was at the base of the lesser trochanter. Hence, bone ingrowth of the proximal sleeve was not obtained, resulting in stem loosening. In cases of a hypoplastic femoral neck, a tapered cone-type stem is recommended, whereas in those of a Dorr type C femur with significant medullary canal enlargement due to osteoporosis, a cemented stem is recommended.

(b) Intraoperative fractures

The incidence of intraoperative fractures, including small cracks, in cementless THA with subtrochanteric shortening osteotomy is higher than that in conventional THA, ranging from 5% to 46% in reported cases. This may be due to the hypoplastic femur and mismatch of the proximal and distal medullary canal geometry caused by the shortening osteotomy. Prophylactic wire clamping has been reported in some studies.

(c) Dislocation

The acetabulum is hypoplastic, resulting in a smaller cup and head size. Both the ROM and jumping distance of the implant are reduced, resulting in decreased resistance to dislocation. Soft-tissue resistance to dislocation is also reduced in patients who require complete dissection of the surrounding joint capsule. In some cases, the stem size decreases owing to the narrow medullary canal, and the femoral offset may be insufficient. The dislocation rate of THA with subtrochanteric shortening osteotomy was reported to be 5.88% (95% confidence interval [CI]: 4.22–7.80%) in a meta-analysis, which was higher than that of conventional primary THA [22]. The dislocation rate is expected to decrease in the future with the use of larger diameter heads with improved polyethylene liners, stems whose femoral offset can be adjusted independently of the medullary canal shape according to modularity, and navigation surgery.

(d) Nerve palsy

Although the amount of leg lengthening allowed to prevent postoperative nerve palsy in THA is unknown

[24], many studies have set the amount of femoral shortening to prevent leg lengthening as ≥3–4 cm [15]. Sonohata et al. have reported that nerve palsy occurred in four of six THA cases performed without subtrochanteric shortening osteotomy, but not in three cases with subtrochanteric shortening osteotomy. The leg extension was 4.5–6.5 cm in patients with neuropathy and 3.0–5.5 cm in those without neuropathy [25, 26]. The incidence of nerve palsy in patients receiving THA with subtrochanteric shortening osteotomy was reportedly 2.63% (95% CI: 1.60–3.87%) in a meta-analysis [22].

(e) Non-union at the osteotomy site

Bone healing of shortened osteotomies is determined by the rotational stability of bone fragments based on stem fixation and osteotomy design, use of metal plates or not, and method of bone grafting (bone plate or bone chips). Thus, the results of previous reports vary. The non-union rate of THA with subtrochanteric shortening osteotomy was reportedly 3.79% (95% CI: 2.60–5.20%) [22]. The major osteotomy designs are as follows: horizontal, oblique, V-shaped, and step cut, with each method having its own advantages and disadvantages. However, regardless of the method used, obtaining intraoperative rotational stability at the osteotomy site is important. In the case of cemented stems, osteosynthesis failure and prolonged fusion due to cement intervention in osteotomy associated with cement pressurization may occur. Inoue et al. have reported that the time to bone union was 9.8 months for cemented stems and 5.0 months for cementless stems, indicating that bone union was achieved longer with cemented fixation [4]. Kawai et al. have reported that the non-union of the osteotomy was unrelated to the use of cement and that for shorter required femoral resection lengths, that is smaller bone plates, the risk of delayed union was higher [5].

Although THA for a Crowe group IV dislocated hip is one of the most difficult surgical procedures, subtrochanteric femoral shortening osteotomy is advantageous as it allows both optimal reconstruction of the hip abductor mechanism and avoidance of excessive elongation of the leg. In selecting the stem brand, confirming the fit of the stem, stability of the osteotomy site, femoral anteversion and offset, and leg length in preoperative planning is crucial.

References

1. Anwar MM, et al. Total hip arthroplasty in the neglected congenital dislocation of the hip: a five- to 14-year follow-up study. Clin Orthop Relat Res. 1993;295:127–34.
2. Akiyama H, et al. Cemented total hip arthroplasty with subtrochanteric femoral shortening transverse osteotomy for severely dislocated hips. J Orthop Sci. 2011;16:270–7.
3. Hasegawa Y, et al. Total hip arthroplasty for Crowe type IV developmental dysplasia. J Arthroplast. 2012;27:1629–35.
4. Inoue D, et al. Comparison of mid-term clinical results between cementless and cemented femoral stems in total hip arthroplasty with femoral. Arch Orthop Trauma Surg. 2021;141:1057–64.
5. Kawai T, et al. Total hip arthroplasty combined with subtrochanteric transverse shortening osteotomy: factors associated with delayed union at the osteotomy site. J Am Acad Orthop Surg Glob Res Rev. 2020;4:e20.00056.
6. Kawai T, et al. Cemented total hip arthroplasty with transverse subtrochanteric shortening osteotomy for Crowe group IV dislocated hip. J Arthroplast. 2011;26:229–35.
7. Makita H, et al. Results on total hip arthroplasties with femoral shortening for Crowe's group IV dislocated hips. J Arthroplast. 2007;22:32–8.
8. Nagoya S, et al. Cementless total hip replacement with subtrochanteric femoral shortening for severe developmental dysplasia of the hip. Surg Br. 2009;91:1142–7.
9. Oe K, et al. Subtrochanteric shortening osteotomy combined with cemented total hip arthroplasty for Crowe group IV hips. Arch Orthop Trauma Surg. 2013;133:1763–70.
10. Ogawa H, et al. Subtrochanteric transverse shortening osteotomy in cementless total hip arthroplasty achieved using a modular stem. Orthopedics. 2011;34:170.
11. Oinuma K, et al. Total hip arthroplasty with subtrochanteric shortening osteotomy for Crowe grade 4 dysplasia using the direct anterior approach. J Arthroplast. 2014;29:626–9.
12. Onodera S, et al. Cementless total hip arthroplasty using a modular S-ROM prosthesis combined with corrective proximal femoral osteotomy. J Arthroplast. 2006;21:664–9.
13. Sonohata M, et al. Total hip arthroplasty with femoral subtrochanteric osteotomy after Schanz osteotomy. J Orthop Sci. 2016;21:469–74.
14. Sonohata M, et al. Total hip arthroplasty combined with double-chevron subtrochanteric osteotomy. J Orthop Sci. 2012;17:382–9.
15. Takao M, et al. Cementless modular total hip arthroplasty with subtrochanteric shortening osteotomy for hips with developmental dysplasia. J Bone Joint Surg Am. 2011;93:548–55.
16. Hattori T, et al. The epidemiology of developmental dysplasia of the hip in Japan: findings from a nationwide multi-center survey. J Orthop Sci. 2017;22:121–6.
17. Sugano N, et al. Polyethylene sockets and alumina ceramic heads in cemented total hip arthroplasty. A ten-year study. J Bone Joint Surg Br. 1995;77:548–56.
18. Takao M, et al. The results of a press-fit-only technique for acetabular fixation in hip dysplasia. J Arthroplast. 2011;26:562–8.
19. Sugano N, et al. Predicting the position of the femoral head center. J Anthroplasty. 1999;14(1):102–7.
20. Fujishiro T, et al. Predicting leg-length change after total hip arthroplasty by measuring preoperative hip flexion under general anaesthesia. Orthop Surg (Hong Kong). 2012;20:327–30.
21. Fujishiro T, et al. Leg length change in total hip arthroplasty with subtrochanteric femoral shortening osteotomy for Crowe type IV developmental. J Arthroplast. 2012;27:1019–22.
22. Li C, et al. Comparison of transverse and modified subtrochanteric femoral shortening osteotomy in total hip arthroplasty for developmental dysplasia of the hip: a meta-analysis. BMC Musculoskelet Disord. 2014;15:331.
23. Hotokebuchi T, et al. A new device for a V-shaped subtrochanteric osteotomy combined with total hip arthroplasty. J Arthroplast. 2006;21:135–7.
24. Farrell CM, et al. Motor nerve palsy following primary total hip arthroplasty. J Bone Joint Surg Am. 2005;87:2619–25.
25. Sonohata M, et al. Nerve palsy after total hip arthroplasty without subtrochanteric femoral shortening osteotomy for a completely dislocated hip joint. Open Orthop J. 2016;10:785–92.
26. Kawano S, et al. Risk factors for the development of nerve palsy following primary total hip arthroplasty. Open Orthop J. 2018;12:164–72.

Ankylosed Hip Joint

75

Motoki Sonohata

1 Indications for Total Hip Arthroplasty for Ankylosed Hip Joints

Ankylosed hip joints are painless despite the loss of hip mobility. Therefore, hip pain is not an indication for THA. In ankylosed hip joints, THA is indicated for low back pain, ipsilateral knee pain, and contralateral hip pain caused by the ankylosing hip. In other words, the most common indication for THA of the ankylosed hip joint is adjacent joint dysfunction. Coxitis knee [1] and hip-spine syndrome [2] are well-known adjacent joint disorders of the ankylosed hip (Fig. 75.1). THA or TKA (total knee arthroplasty) for ipsilateral knee pain or contralateral hip pain caused by an ankylosed hip joint should be performed with caution. TKA of the ipsilateral knee joint or THA of the contralateral hip joint may cause implant failure due to the unusual load applied. Therefore, THA of the ankylosed hip joint is prioritized for adjacent joint disorders of the ankylosed hip joint.

In addition, muscle atrophy and fatty degeneration of the muscles around the ankylosed hip joint due to long-term immobilization are significant, requiring a long period to recover muscle strength after THA, and complete recovery is difficult. In THA for ankylosed hip joints, 26–33% of patients are able to walk unassisted at an average of 8–13 years after surgery [3, 4]. It is well understood that "an immobile but functional joint can be replaced by a movable but nonfunctional joint [5]. The indications for THA in patients seeking mobility of the hip joint should be carefully evaluated.

2 Preoperative Evaluation

The preoperative evaluation of an ankylosed hip should include not only plain radiographs but also CT or MRI, with CT providing a more detailed assessment of the bony morphology of the ankylosed hip and quantitative assessment of

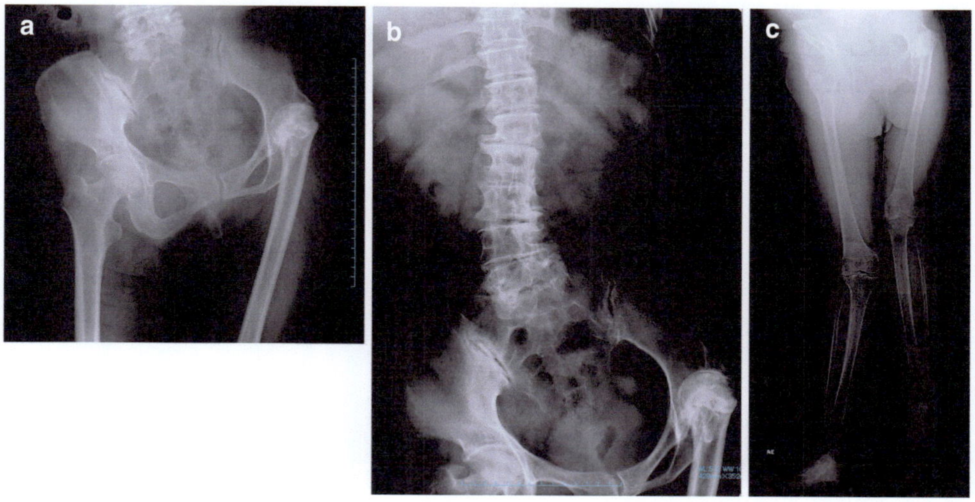

Fig. 75.1 Adjacent joint disorder. 73-year-old woman. After right hip fusion surgery. (**a**) Left hip joint ankysosis in adduction; (**b**) Hip-spine syndrome; (**c**) Coxitis knee

M. Sonohata (✉)
Saga Central Hospital, Saga, Japan

Y. Nakashima (ed.), *Advances in Total Hip Arthroplasty*, https://doi.org/10.1007/978-981-97-2461-1_75

the muscles ankylosed hip joint and MRI providing a more qualitative assessment of the ankylosed hip joint (Fig. 75.2).

The causes of ankylosis of the hip joint include osteoarthritis, rheumatoid arthritis, or pyogenic arthritis. The most important preoperative evaluation is fixed limb position. In cases of arthrodesis, detailed assessment of the greater trochanter is required; osteotomized? Used for bone graft?, and so on. THA combined with corrective subtrochanteric osteotomy is required in patients with femoral deformity after multiple hip surgeries.

There are many cases of ankylosis in a significant malaligned position, regardless of whether arthrodesis is performed. Therefore, the THA approach for ankylosed hip joints should be considered carefully (Fig. 75.3).

3 Surgical Techniques

(a) Approach

Because there are remarkable adhesions and anatomic abnormalities in ankylosed hip joints regardless of whether arthrodesis is performed, a widespread surgical approach is required.

The lateral approach with osteotomy of the greater trochanter is often chosen; however, a posterolateral approach has also been reported in some cases.

In most cases, atrophy, fibrosis, and fatty degeneration of the gluteus muscle is observed; therefore, a maximum atraumatic operation is needed. Atrophic, fibrotic, and fatty degeneration of the gluteus muscle can easily be torn or

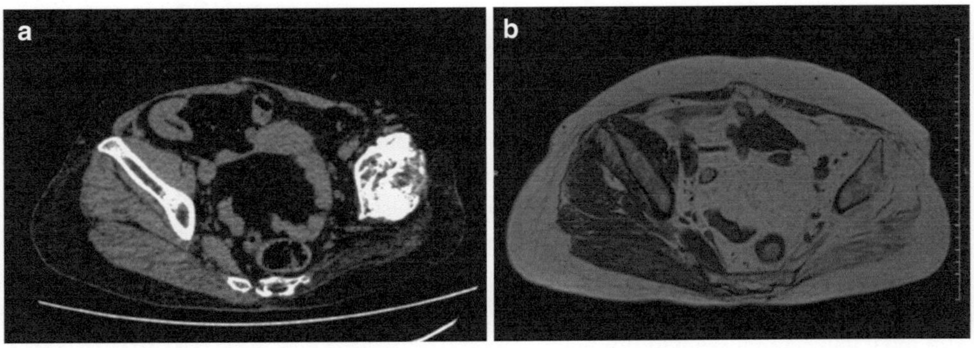

Fig. 75.2 Preoperative CT of the case shown in Fig. 75.1. (**a**) CT shows marked atrophy of the gluteus muscle; (**b**) MRI shows atrophy as well as fattening of the gluteus muscle

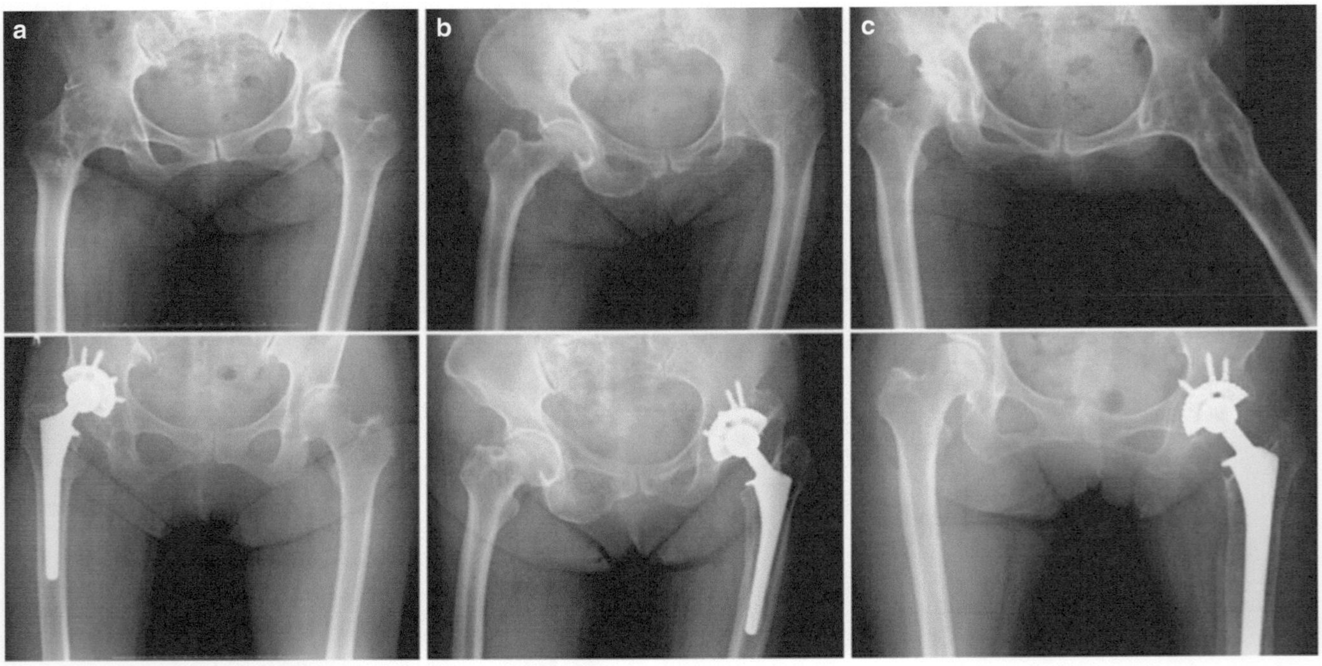

Fig. 75.3 THA for ankylosed hip joint. (**a**) A 63-year-old woman, upper: ankylosis in good position, lower: 2 years after surgery. (**b**) A 76-year-old woman, upper: ankylosis in adduction, lower: 4 years after surgery. (**c**) A 66-year-old woman, upper: ankylosis in abduction, lower: 5 years after surgery

detached from the muscle attachment site by intraoperative manipulation; therefore, great care is needed.

The sciatic nerve requires special attention during surgery. In cases of ankylosis during external rotation, the sciatic nerve is in contact with the greater trochanter and should be thoroughly checked before osteotomy of the fused joint. In a few cases where extra-articular fixation technique was used, the sciatic nerve and vascular canal may be entrapped within the bone during the bone remodeling process.

(b) Osteotomy of ankylosed site

Using the lateral approach, it is relatively easy to identify the site of ankylosis. However, it may be difficult to identify the ankylosis site in cases of external rotational ankylosis using the posterolateral approach.

Osteotomy should be performed with care to protect the sciatic nerve and to maintain the superior lateral part of the femoral neck as much as possible on the pelvic side. In the ankylosed hip joint, there is little subchondral bone in the area where the cup is placed; therefore, it is more difficult to obtain a stronger initial fixation than in the usual primary THA. Therefore, maintaining the superior lateral part of the femoral neck on the pelvic side increases the bony coverage of the cup and serves as a reference point for cup placement.

(c) Implant placement

Even after osteotomy of the ankylosed site, the surgeon cannot achieve good ROM of the hip joint. There is loss of soft tissue extensibility due to long-term ankylosis (more extensive scar tissue is present in post-fusion and post-infection cases). In addition to incision and excision of scar tissue, a full circumferential incision of the joint capsule and dissection of the gluteus maximus muscle at the femoral attachment are often necessary.

On the acetabular side, surgeons should identify the pubis, sciatica, and inferior margin of the acetabulum after osteotomy. As mentioned above, almost no subchondral bone is present in the ankylosed hip joint. Surgeons must carefully ream the cancellous bone directly.

In particular, it is necessary to avoid damaging the posterior and anterior acetabular walls. Although reaming of the acetabulum is usually performed on the medial wall of the pelvis in THA, there are cases in which impingement due to excess bone is feared if the acetabulum is medialized to the medial wall of the pelvis in an ankylosed hip joint. In many cases, it is necessary to place a cup in line with the osteotomy; therefore, the surgical plan should take this into consideration. The upper lateral part of the femoral neck on the pelvic side is one of the reference points for cup placement.

The procedure on the femoral side is similar to that of the usual THA, but it should be noted that there are many cases of cortical atrophy.

(d) Heterotopic ossification

In THA for ankylosing groins, many bone fragments are produced in the operative field, and postoperative heterotopic ossification is a concern. Prophylactic irradiation is recommended in some cases, but has not been performed in Japan. The efficacy of nonsteroidal anti-inflammatory drugs has been reported; however, long-term use of these drugs should be dealt with caution because of their side effects. In Japan, disodium etidronate is indicated to prevent heterotopic ossification in the early and advanced stages of hip arthroplasty. Bone fragments should be carefully removed from the surgical field and the amount of washing water in the surgical field should be increased.

4 Clinical Results

Ankylosed hip joints rarely improve to a normal range of motion immediately after THA. The long-term results of THA for ankylosis have been reported to be excellent, but less than half of patients are able to walk without walking aids after surgery [3, 5]. The most frequently improved outcome is chronic low back pain, followed by knee pain [6].

References

1. Smillie IS. Angular deformity, disease of the knee joint. 2nd ed. Edinburgh: Churchill Livingstone; 1974.
2. Offierski CM, et al. Hip-spine syndrome. Spine (Phila Pa 1976). 1983;8:316–21.
3. Kitajima M, et al. Ten-year results of cementless total hip arthroplasty for ankylosing hips. Hip Joint. 2017;43:306–9. Japanese
4. Reikerås O, et al. Total hip arthroplasty for arthrodesed hips. 5- to 13-year results. J Arthroplast. 1995;10:529–31.
5. Callaghan JJ, et al. Chapter 73 conversion of hip arthrodesis to total hip arthroplasty. In: The adult hip: hip arthroplasty surgery, vol. 2. 3rd ed. Philadelphia: Wolters Kluwer; 2016. p. 949–54.
6. Fernandez-Fairen M, et al. Is total hip arthroplasty after hip arthrodesis as good as primary arthroplasty? Clin Orthop Relat Res. 2011;469:1971–83.

THA After Osteotomy

76

Goro Motomura

1 Surgical Planning and Implant Selection

(a) Toal hip arthroplasty (THA) after pelvic osteotomy

As in conventional THA, the bony coverage and height of the acetabular cup are considered when determining the implant placement position. Computed tomography (CT) is recommended because of the possibility of an anteroposterior acetabular wall defect. In cases where the height of the joint is elevated, if the original acetabular position is maintained, a bulky bone graft may be required at the loading site, which may cause problems with the fixation of the acetabular cup. The author uses the cup-center edge (CE) angle (0° or greater) [1] and the distance from the inferior edge of the tear drop to the center of the head (approximately within 30 mm) [2–4] as guidelines for placement (Fig. 76.1).

In cases with sclerotic changes in the acetabulum, it may be difficult to obtain a scratch fit with some types of cementless cups; therefore, a surface with a high coefficient of friction, such as three-dimensional (3D) porous cups, is useful. In cases where the cup is placed medially, the stem offset should be considered when selecting the model.

(b) THA after proximal femoral osteotomy

In addition to changes in the proximal femoral morphology associated with osteotomy, the hip is often associated with external rotation contracture and limitation of hip extension (Fig. 76.2), limiting preoperative planning based on X-rays alone. A CT scan is essential, and a 3D template is also useful.

After varus or valgus osteotomy, in which the bone fragment moves only in the coronal plane, the calcar portion can be formed according to the stem shape, and the conventional fit and fill or tapered-wedge type can be placed without alignment failure (Figs. 76.3 and 76.4). When removal of the fixation material used at the time of osteotomy is required during THA, a stem length longer than that of the most distal screw hole should be used to prevent fracture complications (Fig. 76.3). After valgus osteotomy, lateralization of the greater trochanter relative to the femoral diaphysis has been achieved. Therefore, care should be taken when using a high-offset stem because it may cause a sense of overhang in the greater trochanter (Fig. 76.4).

After transtrochanteric rotational osteotomy, the morphology of the proximal femur is significantly altered; in particular, femoral neck anteversion is not easily identified, even with CT. Although the conventional fit and fill or tapered-wedge type stem can be used in most cases, anteversion of the stem is not possible in some cases. Therefore, preoperative planning using a 3D template is useful (Fig. 76.5). Additionally, it is recommended to prepare a stem with freely adjustable anteversion as a backup. Since intraoperative and postoperative bleeding are often more frequent than in other THA procedures, autologous blood should be prepared, if possible.

Leg shortening of 10–20 mm is common after varus or rotational osteotomies, and leg length correction should also be considered. In preoperative planning, it is necessary to consider whether the leg length can be compensated for by adjusting the stem and neck and to select a model and prepare a back-up system.

G. Motomura (✉)
Department of Artificial Joints and Biomaterials, Graduate School of Medical Sciences, Kyushu University, Fukuoka, Japan
e-mail: motomura.goro.014@m.kyushu-u.ac.jp

Y. Nakashima (ed.), *Advances in Total Hip Arthroplasty*, https://doi.org/10.1007/978-981-97-2461-1_76

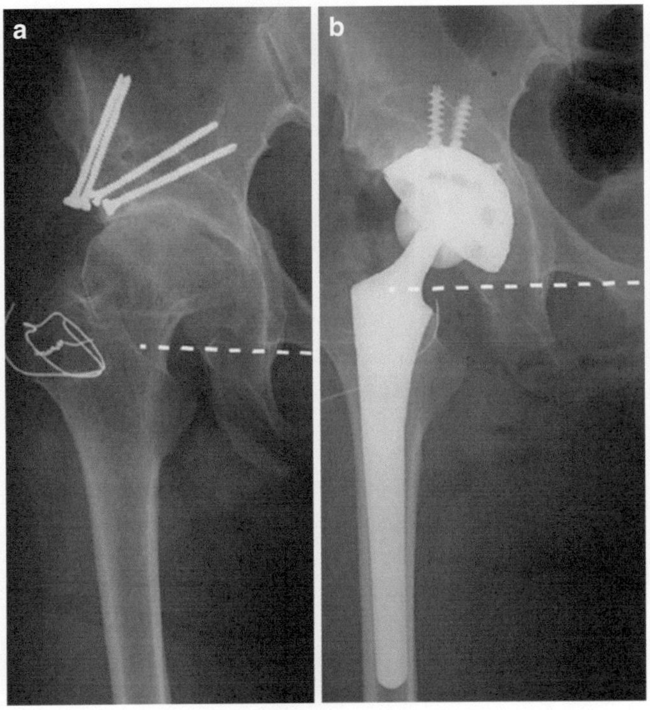

Fig. 76.1 A case of total hip arthroplasty after acetabular translocation. (**a**) Preoperative radiograph. The height of the joint is slightly elevated (dotted line connecting the lower edge of the tear drop). (**b**) Postoperative radiograph. The cup is placed at an acceptable height

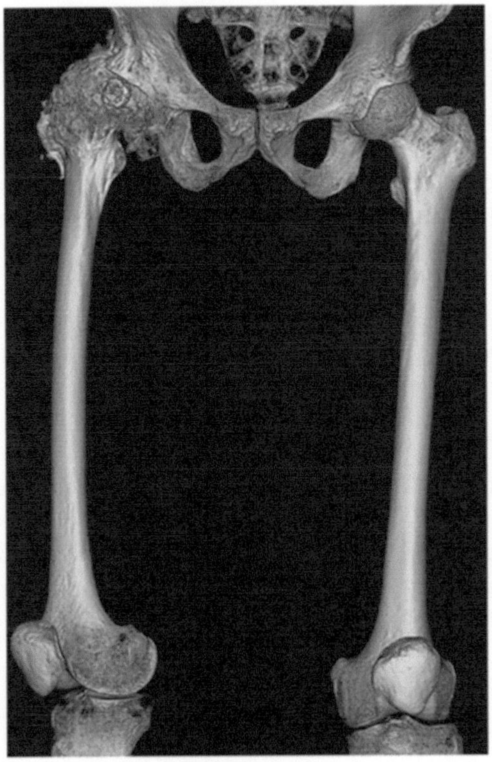

Fig. 76.2 Three-dimensional computed tomography of the frontal pelvis after transtrochanteric rotational osteotomy of the right femoral head. The right lower limb was contracted in external rotation

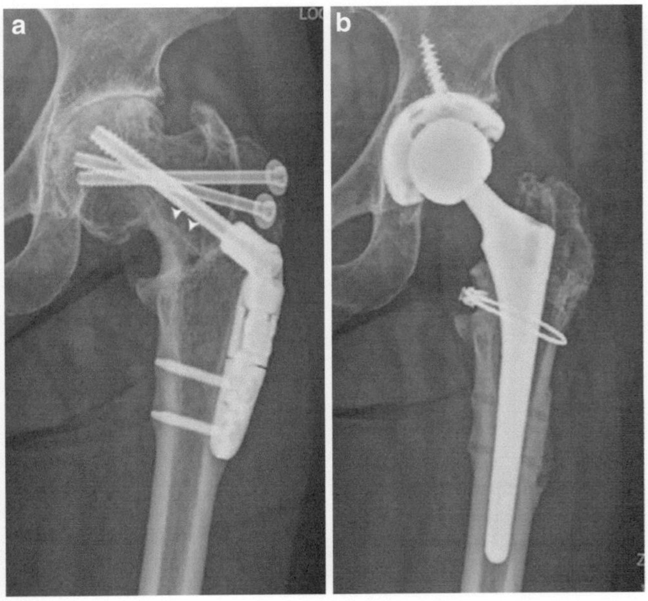

Fig. 76.3 A typical case of total hip arthroplasty after varus osteotomy. (**a**) Preoperative radiograph. A sclerotic lesion was observed in the calcar area (arrowhead). (**b**) Postoperative radiograph. The sclerotic lesion was resected according to stem shape. The stem length exceeded the most distal screw hole. Prophylactic wiring with cable was performed

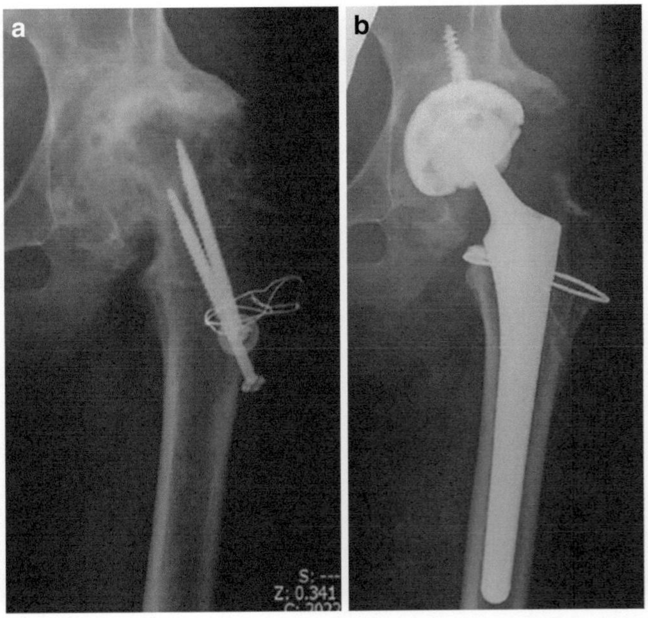

Fig. 76.4 A typical case of total hip arthroplasty after valgus osteotomy. (**a**) Preoperative radiograph. Lateralization of the greater trochanter, relative to the femoral diaphysis, was observed. (**b**) Postoperative radiograph. Prophylactic wiring with cable was performed

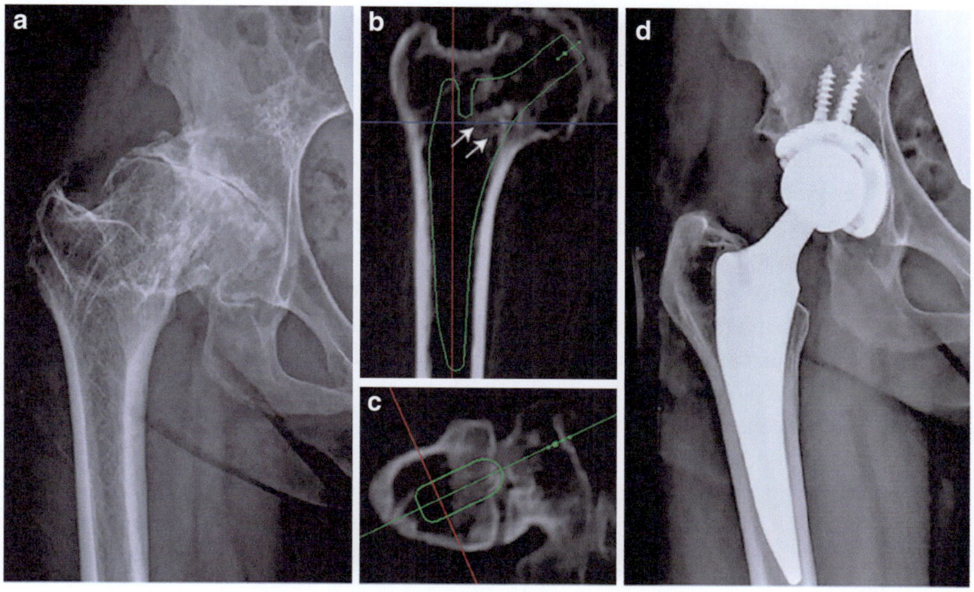

Fig. 76.5 A typical case of total hip arthroplasty after anterior rotational osteotomy of the femoral head. (**a**) Preoperative X-ray image. (**b**) Coronal section of the three-dimensional (3D) template. A sclerotic lesion (arrow) was observed in the calcar area. (**c**) Axial section of the 3D template (neck level). Stem anteversion was achieved. (**d**) Postoperative X-ray image

2 Tips for Surgery

Surgery is performed in the true lateral position using the posterior approach. A skin incision is made through a lateral longitudinal incision through the greater trochanter, and any available skin incision from prior surgery is used. In patients with a previous surgery, the subfascial area is widely adherent, and the view is different from the familiar view. It is easier to manipulate if the adhesions are separated distally over the vastus lateralis fascia, anteriorly and superiorly over the gluteus medius, and posteriorly over the short external rotator muscles or capsule. We then move on to the dislocation maneuver. The hip joint is placed in extension internal rotation, and if the short external rotator muscle can be identified, the posterior joint is exposed as usual for THA. If it cannot be identified, the posterior part of the joint is exposed under the periosteum along the intertrochanteric ridge. In cases where orientation is difficult, such as after rotational osteotomy, it is recommended to use the lesser trochanter as a landmark. If external rotation contracture is severe, the location of the sciatic nerve needs attention, and it is important to keep in mind subperiosteal expansion. After osteotomy of the pelvic side, the femoral head may not be exposed because the posterior wall of the acetabulum may cover it. In such cases, the femoral head can be visualized by removing approximately 1 cm of the posterior wall with a flat chisel. If the deformity of the hip joint is severe, its dislocation may be difficult. In such cases, a small elevatorium is inserted into the joint space, and the head is lifted from the inside by pushing the elevatorium down. When removal of the fixation material used at the time of osteotomy is required during surgery, the femur should be prepared in advance for dislocation before removal of the materials to avoid applying strong torsional force to the femur through the screw hole. If dislocation is not possible, the neck is osteotomized, and the femoral head is resected piece by piece.

(a) THA after pelvic osteotomy

There is no major difference compared with the usual THA. The procedure is first performed on the femoral side first and the anteversion angle of the stem is confirmed before proceeding to the acetabular side. The center of the cup is confirmed according to the preoperative plan and the cup is placed according to the procedure. Preoperative and intraoperative radiographic confirmations are helpful. If a good press fit of the acetabular cup cannot be achieved, additional screw fixation is used.

(b) THA after proximal femoral osteotomy

(i) After transtrochanteric rotational osteotomy

After transtrochanteric rotational osteotomy, the appearance of the osteotomy surface of the femoral neck is completely different from normal, and it is difficult to identify the insertion point of the stem [5]. First, the femoral axis is confirmed by inserting elevatorium on both sides of the diaphysis. Rasping is performed after confirming the bone axis with a starter reamer; however, it is often difficult to smoothly insert the starter reamer because of hardening of the medullary cavity. A hole is made with a round chisel along the axis of the femur, a curette is inserted and tapped lightly with a hammer to confirm that the reamer continued to the medullary cavity of the diaphysis, and a starter reamer is inserted to confirm the overall axis of the femur. When rasping, it is necessary to proceed with it while resecting the sclerotic bone in the medulla to create an anteversion of the stem, rather than inserting the rasp so that the rasp can enter [5]. In cases where the rasp is difficult to insert in the anteversion position, a stem that allows the anteversion to be set freely is used, as described above. In the osteotomy plane of the femoral neck, the bone overhangs anteriorly because of the rotation of the proximal osteotomied bone (Fig. 76.6), which causes impingement during flexion and internal rotation and should be resected. With a rasp inserted, the rasp line should be used as a guide for resection (Fig. 76.6) [3]. The thickened and degenerated anterior capsule can also be resected, and the findings of impingement during flexion and internal rotation can be improved.

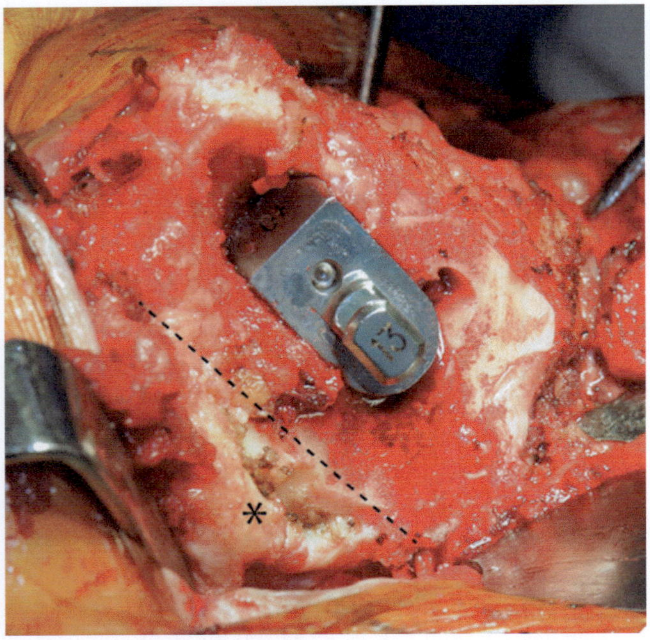

Fig. 76.6 Femoral neck osteotomy plane of total hip arthroplasty after anterior rotational osteotomy of the right femoral head (top is posterior, left is lateral, with the rasp inserted). Anteriorly, the bone overhang (asterisk) is resected using the rasp line as a guide (dotted line)

(ii) After valgus/varus osteotomy

After valgus osteotomy, the difference from the normal proximal femoral morphology should be considered because of lateralization of the greater trochanter (Fig. 76.4). As in rotational osteotomy, the bony axis of the metaphysis should be confirmed. Partial plastic resection of the calcar area is often necessary to place the stem in a straight position. In cases of varus osteotomy, the medial calcar area is well remodeled; therefore, we can use a circular chisel or Luer forceps to resect a portion of the sclerotized bone for intermediate placement of the stem.

(iii) Intraoperative peri-stem fracture handling and prevention

During THA after osteotomy of the proximal femur, more attention should be paied to fracture complications than usual. Fracture fixation with a cable is often sufficient. If a fracture is identified intraoperatively, good fixation can be achieved by fastening two or three cables. The start of loading should be delayed, as necessary, and postoperative radiographs should be periodically checked to ensure the presence of stem sinking. If the risk of fracture is judged to be high considering bone quality and other factors, one or two additional cables are added prophylactically (Figs. 76.3 and 76.4).

3 Clinical Results

(a) THA after pelvic osteotomy

Many reports from Japan have used cementless cups (Table 76.1) [2, 6–10]. Many reports indicate that THA

Table 76.1 Results of total hip arthroplasty after pelvic osteotomy

First author (year)	Number (hip)	Pelvic osteotomy	Follow-up (year)	Cementless/cement cup (hip)	Bulk bone grafting	Cup position	Complication (hip)/ Revision (hip)
Komiyama (2021)	37	Transpositional osteotomy of the acetabulum	5.7	37/0	None	Superolateral	None
Osawa (2016)	52	Eccentric RAO: 36 RAO: 16	5.4	44/8	11	Lateral	Dislocation (1)/ none
Tamaki (2016)	34	RAO: 21 Chiari: 13	RAO: 3.8 Chiari: 3.5	34/0	RAO: 7 Chiari: 2	N.A.	RAO: Cup migration (2)/(1) Chiari: Dislocation (1)/ none
Ito (2015)	44	RAO	4.6	44/0	None	Lateral	None
Fukui (2015)	22	RAO	8.2	22/0	None	Superolateral	None
Yuasa (2015)	24	RAO	7.0	21/3	None	N.A.	Infection (1)/(4)

RAO rotational acetabular osteotomy, *N.A.* not available

Table 76.2 Results of total hip arthroplasty after proximal femoral osteotomy

First author (year)	Number (hip)	Type of osteotomy	Follow-up (year)	Cementless/cement stem (hip)	Complication (hip)	Reason for revision surgery(hip)
Utsunomiya (2017)	24	TRO	8.3	24/0	Dislocation(2), infection(1)	None
Osawa (2017)	34	TRO	10.5	28/6	Dislocation(1), fracture(1)	Stem loosening(1), cup loosening(1), polyethylene wear (1), fracture(1)
Ohishi (2016)	66	Valgus: 42 Varus: 24	7.3	66/0	Fracture(4)	Fracture(2)
Suzuki (2007)	30	Valgus	7.2	30/0	Fracture(1)	Cup loosening(1)
Iwase (1999)	30	Valgus	7.0	18/12	None	Stem loosening(3), cup loosening(2)

TRO transtrochanteric rotational osteotomy

after pelvic osteotomy is inferior to conventional THA in terms of operative time and intraoperative blood loss, but the clinical scores are comparable [2, 7–10], however, some reports indicate that THA after pelvic osteotomy has a significantly inferior clinical score [6]. Regarding the position of the cup, many reports indicate that the cup center is shifted superolaterally or laterally [2, 6, 8, 9], but there is no clear influence of the cup position on the clinical outcome in the short-to-medium term. However, there are some reports on selective bulk bone grafting to achieve the original acetabular position [6, 7], but the number of cases is too less to evaluate the advantages and disadvantages of this technique, and future studies are needed.

(b) THA after proximal femoral osteotomy

Many studies from Japan have used cementless stems (Table 76.2) [5, 11–14]. The operative time for THA after rotational osteotomy was longer, and intraoperative blood loss was higher than that for primary THA without previous osteotomy, suggesting that THA after rotational osteotomy is a difficult procedure [5, 11]. Although there are reports of inferior or equivalent clinical outcomes, a systematic review reported that preious rotational osteotomy did not affect the clinical outcome of THA [15]. Although generally good results have been reported in THA after varus or valgus osteotomy [12–14], it has been suggested that complications of fracture should be noted [12].

References

1. Fujii M, et al. Minimum lateral bone coverage required for securing fixation of cementless acetabular components in hip dysplasia. Biomed Res Int. 2017;2017:4937151.
2. Komiyama K, et al. Total hip arthroplasty after periacetabular osteotomy versus primary total hip arthroplasty: a propensity-matched cohort study. Arch Orthop Trauma Surg. 2021;141:1411–7.
3. Komiyama K, et al. Does high hip center affect dislocation after total hip arthroplasty for developmental dysplasia of the hip? Int Orthop. 2019;43:2057–63.
4. Komiyama K, et al. Does high hip center decrease range of motion in total hip arthroplasty? Computer simulation study. J Arthroplast. 2016;31:2342–7.
5. Utsunomiya T, et al. The results of total hip arthroplasty after sugioka transtrochanteric anterior rotational osteotomy for osteonecrosis. J Arthroplast. 2017;32:2768–73.
6. Osawa Y, et al. Significantly poor outcomes of total hip arthroplasty after failed periacetabular osteotomy. J Arthroplast. 2016;31:1904–9.
7. Tamaki T, et al. Total hip arthroplasty after previous acetabular osteotomy: comparison of three types of acetabular osteotomy. J Arthroplast. 2016;31:172–5.
8. Ito H, et al. Total hip arthroplasty after rotational acetabular osteotomy. J Arthroplast. 2015;30:403–6.
9. Fukui K, et al. Does rotational acetabular osteotomy affect subsequent total hip arthroplasty? Arch Orthop Trauma Surg. 2015;135:407–15.
10. Yuasa T, et al. Total hip arthroplasty after previous rotational acetabular osteotomy. Eur J Orthop Surg Traumatol. 2015;25:1057–60.
11. Osawa Y, et al. Total hip arthroplasty after transtrochanteric rotational osteotomy for osteonecrosis of the femoral head: a mean 10-year follow-up. J Arthroplast. 2017;32:3088–92.
12. Ohishi M, et al. Cementless total hip arthroplasty for patients previously treated with femoral osteotomy for hip dysplasia: the incidence of periprosthetic fracture. Int Orthop. 2016;40:1601–6.
13. Suzuki K, et al. Cementless total hip replacement after previous intertrochanteric valgus osteotomy for advanced osteoarthritis. J Bone Joint Surg. 2007;89:1155–7.
14. Iwase T, et al. Total hip arthroplasty after failed intertrochanteric valgus osteotomy for advanced osteoarthrosis. Clin Orthop Relat Res. 1999;364:175–81.
15. Shigemura T, et al. Total hip arthroplasty after failed transtrochanteric rotational osteotomy for osteonecrosis of the femoral head: a systematic review and meta-analysis. Orthop Traumatol Surg Res. 2018;104:1163–70.

Posterior Pelvic Tilt Cases (Spinal and Pelvic Alignment)

77

Toshiyuki Kawai

1 Pathologies of the Hip Joint Following Spinal Degeneration

Osteoarthritis of the hip results in severe degenerative changes to the spinal column, which often causes disruption to the sagittal alignment requiring corrective surgery. The risk of dislocation after total hip arthroplasty (THA) is higher in patients undergoing both THA and spinal fusion. Total hip arthroplasty (THA) dislocation is common in patients undergoing both THA and spinal fusion. A large survey indicated that performing spinal fusion followed by THA results in a higher dislocation rate than performing THA before spinal fusion [1]. Increasing the interval between THA and spinal fusions decreases the dislocation rate [1]. Therefore, if the situation permits, the risk of dislocation may be reduced by performing THA first followed by spinal fusion after an extended period. There is an increase in the risk of hip joint degeneration and the need for THA after long fusion of the spine [2]. Thus, the number of complicated THA cases after spinal fusion is likely to increase in the future based on the recent increase in the number of long fusion cases.

Patients with posterior pelvic tilting often have a flat back. In addition, they tend to have a stiff pelvis with decreased mobility. These conditions are associated with not only a higher dislocation rate but also a lower gait score and functional score after THA [3]. Therefore, a detailed preoperative evaluation is necessary to manage the risk of THA.

2 Spinal Alignment Evaluation

(a) Required measurements of spinal alignment

Various spinal alignment indices contribute to the risk of hip dislocation. As such, routine measurements of spinal alignment are recommended in the preoperative management of THA. We recommend that the following spinal alignment measurements should be included in the THA preoperative examination: (1) frontal and lateral views of the patient in the standing position; (2) a seated lumbar view; and (3) a lateral view of the pelvis. Using the standing lateral view, the following measurements were taken: sacral slope (SS); pelvic tilt; pelvic incidence (PI); lumbar lordosis (LL); the sagittal vertical axis; and T1-pelvis angle (TPA) (Fig. 77.1). With the patient in the sitting position, lateral lumbar pelvic imaging was used to measure the SS. The SS change from the standing to sitting position was calculated.

(b) Spinal alignment parameters affecting the risk of dislocation.

High preoperative TPA [4], high preoperative pelvic-incidence lumbar lordosis (PI-LL) mismatch [4], and low SS change between the standing and sitting positions [5, 6]) are

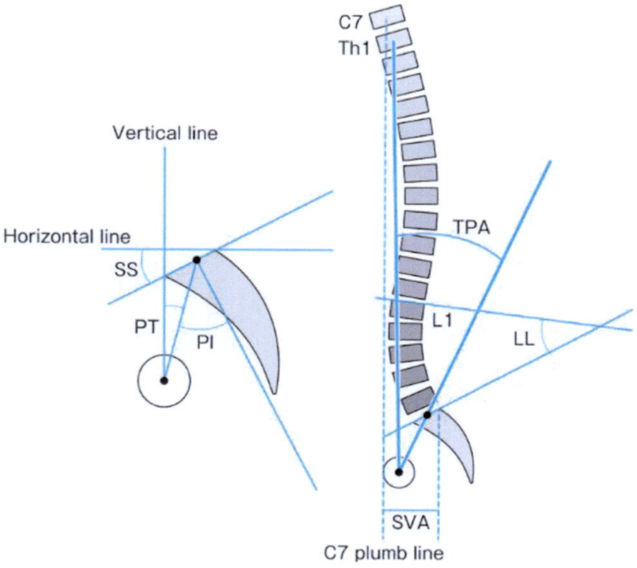

Fig. 77.1 Main spinal and pelvic parameter measurement methods

T. Kawai (✉)
Kyoto University, Kyoto, Japan
e-mail: kawait@kuhp.kyoto-u.ac.jp

Y. Nakashima (ed.), *Advances in Total Hip Arthroplasty*, https://doi.org/10.1007/978-981-97-2461-1_77

associated with a significantly increased dislocation rate. In addition, it has been pointed out that the risk of anterior implant impingement is increased in cases with a small PI-LL mismatch. Furthermore, the risk of posterior implant impingement is increased in cases with large PI-LL mismatch [7]. In the standing position, patients with an anterior pelvic plane tilted more than 10° posteriorly to the vertical plane are at risk for anterior dislocation [8]. Approximately one in three patients undergoing THA has a posterior pelvic tilt of 10° or more. Therefore, radiographic measurements made in the standing position should be routinely performed. One in seven patients undergoing THA had a preoperative PI-LL mismatch of >20° [8]; these patients have a high risk of dislocation. Routine assessment of spinal-pelvic alignment should be performed to identify latent high-risk patients and to reduce the incidence of so-called "repeated dislocations despite apparently good implant positioning."

Patients who have undergone lumbar fusion have a high risk of anterior impingement and posterior dislocation during squatting because of the small increase in cup anteversion from the standing to sitting position [9]. Patients who have undergone three or more vertebrae levels of lumbar fusions have a higher risk of dislocation than those who have undergone lumbar fusion at one or two levels [10]. After spinal fusion, it is necessary to maintain adequate pelvic mobility during hip flexion and extension.

(c) Changes in spinal alignment after THA surgery.

The degree of posterior pelvic tilt has improved after THA in numerous cases [11]. Improvement in sagittal alignment after THA tends to be more pronounced in patients with severe hip deformities [12]. However, 5 years after THA, approximately 8% of patients have a posterior pelvic tilt of 20°or more in the standing position compared to the supine position [13]. In addition, patients with a large PI have shown only a slight improvement in pelvic tilt, after THA [14].

The identification of large differences in leg length and the presence of scoliosis preoperatively is necessary to predict the extent to which scoliosis will improve following leg length correction. The surgeons can palpate the spinous process and iliac crest, with the patient in a seated position to confirm the compensatory nature of the spine. A seated anteroposterior view and lateral bending view of the entire spine can also be used to estimate the degree of postoperative improvement in scoliosis.

3 Tips and Points of Surgery

(a) Preoperative planning
   (i) Reference plane of measurement

Lewinnek's concept of a safe zone plays a key role in implant placements. This concept states that the cup component should be kept within an abduction angle of 40° ± 10° and an anteversion angle of 15° ± 10°. In addition, cup placement outside this range will result in a high dislocation rate [15]. Lewinnek's safe zone, includes the radiographic anteversion that was calculated from the ratio of the cup major to minor diameters on an anteroposterior radiograph taken orthogonally to the anterior pelvic plane [16]. Most actual dislocators have acetabular implants and are placed within the Lewinnek safe zone [17]. This suggests that there are limitations in the current knowledge of the dislocation risk based on the angle of placement relative to the anatomical plane. Although previous studies have discussed the validity of the Lewinnek safe zone concept, angular measurements are often made on a functional pelvic plate orthogonal to the CT table (Fig. 77.2), and only approximately 10% of the papers have discussed the safe zone based on the correct plane [18]. It is necessary to be aware of the plane in which measurements were made before interpreting the results.

Since the Lewinnek safe zone is an anatomical measurement index, it is important not to overestimate the validity of this concept when surgery is planned for a patient whose anterior pelvic plate is largely tilted. Therefore, the sagittal tilt of the pelvis in the sitting and standing positions should be con-

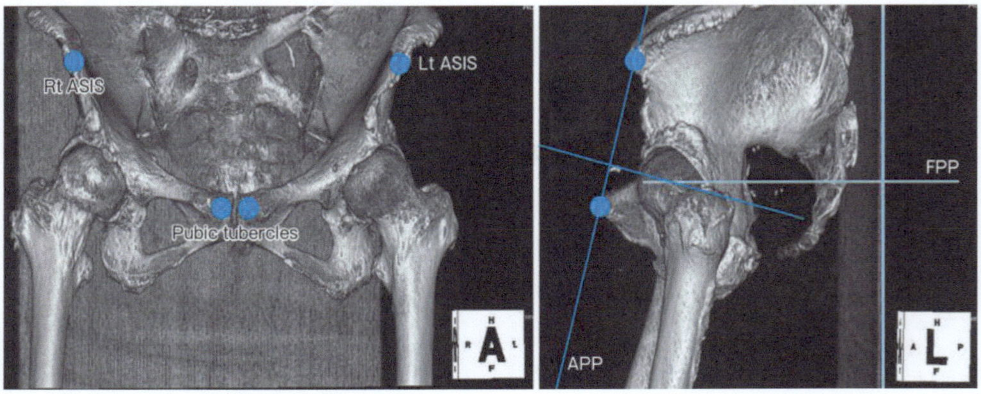

Fig. 77.2 Reference plane for measurement of the front opening. *ASIS* anterior superior iliac spine, *APP* anterior pelvic plane, *FPP* functional pelvic plane

sidered, and placement planning should be based on functional plane measurements in both the sitting and standing positions (Fig. 77.3).

(ii) Calculation of cup installation angle

In patients with a PI-LL mismatch of 20° or more and a change in SS of 10° or less from the standing to sitting position, the risk of dislocation is considered to be high. Therefore, the cup should be placed with consideration given to the pelvic tilt angle in the standing and sitting positions. The cup should be adjusted to limit the anteversion when the pelvis is tilted posteriorly in the standing position. The anterior cup opening should be kept less than 30° in the functional pelvic plane in the standing position [19].

For every 1° posterior tilt of the pelvis, the anterior opening of the acetabular component increases by approximately 0.7° [20]. This relationship can be used as a guideline to determine the degree to which the anterior opening in the supine position changes when the patient is standing or sitting.

Therefore, the risk of posterior dislocation should also be evaluated. A cup anteversion of more than 30° to the functional pelvic plane in the sitting position should be maintained to prevent posterior dislocation [19]. To avoid anterior dislocation, the dislocation risk is reduced by maintaining a cup anteversion of less than 30° relative to the functional pelvic plane in the standing position [19]. Using the SS and cup anteversion noted on CT with the patient in the supine position as points of reference, changes in anteversion can be estimated in the sitting and standing positions. Optimal anteversion should be determined so that the estimated anteversion in the sitting position is more than 30° and that in the standing position is less than 30°. In the case shown in Fig. 77.4, when the cup was placed according to the acetabular anteversion, the estimated cup anteversion was 73.8° in the sitting position and 51.7° in the standing position, which means that the patient had a high risk of anterior dislocation. To reduce the estimated cup anteversion in the standing position to less than 30°, a cup anteversion of less than 13.3° to the FPP in the supine position can be achieved using the

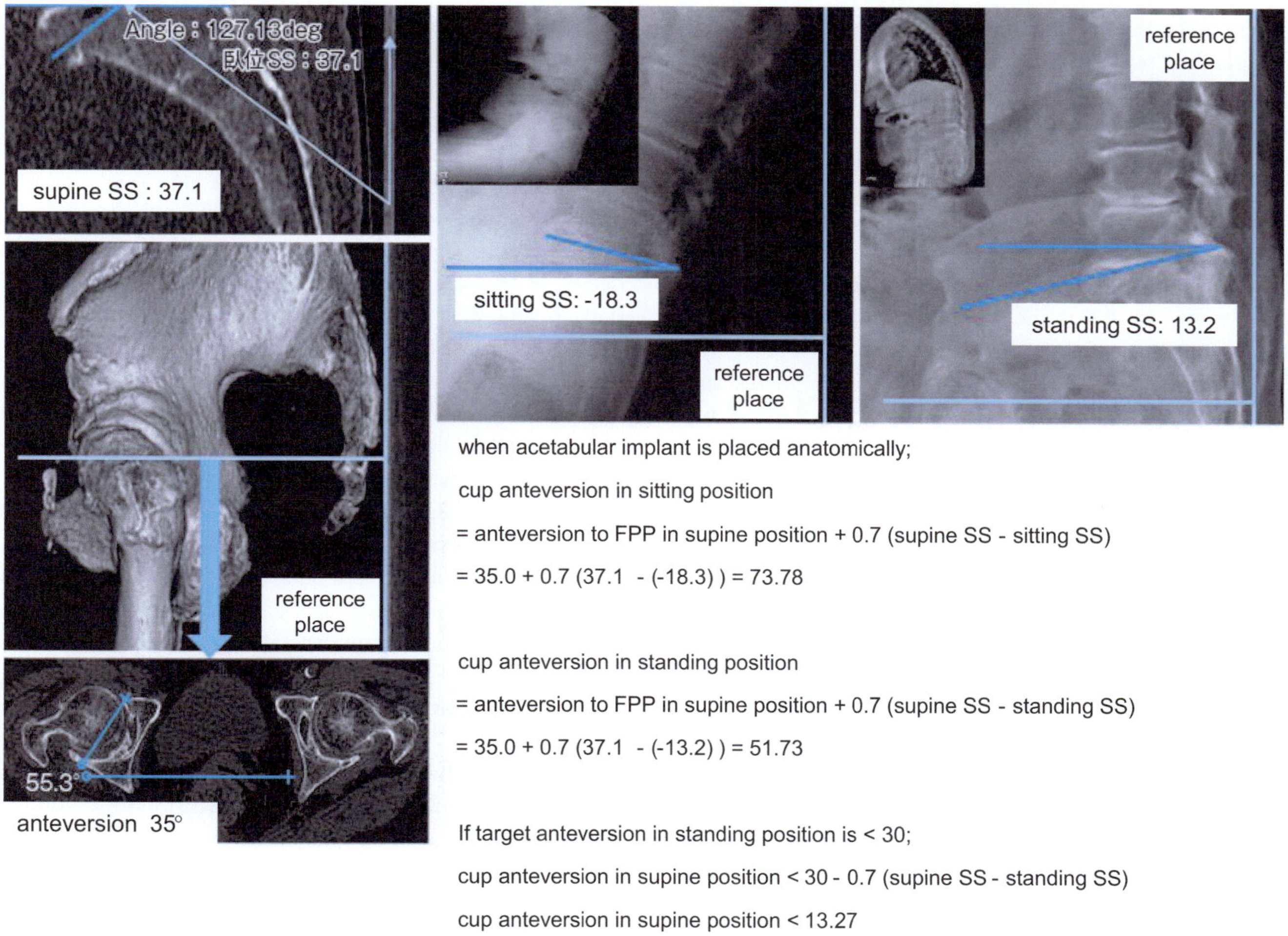

Fig. 77.3 Calculation of functional anteversion in supine, sitting and standing position, based on the Sacral slope (SS) in each position

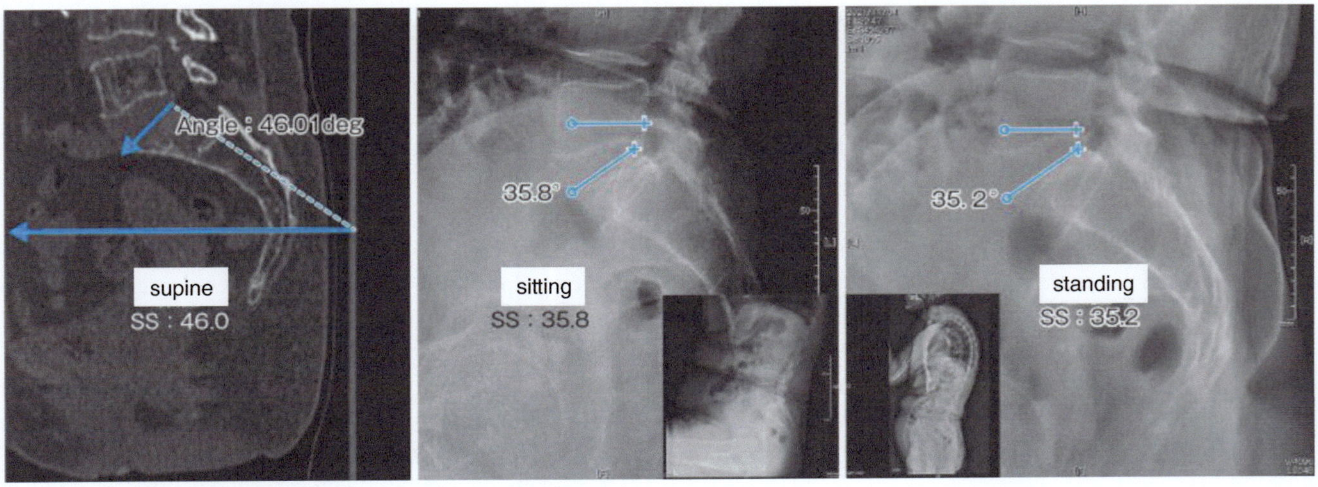

Fig. 77.4 Calculation of the assumed anterior spread in each body position based on the change in sacral slope in the supine, seated, and standing positions

following formula for estimated cup anteversion in the standing position:

Estimated cup anteversion in the standing position = cup anteversion to FPP in the supine position +0.7 × (SS in the standing position − SS in the supine position) < 30. This formula can be used in patients with a high risk of anterior dislocation.

If the risk of posterior dislocation is considered high, a cup anteversion of >30° in the sitting position is desirable.

The estimated cup anteversion in the sitting position = cup anteversion to FPP in the supine position +0.7 × (SS in the sitting position − SS in the supine position) > 30.

The range of optimal anteversion to FPP in the supine position derived from these inequalities is 0.7 × (SS in sitting–SS in standing position). This finding suggests that in cases in which the SS in the sitting position is larger than the SS in the standing position (Fig. 77.3), there is no optimal anteversion to FPP in the supine position, which can avoid both the risk of anterior dislocation and the risk of posterior dislocation.

Attempts to reduce the anteversion of the acetabular component based on the above calculations, which are aimed at reducing the risk of anterior dislocation (Fig. 77.5), may result in an overhang of the anterior edge of the cup from the acetabular bone. If the maximum value of this overhang exceeds 12 mm, the risk of iliopsoas impingement is significantly increased (Fig. 77.6) [21]. Therefore, implant placement largely reduces the anteversion of the cup compared with the anteversion of the acetabular bone, which is associated with a risk of iliopsoas impingement. In cases where the acetabular implant was expected to protrude more than 12 mm when planning for an anteversion of 30° in the standing anteroposterior pelvic plane, a dual-mobility cup was used instead of adjusting the cup anteversion in an attempt to reduce the dislocation risk.

(b) Implant selection -indication for dual-mobility cup

The dual-mobility cup has a mobile polyethylene liner, a larger oscillation angle than the conventional fixed polyethylene liner THA, and a larger jumping distance when impingement occurs, resulting in decreased risk of dislocation. The dislocation rate is approximately one-sixth of fixed polyethylene THA [22]. The dual-mobility cup has a 10-year survival rate of 98.6% [23] and a 15-year survival rate of 98.4% [24]. Polyethylene wear is comparable to that of conventional low-friction arthroplasty [25]. However, many types of dual-mobility arthroplasty have a modular shell and a metal liner. These dual-mobility cups may increase the concentration of metal ions in the blood [26]. Therefore, it is safer to use a dual-mobility cup only in cases with a high risk of dislocation.

Currently, we use dual-mobility cups in patients who have undergone thoracolumbar fusions of 7 or more vertebrae, the fusions of 3 or more vertebrae including the sacrum, in patients with a severe mismatch of PI-LL > 20°, or a change of less than 10° in SS from the standing to sitting position.

(c) Determination of the intraoperative cup placement angle

As mentioned above, it is recommended to use points of reference to calculate the optimal anteversion based on the pelvic tilt in the standing and sitting positions in each case and to place the cup according to these calculations. It is possible to achieve anteversion close to the planned placement by using the direction of the transverse ligament as a reference or by calculating the estimated protrusion distance of the cup from the anterior and posterior acetabular walls on

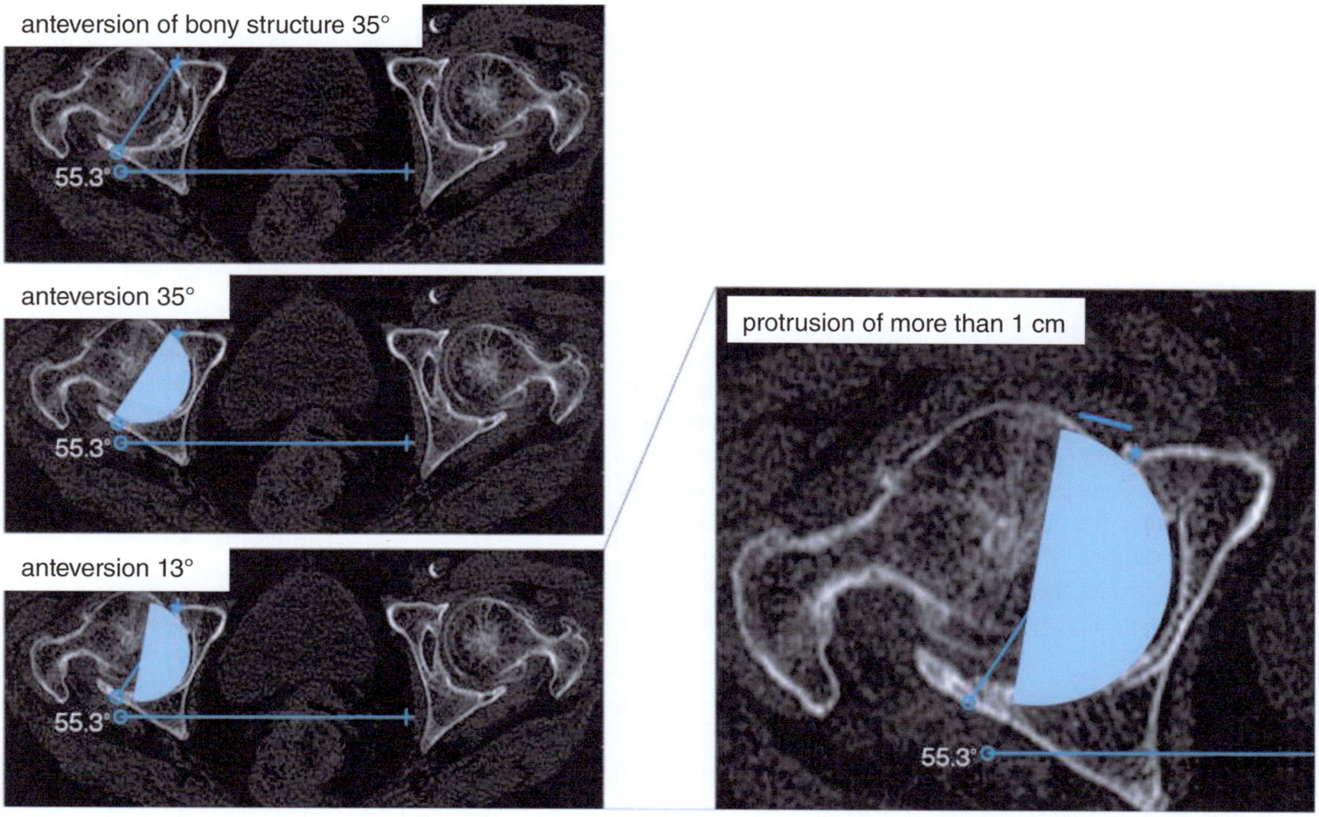

Fig. 77.5 Installation plan based on the calculated cup installation angle to avoid dislocation risk. If navigation is not used, the cup protrusion and osteophyte protrusion distances can be calculated in advance for more accurate intraoperative cup placement

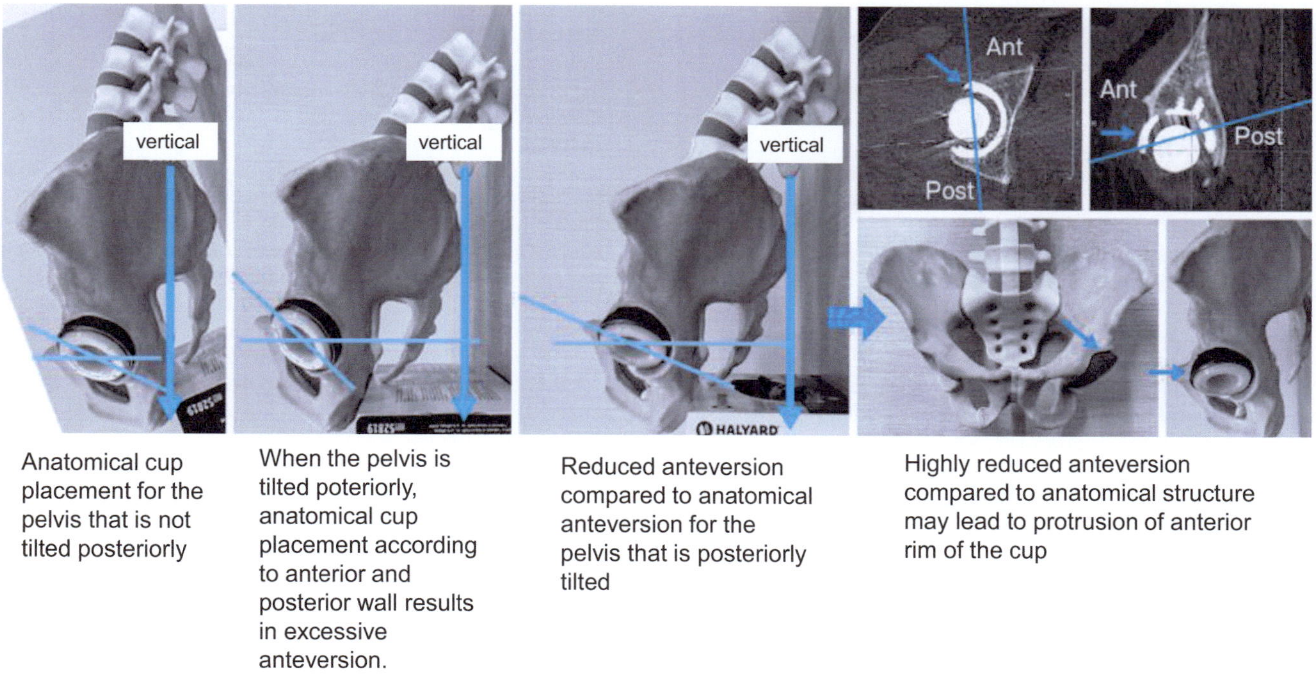

Fig. 77.6 When the cup is placed with a reduced anterior cup opening in a case of posterior pelvic tilt, the cup protrudes anteriorly and upward, which may result in irritation of the iliopsoas muscle. *Ant* anterior, *Post* posterior

the CT slice before surgery, as shown in Fig. 77.5. Since various navigation devices are now available, using them to place the cup according to the pre-calculated anteversion is also a promising option.

4 The Need for Collaboration with Spine Surgeons

Based on preoperative standing and sitting imaging, cup placement can be performed with a reduced risk of dislocation. However, optimal cup placement is calculated based on alignment at the time of surgery. There are cases in which spinal degeneration progress and pelvic tilt changes significantly after THA surgery and cases in which spinal alignment abnormalities present at the time of THA are corrected by subsequent spinal surgery. Therefore, it is necessary to share information regarding the prognosis of postoperative spinal degeneration and alignment correction plans with spine surgeons.

References

1. Malkani AL, et al. Total hip arthroplasty in patients with previous lumbar fusion surgery: are there more dislocations and revisions? J Arthroplast. 2018;33:1189–93.
2. Kawai T, et al. Number of levels of spinal fusion associated with the rate of joint-space narrowing in the hip. J Bone Joint Surg Am. 2021;103:953–60.
3. Ochi H, et al. Sagittal spinopelvic alignment predicts hip function after total hip arthroplasty. Gait Posture. 2017;52:293–300.
4. DelSole EM, et al. Total hip arthroplasty in the spinal deformity population: does degree of sagittal deformity affect rates of safe zone placement, instability, or revision? J Arthroplast. 2017;32:1910–7.
5. Esposito CI, et al. Total hip arthroplasty patients with fixed spinopelvic alignment are at higher risk of hip dislocation. J Arthroplast. 2018;33:1449–54.
6. Stefl M, et al. Spinopelvic mobility and acetabular component position for total hip arthroplasty. Bone Joint J. 2017;99-B(1 Supple A):37–45.
7. Hagiwara S, et al. Impact of spinal alignment and stiffness on impingement after total hip arthroplasty: a radiographic study of pre- and post-operative spinopelvic alignment. Eur Spine J. 2021;30:2443–9.
8. Vigdorchik JM, et al. Prevalence of risk factors for adverse spinopelvic mobility among patients undergoing total hip arthroplasty. J Arthroplast. 2021;36:2371–8.
9. Lazennec JY, et al. What is the impact of a spinal fusion on acetabular implant orientation in functional standing and sitting positions? J Arthroplast. 2017;32:3184–90.
10. Buckland AJ, et al. Dislocation of a primary total hip arthroplasty is more common in patients with a lumbar spinal fusion. Bone Joint J. 2017;99-B:585–91.
11. Kanto M, et al. Influence of spinopelvic alignment on pelvic tilt after total hip arthroplasty. Orthop Surg. 2019;11:438–42.
12. Weng W, et al. The effect of total hip arthroplasty on sagittal spinal-pelvic-leg alignment and low back pain in patients with severe hip osteoarthritis. Eur Spine J. 2016;25:3608–14.
13. Suzuki H, et al. Postural and chronological change in pelvic tilt five years after total hip arthroplasty in patients with developmental dysplasia of the hip. J Arthroplast. 2016;31:317–22.
14. Kim Y, et al. How do global sagittal alignment and posture change after total hip arthroplasty? Int Orthop. 2020;44:267–73.
15. Lewinnek GE, et al. Dislocations after total hip-replacement arthroplasties. J Bone Joint Surg Am. 1978;60:217–20.
16. Murray DW. The definition and measurement of acetabular orientation. J Bone Joint Surg Br. 1993;75:228–32.
17. Abdel MP, et al. What safe zone? The vast majority of dislocated THAs are within the Lewinnek safe zone for acetabular component position. Clin Orthop Relat Res. 2016;474:386–91.
18. Burapachaisri A, et al. Lewinnek safe zone references are frequently misquoted. Arthroplast Today. 2020;6:945–53.
19. Vigdorchik JM, et al. Otto Aufranc award: a simple hip-spine classification for total hip arthroplasty: validation and a large multicentre series. Bone Joint J. 2021;103-B(7 supple B):17–24.
20. Lembeck B, et al. Pelvic tilt makes acetabular cup navigation inaccurate. Acta Orthop. 2005;76:517–23.
21. Cyteval C, et al. Iliopsoas impingement on the acetabular component: radiologic and computed tomography findings of a rare hip prosthesis complication in eight cases. J Comput Assist Tomogr. 2003;27:183–38.
22. Romagnoli M, et al. The efficacy of dual-mobility cups in preventing dislocation after total hip arthroplasty: a systematic review and meta-analysis of comparative studies. Int Orthop. 2019;43:1071–82.
23. Fessy M-H, et al. Midterm clinical and radiographic outcomes of a contemporary monoblock dual-mobility cup in uncemented total hip arthroplasty. J Arthroplast. 2019;34:2983–91.
24. Puch J-M, et al. Dual-mobility cup in total hip arthroplasty in patients less than fifty five years and over ten years of follow-up: a prospective and comparative series. Int Orthop. 2017;41:475–80.
25. Prudhon JL, Verdier R, Caton JH. Low friction arthroplasty and dual mobility cup: a new gold standard. Int Orthop. 2017;41:563–71.
26. Nam D, et al. Is a modular dual mobility acetabulum a viable option for the young, active total hip arthroplasty patient? Bone Joint J. 2019;101-B:365–71.

Coxa Vara

78

Yoko Miura

1 Coxa Vara

Coxa vara is a deformity caused by congenital dislocation of the hip joint or Legg-Calvé-Perthes disease in childhood that disrupts blood flow to the femoral head, leading to temporary necrosis of the bone and resulting in flattening and shortening of the femoral head, elongation of the greater trochanter, and abnormal acetabular morphology [1, 2] (Fig. 78.1a).

Total hip arthroplasty (THA) for adult patients with coxa vara is often more difficult than that for general hip osteoarthritis for the following reasons [3]:

(a) Narrow surgical space

The working space is often narrow despite neck osteotomy owing to contracted soft tissue and a shortened femoral head, which makes it more difficult to approach the acetabulum. Sufficient capsular and soft tissue release is often necessary to achieve adequate exposure.

(b) Abnormal femoral anteversion

The proximal femur often has excessive anteversion (Fig. 78.1b) for which correction is often required to enable femoral implant insertion.

(c) Imbalance between metaphyseal and diaphyseal medullary canals

The metaphyseal medullary canal is relatively wider than that of the diaphysis (Fig. 78.1a), making it difficult to select an adequate femoral component that fits the meta- and diaphyseal medullary canals.

(d) Leg length discrepancy

Due to shortening of the femoral head and neck, a severe leg length discrepancy is common (Fig. 78.1a). Owing to soft tissue contracture, correcting the leg length discrepancy is often difficult.

(e) Soft tissue contracture

The joint capsule is often thickened, and the surrounding soft tissues contracted. Sufficient soft tissue release is required to approach the acetabulum and the femur.

(f) Childhood surgical history

THA becomes more complicated in patients who underwent surgical procedures in childhood, leading to increased joint and soft tissue contractures and an altered bone morphology.

Y. Miura (✉)
Funabashi Orthopedic Hospital, Funabashi, Chiba, Japan

Y. Nakashima (ed.), *Advances in Total Hip Arthroplasty*, https://doi.org/10.1007/978-981-97-2461-1_78

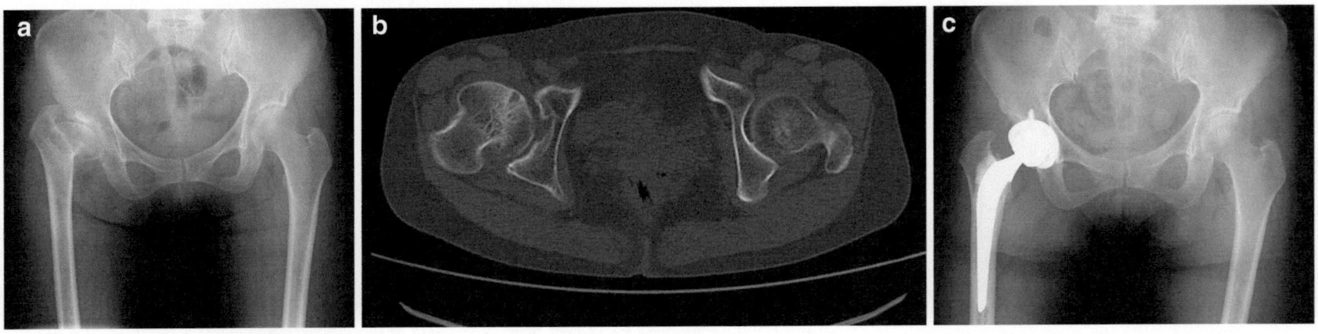

Fig. 78.1 Images of the right hip of a 55-year-old woman. (**a**) Preoperative standing simple radiograph showing a leg length difference of approximately 3 cm compared with the healthy side. (**b**) Preoperative computed tomography image showing hyperanteversion of the femoral neck, shortening of the neck, and flattening of the femoral head versus the healthy side. (**c**) Postoperative standing simple radiograph showing a cemented stem used to correct the anterior torsion

2 Indications for Surgery

The presence of knee and low back pain that is refractory to conservative treatment due to leg length discrepancy as well as end-stage hip osteoarthritis affects patient quality of life and function.

3 Implant Selection

On the femoral side, it is necessary to select a component that allows version correction and fits the imbalance between the metaphyseal and diaphyseal medullary canals.

Good results using standard cementless stems have also been reported [4]. On the other hand, there are some reports that the incidence of femoral fracture in THA for patients with coxa vara is slightly higher than that for general patients, and most of these fractures occur when monoblock cementless stems are used [5].

Modular femoral components easily adjust neck length, offset, and anteversion and can easily fit with the imbalance between the diaphyseal and metaphyseal medullary canals. Excellent outcomes have been reported for modular femoral components [6, 7]. However, concern persists regarding taper fretting, corrosion, and stem fracture of the modular femoral components [8].

Cemented femoral components are also good options because they can be used to easily achieve version correction versus cementless stems and do not create concern about mismatch between the metaphyseal and diaphyseal medullary canals (Fig. 78.1c).

On the acetabular side, cementless or cemented components are acceptable if good fixation is achieved.

4 Preoperative Planning and Surgical Techniques

(a) Preoperative templating

Because multiplanar deformities often present on the femoral and acetabular sides in cases of coxa vara, such as leg length discrepancy, abnormal femoral anteversion, and mismatch between the metaphyseal and diaphyseal medullary canals, it is important to complete proper preoperative planning using radiographs and computed tomography.

(b) Acetabular exposure

If reaming is performed without sufficient acetabular exposure, cup malpositioning and insufficient fixation may occur. The surgical space should be sufficiently exposed to enable soft tissue release and easy reaming.

(c) Cup insertion

In cases of coxa vara, the acetabulum is often shallow, enlarged, and retroverted [9]. When a bone defect is present, a bone graft should be positioned above the cup. When contracture is severe and reduction is expected to be difficult, the use of a smaller head may be considered.

(d) Femoral preparation

A short and flattened head often leads to a lower-level neck osteotomy. Effort should be made to preserve as much of the hard cortical bone of the proximal femur as possible to prevent postoperative periprosthetic femoral fractures.

If the contracture is severe, circumferential capsular and soft tissue release can improve femoral mobility and facilitate femoral access.

(e) Reduction

In cases of severe leg length discrepancy or severe joint contracture, hip joint reduction is often difficult. Stretching of the hip joint and use of a reduction device can facilitate the reduction [10].

5 Clinical Results

Although many reports have shown that the clinical outcomes of THA of patients with coxa vara are as good as those of patients with general hip osteoarthritis [4, 6, 7, 11, 12], there are also some reports of high complication rates and inferior clinical outcomes [5, 13].

(a) Complications

(i) Femoral fracture

In cases of coxa vara, the femoral morphology may lead to a high frequency of periprosthetic femoral fractures [13, 14]. Most fractures occur with the usual monoblock cementless stems, and modular femoral components reportedly reduce the fracture risk [14].

(ii) Neurological deficit

The incidence of neurological injury (mainly in the sciatic nerve) is reportedly approximately 3%, higher than that of general THA (<1%) [3, 13, 14]. A history of surgery in childhood and aggressive leg lengthening may be a risk factor [11, 14].

(b) Leg length correction

Strict leg length correction may cause low back pain in patients with severe deformities of the lumbar spine, and excessive leg lengthening may not be necessary [15].

Postoperative rehabilitation often improves pelvic tilt and range of motion, and the subjective leg length discrepancy gradually improves.

(c) Mid- to long-term results

Compared with the preoperative period, the functional score improved significantly, and many reports indicated that the results are comparable to those of conventional THA [3, 4]. On the other hand, the revision rate at the mid- to long-term time points is 0–21%, although reports vary widely [14].

Many reports indicate that the THA results for patients with coxa vara as well as those of THA with general osteoarthritis are satisfactory.

6 Prospects

THA in patients with coxa vara is challenging. Patients are often younger than general THA patients and require a more accurate and appropriate implant placement. Careful preoperative planning, informed consent, and follow-up with postoperative rehabilitation are important.

References

1. Herring JA. The treatment of Legg-Calvé-Perthes disease. A critical review of the literature. J Bone Joint Surg Am. 1994;76:448–58.
2. Stulberg SD, et al. The natural history of Legg-Calvé-Perthes disease. J Bone Joint Surg Am. 1981;63:1095–108.
3. Traina F, et al. Long-term results of total hip replacement in patients with Legg-Calvé-Perthes disease. J Bone Joint Surg Am. 2011;93:e25.
4. Luo ZY, et al. Monobloc implants in cementless total hip arthroplasty in patients with Legg-Calve-Perthes disease: a long-term follow-up. BMC Musculoskelet Disord. 2017;18:1–10.
5. Sansanovicz D, et al. Cementless total hip arthroplasty in patients with osteoarthrosis secondary to Legg-Calvé-Perthes disease compared with primary osteoarthrosis: a case-control study.
6. Seufert CR, et al. Treatment of arthritis associated with Legg-Calve-Perthes disease with modular total hip arthroplasty. J Arthroplast. 2015;30:1743–6.
7. Takahashi E, et al. Satisfactory outcomes in patients operated with primary total hip arthroplasty for Perthes-like deformities: results from a surgical technique utilizing a conical stem, an elevated hip center, and no shortening femoral osteotomy. Arthroplast Today. 2021;7:29–36.
8. Krishnan H, et al. Modular neck femoral stems. Bone Joint J. 2013;95:1011–21.
9. Sankar WN, et al. The development of acetabular retroversion in children with Legg-Calvé-Perthes disease. J Pediatr Orthop. 2008;28:440–3.
10. Oinuma K, et al. A new anterior total hip arthroplasty reduction technique for patients who require leg lengthening. Hip Int. 2020;30:587–91.
11. Anthony CA, et al. Total hip arthroplasty in patients with osteoarthritis associated with Legg-Calve-Perthes disease: perioperative complications and patient-reported outcomes. J Arthroplast. 2021;36:2518–22.
12. Engeæter LB, et al. Low revision rate after total hip arthroplasty in patients with pediatric hip diseases. Acta Orthop. 2012;83:436–41.
13. Baghdadi YMK, et al. Total hip arthroplasty for the sequelae of Legg-Calvé-Perthes disease. Clin Orthop Relat Res. 2013;471:2980–6.
14. Hanna SA, et al. Systematic review of the outcome of total hip arthroplasty in patients with sequelae of Legg-Calvé-Perthes disease. Arch Orthop Trauma Surg. 2017;137:1149–54.
15. Koga D, et al. The effect of preoperative lateral flexibility of the lumbar spine on perceived leg length discrepancy after total hip arthroplasty. Med Dent Sci. 2009;56:69–77.

Dorr Classification Type C

79

Takuya Nakamura

1 Historical Background of Studies on the Geometry and Bone Quality of the Proximal Femur

In the late 1970s, cementless stems with osteogenic fixation capacities similar to the cementless stems of today began to be used because of concerns about cemented stem laxity and the concept of "cement disease" which was prevalent in the 1970s. The average age of patients who received cementless stems in North America at the end of 1980 was 45–55 years old [1]. At that time, it was believed that (1) morphological compatibility between the stem and supporting bone and (2) good bone quality of the femur were important for good fixation of the cementless stem. Further research in this field is urgently needed. Noble et al. used the canal flare index to classify proximal femurs by shape into champagne flute, normal, and stovepipe shapes (Fig. 79.1a) [1]. Dorr et al. studied the biochemical and histological characteristics of different proximal femur shapes and proposed a radiographic classification that also evaluated the bone quality of the femur [2]. This classification is still frequently used, particularly cementless stems.

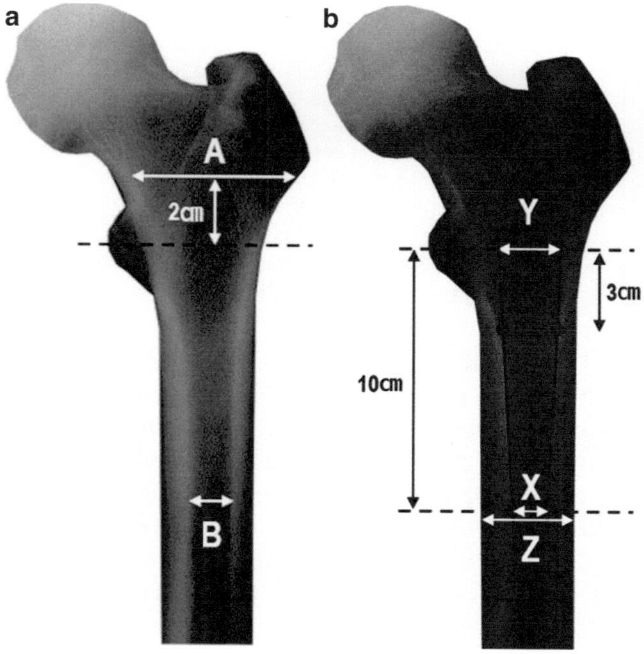

Fig. 79.1 Quantitative indices of the proximal femur. (**a**) Canal flare index (1) is defined as the ratio of the intracortical width of the femur at a point 20 mm proximal to the lesser trochanter (*A*) and canal isthmus (*B*). Less than 3.0 is classified as "stovepipe;" 3.0 to 4.7 "normal;" and 4.7 to 6.5 "champagne flute." (**b**) Cortical index: Z-X/Z (calculated using frontal and Lauenstein lateral radiographs) and canal to calcar isthmus ratio: X/Y (calculated on frontal radiographs) (2). X, intramedullary. Canal diameter at a point 10 cm distal to the mid-lesser trochanter. Y, intramedullary. Canal diameter at the level of the mid-lesser trochanter; Z, femoral diaphyseal diameter at a point 10 cm distal to the mid-lesser trochanter

T. Nakamura (✉)
Department of Orthopedic Surgery, Toyama Nishi General Hospital, Toyama, Japan

Y. Nakashima (ed.), *Advances in Total Hip Arthroplasty*, https://doi.org/10.1007/978-981-97-2461-1_79

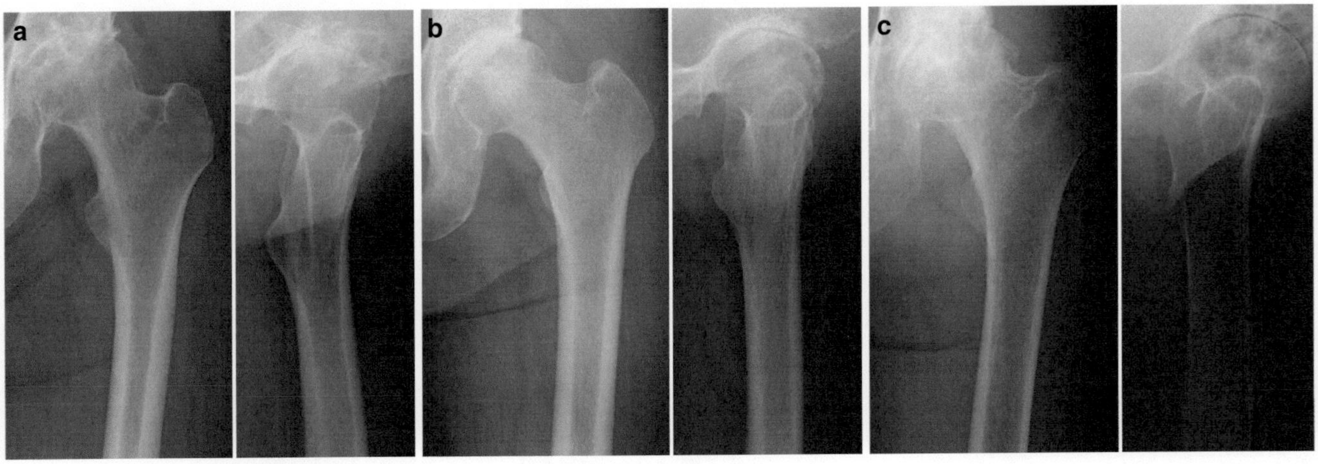

Fig. 79.2 Dorr classification. (**a**) Dorr type A features. Thick cortical bone in the frontal view. Large posterior cortical bone in lateral view. Medial and posterior cortical bone thickens immediately below the lesser trochanter. Narrow metaphyseal medullary cavity. Cortical bone is dense with well-defined margins. Funnel-shaped. (**b**) Dorr type B features. The medial and posterior cortical bones are further reduced. The proximal portion of the posterior cortical bone has thinned or is absent, and the medullary cavity is enlarged. Irregular posterior endosteum of cortical bone. Also funnel-shaped. (**c**) Dorr type C features. The medial and posterior cortical bones are almost completely absent. The anterior and posterior cortical bones are sometimes dramatically thin and opaque in the lateral view. The lateral view shows a very wide medullary cavity. Stovepipe shaped

2 Dorr Classification (Fig. 79.2)

The Dorr classification, which is used to evaluate the bone quality of the proximal femur, is based on frontal and Lauenstein lateral radiographs (Fig. 79.2). Basically, bone quality can be classified into types A, B, and C based on radiographic appearance, but a relationship with quantitative indices (Fig. 79.1b) has also been reported [2]. Dorr et al. stated that, for implant fixations, "type A can tolerate fixations well with or without cement, type B has no issues with fixation, and type C is not a suitable environment for fixation" [2], a guideline which is still used today. However, the Dorr classification has intra- and inter-rater disagreement, and it has been reported that the agreement rate can be improved by measuring quantitative indices [3].

3 Support for Dorr Type C

Most cases of femoral neck fractures that require artificial head insertion occur in Dorr type C femurs. The incidence of femoral neck fractures has stopped increasing in recent years in Western countries but is still increasing in Asia [4]. The number of type C total hip arthroplasty (THA) cases is expected to gradually increase as the global population continues to age. The choice of implant for patients with type C femurs has always been a matter of debate, and there are still many unresolved issues that require further investigation.

(a) Cementless stems in patients with Dorr type C femurs

Type C femurs are associated with bone fragility, and it is difficult to obtain a morphological match with stems [2]. Cementless stems are associated with concerns regarding strong initial fixations and bony fixations, and the use of cementless stems for patients with type C femurs is generally not recommended due to concerns about implant laxity and peri-implant fractures [5, 6]. However, there are increasing reports that cementless titanium alloy stems with modern plasma spray coatings or hydroxyapatite surface treatments can be used in patients with type C femurs [7–10]. This has led some researchers to believe that the Dorr classification may no longer be appropriate for determining the type of implant to be used, particularly for patients with type C femurs. Some believe that type C can be further divided into two subtypes, and the subtype with extremely weak bone quality would be better suited for modern implants. However, it should be noted that many reports with good results on these type C implants are based on a small number of cases, and there have also been reports with poor results using the same implants. In addition, it has been reported that a high rate of postoperative peri-implant fractures occurred in patients with type C when cementless short stems were used, although there were no differences in the results between young and elderly patients [11]. This suggests that the Dorr classification is still useful for implant selection in modern times and is more significant than age-based classifications (Fig. 79.3).

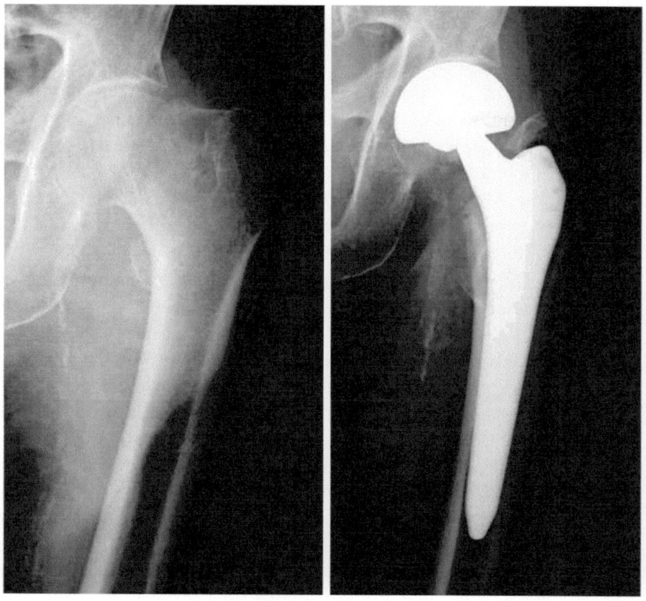

Fig. 79.3 102-year-old woman, Dorr type B. A Zweymüller-type cementless stem was used. Postoperative fixation was good

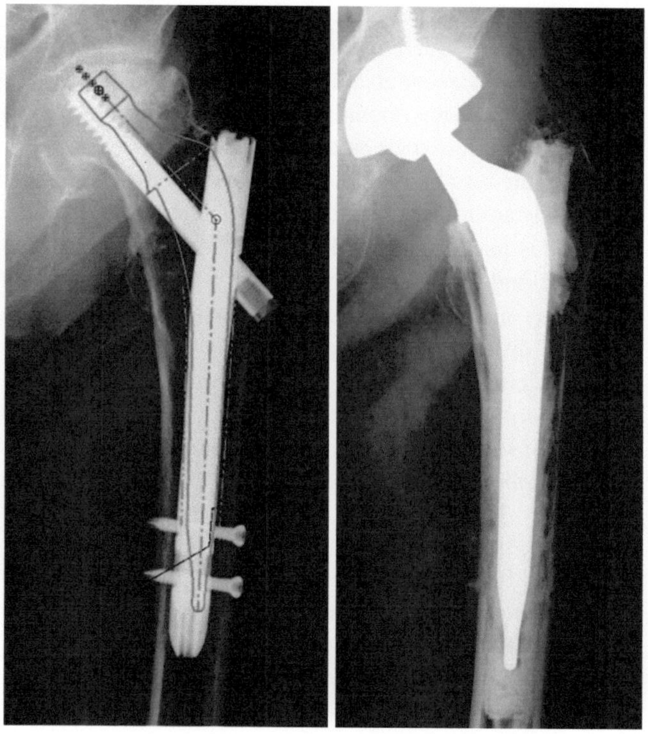

Fig. 79.4 In cases with a wide medullary cavity due to bone fragility, even the largest cement stem can be undersized, and the cement mantle may be thickened

It is not clear what shapes of cementless stems are best for type C femurs, but proximally fixed stems tend to concentrate stresses more proximally in the stovepipe shape with a wider medullary cavity. Although a stem that transmits load proximally is considered ideal to avoid postoperative stress sealing, excessive proximal load bearing in fragile bones is associated with a risk of peri-implant fractures. In addition, postoperative stress shielding cannot be avoided even with a proximally fixed stem in type C femurs with poor bone quality [12]. It is believed that fixation at the diaphysis, such as with the Zweymüller-type stem, provides more stability for type C femurs, but postoperative stem subsidence and peri-implant fractures have been reported under these conditions [13]. In any case, forceful insertion of a proximal fixation-type stem can cause fracture of the calcar area, whereas forceful insertion of a distal fixation-type stem can induce fracture in the metaphysis. Gentle hammering during insertion should be considered, and excessive force should be avoided by hitting the hammer using a steady rhythm. In addition, at the end of broach insertion, the broach could be inserted more deeply and stabilized by tapping the broach directly with the stem impactor in small increments. This helps prevent postoperative sinking of the stem and peri-implant fractures.

(b) Cement stem in Dorr type C

In Dorr type C femurs with poor bone quality, the use of cemented stems is generally recommended due to the uncertain results with cementless stems [5]. In contrast, the stovepipe shape does not correlate closely with cemented stem laxity, but poor bone quality and thin cortical bone correlate with laxity even in cemented stems [14], suggesting that the bone quality of the proximal femur affects the outcome more than the shape in type C femurs, so evaluation of bone quality is important. In addition, in cases with extreme bone atrophy, the medullary cavity may be enlarged and the largest cemented stem may be undersized, resulting in a thick cement mantle (Fig. 79.4) [15]. The generally recommended mantle size is 1–2 mm, and an undersized stem may cause premature laxity [16].

Some surgeons dislike femoral cementation due to technical difficulties and long operative times. Furthermore, cementation tends to be avoided because it is associated with bone-cement implantation syndrome, which is a fatal complication [17]. However, there is a report that shows no difference in mortality between cementless and cemented fixation in elderly patients [5], and future studies are needed regarding mortality in patients with type C femurs. Although the number of surgeons with no experience in cementation has been increasing recently, it is essential for all to master the cementation technique if possible.

(c) Operator factors and impact on outcomes in patients with Dorr type C femurs

Both cementless and cemented stems have different intraoperative and postoperative fracture rates depending on stem shape, and it is recommended that stems with high peri-implant fracture rates should not be used in elderly patients with fragile bones, such as those classified as type C [18]. However, the results of THA in patients with type C femurs vary widely among reports, both for cementless and cemented stems. This may be due to the familiarity of surgeons with cemented and cementless insertions [19]. Familiarity with stems may also be a factor. In addition to differences in implants, differences in surgeons has also been shown to impact outcomes due to not only the proficiency of the surgeon in terms of insertion techniques, but also an understanding of stem characteristics and habitual use of stems, as well as familiarity with the device itself.

(d) Others

In patients Dorr type C femurs, a large stem tends to be inserted due to the presence of wide medullary cavities. Generally, the neck of the stem becomes proportionally longer as the stem size increases, and there is a concern that the neck may become longer than the original neck length. Care should be taken when selecting a stem to ensure that it does not have an excessively large offset or overly long leg length.

In addition, regardless of implant selection, poor bone quality can lead to poor long-term outcomes in both cementless and cemented cases; therefore, osteoporosis must also be treated appropriately to improve long-term outcomes [20].

References

1. Noble PC, et al. The anatomic basis of femoral component design. Clin Orthop. 1988;235:148–65.
2. Dorr LD, et al. Structural and cellular assessment of bone quality of the proximal femur. Bone. 1993;14:231–42.
3. Nakaya R, et al. Reproducibility of the Dorr classification and its quantitative indices on plain radiographs. Orthop Traumatol Surg Res. 2019;105:17–21.
4. Cooper C, et al. Secular trends in the incidence of hip and other osteoporotic fractures. Osteoporos Int. 2011;22:1277–88.
5. Jämsen E, et al. High early failure rate after cementless hip replacement in the octogenarian. Clin Orthop. 2014;472:2779–89.
6. Thomsen MN, et al. Fracture load for periprosthetic femoral fractures in cemented versus uncemented hip stems: an experimental in vitro study. Orthopedics. 2008;31:653.
7. Dalury DF, et al. Modern proximally tapered uncemented stems can be safely used in Dorr type C femoral bone. J Arthroplast. 2012;27:1014–8.
8. Reitman RD, et al. Thirteen year results of total hip arthroplasty using a tapered titanium femoral component inserted without cement in patients with type C bone. J Arthroplast. 2003;18:116–21.
9. Meding JB, et al. High survival of uncemented proximally porous-coated titanium alloy femoral stems in osteoporotic bone. Clin Orthop Relat Res. 2010;468:441–7.
10. Kelly SJ, et al. Use of a hydroxyapatite-coated stem in patients with Dorr type C femoral bone. Clin Orthop Relat Res. 2007;465:112–6.
11. Gkagkalis G, et al. Cementless short-stem total hip arthroplasty in the elderly patient—is it a safe option? A prospective multicentre observational study. BMC Geriatr. 2019;19:112.
12. Nakamura T, et al. Stress differences when a tapered wedge stem with short stem length and small thermal spray range is inserted into femurs of different geometries. Hip Joint. 2014;40:620–3. Japanese
13. Taylor F, et al. Hemiarthroplasty of the hip with and without cement: a randomized clinical trial. J Bone Joint Surg Am. 2012;94:577–83.
14. Nixon M, et al. Does bone quality predict loosening of cemented total hip replacements? J Bone Joint Surg Br. 2007;89:1303–8.
15. Mears SC. Management of severe osteoporosis in primary total hip arthroplasty. Curr Transl Geriatr and Exp Gerontol Rep. 2013;2:99–104.
16. Cristofolini L, et al. Effect of undersizing on the long-term stability of the Exeter hip stem: a comparative in vitro study. Clin Biomech (Bristol, Avon). 2010;25:899–908.
17. Donaldson AJ, et al. Bone cement implantation syndrome. Br J Anaesth. 2009;102:12–22.
18. Carli AV, et al. Periprosthetic femoral fractures and trying to avoid them: what is the contribution of femoral component design to the increased risk of periprosthetic femoral fracture? Bone Joint J. 2017;99-B(1 Supple A):50–9.
19. Stihsen C, et al. Cementless total hip arthroplasty in octogenarians. J Arthroplast. 2017;32:1923–9.
20. Bhandari M, et al. Effect of bisphosphonates on periprosthetic bone mineral density after total joint arthroplasty. A meta-analysis. J Bone Joint Surg Am. 2005;87:293–301.

Bilateral Same-Day Surgery

80

Atsutoshi Maki and Hiroyoshi Iwaki

1 Bilateral Total Hip Arthroplasty

In Japan, the prevalence of osteoarthritis of the hip is 1.0–4.3%, according to the guidelines for the treatment of osteoarthritis of the hip [1]. A small number of patients have bilateral hip disorders, such as rheumatoid arthritis and osteonecrosis. Total hip arthroplasty (THA) is one of the most effective surgical procedures in orthopedics; hence, the demand for bilateral THA is increasing.

Bilateral THA is planned based on two methods: bilateral simultaneous THA, which is performed on the same day, and bilateral staged THA. Bilateral simultaneous THA was first reported by Chanley et al. in 1971 [2], and since then, numerous studies and discussions about it have followed [3, 4]. The expected advantages of this method include completion of treatment with a single anesthetic and shortened hospital stay. In fact, bilateral simultaneous THA has a shorter hospitalization period than bilateral staged THA, as the latter combines the two hospitalization periods [5]. Furthermore, the difference becomes greater when the overall treatment time, including the waiting period prior to surgery, is compared in addition to the duration of hospitalization itself, which leads to reduction in medical costs [6]. However, the increased invasiveness of a single surgery may lead to significant blood loss and infection, as well as serious postoperative complications such as thromboembolism and cardiovascular events. Ultimately, the benefits of bilateral simultaneous THA are not paramount, and its effectiveness should be debated since it does not result in a definite increase in complications compared to bilateral staged THA. We discuss the complications based on previous studies, indications for treatment considering complications, surgical techniques specific to bilateral simultaneous THA, and our own experience.

2 Complications Based on Previous Studies

One adverse event of concern in bilateral simultaneous THA is the amount of blood loss, which is not greater than that in bilateral graded THA [7]. However, it is believed that the blood loss volume is increased compared with that in unilateral THA, which requires a greater amount of blood transfusion accordingly [8]. Moreover, the corrected operative time in bilateral simultaneous THA is neither prolonged nor shorter than that in bilateral staged THA [9], but it is longer than that in unilateral THA. Although the time required for a single operation may be a risk factor for deep vein thrombosis [10], there is no difference in the incidence of deep vein thrombosis between bilateral simultaneous THA and bilateral staged THA [11]. In addition, complications that usually occur with THA, such as dislocation, infection, and fracture, were not significantly different between the two methods [12]. The major causes of postoperative death are pulmonary embolism and cardiovascular events [13], which are not more common in bilateral simultaneous THA [14]. However, the incidence of these major complications has increased in institutions that are not experiences in performing THA [15].

3 Criteria for Indication in Our Hospital

The indication for unilateral THA is painful hip arthroplasty refractory to conservative treatment and is same for bilateral simultaneous THA. However, since the benefits of bilateral simultaneous THA were determined, the need for a stricter indication criteria was proposed [16]. In fact, a database analysis showed that bilateral simultaneous THA selected a

A. Maki (✉) · H. Iwaki
Osaka Orthopedic Hospital, Osaka, Japan

Y. Nakashima (ed.), *Advances in Total Hip Arthroplasty*, https://doi.org/10.1007/978-981-97-2461-1_80

larger proportion of male patients than bilateral staged THA, which also suggested that younger patients were selected [15]. Therefore, we added the "American Society of Anesthesiologists Physical Status Classification System (ASA) Class 3 or higher" to the exclusion criteria for bilateral simultaneous THA [17]. The ASA classification is a well-known scoring system for predicting postoperative prognosis. Sex, age, and body mass index (BMI) were not included in the exclusion criteria. However, the need to shorten the treatment period, which is the basis for selecting bilateral simultaneous THA instead of bilateral staged THA, should be determined by considering age and activity.

4 Surgical Procedure in Our Hospital

The same technique can be used for THA and conventional hip arthroplasty. However, since it is necessary to operate on both hips during a single anesthesia session, there is a need to shorten the operation time. From this point of view, supine surgery is more compatible with bilateral simultaneous THA than lateral THA because it does not require positional changes, and the surgeon can perform surgery on the contralateral hip at the time of wound closure [18].

The order of surgery was determined by the hip side which had greater leg shortening. The goal was to place the implant in the primary labral position; however, in patients with severe deformities or contractures, the cup may be placed higher or leg shortening may be maintained. To avoid excessive postoperative leg length discrepancy in such cases, we operated on the hip side with greater leg shortening first and corrected the contralateral hip to some extent while confirming the position and stability of the primary labrum (Fig. 80.1).

Bilateral simultaneous THA can be performed in the same way as unilateral THA if blood transfusion and collection devices are available and if a surgeon skilled in conventional THA performs the procedure.

5 Example of a Self-Experiment

We describe the autopsy of 181 patients (362 hips) who underwent bilateral simultaneous THA between 2016 and 2021. This was a retrospective study. All procedures were performed by the same surgeon who was skilled in THA, and the surgical technique used was the anterolateral supine approach [19]. Gait training was started the day after surgery, and a clinical path was used with a 3-week hospitalization period. The indication for blood transfusion was a hemoglobin level of 7.0 g/dL or less.

The mean operative time was 115.0 min, and the mean intraoperative blood loss was 226.6 mL. Patients were ambulatory on an average of 2.3 days after surgery. The mean length of hospital stay was 27.5 days, and 6 patients (3.3%) deviated from the clinical path. Complications included dislocation in 2 patients (0.6%), fracture in 2 patients (0.6%), deep vein thrombosis in 14 patients (7.7%), deep infection in 3 patients (0.8%), and gastrointestinal bleeding in 1 patient (0.6%). No serious complications occurred, such as pulmonary embolism or cardiovascular events. Three patients (1.7%) required blood transfusion. These results indicate that the expected benefits of bilateral same-day THA were achieved without an apparent increase in complications (Table 80.1) [20, 21].

Among the 181 patients included in the study, 23 patients had 400–500 mL of autologous blood in their blood stores, regardless of patient factors. We performed propensity score matching in a prospective cohort using blood storage as the starting point, and examined the association between blood storage and transfusion. The analysis showed no difference in transfusion frequency between the two groups, and transfusion could not be avoided by blood storage (Table 80.2). Therefore, after the study, bilateral simultaneous THA was performed in almost all patients without blood storage and no adverse effects occurred.

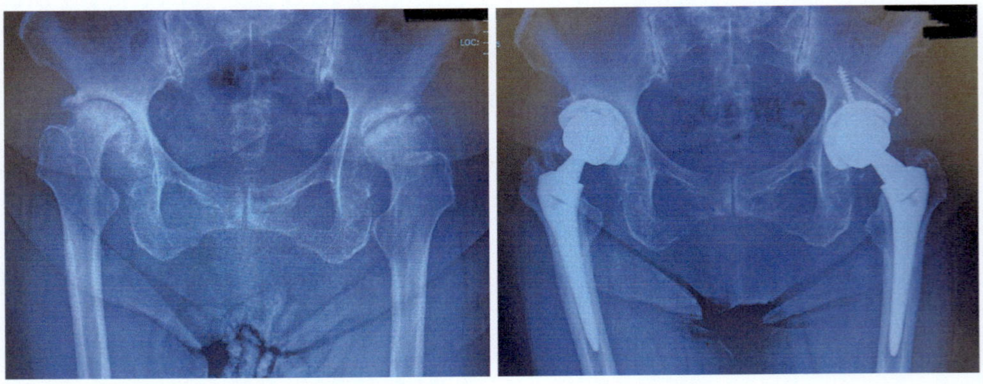

Fig. 80.1 Preoperative and postoperative radiographs. Surgery was performed starting from the right hip with significant leg shortening. The implant was placed at a high level as planned. In the left hip, joint stability was achieved while avoiding excessive leg length differences by using the previously operated right hip as an indicator

Table 80.1 Survey results

Mean (standard deviation)	
Number of cases (persons)	181
Sex (male/female)	22/159
Age (years)	65.7 (9.5)
BMI (kg/m^2)	23.6 (4.6)
Crowe classification	I: 129 II: 35 III: 17 IV: 0
Preoperative hemoglobin level (g/dl)	12.9 (1.4)
Operation time (minutes)	115.0 (23.0)
Intraoperative blood loss (ml)	226.6 (153.9)
Postoperative leg length difference (mm)	2.1 (5.5)
Number of days to acquire walking ability (days)	2.3 (3.8)
Length of hospital stay (days)	27.5 (15.3)

(%)	
Complication	**Incidence**
Dislocation	2 (0.6) <1.0–5.0%>
Fracture	2 (0.6)
Deep vein thrombosis	14 (7.7) <20.0–30.0%>
Deep infection	3 (0.8) <1.0–5.0%>
Gastrointestinal bleeding	1 (0.5)
Pulmonary embolism	0 (0) <0.5–1.0%>
Cardiovascular events	0 (0)
Death	0 (0)

< > indicates complication rates obtained from the guidelines for the treatment of osteoarthritis of the hip joint
Patients were weaned early postoperatively and discharged without deviation from the clinical pathway. Benefits of bilateral simultaneous THA without apparent increased complications compared to conventional THA studied in the past

Table 80.2 Association between blood storage and transfusion

(Standard deviation)		
Before matching	Blood-storage group	Non-storage group
Number of cases	23	158
Sex (male/female)	3/20	19/139
Age (years)	63.0 (7.9)	66.0 (9.7)
BMI (kg/m^2)	23.0(4.7)	23.7(4.5)
Crowe classification	I,II:19 III, IV:4	I,II:144 III, IV:14
Preoperative hemoglobin level (g/dl)	12.9 (1.5)	12.5 (1.2)
Operation time (minutes)	112.4 (17.9)	114.0 (20.3)
Intraoperative blood loss (ml)	217.0 (149.3)	284.2 (166.7)

After matching			p value
Frequency of blood transfusion	0/23	1/76	p = 0.77

Propensity scores were estimated using logistic regression (area under the curve [AUC] = 0.68) from covariates (age, BMI, Crowe classification, and preoperative hemoglobin level) observed before blood storage. Treatment effects were estimated by a one-sided test of independence using Fisher's direct establishment method

Bilateral simultaneous THA can be a useful treatment strategy if the institution is well-equipped and skilled in performing THA, including transfusion preparation and complication management.

References

1. Nakashima Y, et al. [Clinical Question 1: What is the prevalence of osteoarthritis of the hip in Japan? Osteoarthritis of the Hip Joint Clinical Practice Guideline 2016.] Nankodo, Tokyo, p10, 2016. Japanese.
2. Jaffe WL, et al. Bilateral Charnley low-friction arthroplasty as a single operative procedure: a report of fifty cases. Bull Hosp Joint Dis. 1971;32:198–214.
3. Saito S, et al. One-versus two-stage bilateral total hip arthroplasty. Orthopedics 2010;33.
4. Yoshii H. Comparison of patient satisfaction after unilateral or simultaneous bilateral total hip arthroplasty through a direct anterior approach: evaluation using the Japanese Orthopaedic Association Hip Disease Evaluation Questionnaire. J Orthop Sci. 2016;21:332–5.
5. Ritter MA, et al. Bilateral total hip arthroplasty: a single procedure. Clin Orthop Relat Res. 1980;149:185–90.
6. Berend KR, et al. Simultaneous versus staged cementless bilateral total hip arthroplasty: perioperative risk comparison. J Arthroplasty. 2007;22:111–5.
7. Salvati EA, et al. Bilateral total hip-replacement arthroplasty in one stage. J Bone Joint Surg Am. 1978;60:640–4.
8. Alfaro-Adrián J, et al. One- or two-stage bilateral total hip replacement. J Arthroplasty. 1999;14:439–45.
9. Parvizi J, et al. Bilateral total hip arthroplasty: one-stage versus two-stage procedures. Clin Orthop Relat Res. 2006;453:137–41.
10. Macaulay W, et al. Single-stage bilateral total hip arthroplasty. J Am Acad Orthop Surg. 2002;10:217–21.
11. Eggli S, et al. Bilateral total hip arthroplasty: one stage versus two stage procedure. Clin Orthop Relat Res. 1996;328:108–18.
12. Lorenze M, et al. A comparison of the cost effectiveness one-stage versus two-stage bilateral total hip replacement. Orthopedics. 1998;21:1249.
13. Blom A, et al. Early death following primary total hip arthroplasty: 1,727 procedures with mechanical thrombo-prophylaxis. Acta Orthop. 2006;77:347–50.
14. Shao H, et al. Bilateral total hip arthroplasty: 1-Stage or 2-Stage? A meta-analysis. J Arthroplasty. 2017;32:689–95.
15. Partridge TCJ, et al. Simultaneous or staged bilateral total hip arthroplasty? An analysis of comparisons in 14,460 patients using national data. Arthroplasty. 2020;35:166–71.
16. Ritter MA, et al. One-staged contralateral or ipsilateral total hip and total knee arthroplasty. J Arthroplasty. 2002;17:528–32.
17. Saklad M, et al. Grading of patients for surgical procedures. Anesthesiology. 1941;2:281–4.
18. Tamaki T, et al. Perioperative comparison rate of one-stage bilateral total hip arthroplasty using the direct anterior approach. J Orthop Sci. 2016;21:658–61.
19. Pfluger G, et al. Minimally invasive total hip replacement via the anterolateral approach in the supine position. Int Orthop. 2007;31:S7–11.
20. Nakashima, et al. [Y Osteoarthritis Clinical Practice Guideline 2016.] Nanko-do, Tokyo, p. 53, 2016. Japanese.
21. Akagi M, et al. [Guidelines for the prevention of symptomatic venous thromboembolism 2017.] Nankodo, Tokyo, p. 48, 2017. Japanese.

Part XV

Management for Complications

Pain Management

81

Ryo Sugama

Postoperative pain after total hip arthroplasty (THA) is not as severe as that after total knee arthroplasty (TKA) and has not received much attention until recently. However, with the increasing demand for minimally invasive surgery, rapid recovery, and early return to society, the concept of perioperative pain control in THA has changed.

Nearly 30 years have passed since the concept of multimodal pain management was first proposed. Its basic concept is to administer various drugs to various sites using various methods and at various times, but it has only recently been used for pain management in actual joint replacement surgery. Pain management after THA includes epidural block, patient-controlled analgesia (PCA), lumbar plexus block, femoral nerve block (FNB), lateral femoral cutaneous nerve block, intra-articular injection, and periarticular multimodal drug injection. Periarticular multimodal drug injection, the intraoperative injection of a large amount of injectable solution containing several drugs directly into the periarticular area with the main purpose of pain control, has recently attracted considerable attention and research focus. Since the publication of two papers on TKA in 2006 [1, 2], periarticular multimodal drug injection has become the mainstay of postoperative pain management after THA/TKA in Europe and the United States, but it remains not well recognized in Japan, as the fixed concept that "surgery is painful" is an obstacle to its widespread use. However, perioperative pain management should be pursued in Japan to increase patient satisfaction and has attracted attention in recent years.

1 Periarticular Multimodal Drug Injection in THA

Busch et al. reported the analgesic effect of periarticular injection of a solution containing ropivacaine and morphine in patients with THA and reported that the amount of analgesic use was low in the injection group up to 24 h postoperative [3]. Andersen et al. reported that analgesic use was lower up to 4 days postoperative in the injection group due to the periarticular injection of a solution centered on ropivacaine [4]. Andersen et al. used an epidural block as a comparator and reported that analgesic use was low at 96 h postoperative and that pain scores were low until the second postoperative week in a periarticular multimodal drug injection centered on ropivacaine [5]. In a prospective randomized controlled trial in Japan, Nakai et al. [6] evaluated the efficacy of periarticular multimodal drug injection for postoperative THA pain and found it superior in pain relief on the day of surgery. This study is noteworthy in that the researchers used a drug solution similar to that reported from Europe and the United States (morphine 10 mg, bupivacaine 100 mg, adrenaline 0.3 mg, and saline totaling 30 mL), used diclofenac suppository as a rescue, mentioned the incidence of side effects, and evaluated postoperative the incidence of side effects. The postoperative rehabilitation progress was also evaluated.

The authors also focused on periarticular multimodal drug injection and reported its usefulness in Japanese patients. The following is an introduction to the periarticular multimodal drug injection that the authors have been using. We believe that this article will help confirm and reaffirm that periarticular multimodal drug injection is a technique that can be easily performed by anyone.

R. Sugama (✉)
Osaka Metropolitan University, Osaka, Japan
e-mail: sugama@omu.ac.jp

Y. Nakashima (ed.), *Advances in Total Hip Arthroplasty*, https://doi.org/10.1007/978-981-97-2461-1_81

2 Menu and Technique of Periarticular Multimodal Drug Injection

Table 81.1 shows the components of cocktail injections used by the authors, which included local anesthetics, narcotics, steroids, adrenaline, nonsteroidal anti-inflammatory drugs (NSAIDs), and saline totaling approximately 45 mL. The injection sites were the subcutaneous tissue before skin incision, anterior capsule after implantation, iliopsoas tendon, anterior synovium, gluteus maximus attachment site, posterior capsule, external rotator muscles, abductor muscles, thigh fascia, and the subcutaneous tissue. After implantation, we injected the anterior capsule and thigh fascia. In the subcutaneous region, we minimized injections into shallow areas due to skin necrosis. One must carefully inject small amounts into all sites. Approximately 45 mL of the drug solution was divided into two 20-mL syringes and 1–2 mL was injected into each site using a 23-G catheter needle (approximately 7 mL before the skin incision, 15 mL near the anterior capsule, and the remaining 23 mL near the femoral fascia). To reduce the complications of nausea and vomiting, it is also important to use an antiemetic agent and thoroughly wash the surgical field after injecting the cocktail.

3 Effect of Periarticular Multimodal Drug Injection

A study of 43 hips is presented [7]. Patients were randomly divided into a periarticular multimodal drug injection group (cocktail group) and an epidural block group (Epi group). In the cocktail group, the cocktail injection (Table 81.1) was injected into the periarticular soft tissues before the skin incision and after the implantation; all other surgical procedures were identical. The Epi group underwent epidural catheter placement for 3 days postoperative. Postoperative pain control consisted of iv PCA with fentanyl for up to 24 h after surgery. The two groups were compared using the visual analog scale (VAS) for pain assessment and on the dose of fentanyl used for PCA. The VAS score for postoperative pain on postoperative day 1 was significantly lower in the cocktail group (7.5 ± 9.4) than the Epi group (17.8 ± 16.1) ($p < 0.05$). Figure 81.1 shows the change in VAS scores over time, indicating that periarticular multimodal drug injection is effective on the first postoperative day and reduces overall pain throughout the perioperative period. The amount of fentanyl used in the cocktail was significantly lower than that used in the postoperative period. The amount of fentanyl used was significantly lower in the cocktail group (0.24 ± 0.23 mg) than in the Epi group (0.39 ± 0.19 mg) ($p < 0.05$).

Table 81.1 Components of periarticular multimodal drug injection

Periarticular multimodal drug injection (approximately 45 mL)	
Ropivacaine	300 mg
Morphine	10 mg
Methylprednisolone	40 mg
Adrenaline	300 μg
Ketoprofen	50 mg

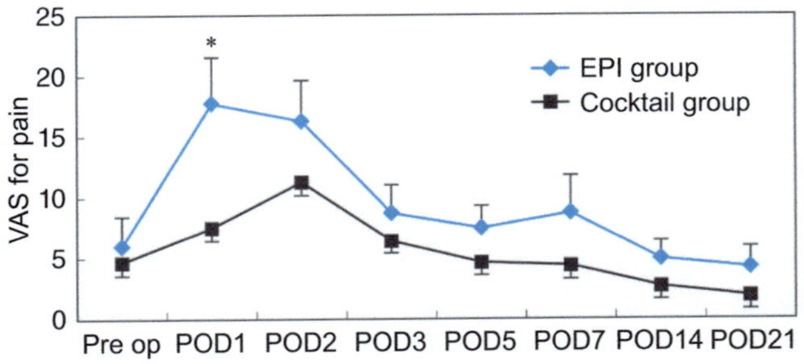

Fig. 81.1 Change in postoperative pain over time. The mean visual analog scale score for pain on postoperative day 1 was significantly lower in the cocktail group. The periarticular multimodal drug injection effectively reduced postoperative pain throughout the perioperative period

Hashimoto et al. reported the effect of steroid-containing periarticular multimodal drug injection on blood tests [8]. The drug solution used in this study was 41 mL, containing 100 mg of levobupivacaine, 40 mg of triamcinolone acetonide, and saline. The leukocyte counts were higher and creatine kinase, C-reactive protein, and D-dimer levels were lower in the cocktail group, suggesting that the periarticular multimodal drug injection may reduce the incidence of deep vein thrombosis. The patients in the cocktail group also showed a decreased mean VAS score for pain.

4 Which Drug Works?

The results of the studies described thus far indicate that periarticular multimodal drug injections effectively control postoperative pain. However, the drug responsible for this effect remains unclear.

(a) Are steroids necessary?

To investigate the need for steroids, we compared the two groups of 68 hips (30 in the steroid group, 38 in the no steroid group) [9]. VAS scores for pain did not differ significantly between the two groups from postoperative days 1 to 21; however, from postoperative days 4–10, VAS scores for pain tended to be lower in the steroid group (day 5:0.5 ± 1.6 vs. 1.4 ± 4.1, $p = 0.260$). The mean dose of fentanyl in the PCA was 0.15 ± 0.22 mg in the steroid group and 0.18 ± 0.21 mg in the no steroid group, with no significant intergroup difference ($p = 0.6025$). There was no significant difference in the start date of walking with parallel bars, a walker, a cane, or stairs, but the mean timing of active straight leg raising was significantly earlier in the steroid group (1.9 ± 1.7 days) than in the no steroid group (4.2 ± 5.1 days) ($p < 0.05$). There were no cases of infection in either group.

Kurosaka et al. reported the effect of periarticular cocktail injection with or without steroids on postoperative pain in 187 THA procedures [10]. A notable feature of this study was the use of the area under the curve (AUC) as the sum of pain. The AUC of the VAS score for pain from the immediate postoperative period to 24 h postoperative was 550 ± 362 versus 392 ± 320 in the no steroid versus steroid group, respectively. The steroid group also reported significantly lower VAS scores at 12, 15, 18, 21, 24, and 48 h postoperative. The authors concluded that the addition of steroids to periarticular multimodal drug injections for postoperative analgesia is beneficial.

(b) Are opioids necessary?

Iwakiri et al. investigated the effects of postoperative pain with or without morphine in 100 THA cases [11]. The drug solution contained local anesthetics, steroids, NSAIDs, and adrenaline, and patients were randomly divided into groups according to the presence or absence of morphine. The morphine dose was determined based on 0.1 mg/kg a body weight. No intergroup difference in mean VAS scores for pain was noted. The nausea numerical rating scale (NRS) score and amount of antiemetic drug required were significantly higher in the group treated with morphine. Accordingly, the author concluded that morphine is unnecessary for periarticular multimodal drug injection.

We present a report on TKA supporting the use of opioids in 42 knees randomly divided into a morphine-containing group (morphine group) and a morphine-free group (control group) [12]. In this study, 5 mg of morphine was used. On the first postoperative day, the resting VAS score for pain was 2.4 ± 6.2 in the morphine group and 10.5 ± 17.7 in the control group, showing no significant difference. However, the VAS score for pain during movement was significantly lower in the morphine versus control group (11.1 ± 13.4 and 28.0 ± 20.4) ($p < 0.05$). This suggests that opioids are useful in periarticular multimodal drug injections.

Furthermore, there are reports on the significance of the local administration of opioids, which are believed to act on the central nervous system [13]. In 32 cases of bilateral simultaneous TKA, a morphine-containing drug solution was randomly injected on one side (morphine side) versus a non-morphine-containing drug solution on the opposite side (control side). The dose of morphine used in this study was 5 mg. The resting VAS score for pain did not differ on postoperative day 1 but was significantly lower on the morphine side on postoperative days 2, 5, and 10 ($p < 0.05$). The VAS score for pain during movement was significantly lower on the morphine side on postoperative days 3, 4, and 5 ($p < 0.05$). A greater percentage of patients rated the morphine versus control side as better. The results indicated that the morphine side tended to have a lower VAS score for pain, and a higher percentage of patients rated the morphine versus control side as better after bilateral simultaneous TKA of the same individuals. These results indicate that opioids in periarticular multimodal drug injections have local and analgesic effects.

(c) Are NSAIDs necessary?

Kim et al. reported a difference in the efficacy of cocktail injection with versus without ketorolac, an NSAID, in patients with TKA and found that its addition produced significantly stronger analgesia [14]. Ketorolac is not approved in Japan; therefore, the authors used ketoprofen as a substitute. We believe that ketoprofen has a similar effect, although the degree of the difference requires investigation.

Periarticular multimodal drug injection of multiple drugs is a pharmacologically effective and versatile analgesic treatment. Interactions between multiple drugs are also expected to result in stronger pain suppression. Steroids, opioids, and NSAIDs are important drugs used in periarticular multimodal drug injections.

5 When to Administer the Drug?

Kurosaka et al. investigated the timing of periarticular multimodal drug injections in 140 THA patients [15]. Compared to the late-stage group, in which injections were administered after implantation, the early-stage group, in which injections were administered before arthrotomy, showed a lower VAS score for pain in the recovery room (46 ± 30 vs. 30 ± 28, respectively). Thus, preemptive analgesia may be important for preventing peripheral and central nervous system sensitization (hyperalgesia).

6 Future Prospects

The most important question is not whether cocktail therapy is effective but rather how to provide painless surgery and increase patient satisfaction. Periarticular multimodal drug injections and other techniques should be used to achieve this goal. Tetsunaga et al. reported the use of periarticular multimodal drug injections in combination with continuous FNB [16]. They compared the analgesic effects of a continuous FNB, periarticular multimodal drug injection (ropivacaine 225 mg, morphine 10 mg, adrenaline 0.5 mg, and saline totaling 50 mL) alone and in combination. The VAS score for pain was lower in the patients receiving the combination versus single treatment. Methods of perioperative pain control other than periarticular multimodal drug injection are also available. Aoyama et al. compared continuous quadratus lumborum block with continuous FNB and found that patients treated with the latter had significantly lower VAS scores for pain at 6 and 24 h postoperative [17]. Nishio et al. investigated whether the addition of sciatic nerve block to continuous FNB increased the analgesic effect [18]. In the recovery room, no intergroup difference in pain VAS score was noted, but the NRS for pain was lower at 6 and 12 h postoperative in the group to which sciatic nerve block was added, suggesting that the combination of continuous FNB and sciatic nerve block is useful in terms of analgesia. Fujino et al. reported the usefulness of a new block, the continuous pericapsular nerve group block [19]. Takeda et al. reported the usefulness of intravenous acetaminophen administration [20]. They compared patients who underwent THA (97 cases) in an intravenous acetaminophen versus control group. In the intravenous acetaminophen group, acetaminophen 1000 mg was administered every 6 h for 24 h postoperative. The NRS for pain at 24 h was lower in the intravenous acetaminophen group, and the amount of fentanyl used in iv PCA at 24 h postoperatively was also lower. However, the excessive pursuit of postoperative pain control alone may cause respiratory depression; therefore, great care must be taken. Yoshida et al. reported that respiratory depression occurred in 22% of cases (5/23 patients) following periarticular multimodal drug injection, including opioids, immediately after surgery through the first postoperative day [21]. In addition, when narcotics are included, their dependence may lead to drug abuse, which has become a problem in Europe and the United States in recent years, and there is a risk that this may lead to new social problems. We believe that rapid progress has been made in the management of postoperative pain in Japan over the past decade. However, the reality is that it has not yet reached the level in Western countries. We should strive to accumulate evidence in Japan to avoid repeating the failures of pain control in Europe and the United States, such as the abuse of narcotics.

References

1. Busch CA, et al. Efficacy of periarticular multimodal drug injection in total knee arthroplasty: a randomized trial. J Bone Joint Surg Am. 2006;88:959–63.
2. Vendittoli PA, et al. A multimodal analgesia protocol for total knee arthroplasty. A randomized, controlled study. J Bone Joint Surg Am. 2006;88:282–9.
3. Busch CA, et al. The efficacy of periarticular multimodal drug infiltration in total hip arthroplasty. Clin Orthop Relat Res. 2010;468:2152–9.
4. Andersen LJ, et al. Postoperative analgesia in total hip arthroplasty: a randomized double-blinded, placebo-controlled study on perioperative and postoperative ropivacaine, ketorolac, and adrenaline wound infiltration. Acta Orthop. 2007;78:187–92.
5. Andersen KV, et al. Reduced hospital stay and narcotic consumption and improved mobilization with local and intraarticular infiltration after hip arthroplasty: a randomized clinical trial of an intraarticular technique versus epidural infusion in 80 patients. Acta Orthop. 2007;78:180–6.
6. Nakai T, et al. A study of the usefulness of a periarticular multimodal drug cocktail injection for pain management after total hip arthroplasty. J Orthop. 2013;10:5–7.
7. Sugama R, et al. [Pain control after hip arthroplasty by periarticular multimodal drug injection.] Hip Joint 2011;37: 659–61. Japanese.
8. Hashimoto A, et al. Periarticular analgesic injection containing a corticosteroid after total hip arthroplasty may prevent deep venous thrombosis: a retrospective comparative cohort study. BMC Musculoskelet Disord. 2021;22:19.
9. Sugama R, et al. [Significance of steroids in periarticular multimodal drug injection after total hip arthroplasty.] Hip Joint 2013;39:877–80. Japanese.
10. Kurosaka K, et al. Addition of corticosteroid to periarticular injections reduces postoperative pain following total hip arthroplasty under general anaesthesia: a double-blind randomized controlled trial. Bone Joint J. 2020;102-B:1297–302.
11. Iwakiri K, et al. Effect of periarticular morphine injection for total hip arthroplasty: a randomized, double-blind trial. Hip Int. 2019;29:245–52.
12. Sugama R, et al. [Significance of opioids in periarticular multimodal drug injection in total knee arthroplasty.] JOSKAS 2017;42:226–7. Japanese.
13. Sugama R, et al. [Significance of opioids in periarticular multimodal drug injection in total knee arthroplasty – a study in bilateral TKA cases.] JOSKAS 2018; 43:426–7. Japanese.
14. Kim TW, et al. Which analgesic mixture is appropriate for periarticular injection after total knee arthroplasty? Prospective, randomized, double-blind study. Knee Surg Sports Traumatol Arthrosc. 2015;23:838–45.
15. Kurosaka K, et al. Comparison of early-stage and late-stage periarticular injection for pain relief after total hip arthroplasty: a double-blind randomized controlled trial. J Arthroplast. 2020;35:1275–80.
16. Tetsunaga T, et al. Combination therapy with continuous three-in-one femoral nerve block and periarticular multimodal drug infiltration after total hip arthroplasty. Pain Res Manage. 2016;2016:1425201.
17. Aoyama Y, et al. Continuous quadratus lumborum block and femoral nerve block for total hip arthroplasty: a randomized study. J Anesth. 2020;34:413–20.
18. Nishio S, et al. Comparison of continuous femoral nerve block with and without combined sciatic nerve block after total hip arthroplasty: a prospective randomized study. Orthop Rev (Pavia). 2017;9:7063.
19. Fujino T, et al. Continuous pericapsular nerve group block for postoperative pain management in total hip arthroplasty: report of two cases. Rep. 2021;7:22.
20. Takeda Y, et al. Evaluating the effect of intravenous acetaminophen in multimodal analgesia after total hip arthroplasty: a randomized controlled trial. J Arthroplast. 2019;34:1155–61.
21. Yoshida K, et al. Respiratory depression after intraoperative opioid-containing periarticular multimodal drug injection for total hip and total knee arthroplasty. J Clin Anesth. 2019;54:147–8.

Bleeding Control

82

Takuma Yamasaki

Bleeding is an unavoidable complication of total hip arthroplasty (THA); however, excessive bleeding is a severe perioperative complication. THA involves the manipulation of bones, such as reaming the acetabulum or femur, which may result in excessive blood loss. Conventional methods for blood loss control include managing hypotensive anesthesia and autologous or allogeneic blood transfusions to compensate for blood loss based on the expected blood loss. However, allogeneic blood transfusion should be avoided as much as possible because of concerns of serious complications, such as transfusion-related infections, including infections of hepatitis viruses and human immunodeficiency virus (HIV), graft-versus-host disease, production of abnormal irregular antibodies due to allogeneic immunity, and transfusion-related circulatory overload. Current hemorrhage control measures recommend patient blood management (PBM), which focuses on the resolution of anemia, prevention of blood loss, and appropriate blood transfusion. Autologous blood transfusions or hemostatic measures should be considered to minimize allogeneic blood transfusion as well as ensure smooth and safe surgical and perioperative management [1]. PBM should be planned for preoperative, intraoperative, and postoperative periods. It should include increasing preoperative hemoglobin levels and optimizing hemostatic coagulation, improving surgical techniques and hemostatic procedures, reducing intraoperative and postoperative blood loss through improved anesthetic management, and using limited and restrictive blood products based on the indication [2].

The three pillars of preoperative PBM are the following: (1) preoperative evaluation of anemia and iron deficiency and optimizing blood hemoglobin levels using iron and folic acid supplementation, if indicated; (2) identification and management of the risk of bleeding, including antithrombotic therapy; and (3) evaluation of the cardiopulmonary reserve to tolerate anemia. The three pillars of intraoperative PBM are the following: (1) determination of the timing of surgery to optimize erythropoiesis and erythrocyte volume; (2) minimizing intraoperative blood loss and hemostasis using meticulous surgical techniques and anesthetic management, use of autotransfusion, patient positioning and warming to avoid coagulopathy, and pharmacotherapy (including antifibrinolytic agents), and (3) optimization of the cardiopulmonary function in cases of anemia, including ventilation and oxygenation, and using limited transfusions, even if allogeneic blood is used. In the postoperative period, the following three points should be considered: (1) management of anemia and iron deficiency and attention to the prescribed drugs and drug-drug interactions; (2) monitoring and management of postoperative bleeding, patient temperature, and diagnosis and early treatment of infectious diseases to avoid wasting anemia; and (3) ensuring adequate oxygen supply and reducing oxygen consumption to improve postoperative anemia [2].

Of the bleeding control measures described above, this section describes the effects of intraoperative drainage and tranexamic acid on hemostasis based on the evidence available to date.

(a) Drain placement

Drainage after THA is commonly used to prevent wound hematoma and swelling, which promotes wound healing and prevents postoperative infections (prophylactic drainage). Additionally, a drain can help estimate the volume of early postoperative bleeding (informational drainage).

During drainage, a tubular drain is used for closed drainage (Fig. 82.1). It is inserted through the fascia to an opening on the skin at the time of wound closure to prevent wound infection caused by the drain as much as possible (Fig. 82.2). The length of the tube should be adjusted such that the tip of the drain is placed within the wound to produce the most effective suction force (Fig. 82.3).

T. Yamasaki (✉)
National Hospital Organization, Kure Medical Center, Hiroshima, Japan

Y. Nakashima (ed.), *Advances in Total Hip Arthroplasty*, https://doi.org/10.1007/978-981-97-2461-1_82

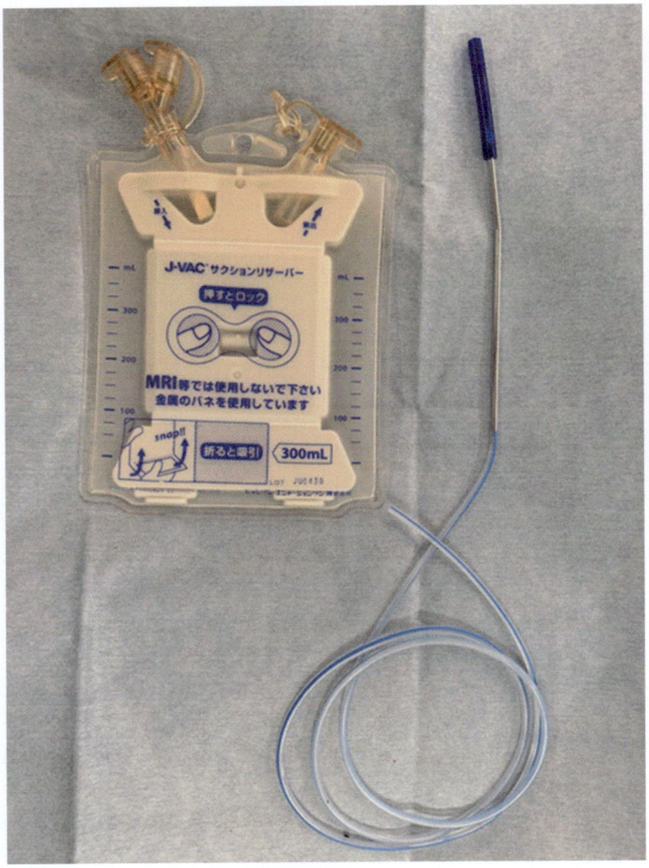

Fig. 82.1 Tubular drain and reservoir

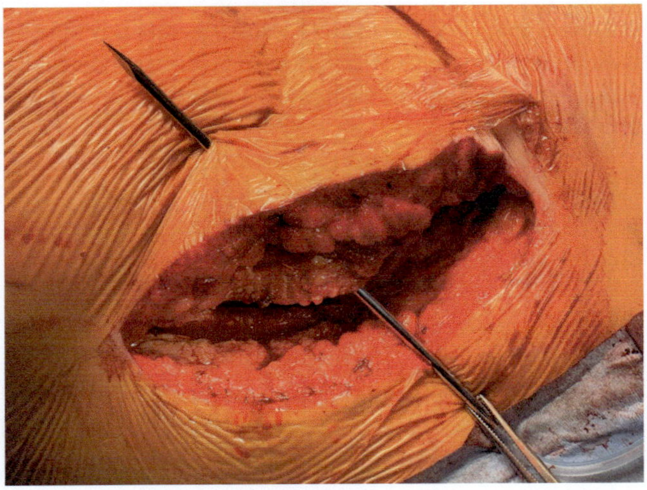

Fig. 82.2 Direction of drain insertion. Penetrate the fascia from within the wound and bring it outside the wound through the skin

There are two methods of drain placement: one is to release the drain immediately after the surgery and the other is to clamp the drain for 30 min to 2 h after the surgery [3]. A comparison of postoperative blood loss between the two methods revealed that the clamping method significantly reduced the blood loss with further reduction in the volume of blood loss when the drain was clamped for 1–2 h rather than 30 min [3].

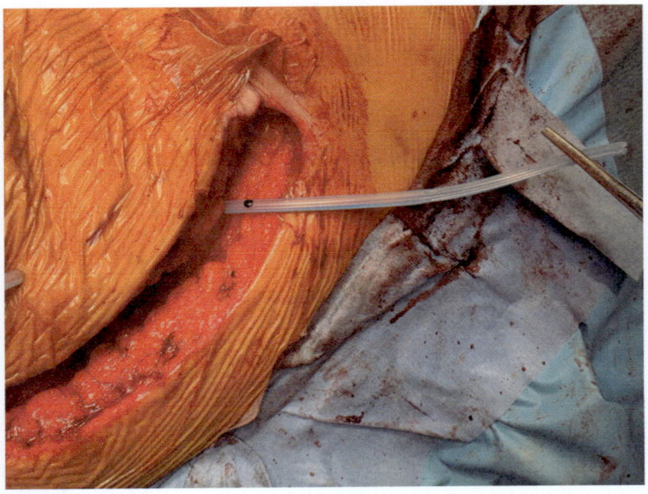

Fig. 82.3 Adjust the drain so that the tip of the drain is where the most suction force can be achieved

A disadvantage of a drain is that using it for long periods can result in wound infection. Additionally, in clinical practice, drains result in restricted patient movements as well as an additional workload for the nursing team in terms of drain management. In a comparative study of THA with and without drains, no significant difference was observed in intraoperative and postoperative blood loss or another report mentioned that blood loss was greater in the group with drain placement [4]. Furthermore, no significant differences were noted between the groups in postoperative changes in thigh circumference, pain on the visual analog scale, and rehabilitation, such as walking with a walker or cane [5]. However, patients without drains were able to leave the bed and get the balloon catheter removed significantly earlier than patients with drains [6]. Drains constitute a risk factor for postoperative heterotopic ossification [7]. Consequently, the usefulness of drains, which was taken for granted, is becoming debatable.

Systematic reviews on the use of drains have reported the effectiveness of drains in terms of wound hematomas and exudate; however, this data remains controversial [8–10]. No significant differences were reported in other perioperative complications between patients with and without drains; however, the volume of blood loss, duration of hospitalization, and the number of allogeneic blood transfusions were lower in patients without drains [11–13]. In summary, the use of a drain should be considered in each patient based on each patient's bleeding tendencies and the estimated postoperative blood loss, which is based on the volume of intraoperative bleeding.

Fig. 82.4 Formula for calculating the expected total blood loss. *Ht* Hematocrit

Circulating blood volume = k1 x height (m)3 + k2 x weight (kg) + k3

Male: k1 = 0.3669, k2 = 0.03219, k3 = 0.6041

Female: k1 = 0.3561, k2 = 0.03308, k3 = 0.1833

Estimated total blood loss

= circulating blood volume x (preoperative Ht value - postoperative Ht value) / average of pre-and postoperative Ht values

(b) Tranexamic acid

The total postoperative blood loss following THA usually includes intraoperative and postoperative blood loss as well as the drainage volume. However, hidden blood loss (HBL), such as the volume of blood that remains in the intra-articular space or leaks into the periarticular space, must also be included. HBL is estimated to account for 26–60% of the total blood loss (Fig. 82.4) [14, 15]. Reducing HBL not only decreases the total blood loss but also promotes postoperative recovery of hip function [15, 16].

In recent years, the efficacy of tranexamic acid has been reported in the management of surgical hemorrhage. Following primary hemostasis via platelet aggregation, fibrin is stabilized by the coagulation system for complete hemostasis; subsequently, it is degraded by the fibrinolytic system. Tranexamic acid binds to the lysine-binding site of plasminogen in the fibrinolytic system and inhibits the binding of plasminogen to fibrin via lysine residues. The inhibition of plasminogen activation prevents fibrin degradation, which produces a hemostatic effect. Tranexamic acid also reduces the volume of intraoperative and postoperative bleeding in THA and has been demonstrated to reduce HBL [17]. Tranexamic acid is generally administered as an intravenous infusion immediately before the surgery or as an intra-articular injection before wound closure. The dose of tranexamic acid used varies from 15 mg/kg to 3 g; however, 1 g is the usual dose across different methods [18]. The author uses 1 g tranexamic acid for intravenous infusion immediately before the surgery and 1 g for intra-articular injection before wound closure along with an analgesic (ropivacaine hydrochloride 150 mg + ketoprofen 50 mg) (Figs. 82.5 and 82.6). Many studies have compared intravenous and intra-articular tranexamic acid as well as the overall efficacy of combining them. In a meta-analysis comprising 26 randomized controlled trials that compared intravenous, intra-articular, and the combination of intra-articular and intravenous tranexamic acid in arthroplasty, no significant differences were observed in the total blood loss and drainage volumes, postoperative hemoglobin levels, changes in hemoglobin levels, duration of hospitalization, blood transfusion rate, or the incidence of venous thromboembolism. No significant differences were noted in the total blood loss volume, change in hemoglobin level, and transfusion rate between patients who were administered a single agent and those who were administered both intravenous and intra-articular injections; however, no significant differences were observed in the drainage volume, postoperative hemoglobin

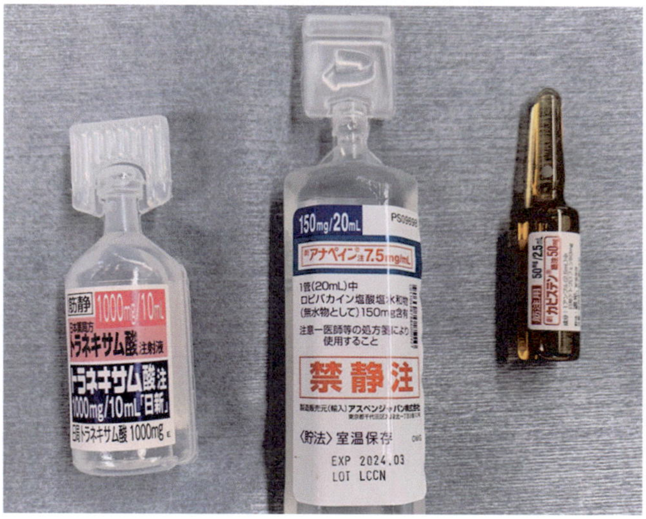

Fig. 82.5 Drugs used in intra-articular injections (tranexamic acid (left) + ropivacaine hydrochloride (center) + Right: ketoprofen (right))

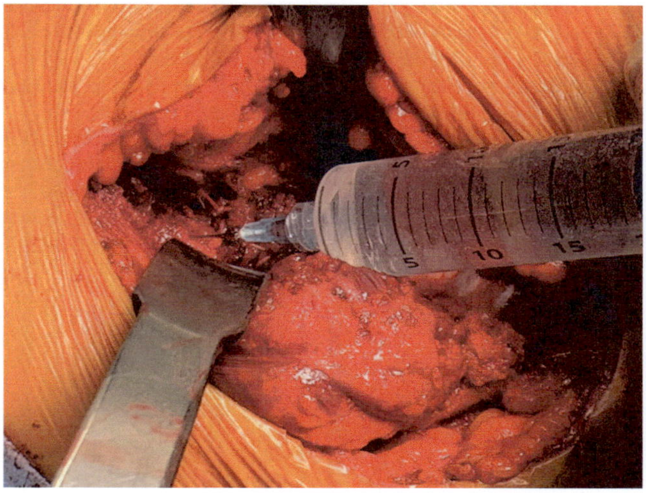

Fig. 82.6 Intra-articular injection of tranexamic acid before wound closure

level, and incidence of venous thromboembolism between them [19]. The incidence of venous thromboembolism, which is a complication of concern, has not been noted to increase with tranexamic acid [20, 21].

A report on the frequency of postoperative complications associated with tranexamic acid and drainage revealed that tranexamic acid was associated with a significantly reduced intraoperative blood loss and estimated total blood loss with or without drainage. Additionally, the incidence of venous thromboembolism or postoperative pain was not affected by either tranexamic acid or drainage [22].

Therefore, intravenous and intra-articular tranexamic acid injections without a drain are considered the most effective measures for preventing intraoperative bleeding. Tranexamic acid and drains should be considered after comprehensively considering the patient's preoperative general condition, including anemia and its severity, bleeding tendencies, thrombus formation capacity, surgical technique, pelvis and femur bone quality, scheduled operation duration, and the skills of the surgeon. Preoperative blood transfusion should be avoided as much as possible; however, preoperative blood storage and postoperative autotransfusions must be considered.

References

1. Kino S. [Patient blood management (PBM).] J Clin Exp Med 2012;243:273–8. Japanese.
2. Yamaura K, et al. [Concept of patient blood management (PBM).] J Jpn Soc Clin Anesth 2019;39:551–554. Japanese.
3. Yoshimitsu K, et al. [Investigation of drain clamping methods in total hip arthroplasty comparison of drain clamping time.] Nichi Jinko Kansetsu Kaishi 2015;45:259–60. Japanese.
4. Nakamura A, et al. [A review of the need for drains in total hip arthroplasty.] Hip Joint 2020;46:S46–8. Japanese.
5. Yamada J, et al. [Influence of the presence or absence of drain implantation on thigh circumference and rehabilitation during THA.] Nichi Jinko Kansetsu Kaishi 2015;45:401–2. Japanese.
6. Miyama Y, et al. [The effect of drainage on ADL and pain in the early postoperative period after THA surgery.] Hip Joint 2016;42:S33–5. Japanese.
7. Di Benedetto P, et al. Heterotopic ossification in primary total hip arthroplasty: which is the role of drainage? Acta Biomed. 2019;90(Suppl 1):92–7.
8. Kelly EG, et al. Systematic review and meta-analysis of closed suction drainage versus non-drainage in primary hip arthroplasty. Surg Technol Int. 2014;24:295–301.
9. Chen ZY, et al. Is wound drainage necessary in hip arthroplasty? A meta-analysis of randomized controlled trials. Eur J Orthop Surg Traumatol. 2014;24:939–46.
10. Zeng WN, et al. Comparison between drainage and non-drainage after total hip arthroplasty in Chinese subjects. Orthop Surg. 2014;6:28–32.
11. Parker MJ, et al. Closed suction drainage for hip and knee arthroplasty: a meta-analysis. J Bone Joint Surg Am. 2004;86:1146–52.
12. Zhou XD, et al. Do we really need closed suction drainage in total hip arthroplasty? A meta-analysis Int Orthop. 2013;37:2109–18.
13. Long H, et al. The usage of drainage after primary total hip or knee arthroplasty: best evidence selection and risk of bias considerations. BMC Musculoskelet Disord. 2021;22:1028.
14. Matsuya H, et al. [A study of hidden blood loss in hip arthroplasty in our department.] Arthroplasty Regist Japan 2020; 50:319–20. Japanese.
15. Miao K, et al. Hidden blood loss and its influential factors after total hip arthroplasty. J Orthop Surg Res. 2015;10:36.
16. Okamoto M, et al. [Preoperative administration of tranexamic acid for hip arthroplasty reduces hospital stay.] Arthroplasty Regist Japan 2020; 50:181–2. Japanese.
17. Okamoto M, et al. [Preoperative administration of tranexamic acid reduces hidden blood loss in total hip arthroplasty.] Hip Joint 2018;44:85–7. Japanese.
18. Yamaguchi A, et al. Dose optimization of topical tranexamic acid for primary total hip arthroplasty: a prospective cohort study. J Orthop Sci. 2019;24:275–9.
19. Sun Q, et al. Comparison of intravenous, topical or combined routes of tranexamic acid administration in patients undergoing total knee and hip arthroplasty: a meta-analysis of randomised controlled trials. BMJ Open. 2019;9:e024350.
20. Zeng Y, et al. Intravenous combined with topical administration of tranexamic acid in primary total hip arthroplasty: a randomized controlled trial. Orthop Surg. 2017;9:174–9.
21. Zeng Y, et al. Tranexamic acid administration in primary total hip arthroplasty: a randomized controlled trial of intravenous combined with topical versus single-dose intravenous administration. J Bone Joint Surg Am. 2019;98:983–91.
22. Naito Y, et al. [Evaluation of the usefulness of transvenous tranexamic acid administration and drain placement in hip arthroplasty.] Cent Jpn J Orthop Traumat 2021;64:195–6. Japanese.

Deep Vein Thrombosis

83

Kan Sasaki and Masaji Ishii

1 Epidemiology

In Japan, deep vein thrombosis (DVT) and pulmonary thromboembolism (PTE) were originally considered rare diseases and were not highly recognized; however, their prevalence rates have been increasing in recent years due to the aging of the population and improvement in diagnosis rates [1]. DVT and PTE frequently occur in individuals with reduced activity in disaster shelters and have attracted media coverage and attention as economy class syndromes. They are now recognized not only by medical professionals but also by the public. The number of diagnosed cases of PTE in Japan in 1996 was estimated to be 28 per million population, according to an epidemiological study by Kumasaka et al. [2], whereas epidemiological studies by Sakuma et al. [3] and Nakamura et al. [1] estimated the number of cases to be 62 and 126 per million population, respectively, in 2006 and 2011, indicating a four- to sixfold increase over 15 years. The incidence of PTE in Japan in 2006 was about one-fifth to one-eighth of that in the United States [1, 3] compared with approximately 500–600 cases per million people in the United States [4]. Symptomatic PTE is sometimes fatal and is an important perioperative complication not only in total hip arthroplasty (THA) but also in many orthopedic surgeries and other medical specialties. As PTE is caused by DVT, both are collectively called venous thromboembolism (VTE). THA is one of the most common orthopedic surgeries with a high risk of VTE. Therefore, it is essential to take measures to prevent VTE, especially to avoid fatal PTE. In Japan, perioperative fatal PTE has been decreasing since the publication of guidelines for the prevention of VTE in 2004 by a joint committee of related societies and the establishment of a PTE prevention and management fee through health insurance [5]. The joint guidelines were developed based on the American College of Chest Physicians (ACCP) VTE guidelines, sixth edition; however, the Japanese Orthopaedic Association Venous Thromboembolism Prevention Revision Committee revised the guidelines in 2008 to respond to changes in overseas and domestic situations over the years. The most recent revision of the guidelines in 2017 was based on the ninth edition of the ACCP guidelines and the guidelines developed by the American Academy of Orthopaedic Surgeons (AAOS) in 2011, and the major change was that the prevention target was changed from asymptomatic VTE to symptomatic VTE [6]. This section describes the prevention of VTE after THA, focusing on the 2017 guidelines.

2 Frequency of Occurrence

The 2012 ACCP guidelines, ninth edition [7], stated that the incidence of asymptomatic DVT after THA surgery was 40–60% and that the central type was more common than total knee arthroplasty (TKA). However, the 2011 AAOS guidelines [8] reported that the incidence of symptomatic DVT after THA and TKA surgery was 0.7% and that of symptomatic PTE was 0.4% (42,390 patients with THA and TKA, but the exact frequency of no prophylaxis is unknown because only 3 of 25 patients in the placebo or no drug prophylaxis group were treated), and fatal PTE was estimated to be 10% of these cases. In Japan, Fujita et al. [9] reported that DVT was observed in 22.6% of 164 THA cases, of which 9.8% were of the proximal type and 0.7% were symptomatic PTE. According to the results of the Japanese Society of Anesthesiologists Perioperative Pulmonary Thromboembolism (JSA-PTE) survey, the incidence of perioperative PTE in the "hip/extremity" is the second highest after "open chest + open abdomen" at 0.6% [10].

K. Sasaki (✉) · M. Ishii
Yamagata Saisei Hospital, Yamagata, Japan
e-mail: s-kan@msb.biglobe.ne.jp

Y. Nakashima (ed.), *Advances in Total Hip Arthroplasty*, https://doi.org/10.1007/978-981-97-2461-1_83

3 Symptomatic VTE

Symptomatic VTE can be classified into symptomatic and asymptomatic, with symptomatic PTE being the most clinically important. The most common types are the extensive type, in which two or more pulmonary lobar arteries are occluded, and subextensive type, in which one or more regional arteries and less than two pulmonary arteries are occluded; the peripheral type, in which small arteries below the regional artery are occluded, is often asymptomatic [11]. According to the 2008 Joint Study Group guidelines [12] prepared mainly by the Japanese Circulation Society, more than 90% of emboli in acute PTE originate from the lower extremities or pelvic veins, and 61 of 62 autopsy cases of acute extensive PTE are of the inferior leg type [13]. It has been reported that approximately 13% of thrombi formed in the inferior leg muscular veins extend to the central type [14]. Thrombi in the femoral and iliac veins are called secondary thrombi and are assumed to be easily released and embolized because of their weak adherence to the vein wall [11]. These pathological theories are the basis for the importance of preventing lower-extremity DVT to prevent the development of symptomatic PTE; however, recent studies have shown that the central extension of peripheral DVT is as low as 3–3.7% and the incidence of PTE is also low [15, 16]. Furthermore, anticoagulation therapy for patients with peripheral DVT was reported to be ineffective in preventing the development of symptomatic PTE and to increase bleeding complications [17]. The anticoagulation is recommend only for high-risk patients with peripheral DVT, not all patients with peripheral DVT according to the latest 2017 joint study group guideline and JOS guideline for symptomatic DVT [6, 12].

4 Risk Factors

The Joint Study Group guidelines state that THA, TKA, Hip fracture surgeries pelvic osteotomy, lower-extremity malignancy surgery, and multiple traumas/pelvic fracture are at high risk, and surgeons who are responsible for hip surgery should always keep in mind the prevention of VTE not only in THA but also in their daily practice.

- History of VTE

According to the 2011 AAOS guidelines, a history of symptomatic VTE is a strong risk factor for symptomatic VTE after THA or TKA surgery [7]. Because a history of VTE is also a significant risk factor with an odds ratio of 10.3 in the JSA-PTE survey results [18], it should be confirmed before surgery.

- Age

According to the results of the JSA-PTE study, the incidence of perioperative PTE by age group increased with increasing age, and the relative risks as the control of the age group of 19–65 years were 1.9 for those aged 66–85 years and 2.7 for those aged ≥86 years, indicating that the incidence rate increased with age [10].

- Obesity

In a US study, obese women with a BMI of ≥29 kg/m^2 had a higher incidence of PTE than obese women with a BMI of <21 kg/m^2, and the odds ratio was 2.9 [19]. In the JSA-PTE study results, the odds ratio was 1.7 for a BMI of ≥25 kg/m^2, and obesity was considered a risk factor in Japanese studies [18].

- Prolonged bed rest

The results of the JSA-PTE survey showed that prolonged lying in bed for more than 4 days was a significant risk factor for PTE [18].

- Paralysis

Lower leg paralysis leads to decreased active movement of the lower leg muscles and is considered a risk factor for VTE. The incidence of symptomatic VTE has been reported to be higher on the paralyzed side than on the healthy side in patients with stroke [20].

- Congenital thrombophilia

It has been reported that patients with antithrombin deficiency, protein C deficiency, and protein S deficiency have a higher risk of VTE than healthy individuals [21]. Although screening of all patients is impractical, it is necessary to recognize that the risk of VTE is high in patients with known congenital predispositions.

- Antiphospholipid antibody syndrome.

Antiphospholipid antibody syndrome is an autoimmune disease with an acquired thrombophilic predisposition. The binding of antiphospholipid antibodies to phospholipids, which are necessary for the activation of protein C, promotes thrombus formation, resulting in the risk of VTE [22].

- Malignancy

The release of clots from tumor cells and direct damage to venous endothelial cells by malignant cells are believed to cause thrombus formation, and malignant tumors are considered a significant risk factor for PTE in the results of the JSA-PTE study [18]. Furthermore, the incidence of VTE in patients after chemotherapy for malignant tumors is 12.6%, and chemotherapy for malignant tumors is a risk factor for VTE [23].

5 Prevention

(a) Physical prophylaxis

The 2011 AAOS guidelines recommend early weaning and early ambulation for the prevention of pre-VTE after THA and TKA [7] (consensus). Although there is little expectation that more high-quality studies will be reported in the future, these are safe and low-cost methods of prevention. If early walking is difficult, active ankle joint exercises in bed are an effective preventive method for increasing venous blood flow [24].

(b) Elastic stockings

Elastic stockings compress superficial veins to decrease the total venous cross-sectional area and increase the blood flow velocity in deep veins to prevent VTE. Although this is simple, the single preventive effect is not high in high-risk surgery [25], and the risk of peroneal nerve palsy has been reported.

(c) Intermittent pneumatic compression

Intermittent pneumatic compression devices (IPCDs) include intermittent pneumatic compression (IPC), which injects air intermittently into a cuff wrapped around the lower extremity to return blood flow to the vein, and a venous foot pump (VFP), which applies pressure rapidly to the sole of the foot to return blood flow to the vein. The 2011 AAOS prevention guidelines recommend drug prophylaxis and/or mechanical compression prophylaxis (moderate) and IPC in the presence of abnormal bleeding and/or active liver disease (consensus) [8]. The 2012 ACCP guidelines (ninth edition) recommend IPC or no drug therapy in patients at high risk of bleeding after THA, TKA, and HFS [7].

(d) Pharmacotherapy

Anticoagulant therapy has been widely used for the prevention of VTE in Europe and the United States. Warfarin was used in the 1960s, and low dose unfractionated heparin (LDUH) was used in the 1970s. Fondaparinux was developed in 2000, and several new anticoagulants have been launched since then.

The eighth edition of the ACCP guidelines was written for the prevention of all VTEs, including asymptomatic VTE; however, there has been a change in policy since the ninth edition, and the guidelines are now written for the prevention of symptomatic and fatal PTE and symptomatic DVT. The eighth edition described the routine use of either low molecular weight heparin (LMWH), fondaparinux, or warfarin (grade 1A), and for patients at high risk of bleeding, VFP or IPC should be used until the bleeding risk is reduced and then switched or added to an anticoagulant [26]. The ninth edition recommended the use of LMWH, fondaparinux, apixaban, dabigatran, reversaloxaban, LDUH, warfarin, or aspirin; however, the level of evidence decreased to grade 1B [7]. Furthermore, for patients at high risk of bleeding, IPCD use is recommended instead of anticoagulants, unlike in the eighth edition. This is because there is insufficient evidence for the prophylactic administration of anticoagulants when the endpoint is the difference in the frequency of symptomatic and fatal PTE and symptomatic DVT. The 2011 AAOS guidelines also aim to prevent symptomatic VTE in THA and TKA but recommend the use of pharmacological and/or physical prophylaxis in general cases (moderate) [7]. Recent studies have reported that the efficacy and safety of aspirin for the prevention of symptomatic VTE are comparable to those of other anticoagulants, and aspirin has attracted attention because of its cost advantage [27, 28].

(e) Positioning of anticoagulation therapy in Japan

The 2008 Japanese Orthopaedic Association Guidelines for the Prevention of Pulmonary Thrombosis/Venous Thrombosis state that anticoagulant therapy may be considered for high- and highest-risk patients in accordance with the first edition [29]. Currently, five anticoagulants are covered by insurance in Japan: warfarin, unfractionated heparin, enoxaparin, fondaparinux, and edoxaban. Although the safety of enoxaparin and fondaparinux for major bleeding has been confirmed, there have been no comparative studies with symptomatic VTE as an endpoint. Therefore, the 2017 Japanese Orthopaedic Association guidelines for the prevention of symptomatic VTE recommend that patients with THA with a history of highest-risk VTE should be treated with a combination of physical therapy and pharmacological prophylaxis, and the indication for pharmacological therapy in patients deemed to be at high bleeding risk should be carefully considered (grade B) [6]. In general, the choice of prophylaxis should be based on the balance between the risk of VTE and that of bleeding with pharmacological prophylaxis (grade A). Although studies evaluating the effect of prophylactic anticoagulants on symptomatic VTE in Japan are limited, Mihara et al. [30] reported no difference in symptomatic VTE between the two groups in 3295 patients with THA, with aspirin in the low-risk group and a 5-day combination of an Xa inhibitor and aspirin in the high-risk group. Further verification of the efficacy and safety of prophylactic administration of coagulant agents is needed because the incidence of symptomatic/fatal PTE, which is of greatest concern, is lower in Japan than in the West. The JSA-PTE study showed that since 2008, when anticoagulation therapy, as represented by direct oral anticoagulant (DOAC), was introduced in Japan, the mortality rate from symptomatic PTE has gradually decreased. Newer anticoagulant drugs may contribute to a decrease in mortality [31]. Ideally, the timing, volume, and duration of anticoagulant initiation should be tailored to the patient based on an individual VTE/bleeding risk assessment and informed consent.

6 Screening

Screening includes imaging, such as ultrasound and contrast-enhanced CT, and blood tests, such as D-dimer level. Although widespread screening has been considered ideal for the early detection of VTE and early intervention to prevent symptomatic PE, the 2011 AAOS guidelines strongly recommend against routine screening by lower-extremity venous ultrasonography in postoperative THA and TKA (grade A) [8]. The 2012 ACCP guidelines (ninth edition) also strongly recommend routine screening by lower-extremity venous ultrasonography in postoperative THA and TKA (grade B) [7]. This is because routine leg vein echoes are not recommended for postoperative THA or TKA. The reasons for this are that anticoagulation of patients detected by routine ultrasonography of the lower extremities failed to reduce the incidence of symptomatic VTE after discharge and anticoagulation of asymptomatic DVT was associated with an increased risk of major bleeding events [32, 33]. In response to this, the 2017 Japanese Orthopaedic Association guidelines for the prevention of symptomatic VTE also do not recommend routine venous ultrasonographic screening for DVT in THA, TKA, and HFS (grade A). Similarly, D-dimer screening is not recommended. However, DVT by venous ultrasonography or D-dimer or PTE by contrast-enhanced CT should be considered (grade C) in patients at high risk of symptomatic VTE or fatal PTE, such as patients with pelvic fractures, multiple traumas, extensive resection of malignant tumors, and a history of symptomatic PTE. The outcomes of the studies on which the AAOS and ACCP guidelines are based are proximal DVT or symptomatic PTE at 30–90 days postoperatively after hospital discharge, and the effectiveness of venous echo screening in preventing symptomatic VTE in the early postoperative period has not been demonstrated [32, 33]. The results of the JSA-PTE study showed that perioperative PTE occurred between the intraoperative period and the seventh postoperative day, accounting for more than 80% of cases [34]. Screening by venous ultrasonography may also be considered an effective option to estimate the real-time status of the patients in the early postoperative period.

Current VTE prophylaxis in patients undergoing THA should be practiced according to the 2017 Japanese Orthopaedic Association Guidelines for the Prevention of Symptomatic Venous Thromboembolism against the background of the ACCP Guidelines Ninth Edition and 2011 AAOS Guidelines. As emphasized in the guidelines, it is important to understand that the purpose of the guidelines is not to establish uniform prophylaxis but to provide information for selecting the most appropriate prophylaxis for each individual case. However, the incidence of perioperative fatal PTE has not been eliminated, and the risk and prophylaxis may change as the surgical technique and patients change. The importance of physical prophylaxis with IPCD, early bed release, and early ambulation has increased, whereas the role of prophylactic anticoagulation has declined slightly.

Each facility is required to thoroughly discuss prevention methods according to the patient's risk assessment, implement appropriate practices, and carefully explain them to the patient.

References

1. Nakamura M, et al. Current management of venous thromboembolism in Japan: current epidemiology and advances in anticoagulant therapy. J Cardiol. 2015;66:451–9.
2. Kumasaka N, et al. Incidence of pulmonary thromboembolism in Japan. Jpn Circ J. 1999;63:439–41.
3. Sakuma M, et al. Venous thromboembolism: deep vein thrombosis with pulmonary embolism, deep vein thrombosis alone, and pulmonary embolism alone. Circ J. 2009;73:305–9.
4. Heit JA. Epidemiology of venous thromboembolism. Nat Rev Cardiol. 2015;12:464–74.
5. Kuroiwa M, et al. Changes in the incidence, case fatality rate, and characteristics of symptomatic perioperative pulmonary thromboembolism in Japan: results of the 2002-2011 Japanese Society of Anesthesiologists Perioperative Pulmonary Thromboembolism (JSA-PTE) Study. J Anesth. 2015;29:433–41.
6. Japanese Orthopaedic Association Corporation Clinical Practice Guidelines Committee/Japanese Orthopaedic Association Guidelines for the Prevention of Symptomatic Venous Thromboembolism (ed.). [Japanese Orthopaedic Association Guideline for Prevention of Symptomatic Venous Thromboembolism 2017], Nankodo, Tokyo, 2017. Japanese.
7. Falck-Ytter Y, et al. Prevention of VTE in orthopedic surgery patients: antithrombotic therapy and prevention of thrombosis, 9th ed: American College of Chest Physicians Evidence-Based Clinical Practice Guidelines. Chest. 2012;141(2 Suppl):e278–325.
8. Jocobs JJ, et al. Preventing venous thromboembolic disease in patients undergoing elective hip and knee arthroplasty evidenced based guideline and evidence report. American Academy of Orthopaedic Surgeons 2011.
9. Fujita S, et al. Deep venous thrombosis after total hip or total knee arthroplasty in patients in Japan. Clin Orthop Relat Res. 2000;375:168–74.

10. Kuroiwa M, et al. [Characteristics of perioperative pulmonary thromboembolism in Japan and issues related to prevention: results of the JSA-PTE survey.] Jpn Soc Clin Anesth 2015;35:217–24. Japanese.
11. Kunieda T, et al. [Diagnosis and treatment of acute pulmonary thromboembolism.] Jpn J Med Pharm Sci 1998;39:45–53. Japanese.
12. Japanese Circulation Society, et al. [Guidelines for the diagnosis and treatment of cardiovascular disease (report of the 2008 Joint Study Group). Guidelines for the diagnosis, treatment, and prevention of pulmonary thromboembolism and deep vein thrombosis (revised 2009)], 2009. Japanese.
13. Ro A, et al. [Venous thromboembolism: etiology, diagnosis, treatment and prevention] Clinical Imagiology 2006;22:246–56. Japanese.
14. Labropoulos N, et al. Early thrombus remodeling of isolated calf deep vein thrombosis. Eur J Vasc Endovasc Surg. 2002;23:344–8.
15. Macdonald PS, et al. Short-term natural history of isolated gastrocnemius and soleal vein thrombosis. J Vasc Surg. 2003;37:523–7.
16. Schwarz T, et al. Therapy of isolated calf muscle vein thrombosis: a randomized, controlled study. J Vasc Surg. 2010;52:1246–50.
17. Sales CM, et al. Management of isolated soleal and gastrocnemius vein thrombosis. J Vasc Surg. 2010;52:1251–4.
18. Kuroiwa M, et al. [Perioperative pulmonary thromboembolism in Japan: a survey by the Japanese Society of Anesthesiologists on perioperative pulmonary thromboembolism (2005–2007).] Ther Res 2009;30:575–6. Japanese.
19. Goldhaber SZ, et al. A prospective study of risk factors for pulmonary embolism in women. JAMA. 1997;277:642–5.
20. Turpie AG, et al. Double-blind randomised trial of Org 10172 low-molecular-weight heparinoid in prevention of deep-vein thrombosis in thrombotic stroke. Lancet. 1987;1:523–6.
21. Miyata T, et al. Genetic risk factors for deep vein thrombosis among Japanese: importance of protein S K196E mutation. Int J Hematol. 2002;83:217–23.
22. Kamikura Y, et al. [Risk factors.] Jpn J Clin Med 2003;24:484–90. Japanese.
23. Khorana AA, et al. Incidence and predictors of venous thromboembolism (VTE) among ambulatory high-risk cancer patients undergoing chemotherapy in the United States. Cancer. 2013;119:648–55.
24. Ishii M, et al. [Comparison of hemorrhoidal venous flow improvement based on venous flow velocity for DVT prevention.] Hip Joint 2003; 14:49–62. Japanese.
25. Wells PS, et al. Graduated compression stockings in the prevention of postoperative venous thromboembolism. A meta-analysis. Arch Intern Med. 1994;154:67–72.
26. Hirsh J, et al. Executive summary: American College of Chest Physicians Evidence-Based Clinical Practice Guidelines (8th Edition). Chest. 2008;133(6 Suppl):71S–109S.
27. Muscatelli SR, et al. Non-inferiority of aspirin for venous thromboembolism prophylaxis after hip arthroplasty in a statewide registry. J Arthroplast. 2021;36:2068–75.e2.
28. Matharu GS, et al. Is aspirin as effective as the newer direct oral anticoagulants for venous thromboembolism prophylaxis after total hip and knee arthroplasty? An analysis from the National Joint Registry for England, Wales, Northern Ireland, and the Isle of Man. J Arthroplast. 2020;35:2631–2639.e6.
29. Japanese Orthopaedic Association Corporation Guideline for Prevention of Pulmonary Thromboembolism/Deep Vein Thrombosis (Venous Thromboembolism), Revision Committee (ed.) [Japanese Orthopaedic Association Corporation Guideline for Prevention of Venous Thromboembolism.] Nankodo, Tokyo, 2008. Japanese.
30. Mihara M, et al. Clinical efficacy of risk-stratified prophylaxis with low-dose aspirin for the management of symptomatic venous thromboembolism after total hip arthroplasty. J Orthop Sci. 2020;25:156–60.
31. Baba E. [Perioperative embolism.] LiSA 2021; 28 (Suppl Autumn):21–5. Japanese.
32. Schmidt B, et al. Ultrasound screening for distal vein thrombosis is not beneficial after major orthopedic surgery. A randomized controlled trial. Thromb Haemost. 2003;90:949–54.
33. Robinson KS, et al. Ultrasonographic screening before hospital discharge for deep venous thrombosis after arthroplasty: the post-arthroplasty screening study. A randomized, controlled trial. Ann Intern Med. 1997;127:439–45.
34. Kuroiwa M, et al. [Frequency and characteristics of perioperative pulmonary thromboembolism in Japan based on the results of the 2004 survey on the incidence of perioperative pulmonary embolism.] Jpn J Anesthesiol 2006;55:1031–8. Japanese.

84 Infection

Naomi Kobayashi

1 Prevention

(a) Risk factors

Risk factors for periprosthetic joint infection (PJI) or surgical site infection (SSI) include skin lesions, oral hygiene (dental caries), urinary tract infections, smoking, low body weight, obesity, hepatic dysfunction, renal dysfunction, anemia, diabetes mellitus, and poor nutrition. These factors should be corrected preoperatively if possible. Past trauma or hip surgery is also considered a risk factor. At ICM2018, there were conflicting statements about whether carriers of *Staphylococcus aureus* in the nasal cavity and on the skin should be investigated and if the bacteria should be eradicated. Finally, the answer was inconclusive. However, a recent meta-analysis showed that screening and eradication of *Staphylococcus aureus* has a significant effect on infection prevention [1]. In particular, patients with diabetes are at a higher risk for nasal methicillin-resistant *Staphylococcus aureus* (MRSA) carriage [2].

(b) Perioperative antimicrobial administration/operative field disinfection

The recommended prophylactic antimicrobial treatment for total hip arthroplasty (THA) in adults is 2 g of cefazolin every 4 h (ICM 2018). However, if the patient weighs ≥120 kg, 3 g should be considered. The need for a single or additional dose is controversial, and the addition of vancomycin should be considered in patients with MRSA.

Chlorhexidine gluconate or povidone-iodine with isopropyl alcohol is recommended to disinfect the surgical field (WHO, ICM 2018). Adhesive drapes containing povidone-iodine should be used. However, it should be noted that the risk of contamination is increased by fraying of the povidone-iodine-containing drape from the wound margin [3].

(c) Operating room environment

The number of people in the operating room and the number of times the door is opened and closed correlate with the number of airborne particles in the operating room. Therefore, the number of personnel in the room and number of door openings should be kept to a minimum. However, there is no clear evidence that laminar flow ventilation reduces postoperative infections. Sources of contamination include light handles, suction tube tips, and scalpel blades used for skin incisions, which should be avoided or replaced periodically.

(d) Wash solution/wound management

Several guidelines recommend washing with dilute povidone-iodine solution (0.35%) (WHO, CDC, ICM 2018). A recent meta-analysis has reconfirmed that dilute povidone-iodine solution use reduces the incidence of PJI compared with saline use [4]. On the other hand, there is no clear evidence for the efficacy of pulse rinsing, and it is recommended that antimicrobial-containing rinses should not be used due to the risk of resistance and cost-effectiveness (ICM 2018). Although there is no direct evidence of the association between incidence of PJI/SSI and drain placement, drain placement may be indirectly related to postoperative infections through increasing blood loss and the need for blood transfusion. Silver-containing coatings have been shown to decrease SSI [5].

N. Kobayashi (✉)
Yokohama City University Medical Center, Yokohama, Kanagawa, Japan

Y. Nakashima (ed.), *Advances in Total Hip Arthroplasty*, https://doi.org/10.1007/978-981-97-2461-1_84

2 Diagnosis

(a) Diagnostic criteria/bacterial culture

The diagnostic criteria for PJI proposed in ICM2018 are shown in Table 84.1. The criteria are classified into major and minor criteria. PJI is diagnosed when a score of 6 or more is obtained on any of the major criteria or on the minor criteria. If the score is 3–5, the diagnosis is considered unconfirmed, and additional diagnostic methods such as polymerase chain reaction (PCR) and nuclear medicine imaging are recommended. Especially in culture-negative cases, the diagnosis is often unconfirmed and clinically problematic. The identification rate of the causative organism can be improved by culture methods such as blood cultures, instead of the usual culture on plain medium only [6]. The efficacy of ultrasonography treatment has been confirmed as a means of improving the detection rate of biofilm-forming bacteria adhering to implants [7].

Gram-positive bacteria account for approximately 80% of the causative organisms of PJI, and more than half of these are staphylococcal infections, mainly *S. aureus* and coagulase-negative staphylococci (CNS) [8]. In contrast, various gram-negative bacteria and fungi such as Candida spp. have also been reported as causative organisms, and certain causative organisms are associated with the risk of multiple surgeries and low treatment success rates [9]. Thus, if the causative organism is not known, we initially assume infection by staphylococci; however, we must remember that atypical causative organisms such as gram-negative bacteria and fungi may also be present. Culture-negative cases are more likely to occur when there is no inflammatory response, such as an increased white blood cell count [10].

(b) Various biomarkers/histopathological diagnoses

The first screening tests are serum C-reactive protein (CRP) level and erythrocyte sedimentation rate (ESR). The cutoff values in chronic PJI are 1 mg/dL for CRP and 30 mm/h for ESR. The cumulative sensitivity and specificity of D-dimer for the diagnosis of PJI have been reported to be 0.82 and 0.7, respectively [11]. D-dimer from serum is considered more diagnostically accurate than D-dimer from plasma. The alpha-defensin test and leukocyte esterase test have been included in the diagnostic criteria as biomarkers for synovial fluid, and both have been reported to have high diagnostic accuracy [12]. The leukocyte esterase test is more advantageous in terms of cost and versatility and is easier to use in daily practice than the alpha-defensin test.

Rapid histopathologic diagnosis using frozen sections is one of the few tests that can be performed intraoperatively, and excellent diagnostic accuracy has been reported [13]. The degree of acute inflammatory reaction is determined by counting neutrophil infiltrates.

(c) Supplementary diagnosis

In cases where PJI is suspected but bacterial culture is negative and a definitive diagnosis cannot be made, some type of supplementary diagnosis should be attempted, such as PCR, which has recently become possible with fully automated real-time PCR [14] and next-generation sequencing [15]. These molecular diagnostics are the only way to confirm the presence of bac-

Table 84.1 ICM2018 periprosthetic joint infection diagnostic criteria

Major criteria (at least one of the following two criteria must be met)				Diagnosis
Two positive growths of the same organism using standard culture methods				Infection
Sinus tract with evidence of communication to the joint or visualization of the prosthesis				
Minor criteria	Threshold		Score	Diagnosis
	Acute[a]	Chronic		
Serum CRP (mg/L) or D-dimer (μg/L)	100 Unknown	10 860	2	Combined preoperative and postoperative score: ≥6 infected 3–5 inconclusive[c] <3 not infected
ESR elevation (mm/h)	No role	30	1	
Elevated joint fluid leukocyte count (cells/μL) or leukocyte esterase or α-defensin positive (signal/cutoff)	10,000 ++ +	3000 ++ +	3	
Elevated polymorphonuclear neutrophils (PMN) in joint fluid (%)	90	70	2	
Single positive culture			2	
Positive histopathology			3	
Presence of obvious intraoperative purulence[b]			3	

CRP C-reactive protein, *ESR* erythrocyte sedimentation rate
[a]This criterion has not been validated for acute infections
[b]Not indicated in cases where adverse local tissue reactions are suspected
[c]Next-generation sequencing and other molecular biological diagnostics

teria in culture-negative cases. Other than bacterial culture, various tests of synovial fluid, and biomarkers, diagnostic imaging is another approach to diagnosis. The usefulness of nuclear medicine imaging such as bone scintigraphy, leukocyte scintigraphy, and positron emission tomography (PET) has been widely reported [16]. Nuclear medicine imaging such as PET can be used to identify infected sites in cases where implants are inserted at multiple sites and infected sites are difficult to identify. Specimens are collected from the site of accumulation to improve the positive detection rate of bacterial cultures and histopathological examination [17].

3 Treatment

(a) Debridement, antibiotics, and implant retention (DAIR)

If there an early infection without laxity occurs within 4 weeks postoperatively, implant preservation with cleansed debridement (DAIR) should be attempted as soon as possible. The cumulative success rate of DAIR for THA is 72.2%, especially when performed within 7 days of onset. Furthermore, the replacement of modular components such as polyethylene liners and heads increases the success rate of DAIR [18]. Postoperative antimicrobial therapy, as described below, is key to the success of DAIR and should be continued for at least 6 weeks (ICM 2018).

(b) One-stage revision surgery

One-stage revision can be considered if the causative organism is identified and not resistant, the soft tissue is in good condition, the bone defect is not large, and the patient is in good general health. A meta-analysis of 44 cohort studies showed that the results of one-stage revision were comparable to those of two-stage revision [19]. There are also advantages in terms of patient activity of daily life (ADL) maintenance, early return to work, and health care economics. The primary revision procedure requires thorough cleaning and debridement after removal of the implant and replacement of contaminated machinery and drapes before reinstallation of the new implant.

(c) Two-stage revision

Two-stage revision is indicated when one-stage revision is not indicated due to the presence of resistant organisms, poor skin and soft tissue condition, massive bone defects, or immunocompromised conditions. PJI caused by fungi such as Candida spp. is refractory, and the results of DAIR and one-stage revision are inferior. Therefore, two-stage revision is recommended [20]. The reinfection rate at 2 years after two-stage revision surgery is 4.58%, and the reinfection rate increases to 7.34% by the last follow-up [21]. There is no consensus on the timing of two-stage revision surgery. Therefore, a combination of markers should be used to determine the timing, rather than a single marker.

(d) Antimicrobial therapy

The majority of PJI-causing bacteria are staphylococci, and antimicrobial agents should be selected based on their drug susceptibility and local transferability. The efficacy of antimicrobial therapy with rifampicin against biofilm-forming bacteria has been reported in many cases. In DAIR of PJI with staphylococci as the causative organism, a significantly higher treatment success rate was achieved with rifampicin than without rifampicin, and good results were obtained with quinolones and clindamycin as combination agents [22]. However, the use of rifampicin early in treatment (within 5 days after surgery) is considered a risk factor for treatment failure [22], presumably because the use of rifampicin in the presence of a high bacterial load increases the risk of resistance. A minimum duration of 6 weeks is recommended for DAIR and one-stage revisions (ICM2018). On the other hand, a multicenter randomized controlled trial in two-stage revision surgery showed that continuation of oral antimicrobial treatment for 3 months after surgery improved the rate of quiescent infection [23].

References

1. Ribau AI, et al. Is preoperative Staphylococcus aureus screening and decolonization effective at reducing surgical site infection in patients? A systematic review and meta-analysis with a special focus on elective total joint arthroplasty. J Arthroplasty. 36:752–66.e6.
2. Tateiwa T, et al. Risk factors for preoperative nasal carriage of methicillin-resistant bacteria in total hip and knee arthroplasty patients. Orthop Sci. 2022;27:798–803.
3. Hanada M, et al. Intraoperative bacterial contamination in total hip and knee arthroplasty is associated with operative duration and peeling of the iodine-containing drape from skin. Eur J Orthop Surg Traumatol. 2020;30:917–21.
4. Kobayashi N, et al. Effectiveness of diluted povidone-iodine lavage for preventing periprosthetic joint infection: an updated systematic review and meta-analysis. J Orthop Surg Res. 2021;16:569.
5. Rothfusz CA, et al. Wound dressings for hip and knee total joint arthroplasty: a narrative review. JBJS Rev. 2021;9:e20.00301.
6. Watanabe S, et al. Differences in diagnostic properties between standard and enrichment culture techniques used in periprosthetic joint infections. J Arthroplast. 2020;35:235–40.

7. Hameister R, et al. What is the role of diagnostic and therapeutic sonication in periprosthetic joint infections? J Arthroplast. 2018;33:2575–81.
8. Kheir MM, et al. Culturing periprosthetic joint infection: number of samples, growth duration, and organisms. J Arthroplast. 2018;33:3531–3536.e1.
9. Cunningham DJ, et al. Specific infectious organisms associated with poor outcomes in treatment for hip periprosthetic infection. J Arthroplast. 2017;32:1984–1990.e5.
10. Watanabe S, et al. Clinical characteristics and risk factors for culture-negative periprosthetic joint infections. J Orthop Surg Res. 2021;16:1–9.
11. Lu G, et al. D-dimer in the diagnosis of periprosthetic joint infection: a systematic review and meta-analysis. J Orthop Surg Res. 2020;15:265.
12. Chen Y, et al. Reliability of synovial fluid alpha-defensin and leukocyte esterase in diagnosing periprosthetic joint infection (PJI): a systematic review and meta-analysis. J Orthop Surg Res. 2019;14:453.
13. Tsaras G, et al. Utility of intraoperative frozen section histopathology in the diagnosis of periprosthetic joint infection: a systematic review and meta-analysis. J Bone Joint Surg Am. 2012;94:1700–11.
14. Yang F, et al. An automated real-time PCR assay for synovial fluid improves the preoperative etiological diagnosis of periprosthetic joint infection and septic arthritis. J Orthop Res. 2021;39:348–55.
15. Tarabichi M, et al. Diagnosis of periprosthetic joint infection: the potential of next-generation sequencing. J Bone Joint Surg Am. 2018;100:147–54.
16. Verberne SJ, et al. The accuracy of imaging techniques in the assessment of periprosthetic hip infection: a systematic review and meta-analysis. J Bone Joint Surg Am. 2016;98:1638–45.
17. Choe H, et al. Use of 18F-fluoride PET to determine the appropriate tissue sampling region for improved sensitivity of tissue examinations in cases of suspected periprosthetic infection after total hip arthroplasty. Acta Orthop. 2011;82:427–32.
18. Tsang S-TJ, et al. Outcomes following debridement, antibiotics and implant retention in the management of periprosthetic infections of the hip: a review of cohort studies. Bone Joint J. 2017;99-B:1458–66.
19. Kunutsor SK, et al. One- and two-stage surgical revision of peri-prosthetic joint infection of the hip: a pooled individual participant data analysis of 44 cohort studies. Eur J Epidemiol. 2018;33:933–46.
20. Kuiper JWP, et al. 2-stage revision recommended for the treatment of fungal hip and knee prosthetic joint infections. Acta Orthop. 2013;84:517–23.
21. Otten MR, et al. Two-stage reimplantation of a prosthetic hip infection: systematic review of long-term reinfection and pathogen outcomes. J Arthroplast. 2021;36:2630–41.
22. Beldman M, et al. If, when, and how to use rifampin in acute staphylococcal periprosthetic joint infections, a multicenter observational study. Infect Dis. 2021;73:1634–41.
23. Frank JM, et al. The mark Coventry, MD, award: Oral antibiotics reduce reinfection after two-stage exchange: a multicenter, randomized controlled trial. Clin Orthop Relat Res. 2017;475:56–61.

85 Adverse Reactions to Metal Debris (ARMD)

Takashi Nishii

1 Pathogenesis of ARMD

Adverse reactions to metal debris (ARMD) and adverse local tissue reactions (ALTR) are mainly observed in total hip arthroplasty (THA) and hip resurfacing arthroplasty with metal-on-metal (MoM) articulations. Metal wear particles and metal ions cause the formation of non-infectious, non-neoplastic pseudotumor lesions and extensive fluid correction around the joint, which if allowed to progress, can lead to soft tissue damage in the joint capsule, muscle tissue, and bone tissue around the implant [1]. Similar pathologies can occur from corrosion and fretting (taper corrosion, trunnionosis) at the junctions between the metal head and stem neck, including THA with a metal head and polyethylene liner (metal-on-polyethylene: MoP) articulations (Fig. 85.1) [2, 3]. Although metal allergy disorders, such as McKee-Farrar-type MoM THA and disorders of the junctions between the metal head and stem neck have been reported before the 1990s [4], ARMD-related disorders have gained great attention as the clinical use of MoM THA/resurfacing hip prostheses using large-diameter femoral heads rapidly became widespread in the late 1990s [1]. In the pathogenesis of ARMD, cobalt and chromium ions have been implicated in stimulating the migration of T lymphocytes and the activation of macrophages [5]. ARMD is rare in THAs without cobalt-chromium metal for articulation or stem neck junctions [6].

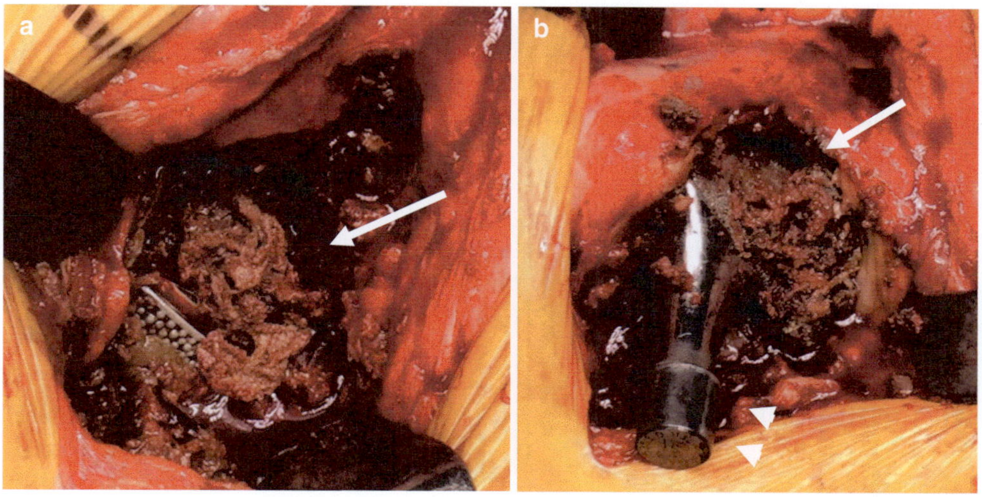

Fig. 85.1 Reoperation for ARMD of MoM THA with a large femoral head. (**a**) The joint is filled with dark red tissues (arrow), and (**b**) osteolysis is observed around the cup and stem (arrow). Corrosion of the stem neck joint was observed when the head was removed (arrowheads)

T. Nishii (✉)
Osaka Habikino Medical Center, Osaka, Japan

Y. Nakashima (ed.), *Advances in Total Hip Arthroplasty*, https://doi.org/10.1007/978-981-97-2461-1_85

Table 85.1 Frequency of adverse reactions to metal debris for various hip prostheses

First reporter	Number of cases	Survey method	Frequency	Frequency of occurrence or revision
Sugano [7]	13,567	Questionnaire for domestic hospitals	Frequency of occurrence Revision frequency	MoM[a] resurfacing type: 0.5% MoM THA: 1.2% MoM resurfacing type: 0.% MoM THA: 0.7%
Hussey [2]	1352	Symptoms/blood metal ion measurement in cases of osteolysis image, etc.	Frequency of occurrence	MoP[a] THA: 3.2%
Persson [8]	2409	Histological findings at revision	Revision frequency	MoP THA: 0.5% (average 7 years after surgery)
de Steiger [6]	370,685	Registry (AOANJRR[c])	Cumulative revision frequency	Non-MoM modern THA[b]: 0.1% (17 years after surgery) Large diameter MoM THA with diameter of 36 mm or larger: 15.5% (17 years after surgery) Small diameter MoM THA with 32 mm or smaller head: 0.9% (9 years after surgery)

[a]*MoM* metal-to-metal bearing, *MoP* metal-to-polyethylene bearing, *THA* total hip arthroplasty
[b]Non-MoM modern THA: THA using metal or ceramic heads vs. cross-linked polyethylene bearings, and ceramic vs. ceramic bearings
[c]*AOANJRR* Australian Orthopaedic Association National Joint Replacement Registry

2 Frequency of ARMD and Revision (Table 85.1)

The frequency of ARMD is influenced by the method of investigation and diagnosis, as some cases are asymptomatic and cannot be diagnosed using simple radiological evaluation alone. However, there is no large-scale survey using imaging tests with a high sensitivity for ARMD detection, such as magnetic resonance imaging (MRI). In a large-scale questionnaire survey in Japan, the frequency of MoM THA was higher than that of MoM hip resurfacing arthroplasty. The frequency of ARMD varies from 0% to 3.1% among THAs depending on the implant brands [7]. In a study using an overseas registry, the frequency of revision due to ARMD was higher in cases with a head diameter ≥ 36 mm [6]. For MoP THA, relatively high frequencies of ARMD and revision have also been reported in specific brands and cases with head diameter of 36 mm or larger [2, 6, 8].

3 Diagnosis of ARMD

(a) Symptoms

Symptoms of ARMD may vary depending on the type, location, and extent of the lesion, and include pain, discomfort, mobility problems, swelling, and edema/numbness of the lower extremities. Since symptomatic patients are more likely to have ARMD than asymptomatic patients [9], it is important to monitor patients' symptoms during follow-up. However, it has been reported that asymptomatic patients are common, and it is important to pay attention to minor symptom and radiographic changes in patients who have been treated with an implant model with a high incidence of ARMD or who have high patient risk factors.

(b) Characteristics of each test method

(i) Plain radiograph

Although direct diagnosis of soft tissue lesions is difficult, intraosseous lesions, such as radiolucent areas, osteolysis around the cup and proximal portion of the stem, and implant migration can be indicators of advanced ARMD. Since these minor changes are difficult to detect, careful comparison of these radiographic changes over time is recommended (Fig. 85.2).

(ii) MRI

MRI is most useful for lesion delineation owing to the high signal contrast among the lesion, adjacent soft tissue, and bone. MRI also allows the identification of three-dimensional location and extension of lesions. Metal artifact reduction sequences (MARS) are particularly useful for the assessment of large-diameter cobalt-chromium metallic heads. In addition to conventional MARS techniques, such as the use of a wider receiver bandwidth, fast spin–echo, and fat-suppressed short tau inversion recovery (STIR) sequence, advanced sequences for MARS, such as multiacquisition variable-

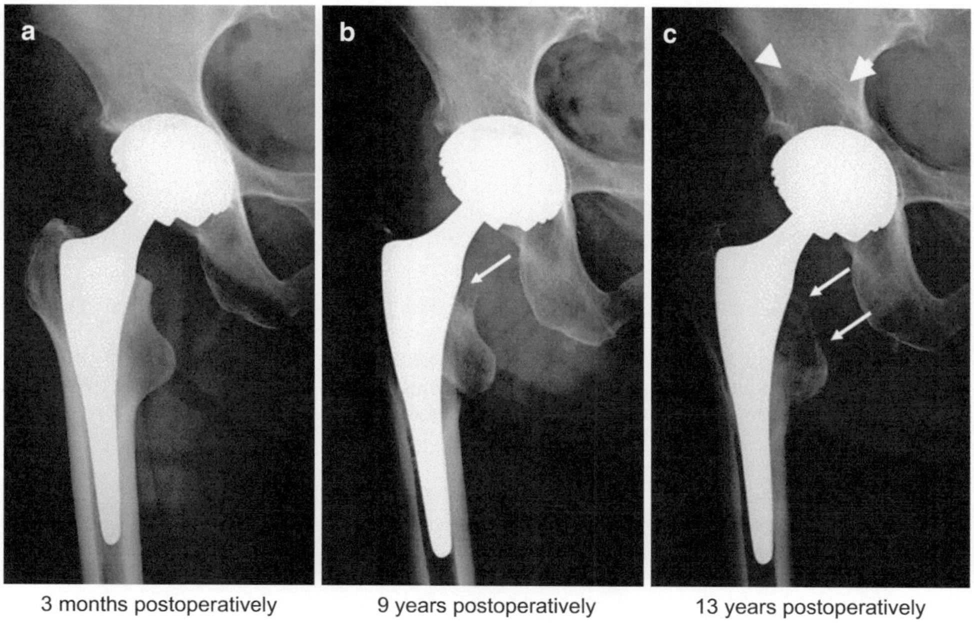

Fig. 85.2 (**a**–**c**) ARMD case after MoM THA with a large femoral head. (**a**) There were no abnormal findings at 3 months postoperatively. (**b**) Osteolysis of the proximal medial part of the stem (arrow) was observed on radiographs 9 years postoperatively. (**c**) Enlarged osteolysis of the proximal medial part of the stem (arrows) and osteolysis around the cup (arrowheads) were observed 13 years postoperatively

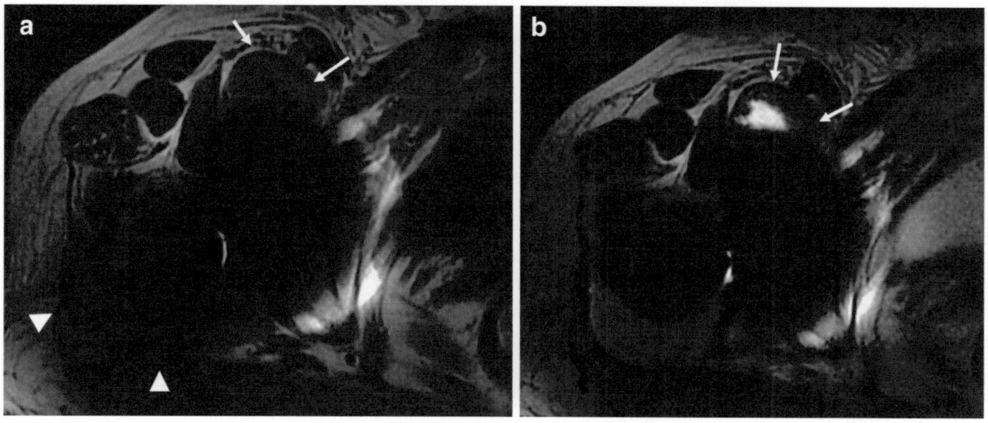

Fig. 85.3 MRI cross-sectional images 8 years after MoM THA with a large femoral head. (**a**) T1-weighted image shows erythematous lesions in the anterior (arrows) and posterior (arrowheads) regions of the articulation with a slightly higher signal intensity than that of the bladder effusion due to the metallic component in the lesion. (**b**) T2-weighted image shows an anterior lesion with a thick erythematous wall (arrows)

resonance image combination (MAVRIC) and slice-encoding for metal artifact correction (SEMAC), have been introduced in Japan in recent years.

The lesions were evaluated by type classification (cystic type with uniform liquid component, solid type with high solid component, and cystic-solid type in between), thickness of the cyst wall, size, location (anterior, lateral, posterior), and continuity with the joint (Fig. 85.3). It has been shown that compared to other patients, those with anterior, solid type, enlarged lesions, and thick cystic wall lesions are likely to have more severe symptoms and require revision surgery more frequently [10, 11].

(iii) CT

Although the image contrast between the soft tissue and the lesion is not high, lesions can often be identified from the enlarged shadows of the gluteus medius and iliopsoas muscle compared with the contralateral side, and from abnormalities in the location of the muscles (Fig. 85.4). In computed tomography (CT) imaging, methods that reduce metal artifacts are useful. High tube voltage, thin slice width imaging, the dual energy method (different processing between low- and high-voltage images), and the use of specialized metallic artifact reduction software are effective.

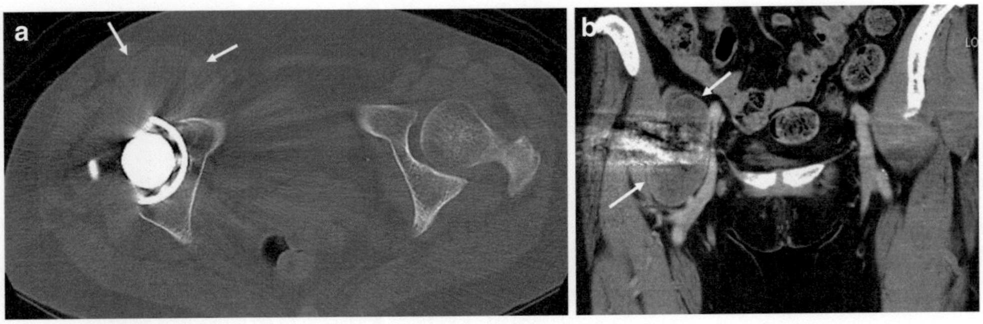

Fig. 85.4 CT images of right MoM THA 14 years postoperatively. (**a**) The iliopsoas muscle is clearly enlarged anterior to the right articulation (arrows) compared with the contralateral muscle. (**b**) Coronal reconstruction image shows a lesion extending cephaloposteriorly along the iliopsoas muscle (arrows)

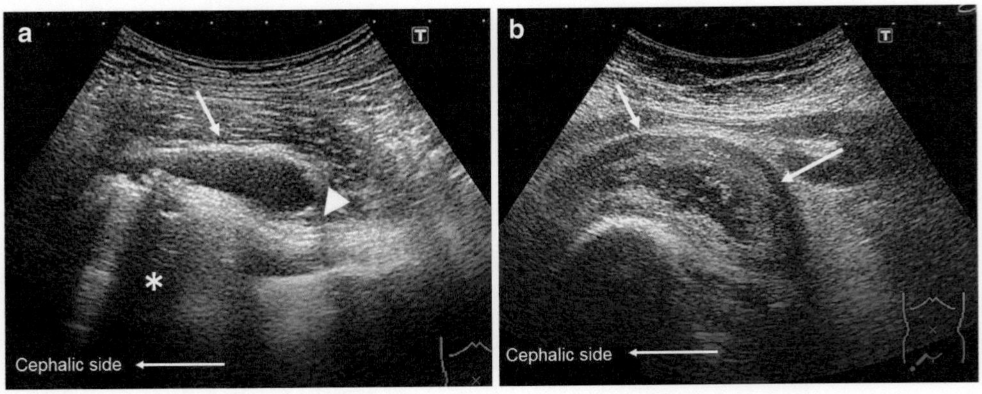

Fig. 85.5 (**a**) Ultrasound images obtained 2 years after MoP THA. Cystic lesion (arrow) extending anteriorly (*femoral head) with intra-articular traffic (arrowhead). (**b**) Ultrasound image obtained 2 years after the MoM THA. A cystic lesion with a thick cystic wall (arrows) is observed along the iliopsoas muscle

CT is particularly important for evaluating the location and size of osteolysis, the angle and position of implant placement, and peri-implant bone preservation in anticipation of surgical treatment.

(iv) Ultrasound

Ultrasound examination has the advantage of being able to evaluate areas close to the articulation of the THA without the influence of metallic artifacts, such as those seen on MRI and CT. The probe was applied anteriorly and laterally to the hip joint to search for fluid retention and cystic lesions around the articulation and greater trochanter (Fig. 85.5). It is possible to classify the cystic/solid type of lesion, evaluate thickening of the cystic wall, and evaluate its traffic with the joint. Ultrasound has been reported to have a diagnostic accuracy comparable to that of MRI, and many experts recommend it as the first screening imaging test for ARMD lesions [12]. However, using ultrasound is difficult when evaluating lesions that extend behind the articulation covered with thick gluteus muscles or proximal to the articulation. The accuracy of ultrasound evaluation may also be affected by the skill of the examiner. Ultrasound is considered an adjunctive diagnostic method next to MRI and CT, and is useful for periodic observation of lesion progression and confirmation of the position of the puncture needle during lesion and joint puncture.

(v) Blood metal ion measurement

In the announcements of alerts and recommendations by the UK Medicines and Healthcare Products Regulatory Agency (MHRA) and the US Food and Drug Administration (FDA) for MoM implant cases, measurement of blood metal ion concentrations is regarded as the primary screening method for ARMD. Previously, an abnormal blood cobalt and chromium concentration of 7 μg/L or higher was considered an indicator of ARMD and the need for surgical intervention [13]. However, in recent years, various points have been raised, including the need to set different cut-off values for different implant models and the impossibility of setting clear cut-off values [14, 15]. The significance of blood metal ion measurements and their effective utilization have not yet been determined. In Japan, blood metal ion concentration testing is not covered by insurance and is not considered a standard diagnostic test because of the lack of facilities to perform practical analysis.

(vi) Joint and lesion puncture

Fluid is collected from the joint or lesion under ultrasound guidance, and microscopic and culture examinations are performed to differentiate other diseases, such as infection and hematoma related to iliopsoas muscle impingement [16].

A low-viscosity, yellowish, or grey turbid fluid with metal abrasion particles is often collected from ARMD lesions. However, it is difficult to determine the presence of ARMD based on the appearance and color of the aspirate alone since relatively clear or bloody fluid may also be present.

(vii) Histopathological examination

Typical histopathological findings of these lesions include aggregations of perivascular lymphocytes and plasma cells called aseptic lymphocyte-dominated vasculitis-associated lesions (ALVAL), scattered metal debris-containing macrophages, necrosis of superficial synovial cells, and accumulation of fibrin [17, 18]. As a histological evaluation of metal hypersensitivity that is not dependent on excessive metal wear particles, the ALVAL scoring system, which points to the aggregation of synovial surface layer cells, macrophages, and lymphocytes and the state of tissue structure, has also been proposed [18].

(c) Differential diagnosis

When pseudotumors, fluid collection, or osteolysis are observed on imaging examinations, it is important to distinguish infection, neoplastic lesions, or aseptic loosening of the implant from the viewpoint of treatment selection [16]. It is often difficult to distinguish infections caused by attenuated bacteria, such as coagulase-negative staphylococci (CNS). In addition to blood biochemical and culture tests of joint fluid and effluent, it is also necessary to carefully examine the pathology and culture of periarticular tissues during revision surgery.

4 Management of MoM Cases

More careful follow-up is required for MoM THA and resurfaced hip arthroplasty cases, which have a higher risk of ARMD than cases of other bearing materials. The 2017 update of the MHRA's alert statement [13] recommends blood metal ion measurement in all patients and imaging examinations, such as MRI and ultrasound, depending on the femoral head diameter and presence of symptoms. Moreover, in Japan, the Japanese guidelines for the treatment of metal-on-metal total hip arthroplasty complications (hereafter referred to as the Japanese MoM guideline) [10] recommend MRI examination in cases with abnormal radiographic findings or clinical symptoms, and in cases using implant models that are considered to have a high incidence of ARMD (Fig. 85.6). Annual follow-up is recommended, even in the absence of abnormal findings or clinical symptoms.

In THA with a polyethylene liner, the risk of trunnionosis is higher in cases of a combination of titanium stem neck and cobalt-chromium head than in cases of a combination of similar metals or ceramic head because of the involvement of dissimilar metal contact corrosion.

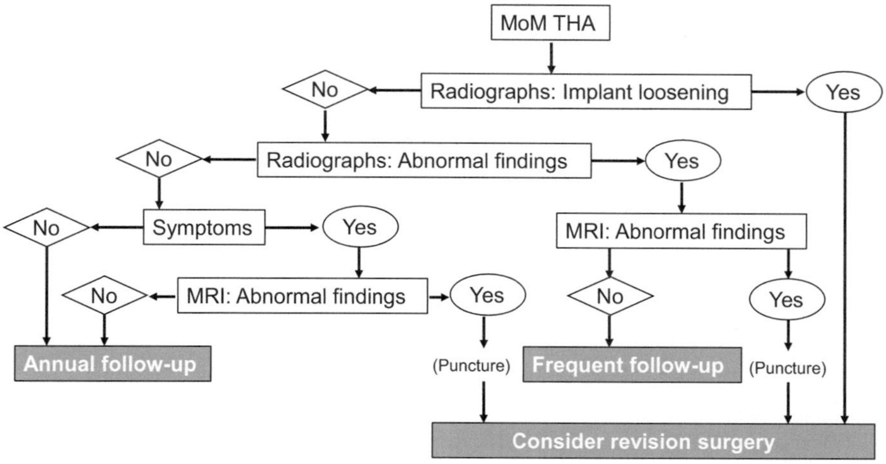

Fig. 85.6 Algorithm of medical treatment after MoM THA surgery. Note 1: For implant models with a high incidence of ARMD (MicroPort Conserve Plus, Corin Cormet, Depuy Ultamet), MRI is recommended. Note 2: MRI with metal artifact reduction sequences (MARS) is recommended. Abnormal MRI findings include the presence of a pseudotumor larger than 5 cm, a tendency toward enlargement, or edematous images of bone or muscle in the vicinity of the metal. (Cited from the Committee to Investigate Complications of Metal-on-Metal Hip Prostheses of the Japanese Society for Replacement Arthroplasty. 2016. Available at: https://jsra.info/pdf/Metal-on-Metal.pdf)

5 Response to ARMD Cases

(a) Natural history and conservative treatment

The natural history of ARMD lesions can vary depending on the implant model used and nature of the lesion. MoM THA with small-diameter heads, cystic type with uniform fluid component, and small ARMD lesions show a non-progressive or shrinking trend, whereas solid type with massive internal components and lesions with thick cystic walls tend to increase in size and are reoperated more frequently [10, 19, 20]. In cases of MoM implant models with a high revision frequency, large solid mass lesions, and bone lesions, such as lysis, careful follow-up is needed, and surgical treatment should be considered before excessive progression of bone and soft tissue damage occurs. Conservative treatments, such as drainage and pharmacological treatment, have not been proven to be effective.

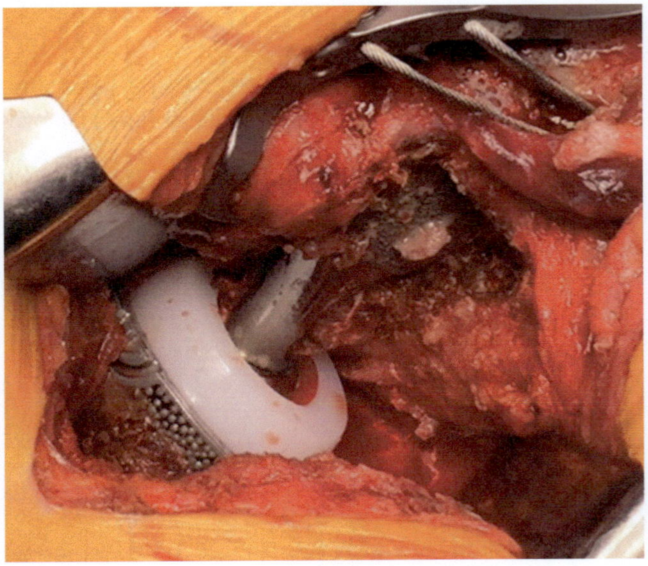

Fig. 85.7 In the case of Fig. 85.1, the cup and stem were preserved, and the head was replaced with a dual mobility system

(b) Surgical treatment

Postoperative outcomes are poor in patients with advanced ARMD who undergo revision surgery, and early surgical treatment is recommended for those at high risk of progression [21]. The Japanese MoM guideline [10] lists the following findings as lesions that frequently require revision surgery: (1) maximum diameter of 50 mm or more; (2) solid or mixed contents; (3) lumpy or irregular shape; and (4) pseudotumor wall thickness of 2 mm or more.

In the surgical treatment of MoM, the entire cup or liner should be replaced with bearings with a metallic head-to-polyethylene liner (MoP) or ceramic head-to-polyethylene liner (CoP). However, when the cup integrated with the metal liner was firmly fixed to the surrounding bone and the condition of the surrounding bone was poor due to osteolysis, there are concerns regarding the fixation of the cup at the time of revision. A dual mobility system with a double structure of an inner head and an external polyethylene head was used while preserving the cup [22] (Fig. 85.7). Although the dual-mobility system is effective in reducing the risk of dislocation after revision, there are concerns regarding the risk of corrosion of the metal inner head and stem trunnion and the lack of reports on medium- to long-term outcomes in several cases.

If the fretting/corrosion of the trunnion is severe at the time of revision in patients with ARMD, the stem should be revised because the recurrence rate increases when the metallic head is reinterlocked. However, the removal of a firmly adhered stem is a highly invasive operation, and there are concerns about the risk of expansion of surrounding bone damage; therefore, a titanium stem can be preserved and replaced with a ceramic head by fitting a titanium sleeve adaptor to the stem neck [23].

Results of revision surgery for ARMD are not favorable, and postoperative dislocation, component loosening, infection, and recurrence of ARMD are reported to be frequent [23, 24]. The use of a large-diameter head or dual-mobility system at the time of revision surgery is one of the most effective countermeasures against dislocation, and the use of a constrained liner is also considered for repeated dislocations. If the ARMD lesion is enlarged and the joint supportive function is impaired due to necrosis of soft tissues, such as the abductor muscles, these measures may not be effective, and reconstruction of the abduction function by shifting part of the gluteus maximus muscle and the tensor fascia femoris muscle [25] may be necessary.

References

1. Langton DJ, et al. Early failure of metal-on-metal bearings in hip resurfacing and large-diameter total hip replacement: a consequence of excess wear. J Bone Joint Surg Br. 2010;92:38–46.
2. Hussey DK, et al. Ten-year cross-sectional study of mechanically assisted crevice corrosion in 1352 consecutive patients with metal-on-polyethylene total hip arthroplasty. J Arthroplast. 2017;32:2546–51.
3. Nakamura M, et al. Failure rates of Asian-type anatomic medullary locking stemmed metal-on-metal total hip replacement: a cause for adverse tissue reactions to metal debris (ARMD). J Orthop Sci. 2016;21:779–85.
4. Jones DA, et al. Cobalt toxicity after McKee hip arthroplasty. J Bone Joint Surg Br. 1975;57:289–96.

5. Hasegawa M, et al. Immune response in adverse reactions to metal debris following metal-on-metal total hip arthroplasty. BMC Musculoskelet Disord. 2016;17:221.
6. de Steiger RN, et al. What is the risk of THA revision for ARMD in patients with non-metal-on-metal bearings? A study from the Australian National Joint Replacement Registry. Clin Orthop Relat Res. 2020;478:1244–53.
7. Sugano N, et al. Nationwide investigation into adverse tissue reactions to metal debris after metal-on-metal total hip arthroplasty in Japan. Orthop Sci. 2014;19:85–9.
8. Persson A, et al. Revision for symptomatic pseudotumor after primary metal-on-polyethylene total hip arthroplasty with a standard femoral stem. J Bone Joint Surg Am. 2018;100:942–9.
9. Galea VP, et al. What is the clinical presentation of adverse local tissue reaction in metal-on-metal hip arthroplasty? An MRI study. Clin Orthop Relat Res. 2019;477:353–60.
10. Japanese Society for Replacement Arthroplasty Metal-on-Metal Total Hip Arthroplasty Complication Investigation Committee. [Guidelines for the treatment of metal-on-metal total hip replacement complications (recommendations).] 2016. Available at: https://jsra.info/pdf/Metal-on-Metal.pdf. Japanese.
11. Hauptfleisch J, et al. A MRI classification of periprosthetic soft tissue masses (pseudotumours) associated with metal-on-metal resurfacing hip arthroplasty. Skeletal Radiol. 2012;41:149–55.
12. Matharu GS, et al. What is appropriate surveillance for metal-on-metal hip arthroplasty patients? Acta Orthop. 2018;89:29–39.
13. Medicines and Healthcare products Regulatory Agency. all metal-on-metal (MoM) hip replacements: updated advice for follow-up of patients. 2017:1–7. Available at: https://www.gov.uk/drug-device-alerts/all-metal-on-metal-mom-hip-replacements-updated-advice-for-follow-up-of-patients
14. Matharu GS, et al. Blood metal ion thresholds to identify patients with metal-on-metal hip implants at risk of adverse reactions to metal debris: an external multicenter validation study of Birmingham Hip Resurfacing and Corail-Pinnacle implants. J Bone Joint Surg Am. 2017;99:1532–9.
15. Matharu GS, et al. No threshold exists for recommending revision surgery in metal-on-metal hip arthroplasty patients with adverse reactions to metal debris: a retrospective cohort study of 346 revisions. J Arthroplast. 2019;34:1483–91.
16. Browne JA, et al. Failed metal-on-metal hip arthroplasties: a spectrum of clinical presentations and operative findings. Clin Orthop Relat Res. 2010;468:2313–20.
17. Davies AP, et al. An unusual lymphocytic perivascular infiltration in tissues around contemporary metal-on-metal joint replacements. J Bone Joint Surg Am. 2005;87:18–27.
18. Campbell P, et al. Histological features of pseudotumor-like tissues from metal-on-metal hips. Clin Orthop Relat Res. 2010;468:2321–7.
19. Hasegawa M, et al. Longitudinal magnetic resonance imaging of pseudotumors following metal-on-metal total hip arthroplasty. J Arthroplast. 2014;29:2236–8.
20. Kwon Y-M, et al. What is the natural history of "asymptomatic" pseudotumours in metal-on-metal hip arthroplasty? Minimum 4-year metal artifact reduction sequence magnetic resonance imaging longitudinal study. J Arthroplast. 2016;31(9 Suppl):121–6.
21. Grammatopolous G, et al. Hip resurfacings revised for inflammatory pseudotumours have a poor outcome. J Bone Joint Surg Br. 2009;91:1019–24.
22. Colacchio ND, et al. Dual mobility for monoblock metal-on-metal revision-Is it safe? J Arthroplast. 2020;35:508–12.
23. Waterson HB, et al. Revision for adverse local tissue reaction following metal-on-polyethylene total hip arthroplasty is associated with a high risk of early major complications. Bone Joint J. 2018;100-B:720–4.
24. Liow MH, et al. Preoperative risk factors associated with poor outcomes of revision surgery for "pseudotumors" in patients with metal-on-metal hip arthroplasty. J Arthroplast. 2016;31:2835–42.
25. Whiteside LA. Surgical technique: gluteus maximus and tensor fascia lata transfer for primary deficiency of the abductors of the hip. Clin Orthop Relat Res. 2014;472:645–53.

Periprosthetic Fracture

86

Satoshi Ikemura

1 Epidemiology

Total hip arthroplasty (THA) is an orthopedic technique which relieves pain and improves hip function in the early postoperative period and is commonly regarded as the greatest invention in the field of orthopedic surgery in the twentieth century [1]. However, the incidence of periprosthetic femoral fractures (PFF), particularly peri-stem fractures, has been increasing in the aging population [2, 3]. According to the UK registry, the incidence of PFF increased by 2.2% during the 3-year period from 2012 to 2015 [2], while the US registry predicted an increase of 4.6% every 10 years for the next 30 years [3]. The mortality rate within 6 months after revision for PFF following THA was reported to be 7.3%, compared with the mortality rate of 0.9% for elective revision THA. PFF is therefore a serious complication that should be avoided [4].

2 Frequency

One systematic review comparing the frequency of PFF between cemented and cementless stems showed that cemented stems had a significantly lower incidence of PFF, of 0.31% (790/255,958 joints), compared to the incidence of 1.03% (901/86,761 joints) for cementless stems [5]. In the Nordic Arthroplasty Registry Association Database, the incidence of PFF was low for both cemented and cementless stems, but had an incidence of 0.07% for cemented stems compared with that of 0.47% for cementless stems [6].

In cementless stems, the incidence of PFF in fit-and-fill stems is 1.15% (627/54,742 joints) [5]. In contrast, 5.6% (58/1, 039 joints) of PFF in fit-and-fill stems were reported, of which 38 (3.7%) had intraoperative fractures and 20 (1.9%) had postoperative fractures [7].

The incidence of PFF in tapered wedge stems was found to be 0.3% (6/2220 joints) [8] and 2.0% (17/850 joints) [9] in two prior studies, which is lower than that in fit-and-fill stems. The difference in the number of steps and the difference in the fixation points (fit-and-fill: four points, taper wedge: two points) are assumed to be related to the difference in the incidence of PFF. However, there have been very few reports comparing the two systems; therefore, we compared and examined the results in our institute.

In our study, we included 753 and 526 joints that underwent an initial THA using fit-and-fill stems (Fig. 86.1a) and tapered wedge stems (Fig. 86.1b), respectively, in our department. The incidence of PFF in the tapered wedge stem group was 0.8% (4/526 joints), which was significantly lower than that in the fit-and-fill group (2.4% [18/753 joints]). In the fit-and-fill group, 11 (1.5%) fractures developed postoperatively and 7 (0.9%) intraoperatively, while all 4 fractures in the tapered wedge stem group were postoperative. Reoperation was required in only 5 joints (0.7%) in the fit-and-fill group. In our study, the incidence of PFF in the tapered wedge stem group was found to be lower than that in the fit-and-fill group.

S. Ikemura (✉)
Kyushu University, Fukuoka, Japan

Y. Nakashima (ed.), *Advances in Total Hip Arthroplasty*, https://doi.org/10.1007/978-981-97-2461-1_86

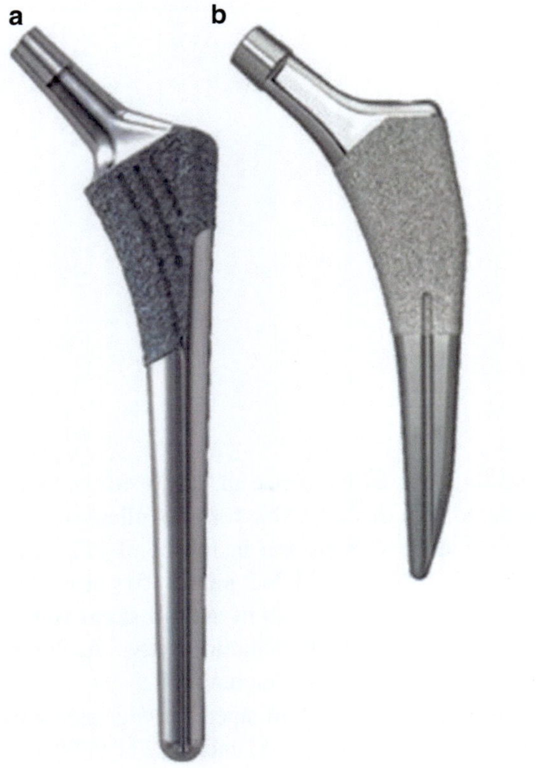

Fig. 86.1 (**a**) A Fit-and-fill stem (Kyocera 910 PerFix). (**b**) A Tapered wedge stem (Kyocera Initia)

3 Risk Factors

The risk factors for PFF include female sex, advanced age, osteoporosis, and low BMI [5, 7, 10, 11]. A prior report in which 23 (71.9%) of 32 of the 5313 joints with PFF were female [10]; this report showed that the risk of PFF was 2.9 times higher in patients aged 70 years or older [11]. Diseases and femoral shape include developmental hip dysplasia and a wide medullary cavity shape (Dorr C) [10, 12]. In addition, after femoral osteotomy, the location and axis of the medullary cavity are difficult to identify because of deformity of the proximal femur, bone thickening in the calcar area, and the presence of osteosclerotic areas in the medullary cavity, performing THA more difficult than normal THA and increasing the risk of PFF [13, 14] (Fig. 86.2).

Surgical factors include cementless stem, short stem, and approaches [5, 6, 9, 15]. In a report from Japan comparing the PFF of THA using a standard tapered wedge stem or a short stem, all PFFs observed in 17 joints (2.0%) were short stems, and multivariate analysis showed that the short stem was an independent influencing factor of PFF [9]. Regarding the approach, anterior approach, which is a more technically difficult femoral manipulation than posterior approach, was a risk factor for PFF [15]. In addition, frequent use of a hammer and striking force during stem insertion have been reported as risk factors for PFF [16].

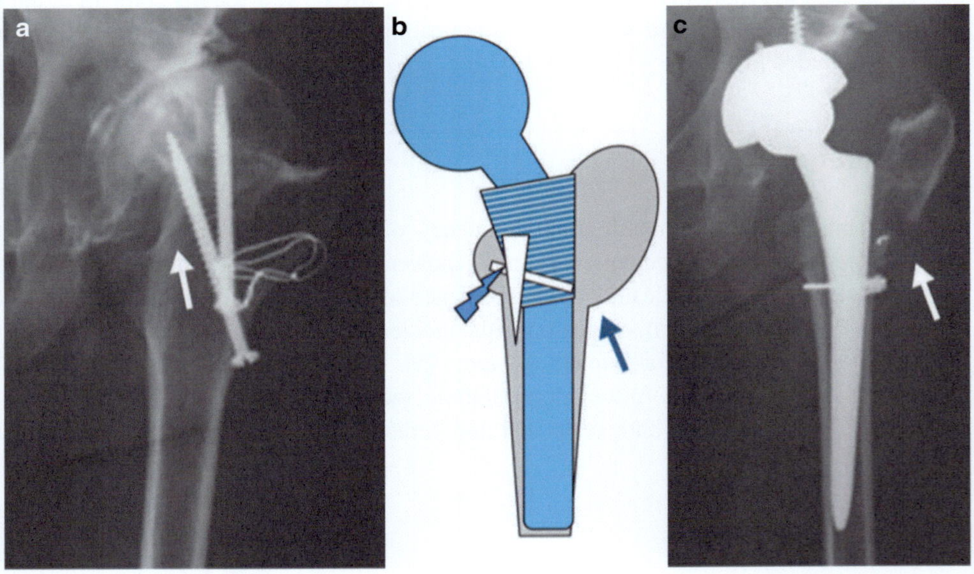

Fig. 86.2 Images of a patient after transtrochanteric valgus osteotomy. (**a**) Steepening, thickening and hardening of the calcar section (arrow). (**b**) Fracture risk in the calcar area (white area). The stem tends to enter externally, and the inferior end of the greater trochanter is thin (arrow). (**c**) Postoperative PFF (Vancouver A_G) (arrow)

Fig. 86.3 Vancouver classification. Type A: Fracture around the trochchanteric area (A_G greater trochanter, A_L lesser trochanter). Type B_1: Fracture around or just below the stem, with no loosening in the stem. Type B_2: Fracture around or just below the stem with loosening of the stem, but no bone loss. Type B_3: Fracture around or just below the stem with loosening of the stem, poor bone quality proximal to the stem, and comminuted fracture. Type C: A distal stem fracture

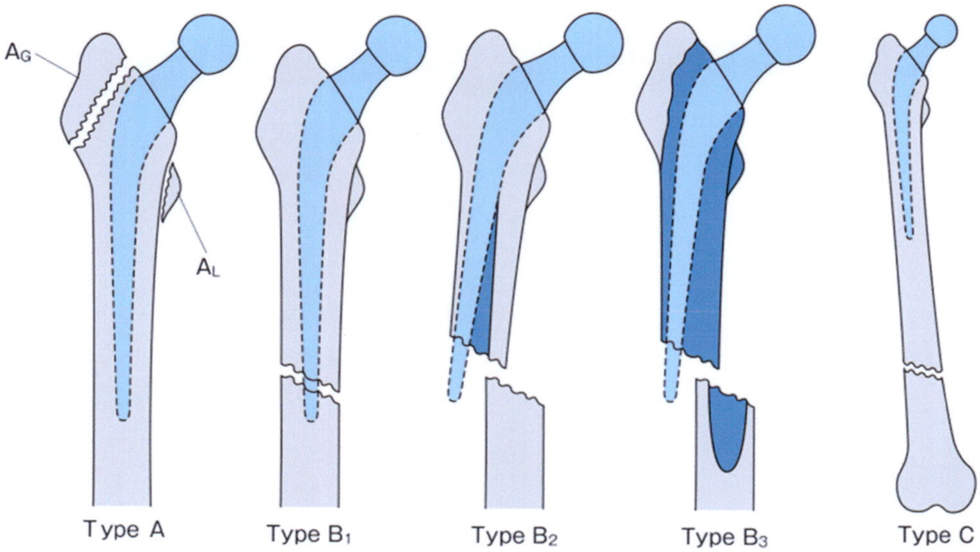

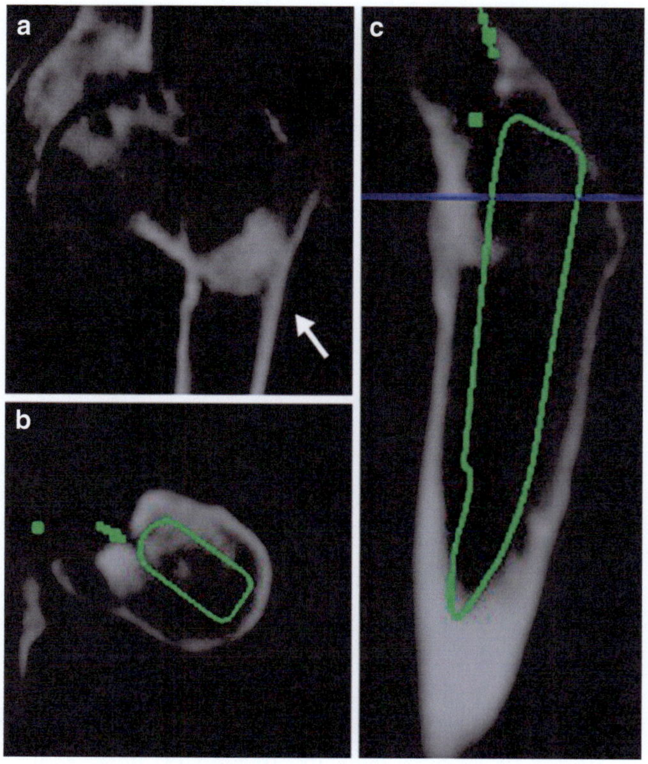

Fig. 86.4 Post-transtrochanteric curved varus osteotomy (CT-MPR image). (**a**) Coronal plane. Osteosclerotic image of the medullary cavity (arrow). Preparation of a surgical bur for excavation of the sclerotic area. (**b**) Axial plane. (**c**) Sagittal plane. 3D templating enables realistic planning, including the implant size

4 Classification

The Vancouver classification is the most frequently used [17] (Fig. 86.3).

In PFF using a proximally fixed cementless stem, a modified Vancouver classification that subdivides the fracture morphology of the proximal femur has recently been proposed as fractures of the calcar area affect stem stability [7].

5 Response Method

(a) Prevention

(i) Pharmacotherapy, preoperative planning, and implant preparation

In a prospective study in Japan, bone mineral density around the stem was shown to increase by 7.3% in the group treated with subcutaneous denosumab injection, an osteoporosis drug in Gruen's zone 7 [18] after THA surgery, which was higher than that in the control group taking oral vitamin D only [19]. In the current situation, where the target age for THA has increased with the extension of healthy life expectancy, denosumab is expected to contribute to the prevention of PFF in elderly patients with osteoporosis.

In cases with femoral deformities, such as following femoral osteotomy, careful preoperative planning and preparation are especially necessary. Preoperative planning should be performed three-dimensionally with CT (Fig. 86.4), and it is important to prepare backup implants, such as distally fixed cementless and cemented stems, and osteosynthetic materials, such as wires and plates, in cases of PFF occurring during surgery.

(ii) Surgical Techniques

Particularly in cases with a femoral deformity, the femoral axis should be confirmed under direct visualization for correct reaming and broaching in the correct

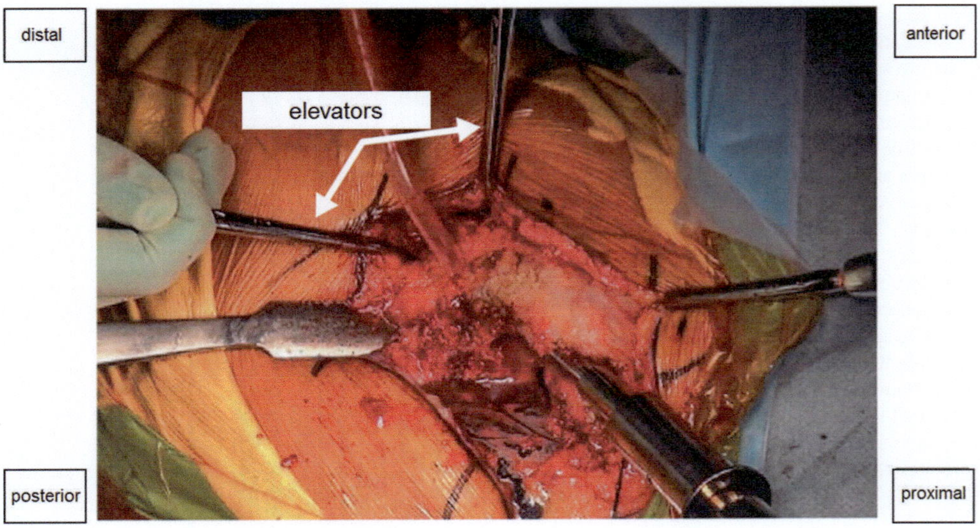

Fig. 86.5 Posterior approach, left hip joint. Reaming while checking the femoral axis in the elevators

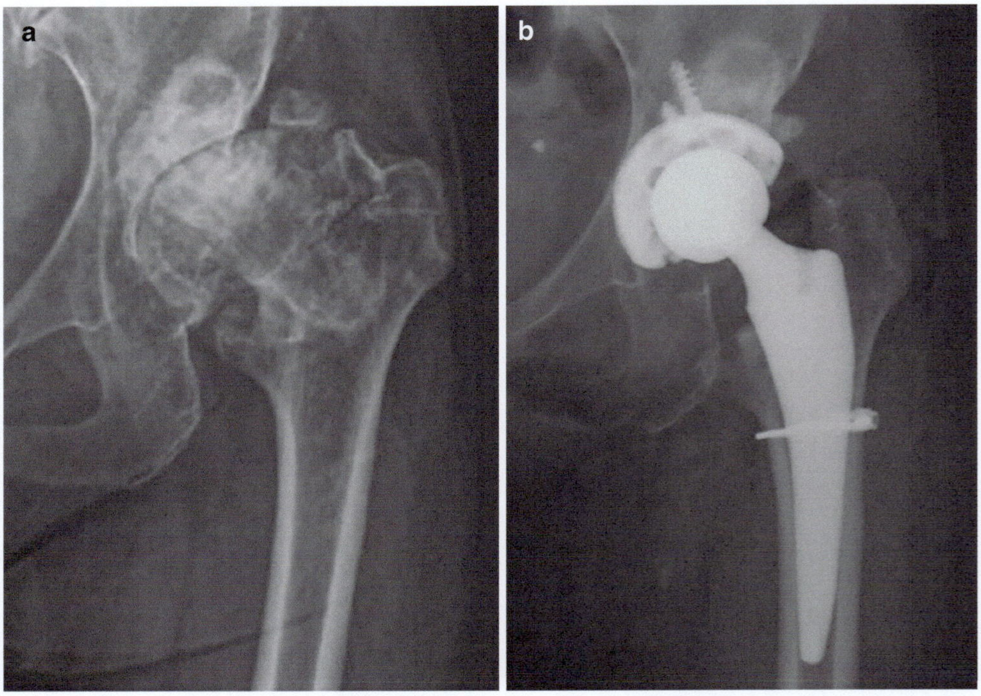

Fig. 86.6 (**a**) Preoperative anteroposterior radiograph (Fig. 86.4 case). (**b**) Postoperative anteroposterior radiograph. After prophylactic wiring under the lesser trochanter, a tapered wedge stem (Kyocera Initia) is inserted

medullary cavity direction. The femoral axis was confirmed by inserting the elevators at both sides of the femoral diaphysis using a longer-than-usual skin incision (Fig. 86.5).

When a proximally fixed cementless stem, such as a fit-and-fill or taper wedge, is used in patients with femoral bone fragility or proximal bone deformity, prophylactic wiring may be performed under the lesser trochanter prior to stem insertion to prevent fractures at the calcar site (Fig. 86.6).

(b) Intraoperative PFF

When treating major trochanteric fractures (Vancouver type A_G), fixation is performed using a cable grip system or tension-band wiring. For fractures localized to the calcar, wiring was performed at the calcar or below the lesser trochanter. In the case of Vancouver type B fractures, a distal fixation stem or a transverse fixation stem should be combined with plate and wiring. In cases of high intraoperative PFF risk, such as femoral bone fragility and deformities, it is important to have a backup implant, as described above. In all cases, the postoperative weight-bearing walking should be limited, and the timing of stating weight-bearing walking should be carefully determined using radiographic images.

(c) Postoperative PFF

Vancouver type A fractures are indicated for conservative treatment or plate-and-wire fixation. Osteosynthesis is indicated for fractures without stem

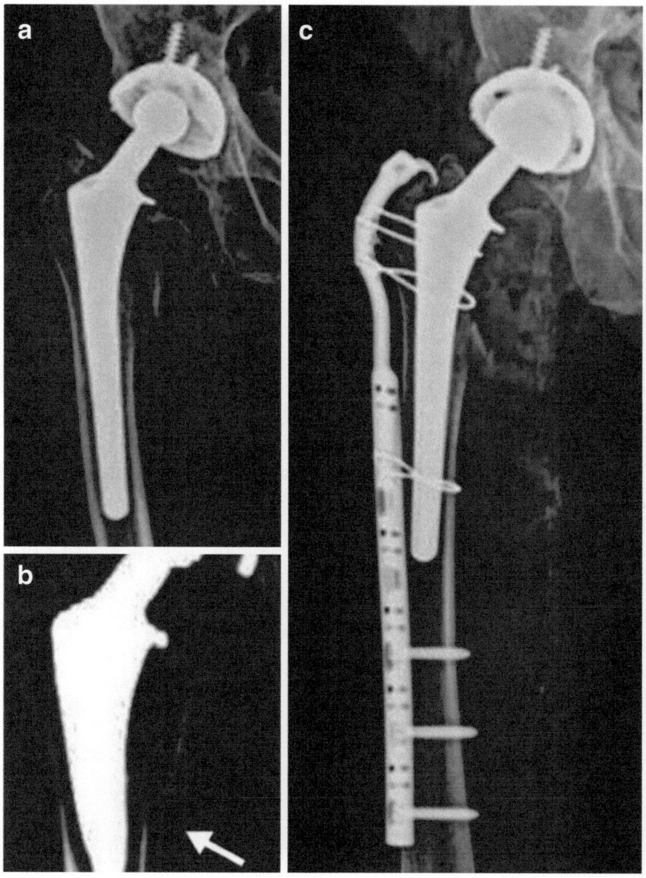

Fig. 86.7 23 years after THA surgery (Kyocera PerFix). (**a**) After fall injury; a fracture line below the lesser trochanter and liner wear (head displaced externally and superiorly) are observed. (**b**) CT-MPR (coronal plane) of a joint diagnosed as Vancouver Type B. Classification of B_1 and B_2 is not possible on imaging. (**c**) Preparation for both osteosynthesis and revision. No loosening of the stem was observed (Type B_1), and osteosynthesis was performed using a long plate. The liner and head were replaced with a 32 mm diameter at the same time

loosening (types B_1 and C) (Fig. 86.7), while Vancouver types B_2 and B_3 with loosening are indicated for revision (Fig. 86.8).

It may be difficult to determine the presence or absence of stem loosening, even with CT, except in the early postoperative period or in cases of stem subsidence. If there is any doubt, the patient should be prepared to undergo both osteosynthesis and revision surgery.

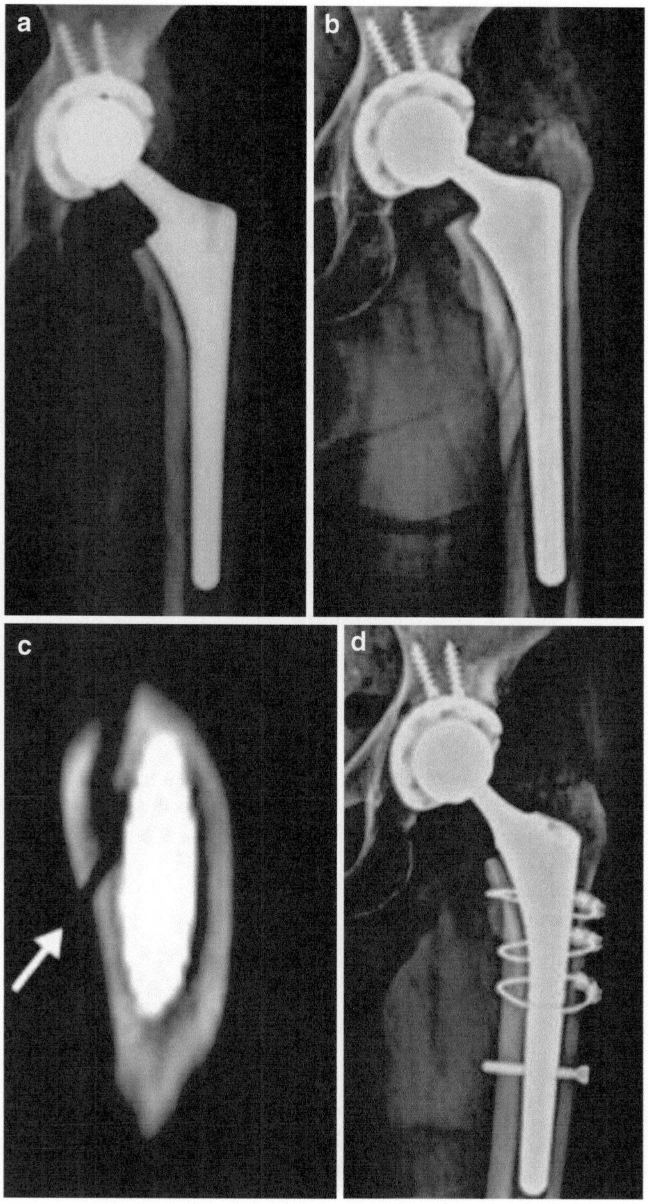

Fig. 86.8 Postoperative PFF (Vancouver type B_2). (**a**) Immediately after surgery (Kyocera 910 PerFix). (**b**) Two weeks postoperatively. Stem subsidence can be seen. (**c**) Fracture line around the stem in CT coronal plane (arrow). (**d**) A radiograph after revision (Kyocera S-LOCK + wiring × 3)

References

1. Learmonth ID, et al. The operation of the century: total hip replacement. Lancet. 2007;370:1508–19.
2. National Joint Registry for England, Wales and Northern Ireland, and the Isle of Man. 12th Annual Report. 2015. http://www.njrcenter.org.uk
3. Pivec R, et al. Incidence and future projections of periprosthetic femoral fracture following primary total hip arthroplasty: an analysis of international registry data. J Long-Term Eff Med Implants. 2015;25:269–75.
4. Young SW, et al. Functional outcome of femoral peri-prosthetic fracture and revision hip arthroplasty: a matched-pair study from the New Zealand registry. Acta Orthop. 2008;79:483–8.
5. Carli AV. Periprosthetic femoral fractures and trying to avoid them: what is the contribution of femoral component design to the increased risk of periprosthetic femoral fracture? Bone Joint J. 2017;99 -B(1 Supple A):50–9.
6. Thien TM, et al. Periprosthetic femoral fracture within two years after total hip replacement: analysis of 437,629 operations in the nordic arthroplasty register association database. J Bone Joint Surg Am. 2014;96:e167.
7. Capello WN, et al. Periprosthetic fractures around a cementless hydroxyapatite-coated implant: a new fracture pattern is described. Clin Orthop Relat Res. 2014;472:604–10.
8. Cooper HJ, et al. Early postoperative periprosthetic femur fracture in the presence of a noncemented tapered wedge femoral stem. HSS J. 2010;6:150–4.
9. Tamaki T, et al. Cementless tapered-wedge stem length affects the risk of periprosthetic femoral fractures in direct anterior total hip arthroplasty. J Arthroplast. 2018;33:805–9.
10. Sheth NP, et al. Operative treatment of early peri-prosthetic femur fractures following primary total hip arthroplasty. J Arthroplast. 2013;28:286–91.
11. Cook RE, et al. Risk factors for periprosthetic fractures of the hip: a survivorship analysis. Clin Orthop Relat Res. 2008;466:1652–6.
12. Herndon CL, et al. Risk factors for periprosthetic femur fracture and the influence of femoral fixation using the mini-anterolateral approach in primary total hip arthroplasty. J Arthroplast. 2020;35:774–8.
13. Ohishi M, et al. Cementless total hip arthroplasty for patients previously treated with femoral osteotomy for hip dysplasia: the incidence of periprosthetic fracture. Int Orthop. 2016;40:1601–6.
14. Utsunomiya T, et al. The results of total hip arthroplasty after Sugioka transtrochanteric anterior rotational osteotomy for osteonecrosis. J Arthroplast. 2017;32:2768–73.
15. Miettinen SSA, et al. Risk factors for intraoperative calcar fracture in cementless total hip arthroplasty. Acta Orthop. 2016;87:113–9.
16. Sakai R, et al. Hammering sound frequency analysis and prevention of intraoperative periprosthetic fractures during total hip arthroplasty. Hip Int. 2011;21:718–23.
17. Duncan CP, et al. Fractures of the femur after hip replacement. Instr Course Lect. 1995;44:293–304.
18. Gruen TA, et al. "Modes of failure" of cemented stem-type femoral components: a radiographic analysis of loosening. Clin Orthop Relat Res. 1979;141:17–27.
19. Nagoya S, et al. Restoration of proximal periprosthetic bone loss by denosumab in cementless total hip arthroplasty. Eur J Orthop Surg Traumatol. 2018;28:1601–7.

Nerve Paralysis

87

Tetsuro Ishimatsu

1 Femoral Nerve Palsy

(a) Point

Femoral nerve palsy, which manifests as quadriceps weakness with or without sensory loss, has been reported as one of the intraoperative complications following total hip arthroplasty (THA). It has a negative impact on the postoperative rehabilitation of patients and results in reduced patient satisfaction [1, 2]. The incidence of femoral nerve palsy following THA was reported to be 0.64% for anterolateral approach THA, 0.40% for anterior approach THA, 0.045% for posterior approach THA, and 0.026% for lateral approach THA among 17,350 patients who underwent initial THA [3]. In a study using magnetic resonance imaging, the distance between the femoral nerve and the acetabular rim was significantly shorter in the supine approach than in the lateral approach [4]. Therefore, more attention must be paid to femoral nerve palsy following THA with the supine approach than with the lateral supine approach.

(b) Causes

Several reports have highlighted the anterior retractor, which is placed in the osseous wall of the anterior acetabulum to pull the iliopsoas muscle, to gain an adequate view of the anterior margin of the acetabulum [2, 5–8] (Fig. 87.1). In particular, a report validated the effect of the anterior retractor on the femoral nerve using a motor-evoked potential (Neuromaster G1, Nihon Kohden), which enables intraoperative measurement of nerve potentials. Compared with the amplitude of the femoral nerve preoperatively, the amplitude of the femoral nerve after the placement of the anterior retraction on the acetabulum was reduced significantly in 17 of 22 patients (77%) who underwent an anterior approach THA and in 29 of 32 patients (93%) who underwent an anterolateral approach THA [5, 6]. The reduced amplitude of the femoral nerve improved significantly after THA, and there were no cases of paralysis. Therefore, the cause could be the anterior retractor itself compressing the femoral nerve through the iliopsoas muscle bulk [5, 6].

(c) Method of response

In a cadaveric study, the shortest distance between the acetabular rim and femoral nerve was reported as an average of 16.6 mm at the acetabular rim at 90°anteriorly, based on the line connecting the superior anterior iliac spine and the center of the acetabulum [7]. In other words, the anterior retractor should be properly positioned on the acetabular rim at 90°anteriorly to avoid femoral nerve palsy. Furthermore, there have been reports of a positive correlation between the thickness of the iliopsoas muscle and the distance from the acetabular rim to the femoral nerve [7]. In patients with small iliopsoas muscles, anterolateral and anterior THA approaches may lead to femoral nerve palsy. As a result, an intermittent release of the anterior retractor and slight flexion of the hip joint may be useful in reducing pressure on the femoral nerve in these patients [9, 10]. Furthermore, preoperative physical therapy using an ergometer may be useful to increase the thickness of the iliopsoas muscle [11].

T. Ishimatsu (✉)
Fukuoka University Faculty of Medicine, Fukuoka, Japan

Y. Nakashima (ed.), *Advances in Total Hip Arthroplasty*, https://doi.org/10.1007/978-981-97-2461-1_87

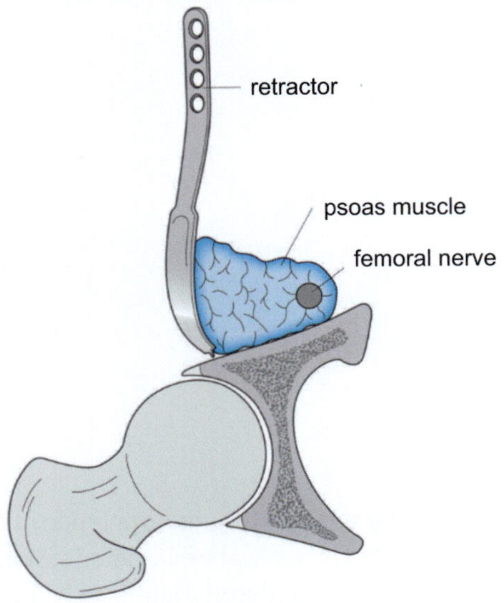

Fig. 87.1 Right hip joint axial dissection (Ishimatsu T. Orthopedic surgery. 2019; 70:470, cited from Fig. 1)

2 Sciatic Nerve Palsy

(a) Point

The majority of sciatic nerve palsy cases following THA occurs with posterior and lateral approaches, and the frequency following the initial THA ranges from 0.08% to 3.7% [12–14].

(b) Causes

In THA, reported causes of sciatic nerve palsy include direct compression of the sciatic nerve by the posterior retractor or a postoperative hematoma, traction associated with leg lengthening, and burns associated with bone cement [15, 16]. However, these remain unelucidated, and acetabular dysplasia and a history of hip arthroplasty have been reported as risk factors [17].

(c) Method of response

The compression of the sciatic nerve associated with the posterior retractor should be noted. In a cadaver-based study, the sciatic nerve was significantly closer to the femoral neck as the hip approached 90° internal rotation and 0° extension to 90° flexion after hip dislocation [18]. In other words, the hip should be in extension when the retractor is inserted under the femoral neck during osteotomy of the femoral neck to prevent sciatic nerve palsy [18].

Regarding sciatic nerve palsy and the amount of leg lengthening after THA, several studies have demonstrated that leg lengthening of 3–5 cm or more is associated with a high risk of nerve injury [12, 19, 20]. Contrarily, in 1284 patients who underwent THA, leg lengthening of up to 5.8 cm did not cause nerve palsy [16]. As a result, the association between sciatic nerve palsy and the amount of leg lengthening is controversial. Furthermore, when sciatic nerve palsy occurred after THA, surgery for shortening the leg extension the day after surgery resulted in nerve palsy improvement 1 week later [12]. Furthermore, the sciatic nerve potential was checked intraoperatively; if it was low, the leg extension was immediately shortened to recover the nerve potential [21]. Therefore, traction associated with leg extension may increase the risk of sciatic nerve palsy, at least in THA. Leg lengthening was categorized using a 2-cm threshold in a systematic review [17]; intraoperative nerve monitoring and wake-up tests should be performed to confirm whether sciatic nerve palsy occurs when leg lengthening of at least 2 cm is performed.

Interfascicular neurolysis has been reportedly used for incompletely recovered sciatic nerve palsy after THA. In a patient who developed secondary hip arthrosis following acetabular fracture surgery, sciatic nerve palsy developed after THA; interfascicular neurolysis was performed 1 year after surgery, and the patient's symptoms improved [22]. Additionally, it was performed 6 months after THA (average, 28 months) in 12 patients with sciatic nerve palsy following THA, and all patients demonstrated significant recovery of motor function [23]. However, there is currently limited evidence regarding the best treatment for patients who have difficulty fully recovering from sciatic nerve palsy after THA [21].

3 Lateral Femoral Cutaneous Nerve Palsy

(a) Point

The lateral femorocutaneous nerve (LFCN) is a sensory nerve that innervates the anterolateral thigh. Damage to this nerve can result in sensory insensitivity, pain, and abnormal sensations in this region. As a result, despite the improvement in hip function following THA, some studies indicate that patients with LFCN palsy have a decreased quality of life [24]. The incidence of LFCN after anterior approach THA ranges from 0% to 81% [25–27]. In a prospective study on 30 patients who underwent bilateral THA in Japan and were randomly assigned to undergoing anterior or anterolateral approach THA on the left or right side, LFCN paralysis occurred 3 months after surgery in 7 (23%) patients who underwent anterior approach THA and in none who underwent anterolateral approach THA [27]. This demonstrates that the risk of LFCN palsy varies depending on THA approach.

(b) Causes

Although the cause remains unclear, anterior approach THA divides the intermuscular space between the sartorius and the tensor fascia femoris muscles, and the LFCN is distributed in this area [26, 28]. As a result, the insertion of the retractor between these muscles and the traction on the fascia of the tensor fascia femoris muscle might simultaneously stretches the branches of the LFCN, resulting in paralysis [29]. The risk factor is a minor preoperative femoral offset, which is considered to require a stronger traction force on the fascia of the tensor fascia femoris muscle [28].

(c) Method of response

The skin incision is supposed to be positioned laterally from the fascia of the tensor fascia lata femoris muscle [29]. A cadaveric study revealed three different branching patterns of LFCN: sartorius type (36%), posterior type (32%), and fan type (32%) [30]. Among them, the fan type, in which multiple branches of the LFCN run across the lateral side of the sartorius and tensor fascia lata muscles, was reported to have a high risk of LFCN injury [31]. Consequently, a report using echography to confirm that the LFCN runs before THA, concluded that if a fan type is observed, the skin incision should be positioned medially to prevent direct injury [31]. The frequency of LFCN paralysis could be reduced by shortening the operative time and decreasing the traction time of the fascia of the tensor fascia lata femoris muscle [32].

Regarding the postoperative recovery of LFCN palsy, one study found that of 7 patients who had LFCN palsy after anterior approach THA, all patients experienced rapid symptomatic improvement until 1 year postoperatively [27]. Contrarily, of 12 patients with LFCN palsy after anterior approach THA, 50% recovered at 1 year, and 83% recovered at 2 years after surgery [32], indicating that recovery may take time.

References

1. Oldenburg M, et al. The frequency, prognosis and significance of nerve injuries in total hip arthroplasty. Int Orthop. 1997;21:1–3.
2. Weale AE, et al. Nerve injury after posterior and direct lateral approaches for hip replacement. A clinical and electrophysiological study. J Bone Joint Surg Br. 1996;78:899–902.
3. Fleischman AN, et al. Femoral nerve palsy following total hip arthroplasty: incidence and course of recovery. J Arthroplast. 2018;33:1194–9.
4. Takada R, et al. Does surgical body position influence the risk for neurovascular injury in total hip arthroplasty? Magnetic resonance imaging study. Orthop Traumatol Surg Res. 2021;107:102817.
5. Ishimatsu T, et al. Motor-evoked potential analysis of femoral nerve status during the direct anterior approach for total hip arthroplasty. J Bone Joint Surg Am. 2018;100:572–7.
6. Kinoshita S, et al. Femoral nerve status during the anterolateral approach for total hip arthroplasty: motor-evoked potential analysis and an influencing factor. J Orthop Sci. 2021;S0949-2658(21):00334-1.
7. Yoshino K, et al. Anatomical implications regarding femoral nerve palsy during a direct anterior approach to total hip arthroplasty: a cadaveric study. J Bone Joint Surg Am. 2020;102:137–42.
8. Hoshino C, et al. Femoral nerve palsy following primary total hip arthroplasty with a direct anterior approach. PLoS One. 2019;14:e0217068.
9. Kim CW. Scientific basis of minimally invasive spine surgery: prevention of multifidus muscle injury during posterior lumbar surgery. Spine (Phila Pa 1976). 2010;35(26 Suppl):S281–6.
10. Wang YC, et al. Dynamic effects of axial loading on the lumbar spine during magnetic resonance imaging in patients with suspected spinal stenosis. J Formos Med Assoc. 2008;107:334–9.
11. Osawa Y, et al. Effects of 16-week high-intensity interval training using upper and lower body ergometers on aerobic fitness and morphological changes in healthy men: a preliminary study. Open Access J Sports Med. 2014;5:257–65.
12. Edwards BN, et al. Contributory factors and etiology of sciatic nerve palsy in total hip arthroplasty. Clin Orthop Relat Res. 1987;218:136–41.
13. Nercessian OA, et al. The use of somatosensory evoked potential during revision or reoperation for total hip arthroplasty. Clin Orthop Relat Res. 1989;243:138–42.
14. Navarro RA, et al. Surgical approach and nerve palsy in total hip arthroplasty. J Arthroplast. 1995;10:1–5.
15. Makita H, et al. The results on total hip arthroplasties with femoral shortening for Crowe's group IV dislocated hips. J Arthroplast. 2007;22:32–8.
16. Nercessian OA, et al. Postoperative sciatic and femoral nerve palsy with reference to leg lengthening and medialization/lateralization of the hip joint following total hip arthroplasty. Clin Orthop Relat Res. 1994;304:165–71.
17. De Fine M, et al. Sciatic nerve palsy following total hip replacement: are patients personal characteristics more important than limb lengthening? Systematic review. Biomed Res Int. 2017;2017:8361071.
18. Kanawati AJ, et al. The change in position of the sciatic nerve during the posterior approach to the hip. Bone Joint J. 2015;97-B:1056–62.
19. Higuchi Y, et al. Leg lengthening of more than 5 cm is a risk factor for sciatic nerve injury after total hip arthroplasty for adult hip dislocation. Nagoya J Med Sci. 2015;77:455–63.
20. Sakai T, et al. Sciatic nerve palsy after cementless total hip arthroplasty. Treatment by modular neck and calcar shortening: a case report. J Orthop Sci. 2002;7:400–2.
21. Shemesh SS, et al. Novel technique for intraoperative sciatic nerve assessment in complex primary total hip arthroplasty: a pilot study. Hip Int. 2018;28:210–7.
22. Kanda A, et al. Treatment of postoperative sciatic nerve palsy after total hip arthroplasty for postoperative acetabular fracture: a case report. Ann Med Surg (Lond). 2016;11:39–41.
23. Regev GJ, et al. Neurolysis for the treatment of sciatic nerve palsy associated with total hip arthroplasty. Bone Joint J. 2015;97-B:1345–9.

24. Homma Y, et al. Lateral femoral cutaneous nerve injury with a direct anterior approach for total hip arthroplasty. Int Orthop. 2016;40:1587–93.
25. Restrepo C, et al. Prospective randomized study of two surgical approaches for total hip arthroplasty. J Arthroplast. 2010;25:671–679.e1.
26. Goulding K, et al. Incidence of lateral femoral cutaneous nerve neuropraxia after anterior approach hip arthroplasty. Clin Orthop Relat Res. 2010;468:2397–404.
27. Takada R, et al. Direct anterior versus anterolateral approach in one-stage supine total hip arthroplasty: focused on nerve injury: a prospective, randomized, controlled trial. J Orthop Sci. 2018;23:783–7.
28. Ozaki Y, et al. Small femoral offset is a risk factor for lateral femoral cutaneous nerve injury during total hip arthroplasty using a direct anterior approach. Orthop Traumatol Surg Res. 2016;102:1043–7.
29. Ropars M, et al. Anatomical study of the lateral femoral cutaneous nerve with special reference to minimally invasive anterior approach for total hip replacement. Surg Radiol Anat. 2009;31:199–204.
30. Rudin D, et al. The anatomical course of the lateral femoral cutaneous nerve with special attention to the anterior approach to the hip joint. J Bone Joint Surg Am. 2016;98:561–7.
31. Ozaki Y, et al. Preoperative ultrasound to identify the distribution of the lateral femoral cutaneous nerve in total hip arthroplasty using a direct anterior approach. SICOT J. 2018;4:42.
32. Bhargava T, et al. Lateral femoral cutaneous nerve impairment after direct anterior approach for total hip arthroplasty. Orthopedics. 2010;33:472.

Instability

88

Ayumi Kaneuji

1 Frequency of Dislocation

In recent years, total hip arthroplasty (THA) outcomes have improved dramatically and incidence of aseptic loosening has decreased. On the other hand, dislocation is cited as a cause of revision surgery and revision rate by dislocation was reported to be the most common cause at 17.3% in a survey of more than 250,000 revision THAs performed in the United States from 2009 to 2013 [1].

The frequency of dislocation after THA is 1–5%, according to the "Guidelines for the Treatment of osteoarthritis (OA) of the hip joint" (Guideline) [2]. The frequency of dislocation after THA for idiopathic osteonecrosis of the femoral head (ONFH) was reported to be 5.2% by the Ministry of Health, Labor, and Welfare Intractable Disease Project Research Group [3]. In addition, it has been reported that 90.9% of cases result in frequent dislocations if there is a dislocation 1 year or more after surgery, suggesting that there is a high possibility of the patient requiring revision replacement [4]. As the number of THA cases increases, the number of patients with postoperative dislocations also increases, resulting in an increase in the number of patients requiring revision THA. Risk factors for dislocation, inspection points, countermeasures, and treatments are described below.

2 Risk Factors

Multiple factors contribute to the development of dislocations. Patient factors include abnormal bone morphology, osteophytes, muscle weakness, poor posture, excessive range of motion, and cognitive dysfunction. In addition, preoperative patient education is effective in reducing the dislocation rate, described as grade C in the Guidelines [2]. There have been reports of postoperative dislocation in patients with myasthenia gravis, a neuromuscular disorder, despite good alignment of the stem and cup [5].

The mechanisms of dislocation are bone-to-implant impingement and implant-to-implant impingement, as well as bone- or soft tissue-to-implant impingement.

Regarding implant-to-implant impingement, a four-dimensional motion analysis showed that dislocation occurred in impingement by the liner and stem neck [6].

In addition, intraoperative confirmation showed that the tendency of anterior dislocation was often due to implant impingement at the posterior area, whereas the tendency of posterior dislocation was often due to upper lateral impingement or soft tissue traction [7].

The angle of implant placement is important to avoiding such impingement, and appropriate implant placement is described as effective in reducing the dislocation rate in the Guidelines at grade C 2.

In terms of the angle of implant placement, an anterior cup opening angle of 20° to 30° is a safe zone [8] and dislocation is more common in a posterior approach with a cup opening angle of 20° or less [9]. However, combined anteversion, which is the sum of the anteversion angle on the stem side and the cup anterior opening angle, is important [10, 11].

Cases with combined anteversion other than 50° ± 10° are 5.8 times more likely to dislocate than cases with 50° ± 10° [12]. In addition to combined anteversion, computer simulations have shown that anterior offset recovery is important for the prevention of posterior dislocation to avoid bone impingement [13].

Thus, it is important to confirm the implant placement angle to avoid impingement of the implant and bone and remove osteophytes that may cause impingement during surgery. In addition, obtaining appropriate muscle tension is an important factor, since the range of motion (ROM) is wide even without impingement and may occur alongside muscle traction [14].

Furthermore, a hip center of 23.9 mm or more above the tear-drop line is a risk factor for dislocation, in addition to

A. Kaneuji (✉)
Kanazawa Medical University, Kahoku, Ishikawa, Japan
e-mail: kaneuji@kanazawa-med.ac.jp

Y. Nakashima (ed.), *Advances in Total Hip Arthroplasty*, https://doi.org/10.1007/978-981-97-2461-1_88

advanced age, small head diameter, and Crowe's classification [15].

Pelvic movement is associated with impingement. In the posterior dislocation group, the posterior pelvic tilt was smaller when patients moved from the standing to the sitting position. This situation is considered to cause anterior impingement and tends to cause posterior dislocation [16].

A similar situation can occur after spinal fusion, and spinopelvic imbalance is a factor in late dislocation [17]. Furthermore, long fusion, in which the spine is deformed and fixed to the pelvis, is a risk factor for dislocation after THA [18, 19]. This is thought to be so because spine and pelvic fusion changes cup alignment, reduces posterior pelvic tilt in the sitting position, and causes the trunk to move forward during standing movements, which may lead to anterior impingement.

Furthermore, doctors with less experience with THA have a higher dislocation rate, resulting in an increase in revision surgery [20].

3 Inspection Points and Countermeasures

The following subsections describe the items to be checked at four stages, from preoperative planning to the revision procedure, and the countermeasures to any problems.

(a) Preoperative planning

As evidence of implants and surgical approaches described in Guidelines [2], it is reported in grade B that the dislocation rate decreases with an implant head of 32 mm or more and it is effective in reducing the dislocation rate by repairing the posterior soft tissue (capsule and external rotator muscles) in the posterior approach.

In grade C, the anterior and lateral approaches reduce the dislocation rate compared to the posterior approach.

The direct anterior approach has a very low dislocation rate in the first month after THA [21].

Intraoperative instability is expected in patients with weak soft tissue tension, diseases with a wide ROM, rheumatoid arthritis, rapidly destructive coxarthrosis, ONFH, subchondral insufficiency fractures of the femoral head, femoral neck fractures, abnormal proximal femoral morphology (post-trauma, abnormal anteversion, and post-femoral osteotomy), and post-acetabular osteotomy. In addition, older people are more likely to experience dislocation than younger people due to muscle weakness [22].

The two main causes of dislocation due to muscle hypotension are vertical (leg length) and horizontal (offset) problems. The leg lengthening improves resistance to dislocation; however, if the leg extends more than 15 mm from the opposite side, lameness persists, resulting in lower patient satisfaction.

In most cases, a leg length discrepancy of 10 mm or less compared with the contralateral side is acceptable [23]. Recently, it has been suggested that the capsule and ligaments around the hip joint function in instability resistance [24]. Therefore, soft tissue-sparing or repaired THA is effective in preventing dislocation in cases without leg lengthening.

An increase in offset of 3 mm or more compared to the normal side decreases dislocation [25], whereas an increase in offset increases pain [26]. Therefore, excessive increase of the offset compared to the contralateral side or preoperatively is not advisable. In terms of resistance to dislocation, a decreased offset is disadvantageous for muscle tension and impingement. Preoperative planning is useful for determining the appropriate leg length and offset before surgery. It is better to select a stem with optimal leg length and offset using a two-dimensional or three-dimensional template for surgical preparation. Specifically, a stem with a large offset should not be selected for patients with a short femoral neck, whereas a stem with a large offset should be selected for those with a long neck. To make an intraoperative decision, it would be useful to prepare a stem that allows a high offset to be selected intraoperatively. In addition, if the anteversion of the femoral neck is excessively large or small, it can be treated by selecting a modular-type stem whose rotation angle can be determined intraoperatively or a cemented stem that makes it relatively easy to control rotation. In addition, dislocation due to changes in the posterior pelvic tilt between the standing and lying positions, which is common in older persons, may be a concern, and anteroposterior radiographs of the hip joint should be taken in the standing and lying positions to plan the cup placement angle considering the posterior pelvic tilt. Moreover, as mentioned above, patients with lumbosacral spine and pelvic fusion are prone to dislocation because of decreased mobility of the lumbar spine and pelvis resulting from positional changes. Lateral functional imaging of the lumbar spine in a sitting position may be helpful for cup positioning.

The use of a large-diameter head with a large oscillation angle and a dual mobility system (DM) should be considered in patients with a low safe ROM for impingement. A systematic review found that intra-prosthetic dislocation and extra-articular dislocation, respectively, occurred in 1.1% and 0.46% of 10,783 first THAs with DM, with all revisions as endpoints. The average rate of dislocation resistance was 98.0% at 8.5 years, and it was reported to be useful for dislocation prevention [27].

Good resistance to dislocations has also been reported in Japan [28–30].

DM has been shown to be resistant to dislocation in revision THAs [27]. The use of navigation systems has been shown to reduce the dislocation rate by accurately reproducing the cup installation angle [31].

(b) Intraoperative confirmation

Intraoperative instability can be caused by a combination of poor soft tissue tension, impingement, and poor alignment or placement of the implant.

After placing the cup and stem, test repositioning is performed to check whether the femoral head dislocates anteriorly or posteriorly. First, soft tissue tension is checked by the shuck test, in which the surgeon pulls the affected limb distally at 0° internal/external rotation/flexion/extension and 5–15° abduction.

Usually, tension should be sufficient to pull the head slightly away from the liner. If the muscle is pulled down too far, the tension is considered weak and stability is evaluated in the dislocated position. However, the shuck test varies depending on the strength of the examiner's muscles, direction and angle of traction, and surgical approach. A certain amount of experience is required to determine the appropriate tension.

The method and acceptable angle for checking dislocations differ among doctors; however, in our department, which uses a posterior approach in the lateral decubitus position, posterior dislocation is checked in maximum flexion, flexion 70°, and adduction 20° with internal rotation 45°, and anterior dislocation is checked in extension 10–20° plus adduction 10° with maximum external rotation.

The following are the precautions and countermeasures to be taken, depending on the situation in the anterior/posterior of dislocation.

(i) When dislocation occurs both in the anterior and posterior directions (Fig. 88.1a)

In this case, the leg length is too short or the offset is insufficient.

(ii) Posterior dislocation (Fig. 88.1b)

When posterior dislocation occurs despite relatively good muscle tension, the first thing to check is whether there is an osteophyte anterior to the acetabulum or whether a thickened capsule remains in the anterior superior part of the acetabulum. An osteophyte may also be present on the anterior femoral neck. Both can cause bone or soft tissue impingement. The surgeon resects osteophyte and thick capsule appropriately, and the dislocation is rechecked. If dislocation is still present, intraoperative radiography of the hip is recommended. The radiographs allow us to confirm the alignment of the cup (inclination and anterior opening angles), the superior and medial position of the cup, as well as the internal or external insertion and offset of the stem and leg length. If there is little anteversion of the stem that can be confirmed by the naked eye or if the anterior opening angle of the cup is small in the X-ray image, the cup is likely to be posteriorly dislocated, and it is recommended to replace the cup or stem. An extremely steep cup inclination angle may also trigger dislocations. Conversely, if the offset of the stem is shorter than that preoperatively, the stem should be changed to a high-offset stem. Combined anteversion of the cup and stem should be between 40° and 60°. A posteriorly elevated liner can also be used. Implant-implant or implant-bone impingement may occur in patients with a very wide ROM of the hip joint. In these cases, we attempt a large-diameter head with the greatest possible ROM or a dual-mobility system.

(iii) Anterior dislocation (Fig. 88.1c)

Anterior dislocation is often caused by a large anterior cup opening angle. Anterior instability is likely to occur if the anterior version of the stem is in excess. If the offset is small, impingement can be avoided by securing a large offset. Anterior dislocation may occur when the lesser trochanter impinges on the sciatic bone. If the cause of anterior dislocation is unknown, it is recommended to confirm the alignment of the implant and offset of the stem by radiography.

(c) Confirmation of the cause of postoperative dislocation

Even if the intraoperative checks described above are performed, postoperative dislocation may occur, depending on various causes. When dislocation occurs, it is important to check the posture and direction of the dislocation; if the hip motion is in flexion or internal rotation, it is usually a posterior dislocation. A posterior approach does not necessarily occur in posterior dislocation, and an anterior approach does not necessarily occur in anterior dislocation. After confirming the direction of the dislocation, a doctor attempts to reduce the dislocation.

The procedure is usually performed under fluoroscopic guidance. In addition, it is often difficult to perform repositioning when the patient is conscious due to pain; therefore, it is often performed under sedation using intravenous anesthesia such as thiopental.

Posterior dislocation can often be repositioned by applying downward traction to the hip in flexion under fluoroscopic guidance and external rotation after the implant head of the bone is lowered below the cup. Anterior dislocation is usually repositioned by placing a large pillow under the lumbar region and applying downward traction to the hip during hyperextension and external rotation. Both techniques are often performed in the supine position; however, it is sometimes difficult to identify the dislocated head position except if the patient is placed in the lateral decubitus position.

If the femoral head cannot be repositioned under intravenous anesthesia, it can often be repositioned by completely weakening the affected limb under general or lumbar anesthesia.

Patients should be evaluated for easy repositioning or excessive downward traction after repositioning. A doctor should check whether there is a tendency for dislocation in flexion, internal rotation, extension, or external rotation under fluoroscopy. If the muscle tension is weak, we check for the causes of muscle weakness, such as aging, disease, or leg shortening, and if necessary, recommend muscle strength training. If impingement is suspected, the prohibited movement is explained to the patient. In addition, the use of orthotics or devices to prevent dislocations should be considered [32].

To investigate the cause of a dislocation, it is necessary to check the alignment of the implant; although cup alignment can be measured on radiographs, the anteversion of the stem is difficult to evaluate. In this case, the anterior cup opening angle and stem anteversion are measured by computed tomography (CT).

The anteversion of the stem should be measured from the femoral condyle; however, it should be kept in mind that combined anteversion alone cannot predict a patient's susceptibility to a dislocation because the tendency toward internal or external rotation naturally depends on the patient.

(d) Method of revision surgery

When a patient experiences recurrent dislocation, a revision surgery may be necessary. First, the problem is investigated by checking the radiographic and CT images for leg length (compared to the preoperative length and immediate postoperative length to see if there

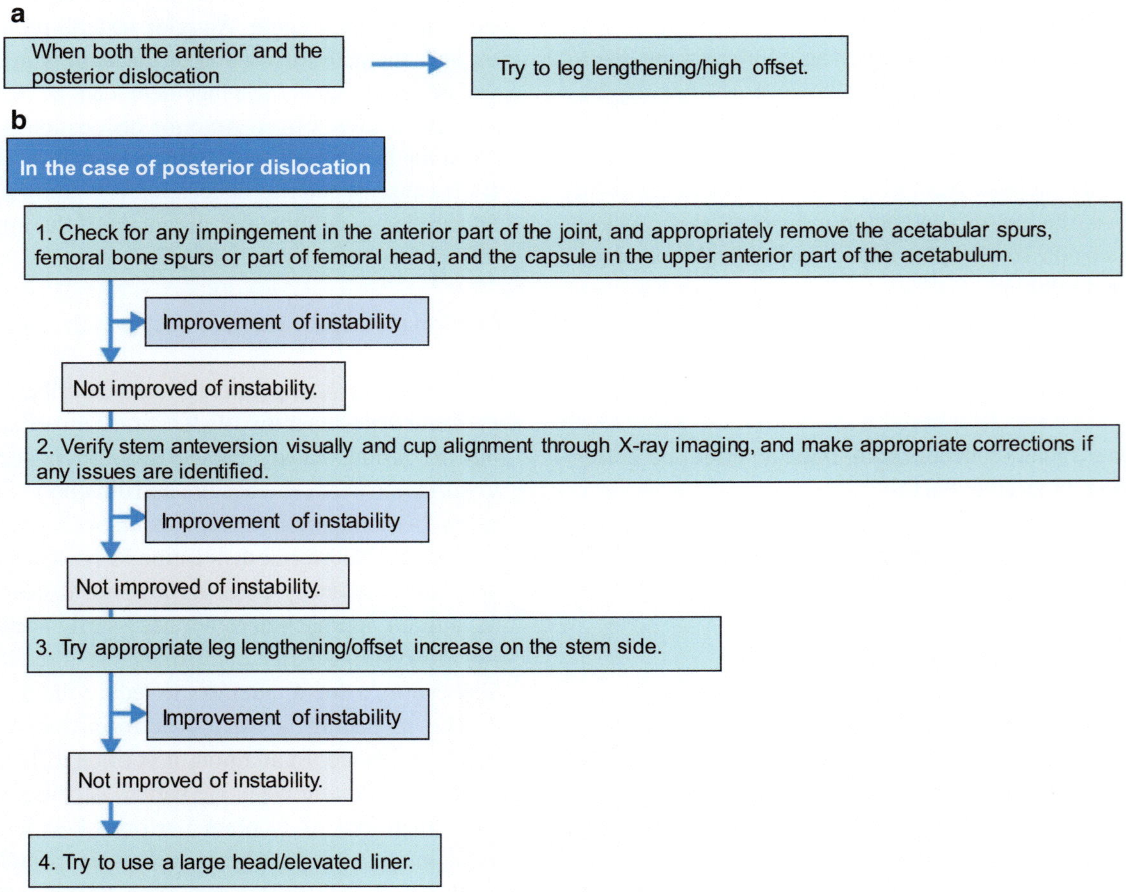

Fig. 88.1 Intraoperative flow for the instability

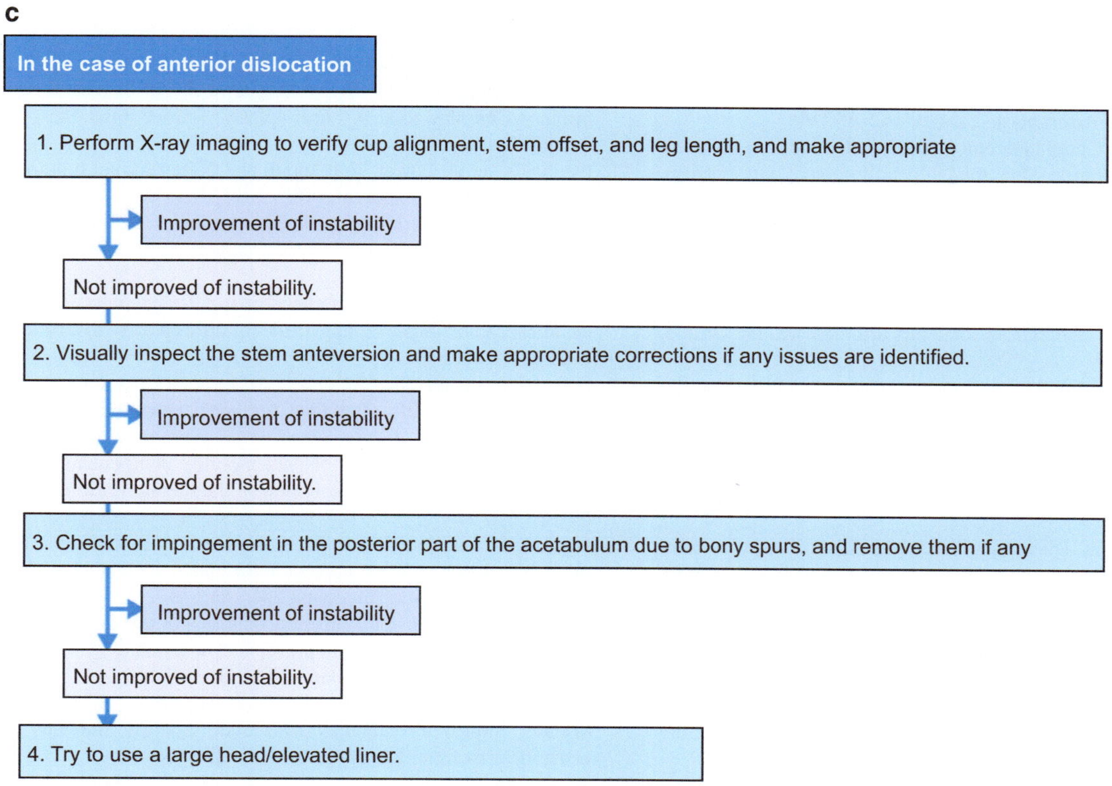

Fig. 88.1 (continued)

was any shortening due to stem subsidence) and cup and stem alignment, as described above. If it is possible to lengthen the femoral head in cases of leg shortening, this should be considered. Furthermore, it is effective to increase the diameter of the implant head. In such cases, it is common to change the flat liner to an elevated liner.

Next, in the case of malalignment of the cup or stem, many doctors would like to avoid the removal and revision of implants that have ingrown into the bone as much as possible. In the case of slight malalignment, it may be possible to avoid subsequent redislocation by changing to a large-diameter head. However, in extreme cases of malalignment, revision with metal implants should be considered. In principle, cup alignment abnormalities should be addressed on the cup side and the stem side should be addressed on the stem side. However, because the removal of a cementless stem can sometimes result in a femoral bone fracture, only a cup-side revision is recommended. Tools for removing cementless cups are available from various implant companies. The cemented stem can be easily removed in the case of a polished stem; therefore, it is recommended that the stem should be removed and replaced.

A multicenter study found that 16 of 88 (18.2%) patients who underwent a revision procedure due to dislocation had a further dislocation, and eight of these patients underwent a re-revision [33].

It is important to ensure that the approach, surgical technique, preoperative planning, and implant selection are performed with the aim of achieving THA without dislocations.

References

1. Gwam CU, et al. Current epidemiology of revision total hip arthroplasty in the United States: a National Inpatient Sample 2009 to 2013. J Arthroplast. 2017;32:2088–92.
2. Clinical Practice Guidelines Committee/Osteoarthritis of the Hip Joint Practice Guidelines Development Committee of Japanese Orthopedic Association Corporation (ed.) [Osteoarthritis of the Hip Joint Clinical Practice Guidelines 2016] Revised 2nd Ed. Nankodo, Tokyo, 2016. Japanese.
3. Kobayashi S, et al. Nationwide multicenter follow-up cohort study of hip arthroplasties performed for osteonecrosis of the femoral head. Int Orthop. 2018;42:1661–8.
4. Itokawa T, et al. Late dislocation is associated with recurrence after total hip arthroplasty. Int Orthop. 2013;37:1457–63.
5. Murotani Y, et al. Unexpected dislocation following accurate total hip arthroplasty caused by excessive hip joint laxity during myasthenic crisis: a case report. J Med Case Rep. 2018;12:331.
6. Miki H, et al. Detecting the cause of dislocation after total hip arthroplasty by patient-specific four-dimensional motion analysis. Clin Biomech (Bristol, Avon). 2013;28:182–6.
7. Tanino H, et al. An in vivo model for intraoperative assessment of impingement and dislocation in total hip arthroplasty. J Arthroplast. 2008;23:714–20.

8. Masaoka T, et al. Study of hip joint dislocation after total hip arthroplasty. Int Orthop. 2006;30:26–30.
9. Nishii T, et al. Influence of component positions on dislocation: computed tomographic evaluations in a consecutive series of total hip arthroplasty. J Arthroplast. 2004;19:162–6.
10. Komeno M, et al. Computed tomographic evaluation of component position on dislocation after total hip arthroplasty. Orthopedics. 2006;29:1104–8.
11. Fujishiro T, et al. The effect of acetabular and femoral component version on dislocation in primary total hip arthroplasty. Int Orthop. 2016;40:697–702.
12. Nakashima Y, et al. Combined anteversion technique reduced the dislocation in cementless total hip arthroplasty. Int Orthop. 2014;38:27–32.
13. Shoji T, et al. Factors affecting impingement and dislocation after total hip arthroplasty – computer simulation analysis. Biomech (Bristol, Avon). 2020;80:105151.
14. Higa M, et al. Effect of acetabular component anteversion on dislocation mechanisms in total hip arthroplasty. J Arthroplast. 2011;44:1810–3.
15. Komiyama K, et al. Does high hip center affect dislocation after total hip arthroplasty for developmental dysplasia of the hip? Int Orthop. 2019;43:2057–63.
16. Snijders TE, et al. The effect of functional pelvic tilt on the three-dimensional acetabular cup orientation in total hip arthroplasty dislocations. J Arthroplast. 2021;36:2184–2188.e1.
17. Heckmann N, et al. Late dislocation following total hip arthroplasty: spinopelvic imbalance as a causal factor. J Bone Joint Surg Am. 2018;100:1845–53.
18. Furuhashi H, et al. Repeated posterior dislocation of the total hip arthroplasty after spinal corrective long fusion with pelvic fixation. Eur Spine J. 2017;26(Suppl 1):100–6.
19. Furuhashi H, et al. Dislocation rate and its risk factors in total hip arthroplasty with concurrent extensive spinal corrective fusion with pelvic fixation for adult spinal deformity. Eur J Orthop Surg Traumatol. 2021;31:283–90.
20. Kishimoto Y, et al. Dislocation is a leading cause of decreased survival rate in primary total hip arthroplasty performed by low-volume surgeons: long-term retrospective cohort study. J Arthroplast. 2020;35:3237–41.
21. Tamaki T, et al. Epidemiology of dislocation following direct anterior total hip arthroplasty: a minimum 5-year follow-up study. J Arthroplast. 2016;31:2886–8.
22. Berry DJ, et al. The cumulative long-term risk of dislocation after primary Charnley total hip arthroplasty. J Bone Joint Surg Am. 2004;86:9–14.
23. Flecher X, et al. Lower limb length and offset in total hip arthroplasty. Orthop Traumatol Surg Res. 2016;102(1 Suppl):S9–20.
24. Logishetty K, et al. Hip capsule biomechanics after arthroplasty: the effect of implant, approach, and surgical repair. Bone Joint J. 2019;101-B:426–34.
25. Forde B, et al. Restoring femoral offset is the most important technical factor in preventing total hip arthroplasty dislocation. J Orthop. 2018;15:131–3.
26. Liebs TR, et al. The influence of femoral offset on health-related quality of life after total hip replacement. Bone Joint J. 2014;96-B:36–42.
27. Darrith B, et al. Outcomes of dual mobility components in total hip arthroplasty: a systematic review of the literature. Bone Joint J. 2018;100-B:11–9.
28. Homma Y, et al. Benefit and risk in short term after total hip arthroplasty by direct anterior approach combined with dual mobility cup. Eur J Orthop Sur Traumatol. 2016;26:619–24.
29. Homma Y, et al. In total hip arthroplasty via the direct anterior approach, a dual-mobility cup prevents dislocation as effectively in hip fracture as in osteoarthritis. Int Orthop. 2017;41:491–7.
30. Lazennec JY, et al. Total hip arthroplasty in patients with Parkinson disease: improved outcomes with dual mobility implants and cementless fixation. J Arthroplast. 2018;33:1455–61.
31. Takashima K, et al. Does a computed tomography-based navigation system reduce the risk of dislocation after total hip arthroplasty in patients with osteonecrosis of the femoral head? A propensity score analysis. J Artif Organs. 2020;23:247–54.
32. Kawahara S, et al. Digitalized analyses of intraoperative acetabular component position using image-matching technique in total hip arthroplasty. Bone Joint Res. 2020;23:360–7.
33. Yoshimoto K, et al. Redislocation after revision total hip arthroplasty for recurrent dislocation: a multicenter study. Int Orthop. 2017;41:253–8.

Iliopsoas Muscle Impingement

89

Ayano Kuwasawa

1 Outline

Iliopsoas impingement (IPI) is a postoperative complication of total hip arthroplasty (THA), first reported by Postel in 1975 [1]. The main symptom is groin pain induced by movements such as climbing stairs, getting in and out of a car, sneezing, or coughing [2–6]. Onset of groin pain has been reported to vary and can occur at any time after an operation, which should be kept in mind when patients complain of groin pain after THA surgery.

2 Anatomy of the Iliopsoas Muscle (Fig. 89.1)

The iliopsoas muscle is the most powerful flexor of the hip joint and gravity-resisting muscle when maintaining a standing posture. It is called the iliopsoas muscle because it consists of three muscles, the iliopsoas, psoas major, and psoas minor, which function as a single muscle. The psoas minor muscle is a branchial bundle of the psoas major muscle and is located in front of the psoas major muscle, but is is absent in half of the population [7]. The psoas major and psoas minor muscles play the main roles in the iliopsoas muscle. The innervation is by the lumbar plexus and the femoral nerve.

The iliopsoas muscle originates from the iliac fossa and inferior anterior iliac spine, and the psoas major muscle has two heads (shallow and deep), the shallow head at the vertebral body of the 12th thoracic vertebra to the 4th lumbar vertebra, and the deep head at the 12th rib and rib process of all lumbar vertebrae. This is the only muscle in the body connecting the trunk and lower limbs.

Both the iliopsoas and psoas major muscles travel through the anterior surface of the acetabulum and exit the muscle fissure below the inguinal ligament to the anterior surface of the femur, after which they become a common tendon and stop at the obturator minor of the femur. Some iliopsoas muscle fibers attach distally beyond the obturator minor [8]. The iliopsoas tendon at the anterior border of the acetabulum related to IPI varies widely among individuals anatomically, and the dissection results of 53 bodies reported that the iliopsoas tendon was in one bundle in 28.3%, two bundles in 64.2%, and three bundles in 7.5% [9].

A. Kuwasawa (✉)
Saitama Cooperative Hospital, Saitama, Japan

Y. Nakashima (ed.), *Advances in Total Hip Arthroplasty*, https://doi.org/10.1007/978-981-97-2461-1_89

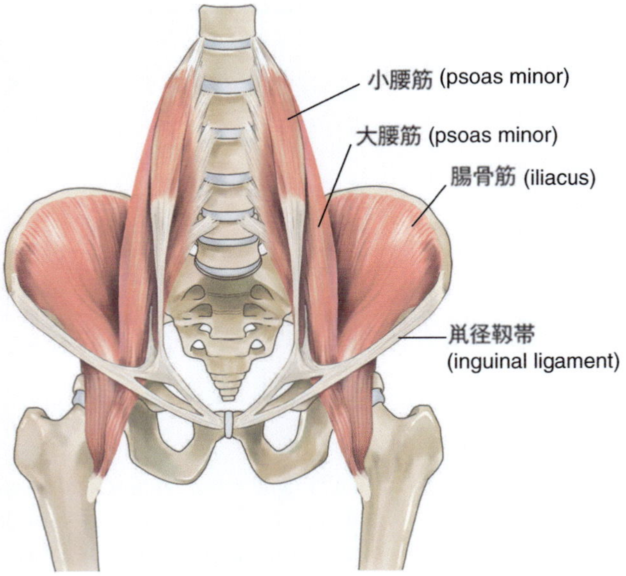

Fig. 89.1 Anatomy of the iliopsoas muscle

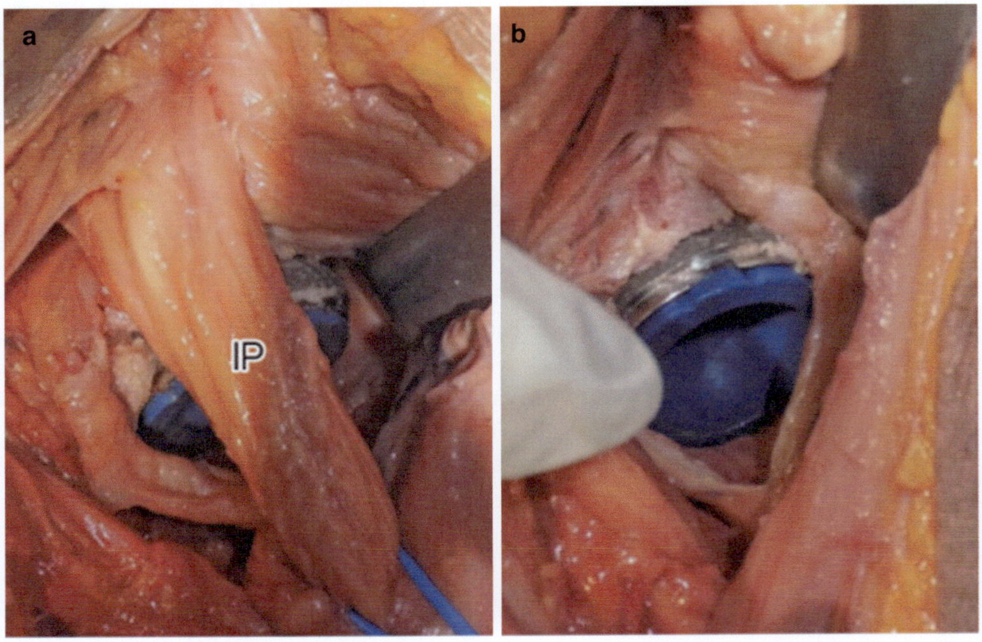

Fig. 89.2 Position of the anterior margin of the acetabulum and the iliopsoas muscle (from cadaver). (**a**) Running of the iliopsoas muscle. (**b**) Anterior margin of the acetabulum on the dorsal surface of the iliopsoas muscle (anteriorly protruding cup)

3 Related Factors (Fig. 89.2)

IPI is caused by the presence of factors that impair the running of the iliopsoas muscle. In particular, protrusion of the anterior margin of the acetabular cup is the most frequently reported factor [5, 10, 11]. The degree of cup protrusion is closely related to the development of IPI, and there are reports that all patients with IPI symptoms have a cup protrusion of 12 mm or more on computed tomography (CT) [11] and that a cup protrusion of 12 mm or more on transverse and 4 mm or more on sagittal images is a risk factor for IPI [5]. The causes of increased protrusion include a small anterior cup opening angle [12, 13] and a progressive pelvic retroversion of 20° or more [12]. It is important to consider cup protrusion when planning THA acetabulum placement.

IPI can also occur in the absence of cup protrusion [2]. Caution should be exercised when the running of the iliopsoas muscle changes significantly after THA, such as with increased leg extension and offset [14] and increased posterior pelvic tilt [12], and when the load on the iliopsoas muscle increases at the anterior acetabular border.

Other factors that have been reported to cause IPI include anterior acetabular wall defects [15], postacetabular rotational osteotomy [16], use of a large diameter cup on the acetabulum [17], use of dual mobility [18], collared stems [19], protruding acetabular cup screws [20], and cement remnants around the cup [21]. Risk factors for development of IPI include anterior intrusion [5], female sex [3], a ratio of acetabular cup to original femoral head diameter greater than 1.1 [3], and postacetabular osteotomy [22].

4 Diagnosis (Figs. 89.3 and 89.4)

Physical examination must include pain in the groin area induced by the hip joint during dynamic extension or automatic hip flexion. In particular, pain is characteristically elicited during automatic flexion from 45° or greater, such as when ascending and descending stairs or getting in and out of a car, which is known as a “car sign” [21].

Radiological findings include radiographic and CT examinations to confirm the protrusion of the acetabular cup, which is the most common cause of IPI. Large protrusions

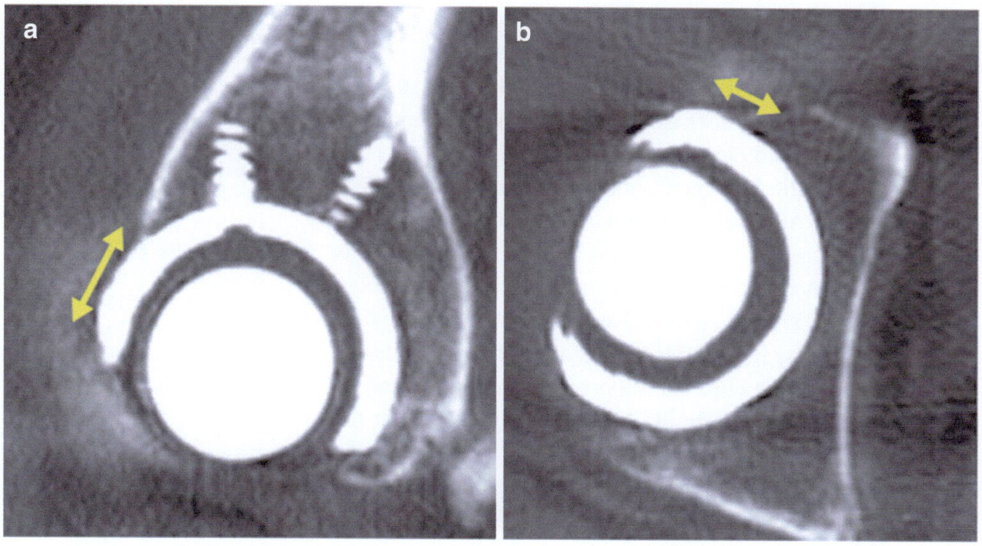

Fig. 89.3 Protruding cup on computed tomography image. (**a**) Sagittal section image. (**b**) Transverse section

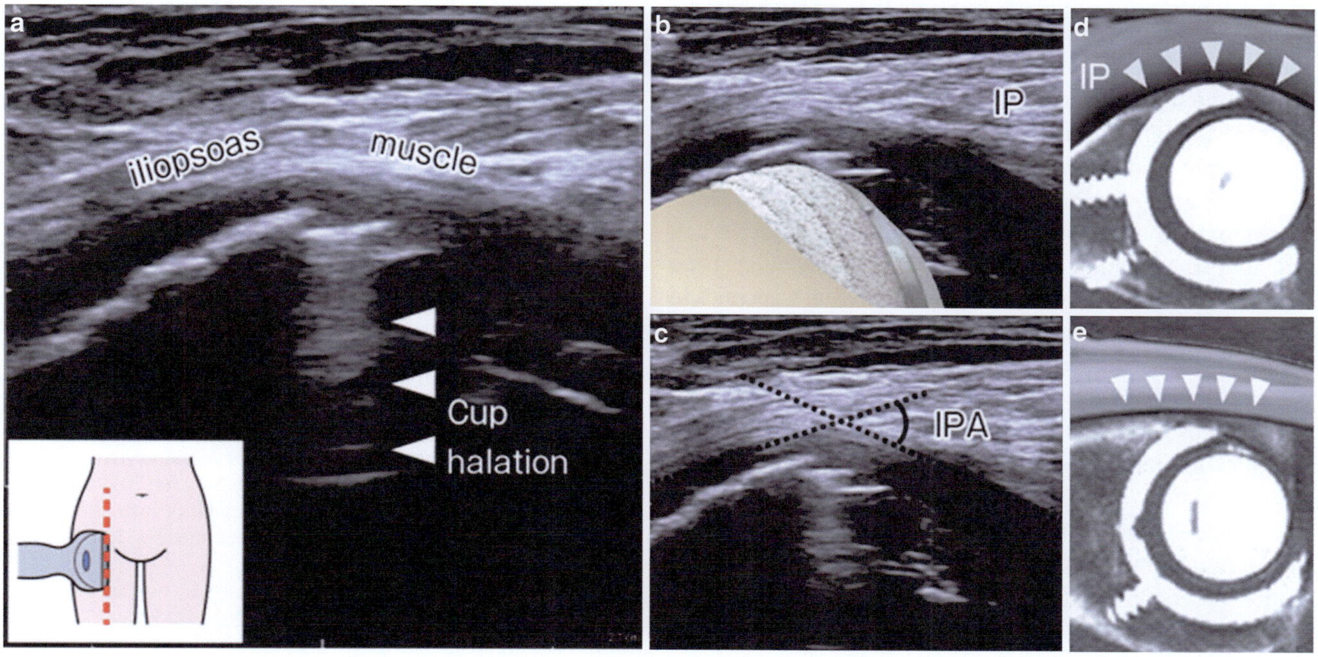

Fig. 89.4 Running of the iliopsoas muscle on ultrasound image. (**a**) Ultrasound image of the iliopsoas muscle (probe applied). (**b**) Collated image of cup protrusion. (**c**) Ilio-psoas angle (IPA). (**d**) Simplified view of iliopsoas muscle running in a case with strong iliopsoas impingement (IPI) (+) IPA. (**e**) Simplified view of iliopsoas muscle running in IPI (−) IPA weak cases

are a strong diagnostic factor for IPI, as it has been reported that many cases of IPI have protrusions of 12 mm or more in diameter [5, 11].

Recently, the use of echo tests has also been recommended as a diagnostic tool. The advantage of this technique is that it can be easily performed in an outpatient setting without exposure to radiation and it can dynamically evaluate the iliopsoas muscle, which is steeply flexed by the anterior margin of the cup and other obstacles, by having the subject perform an automatic hip flexion/extension movement. The iliopsoas muscle on the side of the articular surface in direct contact with the obstructive factor (ilio-psoas angle: IPA) is measured, and it has been reported that cases with a large IPA and steep inflection have more groin pain [23]. When groin pain is severe, block injections using local anesthetics or steroids should be performed; if pain improves, this is the most promising finding for confirming the diagnosis of IPI.

The final decision is made comprehensively by considering each diagnostic factor and differentiating the cause of pain from dislocation, infection, fracture, implant defects such as laxity and migration of the implant, and lumbar spine disease.

5 Treatment

Conservative treatments, including physiotherapy such as stretching, drug therapy such as nonsteroidal anti-inflammatory drugs (NSAIDs), and block injections of local anesthetics and steroids under echo guidance, should be considered for first-line treatment. Although conservative treatment is recommended in all cases, its effectiveness is limited, recurrence is common, and there are reports that surgical treatment is more effective [24]. The most common surgical procedures are iliopsoas tendonotomy [6] and THA revision to change the defective placement of the cup. The iliopsoas tendonotomy can be performed either specularly [4, 25] or directly [14], or at the hip level [4] or at the obturator level [25].

A systematic review comparing specular and direct visualization reported that specularization was less invasive, less risky, and more effective [26].

When dissecting at the hip level, there is an advantage of being able to directly observe the injured area, but care should be taken not to damage the neurovascular bundle, such as the femoral nerve, which is in front of the iliopsoas muscle [27]. It has been reported that the effect of dissection at either site is equivalent [27] and that there is no significant difference in hip flexion force [28].

Although muscle weakness of the iliopsoas muscle is a potential complication of iliopsoas tendonotomy, it has been reported that even if muscle atrophy of the iliopsoas muscle is seen on imaging, return to sports/activities can often still be possible [29, 30] and is not a significant problem in clinical practice. In addition, compared with revision arthroplasty, iliopsoas tendonotomy is less invasive and has a lower complication rate [30], so iliopsoas tendonotomy should be considered first.

On the other hand, in cases where the cup protrudes more than 8 mm from the acetabular rim, the results of iliopsoas tendon dissection tend to be poor, and there have been reports that cup revision was effective [24, 30], so THA revision should be considered to improve poor cup placement in cases where the cup protrusion is large.

References

1. Postel M. [Painful prosthesis. I. Possible causes.] Rev Chir Orthop Reparatrice Appar Mot. 1975;61(Suppl 2):57–61.
2. Eddine TA, et al. [Anterior iliopsoas impingement after total hip arthroplasty: diagnosis and conservative treatment in 9 cases.] Rev Chir Orthop Reparatrice Appar Mot. 2001;87:815–9.
3. Buller LT, et al. Iliopsoas impingement after direct anterior approach total hip arthroplasty: epidemiology, risk factors, and treatment options. J Arthroplast. 2021;36:1772–8.
4. Di Benedetto P, et al. Arthroscopic treatment of iliopsoas impingement syndrome after hip arthroplasty. Acta Biomed. 2019;90(Suppl 1):104–9.
5. Uno T, et al. Risk factors and cup protrusion thresholds for symptomatic iliopsoas impingement after total hip arthroplasty: a retrospective case-control study. J Arthroplast. 2018;33:3288–3296.e1.
6. Guicherd W, et al. Endoscopic or arthroscopic iliopsoas tenotomy for iliopsoas impingement following total hip replacement: a prospective multicenter 64-case series. Orthop Traumatol Surg Res. 2017;103(8S):S207–14.
7. Anderson CN. Iliopsoas: pathology, diagnosis, and treatment. Clin Sports Med. 2016;35:419–33.
8. Tatu L, et al. Descriptive anatomy of the femoral portion of the iliopsoas muscle. Anatomical basis of anterior snapping of the hip. Surg Radiol Anat. 2001;23:371–4.
9. Philippon MJ, et al. Anatomic variance of the iliopsoas tendon. Am J Sports Med. 2014;42:807–11.
10. Chalmers BP, et al. Iliopsoas impingement after primary total hip arthroplasty: operative and nonoperative treatment outcomes. J Bone Joint Surg Am. 2017;99:557–64.
11. Cyteval C, et al. Iliopsoas impingement on the acetabular component: radiologic and computed tomography findings of a rare hip prosthesis complication in eight cases. J Comput Assist Tomogr. 2003;27:183–8.
12. Uno T, et al. Tilt-adjusted cup anteversion in patients with severe backward pelvic tilt is associated with the risk of iliopsoas impingement: a three-dimensional implantation simulation. Clin Orthop Relat Res. 2019;477:2243–54.
13. Weber M, et al. Navigation is equal to estimation by eye and palpation in preventing psoas impingement in THA. Clin Orthop Relat Res. 2017;475:196–203.
14. Heaton K, et al. Surgical release of the iliopsoas tendon for groin pain after total hip arthroplasty. J Arthroplast. 2002;17:779–81.
15. Pozzi L, et al. Reconstruction of the anterior acetabular wall to repair symptomatic defects consecutive to cup malpositioning at total hip arthroplasty. Arthroplast Today. 2021;7:260–263.e0.
16. Nihei K, et al. [High incidence of iliopsoas muscle impingement in THA after RAO]. Hip Joint. 2020;46:739–43. In Japanese.
17. Cobb JP, et al. Why large-head metal-on-metal hip replacements are painful: the anatomical basis of psoas impingement on the femoral head-neck junction. J Bone Joint Surg Br. 2011;93:881–5.
18. Nebergall AK, et al. Analysis of dual mobility liner rim damage using retrieved components and cadaver models. J Arthroplast. 2016;31:1595–602.
19. Brew CJ, et al. Iliopsoas tendonitis caused by overhang of a collared femoral prosthesis. J Arthroplast. 2011;26(504):e17–9.
20. Bricteux S, et al. [Iliopsoas impingement in 12 patients with a total hip arthroplasty]. Rev Chir Orthop Reparatrice Appar Mot. 2001;87:820–5.
21. Jasani V, et al. Pain related to the psoas muscle after total hip replacement. J Bone Joint Surg Br. 2002;84:991–3.
22. Uno T, et al. Association between total hip arthroplasty following periacetabular osteotomy and acetabular component overhang. Eur J Orthop Surg Traumatol. 2020;30:1431–9.
23. Kuwasawa A, et al. [Characteristic echographic findings in iliopsoas muscle impingement syndrome]. Hip Joint. 2020;46:377–83. In Japanese.
24. Dora C, et al. Iliopsoas impingement after total hip replacement: the results of nonoperative management, tenotomy or acetabular revision. J Bone Joint Surg Br. 2007;89:1031–5.
25. Gédouin JE, et al. Technique and results of endoscopic tenotomy in iliopsoas muscle tendinopathy secondary to total hip replace-

ment: a series of 10 cases. Orthop Traumatol Surg Res. 2012;98(4 Suppl):S19–25.
26. O'Connell RS, et al. A systematic review of arthroscopic versus open tenotomy of iliopsoas tendonitis after total hip replacement. Arthroscopy. 2018;34:1332–9.
27. Ilizaliturri VM Jr, et al. Prospective randomized study of 2 different techniques for endoscopic iliopsoas tendon release in the treatment of internal snapping hip syndrome. Arthroscopy. 2009;25:159–63.
28. Ilizaliturri VM Jr, et al. Central compartment release versus lesser trochanter release of the iliopsoas tendon for the treatment of the internal snapping hip: a comparative study. Arthroscopy. 2014;30:790–5.
29. Gouveia K, et al. Iliopsoas tenotomy during hip arthroscopy: a systematic review of postoperative outcomes. Am J Sports Med. 2021;49:817–29.
30. Coulomb R, et al. Arthroscopic iliopsoas tenotomies: a systematic review of surgical technique and outcomes. Hip Int. 2022;32:4–11.

Part XVI

Registry

90 Japanese Orthopaedic Association National Registry System (JOANR)

Haruhiko Akiyama

1 History of the Japan Arthroplasty Register

The National Registry of Artificial Joint Replacement was started in Sweden in 1975 [1] and has since been established in Finland, Norway, Denmark, and other Nordic countries. In this study, we examined the registration system of artificial joints in Japan, which was initiated by the Implant Committee of the Japanese Orthopaedic Association in July 2003. The Registry Subcommittee was established in January 2004 to discuss the registration items, ethical guidelines, confidentiality, etc. In February 2006, 10 institutions that were members of the Registry Subcommittee started a pilot program of total hip arthroplasty and total knee arthroplasty. The pilot registry for hip and knee arthroplasty was started, and after repeated examinations of the registry system, an expanded trial was started in January 2008, covering 130 centers nationwide. In May 2012, the Japan Arthroplasty Register participated in the 1st International Society of Arthroplasty Registries [2] and became an associate. In April 2014, a web-based registration system using the University Hospital Medical Information Network Center (UMIN) was launched, registration of reverse shoulder arthroplasty began in October, and registration of unicompartmental knee arthroplasty and anatomical shoulder arthroplasty surgery began in 2016. The first turning point in the Japan Arthroplasty Register was that, in 2016, participation in the registry system became a prerequisite for accreditation as a teaching and training hospital of the Japanese Orthopaedic Association, resulting in a significant increase in the registration rate and number of registered centers. The next major turning point was the incorporation of the Japan Arthroplasty Register into the Japanese Orthopaedic Association National Registry (JOANR) from April 2021, with registration still mandatory.

H. Akiyama (✉)
Gifu University, Gifu, Japan
e-mail: akiyama.haruhiko.m3@f.gifu-u.ac.jp

2 Significance of the National Registry of Artificial Joint Replacement

In the early 2000s, the importance of the national registry was not well recognized in Japan, and prospective multi-center studies and retrospective studies were mainly conducted. However, the Ministry of Health, Labor, and Welfare (MHLW) and the Pharmaceuticals and Medical Devices Agency (PMDA) have recognized the importance of national registries. For this reason, many national registries were established with the help of the Japan Agency for Medical Research and Development (AMED) through the promotion of the patient registration system in Japan. Currently, its necessity, importance, and significance are widely recognized.

Prostheses are implants that are surgically inserted *in vivo* and must function normally for a long period of time. Considering the current types of implants and their longevity, it is necessary to evaluate them by observing their progress for 10–30 years. Therefore, national registry data should be collected and evaluated over short, medium, and long periods of time. In general, artificial joint data have the following universal and comprehensive significance: the identification of diseases in a country, collection of implant information, evaluation of implant and procedure performance, dissemination of high-performing implants and surgical procedures, elimination of poorly performing implants, investigation of complications, feedback to physicians and hospitals, feedback to implant companies, and international cooperation and the comparison of results and data among countries. The results of the National Registry of Artificial Joint Replacement led to the improvement of surgical techniques at medical institutions and the early detection of implants with poor clinical outcomes [3]. In fact, it was reported that the revision rate decreased in Sweden and that a health economic effect was observed [4]. In addition, data from the prosthesis registry system in Australia and the United Kingdom showed the dangers of metal-on-metal implants [5, 6]. Thus, detailed analyses of registries in

Y. Nakashima (ed.), *Advances in Total Hip Arthroplasty*, https://doi.org/10.1007/978-981-97-2461-1_90

Scandinavia, the United Kingdom, Australia, and other countries are underway. Data on patient factors, such as age, sex, weight, and surgical approach have been published as factors influencing the clinical outcomes of hip prostheses. Implant design is a major influencing factor, and ODEP benchmarking has been conducted to evaluate implants based on registry data [7]. In addition, some countries have published a list of surgical facility results for the evaluation of clinical outcomes. Clinical outcomes have traditionally been evaluated using X-rays and revision procedures; however, in recent years, registries have begun to include patient-reported outcome measures (PROMs), and patient-based evaluations are becoming increasingly common. Thus, THA registries now contain a large amount of information and analyzes outcomes from multiple perspectives.

3 Data from the Japanese Arthroplasty Registry System

According to the 2020 Mid-term Forecast of the Medical Bionics (Artificial Organs) Market and In-depth Analysis of Participating Companies, it was estimated that approximately 260,000 joint replacement operations were performed in Japan from April 2019 to March 2020 (13). Approximately 72,600 artificial hip joints, 78,000 artificial femoral heads, 100,700 artificial knee joints, and 5400 artificial shoulder joints are expected to be implanted [8]. The annual report of the Japanese Arthroplasty Registry System is published on the Japanese Society of Arthroplasty website [9] and the registration data up to the end of March and the end of the fiscal year are cumulatively tabulated. The registration rate of the Japanese Arthroplasty Registry System has increased in recent years, making it possible to calculate the cumulative revision rate of artificial joint implants in Japan. Statistical analyses of various factors and cumulative revision rates are also reported [9].

4 The JOANR

Musculoskeletal diseases targeted by orthopedic surgery are found in a wide range of people, from children to elderly individuals and are one of the main causes of impaired healthy life expectancy in the population. An estimated 1.3 million orthopedic surgeries are done each year and are increasing annually, accompanied by an aging society. Therefore, a comprehensive nationwide registry is required. Orthopedic surgery involves many operations using implants, such as artificial joints and spinal fusion devices, for reconstructive surgery. Information on the status and outcomes of these operations based on real-world evidence from a large-scale registry is extremely important not only to improve national health but also to improve the quality of care, developing new medical devices, and improving health economics. JOANR, as real-world evidence of surgery for musculoskeletal diseases by the Japan Orthopaedic Association, enables various policy responses, such as constructing evidence based on big datasets on surgical treatment for musculoskeletal diseases, investigating the actual conditions of the proposal by the Federation of Surgical Associations of Social Insurance Committee, conducting post-marketing surveillance (PMS), evaluating new medical technologies, and prioritizing musculoskeletal diseases and surgical methods. The JOANR database is a multicenter, retrospective, observational study that began on April 1, 2020 [10]. The database has a two tier structure, with the first consisting of survey items led by the Japanese Orthopaedic Association and the second consisting of survey items led by related societies, such as joint arthroplasty registries. Affiliated societies can use the basic JOANR data by participating in survey items or by linking their registries to JOANR. However, participation in JOANR is subject to the establishment of a secretariat system to ensure the accuracy and security of medical information, including data cleansing and registration data analysis. This is based on the Ministry of Health, Labor, and Welfare's note in 2008 on the points for ensuring reliability in post-marketing database surveys of pharmaceutical products [11]. The data cleaning and analysis in JOANR were conducted in a cloud environment to prevent the registered data from being taken outside the system. In FY2020, approximately 138,000 joint arthroplasty cases were registered in the JOANR, including 61,700 hip arthroplasty cases, 72,600 knee arthroplasty cases, and 3850 shoulder arthroplasty cases; the estimated registration rates were 85.1%, 72.1%, and 71.4%, respectively. In FY2022, the estimated registration rates were 96.1% in THA, 88.8% in TKA, 91.1% in UKA, 94.5% in reverse TSA, respectively.

5 Summary

A national registry of joint arthroplasty is indispensable to understand the status of past and future surgical cases. In Asia, the JOANR the only large database, and it is expected to be utilized for the development of joint arthroplasty in the future.

References

1. The Swedish Knee Arthroplasty Register. Welcome to the Swedish Knee Arthroplasty Register – a National Quality Registry since 1975. https://www.myknee.se/en/. Accessed.
2. International Society of Arthroplasty Registries. Mission Statement. http://www.isarhome.org/. Accessed.

3. von Knoch F, et al. Why do we need a national joint replacement registry in the United States? Am J Orthop (Belle Mead NJ). 2009;38:500–3.
4. Kärrholm J, et al. The Swedish hip arthroplasty register. Annual Report 2015.
5. Australian Orthopedic Association National Joint Replacement Registry. Annual report 2016. 2016. Available at: https://aoanjrr.sahmri.com/annual-reports-2016. Accessed October 29, 2017.
6. National Joint Registry. 12th Annual Report. 2014.
7. Orthopedic Data Evaluation Panel. What is ODEP? https://www.odep.org.uk/Info/Abbreviations.aspx
8. Yano Research Institute Ltd. 2020 Mid-term Forecast of Medical Bionics (Artificial Organs) Market and In-depth Analysis of the Participants. 2021. Japanese. Accessed.
9. Japanese Society for Replacement Arthroplasty. http://jsra.info/jar-report.html. Japanese. Accessed.
10. Japanese Orthopedic Association Corporation. https://www.joanr.org. Japanese. Accessed.
11. Pharmaceutical Safety and Environmental Health Bureau, Ministry of Health, Labor and Welfare of Japan. Notes on reliability assurance in post-marketing database surveys of pharmaceuticals. 2018. https://www.pmda.go.jp/files/000223003.pdf. Japanese. Accessed.